68000 FAMILY ASSEMBLY LANGUAGE

THE PWS SERIES IN ENGINEERING

Anderson, *Thermodynamics*
Askeland, *The Science and Engineering of Materials*, Third Edition
Bolluyt, Stewart, and Oladipupo, *Modeling For Design Using SilverScreen*
Borse, *FORTRAN 77 and Numerical Methods for Engineers*, Second Edition
Bronzino, *Biomedical Engineering and Instrumentation*
Clements, *68000 Family Assembly Language*
Clements, *Microprocessor Systems Design*, Second Edition
Clements, *Principles of Computer Hardware*, Second Edition
Das, *Principles of Foundation Engineering*, Second Edition
Das, *Principles of Geotechnical Engineering*, Third Edition
Das, *Principles of Soil Dynamics*
Fleischer, *Introduction to Engineering Economy*
Gere and Timoshenko, *Mechanics of Materials*, Third Edition
Glover and Sarma, *Power Systems Analysis and Design*, Second Edition
Janna, *Design of Fluid Thermal Systems*
Janna, *Introduction to Fluid Mechanics*, Third Edition
Kassimali, *Structural Analysis*
Keedy, *Introduction to CAD using CADKEY*, Second Edition
Knight, *The Finite Element Method in Mechanical Design*
Logan, *A First Course in the Finite Element Method*, Second Edition
McGill and King, *Engineering Mechanics: Statics*, Second Edition
McGill and King, *Engineering Mechanics: An Introduction to Dynamics*, Second Edition
McGill and King, *Engineering Mechanics: Statics and An Introduction to Dynamics*, Second Edition
Raines, *Software for Mechanics of Materials*
Reed-Hill and Abbaschian, *Physical Metallurgy Principles*, Third Edition
Reynolds, *Unit Operations and Processes in Environmental Engineering*
Sack, *Matrix Structural Analysis*
Schmidt and Wong, *Fundamentals of Surveying*, Third Edition
Segui, *Fundamentals of Structural Steel Design*
Segui, *LRFD Steel Design*
Shen and Kong, *Applied Electromagnetism*, Second Edition
Sule, *Manufacturing Facilities*, Second Edition
Vardeman, *Statistics for Engineering Problem Solving*
Weinman, *FORTRAN for Scientists and Engineers*
Weinman, *VAX FORTRAN*, Second Edition

68000 FAMILY

ASSEMBLY LANGUAGE

ALAN CLEMENTS, BSc, PhD., MBCS
University of Teesside
The Motorola Professor

PWS PUBLISHING COMPANY
Boston

PWS PUBLISHING COMPANY
20 Park Plaza, Boston, MA 02116-4324

I(T)P ™
International Thomson Publishing
The trademark ITP is used under license

PWS Publishing Company is a division of Wadsworth, Inc.

Library of Congress Cataloging-in-Publication Data

Clements, Alan
68000 family assembly language / Alan Clements.
p. cm.
Includes bibliographical references and index.
ISBN 0-534-93275-4
1. Motorola 68000 series microprocessors—Programming.
2. Assembler language (Computer program language) I. Title.
QA76.8.M67C47 1993 93-5102
005.265—dc20 CIP

Printed in the United States of America

5 6 7 8 9 10 — 99 00 01

Sponsoring Editor: *Jonathan Plant*
Assistant Editor: *Mary Thomas*
Editorial Assistant: *Cynthia Harris*
Production Editor: *Monique Calello*
Manufacturing Coordinator: *Ruth Graham*
Interior Designer: *Alan Clements*
Illustrator/Typesetter: *Alan Clements*
Cover Designer: *Robine Andrau/M. Calello*
Cover Photo: *Dominique Sarraute / The Image Bank*
Cover Printer: *John P. Pow Company*
Text Printer and Binder: *RR Donnelley/Harrisonburg*

For my friends

Ruud van der Bijl
Richard H. Eckhouse
Elizabeth, Karen & Grace Prior

PREFACE

Few books begin by providing a justification for their very existence — this one is an exception. Why then should we need to defend a book on assembly language? Consider the following statement: "Virtually all digital computers, from microprocessors to mainframes, execute a low-level machine code, and assembly language is no more than a *symbolic representation of machine code*. Therefore, we can intuitively conclude that assembly language is an essential component of computer science and computer engineering." I shall now explain why the teaching of assembly language is sometimes a controversial aspect of the computer science curriculum, and why I have written this book to support its teaching.

Students are presented with material by their teachers and, in general, they accept the contents of their course without question. In fact, the contents of many a course are the result of a long series of battles fought between various academics. For fifteen years or more, the teaching of assembly language has been supported by its protagonists and denigrated by its antagonists. I shall attempt to summarize the arguments of these two groups.

The protagonist of assembly language argues that its appreciation provides a foundation for the understanding of the digital computer. If you are not familiar with an assembly language you cannot possibly know how a computer works. Clearly, such a body of knowledge must form an essential component of computer science. Three other, and more practical, arguments can be made for learning assembly language. The first is that high-level languages are not always suited to every circumstance. For example, the software required to control a hard disk interface might have to be written in assembly language if high-level languages lack the facilities needed to access the hardware interface. The second reason for teaching assembly language is that a computer's native assembly language is faster than any high-level language running on the same computer. Therefore, programs that have to be run in the minimum time must be written in assembly language. Finally, an assembly language program uses less memory space than its high-level language equivalent.

Sometimes high-level language programmers can't avoid assembly language. Development systems for high-level languages include debugging aids that allow you to step through a program line-by-line in an attempt to locate errors. Some debuggers even allow you to step through the assembly language generated by the compiler. Clearly, you have to understand assembly language in order to use these facilities provided by the debuggers. You might wonder why the high-level language programmer ends up by rooting through assembly language. Many 68000-family programmers write software to run on embedded processors in systems ranging from automobile engines to aircraft control systems. These applications frequently involve an interrupt-driven input/output mechanism and a complex relationship among the applications program, the input/output handlers, and the real-time operating system. In such cases, it is not uncommon to have to perform some of your debugging at the machine code level.

The antagonist argues that assembly language programming flies in the face of history. Computer science has moved away from the real machine that executes machine code to more and more abstract machines that behave in a much more human-like fashion. The high-level language programmer is not concerned with how the real machine operates, and is happy to let the compiler convert his or her program into machine code. Forcing students to study assembly language teaches them all the bad habits (collectively called *hacking*) that computer scientists have been trying to move away from. In any case, you can argue that the manipulation of bits and bytes within a computer is as relevant to today's computer scientist as a knowledge of brain cell chemistry is to a poet. I once worked with a colleague whose proud boast was that he was not only opposed to the teaching of assembly language, but he also thought that even Ada was a little too close to the machine level.

Neither side in the *Great Assembly Language Debate* has a monopoly on the truth, and both parties are partially right. I think that we should teach assembly language, but that our teaching should not be divorced from the world of computer science. The teaching of assembly language should support other areas of computer science. For example, assembly language teaches us about the architecture and organization of a computer. Without this understanding, you cannot design the next generation of computers. The most justified criticism of assembly language teaching is that some courses present the student with little more than a list of assembly language instructions and provide almost no help in writing programs. The principal objectives of this course are:

- To teach the fundamental components of an assembly language: the processor's register set, its instructions, and its addressing modes.
- To teach how assembly language programs are constructed.
- To demonstrate how assembly language can be used to write device drivers to control serial and parallel interfaces.
- To introduce interrupts and exceptions and explain how they are used by the systems programmer.
- To use assembly language to demonstrate the relationship between a high-level language and the underlying machine architecture on which it runs.

One difficulty of teaching assembly language is its location in the curriculum. Because assembly language is called *low-level language* and embodies few intellectually difficult concepts, it is often taught eariy in a computer science course. Yet, without a knowledge of computer architecture and programming in high-level languages, the student finds it hard to relate low-level programming to the rest of his or her course. As we have already pointed out, a knowledge of assembly language is often required by those debugging complex real-time embedded systems. Perhaps assembly language should be taught as part of real-time programming.

Yet another difficulty in teaching assembly language programming is its dependence on a real architecture. High-level languages like Pascal can be divorced from the hardware of the machines on which they run. An assembly language is the native language of a computer and is therefore not designed with pedagogical objectives in mind. To be more blunt, some assembly languages are similar to human languages in the sense that they are confusing, illogical, ambiguous, and full of irregularities. And that's only the good ones.

The writer of any book on assembly language must select a particular target machine. I have chosen the popular 68000 family of microprocessors. This family, made up of the 68000, 68010, 68020, 68030, and 68040, is widely used in both the industrial and academic worlds. More importantly, the architecture of the 68000 family strongly supports the teaching of several components of a computer science course (e.g., real-time systems and operating systems).

As I have already said, one of the difficulties in teaching assembly language programming is caused by the irregularity of instruction sets. You can overcome this problem by teaching a hypothetical machine that is entirely regular. Such an approach is educationally sound, but runs into two difficulties. The first is that, if it is to be taught practically, the teacher must design a hardware emulator or software simulator on which students can test their programs. The second problem with a hypothetical machine is that it doesn't always prepare students well for the real world. Moreover, the student cannot readily access many of the books written about real machines.

My approach to assembly language programming is to compromise between choosing a real and a hypothetical machine. We start by teaching a simplified version of the 68000 assembly language and ignore many features that confuse the fundamental issues. Once the simplified model has been digested, we can introduce the factors that make the 68000 a much more complex processor. Some of these factors are concerned with small details that complicate the life of the assembly language programmer. Some of them are extensions to the simplified model of the 68000 that transform it into a very powerful microprocessor. Since the introduction of the 68000 in 1984, new members of the 68000 family have been designed. We look at the 68000's more powerful successor — the 68020. Although we don't cover the 68030 and the 68040, the architectures of these processors are very similar to the 68020. The only real architectural difference between the 68020 and the 68030/40 is the inclusion of on-chip memory management — a topic beyond the scope of this text.

A particular difficulty encountered by those teaching assembly language is the way in which its component topics are strongly interrelated. For example, you cannot teach an *instruction set* without first having introduced *addressing modes*, and you cannot teach addressing modes without first having described an instruction set. One way of dealing with this impasse is to introduce a very simple instruction set together with a limited number of addressing modes, and then build on this foundation.

Finally, why did I write this book? A few years ago I wrote a book on microprocessor systems design and the 68000 family, and included a section on assembly language programming for readers already familiar with its basic concepts. I taught assembly language to my own students using a combination of books and notes. I couldn't use my own book because it assumed too much prior knowledge of computer architecture. I couldn't find an alternative book that taught 68000 assembly language and the principles of program design, and yet was readable by my students. I had always wanted to write a book on assembly language but was never able to find the time. One day Jonathan Plant, my editor at PWS Publishing, phoned me and asked me to write an assembly language book. I knew that if I didn't write a book on 68000 assembly language I would regret it, maybe not today or tomorrow, but soon and for the rest of my life. So, I rounded up the usual references and began to write.

Acknowledgments

I would like to thank all those who helped me to write this book and to proofread it. In particular I would like to say a special thank you to Roger Allen and Bill Postlethwaite who helped me remove some of the more embarrassing errors, and to my wife Sue, who gave up her valuable time to proofread the manuscript.

Of all those who have helped me over the past few years, no one has done more than Paul Lambert. Paul developed the 68000 cross-assembler and simulator that we use to teach 68000 assembly language programming to our students at Teesside. Although Paul no longer works for the Computer Center, he has always been ready to correct bugs in the software or to extend it whenever I asked.

I would also like to thank those I worked with at PWS Publishing Company who helped me to produce this book: Jonathan, Mary, Cynthia, and Monique.

While writing this text, I have received a lot of help, feedback, and encouragement from those who have reviewed my work. Some were foolish enough to tell me to drop in if I were ever in the USA — I did.

Terry Glagowski
Washington State University

Donald Gustafson
Texas Tech University

Bill Neumann
Arizona State University

James T. Reinhart
Motorola Inc.

James Resh
Michigan State University

Therrill Valentine

It would be a major miracle if I managed to eliminate all errors in a book of this size — not least because this book is my first venture into desktop publishing. I would be most pleased to hear from any readers who find errors or would like to suggest ways in which the book could be improved. I can be contacted through my publisher, PWS Publishing Company in Boston, or at The University of Teesside, Middlesbrough, England TS1 3BA. My E-mail address on the UK JANET network is: a.clements@uk.ac.tees.

Alan Clements

CONTENTS

CHAPTER 7 Subroutines and Parameters 276

CHAPTER 8 Assemblers 326

CHAPTER 9 Peripheral Programming 362

CHAPTER 1

Data Representation and Computer Arithmetic

Before we look at the digital computer and assembly language proper, we examine how numbers are represented in computers, how they are converted from one base to another, and how they are manipulated within the computer.

We begin by looking at binary codes in general and show how patterns of ones and zeros can represent a wide range of different quantities. The main theme of this chapter is the class of binary codes used to represent *numbers* in digital computers. We look at how numbers are converted from the familiar decimal form to binary form and vice versa. Other topics included here are the ways in which we represent and handle *negative* as well as positive numbers, and how the computer uses floating point arithmetic to deal with very large and very small numbers. Finally, we look at some of the logical operations that can be carried out on binary values. These are the AND, OR, NOT, and EOR operations. We don't wish to cover the algebra of logic, Boolean algebra, in this text. All we need to know is how these logical operations can be used by the assembly language programmer.

Since this chapter provides a coverage of several topics in binary arithmetic, you may wish to skip some of the more advanced topics such as complementary arithmetic and floating point arithmetic until you need the information. Equally, if you are already familiar with binary arithmetic, you may skip ahead to Chapter 2.

1.1 Characters, Words, and Bytes

Because of the ease with which two-state logic elements can be manufactured and because of their remarkably low cost, it was inevitable that computer designers chose the binary number system to represent data within a digital computer. The smallest quantity of information that can be stored and manipulated inside a computer is the *bit* (i.e., **b**inary dig**it**). A bit is unique because it cannot be subdivided into any smaller unit of information. Digital computers store information in their memories in the form of groups of bits called *words*. For our current purposes, we can regard the working definition of a *word* as "the basic unit of information stored in memory and processed by a computer." Unfortunately, the term *word* is employed by computer scientists to mean several different things. The number of bits per word varies from computer to computer. Typical computer wordlengths are: 8-, 16-, 32-, and 64-bits.

A group of eight bits has come to be known as a *byte*. Often a word is spoken of as being two or four bytes long, as its bits can be formed into two or four groups of eight bits, respectively. Throughout this section we will use the term *word* to mean the basic unit of information operated on by a computer. We stress this point because 68000 literature employs the term *word* in a more restrictive fashion to mean a 16-bit value.

An n-bit word can be arranged into 2^n unique bit patterns and may represent many things, because there is no intrinsic meaning associated with a pattern of 1s and 0s. For example, the 8-bit values 11001010 and 00001101 do not mean anything. The actual meaning of a particular pattern of bits is the meaning given to it by the programmer. As Humpty Dumpty said to Alice, "A word means exactly what I choose it to mean, nothing more and nothing less."

The computer itself cannot determine the meaning of the word, but simply treats it in the way the programmer dictates. For example, a programmer could read the name of a person into the computer and then perform an arithmetic operation on the pattern of bits representing the letters of the name (say, multiply it by two). The computer would happily carry out the operation, although the result would be quite meaningless. If this is not clear, consider the following example. Suppose that in a certain Chinese restaurant, 46 represents bamboo shoots and 27 represents egg fried rice. If I were to ask for a portion of 73, I would be most *unlikely* to get egg fried rice with bamboo shoots! The following are some entities a word may represent.

- **An Instruction** An instruction or *operation code (op-code)* defines an action that is to be performed by the CPU and is represented by a single word (or by a sequence of words). The relationship between the bit-pattern of the instruction and what it does is arbitrary and is determined by the designer of the instruction set. A particular sequence of bits that means add A to B to one computer might have an entirely different meaning to another.

- **A Numeric Quantity** A word, either alone or as part of a sequence of words, may represent a numerical quantity. Numbers can be represented in one of many formats: BCD integer, unsigned binary integer, signed binary integer, BCD floating point, binary floating point, complex integer, complex floating point, double precision integer, etc.

  The meaning of some of these terms and the way in which the computer carries out its operations in the number system represented by the term will be examined later.

- **A Character** There are many applications of computers in which text is input, processed, and the results printed. The most obvious and spectacular example is the word processor. Programs themselves are frequently in text form when they are first submitted to the computer. The alpha-numeric characters of the English alphabet (A to Z, a to z, 0 to 9) and the symbols *, -, +, !, ?, etc. are assigned binary patterns so that they can be stored and manipulated within the computer. Fortunately, one particular code, called the ASCII code (American Standard Code for Information Interchange), is now in widespread use throughout the computer industry. This is also known as the ISO 7-bit character code, and represents a character by 7 bits, allowing a maximum of $2^7 = 128$ different characters. Of these 128 characters, 96 are the normal printing characters (including both upper- and lowercases). The remaining 32 characters are non-printing. These include carriage return, backspace, line feed, etc. Table 1.1 defines the relationship between the bits of the ISO/ASCII code and the character they represent.

Table 1.1 The ASCII character code

<table>
<tr><th></th><th></th><th>0
000</th><th>1
001</th><th>2
010</th><th>3
011</th><th>4
100</th><th>5
101</th><th>6
110</th><th>7
111</th></tr>
<tr><td>0</td><td>0000</td><td>NUL</td><td>DLC</td><td>SP</td><td>0</td><td>@</td><td>P</td><td>‘</td><td>p</td></tr>
<tr><td>1</td><td>0001</td><td>SOH</td><td>DC1</td><td>!</td><td>1</td><td>A</td><td>Q</td><td>a</td><td>q</td></tr>
<tr><td>2</td><td>0010</td><td>STX</td><td>DC2</td><td>"</td><td>2</td><td>B</td><td>R</td><td>b</td><td>r</td></tr>
<tr><td>3</td><td>0011</td><td>ETX</td><td>DC3</td><td>#</td><td>3</td><td>C</td><td>S</td><td>c</td><td>s</td></tr>
<tr><td>4</td><td>0100</td><td>EOT</td><td>DC4</td><td>$</td><td>4</td><td>D</td><td>T</td><td>d</td><td>t</td></tr>
<tr><td>5</td><td>0101</td><td>ENQ</td><td>NAK</td><td>%</td><td>5</td><td>E</td><td>U</td><td>e</td><td>u</td></tr>
<tr><td>6</td><td>0110</td><td>ACK</td><td>SYN</td><td>&</td><td>6</td><td>F</td><td>V</td><td>f</td><td>v</td></tr>
<tr><td>7</td><td>0111</td><td>BEL</td><td>ETB</td><td>'</td><td>7</td><td>G</td><td>W</td><td>g</td><td>w</td></tr>
<tr><td>8</td><td>1000</td><td>BS</td><td>CAN</td><td>(</td><td>8</td><td>H</td><td>X</td><td>h</td><td>x</td></tr>
<tr><td>9</td><td>1001</td><td>HT</td><td>EM</td><td>)</td><td>9</td><td>I</td><td>Y</td><td>i</td><td>y</td></tr>
<tr><td>A</td><td>1010</td><td>LF</td><td>SUB</td><td>*</td><td>:</td><td>J</td><td>Z</td><td>j</td><td>z</td></tr>
<tr><td>B</td><td>1011</td><td>VT</td><td>ESC</td><td>+</td><td>;</td><td>K</td><td>[</td><td>k</td><td>{</td></tr>
<tr><td>C</td><td>1100</td><td>FF</td><td>FS</td><td>,</td><td><</td><td>L</td><td>\</td><td>l</td><td>|</td></tr>
<tr><td>D</td><td>1101</td><td>CR</td><td>GS</td><td>-</td><td>=</td><td>M</td><td>]</td><td>m</td><td>}</td></tr>
<tr><td>E</td><td>1110</td><td>SO</td><td>RS</td><td>.</td><td>></td><td>N</td><td>^</td><td>n</td><td>~</td></tr>
<tr><td>F</td><td>1111</td><td>SI</td><td>US</td><td>/</td><td>?</td><td>O</td><td>_</td><td>o</td><td>DEL</td></tr>
</table>

We obtain the 7-bit binary code for a character by reading the most-significant three bits in the column in which it appears and the least-significant four bits in the row in which it appears. Thus, the character W is in the 101 column and the 0111 row, so that its ASCII code is 1010111 (or 57 in hexadecimal form). Similarly, the code for a carriage return, CR, is 0001101.

- **A Picture Element** One of the many entities that have to be digitally encoded is the *picture* or *graphical display*. Pictures vary widely in their complexity and there are a correspondingly large number of ways of representing pictorial information. For example, pictures can be *parameterized* and stored as a set of numeric values. When the picture is to be displayed or printed, it is recreated from its parameters. These parameters are the ingredients of a picture and correspond to lines, arcs, and polygons and their positions within the picture.

Another way of storing pictorial information is to employ symbols that can be put together to make a picture. Such an approach is popular with small microprocessor systems and generally employs an 8-bit code to represent the symbols from which the picture is to be constructed. Teletext, Ceefax, and Oracle pictures are all drawn in this way. Figure 1.1 illustrates a typical set of graphics symbols.

Figure 1.1 Representing graphic symbols by codes

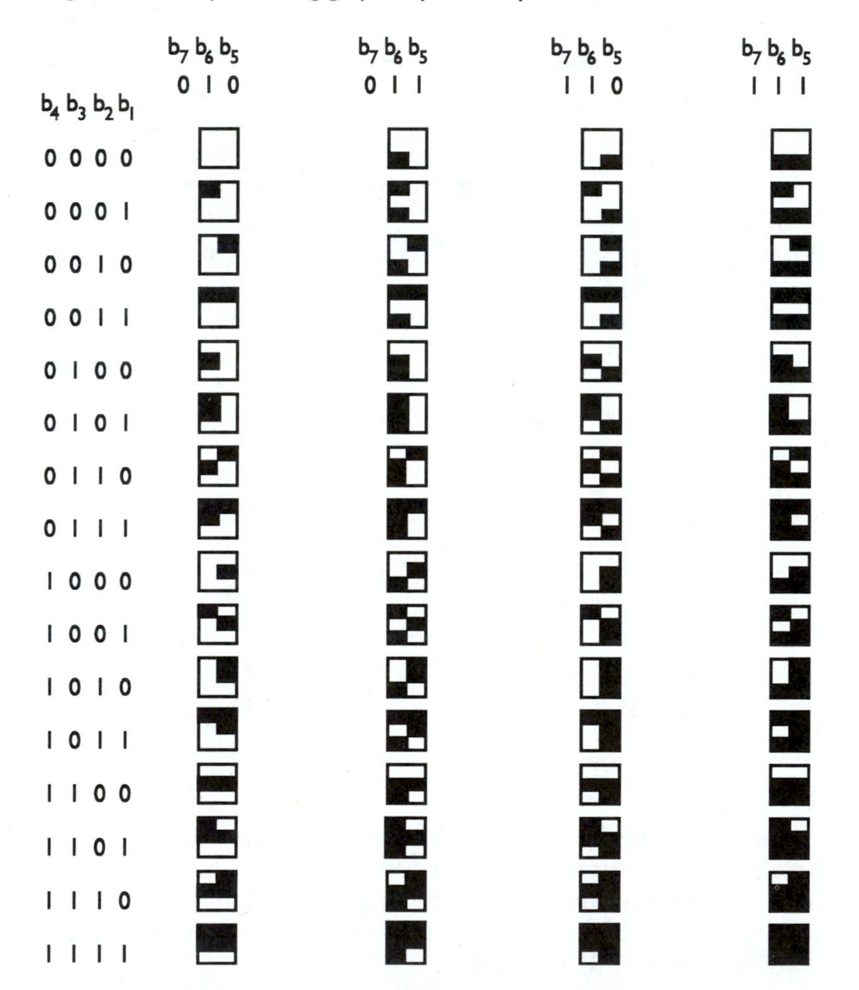

Complex pictures such as photographs cannot readily be reduced to a few fairly crude symbols. Often, the only way such a picture can be stored is as a *bit-map*. Any picture can be transformed into a rectangular array of pixels or picture elements. By analogy with the bit, a pixel is the smallest unit of information of which a picture is composed. Unlike a bit that can have only one of two values, the pixel can have attributes such as color. Figure 1.2 illustrates how an image is made up of pixels. Note that if the pixels are small enough, the image appears to be made of continuous lines and curves, rather than individual dots.

Figure 1.2 Representing an image as pixels

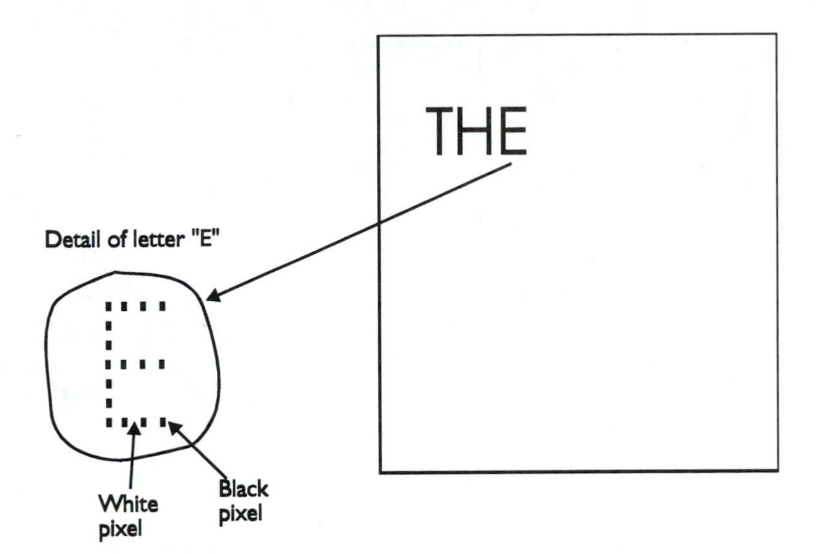

Consider an A4 size picture (approximately 210mm by 296mm). If we wish to store a reasonably high definition A4 picture, we must use approximately 12 pixels/mm in both the horizontal and vertical axes. That is, one square millimeter is made up of 12 x 12 = 144 pixels, and therefore the entire picture is composed of 210 x 296 x 144 = 8,951,040 pixels. This picture represents over 1 Mbyte of storage. If the picture were colored and each pixel could have one of 256 different colors, the total storage requirement would be over 8 Mbytes. From these parameters it should now be clear why high quality computer graphics require such expensive equipment. Typical high quality color video displays have a resolution of 1024 by 768 (i.e., approximately 2^{20} pixels) per frame. There are, however, rather complex techniques for compressing the amount of storage required by a picture. Such techniques operate by locating areas of a constant color and intensity and storing the shape and location of the area and its color. Encoding pictures in this way is similar to the ideas behind parameterizing that we discussed earlier.

- **A Signal** Time-varying analog signals such as those representing music or television images can be captured by a computer, stored in memory, and processed. Perhaps the most spectacular example of this is the CD player. The rapidly varying signal at the output of a microphone placed in front of an orchestra can be turned into a form suitable for processing in a computer by means of an *analog to digital converter* (ADC). Figure 1.3 illustrates the action of an ADC. The analog signal is periodically *sampled* by the ADC and the instantaneous amplitude of the signal recorded as a string of 1s and 0s. The rate at which the analog voltage is sampled (i.e., measured) and the number of bits in the string of 1s and 0s determine how precisely the signal is represented. Once the signal has been captured by the computer it can be manipulated

Figure 1.3 The analog to digital converter, ADC

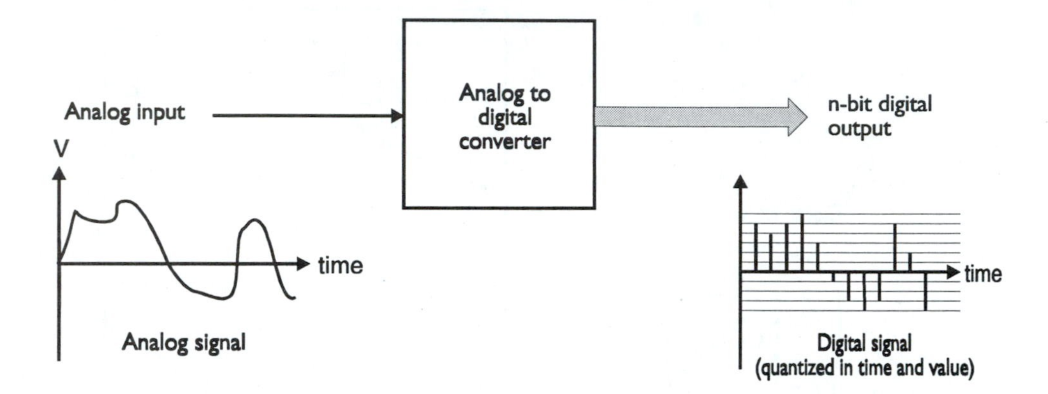

exactly like any other digital quantity. The sequence of binary values representing the analog signal can be fed to a *digital to analog converter* (DAC) and a new sound synthesized. We are now going to look at how numbers can be represented by binary patterns of 1s and 0s.

1.2 Number Bases

We represent numbers in the decimal system by means of *positional notation.* By positional we mean that the value or *weight* of a digit depends on its *location* within a number. As each digit moves one place left, it is multiplied by ten (the base or radix), and as it moves one place right it is divided by ten. Thus, the 9 in 95 is worth ten times the 9 in 59. So, the value 2455.143 is a number represented by the positional notation that we use automatically in everyday life. If this seems obvious and not worthy of mention, consider the Romans. In spite of all the Romans' achievements (e.g., writing Latin textbooks), their mathematics was terribly cumbersome. Because they did not use a positional system, each new large number had to have its own special symbol. Their number system was one of *give and take* so that if X = 10 and I = 1, then XI = 11 (i.e., 10 + 1) and IX = 9 (i.e., 10 - 1). In fact, the Romans did not use Roman numerals in their calculations — they used an abacus.

A number, N, when expressed in positional notation in the base b is written: $a_n a_{n-1} a_{n-2} \ldots a_1 a_0 . a_{-1} a_2 \ldots a_{-m}$ and is defined as:

$$N = a_n . b^n + a_{n-1} . b^{n-1} + \ldots + a_1 . b^1 + a_0 . b^0 + a_{-1} . b^{-1} + a_{-2} . b^{-2} + \ldots + a_{-m} . b^{-m}$$

$$= \sum_{i=-m}^{n} a_i b^i$$

The **a**'s in this equation are called *digits* and may have one of **b** possible values. For example, in the base b = 10, a digit may take any one of the values 0, 1, 2, 3, 4, 5, 6, 7, 8, or 9. Positional notation employs the *radix point* to separate the integer and fractional parts of the number. In decimal arithmetic (b = 10) we speak of the *decimal point* and in binary arithmetic (b = 2) we speak of the *binary point*.

Now let's look at some examples of how the above formula works. The decimal number 1982 is equal to $1 \times 10^3 + 9 \times 10^2 + 8 \times 10^1 + 2 \times 10^0$. Similarly, 12.34 is equal to $1 \times 10^1 + 2 \times 10^0 + 3 \times 10^{-1} + 4 \times 10^{-2}$. The value of the binary number 10110.11 is given by $1 \times 2^4 + 0 \times 2^3 + 1 \times 2^2 + 1 \times 2^1 + 0 \times 2^0 + 1 \times 2^{-1} + 1 \times 2^{-2}$, which is equal to the decimal value 16 + 4 + 2 + 0.5 + 0.25 = 22.75. If we decided to adopt base seven, the number 123 would be equal to the decimal value $1 \times 7^2 + 2 \times 7^1 + 3 \times 7^0 = 49 + 14 + 3 = 66$.

To be more precise in the use of our terminology, we should make it clear that we are talking about *natural* positional numbers here. The natural numbers have positional weights of 1, 10, 100, 1000, ... (decimal) or 1, 2, 4, 8, 16, 32, ... (binary). It is, in fact, perfectly possible to have weightings that are not successive powers of an integer. For example, we can choose a binary weighting of 2, 4, 4, 2 which means that the number 1010 is interpreted as 1x2 + 0x4 + 1x4 + 0x2 = 6.

Currently, those involved with computers are interested in four bases: *decimal, binary, octal,* and *hexadecimal.* The set of digits employed by each of these bases is provided in Table 1.2.

Table 1.2 Number bases and digit sets

Name of base	Base	Set of digits
Decimal	b = 10	a = {0,1,2,3,4,5,6,7,8,9}
Binary	b = 2	a = {0,1}
Octal	b = 8	a = {0,1,2,3,4,5,6,7}
Hexadecimal	b = 16	a = {0,1,2,3,4,5,6,7,8,9,A,B,C,D,E,F,}

People normally work in decimal and computers in binary. We shall see later that the purpose of the octal and hexadecimal systems is as an aid to human memory. It is almost impossible to remember long strings of binary digits. By converting them to the octal or hexadecimal bases (a very easy task), the shorter octal or hexadecimal numbers can be more readily committed to memory. Furthermore, as octal and hexadecimal numbers are more compact than binary numbers (1 octal digit = 3 binary digits and 1 hexadecimal digit = 4 binary digits), they are used in computer texts and core-dumps. The latter term refers to a print-out of part of the computer's memory, an operation normally performed as a diagnostic aid when all else has failed. For example, the eight-bit binary number 10001001 is equivalent to the hexadecimal number 89. Clearly, 89_{16} is easier to remember than

10001001_2. Since the octal base is hardly ever used by 68000 programmers, we shall not consider it further. In this text we usually employ a subscript to indicate the base — although we omit it when the base is entirely obvious.

There are occasions when binary numbers offer advantages over other forms of representation. Suppose a computer-controlled chemical plant has three heaters, three valves, and two pumps, that are designated H1,H2,H3, V1,V2,V3, P1,P2, respectively. An eight-bit word from the computer is fed to an interface unit that converts the binary ones and zeros into electrical signals that switch on (logical one), or switch off (logical zero), the corresponding device. For example, the binary word 01010011 has the effect described in Table 1.3 when presented to the control unit.

By inspecting the binary value of the control word, the status of all devices is immediately apparent. If the output had been represented in decimal (83), or hexadecimal (53), the relationship between the number and its intended action would not be so obvious.

Table 1.3 Decoding the binary string 01010011

Device	H1	H2	H3	V1	V2	V3	P1	P2
Bit	0	1	0	1	0	0	1	1
Status	off	on	off	on	off	off	on	on

How Many Bits Does It Take To Represent a Decimal Number? If we are going to represent decimal numbers in binary form, we need to know how many bits are required to express, say, an n-digit decimal number. Suppose we require m → n bits to represent the largest n-digit decimal number, which is, of course, 99...999 or $10^n - 1$.

We require the largest binary number in m bits (i.e., $11 \ldots 111_2$) to be equal to or greater than the largest decimal number in n bits (i.e., 99 ... 999). That is,

$$10^n - 1 \leq 2^m - 1 \qquad \text{i.e., } 10^n \leq 2^m$$

Taking logarithms to base ten we get:

$$\log_{10}10^n \leq \log_{10}2^m$$

$$n\log_{10}10 \leq m\log_{10}2$$

$$n \leq m\log_{10}2 \qquad \text{or } n \leq 0.30103m$$

$$m \leq 3.322n$$

≥

In other words, it takes approximately 3.3n bits to represent an n-bit decimal number. For example, if we wish to represent decimal numbers up to 1,000,000 in binary, we must use at least 6 x 3.3 bits, which indicates a 20-bit wordlength.

If there is one point that we would like to emphasize here, it is that the rules of arithmetic are the same in base *x* as they are in base *y*. In other words, all the rules we learned for base ten arithmetic can be applied to base two, base 16, or even base five arithmetic. For example, the base five numbers 123 and 221 represent $1x5^2 + 2x5^1 + 3x5^0 = 38_{10}$, and $2x5^2 + 2x5^1 + 1x5^0 = 61_{10}$, respectively. If we add 123_5 to 221_5 we get 344_5, which is equal to the decimal number $3x5^2 + 4x5^1 + 4x5^0 = 99_{10}$. Adding the decimal numbers 38_{10} and 61_{10} also gives us 99_{10}.

Now that we've looked at the structure of binary and decimal numbers, the next step is to consider how to convert a number in one base into another.

Number Base Conversion

We sometimes have to convert numbers from one base to another by means of a pencil-and-paper method. This statement is particularly true when working with microprocessors at the assembly language or the machine code level. In general, computer users need not concern themselves with conversion between number bases, as the computer has software to convert a decimal input into its own internal binary representation of the input. Once the computer has done its job, it converts the binary results into decimal form before printing them.

A knowledge of the effect of number bases on arithmetic operations is sometimes quite vital, as, for example, even the simplest of decimal fractions (e.g., $1/10_{10} = 0.1_{10}$) have no exact binary equivalent. Suppose the computer were asked to continue to add 0.1_{10} to zero and stop when the result reached one. The computer might never stop, because the decimal value 0.1_{10} cannot be represented exactly by a *binary* number. That is, the sum of ten binary representations of 0.1_{10} is never exactly 1. It may be 1.00000000001 or 0.99999999999, which is almost as good as 1, but it is not the same as 1, and a test for equality with 1 will always fail.

Conversion of Integers

In this section we are going to demonstrate how integers are converted from one base to another.

Decimal to Binary To convert a decimal integer to binary, divide the number successively by 2, and after each division, record the remainder which is either 1 or 0. The process is terminated only when the result of the division is 0 remainder 1. Note that in all the following conversions *R* is the remainder after a division.

For example, 123_{10} becomes:

```
123 ÷ 2 = 61     R = 1
 61 ÷ 2 = 30     R = 1
 30 ÷ 2 = 15     R = 0
 15 ÷ 2 =  7     R = 1
  7 ÷ 2 =  3     R = 1
  3 ÷ 2 =  1     R = 1
  1 ÷ 2 =  0     R = 1
```

The result is read from the most-significant bit (the last remainder) upward to give $123_{10} = 1111011_2$.

Decimal to Hexadecimal Decimal numbers are converted from decimal into hexadecimal form in exactly the same way that decimal numbers are converted into binary form. However, in this case the remainder lies in the decimal range 0 to 15, corresponding to the hexadecimal range 0 to F.

For example, 53241_{10} becomes:

$$53241 \div 16 = 3327 \quad R = 9$$
$$3327 \div 16 = 207 \quad R = 15_{10} = F_{16}$$
$$207 \div 16 = 12 \quad R = 15_{10} = F_{16}$$
$$12 \div 16 = 0 \quad R = 12_{10} = C_{16}$$

Therefore, $53241_{10} = CFF9_{16}$.

Binary to Decimal It is possible to convert a binary number to decimal by adding together all its "1" bits weighted by the requisite powers of two. This technique is suitable for small binary numbers up to about seven or eight bits. For example, 1010111_2 is represented by:

```
64  32  16  8  4  2  1        64
 1   0   1  0  1  1  1  =     16
                               4
                               2
                             + 1
                              87
```

A more methodical technique is based on a *recursive* algorithm as follows. Take the left-most non-zero bit, double it, and add it to the bit on its right. Now take this result, double it, and add it to the next bit on the right. Continue in this way until the least-significant bit has been added in. This recursive procedure may be expressed mathematically as: $(a_0 + 2(a_1 + 2(a_2 + ...)))$, where the least-significant bit of the binary number is a_0. For example, 1010111_2 becomes:

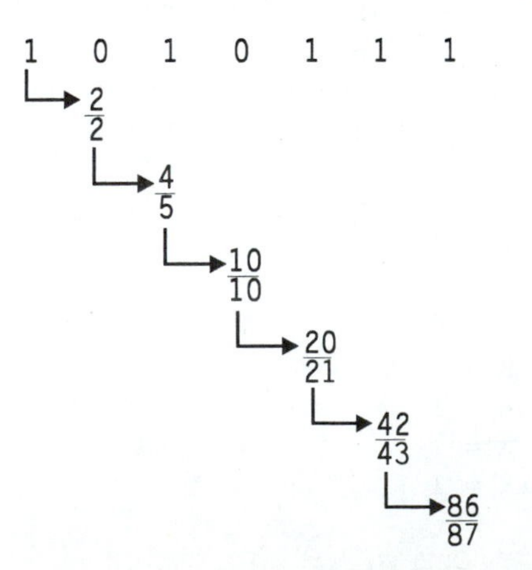

Therefore, $1010111_2 = 87_{10}$.

Hexadecimal to Decimal This method is identical to the procedure for binary, except that 16 is used as a multiplier. For example, $1AC_{16}$ becomes:

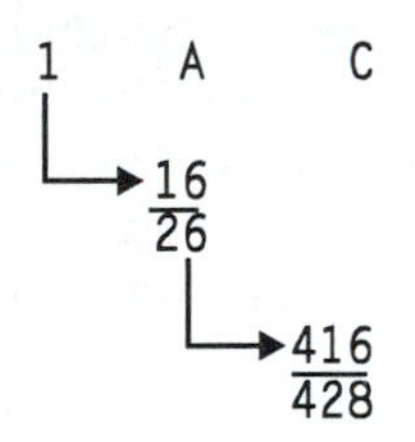

Therefore, $1AC_{16} = 428_{10}$.

Conversions between Binary and Hexadecimal

In much of this book, binary numbers are represented in hexadecimal form. Although some texts favor the octal format, octal numbers are ill-fitted to the representation of 8- or 16-bit binary values. We shall use hexadecimal representations of binary numbers simply because of the ease with which conversions may be made between binary and hexadecimal numbers.

Binary to Hexadecimal The binary number is formed into groups of four bits starting at the decimal point. Each group is replaced by a hexadecimal digit from 0 to 9, A, B, C, D, E, F.

For example, 011001011101_2 becomes 0110 0101 1101.
6 5 D

Therefore, $011001011101_2 = 65D_{16}$.

Hexadecimal to Binary Each hexadecimal digit is replaced by its four-bit binary equivalent.

For example, $AB4C_{16}$ becomes A B 4 C
1010 1011 0100 1100

Therefore, $AB4C_{16} = 1010101101001100_2$.

Conversion of Fractions

The conversion of fractions from one base to another is carried out in a similar way to the conversion of integers, although it's rather more tedious to manipulate fractions manually. Fortunately, you rarely have to perform the pencil-and-paper conversion of fractions outside the classroom. One way of effectively abolishing fractions is to treat all fractions as integers scaled by an appropriate factor. For example, the binary fraction 0.10101 is equal to the integer 10101 divided by 2^5 (i.e., 32), so that, for example, 0.10101 is the same as $10101/2^5 = 21/32 = 0.65625$.

Converting Binary Fractions to Decimal Fractions The algorithm for converting binary fractions to their decimal equivalent is based on the fact that a bit in one column is worth half the value of a bit in the column on its left. Starting at the right-most non-zero bit, take that bit and halve it. Now add the result to the next bit on its left. Halve this result and add it to the next bit on the left. Continue until the binary point is reached.

For example, consider the conversion of 0.01101_2 into decimal form.

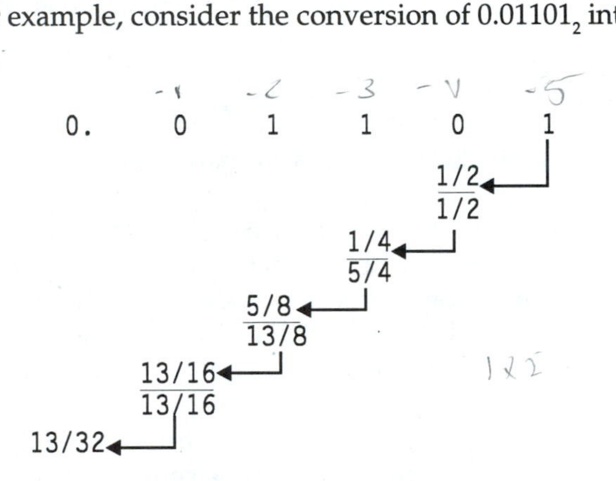

Therefore, $0.01101_2 = 13/32 = 0.40625_{10}$.

Converting Decimal Fractions to Binary Fractions The decimal fraction is multiplied by two and the integer part noted. The integer, which will be either 1 or 0, is then stripped from the number to leave a fractional part. The new fraction is multiplied by two and the integer part noted. We continue in this way until the process ends, or a sufficient degree of precision has been achieved. The binary fraction is formed by reading the integer parts from the top to the bottom as the following example illustrates.

For example, 0.6875_{10} becomes:

```
0.6875 x 2  →  1.3750
0.3750 x 2  →  0.7500
0.7500 x 2  →  1.5000
0.5000 x 2  →  1.0000
0.0000 x 2  →  0.0000  ends the process
```

Therefore, $0.6875_{10} = 0.1011_2$. Now consider the conversion of 0.1_{10} into a binary fraction.

```
0.1000 x 2  →  0.2000
0.2000 x 2  →  0.4000
0.4000 x 2  →  0.8000
0.8000 x 2  →  1.6000
0.6000 x 2  →  1.2000
0.2000 x 2  →  0.4000
0.4000 x 2  →  0.8000
0.8000 x 2  →  1.6000
0.6000 x 2  →  1.2000
0.2000 x 2  →  0.4000
```
etc.

Therefore, $0.1_{10} = 0.0001100110_2$ to ten binary places. As we pointed out before, 0.1_{10} cannot be expressed *exactly* in terms of binary fractions. However, if you wish to convert a fractional decimal number into its binary equivalent, you should select sufficient significant figures to make the precision of the binary value the same as that of the decimal value.

Converting between Hexadecimal Fractions and Decimal Fractions Hexadecimal fractions can be converted into decimal fractions by means of the same algorithm used for binary conversions. All we have to change is the base (i.e., 2 to 16). Consider the conversion of 0.123_{16} into a decimal fraction.

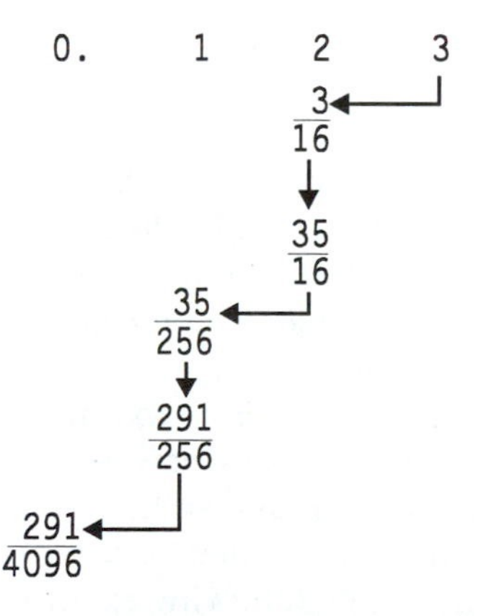

Binary to Hexadecimal Fraction Conversion and Vice Versa The conversion of binary fractions to hexadecimal format is as easy as the corresponding integer conversions. The only point worth mentioning is that when binary digits are split into groups of four, we start grouping bits at the binary point and move to the *right*. Any group of digits remaining on the right containing fewer than four bits must be made up to four bits by the addition of zeros to the right of the least-significant bit. The following examples illustrate this point.

Binary to hexadecimal $0.10101100_2 \rightarrow 0.1010\ 1100_2$
$\rightarrow 0.AC_{16}$

Binary to hexadecimal 0.101011001_2 → $0.1010\ 1100\ 1(000)_2$
→ $0.AC8_{16}$

Hexadecimal to binary $0.ABC_{16}$ → 0.1010 1011 1100
→ 0.101010111100_2

Numbers containing an integer part and a fractional part (e.g., 110101.11010_2 or 123.125_{10}) are converted from one base to another in two stages. The integer part is first converted and then the fractional part (using any appropriate conversion technique). For example, consider the conversion of the decimal value 129.2 into binary form. Assume that the result is a 16-bit value with 8 bits before the binary point and 8 bits after it. The integer part, 129, is given by 10000001, and the fractional part by .00110011 (in 8 bits). Therefore, 129.2 is expressed as 1000000100110011 in 16 bits.

Special Purpose Codes

Throughout this book a group of binary digits generally represents one of three things: a numerical quantity, an instruction, or a character. However, in the world of computing and digital systems many different codes are widely used, each one best suited to the particular job for which it was designed.

BCD Codes

A particularly common code is called BCD or *Binary Coded Decimal*. In theory BCD is a case of having your cake and eating it. Computer designers are forced to rely on two-state logic elements on purely economic grounds. This, in turn, leads to the world of binary arithmetic and the consequent problems of converting between binary and decimal representations of numeric quantities. Binary coded decimal numbers accept the inevitability of two-state logic by coding the individual decimal digits into groups of four bits. Table 1.4 shows how the ten digits, 0 to 9, are represented in BCD, and how a decimal number is converted to a BCD form.

BCD arithmetic is identical to decimal arithmetic and differs only in the way in which the ten digits are represented. The following example demonstrates how a BCD addition is carried out.

```
 1942     0001 1001 0100 0010
+2379    +0010 0011 0111 1001
 4321     0100 0011 0010 0001
```

Although BCD seems a good idea because it makes decimal to binary conversion easy, it suffers from two disadvantages. The first is that BCD arithmetic is more complex than binary arithmetic, simply because the binary tables (i.e., addition, subtraction, multiplication, and division) are exceedingly small and may be

Table 1.4 The BCD code

Decimal	BCD
0	0000
1	0001
2	0010
3	0011
4	0100
5	0101
6	0110
7	0111
8	1000
9	1001

To convert a decimal number into its BCD equivalent, you just translate each decimal digit into its appropriate 4-bit BCD code. For example, the decimal number 1942 is encoded as 0001 1001 0100 0010. After compaction, this is written 0001100101000010.

implemented in hardware by a few gates. On the other hand, the decimal tables involve all combinations of the digits 0 to 9.

The major disadvantage of BCD lies in its inefficient use of storage. A BCD digit requires four bits of storage but only ten symbols are mapped onto ten of the sixteen possible binary codes. Consequently, the binary codes 1010_2 to 1111_2 (10 to 15) are redundant and represent wasted storage. As we demonstrated earlier in this chapter, natural binary numbers require an average of approximately 3.322 bits per decimal digit.

In spite of the disadvantages of BCD, it is frequently found in applications requiring little storage, such as pocket calculators or digital watches. Microprocessors often have special instructions to aid BCD operations, and some interpreters for the language BASIC perform all numeric operations on BCD numbers. There are, in fact, a number of different ways of representing BCD numbers in addition to the basic BCD code presented above. Each of these codes has desirable properties making it suitable for a particular application. We will look at BCD again when we describe the 68000's special BCD instructions.

1.3 Binary Arithmetic

Now that we've introduced binary numbers and demonstrated how it's possible to convert between binary and decimal formats, the next step is to look at how binary numbers are manipulated. Binary arithmetic follows exactly the same rules as decimal arithmetic and all we have to do to work with binary numbers is to learn the binary tables. These tables are somewhat easier than their decimal equivalents, as Table 1.5 demonstrates.

A remarkable fact about binary arithmetic is that if we did not worry about the carry in addition and the borrow in subtraction, then the operations of addition and subtraction would be identical. Such an arithmetic in which addition and

Table 1.5 The binary tables

Addition	Subtraction	Multiplication
0 + 0 = 0	0 - 0 = 0	0 x 0 = 0
0 + 1 = 1	0 - 1 = 1 borrow 1	0 x 1 = 0
1 + 0 = 1	1 - 0 = 1	1 x 0 = 0
1 + 1 = 0 carry 1	1 - 1 = 0	1 x 1 = 1

subtraction are equivalent does exist and has some important applications; this is called modulo-two arithmetic.

The addition of n-bit numbers is entirely straightforward, except that when adding the two bits in each column, a carry bit from the previous stage on the right must also added in. In the following example, we present the numbers to be added on the left and, on the right, we include the carry bits that must be added in.

```
 00110111     00110111                       55
+01010110     01010110                      +86
              111 11      ← carries
 10001101     10001101                      141
```

Hexadecimal addition is performed in the same way as binary addition. All we have to remember is that if we add together two digits whose sum is greater than 16, we must convert the result into a carry digit whose value is 16 plus the remainder (which is the sum less 16). For example, if we add 9_{16} and B_{16}, we get $14_{16} = 20_{10}$ which is 4 carry 1 (i.e., 4 plus 16). Consider the hexadecimal addition $9A345_{16} + 6701B_{16}$.

```
 9A345        9A345
 6701B        6701B
              11  1    ←  carries
101360       101360
```

Subtraction can also be carried out in a conventional fashion, although computers do not subtract numbers in the same way we do. Negative numbers are not usually represented in a sign plus magnitude form. Instead, computers employ *complementary arithmetic*, as we shall soon see. Consider the following conventional pencil-and-paper subtractions.

```
 01010110                        86
-00101010                       -42
  1 1       ←  borrows           44
 00101100
```

The multiplication of binary numbers can be done by the pencil-and-paper method of shifting and adding, although in practice the computer uses a somewhat modified technique.

```
     01101          13
   x 01010        x 10
     00000         130
    01101
   00000
  01101
 00000
0010000010
```

Signed Numbers

Any real computer must be able to deal with *negative* numbers as well as positive numbers. Before we examine how the computer handles negative numbers, we should consider how we deal with them. I believe that people do not, in fact, actually use negative numbers. They use positive numbers (the '5' in -5 is the same as in +5), and place a negative sign in front of the number to remind them that it must be treated in a special way when it takes part in arithmetic operations. In other words we treat all numbers as positive and use a sign (i.e., + or -) to determine what we have to do with the numbers. For example, consider the following two operations.

```
 8          8
+5   and   -5
13          3
```

In each of these examples the numbers are the same, but the operations we performed on them were different; in the first case we *added* them together and in the second case we *subtracted* them. This technique can be extended to computer arithmetic to give the sign and magnitude representation of a negative number.

Sign and Magnitude Representation An n-bit word can have 2^n possible different values from 0 to 2^n-1. For example, an eight-bit word can represent the numbers 0, 1,..., 254, 255. One way of representing a negative number is to take the most-significant bit and reserve it to indicate the sign of the number. The usual convention is to choose the sign bit to be 0 to represent positive numbers and 1 to represent negative numbers. We can express the value of a sign and magnitude number in the form $(-1)^S \times M$, where S is the sign bit of the number and M is its magnitude. If S = 0, $(-1)^0 = +1$ and the number is positive. Conversely, if S = 1, $(-1)^1 = -1$ and the number is negative. For example, in 8 bits we can interpret the two numbers 00001101 and 10001101 as:

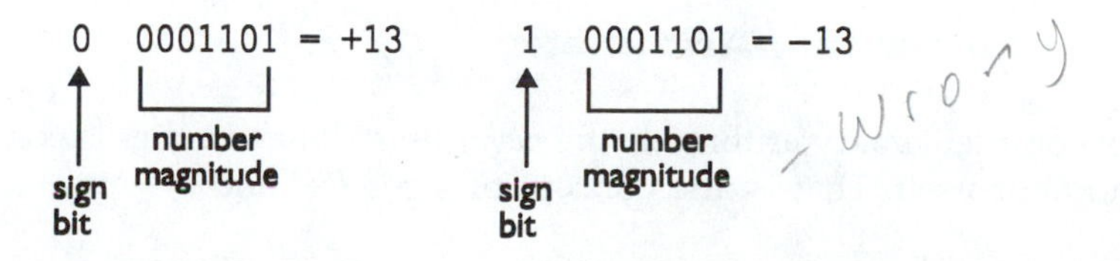

Using a sign and magnitude representation is a perfectly valid way to represent signed numbers, although it is not widely used in integer arithmetic. The range of an n-bit sign and magnitude number is given by:

$$-(2^{n-1} - 1) \text{ to } +(2^{n-1} - 1) \quad \text{or} \quad -2^{n-1} + 1 \text{ to } 2^{n-1} - 1$$

All we have done is to take an n-bit number, use one bit to represent the sign and let the remaining n - 1 bits represent the number. Thus, an eight-bit number represents values from -127_{10} (11111111_2) to $+127_{10}$ (01111111_2). One of the objections to this system is that it has two values for zero:

00000000 = +0 and 10000000 = -0

Let's look at four examples of addition and subtraction in sign and magnitude arithmetic. Remember that the most-significant bit is a sign bit and doesn't take part in the calculation itself. This is in contrast with two's complement arithmetic (see later) in which the sign bit forms an integral part of the number when it is used in calculations. In each example, we perform the calculation by first converting the sign-bit to a plus sign or to a minus sign. Then we carry out the appropriate calculation and, finally, convert the sign of the result into a sign-bit.

	Initial sign and magnitude value	Number with the sign bit converted into a sign	Result with sign converted into a sign bit	Decimal equivalent
1.	001011	+01011		11
	+001110	+01110		+14
		+11001	→ 011001	25
2.	001011	+01011		11
	+100110	-00110		-6
		+00101	→ 000101	+5
3.	001011	+01011		11
	+110110	-10110		-22
		-01011	→ 101011	-11
4.	001011	+01011		11
	-001001	-01001		-9
		+00010	→ 000010	+2

Complementary Arithmetic

In complementary arithmetic the *negativeness* of a number is contained within the number itself. That is, the concept of signs ('+' and '-') may, effectively, be dispensed with. If we add X to Y the operation is that of addition if X is positive and Y is positive, but if Y is negative the end result is that of subtraction (assuming that Y is represented by its negative form). It is important to point out here that complementary arithmetic is used to represent and to manipulate both positive and negative numbers. To demonstrate that there is nothing magical about complementary arithmetic we first examine decimal complements.

Ten's Complement Arithmetic The 10's complement of an n-digit decimal number, N, is defined as $10^n - N$. The 10's complement may also be calculated by subtracting each of the digits of N from 9 and adding 1 to the result. Consider the four-digit decimal number 1234. Its 10's complement is:

```
a.  10^4 - 1234 = 8766        b.   9999
                                  -1234
                                   8765 + 1 = 8766
```

Suppose we were to add this complement to another number, say, 8576. We get:

```
  8576
+ 8766
 17342
```

Now let's examine the effect of subtracting 1234 from 8576 by conventional means.

```
  8576
- 1234
  7342
```

The results of the two operations are identical in the least-significant four digits, but differ in the fifth digit by 10^4. The reason for this is not hard to find. Consider the subtraction of Y from X. We calculate Z = X - Y by adding the 10's complement of Y to X. The 10's complement of Y is defined as 10^4 - Y. Therefore:

$$Z = X + (10^4 - Y) = 10^4 + (X - Y).$$

In other words, we get the desired result, X - Y, together with an unwanted digit in the left-most position. This digit may be discarded. Note that complementing a number twice results in the original number. For example, $-1234 = 10^4 - 1234 = 8766$. Complementing 1234 twice, we get: $-(-1234) = -(8766) = 10^4 - 8766 = 1234$.

Two's Complement Representation The equivalent of 10's complement in binary arithmetic is two's complement. To calculate the two's complement of an n-bit binary number, N, we evaluate $2^n - N$. For example, in 5 bits, if $N = 5 = 00101_2$, then the two's complement of N is given by 2^5 - 00101 = 100000 - 00101 = 11011_2. It is important to note that 11011_2 represents -00101_2 (-5) or +27 depending only on whether we interpret the bit pattern 11011 as a two's complement integer or as an unsigned integer.

Adding the two's complement of N (i.e., 11011) to another binary number automatically carries out the operation of subtraction. In the following demonstration we add 11011 to 01100 (i.e., 12).

```
   01100          12
 + 11011       + (-5)
  100111           7
```

As in the case of the decimal example in 10's complement arithmetic, we get the correct answer together with the $2^n = 2^5$ term which is discarded. Before continuing further, it is worthwhile examining the effect of adding all the combinations of positive and negative values for a pair of 5-bit numbers.

Let X = 9 = 01001_2 and Y = 6 = 00110_2.

-X = 100000 - 01001 = 10111
-Y = 100000 - 00110 = 11010

```
1. +X       01001              2. +X        01001
   +Y     + 00110                 -Y      + 11010
            01111 = 15                     100011 = +3

3. -X       10111              4. -X        10111
   +Y     + 00110                 -Y      + 11010
            11101 = -3                     110001 = -15
```

All four examples give us the result we would expect when the result is interpreted as a two's complement number. However, as Examples 3 and 4 give negative results, they may require a little further explanation. The result in Example 3 is -3. The two's complement representation of -3 is 100000 - 00011 = 11101. Similarly, in Example 4 the two's complement representation of -15 is 100000 - 01111 = 10001.

Example 4 evaluates -X + - Y to give -15 but with the addition of a 2^n term. Because both numbers are negative, we have $(2^n - X) + (2^n - Y) = 2^n + (2^n - X - Y)$. The first term is the redundant 2^n and the second part is the two's complement representation of -X - Y. We can now see that the two's complement system works for all possible combinations of positive and negative numbers.

Calculating Two's Complement Values The two's complement system would not be so attractive if it were not for the ease with which two's complements can be formed. Consider the two's complement of N, which is defined as: $-N \rightarrow 2^n - N$. In this case we are using the symbol $\rightarrow$ to indicate "is represented by."

Suppose we rearrange the equation by subtracting 1 from the 2^n and adding it to the result.

```
-N → (2^n  - 1)   - N + 1
   → 111 ... 1 - N + 1
     └───────┘
      n places
```

For example, in 8 bits (n = 8) we have:

```
-N → (2^8  - 1)
   → 100000000 - N
   → 100000000 - 1 - N + 1  (after rearranging)
   → 11111111  - N + 1
```

The evaluation of the two's complement of N becomes particularly easy because, if a bit of N is 0, subtracting it from 1 gives 1, and if the bit is 1, subtracting it from 1 gives 0. In other words, $1 - N_i = \overline{N_i}$. That is, to form the two's complement of a number we simply invert the bits and add 1. For example, in five bits we have:

$7 = 00111$
$-7 = \overline{00111} + 1 = 11000 + 1 = 11001$

Evaluating two's complement numbers in the above fashion is attractive because it is easy to perform with hardware.

Properties of Two's Complement Numbers

The two's complement system is a true complement system so that +X + (-X) = 0. For example, in five bits +13 = 01101 and -13 = 10011. The sum of +13 and -13 is:

```
 01101
+10011
100000 = 0
```

- There is one unique zero, 00...0.
- If the number is positive the most-significant bit, MSB, is 0, and if it is negative the MSB is 1. Thus, the MSB is a sign bit.
- The range of two's complement numbers in n bits is from $-2^{(n-1)}$ to $+2^{(n-1)} - 1$. For n = 5, this range is from -16 to +15. The total number of *different* numbers is 32 (16 negative, zero, and 15 positive). That is, a five-bit number can uniquely

describe 32 items, and we are free to choose to call these items the natural binary integers 0 to 31, or the signed two's complement numbers -16 to +15.

- The complement of the complement of X is X (i.e., -(-X) = X). In 5 bits +12 = 01100 and -12 = 10011 + 1 = 10100. If we form the two's complement of -12 (i.e., 10100) in the usual fashion by inverting the bits and adding 1, we get: $\overline{10100} + 1 \rightarrow 01011 + 1 \rightarrow 01100$, which is the number we started with.

Let's now see what happens if we violate the range of two's complement numbers. That is, we will carry out an operation whose result falls outside the range of values that can be represented by two's complement numbers. If we choose a five-bit representation, we know that the range of valid signed numbers is -16 to +15. Suppose we first add 5 and 6 and then try 12 and 13.

Case 1		**Case 2**		
5	= 00101	12	= 01100	
+ 6	= 00110	+13	= 01101	
11	01011 = 11_{10}	25	11001	= -7_{10} (as a two's complement number)

In Case 1 we get the expected answer of $+11_{10}$, but in Case 2 we get a negative result because the sign bit is '1'. If the answer were regarded as an unsigned binary number it would be +25, which is, of course, the correct answer. Once the two's complement system has been chosen to represent signed numbers, all answers must be interpreted in this light.

Similarly, adding together two negative numbers whose total is less than -16 also goes out of range. For example, adding $-9_{10} = 10111_2$ to $-12_{10} = 10100_2$, gives:

-9	10111	
-12	+10100	
-21	101011	gives a positive result, $01011_2 = +11_{10}$

Both of these cases represent a condition called *arithmetic overflow*. Arithmetic overflow occurs during a two's complement addition if the result of adding two positive numbers yields a negative result, or if the result of adding two negative numbers yields a positive result. Overflow represents an out of range condition and can result from operations other than addition (e.g., multiplication and division). Arithmetic overflow during an addition can be expressed algebraically. If we let a_{n-1} be the sign bit of A, b_{n-1} be the sign bit of B, and s_{n-1} be the sign bit of the sum of A and B, then:

$$V = \overline{a_{n-1}}\,\overline{b_{n-1}}\,s_{n-1} + a_{n-1}\,b_{n-1}\,\overline{s_{n-1}}$$

If the sign bits of A and B are the same, but the sign bit of the result is different, arithmetic overflow has occurred. Arithmetic overflow is a consequence of two's complement arithmetic, and should not be confused with carry-out, which is the carry bit generated by the addition of the two most-significant bits of the numbers.

An Alternative View of Two's Complement Numbers We have seen that a binary integer, N, lying in the range $0 \leq N < 2^{(n-1)}$, is represented in a negative form by the expression $2^n - N$. We have also seen that this expression can be readily evaluated by inverting the bits of N and adding 1 to the result. Another way of looking at a two's complement number is to regard it as a conventional binary number represented in the positional notation but with the sign of the most-significant bit negative. That is:

$$-N = -d_{n-1}2^{n-1} + d_{n-2}2^{n-2} + \ldots + d_0 2^0,$$

where d_{n-1}, d_{n-2}, ... d_0 are the bits of the two's complement number. Consider the representation of 14, and the two's complement form of -14, in five bits.

$$+14 = 01110$$
$$-14 = 2^n - N = 2^5 - 14 = 32 - 14 = 18 = 10010_2$$
or $$-14 = \overline{01110} + 1 = 10001 + 1 = 10010_2$$

From what we have said earlier, we can regard the two's complement representation of -14 (i.e., 10010) as:

$$-1 \times 2^4 + 0 \times 2^3 + 0 \times 2^2 + 1 \times 2^1 + 0 \times 2^0$$
$$= -16 + (0 + 0 + 2 + 0)$$
$$= -16 + 2 = -14$$

So far we have looked at binary integers and fractions, and at the representation of negative numbers. Now we are going to describe how very large and very small numbers are represented and manipulated by computers.

1.4 Floating Point Numbers

Before we introduce floating point numbers themselves, we are going to show how *fixed-point* numbers are handled by a computer. Consider the following two calculations in decimal arithmetic.

Case 1 Integer arithmetic	**Case 2 Fixed-point arithmetic**
7632135	763.2135
+ 1794821	+ 179.4821
9426956	942.6956

Although Case 1 uses *integer* arithmetic and Case 2 uses *fractional* arithmetic, the calculations are entirely identical. We can extend this principle to computer arithmetic. All the computer programmer has to do is remember where the binary point is assumed to lie. All input to the computer is scaled to match this convention and all output is similarly scaled. The internal operations themselves are carried out as if the numbers were in integer form. This arrangement is called *fixed-point* arithmetic, because the binary point is assumed to remain in the same position. That is, there are always the same number of digits before and after the binary point. The advantage of the fixed-point representation of numbers is that no complex software or hardware is needed to implement it.

A simple example should clarify the idea of fixed-point arithmetic. Consider an eight-bit fixed-point number with the four most-significant bits representing the integer part and the four least-significant bits representing the fractional part. Let's see what happens if we wish to add the two numbers 3.625 and 6.5, and print the result. An input program first converts these numbers to binary form.

$3.625_{10} \rightarrow 11.101_2 \rightarrow 0011.1010_2$ (in 8 bits)

$6.5_{10} \rightarrow 110.1_2 \rightarrow 0110.1000_2$ (in 8 bits)

The computer now regards these numbers as 00111010 and 01101000, respectively. Remember that the binary point is only imaginary. These numbers are added in the normal way to give:

```
 00111010
+01101000
 10100010
```

$10100010_2 = 162_{10}$ (if interpreted as unsigned binary)

The output program takes the result and splits it into an integer part 1010, and a fractional part .0010, and prints the correct answer 10.125. A fixed-point number may be spread over several words to achieve a greater range of values than allowed by a single word. The fixed-point representation of fractional numbers is very useful in some circumstances, particularly for financial calculations. For example, the smallest fractional part might be $0.001 and the largest integer part $999,999. We require a total of 6x4 + 3x4 = 36 bits to represent such a quantity in BCD. In a byte-oriented computer five bytes would be needed for each number.

Fixed-point numbers have their limitations. What about the astrophysicist who is examining the behavior of the sun? An astrophysicist is confronted with quantities such as the mass of the sun (1990000000000000000000000000000000 grams) and the mass of an electron (0.000000000000000000000000000910956 grams).

If astrophysicists were to resort to fixed-point arithmetic, they would need to take an extravagantly large number of bytes to represent a wide range of numbers. A single byte represents numbers in the range 0 to 255, or approximately 0 to ¼ thousand. If the physicist wanted to work with astronomically large and microscopically small numbers, roughly 14 bytes would be required for the integer part of the number and 12 bytes for the fractional part — a 26 byte (208-bit) number. A clue to a way out of our dilemma is to note that both numbers contain a large number of zeros but few *significant* digits.

Representation of Floating Point Numbers

Digital computers often represent and store numbers in a *floating point* format. Just as we represent the decimal number 1234.56 by 0.123456×10^4, the computer handles binary numbers in a similar way. For example, 1101101.1101101 may be represented internally as $0.11011011101101 \times 2^7$ (the 7 is, of course, also stored in a binary format). Floating point notation is sometimes called *scientific notation*. Before looking at floating point numbers in more detail we need to consider the ideas of *range*, *precision*, and *accuracy* that are closely related to the way numbers are represented in floating point format.

Range The range of a number tells us how big or how small it can be. In the example of the astrophysicist we were dealing with numbers as large as 2×10^{33} to those as small as 9×10^{-28}, representing a range of approximately 10^{61}, or 61 decades. The range of numbers represented in a digital computer must be sufficient for the vast majority of calculations that are likely to be performed. If the computer is to be employed in a dedicated application where the range of data to be handled is known to be quite small, then the range of valid numbers may be restricted, simplifying the hardware/software requirements.

Precision The precision of a number is a measure of its *exactness* and corresponds to the number of significant figures. For example, π may be written as 3.142 or 3.141592. The latter case is more precise than the former, because it represents π to one part in 10^7 while the former represents to one part in 10^4.

Accuracy Accuracy has been included here largely to contrast it with precision, a term often incorrectly thought to mean the same as accuracy. Accuracy is a measure of the *correctness* of a quantity. For example, we can say $\pi = 3.141$ or $\pi = 3.241592$. In the former case we have a low precision number that is more accurate than its higher precision neighbor (because the value 3.241592 has a typographic error in the first decimal place). In an ideal world, accuracy and precision would go hand-in-hand. It is up to the computer programmer to design numerical algorithms that preserve the accuracy that the available precision allows. One of the potential hazards of computation is calculations of the form:

$$\frac{A + B}{A - B}$$

Consider the following example.

$$\frac{1234.5687 + 1234.5678}{1234.5687 - 1234.5678} = \frac{2469.1365}{0.0009}$$

When the denominator of the expression is evaluated we are left with 0.0009, a number with only one decimal place of precision. Although the calculation shows eight figures of precision, it may be very inaccurate indeed.

A floating point number is represented in the form: a x r^e, where a is the mantissa (also called an argument), r is the radix or base, and e is the exponent or characteristic. The way in which a computer stores floating point numbers is by dividing the binary sequence representing the number into the two fields illustrated in Figure 1.4. The radix *r* is understood and need not be stored explicitly by the computer. Throughout the remainder of this section the value of the radix in all floating point numbers is assumed to be two. In some computers the radix of the exponent is octal or hexadecimal, so that the mantissa is multiplied by 8^e or 16^e, respectively.

Figure 1.4 Storing a floating point number as exponent and mantissa

Exponent e	Mantissa m	Represents m x 2^e

It is not necessary for a floating point number to occupy a single storage location. Often a number of words are grouped to form the floating point number as described in Figure 1.4. The division between exponent and mantissa need not fall at a word boundary. That is, a mantissa might occupy three bytes and the exponent one byte of a floating point number made up of two consecutive 16-bit words.

Normalization of Floating Point Numbers

By convention the floating point mantissa is always *normalized* (unless it is equal to zero) so that it is expressed in the form 0.1 ... x 2^e. For the moment we are considering positive mantissas only. If the result of a calculation were to yield 0.01 ... x 2^e, the result would be normalized to give 0.1 ... x 2^{e-1}. Similarly, the result 1.01 ... x 2^e would be normalized to 0.101 ... x 2^{e+1}.

By normalizing a mantissa, the greatest advantage is taken of the available precision. For example, the unnormalized 8-bit mantissa 0.00001010 has only four significant bits, while the normalized 8-bit mantissa 0.10100011 has eight significant bits. A positive, floating point, normalized binary mantissa x is of the form:

x = 0.100 ... 0 to 0.11 ... 11

That is, $\frac{1}{2} \leq x < 1$. A special exception has to be made in the case of zero as it cannot, of course, be normalized. A negative, two's complement, floating point mantissa is stored in the form:

x = 1.01 ... 1 to 1.00 ... 0

In this case the negative mantissa, x, is constrained so that $-\frac{1}{2} > x \geq -1$. The floating point number is therefore limited to one of the three ranges described by Figure 1.5. Another representation of floating point mantissas (the IEEE format) uses a *sign and magnitude* representation so that normalized mantissas lie in the

Figure 1.5 Range of valid normalized two's complement mantissas

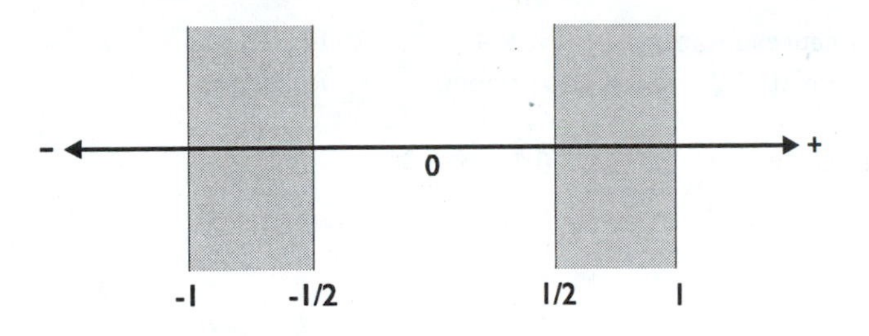

range 1.0000 ... 00 to 1.1111 ... 11 for positive numbers and from -1.0000 ... 00 to -1.1111 ... 11 for negative numbers. That is, a mantissa x is constrained to lie in the ranges: $-2 < x \leq -1$ or $x = 0$ or $1 \leq x < 2$.

Biased Exponents

A floating point representation of numbers must make provision for both positive and negative *numbers*, and positive and negative *exponents*. Consider, for example, the following four decimal numbers.

$+0.123 \times 10^{12}$, -0.756×10^{9}, $+0.176 \times 10^{-3}$, -0.459×10^{-7}

The mantissa of a floating point number is sometimes represented as a two's complement number. The exponent, however, is invariably represented in a biased form. An m-bit exponent permits 2^m possible unsigned integer values from 00 ... 0 to 11 ... 1. Suppose now we re-label these numbers not from 0 to $2^m - 1$, but from -2^{m-1} to $+2^{m-1} - 1$ by subtracting a constant value (or *bias*) of 2^{m-1} from each of the numbers.

What we have is a continuous natural binary series from 0 to N that represents a series of numbers from -B to N - B. For example, if the series (in decimal) were 0,1,2,3,4,5,6,7, we could subtract, say, B = 4 from each number to generate a new series -4,-3,-2,-1,0,1,2,3. We have really invented a new method of representing negative numbers by adding a constant to the most negative number to make it equal to zero. In the example above, we added 4 to each number so that -4 is represented by 0 and -3 by +1, etc.

We create a biased exponent by adding a constant to the true exponent so that the biased exponent is given by $b' = b + B$, where b' is the biased exponent, b the true exponent and B a weighting. The weighting B is frequently either 2^{m-1} or $2^{m-1} - 1$. Consider what happens for the case where $m = 4$ and $B = 2^3 = 8$ (see Table 1.6).

For example, if n = 1010.1111, we normalize it to 0.10101111×2^4. The true exponent is +4 which is stored as a biased exponent of 4 + 8 which is 12 or 1100 in binary form.

The true exponent ranges from -8 to +7, allowing us to represent powers of two from 2^{-8} to 2^{+7}, while the biased exponent ranges from 0 to +15. The advantage of the biased representation of exponents is that the most negative exponent is represented by zero. Conveniently, the floating point value of zero is represented

Table 1.6 Relationship between true and biased exponents

Binary representation of exponent	True exponent	Biased form
0000	-8	0
0001	-7	1
0010	-6	2
0011	-5	3
0100	-4	4
0101	-3	5
0110	-2	6
0111	-1	7
1000	0	8
1001	1	9
1010	2	10
1011	3	11
1100	4	12
1101	5	13
1110	6	14
1111	7	15

by $0.0 \ldots 0 \times 2^{\text{most negative exponent}}$ (see Figure 1.6). The biased exponent system represents zero by a zero mantissa and a zero exponent, as Figure 1.6 demonstrates.

The biased representation of exponents is also called excess n, where n is typically 2^{m-1} or $2^{m-1} - 1$. For example, a 6-bit exponent may be called excess 32 because the stored exponent exceeds the true exponent by 32. In this case, the smallest true exponent that can be represented is -32 and is stored as the value 0. The maximum true exponent that can be represented is 31 and is stored as 63.

A second advantage of the biased exponent representation is that it forms a natural binary sequence. This sequence is monotonic so that increasing the exponent by 1 involves adding one to the binary exponent, and decreasing the exponent by 1 involves subtracting one from the binary exponent. In both cases the binary biased exponent can be considered as behaving like an unsigned binary number.

IEEE Floating Point Format

The Institute of Electronics and Electrical Engineers (IEEE) has produced a standard floating point format for arithmetic operations in mini- and microcomputers (i.e., ANSI/IEEE standard 754-1985). To cater for a number of different applica-

Figure 1.6 Floating point representation of zero

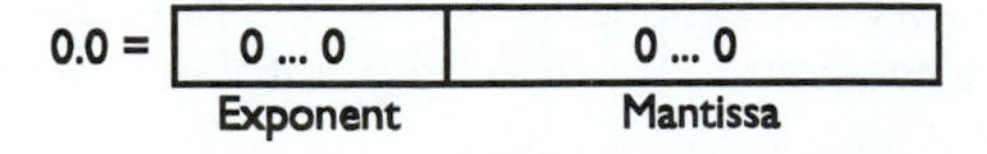

tions, the IEEE has specified three basic formats, called single, double, and quad. Table 1.7 defines the principal features of these three floating point formats.

IEEE floating point numbers are normalized so that their mantissas lie in the range 1 ≤ mantissa < 2. This range corresponds to a mantissa with an integer part equal to 1. An IEEE format floating point number X is defined as:

$$X = -1^{S} \times 2^{E-B} \times 1.F$$

where:

S = sign bit: 0 = positive mantissa, 1 = negative mantissa
E = exponent biased by B
F = fractional mantissa

For example, a single format 32-bit floating point number has a bias of 127 and a 23-bit fractional mantissa. There are two particular points of interest. The first is that a *sign and magnitude* representation has been adopted for the mantissa. If S = 1, the mantissa is negative, and if S = 0, it is positive.

The second point is that the mantissa is always normalized and lies in the range 1.000...00 to 1.111...11. Note that an IEEE floating point number is normalized differently than the floating point numbers we have mentioned earlier. If the mantissa is always normalized, it follows that the leading 1, the integer part, is redundant when the IEEE format floating point number is stored in memory. If we

Table 1.7 Basic IEEE floating point formats

Type	**Single**	**Double**	**Quad**
Field width in bits			
S = sign	1	1	1
E = exponent	8	11	15
L = leading bit	1	1	1
F = fraction	23	52	111
Total width	32	64	128
Exponent			
Maximum exponent	255	2047	32767
Minimum exponent	0	0	0
Bias	127	1023	16383
Normalized numbers (all formats)			
Range of exponents:	(Min E + 1) to (Max E - 1)		
Represented number:	$-1^{S} \times 2^{E\text{-bias}} \times L.F$		

A signed zero is represented by the minimum exponent, L = 0, and F = 0, for all three formats. The maximum exponent has a special function and is used to represent signed infinity for all three formats.

know that a 1 must be located to the left of the fractional mantissa, there is no need to store it. In this way a bit of storage can be saved, permitting the precision of the mantissa to be extended by one bit. The format of the number when stored in memory is given in Figure 1.7.

As an example of the use of the IEEE 32-bit format, consider the representation of the decimal number -2345.125 on a machine having a 16-bit wordlength.

$$-2345.125_{10} = -100100101001.001_2 \quad \text{(as an equivalent binary number)}$$
$$= -1.00100101001001 \times 2^{11} \quad \text{(normalized binary number)}$$

- The mantissa is negative, so the sign bit S is 1.
- The biased exponent is given by +11 + 127 = 138 = 10001010_2.
- The fractional part of the mantissa is .00100101001001000000000 (in 23 bits).

Therefore, the IEEE single format representation of -2345.125 is:

1 10001010 00100101001001000000000

A 32-bit computer would store this number as two consecutive 16-bit words:

1100010100010010 1001001000000000

In order to minimize storage space in a 16-bit memory, floating point numbers are packed so that the sign bit, exponent, and mantissa share part of two or more machine words. When floating point operations are carried out, the numbers are first *unpacked* and the mantissa separated from the exponent. For example, the basic single precision format specifies a 23-bit fractional mantissa, giving a 24-bit mantissa when unpacked and the leading 1 reinserted. If the processor on which the floating point numbers are being processed has a 16-bit word length, the unpacked mantissa occupies 24 bits out of the 32 bits taken up by two words.

If, when a number is unpacked, the number of bits in its exponent and mantissa is allowed to increase to fill the available space, the format is said to be *extended*. By extending the format in this way, the range and precision of the floating point

Figure 1.7 Format of the IEEE 32-bit floating point format

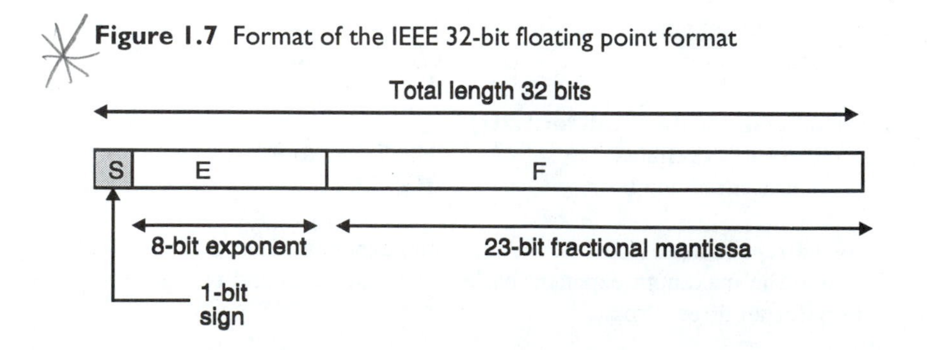

number are considerably increased. For example, a single format number is stored as a 32-bit quantity. When it is unpacked the 23-bit fractional mantissa is increased to 24 bits by including the leading 1 and then the mantissa is extended to 32 bits (either as a single 32-bit word or as two 16-bit words). All calculations are performed using the 32-bit extended precision mantissa. This is particularly helpful when transcendental functions (e.g., sin x, cos x) are evaluated. After a sequence of floating operations have been carried out in the extended format, the floating point number is *packed* and stored in memory in its basic form.

In the 32-bit single IEEE format, the maximum exponent, E_{max}, is +127 and the minimum exponent, E_{min}, is -126. The range of exponents is not +128 to -127 as we might expect. The special value $E_{min} - 1$ (i.e., -127) is used to encode zero, and $E_{max} + 1$ to encode plus or minus infinity or a *NaN*. A NaN is a special entity catered for in the IEEE format and is *Not a Number*. The use of NaNs is covered by the IEEE standard and they permit the manipulation of formats outside the IEEE standard.

Floating Point Arithmetic

Unlike integer and fixed-point number representations, floating point numbers cannot be added in one simple operation. A moment's thought should demonstrate why this is so. Consider an example in decimal arithmetic. Let A = 12345 and B = 567.89. In floating point form these numbers can be represented by:

$$A = 0.12345 \times 10^5 \text{ and } B = 0.56789 \times 10^3$$

If these numbers were to be added by hand, no problems would arise.

```
  12345
+   567.89
  12912.89
```

However, as these numbers are held in a normalized floating point format we have the following problem.

$$\begin{array}{l} 0.12345 \times 10^5 \\ + 0.56789 \times 10^3 \end{array}$$

Addition cannot take place as long as the exponents are different. To perform a floating point addition/subtraction the following steps must be carried out:

1. Identify the number with the smaller exponent.
2. Make the smaller exponent equal to the larger by dividing the mantissa of the smaller number by the same factor by which its exponent was increased.
3. Add (or subtract) the mantissas.
4. If necessary, normalize the result (post-normalization).

In this example $A = 0.12345 \times 10^5$ and $B = 0.56789 \times 10^3$. The exponent of B is smaller than that of A which results in an increase of 2 in B's exponent and a corresponding division of B's mantissa by 10^2 to give 0.0056789×10^5. We can now add A to the denormalized B.

$$
\begin{array}{rl}
A = & 0.1234500 \times 10^5 \\
+\ B = & \underline{0.0056789 \times 10^5} \\
 & 0.1291289 \times 10^5
\end{array}
$$

The result is already in a normalized form and doesn't need post-normalizing. Note that the answer is expressed to a precision of seven significant figures while A and B are each expressed to a precision of five significant figures. If the result were stored in a computer, its mantissa would have to be reduced to five figures after the decimal point (because we are working with five-digit mantissas).

When people do arithmetic they often resort to what may best be called floating precision. If they want greater precision they simply use more digits. Computers use a fixed representation for floating point numbers so that the precision cannot increase as a result of calculation. Consider the following binary example of floating point addition.

$$
\begin{array}{rl}
A = & 0.11001 \times 2^4 \\
B = & 0.10001 \times 2^3
\end{array}
$$

The exponent of B must be increased by 1 and the mantissa of B divided by 2 (i.e., shifted one place right) to make both exponents equal to 4.

$$
\begin{array}{rl}
A = & 0.11001 \quad \times 2^4 \\
B = & \underline{0.010001 \times 2^4} \\
 & 1.000011 \times 2^4
\end{array}
$$

In this case the result has overflowed and must be post-normalized by dividing the mantissa by two and incrementing the exponent.

$$A + B = 1.000011 \times 2^4 = 0.1000011 \times 2^5$$

We have also gained two extra places of precision, forcing us to take some form of action. For example, we can simply truncate the number to six bits to get:

$$A + B = 0.10000 \times 2^5$$

A more formal procedure for the addition of floating point numbers is given in Figure 1.8 as a flowchart. A few points to note about this flowchart are:

- Because (in many implementations) the exponent shares part of a word with the mantissa, it is necessary to separate them before the process of addition can begin. As we pointed out before, this operation is called *unpacking*.

Figure 1.8 Flowchart for floating point addition of $A = a \times 2^{m1}$ and $B = b \times 2^{m2}$

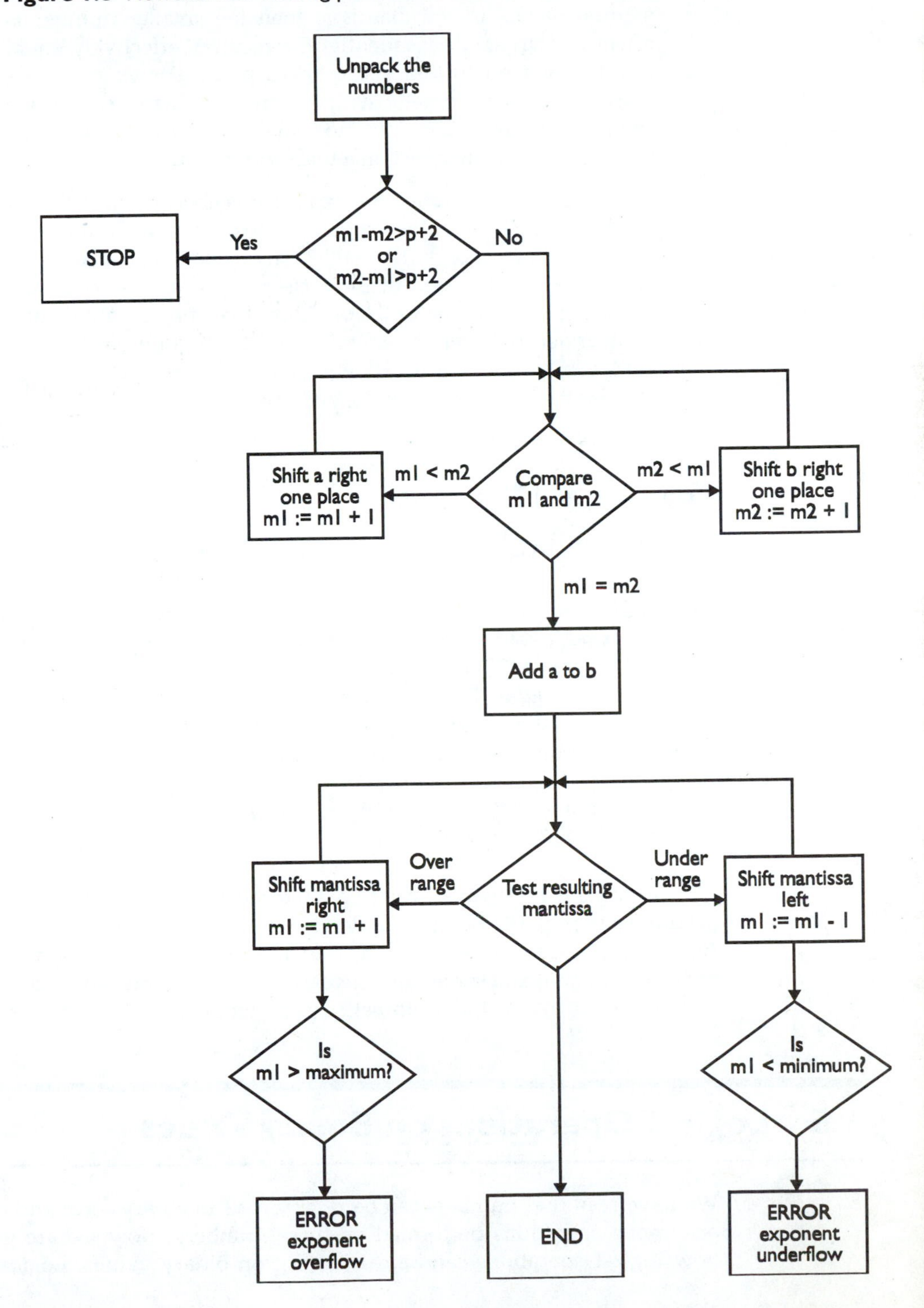

- If the two exponents differ by more than $p + 1$, where p is the number of significant bits in the mantissa, then the smaller number is too small to affect the larger. Consequently, the result is effectively equal to the larger number and no further action takes place. For example, suppose we are using decimal arithmetic with a four-digit mantissa. Clearly, there is no point in adding 0.1234×10^{20} to 0.4567×10^{2}, because adding 0.4567×10^{2} to 0.1234×10^{20} has no effect on a four-digit mantissa.
- During *post-normalization* the exponent is checked to see if it is less than its minimum possible value, or greater than its maximum possible value. This operation corresponds to testing for *exponent underflow* and *exponent overflow*, respectively. Each of these cases represents a condition in which the number is outside the range of values that the computer can handle. Exponent underflow would generally lead to the number being made equal to zero, while exponent overflow would result in an error condition and may require the intervention of the operating system.

Rounding and Truncation

We have seen that some of the operations involved in floating point arithmetic lead to an increase in the number of bits in the mantissa and that some technique is needed to keep the number of bits in the mantissa constant. The simplest technique is called *truncation* and involves nothing more than dropping unwanted bits. For example, 0.1101101 becomes 0.1101 when truncated to four significant bits. A much better technique is *rounding*. If the value of the lost digits is greater than half the least-significant bit of the retained digits, 1 is added to the LSB of the remaining digits. For example, consider rounding to four significant bits the following numbers:

```
a.   0.1101101  →  0.1101 + 1 = 0.1110
b.   0.1101011  →  0.1101
```

Rounding is always preferred to truncation partially because it is more accurate and partially because it gives rise to an *unbiased* error. Truncation always undervalues the result leading to a systematic error, whereas rounding sometimes reduces the result and sometimes increases it. The major disadvantage of rounding is that it requires a further arithmetic operation to be performed on the result.

1.5 Logical Operations on Binary Values

We have seen that numbers can be represented in binary form and conventional arithmetic operations performed on these numbers. Now we are going look at how logical operations can be carried out on binary values. Logical operations regard the values 1 and 0 as *true* and *false*, respectively (although there is no reason why we cannot call 0 true and 1 false). We are not going to provide an extensive coverage of logical operations, but are just going to demonstrate how they affect the assembly language programmer.

AND Operations

The first logical operation is the AND operation. The result of the expression x AND y is true if and only if both bits x and y are true. We can demonstrate this relationship by means of a truth table. A truth table tabulates all possible input values (in this case, x and y) and provides an output for each of the possible pairs of inputs.

```
x y     z=x.y
0 0     0
0 1     0
1 0     0
1 1     1
```

As you can see, both x and y must be 1 for z to be 1. Note that the logical AND operator is represented by a period, so that A AND B is written A.B. When a logical operation is applied to words consisting of m bits, the operation is carried out between pairs of bits. If A and B are two 8-bit values, we calculate A.B by evaluating $a_i.b_i$ for i = 0 to 7. Logical operations, unlike arithmetic operations, do not generate a carry bit from one column to another. If A = 11010110 and B = 01100111, the logical AND of A and B is given by:

```
A     =     11010110
B     =     01100111
A.B = C =   01000110
```

If bit *i* of A and bit *i* of B are both 1, then the corresponding bit of C is also 1. Otherwise it is 0. This operation does not seem very interesting. Let's consider a better example.

```
A     =     11010110
B     =     11110000
A.B = C =   11010000
```

In this case we have ANDed A=11010110 with B=11110000. The effect is to copy the bits of A for which the corresponding bits of B are 1 to C. Any bits of B that are 0 force the corresponding bits of C to zero. You should now see that the AND operator acts like a *mask* and can be used to force one or more bits of a word to zero. For example, if you wish to clear (set to zero) the most-significant bit of a byte, you just AND it with 01111111.

OR Operations

An OR operation on bits x and y is written z = x + y, where z is the logical OR of bits x and y. The logical OR operator is represented by the '+' symbol. Bit z is true if bit x is true or bit y is true, or both bits x and y are true. The truth table for an OR function is as follows:

x	y	z=x+y
0	0	0
0	1	1
1	0	1
1	1	1

If A = 11010110 and B = 01100111, the logical OR of A and B is given by:

```
A   =      11010110
B   =      01100111
A+B = C =  11110111
```

If either bit i of A or bit i of B is 1, then the corresponding bit of C is 1. Otherwise it is 0. As in the case of the AND operation, the OR operation does not seem very interesting. Let's consider a better example.

```
A   =      00010110
B   =      11000000
A+B = C =  11010110
```

In this case we have ORed A = 00010110 with B = 11000000. The effect is to force bits of C to 1 for which the corresponding bits of B are 1. Any bits of B that are 0 cause the corresponding bit of A to be copied to C. The OR operator can be used to force one or more bits of a word to ones. For example, if you wish to set (i.e., make a one) the most-significant bit of a byte, you just OR it with 10000000.

We can use the AND and OR operators to selectively clear and set bits of a word. Suppose we wish to set the least-significant four bits of a byte to 0101. If the byte is X, we can first AND X with 11110101 to clear bits 1 and 3. Then we OR X with 00000101 to set bits 0 and 2.

NOT Operations

The NOT operation *inverts* or *complements* a bit. A 1 becomes a 0, and a 0 becomes a 1. This operator is represented by an overbar. NOT x = $\bar{x}$. We can apply the NOT operation to all the bits of a word. If A = 11010110, $\bar{A}$ = 00101001.

EOR Operations

The AND and OR operators use conventional arithmetic symbols (. and +, respectively), whereas the EOR operator has a special symbol, ⊕. The EOR, exclusive OR, operation is rather like the OR operation, except that the result is true if and only if *just one* of the inputs is true. Some texts use the mnemonic XOR rather than EOR to denote the exclusive OR. Consider the truth table for an EOR operator.

```
x  y    z=x⊕y
0  0    0
0  1    1
1  0    1
1  1    0
```

As you can see, the EOR of two 1's is 0. The EOR operator produces a true output if the inputs are *different*. Now let's look at the effect of an EOR operation on two words.

```
A     =       11010110
B     =       01100111
A⊕B   = C  =  10110001
```

As before, we will look at a more interesting case.

```
A     =       11010110
B     =       00001111
A⊕B   = C  =  11011001
```

In this case, the most-significant four bits of A are EORed with 0 and the result is to copy these bits to C (exactly as if they had been ORed). However, the four least-significant bits of A are EORed with 1 and inverted (complemented) before they are copied to C. The EOR can therefore be employed to selectively invert or *toggle* the bits of a word.

Later we shall see how these logical operations can be used to manipulate the bits of a word in any way we see fit.

Summary

- The smallest unit of data is the bit, which has the value 0 or 1. A group of eight bits is called a *byte*.
- Computers manipulate groups of bits called words. However, some programmers now use the term *word* to mean, specifically, two bytes (i.e., 16 bits).

- A word has no *intrinsic meaning*. A word can be regarded as an instruction, a number, a character, or any other meaning given to it by the programmer.
- The binary number system has digits with two values, 0 and 1. The system is called *positional* because the weighting or value of a digit is determined by its position in a word.
- The hexadecimal system uses the base sixteen and employs the digits 0,1,2,3,4,5,6,7,8,9,A,B,C,D,E,F. Each hexadecimal digit is represented by a 4-bit binary value.
- The rules of arithmetic for decimal, binary, and hexadecimal arithmetic are all identical. Only the base changes.
- The most popular method of representing negative integers is by means of their *two's complement*. The two's complement of N is calculated by inverting all its bits and adding one to the result.
- An n-bit signed two's complement integer can represent positive values in the range 0 to 2^{n-1}-1, and negative values in the range -2^{n-1} to -1. The most-significant bit is 0 if the number is zero or positive, and 1 if it is negative.
- The subtraction of Y - X is carried out by adding the two's complement of X to Y.
- *Arithmetic overflow* occurs in two's complement arithmetic if two positive numbers are added and the result is negative, or if two negative numbers are added and the result is positive.
- You cannot tell whether a number is represented as an unsigned value or as a signed value from looking at its bit pattern. For example, 11111111_2 might be 127 (unsigned) or -1 (signed).
- A 32-bit IEEE floating point number is represented by the expression: $(-1)^S \times 1.F \times 2^{E-127}$, where S is the sign bit, F is the fractional mantissa, and E the stored exponent. The sign bit is 0 for a positive number and 1 for a negative number. The mantissa is normalized in the range 1.00000...00 to 1.00000...11, and the leading one is not stored. The exponent is biased by 127.
- Before floating point numbers can be added or subtracted, their exponents must be made equal. Once the addition/subtraction has been carried out, the result is renormalized if it has gone out of range.
- The following primitive operations may be applied to two bits, x and y:

AND	OR	EOR	NOT
0.0=0	0+0=0	$0\oplus0=0$	$\overline{0}=1$
0.1=0	0+1=1	$0\oplus1=1$	$\overline{1}=0$
1.0=0	1+0=1	$1\oplus0=1$	
1.1=1	1+1=1	$1\oplus1=0$	

- Logical operations may be applied to individual bits or to entire words.

Problems

1. Convert the following decimal integers into their natural binary equivalents.

   a. 15 b. 42 c. 235
   d. 4090 e. 40,900

2. Convert the following natural binary integers into their decimal equivalents.

   a. 110 b. 1110110 c. 11011
   d. 11111110111

3. Convert the following *base five* numbers into their *base nine* equivalents. (For example, $23_5 = 14_9$).

   a. 24 b. 144 c. 1234

4. Convert the following decimal numbers into their binary equivalents, accurate to five binary places.

   a. 1.5 b. 1.1 c. 1/3
   d. 1024.0625 e. 3.141592 f. $1/\sqrt{2}$

5. Convert the following binary numbers into their decimal equivalents.

   a. 1.1 b. 0.001 c. 101.101
   d. 11011.101010 e. 111.111111 f. 10.111101

6. Complete the following table.

Decimal	Binary	Hexadecimal
37		
73		
0.37		
0.73		
	10101010	
	1101101110	
	111011.011101	
	111.1011	
		256
		ABC
		2.56
		AB.C

7. The hexadecimal dump from part of a computer's memory is as follows:

```
0000:    4265   6769   6EFA   47FE   B087   0086   3253   7A29
0010:    C800   E000   0000
```

The dump is made up of rows of groups of four hexadecimal characters. Each row contains up to nine 16-bit groups. The first group of a row (terminated in a colon) is an *address* and defines the first location into which the following groups are to be loaded. For example, the first group in the second row is $0010. The second group is $C800 which is the contents of location $0010. The next group, $E000, is the contents of $0012, and so on. The 22 bytes of data represent the following sequence of items:

a. five consecutive ASCII-encoded characters
b. one unsigned 16-bit integer
c. one two's complement 16-bit integer
d. one unsigned 16-bit fraction
e. one 6-digit natural BCD integer
f. two 32-bit IEEE format floating point numbers

Interpret the hexadecimal dump and convert it into the six items above. For example, convert the floating point numbers to decimal form.

8. Perform the following binary additions:

a.		b.		c.	
	10110		100111		11011011
	00101		111001		10111011
					00101011
					01111111

9. Perform the following hexadecimal additions:

a.		b.		c.	
	42		3357		ABCD
	53		2741		FE10
					123A

10. Suppose that $P = 1234_{16}$, $Q = ABCD_{16}$, and $R = 0F0C_{16}$. Calculate, in hexadecimal arithmetic, the value of the following expressions:

a. P+Q b. P-Q c. P.Q
d. Q-P e. Q/R f. (P+R)(P-R)
g. PQ/(P+Q)

11. Using 8-bit arithmetic throughout, express the following decimal numbers in two's complement form:

a. -4 b. -5 c. 0
d. -25 e. -42 f. -128
g. -127 h. -111

12. Perform the following decimal subtractions in 8-bit two's complement arithmetic. Indicate cases in which arithmetic overflow occurs.

a.	+20 -5	b.	+127 -126	c.	+127 -128
d.	+5 -20	e.	-69 +42	f.	- 20 -111
g.	-127 -2	h.	- 42 + 69	i.	-1 +1

13. Distinguish between the terms *carry* and *overflow* when these are applied to two's complement arithmetic.

14. Explain the meaning of the following terms (in relation to floating point arithmetic).

a.	Exponent	b.	Mantissa
c.	Range	d.	Precision
e.	Packed number	f.	Unpacking
g.	Biased exponent	h.	Exponent overflow
i.	IEEE format	j.	Fractional mantissa

15. When designing a floating point format for a specific application, the programmer must select an appropriate *range* and *precision*. What factors do you think influence the designer's decisions concerning the structure of a floating point number?

16. An IEEE standard 32-bit floating point number is $N = -1^S \times 2^{E-127} \times 1.F$, where S is the sign bit, F is the fractional mantissa, and E the biased exponent.

   a. Convert the decimal number 123.5 into the IEEE format for floating point numbers.

   b. Convert the decimal number 100.25 into the IEEE format for floating point numbers.

   c. Describe the steps that take place when two IEEE floating point numbers are added together. You should start with the two packed floating point numbers and end with the packed sum.

   d. Perform the subtraction of 123.5 - 100.25 using the two IEEE-format binary floating point numbers you obtained for 123.5 and 100.25 in the first parts of this question. You should begin the calculation with the packed floating point representations of these numbers and end with the packed result.

17. Perform the following logical operations on the following binary values:

```
a.  AND   010101          b.  OR    010101
          110011                    110011

c.  EOR   010101          d.  NOT   010101
          110011
```

18. Perform the following logical operations on the following hexadecimal values:

```
a.  AND   ABCD            b.  OR    ABCD
          5CB9                      5CB9

c.  EOR   ABCD            d.  AND   ABCD
          5CB9                      00FF

e.  NOT   ABCD            f.  NOT   1234
```

CHAPTER 2

Introduction to Computer Architecture and Assembly Language

We are now going to look at the *architecture* of a microprocessor. Although there are many definitions of the term architecture, we can regard it as the assembly language programmer's view of the processor. In this chapter, our target architecture is a very simplified model of the 68000 microprocessor. We have decided to take this path because students can soon get lost in the mass of details associated with any real processor. For example, we are going to assume that the 68000 is an 8-bit machine (i.e., it operates on 8-bit operands) and forget that it can operate with 8-, 16-, and 32-bit operands. Once we have digested the simplified 68000, we can introduce its complexities in later chapters. The first part of this chapter shows how a computer is organized and how it executes a program stored in memory. We introduce a notation called *register transfer language* to enable us to describe the computer's internal operations.

We also provide a brief introduction to some of the concepts that underlie an assembly language. An *assembly language* is a language that controls the primitive operations on binary data within a computer. This statement does not devalue the computer, because the end results of these primitive operations can often be most impressive. Remember that the English language itself has just 26 primitive letters that are put together in long sequences according to the rules of English grammar and syntax. No one would imply that English literature is limited just because it is all based on 26 letters. So it is with computers — the fundamental operations are primitive, but their collective effect is anything but.

We have already said that computers perform primitive operations. These operations amount to little more than moving groups of bits from one point in a computer to another; basic arithmetic operations like addition and subtraction; comparing bits with zero; and *branching*. The last operation, branching, means executing one part of a program or another part of the program depending on the outcome of a calculation. One of the reasons that an assembly language often seems so impenetrable to the novice is that you have to learn much fine detail about these primitive operations before you can write meaningful programs. In some ways, learning an assembly language is like studying a foreign language — you have to memorize a lot of vocabulary and grammar before you can have a useful conversation in the language.

2.1 From Problem to Program

Before we launch ourselves into computer architecture and assembly language, we really need to fit these into the scheme of things. Figure 2.1 describes how computer languages can be viewed as a *hierarchical structure*. The outer layer in Figure 2.1 is the *applications language layer* as viewed by the user of the computer. Suppose the computer is playing the game of chess. The applications layer language is the language of chess itself. For example, a valid command at this level might be "move pawn to King four." This language might be represented by the movements of a mouse or by an algebraic chess language entered from a keyboard (e.g., p-k4).

The computer is not really a true chess machine (i.e., it doesn't really have electronic circuits that execute the rules of chess directly). A programmer has analyzed the game of chess, expressed its rules algorithmically, and written a program in a conventional high-level language like Pascal. The second layer down in Figure 2.1 is the *high-level language layer*. At this level the computer appears like a machine that directly executes the high-level language. For example, a high-level language Pascal machine would have electronic circuits that directly implement the statements that make up a Pascal program.

Few computers execute high-level languages directly. Most operate at the *assembly language level* (i.e., machine code level), the third layer down in Figure 2.1. The high-level language written by the programmer to simulate a chess machine is translated into assembly language by a program called a *compiler*. Since the assembly language is the actual *native language* of the computer, the computer can directly execute assembly language. In this book, we concentrate on this level of the

Figure 2.1 Hierarchical organization of a computer

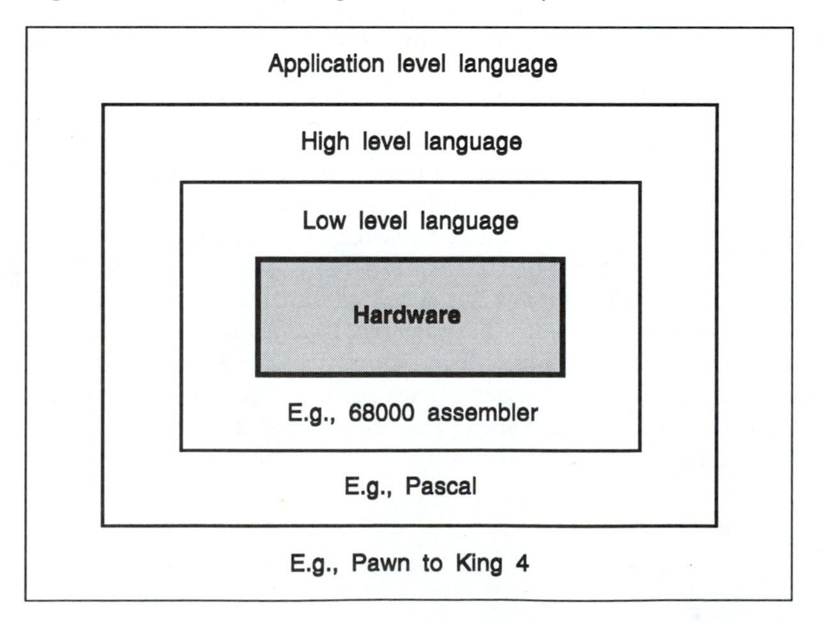

computer and demonstrate how programs can be written in assembly language. Once again, the assembly language programmer's view of a computer is frequently called its *architecture*. To be strictly accurate we should say that a computer executes *machine code* while the programmer writes *assembly language*. An assembly language is a human-readable representation of the binary code executed by a computer.

The innermost layer in Figure 2.1 is separated from the other layers by a heavy line. This line represents the hardware-software barrier. All layers above the line are concerned with *software*. Layers below the line are implemented by the computer's *hardware*. Some computers have additional layers below the hardware-software interface. For example, members of the 68000 family execute each machine code instruction as a series of *microinstructions*. Programmers cannot access the 68000's microinstructions. Only those who designed the 68000 in silicon have control over these microinstructions. A few very specialized machines are microprogrammable, which means that you can define the processor's instruction set. We are now going to look at the structure of a simple computer.

2.2 The CPU and its Memory

Thinking machines come in many different forms, ranging from the discrete logic controller (i.e., sub-computer level) of an automatic drinks vending machine to the human brain itself. In this text we restrict our field of interest to the so-called *Von Neumann machine* that represents today's mainstream computer architecture. John von Neumann was a scientist who, in the 1940s, described the architecture of the computer that was to be the forerunner of today's machines.

The cornerstone of the von Neumann machine is the common memory system that stores both the *sequence of operations* that implement a desired course of action (i.e., the program) and the *data* operated on by the program. Another name for the von Neumann machine is the *stored program computer*. We have to understand the basic principles and implications of the von Neumann machine if we are to discuss the architecture of the 68000 family. But, before we can describe the computer, we need to introduce its memory system.

Although the fine details of the organization of a computer's memory system are complex, the underlying principles are very simple indeed. Figure 2.2 demonstrates how we can regard a memory as a black box with input terminals (the *address* port), input/output terminals (the *data* port), and *control* terminals that make the memory perform a read or a write operation. In the computer world, the term *port* refers to the point at which information enters or leaves a computer (or some part of a computer). The memory is arranged as a sequence of consecutive locations, each specified by a unique *address*. The contents of any given location are referred to as *data*. The signals at the control terminals determine whether the memory is to take part in a read cycle or a write cycle.

In a read cycle, data at the specified address is placed on the data *bus* (the term "bus" refers to the data paths along which signals flow). In a write cycle, data on the data bus is written into the specified address. Any data that was in the location before the write cycle is overwritten and is lost.

The important point to appreciate about a memory is that you supply it with an address and then read data from that location or write data to that location. This data may have any meaning; it could be an instruction, a number, or even the address of another item in memory. By the way, you cannot easily locate data in memory if you do not know its address. For example, suppose the memory contains a series of student records. If you don't know the address of the information about a particular student, you have to read the memory locations, one by one, until you find the required data.

Figure 2.3 illustrates the fundamental components of a von Neumann machine. The *central processor unit*, CPU, is responsible for reading instructions from the memory system and executing them. An instruction is read from memory, executed, and then the next instruction is read, and so on. As you can see from Figure 2.3, there are two signal paths between the CPU and its memory. The *address path* (i.e., address bus) from the CPU to the memory allows the CPU to specify the location within the memory that it would like to access. The bidirectional *data path* between the CPU and memory allows the processor to examine the contents of a memory location in a read cycle and to modify the contents of the

Figure 2.2 The memory

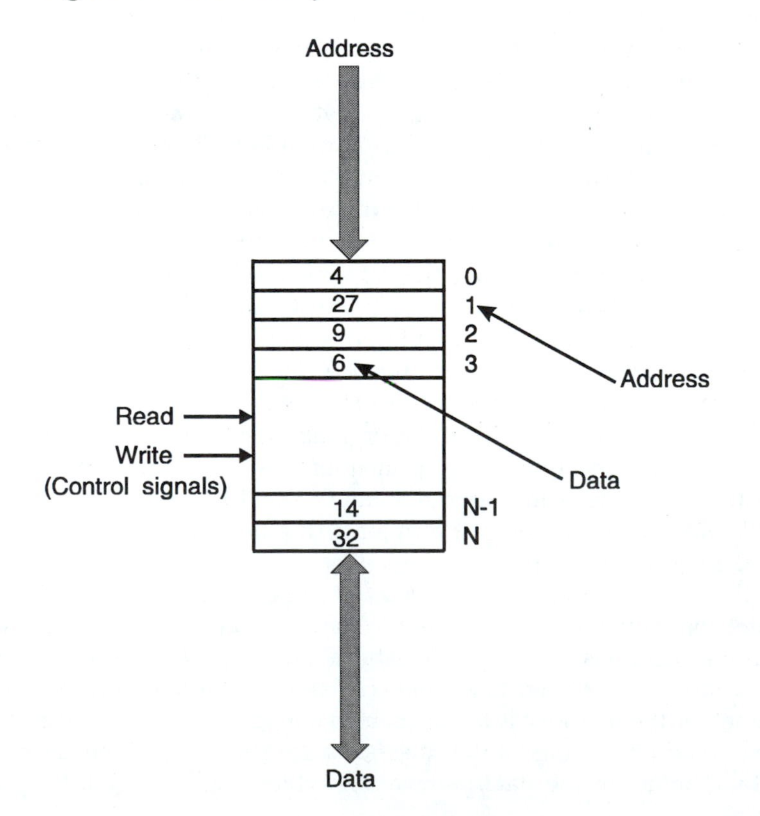

Figure 2.3 The von Neumann machine

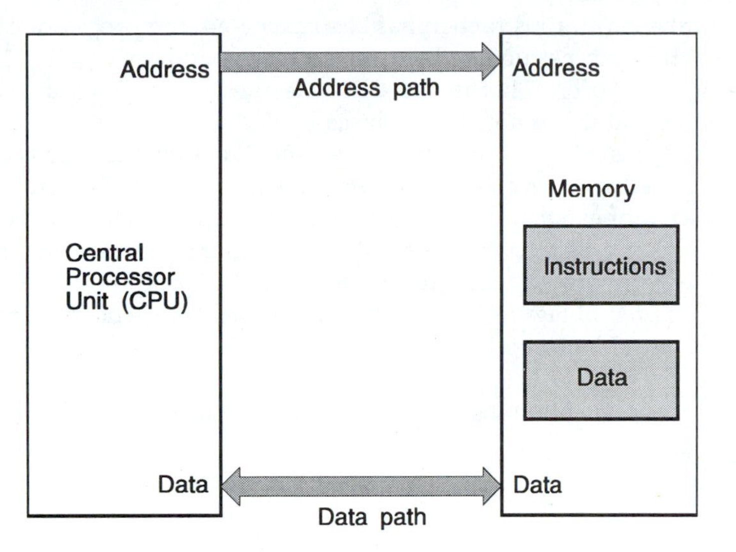

memory location in a write cycle. A bus is said to be bidirectional when data can flow in either direction along the bus (but not in both directions at the same time). Once again we should stress that the memory contains both instructions and the data to be operated on by the instructions. We're now going to describe how the CPU and its memory operate. To simplify our description, we will use *pseudocode*, because it allows us to express an algorithm clearly in a Pascal-like version of plain English. Pseudocode is not a formal language, but a mixture of statements in English together with some of the control structures you would expect to find in a language like Pascal or C. We describe the application of pseudocode to the design of programs in some detail in later chapters. The following fragment of pseudocode describes the action of the von Neumann machine.

```
Module Von_Neumann
      I := 0
      REPEAT
            Get an instruction from memory location I
            Execute the instruction
            I := I + 1
      FOREVER
End Von_Neumann
```

This pseudocode tells us that the processor fetches an instruction from location *I* in the memory, loads it into the CPU, and then executes it. The `REPEAT...FOREVER` construct simply tells us that the sequence of actions between the `REPEAT` and the `FOREVER` are to be repeated over and over again without limit. The variable, *I*, *points* to the location in memory of the next instruction to be executed. That is, *I* is the address of the next instruction. The pseudocode assumes that all programs begin in memory location zero (which is why we assign the value zero to I initially).

After the current instruction has been executed, the processor calculates the address of the next instruction (by means of the operation I := I + 1), fetches it, executes it, and so on. That is, the CPU executes instructions in sequence. We shall soon see that this is not always the case.

For our present purposes, we will assume that an instruction consists of two parts or *fields*, an *operation code* and an *operand address* (Figure 2.4). The *operation code* (or op-code) defines what the instruction does and the *operand address* (or, more simply, address) tells the CPU where the *operand* taking part in the instruction is located. We will normally refer to this as the *address field* of the instruction. The instruction format of Figure 2.4 is not the only possible instruction format — some instructions are able to specify two or more operands.

Figure 2.4 Basic von Neumann machine instruction format

Operation code	Operand address

Pseudocode is particularly useful for describing algorithms in a *top-down fashion*. That is, we can provide a basic overview of the whole process and then expand or *elaborate* the various steps in the process. For example, we can expand the action `Execute the instruction` in the above first-level pseudocode to give:

```
Module Execute the instruction
        Decode the instruction
        IF instruction requires data THEN fetch data from memory END_IF
        Perform operation defined by instruction
        IF instruction requires data to be stored THEN store data in memory END_IF
End Execute the instruction
```

This pseudocode employs the construct `IF condition THEN action END_IF`, that forces the computer to carry out the action if the stated condition is true. The pseudocode tells us that the execution of an instruction requires at least one memory access. It may require two or three memory accesses. The first memory access reads the instruction itself. The second access may be required either to read data from memory for use by the current instruction, or to put data in memory that has previously been created or modified in the CPU by a previous instruction.

Figure 2.5 demonstrates the flow of information between the CPU and its memory when the operation C := A + B is carried out. This operation reads the contents of the memory locations whose addresses are A and B, adds together the data found in these locations, and then deposits the result in memory location C.

Von Neumann machines are frequently said to operate in a two-phase mode, called a *fetch/execute* cycle. Because many instructions require two or more memory accesses per instruction, computer scientists have coined the expression *Von Neumann bottleneck* to indicate that one of the limiting factors of the von Neumann computer is the path between the CPU and the memory. Incidentally, if we had an instruction format that permits you to specify *three* operands, the operation C := A + B would require *four* accesses to memory (one to read the instruction itself, one to read source operand A, one to read source operand B, and one to write back the result to location C).

The four data paths in Figure 2.5 are really the same data path; they have been drawn separately to demonstrate the flow of information during the fetching and execution of the instruction. Note that in Figure 2.5 the addresses A, B, and C are *symbolic addresses* that stand for actual addresses. If we were to employ the actual addresses for these memory locations and write, say, 1002 := 1000 + 1001, it would look very confusing.

We now look at how a von Neumann machine is organized. Figure 2.6 illustrates the address and data paths of part of a hypothetical and highly simplified single-address, stored program machine. In what follows, the term *register* indicates a location within the CPU (as opposed to memory) that stores a unit or *word* of data. There is no conceptual difference between registers and memory locations, as they both perform the same function (i.e., the storage of data). The only difference is that the memory is composed of a large number of sequential locations, while the CPU contains just a few registers. The functional units of this machine are:

MAR The *memory address register* holds the address of the next location in main store to be accessed. The contents of the MAR *point* to the location of information in memory. For example, if the MAR contains the value 12, the CPU is going to access location (i.e., address) number 12 in the memory.

Figure 2.5 Information flow between the CPU and memory when executing C := A + B

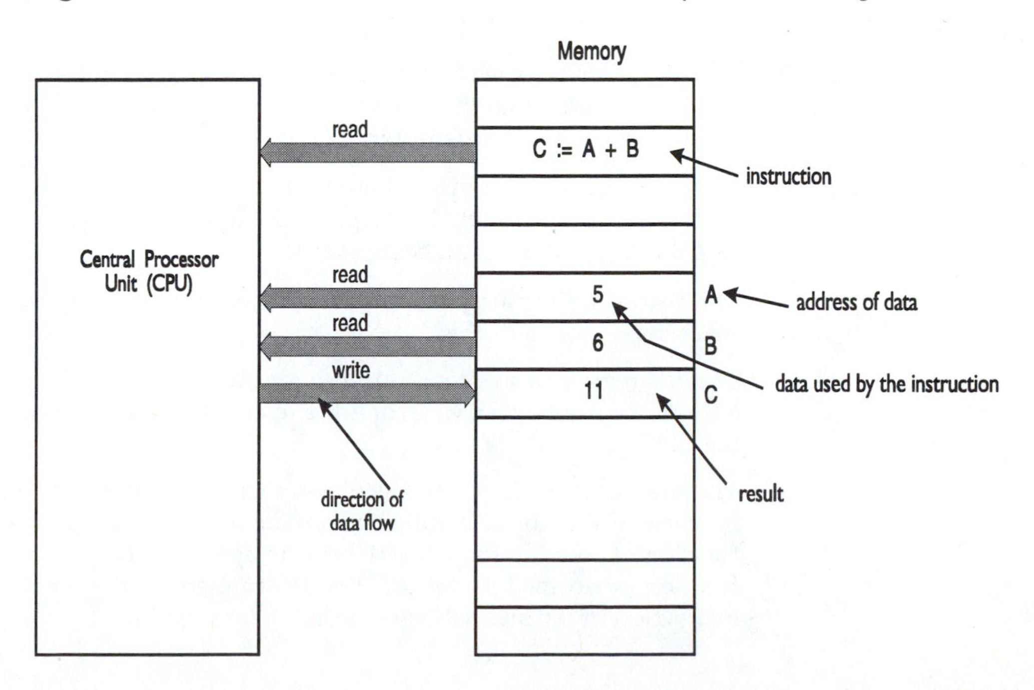

Figure 2.6 Simplified structure of a CPU

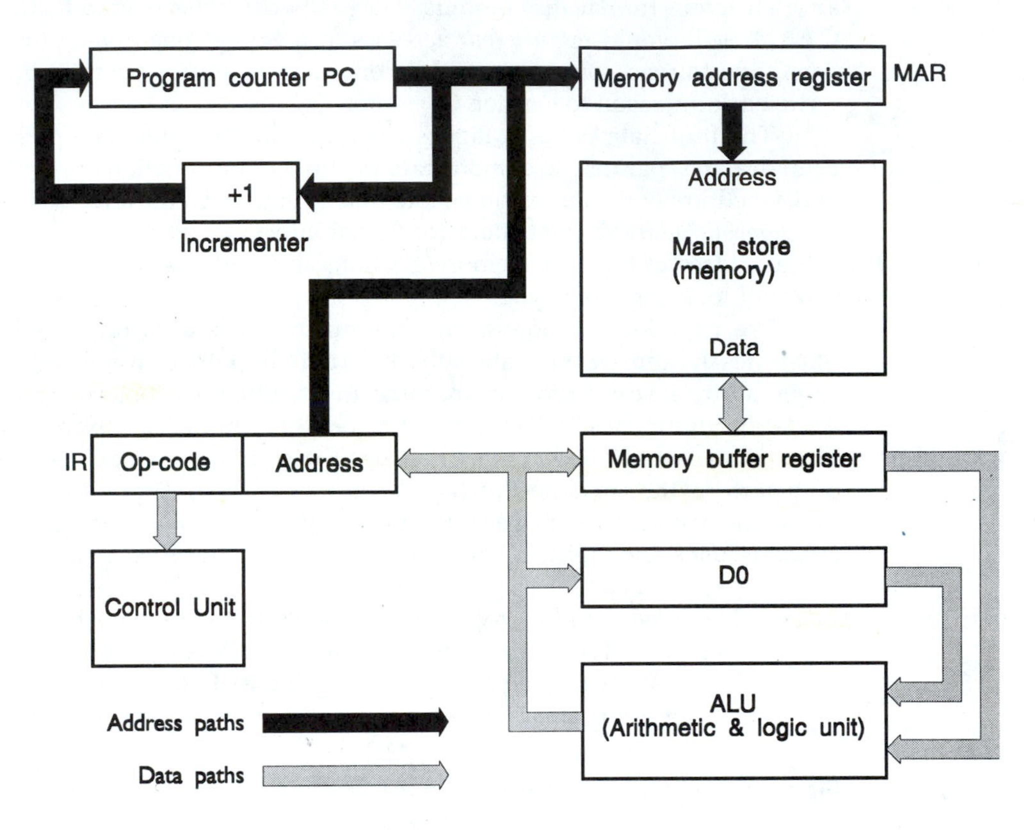

MBR The *memory buffer register* holds the data just read from the main store, or the data to be written into the main store. All information that flows into or out of the memory passes through the MBR.

PC The *program counter* or instruction counter contains the address of the next instruction to be executed. The program counter is used by the CPU to step through the program being executed.

IR The *instruction register* holds the most recently read instruction from the main store.

D0 The *data register* is a general-purpose register that holds data. This data is either to be used by the CPU, or is the result of an operation performed by the CPU.

ALU The *arithmetic and logic unit* calculates a function of one or two inputs. A function of one input is called a *monadic* operation (e.g., NOT x), and a function of two inputs is called a *dyadic* operation (e.g., a+b). The actual function performed by the ALU is determined by the bit pattern of the instruction in the instruction register.

CU The *control unit* interprets the instruction in the IR. That is, the control unit is responsible for converting the bit pattern of an instruction into the sequence of actions necessary to execute the instruction.

Before we explain the operation of the CPU, it is worthwhile developing a notation to help us to describe microprocessor actions clearly and unambiguously. One possible notation is the *register transfer language*, RTL, popularized by Hill and Peterson. Since such a comprehensive language is not necessary for our current purposes, we are going to introduce a simplified RTL notation. The main function of this notation is to describe what computer instructions do and, more importantly, the action of addressing modes. Do not feel that we are unnecessarily complicating matters by introducing a new language, as RTL has a very simple notation and very few rules. RTL makes it much easier to describe the action of assembly language instructions.

As we have just said, information inside a computer is held in registers or memory locations, and the only difference between a register and a memory location is one of detail. Registers are located within the CPU and can be accessed by a short address, because there are usually less than 16 registers. Memory locations are external to the CPU and are accessed by a 16-bit to 32-bit address. Moreover, registers are often dedicated to specific tasks (e.g., the MAR or the MBR).

Square brackets indicate the contents of the register (or the memory location) they enclose. For example, the expression `[MAR]` is interpreted as *the contents of the memory address register*. The expression `[MAR]` = 4 is interpreted as *the contents of the memory address register is equal to the numeric value* 4. A number or a symbolic name on its own represents a literal value. A *literal* is an actual value (sometimes called an immediate value), as opposed to the contents of a register or memory location. For example, "4" means the number four, whereas `[4]` is shorthand for `[M(4)]` and means the contents of memory location 4.

The contents of a location within the system memory (also called *main store* or *immediate access store*, IAS) are indicated by the expression `[M(address)]`, where `address` specifies the location in the memory. For example, `[M(6)]` = 4 is interpreted as *the contents of memory location 6 is the value* 4. That is, if you apply address 6 to the memory and execute a read cycle, the memory provides the value 4 at its data port. This example demonstrates how the notation '( )' is employed to indicate a location within memory. Note that you do not have to specify a location when dealing with a register because each register represents a single location, whereas a memory is an array made up from many memory locations.

The transfer of information between registers (or memory locations) is indicated by the backward arrow, ← . For example, `[PC]` ← 4 is interpreted as *the contents of the program counter are replaced by the number* 4. The symbol '←' is equivalent to the Pascal notation ':='. Consider the high level construct `R` := `P` + `Q`. This high level construct can be written in register transfer language as `[M(R)]` ← `[M(P)]` + `[M(Q)]`, and means *the contents of the memory at address P are added to the contents of the memory at address Q, and their sum is deposited in memory at address R.* Remember that when we described Figure 2.5, we said the high level construct 1002 := 1001 + 1000 was confusing. In RTL we can legitimately write

[M(1002)] := [M(1001)] + [M(1000)] to show that we are adding the contents of two memory locations and putting the result in a third. If we use symbolic notation for the addresses we can write: [M(C)] ← [M(B)] + [M(A)].

Executing an Instruction

We now describe how the computer of Figure 2.6 fetches an instruction from memory and then executes it. The operation of the computer can best be understood by breaking down the action into a series of steps. At the start of the instruction, the program counter contains the address of the current instruction to be executed. The contents of the PC are transferred or copied to the memory address register, MAR. We can represent this action in RTL by:

```
[MAR] ← [PC].
```

Figure 2.7 illustrates the information flow described by this action by means of shaded buses. At this stage, the memory address register is pointing at the next instruction to be executed. Once the program counter has done its job, it is automatically incremented to point to the next instruction in sequence. That is:

Figure 2.7 Fetching an instruction: Step I [MAR] ← [PC]

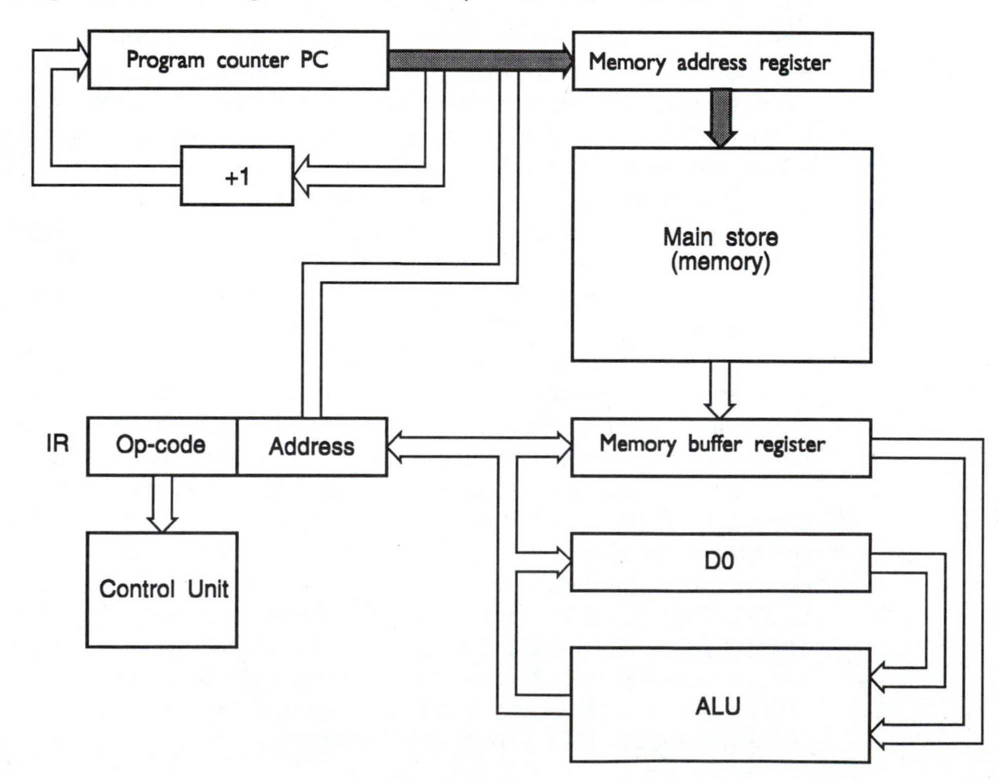

```
[PC] ← [PC] + 1.
```

Note that in the above expression the constant '1' is not enclosed in square brackets, as it is a literal and is not the contents of a register or memory location. Figure 2.8 illustrates the paths used to increment the contents of the program counter.

The next step is to read the contents of the memory location pointed at by the memory address register. The data read from memory is first deposited in a temporary holding register, the MBR (memory buffer register). The read memory operation can be expressed in RTL as:

```
[MBR] ← [M([MAR])].
```

This rather cumbersome expression is read as, *the contents of the memory location pointed at by the contents of the memory address register are copied to the memory buffer register*. Figure 2.9 illustrates this step.

In the final step of the *fetch cycle*, the contents of the MBR are copied to the instruction register IR. This register holds the instruction while it is decoded by the control unit, CU, and acted upon. We can now write the whole fetch cycle in RTL as:

Figure 2.8 Fetching an instruction: Step 2 `[PC] ← [PC] + 1`

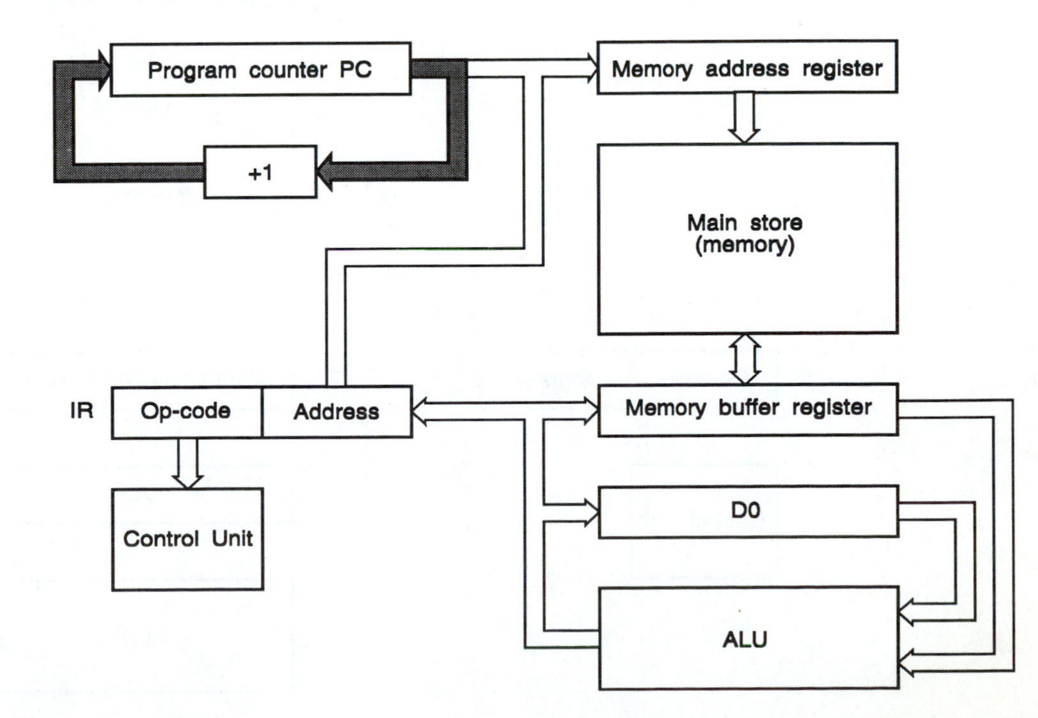

```
FETCH  [MAR] ← [PC]
       [PC]  ← [PC] + 1
       [MBR] ← [M([MAR])]
       [IR]  ← [MBR]
```

Consider how the computer in Figure 2.6 might carry out an addition. We will assume that the instruction just fetched from memory is *add the contents of memory location P to the data register D0*. This instruction has the assembly language form `ADD P`, where P is the address of a memory location containing the number to be added. Note that the instruction need supply only one operand address, P, since the other operand is data register D0 and there is only one data register D0 in the CPU. We shall soon see that the 68000 microprocessor actually has eight general-purpose data registers D0 to D7. Some early computers had a single data register and called it the *accumulator*. The term *accumulator* is historical and comes from the early days of the computer because it implies that this register accumulates the result.

The mnemonic form of the instruction is `ADD P` and its action is defined in RTL as:

```
[D0] ← [D0] + [M(P)]
```

Figure 2.9 Fetching an instruction: Step 3 `[MBR] ← [M([MAR])]`

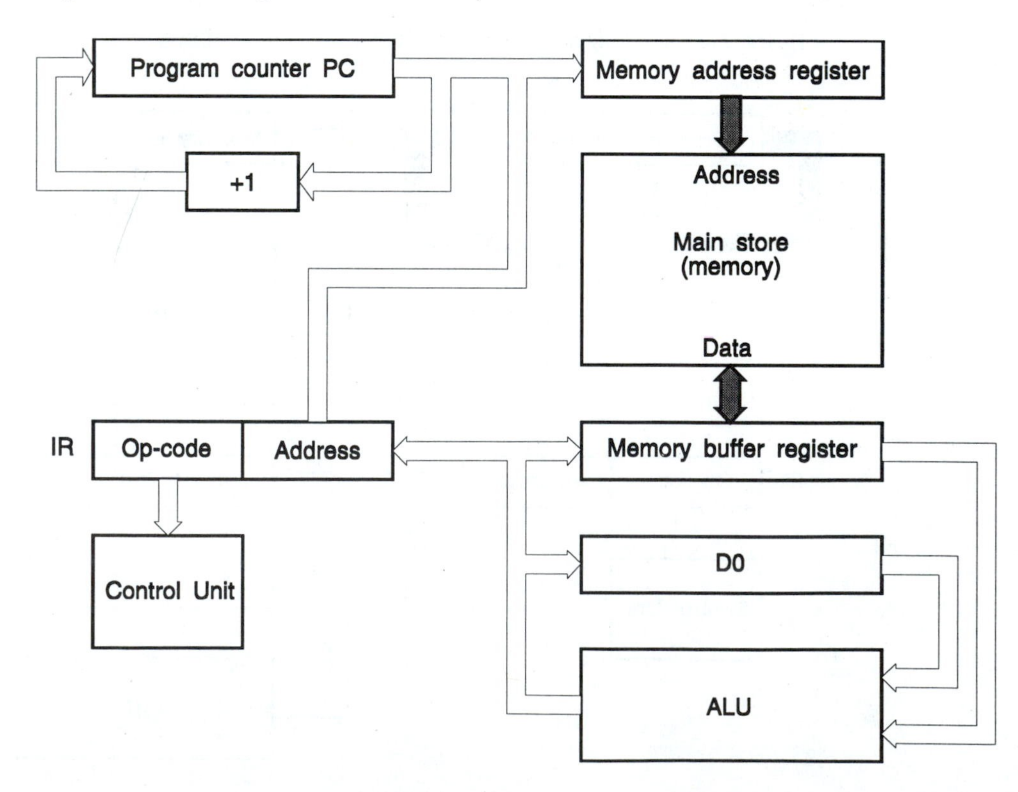

The behavior of the computer during the reading of the instruction and its execution can now be presented in RTL. Note that the first column in the following RTL (i.e., FETCH and ADD) is used to supply a *label field* to identify the corresponding line. We have also added a right-hand column to provide comments in plain English.

```
FETCH  [MAR] ← [PC]                    Send instruction address to MAR
       [PC]  ← [PC] + 1                Point to the next instruction
       [MBR] ← [M([MAR])]              Read the current instruction
       [IR]  ← [MBR]                   Copy the instruction to the IR
ADD    [MAR] ← [IR(address_field)]     Send the operand address to the MAR
       [MBR] ← [M([MAR])]              Read the operand from memory
       [D0]  ← [D0] + [MBR]            Perform the addition
```

The first four RTL operations constitute the *fetch phase* of the cycle in which an instruction is read from memory and the program counter incremented in readiness to fetch the following instruction. The second group of three operations, labelled by ADD, constitutes the *execution phase* of the cycle. Note that the notation [IR(address_field)] means the contents of the address field of the instruction register. All instructions begin with the same fetch cycle and are followed by the execute cycle determined by the instruction fetched from memory in the previous phase. Figure 2.10 illustrates the paths along which data flows during the above execute phase.

Dealing with Literals

The hypothetical computer of Figure 2.6 can execute a linear sequence of operations on data in memory. By *linear* we mean a sequence of instructions in ascending order i, i+1, i+2, i+3, ..., and so on. For example, we can calculate an expression of the form A := B + (D - G) x (P + Q)/J by the following linear sequence of 68000 assembly language instructions.

```
MOVE D,D0
SUB  G,D0          D - G
MOVE D0,Temp
MOVE P,D0
ADD  Q,D0          P + Q
MULU Temp,D0       (P + Q)(D - G)
DIVU J,D0          (P + Q)(D - G)/J
ADD  B,D0          B + (P + Q)(D - G)/J
```

Don't worry about the details of this program, as it is intended only to illustrate a sequence of instructions that process data in a register, D0, or in a memory location, Temp. The column on the right provides a commentary and is ignored by the assembler.

Figure 2.10 Data flow during the execute phase of ADD P

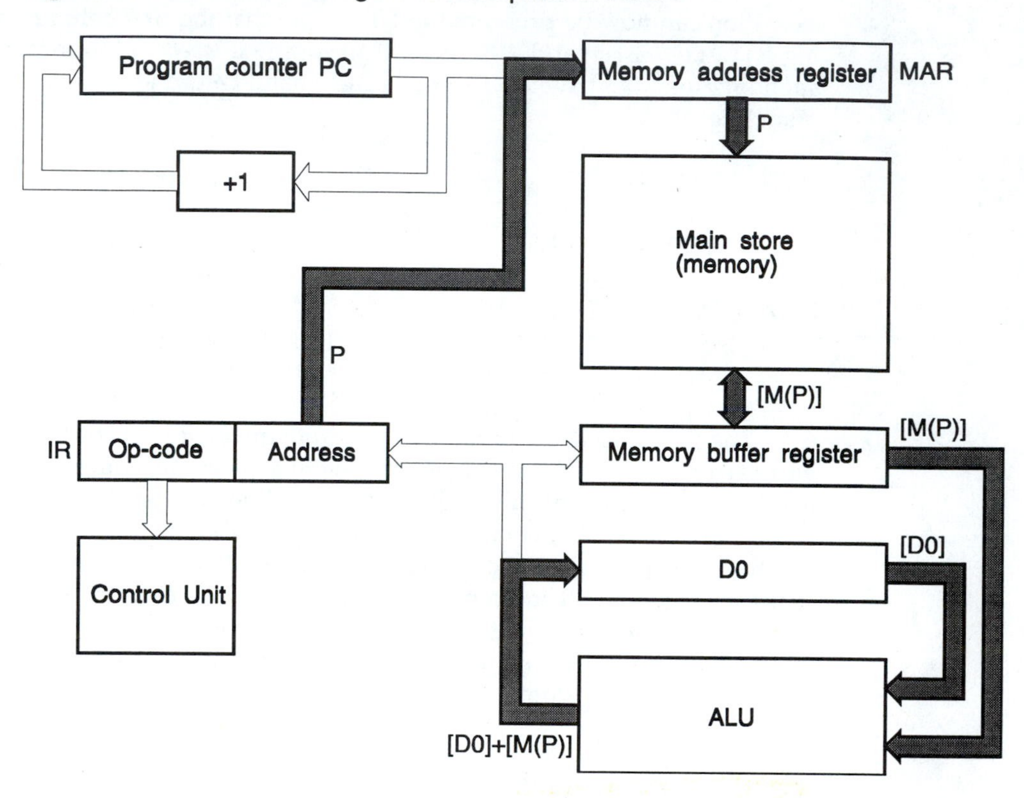

However, the simplified model for a CPU lacks two vital features required to implement a practical computer. First, it cannot operate on *literal* data (i.e., data that is part of the instruction). That is, we can add the contents of memory location 5 to D0, but we cannot add the actual number 5 to the contents of D0. In RTL terms, we can execute:

```
[D0] ← [M(5)]
```

but not `[D0] ← 5`.

Second, the computer of Figure 2.6 cannot execute conditional actions (i.e., the conditional branch instructions required to synthesize high-level language constructs such as `IF...THEN`).

Figure 2.11 shows the additional data paths required to deal with literal operands. Literal operands are handled by routing the address field of the instruction register to the ALU, D0, and MBR registers. A literal is indicated, in 68000 assembly language, by prefixing an operand with the symbol "#" (called *hash* or *pound* in the USA). For example, the high-level language construct `X := Y + 5` might be coded in assembly language as:

```
MOVE Y,D0
ADD  #5,D0
MOVE D0,X
```

In Figure 2.11 the assembly language operation ADD #5,D0 is executed by routing the contents of the address field of the instruction register directly to the ALU, rather than to the MAR. When the addition is performed the actual value "5" is added to the contents of the data register.

Implementing Conditional Behavior

Conditional behavior is implemented by creating an address path between the address field of the instruction register and the program counter, as Figure 2.12 demonstrates. Loading the program counter with a new (i.e., non-sequential) address results in a *jump* or *goto* operation. In order to implement conditional

Figure 2.11 Dealing with literal operands

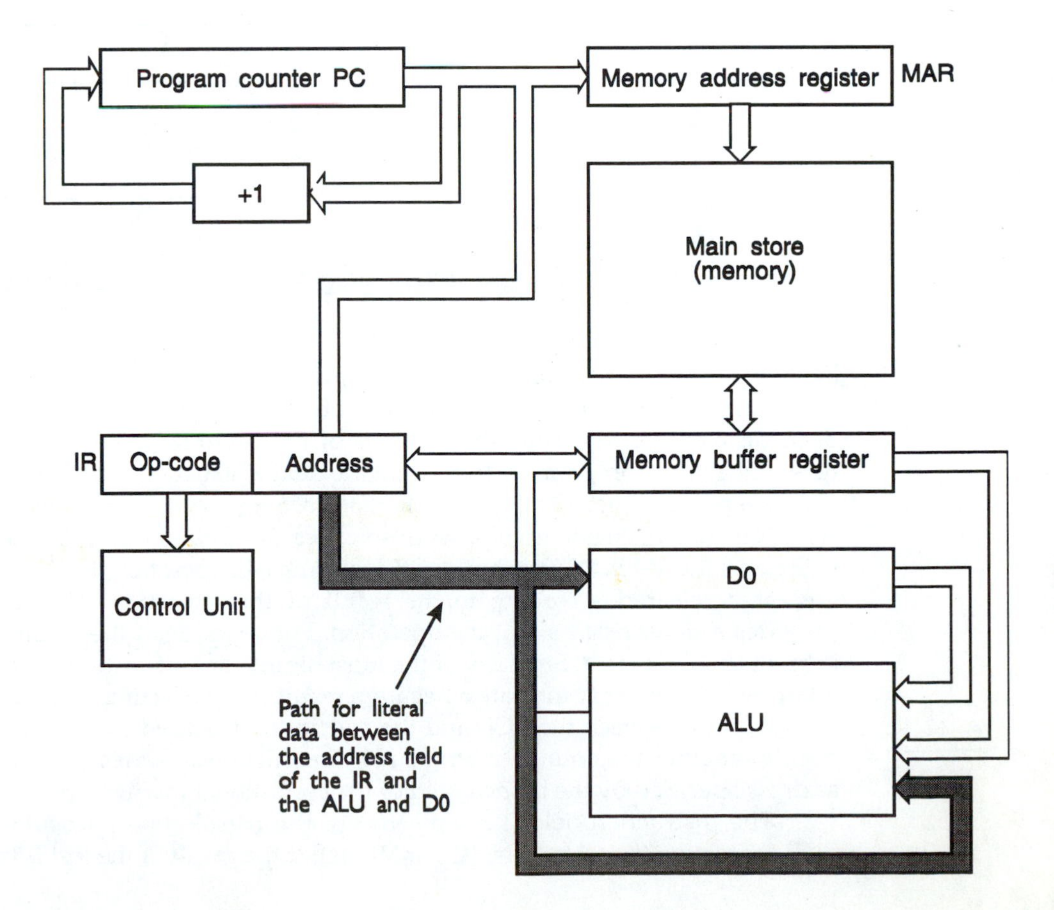

Figure 2.12 Structure of a CPU to handle conditional operations

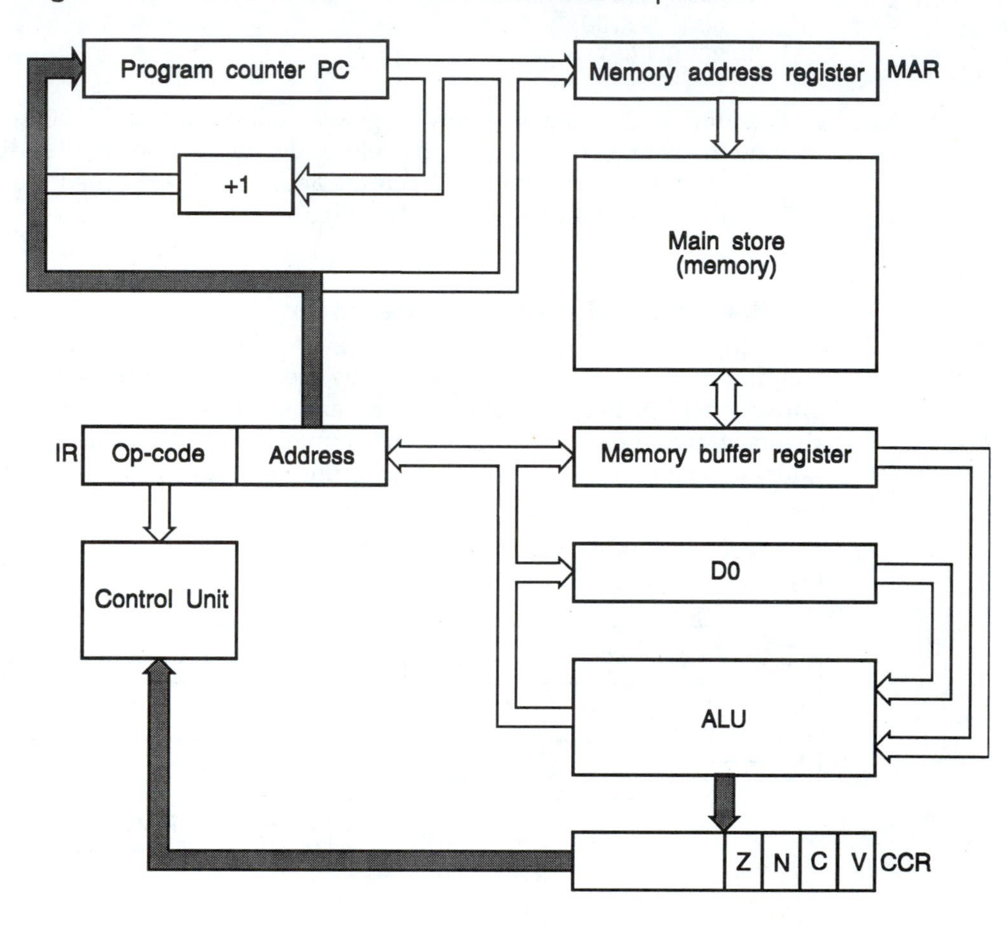

behavior it is necessary to either load the program counter with a *new* value or to continue with the unmodified contents of the PC. That is, the CPU makes a decision whether to modify the contents of the program counter and jump to a new place in the program, or to execute the next instruction as normal.

A special register called the *condition code register*, CCR, is connected to the ALU. The CCR is made up of a set of so called *flag bits* — the Z-bit, the N-bit, the C-bit, and the V-bit. Whenever the CPU executes an operation, the bits of the CCR are set or cleared according to the result of the operation. That is, the CCR provides a *status report* about the operation. For example, if the result is zero, the Z-bit of the CCR is set. Similarly, if the most-significant bit of the result is 1, the N-bit of the CCR is set to indicate a negative result (in two's complement terms).

A path between the CCR and the control unit is used by the control unit to decide whether to continue with the next instruction in series, or to jump to the address specified by the branch field of the conditional instruction.

The instruction field (i.e., op-code) of the IR selects a particular bit in the CCR for testing (i.e., the Z-, N-, C-, or V-bits). If the result of the test is true, the PC is loaded with a new value. Otherwise the contents of the PC are not modified. We can represent a simple conditional operation, `BEQ address` (i.e., branch to location `address` if the zero flag in the CCR is set), in RTL form as:

```
IF [CCR(Z)] = 1 THEN [PC]  ← address.
```

The above line is read as, *If the Z-bit in the condition code register is equal to 1, then load the program counter with the value "address"*.

Remember that when the ALU carries out an operation such as an addition, it modifies a number of *flags* in the condition code register, CCR. The Z-flag is set if the result is zero, the N-flag is set if the most-significant bit is one (i.e., the result is negative in a two's complement sense). The C-flag is made to reflect the value of the carry-out from the most-significant bit of the result. The V-bit is set if arithmetic overflow occurs during the operation. For example, if we add the two 8-bit numbers 11100111 + 11001010, we get a result of 10110001 and Z = 0 (the result is not zero), N = 1 (the most-significant bit is 1), C = 1 (the addition resulted in a carry-out), V = 0 (no arithmetic overflow occurred).

Let's look at an example of conditional behavior. Consider the following high-level language construct:

```
X := P + Q;
IF X = 4 THEN Y := 6
          ELSE Y := X + W;
```

That is, we are going to add variables P and Q to get X, and then set Y to 6 if X is equal to 4. If X is not equal to 4, we set Y to the value of X + W. We will now express this construct in assembly language form suitable for the hypothetical computer of Figure 2.12. Before we can do this, we need to introduce some basic assembly language instructions. We start with just five instructions: `MOVE`, `ADD`, `CMP`, `BEQ`, and `BRA`. These mnemonics are verbalized as: "move, add, compare, branch on zero result, and branch unconditionally," respectively. The action of each of these instructions is described in Table 2.1. Each instruction is described in both RTL and plain English. We will meet all these instructions again when we describe the 68000's instruction set. By the way, the hash symbol, #, in Table 2.1 indicates that the operand is an actual value (i.e., a literal) rather than the name of a register or a memory location.

Now that we've defined some instructions, we can translate the algorithm into assembly language. The following fragment of a program contains instructions on the left and plain English comments on the right.

Assembly language **Comments**

```
MOVE  Q,D0     Load D0 with the contents of location Q
ADD   P,D0     Add the contents of location P to D0
MOVE  D0,X     Put the sum in memory location X
CMP   #4,D0    Compare the contents of D0 with 4
```

```
        BEQ   SIX           If they are the same then go to SIX
        ADD   W,D0          Else add contents of W to D0
        MOVE  D0,Y          Put result in location Y
        BRA   EXIT          Branch to EXIT
SIX     MOVE  #6,D0         Put 6 in D0
        MOVE  D0,Y          Put D0 in memory location Y
EXIT    ....
```

Table 2.1 The definition of some basic assembly language instructions

Instruction	Operation in RTL	Operation in words
MOVE D0,Q	[M(Q)] ← [D0]	Copy the contents of register D0 to memory location Q.
MOVE Q,D0	[D0] ← [M(Q)]	Copy the contents of memory location Q to register D0.
MOVE #Q,D0	[D0] ← Q	Copy the number Q to register D0.
ADD Q,D0	[D0 ← [D0] + [M(Q)]	Add the contents of memory location Q to register D0 and put the result in D0.
ADD D0,Q	[M(Q)] ← [D0] + [M(Q)]	Add the contents of memory location Q to register D0 and put the result in memory location Q.
CMP Q,D0	[D0] - [M(Q)]	Subtract the contents of memory location Q from the contents of register D0. Discard the result. Set up the CCR.
CMP #Q,D0	[D0] - Q	Subtract the number Q from the contents of register D0. Discard the result. Set up the CCR (i.e., update the CCR).
BEQ N	IF Z = 1 THEN [PC] ← N	Branch to location N if the Z-bit of the CCR is set. That is, branch to location N if the result of the last operation yielded zero.
BRA N	[PC] ← N	Unconditional branch to location N. That is, branch always to location N. This instruction is equivalent to the high-level language construct GOTO N.

You read the assembly language program from top to bottom line by line. Each instruction is executed in turn unless a branch instruction (e.g., `BEQ SIX` or `BRA EXIT`) forces a jump to another part of the program. Don't worry if you don't fully follow this program yet. We will return to the design of assembly language programs in the next chapter. At this stage, it is more important that you appreciate the nature of the two-phase fetch/execute cycle and how an instruction is fetched from memory and executed.

We must stress that the comments to the right of the above program do not conform to good programming style. At this stage we are introducing programs solely to explain the action of machine-level instructions. When you come to write real assembly language programs, you should use the comment field to describe the underlying algorithm. For example, it is pointless to use the comment *"Put D0 in memory location Y"* to describe `MOVE D0,Y`, because anyone reading the code will understand the effect of the instruction `MOVE D0,Y`.

We have now described how a simple computer operates. However, we need to describe the rudiments of an assembly language before looking at a real computer, the 68000.

2.3 Introduction to the Assembler

As we have said, *assembly language* is a form of the native language of a computer in which *machine code instructions* are represented by *mnemonics*, and addresses and constants are usually written in *symbolic form* (just as they are in high-level languages). A mnemonic is just an aid to memory. For example, the mnemonic for the instruction that adds together two numbers is `ADD`. A symbolic name is chosen by the programmer to stand for a constant or a variable. A valid symbolic name is made up of a sequence of letters and numbers that begins with a letter (e.g., `Temp1`, `Val4a`). Typical 68000 assemblers limit the number of significant characters in a symbolic name to 8, although they accept longer names. For example, the programmer could use the variables `TempVal23` and `TempVal27`, although the assembler would regard both of these as equal to `TempVal2`.

Figure 2.13 illustrates how the assembler fits into the scheme of things. You create an assembly language program as a *source file* using any suitable text editor. The source file is then submitted to a program called an *assembler* that translates the source file into the target machine's *binary code*. This is usually stored as a file on disk and is also called the *object file*, since binary file and object file are used interchangeably. The assembler might also create an optional *listing file* that can be stored on disk or directed to the printer or screen. The listing file is a text file that contains the assembly language program after assembly plus any error messages.

Assuming that errors were not detected during assembly, the assembler also creates a binary file. The binary file may be stored on disk and later passed to the target machine that is going to execute the program. It may also be fed to a simulator that mimics the target machine and permits the programmer to test the software.

Figure 2.13 The assembler and the files it uses/creates

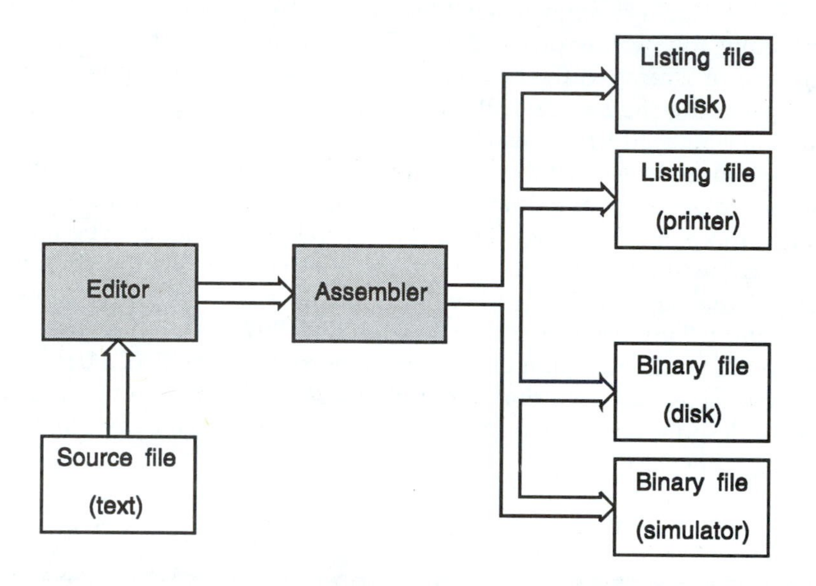

Assembler Directives

Before we can discuss assembly language and the 68000's instruction set, we have to describe some of the conventions adopted when writing assembly language programs. An assembly language is made up of two types of statement: the *executable instruction* and the *assembler directive*. An executable instruction is one of the processor's valid instructions and is translated into the appropriate machine code form by the assembler. We have already met some typical instructions (`ADD`, `SUB`, `MOVE`, `BEQ`).

Assembler directives cannot be translated into machine code; they simply tell the assembler things it need know about the program and its environment. Basically, assembler directives link symbolic names to actual values, allocate storage for data in memory, set up pre-defined constants, and control the assembly process. The assembler directives to be described in this section are: `EQU`, `DC`, `DS`, `ORG`, and `END`. Real assemblers have more directives, but many of these just control the format of the output (i.e., the way in which a program is printed on paper or displayed on a screen) and are not relevant here.

Because assembly language is concerned with the native processor itself, it is strongly dependent on the architecture of the processor. If you learn 68000 assembly language, you will not be able to write, say, 80486 programs with a minimal amount of conversion. This situation contrasts strongly with high-level languages that are independent of the system on which they run. However, the way in which you go about *designing* a 68000 assembly language program is identical to the way in which you go about designing an 80486 assembly language program.

Let's look at an example of a fragment of an assembly language program. Once more, don't worry about understanding this fragment of code, as it is intended only to show what an assembly program looks like.

```
BACK_SP    EQU     $08            ASCII code for backspace
DELETE     EQU     $01            ASCII code for delete
CAR_RET    EQU     $0D            ASCII code for carriage return
           ORG     $004000        Data origin
LINE       DS.B    64             Reserve 64 bytes for line buffer
*
*  Input a character and store it in a buffer
*
           ORG     $001000        Program origin
           LEA     LINE,A2        A2 points to line buffer
NEXT       BSR     GET_DATA       Call subroutine to get input
           CMP.B   #BACK_SP,D1    Test for back space
           BEQ     MOVE_LEFT      If back space then deal with it
           CMP.B   #DELETE,D1     Test for delete
           BEQ     CANCEL         If delete then deal with it
           CMP.B   #CAR_RET,D1    Test for carriage return
           BEQ     EXIT           If carriage return then exit
           MOVE.B  D1,(A2)+       Else store input in memory
           BRA     NEXT           Repeat
MOVE_LEFT  LEA     -1(A2),A2      Move buffer pointer back
           BRA     NEXT           and continue
CANCEL     LEA     LINE,A2        Reset the buffer pointer
           BRA     NEXT           and continue
GET_DATA   MOVE    #5,D0          Read a byte into D1
           TRAP    #15
           RTS
           END
```

The assembly language program is written in four columns (or fields): an optional label that must begin in column 1, an instruction and its operand, and an optional comment field. A word beginning in the left-most column is a programmer-defined *label* and is used by the assembler as a reference to that line. A line beginning with an *asterisk* in its first column is a comment and is entirely ignored by the assembler. The right-most column is an optional comment field and is used by the programmer to document his or her program. This column is ignored by the assembler. Consider the line:

```
NEXT          BSR   GET_DATA          Call subroutine to get input
```

"NEXT" is a label that refers to this line.	"BSR GET_DATA" is the actual instruction to be assembled.	This is the comment field and is entirely ignored by the assembler.

Note that the fragment of code uses the dollar symbol, $, as a prefix to some numbers (e.g., $08 and $004000). A number without a prefix (i.e., a constant) in an assembly language program is treated as a *decimal* value. Prefixing a number with a $ symbol indicates a *hexadecimal* value; prefixing it with a % indicates a *binary* value. Similarly, enclosing a text string in single quotes indicates a sequence of ASCII characters. For example, the values 65, $41, %01000001 and 'A' are all identical as far as the assembler is concerned. These prefixes are necessary because earlier generations of printers were not able to display subscripts and superscripts. By the way, the value $08 is identical to 8, as $8_{10} = 8_{16}$.

The assembler that translates assembly language programs into machine code requires at least one space to separate the label and comment fields from the actual instructions. We can re-write part of the above code in the form below and the assembler will still happily recognize it. Such a format is very difficult to read by the human programmer and should never be used.

```
BACK_SP  EQU $08 ASCII code for backspace
DELETE       EQU $01        ASCII code for delete
CAR_RET EQU             $0D ASCII code for carriage return
                  ORG           $004000  Data origin
LINE          DS.B 64 Reserve 64 bytes for line buffer
*
*  Input a character and store it in a buffer
*
 ORG $001000 Program origin
                LEA    LINE,A2 A2 points to line buffer
NEXT                 BSR       GET_DATA   Call subroutine to get input
 CMP.B  #BACK_SP,D1 Test for back space
            BEQ     MOVE_LEFT If back space then deal with it
       CMP.B #DELETE,D1 Test for delete
```

Before we take a look at some of the features of the program, we will assemble it and produce a listing file of the assembled program. The following listing file contains the original assembly language and the machine code (in hexadecimal form) produced by the assembler. Note that the assembler automatically displays a colon after labels and a semicolon in front of the comment field.

```
 1          00000008  BACK_SP:  EQU    $08         ;ASCII code for backspace
 2          00000001  DELETE:   EQU    $01         ;ASCII code for delete
 3          0000000D  CAR_RET:  EQU    $0D         ;ASCII code for carriage return
 4 00004000                     ORG    $004000     ;Data origin
 5 00004000 00000040  LINE:     DS.B   64          ;Reserve 64 bytes for line buffer
 6                    *
 7                    *  Input a character and store it in a buffer
 8                    *
 9 00001000                     ORG    $001000     ;Program origin
10 00001000 45F84000            LEA    LINE,A2     ;A2 points to line buffer
11 00001004 6100002A  NEXT:     BSR    GET_DATA    ;Call subroutine to get input
12 00001008 0C010008            CMP.B  #BACK_SP,D1 ;Test for back space
13 0000100C 67000016            BEQ    MOVE_LEFT   ;If back space then deal with it
```

```
14 00001010 0C010001              CMP.B    #DELETE,D1    ;Test for delete
15 00001014 67000014              BEQ      CANCEL        ;If delete then deal with it
16 00001018 0C01000D              CMP.B    #CAR_RET,D1   ;Test for carriage return
17 0000101C 6700001A              BEQ      EXIT          ;If carriage return then exit
18 00001020 14C1                  MOVE.B   D1,(A2)+      ;Else store input in memory
19 00001022 60E0                  BRA      NEXT          ;Repeat
20 00001024 45EAFFFF  MOVE_LEFT:  LEA      -1(A2),A2     ;Move buffer pointer back
21 00001028 60DA                  BRA      NEXT          ;and continue
22 0000102A 45F84000  CANCEL:     LEA      LINE,A2       ;Reset the buffer pointer
23 0000102E 60D4                  BRA      NEXT          ;and continue
24 00001030 303C0005  GET_DATA:   MOVE     #5,D0         ;Read a byte into D1
25 00001034 4E4F                  TRAP     #15
26 00001036 4E75                  RTS
27 00001038 4E71      EXIT:       NOP                    ;Continue....
```

Since the 68000 is a real processor we have to say something about the size of the units of data manipulated by the 68000. For example, the 68000 permits operations on data elements of 8, 16 and 32 bits. A 32-bit entity is called a *longword*, a 16-bit entity a *word*, and an 8-bit entity a *byte*. To avoid confusion, longword, word, and byte will always refer to 32-, 16-, and 8-bit values, respectively, throughout this book. We will endeavor to keep things as simple as possible in this chapter and assume that data values are 8-bit quantities and that addresses are 32-bit quantities. We now describe three of the most important assembler directives.

EQU The *equate* directive simply links a *name* to a *value*, making programs much easier to read. For example, it is better to equate the name "CAR_RET" to $0D and use this name in a program, rather than writing $0D and leaving it to the reader to figure out that $0D is the ASCII value for carriage return. Consider the following example of its use.

```
MyData  EQU  128          Define a stack-frame called "MyData" of 128 bytes
        .
        .
        LINK A1,#-MyData  Reserve 128 bytes for local storage
```

Not only is "MyData" more meaningful to the programmer than its numerical value of 128, it's very easy to modify the numeric value by changing the "EQU 128" that appears at the head of the program.

Most 68000 assemblers permit the programmer to employ an *expression* as a valid name. Expressions are evaluated by the assembler to yield a numeric value. For example, you can write the following:

```
FRAME    EQU  128
FRAME2   EQU  FRAME+16
FRAME3   EQU  FRAME2*4
         LEA  FRAME-64,A2
```

In this example, the first assembler directive equates the value 128 to the symbolic name FRAME. The next directive equates the symbolic name FRAME2 to the value FRAME+16. Since FRAME has already been equated to 128, the assembler equates FRAME2 to 128+16, or 144. As you might imagine, the following code is illegal.

```
FRAME   EQU  128
FRAME2  EQU  FRAME3+16
FRAME3  EQU  FRAME*4
```

This code is illegal because the second line attempts to equate FRAME2 to FRAME3 plus 16, but the value of FRAME3 has not yet been evaluated. The following output of a 68000 assembler shows how this error is reported.

```
    1  00000400                          ORG      $400
    2            00000080     FRAME:     EQU      128
    3            00000000     FRAME2:    EQU      FRAME3+16
***** Illegal forward reference - operand 1.
    4            00000200     FRAME3:    EQU      FRAME*4
    5            00000400                END      $400
```

It is sometimes tempting to regard arithmetic operations (e.g., FRAME+16) as part of the executable program. They are not. Arithmetic expressions are evaluated by the assembler and the resulting numbers used in the assembly. Why, then, have assembler writers provided this facility? Consider the following example.

```
Length   EQU  30
Width    EQU  25
Area     EQU  Length*Width
```

There are two reasons for using the above code to represent the value of Area. The first is that the code doesn't have to be modified if ever the values of Length or Width are modified. The second is that the expression Area EQU Length*Width tells the reader how the value Area is arrived at. Whenever the assembly programmer writes Area in a program, the assembler replaces it by the numerical value 750 (i.e., 30 x 25). The following extract from the output of an assembler demonstrates how the expression is treated.

```
    1            0000001E     LENGTH:    EQU      30
    2            00000019     WIDTH:     EQU      25
    3            000002EE     AREA:      EQU      LENGTH*WIDTH
```

DC This directive means *define a constant* and is qualified by .B, .W, or .L, depending on the size of the constant (8, 16, or 32 bits). The constant defined by this directive is loaded in memory at the *current* location. The assembler directive DC.B 3 loads the byte $03 into memory, whereas the directive DC.L 3 loads the longword $00000003 into memory. Constants can be stored in consecutive locations by separating each one with a comma — `DC.B 12,15,32` stores the constants 12, 15, and 32 in three consecutive bytes. Consider the following example.

```
      ORG   $00001000         Start of the data region
First DC.B  10,66             The values 10 and 66 are stored in consecutive bytes
      DC.L  $0A1234           The value $000A1234 is stored as one longword
Date  DC.B  'April 8 1985'    The ASCII characters are stored as 12 bytes
      DC.L  1,2               Two longwords are set up with the values 1 and 2
```

Figure 2.14 uses a memory map to demonstrate the effect of these define-constant directives. By convention, the 68000 memory map is arranged as a word per line. Each word consists of two bytes: an odd byte and an even byte. Low addresses are located at the *top* of the memory map. Addresses and data are often presented in hexadecimal form.

The 68000 microprocessor expects 16-bit words to have even addresses that fall at even *word boundaries*. That is, a 16-bit word should not be stored across a word boundary. For example, you can store the 16-bit value 1234_{16} at memory

Figure 2.14 Example of the effect of DC directives

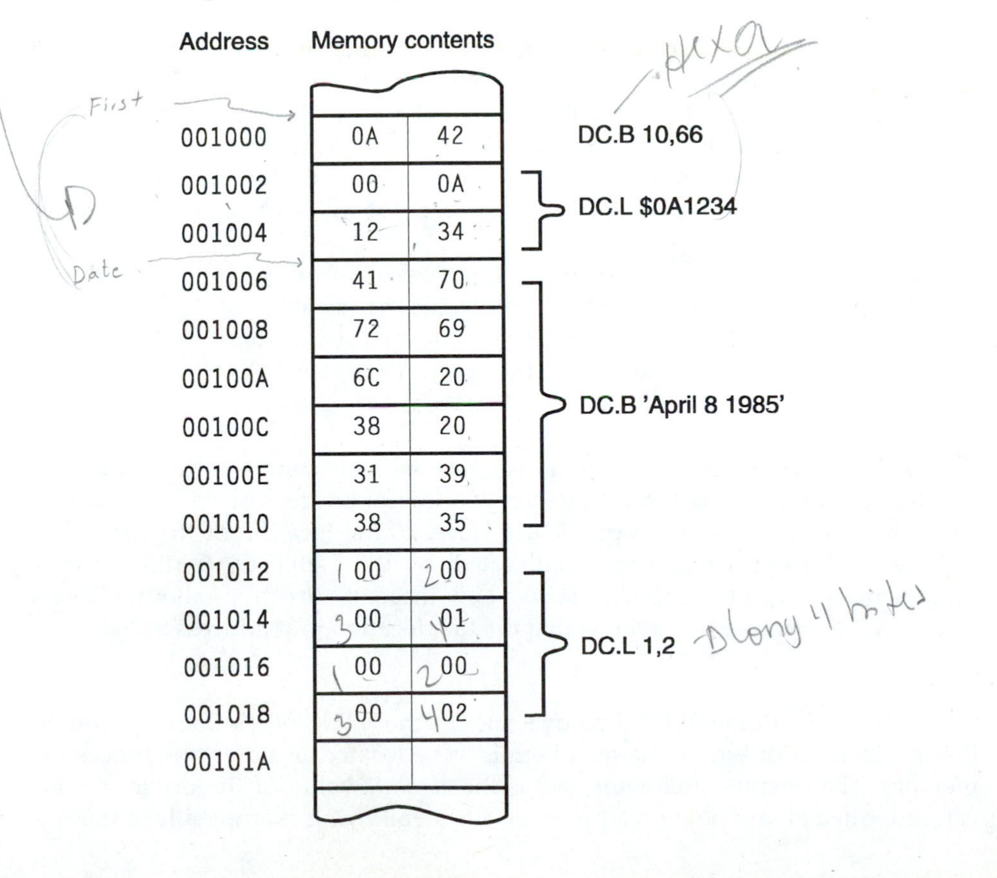

location 1000, but not at memory location 1001. This restriction is an operational detail of the 68000 rather than some fundamental law of computer science. However, you should appreciate that byte boundaries are 0,1,2,3,4,5,..., word boundaries are 0,2,4,6,8,A,..., and longword boundaries are 0,4,8,C,10,14,...

The operand field of a define-constant directive may be a decimal value, a hexadecimal value (prefixed by $), a binary value (prefixed by %), or an ASCII string enclosed by single quotes. Each ASCII character is stored in a single byte. As in the case of the `EQU` directive, we can use expressions that are evaluated by the assembler. Consider the following example.

```
BASE    EQU  57                 Equate "BASE" to $0039
OFFSET  EQU  16                 Equate "OFFSET" to $0010
        DC.L 2*BASE+OFFSET      Store the longword 2*57+16 in memory
```

The assembler directive `DC.L 2*BASE+OFFSET` causes the expression "`2*BASE+OFFSET`" to be calculated and that value to be stored as a longword in memory. That is, the actual value stored in memory is: $2 \times 57+16=130_{10}=82_{16}$. Programmers use expressions as operands for emphasis and clarity. For example, `AREA DC.W 12` is less clear than `AREA DC.W LENGTH*WIDTH`.

DS The *define storage* directive, `DS`, reserves (i.e., allocates) storage locations in memory. It has a similar effect to `DC`, but no information is stored in memory. That is, `DC` is used to set up *constants* that are to be loaded into memory before the program is executed, whereas `DS` reserves memory space for *variables* that are generated by the program at runtime. `DS` is also qualified by `.B`, `.W`, or `.L` and takes the same types of operands as `DC`. Consider the following example.

```
FIRST   DS.B    4      Reserve   4 bytes of memory
TABLE   DS.B  $80      Reserve 128 bytes of memory
ADRS    DS.L    1      Reserve one longword (4 bytes)
VOLTS   DS.W    1      Reserve one word (2 bytes)
LIST    DS.W  256      Reserve 256 words
```

A label in the left-hand column equates the label with the first address of the defined storage. The label references the lowest address of the defined storage value. In the above example, `TABLE` refers to the location of the first of the $80 (i.e., 128_{10}) bytes reserved, or allocated, by the `TABLE DS.B $80` directive. The memory map of Figure 2.15 shows how these `DS` directives allocate storage space. In this example, we assume that the first location is at address $1000.

ORG The *origin* assembler directive sets up the value of the location counter that keeps track of where the next item is to be located in the target processor's memory. The operand following `ORG` is the *absolute* value of the origin. An `ORG` can be located at any point in a program, as the following example illustrates:

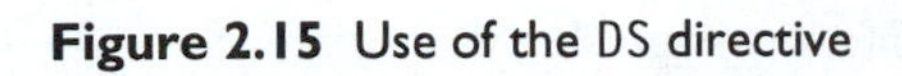

Figure 2.15 Use of the DS directive

Address	Memory contents	
001000		FIRST
001004		TABLE
001083		
001084		ADRS
001088		VOLTS
00108A		LIST
001189		
00118A		

```
          ORG     $001000        Origin for data
TABLE     DS.W    256            Save 256 words for "TABLE"
POINTER_1 DS.L    1              Save one longword for "POINTER_1"
VECTOR_1  DS.L    1              Save one longword for VECTOR_1
INIT      DC.W    0,$FFFF        Store two constants ($0000, $FFFF)
SETUP1    EQU     $03            Equate "SETUP1" to the value 3
SETUP2    EQU     $55            Equate "SETUP2" to the value $55
ACIAC     EQU     $008000        Equate "ACIAC" to the value $8000
RDRF      EQU     0              RDRF = Receiver Data Register Full = 0
PIA       EQU     ACIAC+4        Equate "PIA" to the value $8004
*
          ORG     $018000        Origin for program
ENTRY     LEA     ACIAC,A0       A0 points to the ACIA
          MOVE.B  #SETUP1,(A0)   Write initialization const into ACIA
*
GET_DATA  BTST    #RDRF,(A0)     Any data received?
          BNE     GET_DATA       Repeat until data ready
          MOVE.B  2(A0),D0       Read data from ACIA
          .
          .
          .
```

When this code is presented to an assembler, it produces the following output.

```
1  00001000                          ORG     $001000      ;Origin for data
2  00001000 00000200     TABLE:      DS.W    256          ;Save 256 words for "TABLE"
3  00001200 00000004     POINTER_1:DS.L      1            ;Save one longword for "POINTER_1"
4  00001204 00000004     VECTOR_1:   DS.L    1            ;Save one longword for VECTOR_1
5  00001208 0000FFFF     INIT:       DC.W    0,$FFFF      ;Store two constants ($0000, $FFFF)
6           00000003     SETUP1:     EQU     $03          ;Equate "SETUP1" to the value 3
7           00000055     SETUP2:     EQU     $55          ;Equate "SETUP2" to the value $55
8           00008000     ACIAC:      EQU     $008000      ;Equate "ACIAC" to the value $8000
9           00000000     RDRF:       EQU     0            ;RDRF = Receiver Data Register Full = 0
10          00008004     PIA:        EQU     ACIAC+4      ;Equate "PIA" to the value $8004
11                       *
12 00018000                          ORG     $018000      ;Origin for program
13 00018000 41F900008000 ENTRY:      LEA     ACIAC,A0     ;A0 points to the ACIA
14 00018006 10BC0003                 MOVE.B  #SETUP1,(A0) ;Write initialization const into ACIA
15                       *
16 0001800A 08100000     GET_DATA:   BTST    #RDRF,(A0)   ;Any data received?
17 0001800E 66FA                     BNE     GET_DATA     ;Repeat until data ready
18 00018010 10280002                 MOVE.B  2(A0),D0     ;Read data from ACIA
```

Figure 2.16 provides a memory map for this code. The first occurrence of ORG (i.e., ORG $001000) defines the point at which the following instructions and directives are to be loaded. The three lines after ORG employ the DS directive to define three named storage allocations of 260 words in all. Following these, two words, $0000 and $FFFF, are loaded into memory. The address of the next free location is $00120C. The three EQUs define constants for use in the rest of the program. Thus, whenever the name SETUP1 is used, the assembler replaces it by its defined value, 3. The line PIA EQU ACIAC+4 causes the word PIA to be equated to the value ACIAC (= $008000 + 4 = $008004).

The second ORG (i.e., ORG $018000) in line 12 of the assembled code defines the origin from which the following instructions are loaded. It is not necessary to allocate separate regions of memory for data and instructions in this way — the first instruction would have been located at address $00120C, had the second ORG not been used.

END The *end* directive simply tells the assembler that the end of a program has been reached and that there are no further instructions or directives to be assembled. Most assemblers employ an END directive without parameters. However, the END directive employed by the Teesside 68000 cross-assembler accompanying this text requires a single parameter. This parameter supplies the assembler with the point at which the assembly code is to start executing when it is run. Much of the code used in this text will be located from $400 onwards and the END directive will be END $400.

Figure 2.16 The memory map

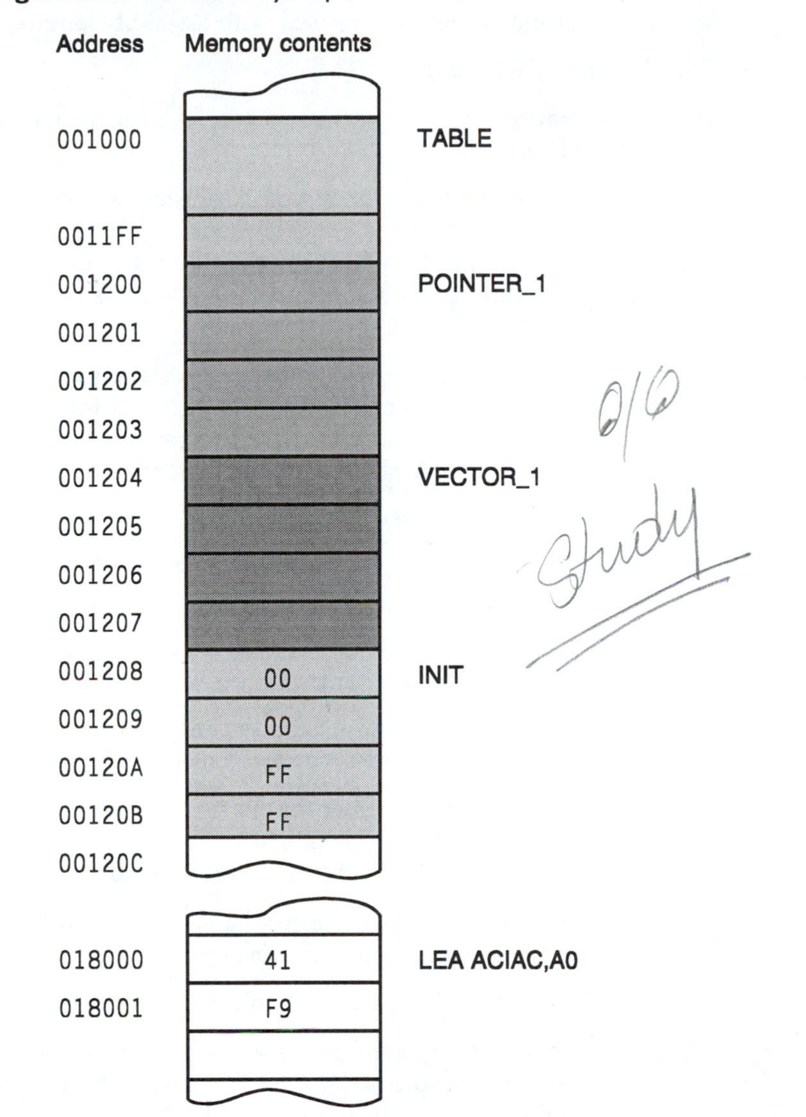

Summary

- A memory is accessed by providing it with an *address* and then writing to or reading from this location. The address defines the location that is to be accessed.
- Register transfer language, RTL, is used to describe the flow of data within a digital system and to define the action of an assembly language instruction.

RTL is nothing more than a notation. RTL is not a computer programming language and should not be confused with assembly language.

- The elements of RTL are:

  a. ***Data movement*** The expression `X ← Y` means that information Y is copied to X.

  b. ***Square brackets*** The contents of a register or a memory location are specified by square brackets. For example, the expression `[X]` means the contents of location X. The symbol X indicates either a register or a memory location. The expression `[X] ← 5` means that the value 5 is written into location X.

  c. ***Round brackets*** Parentheses indicate a location within a memory system. For example, the expression `[M(12)]` is interpreted as the contents of memory location 12. The expression `[M(12)] ← 5` is interpreted as the value 5 is loaded into memory location 12.

- The important registers of a CPU are:

  PC The program counter that contains the address of the next instruction to be executed.

  MAR The memory address register that contains the address of the location in memory being accessed.

  MBR The memory buffer register that holds data about to be written into memory, or data just read from memory.

  IR The instruction register that holds the op-code read from memory during the fetch phase of instruction execution. The IR also has an operand field that holds an operand address or a literal.

  D0 The data register or accumulator that is used to hold temporary data in the CPU during arithmetic and logical operations.

  ALU The arithmetic and logical unit that actually executes instructions.

  CCR The condition code register whose bits are set or cleared after each operation performed by the CPU. These bits are used to implement conditional branches.

- The computer operates in a two-phase fetch/execute mode. In the first phase an instruction is read from memory and decoded. In the second phase it is interpreted or executed.

- Most computer instructions consist of two parts: the op-code and the address of the operand. For example, the instruction `CLR 1000` means clear the contents of memory location 1000. Typical 68000 instructions permit you to specify a data register (one of eight) and an address in memory. For example, the instruction `MOVE 100,D3` copies the contents of memory location 100 to data register D3.

- An assembly language instruction takes one of two types of operand; an *address* or a *literal*. The address specifies the location of the data to be used by the current instruction. The literal is the actual data required by the current instruction. `ADD 1000,D0` means add the contents of memory location 1000 to D0, whereas `ADD #1000,D0` means add the number 1000 to D0.
- A computer executes instructions in order, one by one. Certain instructions can modify this natural sequence. A *conditional branch* can modify the order in which instructions are executed and has the assembly language form `Bcc address` address. If the condition tested by the instruction is false, the next instruction in sequence is executed. If the condition tested is true, the next instruction to be executed is found at the location "`address`."
- An assembly language program consists of two types of statement. The first type includes *executable statements* that are converted into machine code and eventually executed by the processor. The second type includes *assembler directives* that are never executed by the processor, but which tell the assembler something it needs to know to perform its assembly.
- The most important assembler directives are:

  a. `EQU` This equates a symbolic value to a constant.

  b. `DC` This directive defines a constant. That is, it takes a constant and puts it into memory.

  c. `DS` This directive defines storage. That is, it leaves a gap or space in memory for constants and data.

  d. `ORG` This directive resets the origin or starting point of the program or its data. That is, it tells the assembler the point in memory at which to continue loading instructions or data.

Problems

1. Without taking it apart, how can you tell whether a chess computer is:

   a. a chess-engine that directly executes the moves of chess?

   b. a computer programmed in Pascal that implements a program to simulate a chess player?

   c. a computer programmed in its assembly language?

2. Why is it often harder to learn an assembly language than a high-level language?

3. What is a von Neumann machine?

4. A stored program computer locates programs (i.e., instructions) and data in different regions of the same memory. Is this necessary? What are the alternatives?

5. What is the *von Neumann bottleneck* and what do you think can be done to remove it or to reduce its effect on the performance of a computer?

6. When I studied algebra in high school I encountered equations like X = 3Y + 2. Why do you think that such equations lay the foundation for the confusion of future generations of computer programmers?

7. Why do you think that the term *program counter* is a very unfortunate name for the register that holds the address of the next instruction to be executed?

8. What is a *literal* in the context of assembly language?

9. Write an assembly language program to evaluate the expression `B+C(D-G)*(P+Q)/(P-Q)`, where all symbolic names refer to locations in memory. You may assume the existence of the instructions:

```
MOVE D0,Z        MOVE Z,D0     ADD Z,D0      SUB Z,D0
MULU Z,D0        DIVU Z,D0.
```

10. Explain how a computer can make a *decision*.

11. Define the terms:

    a. Assembler b. Native language
    c. Source code d. Object code
    e. List file

12. What is the difference between an *executable instruction* and an *assembler directive*?

13. Write a sequence of assembler directives that will store two longwords, $12345678 and $FF00, in memory starting at location $1000. The first longword is to be called "Test" and the second "Terminal".

14. Draw a memory map to illustrate the following code.

```
        ORG     $1000
        DS.B    4
Mon     DC.W    12
Tue     DS.L    2
Wed     DC.B    1,2,3,4
Thu     DC.B    'A'
```

15. What, if any, are the errors in the following code?

```
        ORG     #1000
X       EQU     5
Y       EQU     $12
        DS.L    12
One     DS.B    1
Two     DC.B    X*Y
Three   DS.B    One+Two
Four    DC.L    Three*Five
Five    DC.B    X+Y
One     DS.B    6
```

16. What would the following code store at address $1018?

```
        ORG     $1000
P       EQU     5
Q       DS.L    6
One     DC.W    P+Q
```

17. What is the difference between the program counter and the location counter?

18. Are the two following fragments of code equivalent?

```
        ORG     $1000
        MOVE.B #2,D0
        MOVE.B D0,Temp

        ORG     $1000
Temp    DC.B    2
```

19. Explain the action of the following sequence of assembler directives.

```
        ORG     $1000
        DC.B    9
        DC.B    '9'
        DC.B    %101
        DC.B    $010
        DC.B    101
        DC.B    '101'
        DC.B    3*'3'+%101-$101
```

CHAPTER 3

Introduction to the 68000

In this chapter we introduce a simplified model of the 68000's architecture and demonstrate some of the basic principles of assembly language programming. At this stage we do not provide a full and exact model of the 68000, since adding too much detail would cause us to lose sight of our basic objectives. We introduce the 68000's register set, some of its instructions and its basic addressing modes. This limited picture of the 68000 does not provide a false impression of the 68000, as we simply hide some of its more powerful features until we have digested its fundamental principles. By the end of this chapter you should be able to write basic 68000 programs.

We also introduce the Motorola's single board computer that enables you to enter 68000 programs and to run them in a *debug mode*. That is, you can monitor the execution of your programs, instruction by instruction, and locate errors. Since not all readers will have access to this or to similar hardware, we include a program that runs on an IBM PC or a compatible that simulates the single board computer. In the next chapter we look at a much more detailed model of the 68000.

3.1 Simplified Model of the 68000

Figure 3.1 describes the 68000's register set. This figure presents a fairly complete picture of the 68000 and we have introduced only three simplifications. These simplifications do not affect what we are going to say about the 68000 in this chapter and help us to avoid long-winded discussion concerning some of the 68000's more esoteric features at this stage. OK, if you really want to know, these simplifications are: the actual 68000 has an 8-bit status byte as well as a CCR (condition code register), a second A7 address register, and only 24 bits from the program counter are connected to the 68000's address bus.

The 68000 has a conventional program counter, PC, that points at the next instruction to be executed. Instead of just a single data register, the 68000 has eight *general-purpose* data registers D0 to D7. It is important to appreciate that these data registers are general in the sense that whatever you can do to D_i, you can also do to D_j. That is, none of the data registers are reserved (by the designers of the 68000) for special applications. Some microprocessors implement special-purpose registers that can be used only in certain ways — not the 68000 family. All eight data registers are 32 bits wide, although, for our current purposes, we will regard them as being just eight bits wide.

Figure 3.1 Simplified model of the 68000's architecture

Figure 3.1 shows how a data register can be divided into two words (D_{00} to D_{15} and D_{16} to D_{31}). When a word operation is applied to a data register, only the low-order word takes place in the operation. Similarly, byte operations can be applied only to bits D_{00} to D_{07} of a 32-bit register.

The eight *address registers*, A0 to A7, are so-called because the information stored in an address register represents the *location* of an item in memory. Suppose that address register A0 contains the number N (i.e., `[A0] = N`). We can say that address register A0 *points* at memory location N. The *pointer* is a key component of all assembly languages (and most high-level languages). In short, address registers are used only to access data in memory. Figure 3.2 illustrates the pointer register. Address register A0 contains the value 1005 and therefore points to memory location 1005. This location holds the data value 57. We can write `[M([A0])] = [M(1005)] = 57`, that is, the contents of the memory location whose

Figure 3.2 The pointer register

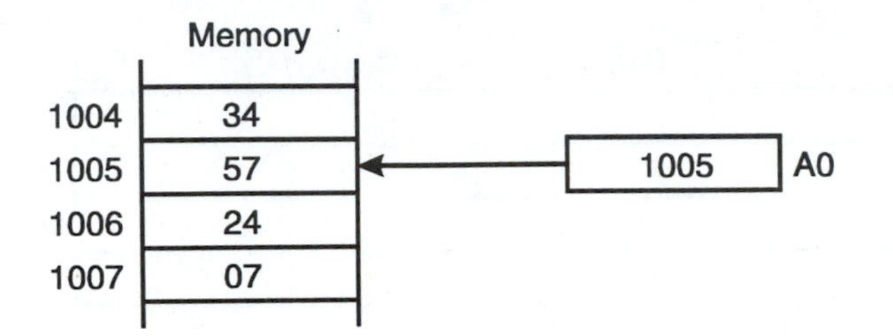

address is given by the contents of A0 is 57. We shall soon discover how memory can be accessed by using an address register as the source of an address.

The 68000's address registers are all 32 bits wide (like data registers), and we will always use them as 32-bit pointers. You cannot perform a byte operation on one of the 68000's address registers, since addresses are 32-bit values. The bits in a data register have an arbitrary meaning and can be arranged in any way the programmer sees fit. For example, a programmer may treat the 32 bits of D0 as four ASCII-encoded characters and the 32 bits of D1 as a 32-bit unsigned integer. The bits of an address register represent a single entity called an address, and an address cannot be subdivided into smaller units.

Address registers A0 to A6 all behave in the same way and you can use them as you like. Whatever you do with A_i, you can also do with A_j, where i and j are in the range 0 to 6. However, although address register A7 behaves like A0 to A6, it has an extra function. It is a *stack pointer* used to keep track of subroutine return addresses. We will describe subroutines shortly. Consequently, the assembly language programmer does not use A7 explicitly, except for certain special applications described in later chapters.

The 68000's *condition code register*, CCR, Figure 3.3, contains the usual N, Z, V, and C flags. We employ the term *usual* because the 68000's CCR is very much like the CCR of most conventional microprocessors. When the 68000 executes an instruction, the CCR is updated to reflect the result of the operation (in fact, some instructions do not update the CCR and leave its bits unchanged). If the result of the operation yields zero, the Z-bit is set to 1. If a carry-out is generated by the most-significant bit of the operand, the C-bit is set. If the operation causes an

Figure 3.3 The condition code register

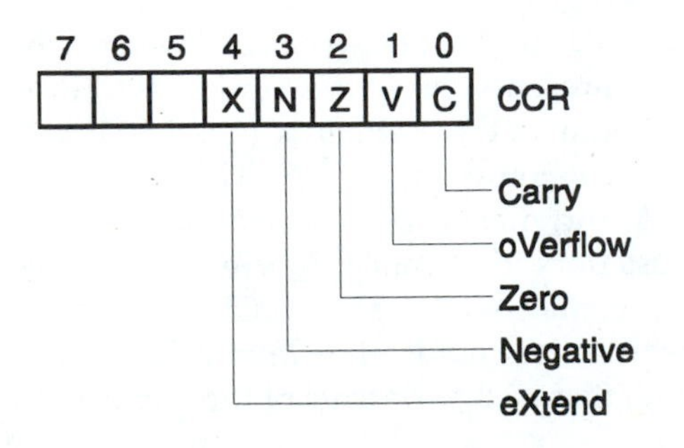

arithmetic overflow (in terms of two's complement arithmetic), the V-bit is set. Finally, if the result is negative (in two's complement terms), the N-bit is set. The CCR also has an X-bit (i.e., the eXtend bit) that we will not consider here. All we need say now is that it is set to the same value as the C-bit after certain operations. Unfortunately, not all microprocessors treat the CCR in the same way. For example, some update it after a data move operation and some don't.

It would be nice to tell you how the bits of the CCR are affected by the 68000's various instructions. Unfortunately it is impossible. Or, at least, it cannot be done in a few words. The relationship between the 68000's instructions and the bits of its CCR is determined by those who designed the 68000 and there are no absolute and universal rules. You have to look at the 68000's instruction set in Appendix B to determine its effect on the CCR. For example, an arithmetic operation such as an addition affects all bits of the CCR. A logical operation like a logical AND does not affect the V-bit, since arithmetic overflow is meaningless when logical operations are performed. An operation like EXG (exchange registers) has no effect on the bits of the CCR.

Now that we have described the 68000's eight data and address registers, the next step is to introduce some of the instructions that operate on data in these registers.

3.2 A Basic Instruction Set

Consider the following three questions: How many instructions do we need to write programs?; How many instructions would we like to have in the best of all possible worlds?; and How many instructions does the 68000 actually implement? The answer to the first question is very simple, although not particularly easy to believe. Computer scientists tell us that all programs can be synthesized by using one single instruction (I've read a paper on the subject. It's weird and convoluted, but it is possible to write an arbitrarily complex program with just one instruction). Saying that you can write a program using an instruction set with one single instruction is rather like saying that you can build a galaxy using only hydrogen atoms — true, but not exactly helpful.

The answer to the second question (how many instructions would you like?) is much debated by computer scientists. On the one hand, a large instruction set provides the programmer with a very powerful toolkit, but it is difficult to use these instructions efficiently. If you have lots of instructions in your instruction set you can probably write a program in many different ways using different sequences of instructions. Indeed, it is difficult to select an optimum sequence of instructions for a given program. In plain English, the programmer is spoiled for choice. On the other hand, a small instruction set may be inefficient but it is often much easier to use. If you have few instructions, the number of ways in which you can express an algorithm is limited. This factor was one of the driving forces behind the RISC (reduced instruction set computer) revolution.

The 68000 has a fairly large instruction set that includes some very powerful instructions. A first encounter with such a rich instruction set can prove quite frightening. So, in this section we are going to cover a tiny fraction of the 68000's instructions. Fortunately, these instructions are sufficient to write programs to perform all the actions we need. The 68000 has instructions that enable us to write programs in a more efficient or concise form.

Although the 68000 has a lot of assembly language level instructions, we can divide all its instructions into a small number of categories. If we think about what the microprocessor (i.e., CPU) actually does, we can identify four fundamental groups of instructions. The most obvious thing we can do with data is to *move* it from one place to another. So, our first group of instructions comprises all those operations that transport data from one location to another. The second thing we can do with data is to *modify* it or to operate on it (for example, by addition or multiplication). The second group of instructions comprises all those operations that act on data. The third group of instructions includes those that *modify the sequence* in which instructions are executed and those that help us examine the result of an operation and then choose between two courses of action depending on the result. This group includes the conditional branches that enable the assembly language programmer to synthesize high-level constructs like loops. Subroutine and procedure calls also belong to this group. The fourth group of instructions includes all the actions that determine the operating mode of the processor (this group is not considered further here).

68000-family dyadic instructions (i.e., with two operands) are described as *register-to-register*, *register-to-memory* or *memory-to-register*. That is, the instructions take two addresses, one of which is a register and the other a memory location (or a register). A few instructions do permit direct *memory-to-memory* operations.

We are now going to look at the first three groups of instructions, starting with data movement. But before we can do this, we must re-introduce a special addressing mode we first met in Chapter 2. The instruction `MOVE.B D3,1234` means *move the contents of data register D3 to memory location 1234*. Very frequently we need to use an actual *value* in an operation rather than the *contents* of a memory location. Suppose we need to put the number 25 in data register D2. The operation `MOVE.B #25,D2` loads the value 25 into register D2. The 68000 assembler employs the special symbol '#' to indicate that the following number is an *actual value* (called a *literal* or an *immediate* value) and is not a reference to a memory location. The assembly language instruction, `MOVE.B #25,D2` is expressed as `[D2]` ← `25` in RTL. Similarly, the assembly language instruction `MOVE.B 25,D2` is expressed as `[D2]` ← `[M(25)]`. Probably the single most common mistake made by those learning to program the 68000 is to forget to prefix the operand by the '#' symbol when dealing with a literal.

By the way, you should not confuse the symbol '#' with the symbols '$' and '%'. The prefix '#' before an operand tells the assembler that the following value is the actual operand and not a memory reference, while the symbols '$' and '%' simply tell the assembler that the following number is in the hexadecimal or binary base, respectively. For example, the operation `MOVE.B #$14,D0` means *load the number* 20_{10} *(i.e.,* 14_{16}*) into data register D0.*

Data Movement Instructions

The 68000's instruction set includes 13 data movement operations. All a data movement instruction does is copy information from one place to another. Hardly exciting, but it has been reported that 70% of the average program consists of data movement operations. The 68000 implements the following data movement instructions: MOVE, MOVEA, MOVE to CCR, MOVE to SR, MOVE from SR, MOVE USP, MOVEM, MOVEQ, MOVEP, LEA, PEA, EXG, SWAP. In this introduction we will look at just the plain vanilla MOVE instruction.

The MOVE instruction copies an 8-, 16-, or 32-bit value from one memory location or register to another memory location or register (i.e, MOVE.B, MOVE.W, MOVE.L, respectively). Most of the 68000's instructions indicate the size of an operand by appending a suffix to the mnemonic. If you don't use a suffix, the default size is normally .W. A few 68000 instructions are intrinsically 8-bit or 32-bit operations. Consider the following examples of the MOVE instruction:

Assembler form	RTL description	Transfer type
MOVE.B D1,D2	[D2] ← [D1]	Register to register
MOVE.B D1,1234	[M(1234)] ← [D1]	Register to memory
MOVE.B 1234,D3	[D3] ← [M(1234)]	Memory to register
MOVE.B 1234,2000	[M(2000)] ← [M(1234)]	Memory to memory

The 68000's MOVE instruction is its most general instruction in the sense that both source and destination operands may be registers or memory locations. This is not true of all other instructions, as most of the 68000's dyadic operations take place between the contents of a memory location and the contents of a register (or between the contents of two registers).

It is important to appreciate that the MOVE instruction is really a *copy* instruction, because the data is copied from its source to its destination. The value of the source is not modified by this instruction. You can move a literal as well as the contents of a memory location:

Assembler form	RTL description	Transfer type
MOVE.B #4,D2	[D2] ← 4	Literal to register
MOVE.B #12,1234	[M(1234)] ← 12	Literal to memory

You can load a literal into a register or a memory location, but not vice versa. An instruction of the form MOVE.B 1234,#12 is entirely meaningless, as you can't move the contents of memory location 1234 into *the number 12* (this is therefore an illegal instruction). Note that executing a MOVE instruction updates the contents of the condition code register even though no actual data processing takes place. For example, executing the instruction MOVE.B #0,D3 (i.e., put the value 0 in register D3) sets the Z-bit of the CCR to 1 to indicate that the result of the operation was zero. Some microprocessors do not update the CCR following a MOVE operation.

Sequence Control Instructions

When I first started writing this introduction to the 68000's instruction set, I wanted to describe arithmetic instructions and the basic instructions that operate on data immediately after I had introduced data move instructions. Although it is perfectly easy to define these instructions without introducing other classes of instructions, it is virtually impossible to demonstrate how they are used in practice. For example, we can add two numbers together by means of an ADD instruction, but we cannot test the result to determine whether it is negative or has gone out of range without introducing instructions that control the flow of a program.

At this stage, we are going to introduce just six of the 68000's fourteen sequence control instructions (i.e., conditional branches). These are: BEQ, BNE, BCC, BCS, BVC, and BVS. Each of these instructions is written in the mnemonic form: B_{cc} address, where the subscript 'cc' defines the condition to be tested and address is the *branch target address*. Programmers almost never use a *numeric* branch target address, as it is difficult for a programmer to calculate the target address of the branch. In any case, a numeric branch address would make a program virtually unreadable. Programmers label the line they wish to jump to and use a symbolic address (the assembler automatically replaces the symbolic address by the actual address).

By the way, we are slightly simplifying matters a little here. The 68000 does not store the actual address of a branch target as part of a branch instruction, but the number of bytes it has to skip forward or backward from the current instruction. We will return to this point when we introduce relative addressing.

Consider the instruction BCC (branch on carry clear):

```
          BCC      Check_5
          MOVE.B   D1,D2
          .
          .
Check_5   MOVE.B   #1,D4
```

The first line, BCC Check_5, means *If the carry bit in the CCR is clear (i.e., [CCR(C)] = 0), then execute the instruction whose address is given by the line labelled with the symbolic name* Check_5. We call the line labelled Check_5 the *branch target address*. Otherwise (i.e., if [CCR(C)] = 1), execute the instruction immediately following the BCC Check_5 (i.e., the instruction MOVE.B D1,D2). Figure 3.4 illustrates the effect of a branch instruction graphically.

We can add comments in RTL to this fragment of code to illustrate its operation.

```
        BCC     Check_5   IF [CCR(C)] = 0 THEN [PC] ← Check_5
        MOVE.B D1,D2                          ELSE execute this instruction
        .
        .
Check_5 .
```

Figure 3.4 The conditional branch

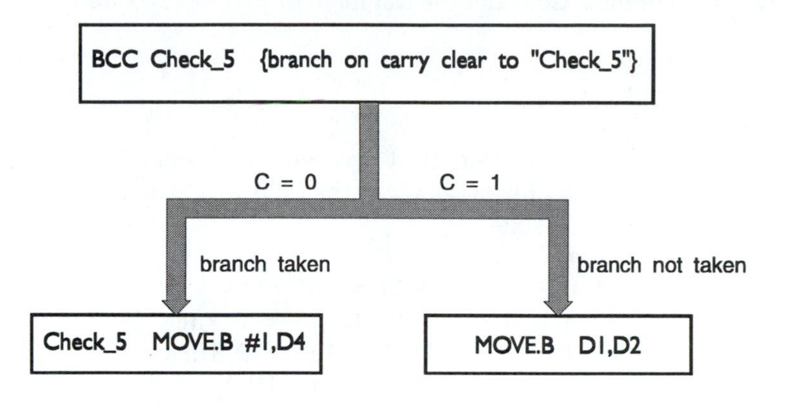

Real programmers would never use RTL when adding comments to their programs. We have added comments in RTL purely to define the effect of the instructions. A programmer might document the above fragment of code in the following way.

```
          BCC     Check_5    IF carry bit is clear THEN Check_5
          MOVE.B D1,D2                                ELSE ......
          .
          .
Check_5   .
```

The meaning of each of the six conditional branch instructions is as follows:

Mnemonic	**Instruction**	**Branch taken on**
BNE	Branch on not equal	[CCR(Z)] = 0
BEQ	Branch on equal	[CCR(Z)] = 1
BCC	Branch on carry clear	[CCR(C)] = 0
BCS	Branch on carry set	[CCR(C)] = 1
BVC	Branch on overflow clear	[CCR(V)] = 0
BVS	Branch on overflow set	[CCR(V)] = 1

Note the use of the rather strange mnemonics BNE and BEQ for the two instructions that test the state of the Z-bit of the CCR. You might expect to see the mnemonics BZC (branch on Z clear) and BZS (branch on Z set). The reason why the actual mnemonics have been chosen is easy to find. If you wish to compare two values you can subtract one value from the other. For example, the contents of data registers D1 and D2 can be compared by means of the instruction SUB D1,D2. If the result of the subtraction is zero, the Z-bit will be set to one, indicating that the values were the same. Therefore, the mnemonic BEQ (branch on equal) is used rather than BZS.

Before we introduce arithmetic operations, we will look at a simple example of a conditional branch. Consider the fragment of high-level code:

```
IF X1 = 0 THEN X1 := Y1
          ELSE X1 := Y2
```

This construct sets X1 to Y1 if X1 was zero; otherwise, it sets X1 to Y2. We can code this in 68000 assembly language in the following way (if you notice an error, don't worry — it's deliberate).

```
     MOVE.B  X1,D0       [D0] ← [M(X1)]
     BEQ     Zero        IF [M(X1)] = 0 THEN [PC] ← Zero
     MOVE.B  Y2,X1       IF [M(X1)] ≠ 0 THEN [M(X1)] := [M(Y2)]
Zero MOVE.B  Y1,X1       [M(X1)] := [M(Y1)]
```

Once again we must point out that the programmer would not use RTL to document a program. Many programmers prefer to document assembly language in something approaching a high-level language. There is little point in noting that `MOVE.B X1,D0` copies X1 into D0, as it is obvious from the instruction. We performed the `MOVE` operation to force the 68000 to update its condition code register. The purpose of documenting an assembly language program is to give the reader the big picture. The following example demonstrates how we might comment the above program.

```
     MOVE.B  X1,D0      IF X1 = 0
     BEQ     Zero          THEN X1 := Y1
     MOVE.B  Y2,X1         ELSE X1 := Y2
Zero MOVE.B  Y1,X1
```

Figure 3.5 illustrates the action of this code diagrammatically. Unfortunately, the above program cannot work. If the result of the test is *false* (i.e., `[M(X1)]` ≠ 0), the `ELSE` part of the construct is executed (i.e., the instruction `MOVE.B Y2,X1`). However, once the instruction has been executed, the next instruction (i.e., the instruction `MOVE.B Y1,X1`) is inadvertently executed. We need to execute the `ELSE` part and skip the `THEN` part of the construct. The program can be readily modified to deal with this problem:

```
     MOVE.B  X1,D0      [D0] ← [M(X1)]
     BEQ     Zero       IF [M(X1)] = 0 THEN [PC] ← Zero
     MOVE.B  Y2,X1      IF [M(X1)] ≠ 0 THEN [M(X1)] := [M(Y2)]
     BRA     Exit       [PC] ← Exit {Skip the 'THEN' part}
Zero MOVE.B  Y1,X1      [M(X1)] := [M(Y1)]
Exit ......
```

After the `MOVE.B Y2,X1` instruction has been executed, an unconditional branch is made to the line we have called '`EXIT`'. An unconditional branch is *always*

Figure 3.5 Coding a conditional branch — incorrectly!

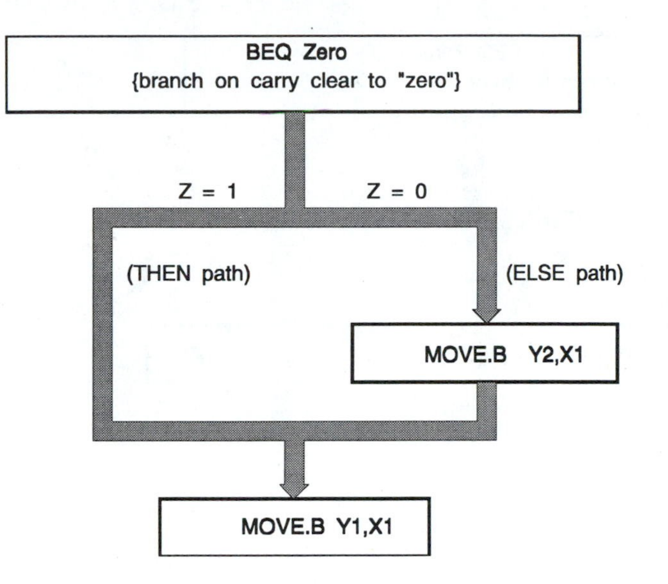

taken and has the assembly language form `BRA Label` (it is equivalent to the high-level construct `GOTO`). Figure 3.6 illustrates this modification diagrammatically.

Before leaving this topic, we will briefly introduce the *subroutine* (we have a lot more to say about subroutines later). In all but the most trivial programs you frequently have to execute a series of operations several times. Consider the following sequence of everyday activities.

```
Get Up
Eat
Go to Work
Eat
Do more Work
Return Home
Eat
Go to Bed
```

The operation `Eat` occurs three times in this sequence. If it were a program, the action `Eat` would have to be replaced by a sequence of instructions every time it appeared. Writing large programs in this way would require an excessive repetition of code. The *subroutine* provides a solution to this problem.

A subroutine is a section or fragment of code that can be *called* or invoked from any point in a program. A subroutine usually performs a single coherent task (e.g., input a character or calculate a mathematical function — it is not simply a random sequence of instructions). Once the subroutine has been executed, a return can be made to the point immediately following the subroutine call. Figure 3.7 illustrates these points.

Figure 3.6 Coding the IF...THEN...ELSE construct

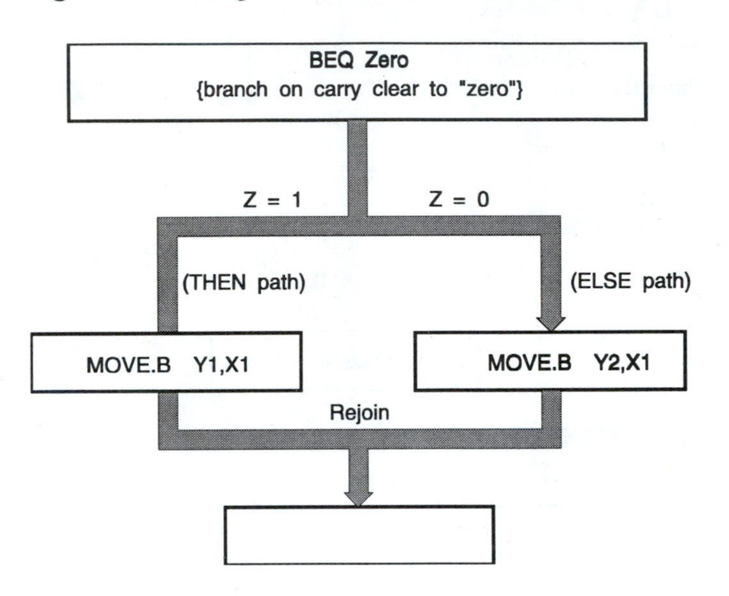

The principal advantage of a subroutine is that it can be called from different places within a program and a correct return always made to its calling point. Consider the example of Figure 3.8 in which the same subroutine is called twice.

Subroutines are delightfully simple to use. All we have to write is `BSR label` and the processor automatically jumps to the line in the program labelled `label`. Consider the following example of a subroutine, `Add12`, that adds the contents of data register D1 to register D2 and subtracts the value 12 from the result (an unrealistically simple subroutine).

```
      BSR    ADD12          Call subroutine 'ADD12'
      .
      .
      .
      BSR    ADD12          Call subroutine 'ADD12'
      .
      .
      .
ADD12 ADD.B D1,D2           D2 := D2 + D1
      SUB.B #12,D2          D2 := D2 - 12
      RTS                   Return from subroutine
```

In this fragment of code the *same* subroutine is called twice, each time returning to the instruction following the subroutine call. Note that the subroutine itself is simply a block of code terminated by the instruction `RTS` (return from subroutine). The effect of an `RTS` instruction is to force the 68000 to load the program counter with the address of the next instruction following the subroutine call. At this point we simply wish to indicate the existence of subroutines. Later we will show how the call and return mechanism works and how you might pass parameters (i.e., data) to the subroutine.

Now that we have introduced the concepts of branching and subroutines, we can put them to good use when we describe the action of arithmetic and logical instructions.

Figure 3.7 The subroutine call

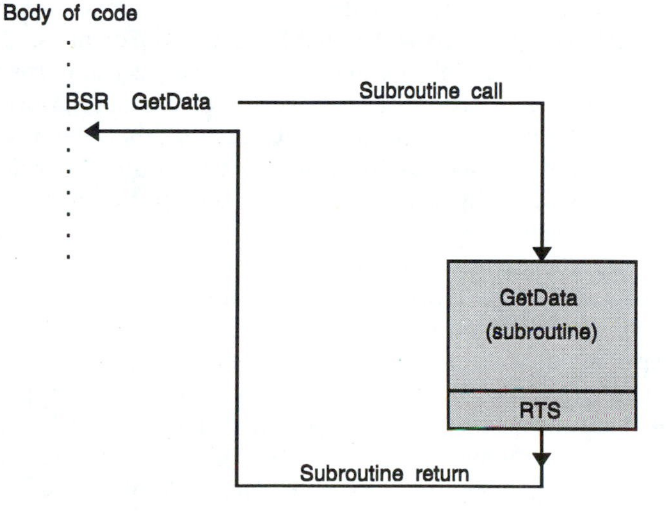

Figure 3.8 The nested subroutine call

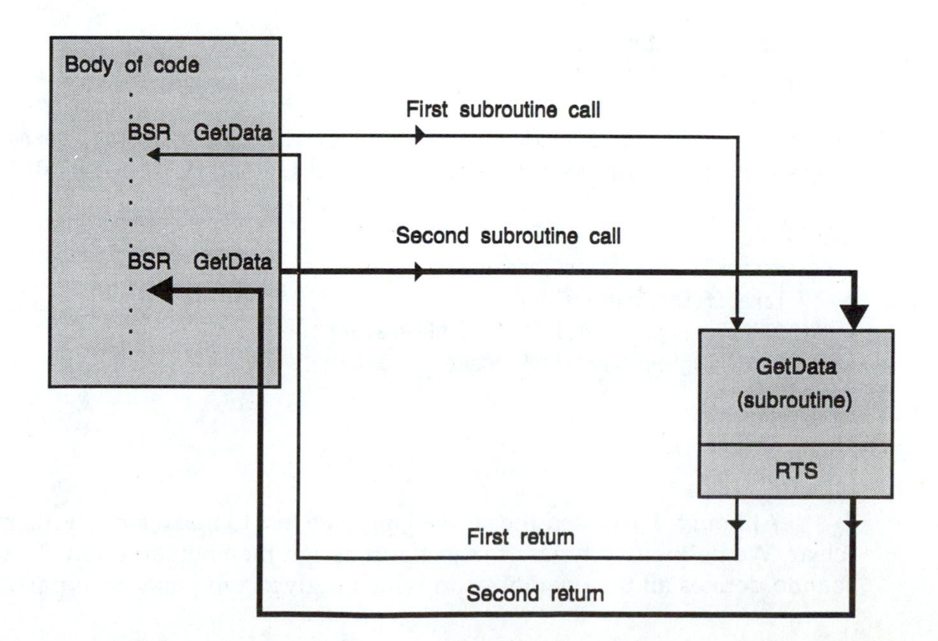

Arithmetic Operations

A computer's arithmetic instructions comprise those that act on data representing numerical values. I know that this is a statement of the apparently obvious. However, you must appreciate that *at the machine level* the processor will do exactly what it is told. It will quite happily multiply the ASCII code for the letter "A" by two if that is what you instruct it to do. The letter "A" is represented by the binary value 01000001, which is the same bit-pattern as the representation of the integer 65. Multiplying the number 65 by two makes more sense than multiplying the code for "A" by two. Before continuing, we will examine the concept of *data typing*.

High-level languages such as Pascal, Modula, and Ada are said to be *strongly typed* which means each data element is defined to be a certain type and may take part only in operations appropriate to its type. For example, a programmer can define a new data type called *vector* as a data element that may take part in only two legal operations: the calculation of a vector sum and the calculation of an inner product. That is, a programmer can apply only two legal operations to data structures declared to be of type vector.

Many computer scientists are deeply suspicious of assembly languages because they are not strongly typed and programmers can perform any operation on any data structure. If you think *strong typing* is an abstract concept, try the following experiment next time you are flying in an aircraft as `type passenger`. Objects of `type passenger` may take part only in operations appropriate to this type (e.g., `FastenSeatBelt`, `EatPlasticFood`, `RemainSeatedUntilToldToMove`). Objects of type `pilot` have a much richer set of operations (e.g., `TakeOff`, `EngageAutopilot`, `ClimbToFlightLevel[n]`, `ReduceEnginePower`, and `OperateRadio`). We will demonstrate how strong typing affects the programmer by continuing this example with an entirely hypothetical high-level language.

```
SmoothOperator: type pilot
AlanClements  : type passenger

Pilot     cando {TakeOff, EngageAutopilot, ClimbToFlightLevel[n], OperateRadio}
Passenger cando {FastenSeatBelt, EatPlasticFood, RemainSeatedUntilToldToMove}

BEGIN
     FastenSeatBelt(AlanClements)
     TakeOff(SmoothOperator)
     ClimbToFlightLevel[100](SmoothOperator)
     ReduceEnginePower(AlanClements)
     .
     .
END
```

Although I invented the above language just to make a point, its meaning is clear. We define two types of object and assign members to them. The construct `cando` defines all the operations in which a given type may take part. When we provide a sequence of executable instructions, you can see that the operation `ReduceEnginePower(AlanClements)` is illegal because `AlanClements` does not belong to the type that gets to fly aircraft. This error would be flagged by the compiler. So, next time you fly, ask if you can go into the cockpit and perform some of the operations normally carried out by type `pilot` and see what happens (but please don't mention me or my book). I have provided this long-winded example to emphasize that it is easy to do things in an assembly language that a high-level language would not permit. Although there are occasions when this freedom from type restrictions makes assembly language programming much easier, you have to be very careful not to perform inappropriate operations on data. One of the most common assembly language programming errors is to perform an operation on a data element inappropriate to its type. We now return to arithmetic operations.

The 68000 has a conventional set of arithmetic operations, all of which are integer operations. Floating point operations are not directly supported by the 68000. At this stage we will look at some of the 68000's basic arithmetic operations and leave the more complex operations until later. The arithmetic instructions we are going to use in this section are: `ADD`, `SUB`, `CLR`, `NEG`, `ASL`, and `ASR`.

`ADD` The `ADD` instruction adds the contents of a source location to the contents of a destination location and deposits the result in the destination location. Either the source or destination must be a data register. Memory to memory additions are not permitted. The following are examples of `ADD` instructions.

Assembler form	**RTL description**	**Transfer type**
`ADD.B D1,D2`	`[D2] ← [D1] + [D2]`	Register to register
`ADD.B D1,1234`	`[M(1234)] ← [D1] + [M(1234)]`	Register to memory
`ADD.B 1234,D3`	`[D3] ← [M(1234)] + [D3]`	Memory to register
`ADD.B #12,D2`	`[D2] ← 12 + [D2]`	Literal to register

An `ADD` instruction affects the CCR's Z-, C-, N-, and V-bits. We later provide examples of the effects of various arithmetic operations on the contents of the CCR. We now look at how an add instruction might be applied and use the occasion to refresh our memories about assembly language directives.

```
        ORG     $400      Data area starts here
V1      DC.B    12        Store the constant 12 in memory and call it the location 'V1'
V2      DC.B    14        Store the constant 14 in memory and call it the location 'V2'
V3      DS.B    1         Reserve one byte of storage at location V3
*
        ORG     $600      Reset the origin to $600 for the program
        MOVE.B  V1,D0     V3 := V1 + V2
        ADD.B   V2,D0
        MOVE.B  D0,V3
        BCS     Error1    IF carry set THEN error
```

The real purpose of this code is to demonstrate how you might use symbolic values rather than numeric values. No programmer would ever use actual numeric addresses (except under very special circumstances).

`SUB` The subtraction operation subtracts the source operand from the destination operand and deposits the result in the destination operand. For example, the assembly language instruction `SUB.B D2,D0` means subtract the contents of data register D2 from data register D0. This operation is expressed in RTL as `[D0(0:7)]` ← `[D0(0:7)] - [D2(0:7)]`. Similarly, `SUB.W D2,D0` is interpreted in RTL as `[D0(0:15)]` ← `[D0(0:15)] - [D2(0:15)]`.

`CLR` The *clear* instruction simply loads the contents of the specified operand with zero. We do not really need an explicit `CLR` instruction, since `MOVE.B #0,D0` is identical to `CLR.B D0`. That is, we can move the literal value zero into a location to achieve the same effect as a clear operation. The `CLR` instruction is implemented simply because it is faster than a move instruction and it is so frequently used to initialize registers and memory locations.

`NEG` The negate instruction subtracts the destination operand from zero and deposits the result at the destination address. That is, `NEG X` has the effect of calculating -X. This is a monadic operation (i.e., it takes a single operand) and has the assembly language form `NEG Di` or `NEG address`. The operand address may be a memory location or a data register. This instruction simply forms the two's complement of an operand. For example, if [D4] = 01101100, executing `NEG.B D4` results in [D4] = 10010100.

`ASL` The arithmetic shift left instruction moves the bits of the specified operand left by the stated number of places. If we take a decimal number (e.g., 123), and shift it one place left we multiply it by ten (i.e., 1230). Shifting a binary number one place left multiplies it by two.

One assembly language form of the `ASL` instruction is: `ASL #n,Di`. The first operand, `#n`, specifies n shifts left (i.e., multiply the operand by 2^n). If [D0] = 00010111, executing `ASL.B #1,D0` results in 00101110. Executing `ASL.B #2,D0` results in 01011100. When you shift the bits of a number by one or more places, three things happen. The bits get shifted; a new bit is shifted into the vacated position at one end; and a bit is shifted out of the other end. When an *arithmetic* shift left takes place, a *zero* is shifted into the right-hand end to replace the vacated bit. The bit shifted out of the left-hand end is not lost but is shifted into the C-bit of the CCR (and also into the X-bit). In a multiple shift, the last bit shifted out goes to the C-bit. Figure 3.9 illustrates the action of an `ASL` instruction.

`ASR` The arithmetic shift right instruction shifts the bits of the specified operand right by the stated number of places. In this case the operand is divided by two for each place shifted right. For example, if [D0] = 00010111, executing the instruction `ASR.B #1,D0` yields 00001011, while executing `ASR.B #2,D0` yields 00000101.

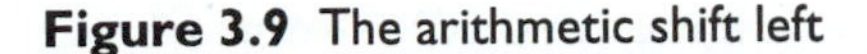

Figure 3.9 The arithmetic shift left

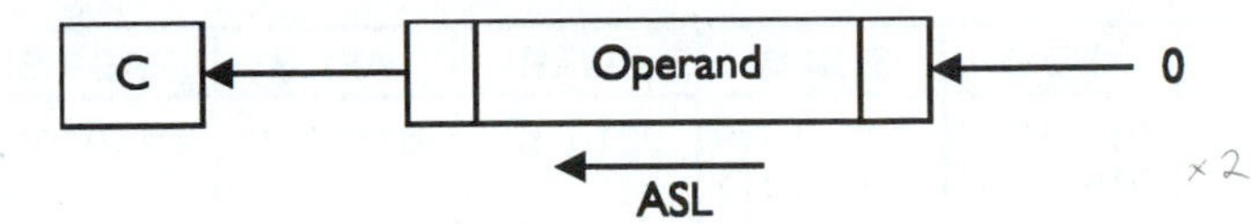

Now, suppose that [D0] = 11010110 and we execute an `ASR.B #1,D0`, what do we get? You might expect to see 01101011. This is not the result that you would get. The actual result is 11101011. An arithmetic shift right instruction shifts all bits one place right but moves the bit shifted out of the most-significant position back into the *most-significant position*. Figure 3.10 illustrates the operation of an `ASR` instruction.

It's not difficult to appreciate how an `ASR` operates when you consider that the 68000 supports two's complement numbers. In two's complement arithmetic the most-significant bit is a sign bit. Thus, the two's complement number 11101010_2 represents -22_{10}. If we shift this value one place right and replicate the sign bit, we get: 11110101_2, equivalent to -11_{10}.

Figure 3.10 The arithmetic shift right

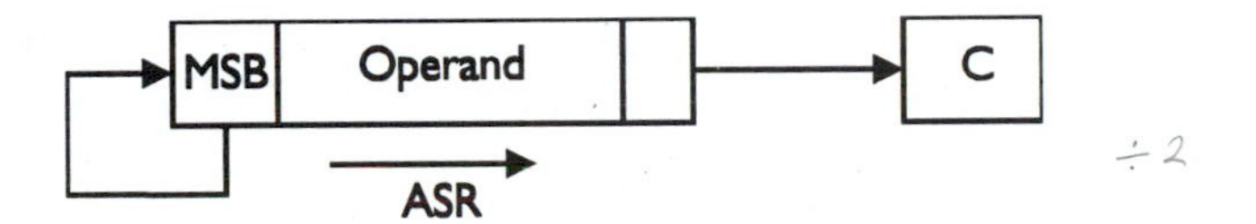

Effect of Arithmetic Operations on the CCR

Let's now see what effect arithmetic operations have on the contents of the CCR. All arithmetic operations modify both the destination operand and the contents of the condition code register. Consider the examples in Figure 3.11 in which we provide the source and destination operands *before* the operation and the destination operand and condition code register *after* the operation.

Note that the X-bit of the CCR is set to the same value as the carry bit during shift and add or subtract operations. The X-bit is not affected by a clear operation and the '-' symbol in the CCR in Figure 3.11 indicates that the state of

Figure 3.11 Effect of arithmetic operations on the CCR — example 1

	ADD.B	SUB.B	CLR.B	ASL.B	ASR.B
Source Destination	00101011 00011010	00011010 00101011	00101011	00101011	00101011
Destination	01000101	00010001	00000000	01010110	00010101
CCR	XNZVC 00000	XNZVC 00000	XNZVC -0100	XNZVC 00000	XNZVC 10001

Figure 3.12 Effect of arithmetic operations on the CCR — example 2

	ADD.B	SUB.B	CLR.B	ASL.B	ASR.B
Source Destination	01101011 01011010	01011010 00011011	00101011	01101011	01101011
Destination	11000101	11000001	00000000	11010110	00110101
CCR	XNZVC 01010	XNZVC 11001	XNZVC -0100	XNZVC 01010	XNZVC 10001

the X-bit is not modified by the CLR operation. These results are not particularly interesting, so let's look at some more interesting data in Figure 3.12.

When adding the two numbers (i.e., $01101011_2 = 107_{10}$ and $01011010 = 90_{10}$), the result (i.e., $11000101_2 = 197_{10}$) is outside the range permitted by two's complement values in 8 bits (i.e., -128 to +127). In this case, the binary result, 11000101, is negative because its sign-bit is 1. The N-bit of the CCR is therefore set to indicate a negative value. The V-bit (overflow bit) is also set to indicate that an out-of-range condition has resulted. In other words, the V-bit is telling us that the N-bit is giving a *false* or misleading reading and that the number is not really negative.

The subtraction in Figure 3.12 is of a positive number from a smaller positive number. In this case, the N-bit of the CCR is set to indicate a negative result. The V-bit is not set because overflow has not occurred. Note that a carry-out is generated due to the nature of two's complement arithmetic and therefore the C-bit (and the X-bit) is set.

Now consider the effect of the ASL instruction. A one is shifted into the most-significant bit position and the sign of the number has changed. Consequently, overflow has occurred and both the V- and N-bits are set. The ASR instruction does not result in overflow or a change in sign. However, a 1 is shifted out of the least-significant bit position and is caught by the C-bit and X-bits. Let's look at another batch of arithmetic operations, Figure 3.13.

Note that the addition of two negative numbers produces a positive result. Consequently, the V-bit is set and the N-bit cleared (as the most-significant bit of the result is a 0). The C-bit is set (since a carry-out was generated).

It is important to appreciate that the result of any arithmetic operation preserves all the information. Even if the result is wrong (due to overflow or to a

Figure 3.13 Effect of arithmetic operations on the CCR — example 3

	ADD.B	SUB.B	CLR.B	ASL.B	ASR.B
Source Destination	10101011 10011010	10101011 11011010	10101011	11101011	11101011
Destination	01000101	00101111	00000000	11010110	11110101
CCR	XNZVC 10011	XNZVC 00000	XNZVC -0100	XNZVC 11001	XNZVC 11001

bit shifted out), the state of the bits in the CCR can be used to recover from the error. Does the 68000 perform unsigned integer addition or signed integer addition using two's complement arithmetic? That was a trick question — there's no difference between signed and unsigned arithmetic. The only difference lies in the way in which you interpret the result of the addition. We will now look at an example, Figure 3.14.

In Figure 3.14 we have performed two pairs of addition; one using unsigned arithmetic and the other using signed arithmetic. Note how, for each pair of operations, the result (i.e., the sum) and the contents of the CCR are identical. This is as it should be — the 68000 does not know whether the data you are adding represents signed or unsigned numbers. The 68000 simply sets N if the most-significant bit is a 1, and V if the sign of the two numbers being added differs from their result. It is up to the programmer to make appropriate use of the contents of the CCR.

The addition of the first pair of numbers produces a correct result when interpreted as an unsigned value, as 197 is within the permitted range of 8 bits, 0 to 255. The addition of the second pair of numbers produces an incorrect result when viewed as a two's complement value, because the permitted range of values is -128 to +127.

The addition of the third pair of unsigned numbers (171 + 138) yields an incorrect value of 53, because the correct sum, 309, is outside the range of unsigned values. Finally, the addition of the fourth pair of signed numbers (-85 + -118) yields an incorrect result of +53 because the result is out of range.

How do we check whether the result is out of range? The answer is simple. If we are using unsigned addition, we check the state of the C-bit, and if we are using signed addition, we check the state of the V-bit. Consider the following two fragments of code.

Case 1 (unsigned arithmetic)

```
          ADD.B  D1,D2
          BCS    Error
            .
            .
Error       .
```

Case 2 (signed arithmetic)

```
          ADD.B  D1,D2
          BVS    Error
            .
            .
Error       .
```

Figure 3.14 Effect of arithmetic operations on the CCR — example 4

	ADD.B unsigned	ADD.B signed	ADD.B unsigned	ADD.B signed
Source Destination	01101011 (107) 01011010 (090)	01101011 (107) 01011010 (090)	10101011 (171) 10001010 (138)	11101011 (-085) 10001010 (-118)
Destination	11000101 (197)	11000101 (-081)	00110101 (053)	00110101 (053)
CCR	XNZVC 01010	XNZVC 01010	XNZVC 00011	XNZVC 00011

In Case 1, unsigned arithmetic, the addition is performed and a *branch on carry set* instruction executed. If the addition causes the carry bit to be set, the result is out of range (i.e., greater than 255) and a branch to the line labelled by 'ERROR' is made. Suitable code at this location must be provided to deal with the problem.

In Case 2, signed arithmetic, addition is performed in exactly the same way as in Case 1. However, the branch to error instruction is made only if two's complement arithmetic overflow takes place (i.e., the V-bit of the CCR is set). As you can see, it is up to the programmer to make the appropriate test after an addition (or any other arithmetic/logical operation). Note that you do not always have to perform such a test. If you know that the result can never go out of range, you do not need to test it.

Logical Operations

The 68000 implements four Boolean operations: AND, OR, EOR, and NOT. In general, logical operations are used to modify one or more bits of an operand. A logical AND masks out bits (i.e., clears them), an OR sets bits, and an EOR toggles them (i.e., causes them to change state). The following instructions illustrate the effect of these logical operations. In each case an immediate operand is used. The low-order byte of D0 before each operation is 11110000_2.

Assembly language	RTL action	Result of operation
AND.B #%10100110,D0	[D0] ← 10100110 . 11110000	[D0] ← 10100000
OR.B #%10100110,D0	[D0] ← 10100110 + 11110000	[D0] ← 11110110
EOR.B #%10100110,D0	[D0] ← 10100110 ⊕ 11110000	[D0] ← 01010110

We now demonstrate how logical instructions can be used. We provide first a simple example and then a more typical example of the use of logical operations. Suppose we wish to set the most-significant bit of the byte in D0, clear its least-significant bit, and change the state of bits 3 and 4.

```
OR.B  #%10000000,D0    Use an OR operation to set the most-significant bit
AND.B #%11111110,D0    Use an AND operation to clear the least-significant bit
EOR.B #%00011000,D0    Use an EOR operation to toggle bits 3 and 4
```

Let's look at an example of logical operations that requires us to test certain bits of a byte and then act on the result. A computer's input port is made up of 8 bits, ABCDEFGH (the least-significant bit is H). The same computer has an 8-bit output port with bits PQRSTUVW (the least-significant bit is W). The input port is located at address $008000 and the output port at address $008001. The memory map of Figure 3.15 describes the structure of these input and output ports.

Note that bit A of the input port is automatically set to 1 if bits B to H of the input are valid, and is set to zero otherwise. The output must be controlled according to the algorithm:

Figure 3.15 The memory-mapped input/output port

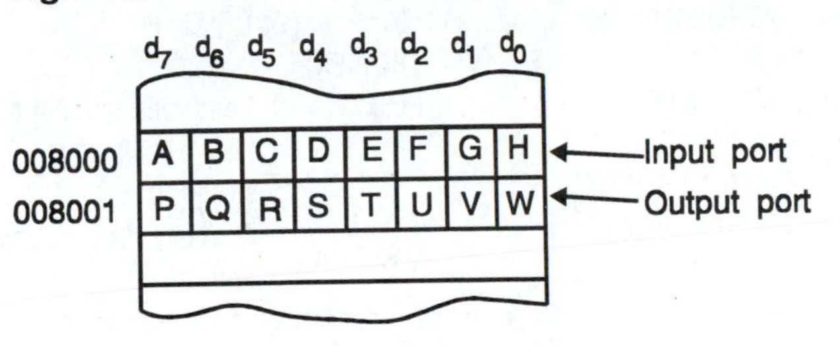


```
IF C = 1 OR F = 0 THEN P := 1; Q := 0 END_IF
IF C = 0 AND F AND H = 1 THEN P := 0 END_IF
```

In this problem we have to test certain bits of the input and then set or clear certain bits of the output. Since we have not yet covered program design, we will present the algorithm in almost plain English.

```
Read the input port UNTIL bit A = 1
Check bit C. IF C is 1 THEN set P to 1 and clear Q to 0 EXIT
             IF C is 0 THEN check bits F and H
                            IF both F and H are 1 THEN clear P to 0 EXIT
Check bit F. IF F is 0 THEN set P to 1 and clear Q to 0 EXIT
EXIT
```

We can code this algorithm in the following way:

```
ReadAgain MOVE.B  $00008000,D0    Read the input port to get bits A to H in D0
          MOVE.B  D0,D1           Save a copy of the input in D1
          AND.B   #%10000000,D0   Clear all bits of input except bit A
          BEQ     ReadAgain       Keep reading the input until bit A is set to 1
          MOVE.B  D1,D0           Restore the input to D0
          AND.B   #%00100000,D0   Clear all bits except bit C
          BEQ     CisZero         IF bit C is not 1 THEN deal with this case
          MOVE.B  $00008001,D2    ELSE Read the current output
          OR.B    #%10000000,D2        Set bit P = 1
          AND.B   #%10111111,D2        Clear bit Q = 0
          MOVE.B  D2,$00008001         Restore the output
          BRA     EXIT                 and EXIT
CisZero   MOVE.B  D1,D0           Restore the input
          AND.B   #%00000101,D0   Clear all bits except F and H
          CMP.B   #%00000101,D0   Test for both F and H = 1
          BNE     TestF           IF H,F are not 1,1 THEN EXIT
          MOVE.B  $00008001,D2                       ELSE Read the current output
          AND.B   #%01111111,D2                           Clear P = 0
          MOVE.B  D2,$00008001                            Restore the output
          BRA     EXIT                                    and EXIT
```

Polling Loop p 276

```
TestF    MOVE.B  D1,D0          Restore the input again
         AND.B   #%00000100,D0  Clear all bits except bit F
         BNE     EXIT           IF F = 1 THEN EXIT
         MOVE.B  $00008001,D2            ELSE F = 0 Read the current output
         OR.B    #%10000000,D2                      Set bit P = 1
         AND.B   #%10111111,D2                      Clear bit Q = 0
         MOVE.B  D2,$00008001                       Restore the output
                                                    and EXIT
EXIT     .....                  Exit point for this fragment of code
```

The program is rather more long-winded than need be. Since we have introduced only a few 68000 instructions, we have to perform some operations in a rather round-about fashion. However, the point of the program is to demonstrate how the AND operation can be used to mask out bits of an operand, and the OR operation can be used to selectively set bits.

3.3 Register Usage

Before we go on to addressing modes, I'd like to make a comment about the way in which data registers are allocated to variables. In the previous problem, we used registers D0, D1, and D2. Why didn't we use registers D5, D6, and D7? A long time ago, when I used to teach 6502 assembly language, life was relatively easy. This first-generation 8-bit processor has only one general-purpose register, called the A register or *accumulator*. All variables and constants used by the 6502 programmer had to be kept in main memory and loaded into the accumulator, processed, and then the result returned to memory. The question of register allocation did not arise. The 68000 has eight general-purpose data registers (i.e., accumulators) and makes life both easier and more difficult for the programmer.

The advantage of multiple data registers is obvious. Frequently accessed data is held in the microprocessor rather than in external memory. Accesses to data registers are not only faster than accesses to external memory, but they require shorter instructions. An instruction requires 32 bits to specify the location of an operand in memory, but only 3 bits to specify one of eight data registers.

The problem posed by multiple data registers is one of *organization*. How does the human programmer or the compiler allocate variables and constants to individual data registers? You get no help from the 68000, as it places no restriction on what data register does what — anything you can do with D0 you can do with data registers D1 to D7. Two problems of register allocation can be identified. The first concerns *ergonomics* and the second *efficiency*. The ergonomics of a program include its readability and its user interface. We want to choose a means of register allocation that makes the life of the programmer easier. Automatic compilers are not concerned with ergonomics, as one name is just as significant to a compiler as another. The efficiency of register allocation is all about the optimum use of registers and external memory. We do not deal with this topic here as it is the province of compiler writers.

Let's look at an example of register allocation. Suppose you have to write a program to calculate the number of hours that someone spends on a certain activity each workday of the week. If the algorithm requires the variables representing the days to be accessed frequently, it might make sense to allocate a data register to each of the five working days. For example, we could allocate D3 to Monday, D7 to Tuesday, D0 to Wednesday, D4 to Thursday, and D1 to Friday. This allocation is acceptable to the 68000 and is entirely correct (i.e., it violates no syntax rule). It is also nonsense. Imagine debugging such a program. A much more logical decision from an ergonomic point of view is to employ the same numbering system as the airlines in their timetables (i.e., Monday = day 1, Tuesday = day 2, ... , Sunday = day 7). In this case, we would allocate D1 to Monday, D2 to Tuesday, ... , and D5 to Friday. Any programmer reading the operation `ADD Hours,D3` would immediately know that the total for 'Wednesday' was being incremented by 'Hours'. By the way, some assemblers permit the programmer to rename registers. For example, you might be able to rename data register D1 Monday. Consequently, the operation `ADD.B #5,Monday` is identical to the operation `ADD.B #5,D1`.

Programmers allocate data registers in their own idiosyncratic ways, but most do employ some methodology. For example, I divide registers into groups of *local* and *global* registers (wherever possible). Local registers are used by a particular procedure (subroutine) and are really *scratchpad* registers. That is, they are allocated to short-term variables and you use them rather like the back of an envelope for jotting down notes. Global registers are used to hold variables required by all parts of the program and the variables in these registers survive throughout the life of the program. For example, I might allocate a global variable called `TotalErrors` to D6. In this case, D6 is used only to hold the variable `TotalErrors` and is never used for any other purpose.

It doesn't really matter what algorithm you choose to allocate variables to registers as long as it's clear and consistent and well documented. Many programmers provide a section of code with a *header* (i.e., a comment block) to show the reader of the program how registers have been allocated. Consider a simple example.

```
*********************************************************************
*
*   GetChar:              Input an ASCII-encoded character into D0
*
*   Input parameters:     None
*   Output parameters:    ASCII character in D0
*                         Error code in D6
*
*   Registers modified: D0, D1, D6
*
*********************************************************************
```

```
GetChar  MOVE.B  ACIAC,D1          REPEAT Read the ACIA's status byte
         BTST.B  #RDRF,D1                 Test the RxDataRegisterFull bit
         BEQ     GetChar           UNTIL  ACIA ready
         MOVE.B  ACIAD,D0          Get received data
         AND.B   #%01111100,D1     Mask error status
         MOVE.B  D1,D6             Copy status to D6
         RTS                       Return
*
**********************************************************************
```

This fragment of code demonstrates how you might lay out a subroutine and document its register usage. Someone who later debugs the program (including yourself) would immediately see that the subroutine returned data in D0 and D6, and that any data that was in D1 before the subroutine was called would be corrupted by the subroutine. While we are on the topic of corruption, it's worth pointing out that a common cause of error in an assembly language program is using a register to hold one variable, reloading it with a second variable (which overwrites or corrupts the first one), and then forgetting that you have lost the first variable. Consider the following fragment of erroneous code.

```
*       D1 = Temp6
*       D2 = Val12
*       D3 = Val5
        MOVE.B  #4,D1      Temp6 := 4
        ADD.B   D3,D1      Temp6 := Temp6 + Val5
        .
        .
        .
        MOVE.B  #6,D1      Calculate a new result for Val13
        ADD.B   D2,D1
        SUB.B   D3,D1
        MOVE.B  D1,Val13
        .
        .
        .
        SUB.B   #18,D1    Temp6 := Temp6 - 18
        .
        .
```

In this fragment of code we use register D1 to represent a variable called `Temp6`. Unfortunately, in the middle of the subroutine we forget that D1 is in use and employ it to perform an intermediate calculation. When the line `SUB.B #18,D1` is executed at the end of this code fragment, D1 contains not `Temp6` as we expect, but `Val13` that has overwritten `Temp6`. Of course, such an error is really a programming blunder, but it's remarkably easy to make when you are writing longish assembler language programs.

Now that we've introduced some of the 68000's instructions, the next step is to look at some of its fundamental addressing modes.

3.4 Addressing Modes

Addressing modes are concerned with the way in which data is *accessed*, rather than the way in which data is *processed*. A microprocessor's addressing modes are as important as its instruction set, since you can't do clever things with data until you've first found it. There's nothing mysterious about addressing modes, and their computer science use is analogous to the way in which we use addressing modes in everyday life. For example, if someone asks you where a friend lives you might say, '61 William Street.' Equally, you might say, 'It's the third house to the left of the bar.' In the first case you specify the *actual* address and in the second case you specify it in *relationship* to a known point (the bar). Computers operate in exactly the same way, but with rather a large number of variations.

In this introduction we are not going to present all variations on the 68000's addressing modes and will concentrate on its fundamental modes: absolute, immediate, and address register indirect. In theory, only one addressing mode is required (i.e., absolute addressing), as all other addressing modes can be synthesized from absolute addressing. In practice, powerful addressing modes increase the efficiency of programs.

Absolute Addressing

In *absolute* addressing the instruction uses the *actual* or *absolute* address of the operand. For example, `CLR.B $234` means *clear the contents of the memory location whose address is* 234_{16}. Symbolic addresses are invariably used by the programmer rather than numeric addresses. For example, and you might write the instruction `CLR.B InputStatus` rather than `CLR.B $234`. The operation `MOVE.B $234,$2000` copies the contents of memory location 234_{16} into location 2000_{16} and uses absolute addressing for both the source and destination operands.

Consider the instruction `MOVE.B D2,$2000` that means copy the contents of data register D2 into memory location 2000_{16}. The destination address (i.e., $2000) is an absolute address, but what is the source address? 68000 literature calls this addressing mode *data register direct*, since the operand comes directly from a data register. It is a reasonable name for this addressing mode, but you could also call it *absolute addressing*. As you know, there is no *fundamental* difference between a data register and a location in memory (it's just that the register lives in an exclusive district on the same chip as the CPU, while the memory location lives on a large estate with millions of similar locations). So, it doesn't really matter whether you specify an operand by its actual location in memory or by its register name. In both cases, you are telling the CPU exactly where to find data.

The following fragment of code employs absolute addressing for all its instructions (or data register direct addressing, if you wish to make the distinction).

```
MOVE.B Input,D0     Read the input
ADD.B  D4,D0        Input := Input + D4
SUB.B  Time,D0      Input := Input - Time
MOVE.B D0,Output    Write the output
```

Immediate Addressing

We have already encountered immediate addressing in which the *actual* operand is supplied with the instruction. For example, the source operand in the instruction `MOVE.B #25,D2` is an immediate operand, because the actual value 25 is loaded into data register D2.

The immediate addressing mode is not actually required to realize a digital computer. For example, we could implement the operation `[D2]` ← `[D2] + 25` by first storing the constant 25 in, say, memory location $2000 and then executing the instruction `ADD.B $2000,D2`. The following fragment of code illustrates this point.

```
Hours DC.B   25          Store the number 25 in memory
      .
      .
      ADD.B  Hours,D2    Add 'Hours' (i.e., 25) to D2
```

Now let's write the same code using immediate addressing.

```
Hours EQU    25          Equate symbolic name Hours to the value 25
      .
      .
      ADD.B  #Hours,D2   Add 'Hours' (i.e., 25) to D2
```

These two fragments of code perform the same action, and yet are profoundly different. In the first case, the assembler directive, `Hours DC.B 25`, stores the value 25 in the memory location whose symbolic name is `Hours`. The instruction `ADD.B Hours,D2` causes the processor to read the contents of memory location `Hours` and to add the value it finds there (i.e., 25) to the contents of register D2. Note that executing this instruction involves two memory references: one to read the instruction itself and one to read the operand in memory. Note also that since the operand `Hours` is a memory location within the data area, its contents can be modified by any data operation while the program is running.

The second fragment of code uses immediate addressing. The instruction `ADD.B #Hours,D2`, requires only one memory reference because the operand, 25, is encoded as part of the instruction. Note that the assembler directive `Hours EQU 25` simply equates a symbolic name to an actual value to help make the program more readable. The constant in this version of the program cannot be changed when the program runs, as it is part of the program and not the data. To change the constant you would actually have to modify the instruction itself.

Once again, let me say that the advantage of immediate addressing is its speed. In Chapter 2 we mentioned a feature of the stored program computer — the so-called von Neumann bottleneck. This bottleneck refers to the congestion on the data highway between the CPU and the memory caused by the need to access memory twice per instruction (once to read the instruction itself and once to access the data used by the instruction). An instruction with an immediate operand requires only one memory access because the operand forms part of the instruction itself — hence the name *immediate*.

Applications of Immediate Addressing

Assembly language programmers employ immediate addressing when using a constant that does not change during the course of a program. Suppose we wish to implement the high-level language construct: `IF 7 < P < 25 THEN X := 6`. This algorithm can be translated into assembly language as:

```
           MOVE.B  P,D0          Get value of P in D0
           SUB.B   #8,D0         IF P < 7 THEN EXIT
           BMI     OutOfRange
           MOVE.B  P,D0          Get P in D0 again
           SUB.B   #25,D0        IF P > 25 THEN EXIT
           BPL     OutOfRange
           MOVE.B  #6,X          As 7 < P < 25  X := 6
OutOfRange ...                   EXIT here
```

Note that we have introduced two new branch instructions, `BPL` (branch on positive) and `BMI` (branch on minus). The `BPL` instruction forces a branch if the number is positive in two's complement terms (i.e., its most-significant bit is zero). Similarly, the `BMI` instruction forces a branch if the number is negative (i.e., its most-significant bit is one).

In the above fragment of code, we load P into data register D0 and then subtract the number 8. If the result is negative, P must be less than or equal to 7 and the `BMI OutOfRange` instruction forces an exit. If the result is positive, P is greater than 7, so we subtract 25 from P. If the result is positive, P is not less than 25 and we use `BPL OutOfRange` to exit. If we do not take the branch, the next instruction (`MOVE.B #6,X`) is executed to implement `X := 6`.

A real programmer would not write the above code. The subtract instruction performs the desired comparison, but it destroys the data being compared. That is, `SUB.B #8,D0` performs the action `[D0]` ← `[D0] - 8`, which destroys the contents of D0. Consequently, we have to reload D0 with P. Most microprocessors have a special *non-destructive* compare instruction that performs a comparison by subtraction but does not modify the destination operand. The 68000 microprocessor has a CMP.B source,destination instruction that subtracts the source from the destination and then sets the bits of the CCR accordingly. We can use the compare instruction to rewrite the code in the form:

```
            MOVE.B  P,D0            Get P in D0
            CMP.B   #8,D0           IF P < 7 THEN EXIT
            BMI     OutOfRange
            CMP.B   #25,D0          IF P > 25 THEN EXIT
            BPL     OutOfRange
            MOVE.B  #6,X            7 < P < 25  X := 6
OutOfRange ...                      EXIT here
```

As you can see, this version does not require P to be reloaded from memory after the first test has been carried out.

Another application of literal addressing is in implementing loops of the form FOR I := 1 TO N. In this case N represents an immediate value that is known at the time the program is written. Suppose we wish to add together the first 10 integers (i.e., 1 + 2 + 3 + ... + 10).

```
            CLR     D1              Total := 0
            MOVE.B  #1,D0           Increment := 1
Next        ADD.B   D0,D1           FOR Count := 1 TO 10
            ADD.B   #1,D0               Increment := Increment + 1
            CMP.B   #11,D0          END_FOR
            BNE     Next
```

The immediate addressing mode is very easy to both understand and use in practice. Unfortunately, one of the most common programming errors is to omit the '#' symbol which informs the assembler that the following operand is immediate. For example, if in the above program we had written CMP.B 11,D0 (instead of MOVE.B #11,D0), the loop would be terminated only when the contents of D0 are equal to the contents of memory location 11.

Address Register Indirect Addressing

Having got so far in this text, you could be forgiven for wondering why the 68000 has eight address registers, A0 to A7, since we have not used them at all. These registers are exploited in conjunction with an addressing mode called *address register indirect*. All this fancy name means is that the *address* of an operand is found in an address register, rather than in the instruction. Address register indirect addressing is indicated to the assembler by enclosing an address register in round parentheses. Note that my RTL uses square brackets for this purpose.

Consider the effect of the instruction CLR.B (A0). The nature of the operation is obvious — set the contents of a certain memory location to zero. But which location? The location to be cleared is the location whose address is in address register A0. In terms of RTL, we can express the instruction as `[M([A0])]` ← 0. That is, clear the contents of the memory location whose address is the contents of address register A0. In Figure 3.16 A0 points at memory location 1000_{16}. Executing the `CLR.B (A0)` instruction does not clear the contents of A0 but the location at which it is *pointing* (i.e., 1000_{16} is loaded with 00_{16}).

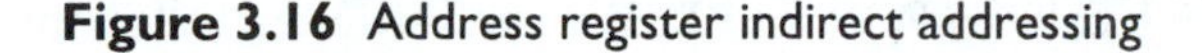

Figure 3.16 Address register indirect addressing

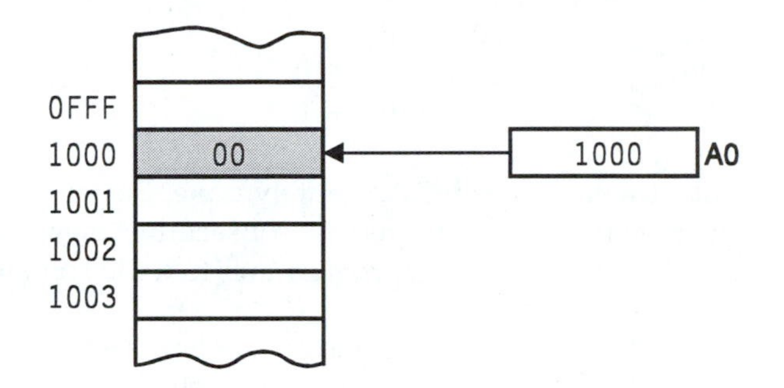

You should now be able to see why this addressing mode is called address register *indirect*. In order to access the operand *two* accesses must be made. The first is to the address register to find the actual address of the operand and the second is to the operand itself.

Address register indirect addressing seems to be a step in the wrong direction, as it increases the number of accesses required to locate an operand. This is not entirely true. Suppose we have to read the contents of memory location 123456_{16} many times. We could first employ the instruction `MOVE.B $123456,D0`. This instruction is, itself, quite long because it requires a 16-bit operation code (i.e., `MOVE.B AbsoluteAddress,DataRegister`) plus a 32-bit source operand address (i.e., 123456_{16}). Suppose we perform the following trick:

```
    MOVEA.L #$123456,A0  Load A0 with the 32-bit value 123456₁₆
*                        Use MOVEA to load an address register
    .
    .
    MOVE.B  (A0),D0      Repeat this section many times
    .
    .
```

The address of the operand, 123456_{16}, is loaded into the address register A0 once by the instruction `MOVEA.L #$123456,A0`. Although we have not emphasized the 68000's ability to handle 16-bit and 32-bit operands, you should appreciate that adding a `.L` suffix to the `MOVE` instruction (instead of a `.B` suffix) causes a 32-bit longword to be moved. Address register A0 can be said to *point* to the location of the operand. The operand itself is accessed by the instruction `MOVE.B (A0),D0`. This instruction does not require an explicit 32-bit operand address and is therefore only 16 bits long. Since the address register is on the chip along with the CPU, little time is lost in reading the contents of A0 to locate the operand in memory. In other words, address register indirect addressing is faster than absolute addressing.

However, this example doesn't reveal the real reason for implementing address register indirect addressing. Up to now we have considered data to be a *constant* or a *variable*, and an address to be a *constant*. Suppose we wish to add three numbers in memory locations P, Q, and R. We can write the following code:

```
MOVE.B  P,D0        D0 := P
ADD.B   Q,D0        D0 := D0 + Q
ADD.B   R,D0        D0 := D0 + R
```

What's the problem with this code? Absolutely none. But now consider the task of adding together 100 numbers stored in consecutive memory locations starting at 2000_{16}. You can't really expect a programmer to write something like:

```
MOVE.B  $2000,D0    Get the first number in D0
ADD.B   $2001,D0    Add the second number
ADD.B   $2002,D0    and the third
ADD.B   $2003,D0    not to mention the fourth
.                   .
.                   .
ADD.B   $2063,D0    Now add in the 100th number
```

The above code is perfectly correct, but no one would ever write it. There has to be a better way of adding together 100 numbers. Suppose we use address register indirect addressing to perform an addition by means of the instruction `ADD.B (A0),D0`. This instruction adds the contents of the memory location pointed at by A0 to data register D0.

Now, suppose we perform a simple operation on the contents of A0. We can do this because the contents of address registers can take part in operations just like the contents of data registers. Imagine that we add 1 to the contents of A0 by means of the instruction `ADDA.L #1,A0` (we use the suffix `.L` because all addresses are longword values, and we use the mnemonic `ADDA` because the 68000 uses `ADDA`, `SUBA`, `CMPA`, and `MOVEA` for operations whose destination is an address register). After adding one to the contents of A0, A0 no longer points to the same location in memory — it points to the *next* location. By operating on A0 we can generate a *variable address* that permits us to step through the 100 numbers in the following way.

```
      CLR.B    D0          Preset the total to zero
      MOVEA.L  #$2000,A0   A0 points to the first number
Next  ADD.B    (A0),D0     REPEAT Get the number pointed at by A0
      ADDA.L   #1,A0              Move A0 to point to the next number
      CMPA.L   #$2064,A0   UNTIL  100 numbers have been added together
      BNE      Next
```

Figure 3.17 illustrates the effect of this fragment of code diagrammatically. Each time the instruction `ADD.B (A0),D0`, is executed, the contents of the memory location pointed at by A0 are added to the contents of data register D0. Once the data from memory has been added to the contents of D0, the instruction `ADDA.L #1,A0` increments A0 by 1 so that it points at the *next* location in memory. This addressing mode permits the programmer to handle many types of data structure (e.g., the table, array, matrix, vector, list, queue, etc.).

We have now covered all the elements of the 68000 microprocessor we need to write programs. In later chapters we add new components enabling us to write more efficient programs and we look at how you go about constructing programs. But before we do this, we are going to introduce the MC68000ECB single board computer and a program that simulates its behavior.

Figure 3.17 Illustration of address register indirect addressing mode

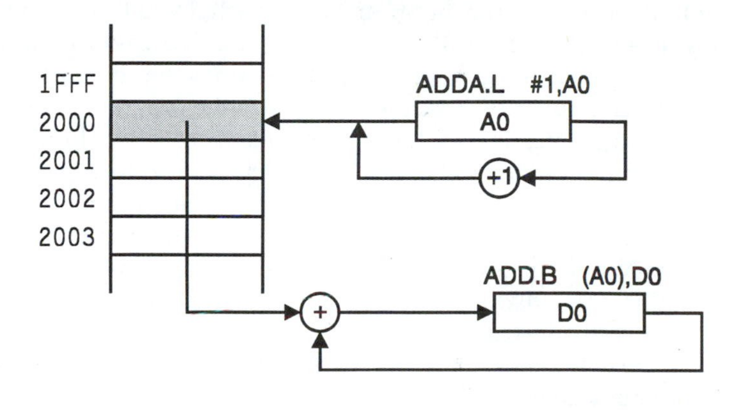

3.5 Running and Debugging 68000 Programs

Do we really need computers to teach computing? It might sound like a silly question, but some students study astrophysics without going into space. Others study particle physics without hands-on experience on a nuclear reactor. So, do we really need access to computers to learn computer science and programming? I suppose that the answer to this question must be no. If the computer executes the instructions of your program and does exactly what it is told, any errors must lie in the program and not the computer. Consequently, all you have to do is to learn how to write correct programs.

The term *formal methods* began to appear in many research papers, books, and computer science syllabuses in the late 1980s and early 1990s. Formal methods comprise various mathematical techniques that enable an algorithm or program to be specified in a mathematical notation and then proved correct. The growing interest in formal methods demonstrates that we are becoming more and more interested in ensuring that a program will work correctly even before we run it. When programs did little more than calculate payrolls and send you a bill for $0.00 (and then threaten to sue for non-payment), we could laugh at these errors. Now that computers directly control fly-by-wire aircraft and the computer can override a pilot's command, the need to write correct programs is vital.

From what we have just said, we can conclude that it is not *necessary* to use computers to help you to write programs. However, it is very *helpful* to test your programs on a computer. Although computers do what you tell them to do, programmers often know what they want a program to do but tell the computer to do something quite different. For example, you might wish to increment the *contents* of D0 by 25 (i.e., `[D0] ← [D0] + 25`) and write `ADD.B 25,D0`. The effect of this instruction is to add the contents of memory location 25 to D0, rather than the actual value 25. You should, of course, have written `ADD.B #25,D0`.

When you make such an error, you might not notice it. Each time you read through the listing of your program you see what you thought you wrote and not what you actually wrote. When the program runs, you obtain an unexpected result. Even worse, if the program itself modifies the contents of location 25, you might get a different result each time you run the program.

One way of dealing with problems like these is to run the program in a so-called *debug mode*. Instead of loading the program and letting it run to completion, you can stop it at any point or you can step through it instruction by instruction. Whenever the program stops (or has completed an instruction) you can examine the contents of the processor's registers and read memory locations. If you had written `ADD.B 25,D0` instead of `ADD.B #25,D0`, you would be able to observe that executing the instruction does not add 25 to D0 (as you had intended).

We are now going to introduce two environments that enable you to debug 68000 assembly language programs. One requires specific 68000-based hardware, and the other is a *simulator* that mimics the behavior of the hardware on a PC.

The MC68000ECB Single Board Computer

When the 68000 microprocessor was first introduced, Motorola marketed a demonstration system aimed at the world of education. If you are launching a new architecture, it makes sense to invest in education, as today's students are tomorrow's users. The MC68000ECB designed by Motorola was a basic, low-cost, single board computer with a 68000, a monitor in ROM called TUTOR (i.e., a simple operating system), 64K bytes of read/write memory, two serial ports, and a combined parallel port and timer. It allows students both to practice writing assembly language programs and to perform experiments in interfacing.

The MC68000ECB was probably largely responsible for making the 68000 so popular with academics who teach computer architecture and organization. Although the MC68000ECB is no longer produced and directly supported my Motorola, a number of similar products have been marketed by independent companies. However, all of them emulate the software of the ECB (i.e., TUTOR) to a greater or lesser extent.

The detailed operation of the ECB is not covered in this text. If you have one, it will already be set up and you will have access to its manuals. All we are interested in is the way in which you can employ it to write and debug programs. We are now going to describe the software that simulates the behavior of the ECB.

The Teesside 68000 Cross-assembler and Simulator

At the University of Teesside we were quick to recognize the advantages of using the 68000 as a vehicle for teaching high-performance computer architectures. Unfortunately, given the large number of students studying computer science, information technology, software engineering, electronics, and computer technology, we could not provide each student with his or her own ECB. Paul Lambert, a member of the Computer Center, designed a software simulator for the ECB to run on the University's network of Prime computers. This software consists of two parts, a cross-assembler that takes a program in 68000 assembly language and produces binary code, and a simulator that mimics the ECB (i.e., it simulates both the 68000 microprocessor and the ECB's monitor, TUTOR).

A student sitting at a terminal is able to write a 68000 program and assemble it. If the program is error-free (i.e., without syntax errors), the student can load it into the simulator and debug it. The simulator behaves like a synthetic 68000 and executes 68000 binary code. Since the functions of the ECB's operating system are built into the simulator, students can examine memory or step through the program as they would in a real 68000 system. The only significant difference between the simulator and the ECB is that the simulator does not operate in real-time, you cannot connect it to an oscilloscope or to a logic analyzer to observe its electrical activity, and you cannot perform a wide range of input/output operations. When we say that the simulator does not run in *real-time*, we mean that the time it takes to execute an instruction is not the same as the time it takes a real 68000 to execute the same instruction. The simulator runs many times slower than a real 68000.

Using the cross-assembler and simulator is easy. Any ASCII text editor can be used to create a source file (i.e., assembly language program). This source file is *cross-assembled* to produce a *binary file* (the 68000 code to be executed) and an optional *listing file*. The term *cross-assembler* indicates that the program is assembled on a machine that is not the same as that which executes the target code — the target machine is a 68000, and the assembler is running on an 80x86 processor. The listing file contains the source program together with the assembled code in hexadecimal format and any error messages. If errors are found during the assembly phase, you must return to the source file and edit it to correct them.

When the program has been assembled and no syntax errors have been made, the simulator can be run using the binary file created by the cross-assembler. If the run is successful and the program behaves as expected, all is well. If the program does not run correctly, the facilities of the simulator can be used to locate the source of the problem. Then you have to return to the source file, edit it, assemble it, and run the simulator again, as Figure 3.18 demonstrates. This entire process is repeated until the program works or you give up and go home.

Figure 3.18 Using the cross-assembler and simulator to design a program

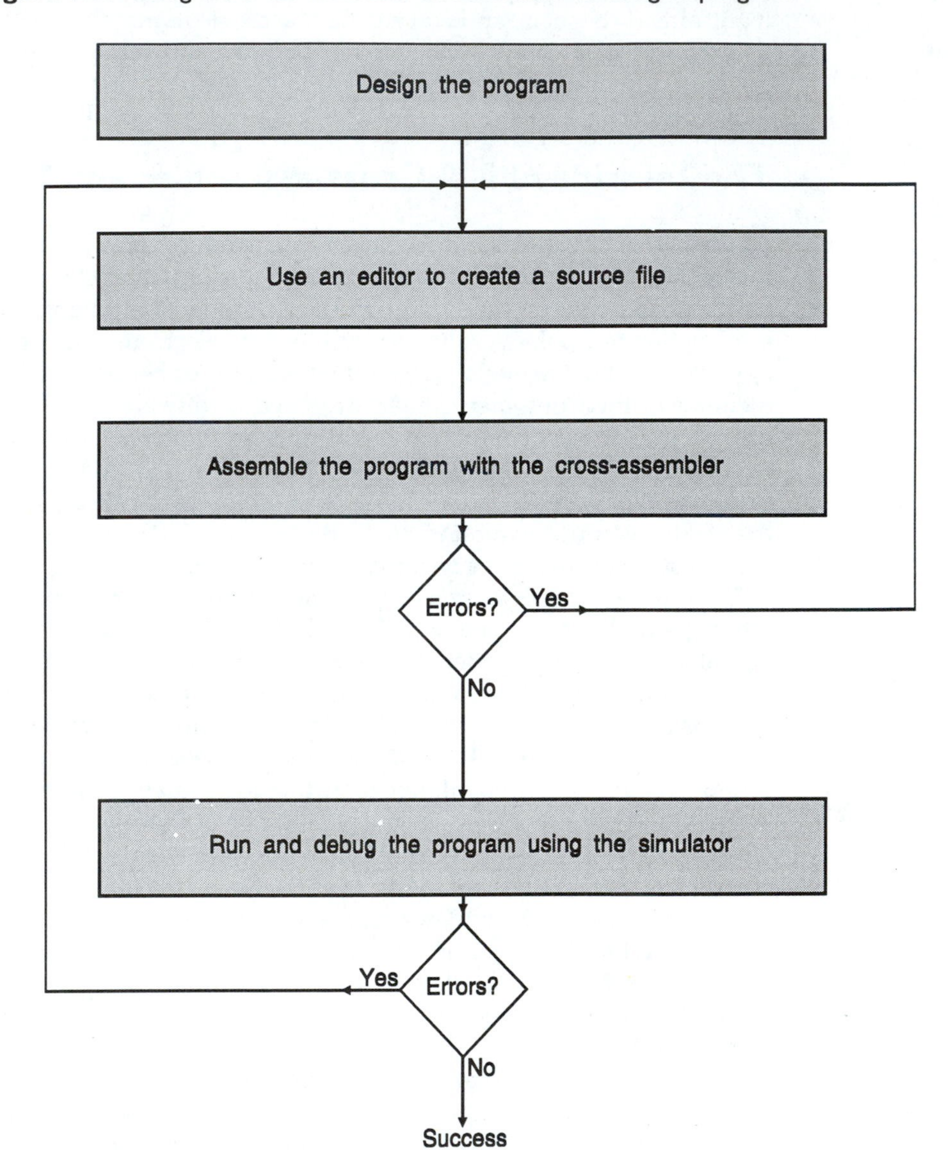

Debugging a program in this way involves two steps. The first is producing a *syntactically correct* program that contains no errors. That is, the instructions and assembler directives all conform to the standard for the assembler. This stage detects many of the typographic errors, illegal instructions and addressing modes, missing subroutines, and the use of the same label more than once.

Syntax errors are best corrected by taking a hard copy of the listing and editing the source file until they have been removed. Once the program is error-free, the simulator can be used to detect *semantic* errors. These are errors of logic in the program design (e.g., using an `ADD A0,D0` instruction when you intended to write an `ADD (A0),D0` instruction). By the way, once the program runs in the manner you expect, the debugging process has not ended. Bugs can still surface long after a program has passed all its initial testing.

The cross-assembler and simulator package proved to be a great success. Toward the end of the 1980s the University became more and more involved with institutions in mainland Europe and students began to buy their own computers. I asked the Computer Center to produce a PC-based version of their software to support our own students with PCs and a course I was teaching in Heraklion, Crete. This was written by Eric Pearson and is the version distributed with this book. We now look at the cross-assembler and simulator in greater detail.

The Cross-assembler

The 68000 cross-assembler runs on any IBM PC or compatible clone. It takes as its input an ASCII-encoded text file and produces a binary file. Any suitable editor can be used to create and edit the source file. The source file must have the extension '`.X68`', in order for the cross-assembler to *recognize* it (e.g., `TEST.X68`). The binary file is the 68000 binary code that was generated by the cross-assembler and has the same name as the source file but with the extension '`.BIN`'. This code will run on an appropriate 68000 microprocessor system. The cross-assembler is executed by entering the command: `X68K <file name>`. For example, to cross-assemble `TEST.X68` you enter `X68K TEST`. Note that it is not necessary to use the full name of the file with its '`.X68`' extension. In what follows, we use a simple program called '`TEST.X68`' to demonstrate the system.

If we wish to produce a listing file as well as the binary file, we need to enter the line: `X68K <file name> -L`. In this case we enter `X68K TEST -L`, which produces two new files: `TEST.BIN` and `TEST.LIS`. The listing file is an ASCII-encoded file. The following fragment of code illustrates a typical listing file.

```
Source file: TEMP.X68
Assembled on: 93-01-08 at: 15:52:34
          by: X68K PC-2.0 Copyright (c) University of Teesside 1989,93
Defaults: ORG $0/FORMAT/OPT A,BRL,CEX,CL,FRL,MC,MD,NOMEX,NOPCO

   1 00001000                    ORG     $001000     ;Origin for data
   2 00001000 00000200  TABLE:     DS.W    256         ;Save 256 words for "TABLE"
   3 00001200 00000004  POINTER_1: DS.L    1           ;Save one longword for "POINTER_1"
   4 00001204 00000004  VECTOR_1:  DS.L    1           ;Save one longword for VECTOR_1
   5 00001208 0000FFFF  INIT:      DC.W    0,$FFFF     ;Store two constants ($0000, $FFFF)
   6          00000003  SETUP1:    EQU     $03         ;Equate "SETUP1" to the value 3
   7          00000055  SETUP2:    EQU     $55         ;Equate "SETUP2" to the value $55
   8          00008000  ACIAC:     EQU     $008000     ;Equate "ACIAC" to the value $8000
   9          00000000  RDRF:      EQU     0           ;RDRF = Receiver Data Register Full = 0
  10          00008004  PIA:       EQU     ACIAC+4     ;Equate "PIA" to the value $8004
```

```
11                          *
12 00018000                          ORG      $018000        ;Origin for program
13 00018000 41F900008000 ENTRY:      LEA      ACIAC,A0       ;A0 points to the ACIA
14 00018006 10BC0003                 MOVE.B   #SETUP1,(A0)   ;Write initialization const into ACIA
15                          *
16 0001800A 08100000     GET_DATA:   BTST     #RDRF,(A0)     ;Any data received?
17 0001800E 66FA                     BNE      GET_DATA       ;Repeat until data ready
18 00018010 10280002                 MOVE.B   2(A0),D0       ;Read data from ACIA
19          00018000                 END      $18000

Lines: 19, Errors: 0, Warnings: 0.
```

If assembler errors are detected, we will be informed. Once assembly is complete and no errors are detected, we can move on to the simulation phase.

The Simulator

The simulator permits us to run 68000 binary code (generated by the cross-assembler) on an IBM PC, as if the PC were an actual 68000. Simulation is not perfect, as it is much slower than running the same code on a true 68000 machine. Furthermore, it does not provide the same *environment* as the target machine on which the code is eventually to run. By *environment*, we mean the operating system and input/output devices. However, the simulator does enable us to test much of the code before it is run on the target machine. Simulation permits us to design and debug software before the target machine has been built. Note that in a practical microprocessor development system, a cross-compiler would probably be used to generate binary code rather than a cross-assembler. For example, you might write the source code in C, rather than assembly language.

The simulator is called by typing: `E68K <filename>`. In our case we type `E68K TEST`. As in the case of the cross-assembler, we do not have to explicitly specify the extension '`.BIN`'.

After starting the simulator with the command '`E68K TEST`', the computer displays the copyright banner, the filename, the low address, high address, and starting point. The starting point is the address of the first executable instruction supplied by the `END` directive in the source file (e.g., `END $400`). The low and high addresses are the region of (synthetic simulator) memory allocated by the simulator to the program.

Once the emulator has been loaded, nothing happens. The computer sits there and waits for a valid command from the keyboard. Here we provide an overview of the simulator's basic facilities. Commands are in the form of two characters followed by any parameters.

HELP The `HELP` instruction tells us what commands the simulator recognizes. This command serves as a reminder and prompts the simulator to tell us what legal commands are available. Typing `HELP` displays the following information.

```
HELP
 .D0 .D1 .D2 .D3 .D4 .D5 .D6 .D7
 .A0 .A1 .A2 .A3 .A4 .A5 .A6 .A7
 .PC .SR .US .SS .C  .N  .V  .X  .Z

 BF - Block fill,                            GT - Go with temp brkpt.
 BM - Block move,                            HE - Help.
 BR - Set/display brkpts,                    LO - Load binary file.
 BS - Block search,                          LOG - LOG screen output.
 DC - Data conversion,                       MD - Memory dump.
 DF - Display registers,                     MM - Memory modify.
 DU - Dump memory to file,                   MS - Memory set.
 GD - Go direct,                             NOBR - Remove brkpts.
 GO - Execute program with brkpts,           TR - Trace program.
                                             QUIT - Return to MS-DOS.

 For more detailed information type: HE command
                                     HE INFO
                                     HE EXCEPTIONS
                                     HE TRAP#15
```

You can obtain more information about a specific command by entering HE <command>. For example, HE DF provides information about the DF function.

The facilities provided by the simulator fall into two groups: those that enable us to set up or to examine the environment and those that enable us to monitor the execution of the program. We first look at the functions that let us look at the environment — the memory and registers.

MD The memory display command allows us to look at the contents of memory. This memory is, of course, the memory of the simulated 68000 processor. If we type MD followed by an address, the simulator displays the contents of 16 consecutive bytes in hexadecimal format each time you hit the enter key.

```
>MD 44E

00044E  41 6C 61 6E 20 43 6C 65 6D 65 6E 74 73 0A 00 00
00045E  00 00 00 00 00 00 00 00 00 00 00 00 00 00 00 00.
```

Entering a period terminates this function. The memory display function is frequently used to examine an area of data. For example, if you have a table of 10 bytes and run a program to put the bytes into ascending order, you can use the MD function to determine whether the program has worked correctly.

The memory display function can also be used to disassemble memory (i.e., the reverse function of an assembler). Binary code is read from memory and converted into 68000 mnemonics. This disassembler is not able to convert addresses into symbolic form and therefore all addresses are expressed numerically.

In the following example we have entered the command MD 400 -DI and the simulator has displayed 16 lines of the program starting from 400. The extension to the MD command, -DI, tells the simulator to disassemble the contents of memory rather than presenting it in hexadecimal form. You don't need to follow this program as it's for demonstration purposes only. Note that the disassembled code presents addresses in hexadecimal form and literals in decimal form.

```
>MD 400 -DI

000400: MOVEA.L  #4096,SP
000406: BSR.L    $00000434
00040A: LEA      $0000044E,A0
000410: BSR.L    $00000442
000414: MOVE.B   D1,(A0)+
000416: CMPI.B   #10,D1
00041A: BNE.S    $00000410
00041C: BSR.L    $00000434
000420: MOVE.B   -(A0),D1
000422: BSR.L    $00000448
000426: CMPA.W   #1102,A0
00042A: BNE.S    $00000420
00042C: BSR.L    $00000434
000430: STOP     #8192
000434: MOVEM.L  A0/D0-D7,-(SP)
000438: MOVEQ    #0,D0
```

We can employ the disassembly function to examine code — we might not have an assembled listing of the source file available. It is rather difficult to follow a disassembled program as the disassembler is not able to bind actual addresses and data values to their symbolic names. For example, all branch and subroutine addresses are given as actual or absolute values.

MM The memory modify function, MM, allows you to modify the contents of a memory location. Entering MM <address> causes the simulator to display the contents of memory location <address>. If you enter new data, it will overwrite the existing data and the simulator will display the contents of the next location. The memory modify command is terminated by entering a period. You might employ this function to set up test data prior to running a program.

DF The display formatted register command, DF, lists the contents of all the 68000's internal registers on the screen. These include D0 to D7, A0 to A7, the supervisor stack pointer, SS, the user stack pointer, US, the status register, SR, and the condition codes. The instruction pointed at by the PC is disassembled and displayed on the screen. Note that this is the next instruction to be executed and not the instruction that has just been executed. The following is an example of the screen you might see after entering DF.

```
PC=000400 SR=2000 SS=00A00000 US=00000000          X=0
A0=00000000 A1=00000000 A2=00000000 A3=00000000 N=0
A4=00000000 A5=00000000 A6=00000000 A7=00A00000 Z=0
D0=00000000 D1=00000000 D2=00000000 D3=00000000 V=0
D4=00000000 D5=00000000 D6=00000000 D7=00000000 C=0
——————>MOVEA.L  #4096,SP
```

Note that the information you see on the screen is the contents of the 68000's synthetic registers in the simulator. These are the values that will be used if the program is run.

Following the register list is the instruction at the address pointed at by the PC (in this example, it is `MOVEA.L #4096,SP`). Note that this is the first instruction in the program because the program counter has been set to 400 hexadecimal. The instruction in the original program was written `MOVEA.L #$1000,A7`. As we said, the simulator displays literal data in decimal form when code is disassembled (hence, 4096 rather than $1000). Note also that address register A7 has been renamed as SP (i.e., stack pointer).

The contents of a register are modified by entering: `.<register> <value>`. For example, the program counter is set to $600 by entering `.PC 600`. Similarly, we can set the contents of D3 to $12345678 by entering `.D3 12345678`.

GO The `GO` command causes the simulator to execute the current program starting from the value in the program counter. Execution continues until:

a. the program reaches a `STOP` instruction

b. the program is interrupted from the keyboard by the ESC key

c. the program reaches a breakpoint (set by the user).

If *nothing* appears to happen after you have typed `GO`, the program is probably either waiting for input from the keyboard or it is in a loop from which it cannot exit. In the latter case, the escape key must be used to stop execution.

TR The trace command, `TR`, is one of the most useful of the simulator commands and permits the execution of a single instruction at a time. After an instruction has been executed, the contents of all registers are displayed on the screen. Entering further carriage returns will step through the program instruction by instruction. You can use the trace command to walk through a program instruction by instruction until you find the cause of an error. You really need a hard copy of the assembled program listing with you when you use the trace function. However, you cannot rely on the trace function alone. Suppose you clear an array of 10,000 bytes in your program by going around a loop made up of two instructions 10,000 times. To trace this loop you have to hit the return button a mere 20,000 times......

BR The breakpoint command, `BR <address>`, places a *marker* in the program at the location specified by `<address>`. When the program is executed, it runs normally until the breakpoint address is reached. The simulator then halts execution and prints the contents of the registers on the screen (i.e., in the same way as a DF or a TR command). If necessary, we can use one or more register modify commands (e.g., .D0 1234) to modify registers at this stage. A breakpoint gives you a snapshot of the state of the processor at any desired point in the execution of a program. It is possible to set more than one breakpoint active at a time.

The breakpoint function overcomes the limitations of the trace function we described above. By putting a breakpoint after the end of a loop that is executed many times, we can execute the code of the loop and then halt the processor at its termination. Then we can revert to the trace mode (if we wish). One of the arts of program debugging is to insert breakpoints at judicious points. If the programmer knows the expected behavior of his or her program, it is usually possible to locate the point at which the actual behavior of the program differs from the intended behavior.

Note that we can resume execution after a breakpoint by entering TR (to get past the breakpoint), followed by GO (to run the program). Entering a GO after a breakpoint will simply rerun the instruction at the breakpoint again.

QU The QUIT command leaves the simulator and returns you to the MS-DOS operating system. Once again, note that if a program is being emulated and the system hangs up for whatever reason, hitting the escape key will force an interrupt and an exit from the simulation mode.

Using the Simulator

We are now going to walk through a demonstration of the cross-assembler and simulator. For the purpose of this example, we will construct a program that takes a sequence of consecutive bytes in memory and searches the sequence for the largest byte, which is to be deposited in data register D1. The sequence is terminated by the value 0. The program is to start at location 400_{16} and the data array at location 1000_{16}. Why did we choose these values? 68000 programmers never locate a program in the range 00000000_{16} to $000003FF_{16}$ (i.e., the first 1024 bytes of memory space), because the 68000 dedicates this region of memory to exception vectors. These vectors are required to implement the 68000's interrupt-handling system and are described further in Chapter 10. Data is frequently located immediately after the program. However, as we might need to examine the data while debugging the program, it is a good idea to locate it at a memorable location, like 1000_{16}.

The instruction STOP #$2700 is a special operation that causes the 68000 to stop processing and to load the value 2700_{16} into its status register. We will meet this instruction again in Chapter 10. At this stage, all you need know is that it offers a convenient way of terminating a program that you are simulating.

The first step is to write the program and enter it into the computer using a text (ASCII) editor.

```
*****************************************************************
* Program to find the largest value in an array of bytes        *
*                                                               *
* The array starts at memory location $1000 and is terminated   *
* by a null byte, $00                                           *
*****************************************************************
*
         ORG      $400                 Program origin
         MOVEA.L  #Data,A0             Use A0 as a pointer
         CLR.B    D0                   D0 will record the largest byte
Next     MOVE.B   (A0),D1              REPEAT: Read a byte
         BEQ      Exit                    IF data := 0 THEN Exit
         CMP.B    D0,D1                   IF New > Old
         BLE      EndTest
         MOVE.B   D1,D0                   THEN Old := New
EndTest  ADDA.L   #1,A0                   Point to next byte in array
         BRA      Next                 END REPEAT
Exit     STOP     #$2700               Halt processor at end of program
*
         ORG      $1000                Data origin
Data     DC.B     12,13,5,6,4,8,4,10,0 Sample data
         END      $400                 Program terminator and entry point
```

The next step is to assemble the program with the cross-assembler. Typing `X68K XX -L` assembles the source file to produce the listing file (`XX.LIS`) and the binary file (`XX.BIN`). We can examine the listing simply by printing the `XX.LIS` file.

```
Source file: XX.X68
Assembled on: 93-03-29 at: 19:14:04
        by: X68K PC-2.1 Copyright (c) University of Teesside 1989,93
Defaults: ORG $0/FORMAT/OPT A,BRL,CEX,CL,FRL,MC,MD,NOMEX,NOPCO

  1                            *****************************************************************
  2                            * Program to find the largest value in an array of bytes        *
  3                            *                                                               *
  4                            * The array starts at memory location $1000 and is terminated   *
  5                            * by a null byte, $00                                           *
  6                            *****************************************************************
  7                            *
  8 00000400                            ORG      $400          ;Program origin
  9 00000400 207C00001000               MOVEA.L  #DATA,A0      ;Use A0 as a pointer
 10 00000406 4200                       CLR.B    D0            ;D0 will record the largest byte
 11 00000408 1210              NEXT:    MOVE.B   (A0),D1       ;REPEAT: Read a byte
 12 0000040A 67000012                   BEQ      EXIT          ;IF data := 0 THEN Exit
 13 0000040E B200                       CMP.B    D0,D1         ;IF New > Old
 14 00000410 6F000004                   BLE      ENDTEST
```

```
15 00000414 1001                  MOVE.B   D1,D0                 ;THEN Old := New
16 00000416 D1FC00000001 ENDTEST: ADDA.L   #1,A0                 ;Point to next byte in array
17 0000041C 60EA                  BRA      NEXT                  ;END REPEAT
18 0000041E 4E722700     EXIT:    STOP     #$2700                ;Halt processor at end of program
19                       *
20 00001000                       ORG      $1000                 ;Data origin
21 00001000 0C0D05060408 DATA:    DC.B     12,13,5,6,4,8,4,10,0  ;Sample data
            040A00
22          00000400              END      $400                  ;Program terminator and entry point

Lines: 22, Errors: 0, Warnings: 0.
```

As no errors occurred during the assembly phase, we know that the program is syntactically correct and therefore we can run it. We do this by entering `E68K XX`. The following sequence demonstrates how a session with the simulator might look. All text in the `Letter Gothic font` was produced by the computer and all text in this font represents my comments.

```
E68K XX
[E68k PC-2.0 Copyright (c) University of Teesside 1989,93]

Address space 0 to ^10485759 (10240 kbytes).

Loading binary file "XX.BIN".
Start address: 000400, Low: 00000400, High: 00001008
```

We begin by examining the data that the program has to search for the highest value. The `MD` (memory display) function lists the data a byte at a time. We enter the command `MD <address>`, using the value 1000 for the address. The simulator expects addresses to be in hexadecimal format.

```
>MD 1000

001000  0C 0D 05 06 04 08 04 0A 00 00 00 00 00 00 00 00
001010  00 00 00 00 00 00 00 00 00 00 00 00 00 00 00 00.
```

After the first line of 16 data bytes was printed, we entered a carriage return and a second line was displayed. A period terminated the `MD` operation. As you can see, the data contains the nine values including the zero terminator. All other data is zero because the simulator clears memory and registers when it is first loaded. Note that a real processor and memory does not do this — you cannot count on your data being initialized to zero before you run a program. We will now run the program and see what happens.

```
>GO
Processor halted at: 000422
```

Well, not a lot happened. All we know is that the program ran — although we have no idea what actually took place. The program was halted at address 000422_{16} by the STOP instruction. We can use the DF (display register) function to examine the contents of the 68000's registers at the point at which the program halted.

```
>DF
PC=000422 SR=2700 SS=00A00000 US=00000000          X=0
A0=00001008 A1=00000000 A2=00000000 A3=00000000 N=0
A4=00000000 A5=00000000 A6=00000000 A7=00A00000 Z=0
D0=0000000D D1=00000000 D2=00000000 D3=00000000 V=0
D4=00000000 D5=00000000 D6=00000000 D7=00000000 C=0
------->ORI.B    #0,D0
```

The contents of data register D0 are $0D_{16}$ (i.e., 13_{10}) which is indeed the largest value in the list of bytes. We can assume that the program is working. Note that the next instruction to be executed is ORI.B #0,D0. The DF function shows the next instruction after the STOP instruction. Since our program does not continue beyond this point, the simulator has read zeros from memory and interpreted them as an ORI.B #0,D0 instruction which has the op-code $0000.

Let's try modifying the data in the array at address 1000_{16}. We can use the MM (memory modify) instruction to do this. You enter MM <address> and the simulator provides you with the byte at that address. You can then either type a carriage return to examine the next location or enter new data. Like most other functions, you can break out of the MM function by entering a period.

```
>MM 1000
001000 0C ?
001001 0D ?
001002 05 ?
001003 06 ? FE
001004 04 ? 01
001005 08 ? FF
001006 04 ?
001007 0A ? 00
001008 00 ? .
```

The new highest value is FF_{16} at location 1005_{16}. Let's rerun the program and use the DF function to see what happens. Before you can rerun the program you must reset the simulator. You can do this by resetting the program counter to $400 (by means of .PC 400) or by quitting the simulator and then re-entering it.

```
>GO
Processor halted at: 000422

>DF
PC=000422 SR=2700 SS=00A00000 US=00000000          X=0
```

```
A0=00001007 A1=00000000 A2=00000000 A3=00000000 N=0
A4=00000000 A5=00000000 A6=00000000 A7=00A00000 Z=0
D0=0000000D D1=00000000 D2=00000000 D3=00000000 V=0
D4=00000000 D5=00000000 D6=00000000 D7=00000000 C=0
------>ORI.B     #0,D0
```

At the end of the run, data register D0 contains the value $0D_{16}$, rather than the expected value of FF_{16}. So, what went wrong? Was the algorithm faulty or have we failed to notice a subtlety? We will now rerun the program in the trace mode to see if we can find the point at which the error appears. The program will be rerun by resetting the contents of the PC, and data registers D0 and D1. We will enter TR to begin the trace mode and a few carriage returns to step through the program.

```
>.PC 400

>.D1 0

>.D0 0

>TR

PC=000406 SR=2700 SS=00A00000 US=00000000          X=0
A0=00001000 A1=00000000 A2=00000000 A3=00000000 N=0
A4=00000000 A5=00000000 A6=00000000 A7=00A00000 Z=0
D0=00000000 D1=00000000 D2=00000000 D3=00000000 V=0
D4=00000000 D5=00000000 D6=00000000 D7=00000000 C=0
------>CLR.B     D0

Trace>

PC=000408 SR=2704 SS=00A00000 US=00000000          X=0
A0=00001000 A1=00000000 A2=00000000 A3=00000000 N=0
A4=00000000 A5=00000000 A6=00000000 A7=00A00000 Z=1
D0=00000000 D1=00000000 D2=00000000 D3=00000000 V=0
D4=00000000 D5=00000000 D6=00000000 D7=00000000 C=0
------>MOVE.B    (A0),D1
Trace>

PC=00040A SR=2700 SS=00A00000 US=00000000          X=0
A0=00001000 A1=00000000 A2=00000000 A3=00000000 N=0
A4=00000000 A5=00000000 A6=00000000 A7=00A00000 Z=0
D0=00000000 D1=0000000C D2=00000000 D3=00000000 V=0
D4=00000000 D5=00000000 D6=00000000 D7=00000000 C=0
------>BEQ.L     $0000041E

Trace>
```

```
PC=00040E SR=2700 SS=00A00000 US=00000000          X=0
A0=00001000 A1=00000000 A2=00000000 A3=00000000 N=0
A4=00000000 A5=00000000 A6=00000000 A7=00A00000 Z=0
D0=00000000 D1=0000000C D2=00000000 D3=00000000 V=0
D4=00000000 D5=00000000 D6=00000000 D7=00000000 C=0
———>CMP.B    D0,D1

Trace>
PC=000410 SR=2700 SS=00A00000 US=00000000          X=0
A0=00001000 A1=00000000 A2=00000000 A3=00000000 N=0
A4=00000000 A5=00000000 A6=00000000 A7=00A00000 Z=0
D0=00000000 D1=0000000C D2=00000000 D3=00000000 V=0
D4=00000000 D5=00000000 D6=00000000 D7=00000000 C=0
———>BLE.L    $00000416
```

At this stage we have loaded the first value in the list, $12 = 0C_{16}$, into register D1 and have compared it to the contents of register D0 (i.e., 0). As we are going to compare the contents of register D1 with the contents of register D0 and branch only if D1 is less than D0, we expect that the branch will not be taken.

```
Trace>
PC=000414 SR=2700 SS=00A00000 US=00000000          X=0
A0=00001000 A1=00000000 A2=00000000 A3=00000000 N=0
A4=00000000 A5=00000000 A6=00000000 A7=00A00000 Z=0
D0=00000000 D1=0000000C D2=00000000 D3=00000000 V=0
D4=00000000 D5=00000000 D6=00000000 D7=00000000 C=0
———>MOVE.B   D1,D0
```

As you can see, we did not take the branch and are now copying the new large value to register D0. We will continue the tracing.

```
Trace>
PC=000416 SR=2700 SS=00A00000 US=00000000          X=0
A0=00001000 A1=00000000 A2=00000000 A3=00000000 N=0
A4=00000000 A5=00000000 A6=00000000 A7=00A00000 Z=0
D0=0000000C D1=0000000C D2=00000000 D3=00000000 V=0
D4=00000000 D5=00000000 D6=00000000 D7=00000000 C=0
———>ADDA.L   #1,A0

Trace>
PC=00041C SR=2700 SS=00A00000 US=00000000          X=0
A0=00001001 A1=00000000 A2=00000000 A3=00000000 N=0
A4=00000000 A5=00000000 A6=00000000 A7=00A00000 Z=0
D0=0000000C D1=0000000C D2=00000000 D3=00000000 V=0
D4=00000000 D5=00000000 D6=00000000 D7=00000000 C=0
———>BRA.S    $00000408
```

```
Trace>

PC=000408 SR=2700 SS=00A00000 US=00000000          X=0
A0=00001001 A1=00000000 A2=00000000 A3=00000000 N=0
A4=00000000 A5=00000000 A6=00000000 A7=00A00000 Z=0
D0=0000000C D1=0000000C D2=00000000 D3=00000000 V=0
D4=00000000 D5=00000000 D6=00000000 D7=00000000 C=0
------>MOVE.B   (A0),D1

Trace>

PC=00040A SR=2700 SS=00A00000 US=00000000          X=0
A0=00001001 A1=00000000 A2=00000000 A3=00000000 N=0
A4=00000000 A5=00000000 A6=00000000 A7=00A00000 Z=0
D0=0000000C D1=0000000D D2=00000000 D3=00000000 V=0
D4=00000000 D5=00000000 D6=00000000 D7=00000000 C=0
------>BEQ.L    $0000041E

Trace>

PC=00040E SR=2700 SS=00A00000 US=00000000          X=0
A0=00001001 A1=00000000 A2=00000000 A3=00000000 N=0
A4=00000000 A5=00000000 A6=00000000 A7=00A00000 Z=0
D0=0000000C D1=0000000D D2=00000000 D3=00000000 V=0
D4=00000000 D5=00000000 D6=00000000 D7=00000000 C=0
------>CMP.B    D0,D1

Trace>

PC=000410 SR=2700 SS=00A00000 US=00000000          X=0
A0=00001001 A1=00000000 A2=00000000 A3=00000000 N=0
A4=00000000 A5=00000000 A6=00000000 A7=00A00000 Z=0
D0=0000000C D1=0000000D D2=00000000 D3=00000000 V=0
D4=00000000 D5=00000000 D6=00000000 D7=00000000 C=0
------>BLE.L    $00000416

Trace>

PC=000414 SR=2700 SS=00A00000 US=00000000          X=0
A0=00001001 A1=00000000 A2=00000000 A3=00000000 N=0
A4=00000000 A5=00000000 A6=00000000 A7=00A00000 Z=0
D0=0000000C D1=0000000D D2=00000000 D3=00000000 V=0
D4=00000000 D5=00000000 D6=00000000 D7=00000000 C=0
------>MOVE.B   D1,D0

Trace>
```

```
PC=000416 SR=2700 SS=00A00000 US=00000000          X=0
A0=00001001 A1=00000000 A2=00000000 A3=00000000 N=0
A4=00000000 A5=00000000 A6=00000000 A7=00A00000 Z=0
D0=0000000D D1=0000000D D2=00000000 D3=00000000 V=0
D4=00000000 D5=00000000 D6=00000000 D7=00000000 C=0
———>ADDA.L    #1,A0
```

At this stage we have gone round the loop a second time and have found that the next value in the sequence, 13 = $0D_{16}$, is larger than the first and have recorded it in data register D0. So far all is well and the program is functioning correctly. But we know that it sticks with a maximum value of $0D_{16}$ and does not detect FF_{16}. Why? Let's continue tracing and see what happens.

```
Trace>

PC=00041C SR=2700 SS=00A00000 US=00000000          X=0
A0=00001002 A1=00000000 A2=00000000 A3=00000000 N=0
A4=00000000 A5=00000000 A6=00000000 A7=00A00000 Z=0
D0=0000000D D1=0000000D D2=00000000 D3=00000000 V=0
D4=00000000 D5=00000000 D6=00000000 D7=00000000 C=0
———>BRA.S     $00000408

Trace>

PC=000408 SR=2700 SS=00A00000 US=00000000          X=0
A0=00001002 A1=00000000 A2=00000000 A3=00000000 N=0
A4=00000000 A5=00000000 A6=00000000 A7=00A00000 Z=0
D0=0000000D D1=0000000D D2=00000000 D3=00000000 V=0
D4=00000000 D5=00000000 D6=00000000 D7=00000000 C=0
———>MOVE.B    (A0),D1

Trace>
PC=00040A SR=2700 SS=00A00000 US=00000000          X=0
A0=00001002 A1=00000000 A2=00000000 A3=00000000 N=0
A4=00000000 A5=00000000 A6=00000000 A7=00A00000 Z=0
D0=0000000D D1=00000005 D2=00000000 D3=00000000 V=0
D4=00000000 D5=00000000 D6=00000000 D7=00000000 C=0
———>BEQ.L     $0000041E

Trace>
PC=00040E SR=2700 SS=00A00000 US=00000000          X=0
A0=00001002 A1=00000000 A2=00000000 A3=00000000 N=0
A4=00000000 A5=00000000 A6=00000000 A7=00A00000 Z=0
D0=0000000D D1=00000005 D2=00000000 D3=00000000 V=0
D4=00000000 D5=00000000 D6=00000000 D7=00000000 C=0
———>CMP.B     D0,D1
```

```
Trace>

PC=000410 SR=2709 SS=00A00000 US=00000000          X=0
A0=00001002 A1=00000000 A2=00000000 A3=00000000 N=1
A4=00000000 A5=00000000 A6=00000000 A7=00A00000 Z=0
D0=0000000D D1=00000005 D2=00000000 D3=00000000 V=0
D4=00000000 D5=00000000 D6=00000000 D7=00000000 C=1
------>BLE.L    $00000416

Trace>

PC=000416 SR=2709 SS=00A00000 US=00000000          X=0
A0=00001002 A1=00000000 A2=00000000 A3=00000000 N=1
A4=00000000 A5=00000000 A6=00000000 A7=00A00000 Z=0
D0=0000000D D1=00000005 D2=00000000 D3=00000000 V=0
D4=00000000 D5=00000000 D6=00000000 D7=00000000 C=1
------>ADDA.L   #1,A0

Trace>

PC=00041C SR=2700 SS=00A00000 US=00000000          X=0
A0=00001003 A1=00000000 A2=00000000 A3=00000000 N=0
A4=00000000 A5=00000000 A6=00000000 A7=00A00000 Z=0
D0=0000000D D1=00000005 D2=00000000 D3=00000000 V=0
D4=00000000 D5=00000000 D6=00000000 D7=00000000 C=0
------>BRA.S    $00000408

Trace>

PC=000408 SR=2700 SS=00A00000 US=00000000          X=0
A0=00001003 A1=00000000 A2=00000000 A3=00000000 N=0
A4=00000000 A5=00000000 A6=00000000 A7=00A00000 Z=0
D0=0000000D D1=00000005 D2=00000000 D3=00000000 V=0
D4=00000000 D5=00000000 D6=00000000 D7=00000000 C=0
------>MOVE.B   (A0),D1

Trace>

PC=00040A SR=2708 SS=00A00000 US=00000000          X=0
A0=00001003 A1=00000000 A2=00000000 A3=00000000 N=1
A4=00000000 A5=00000000 A6=00000000 A7=00A00000 Z=0
D0=0000000D D1=000000FE D2=00000000 D3=00000000 V=0
D4=00000000 D5=00000000 D6=00000000 D7=00000000 C=0
------>BEQ.L    $0000041E

Trace>
```

```
PC=00040E SR=2708 SS=00A00000 US=00000000          X=0
A0=00001003 A1=00000000 A2=00000000 A3=00000000 N=1
A4=00000000 A5=00000000 A6=00000000 A7=00A00000 Z=0
D0=0000000D D1=000000FE D2=00000000 D3=00000000 V=0
D4=00000000 D5=00000000 D6=00000000 D7=00000000 C=0
———>CMP.B    D0,D1

Trace>

PC=000410 SR=2708 SS=00A00000 US=00000000          X=0
A0=00001003 A1=00000000 A2=00000000 A3=00000000 N=1
A4=00000000 A5=00000000 A6=00000000 A7=00A00000 Z=0
D0=0000000D D1=000000FE D2=00000000 D3=00000000 V=0
D4=00000000 D5=00000000 D6=00000000 D7=00000000 C=0
———>BLE.L    $00000416

Trace>

PC=000416 SR=2708 SS=00A00000 US=00000000          X=0
A0=00001003 A1=00000000 A2=00000000 A3=00000000 N=1
A4=00000000 A5=00000000 A6=00000000 A7=00A00000 Z=0
D0=0000000D D1=000000FE D2=00000000 D3=00000000 V=0
D4=00000000 D5=00000000 D6=00000000 D7=00000000 C=0
———>ADDA.L   #1,A0

Trace>.
```

We stopped the trace mode because something strange happened. The `MOVE.B (A0),D0` instruction loaded FE_{16} into data register D1. However, when we compared the contents of registers D0 and D1 using the instruction sequence:

```
    CMP.B D0,D1        Largest - New
and BLE   416          Branch IF Largest > New,
```

we *took the branch*. But surely, the branch shouldn't be taken because the contents of D1 are greater than those of D0, not less? The number in D1 is greater than that in D0 only if we consider the number to be *unsigned*. That is, data register D0 contains the value $0D_{16}$ = 12 and data register D1 contains FE_{16} = 254. The `BLE` instruction is designed to operate with *signed* integers. The data in D1 is actually -2 when interpreted as a signed integer. The instruction `BLE` is the source of the error. We need to edit the source program and replace the line `BLE EndTest` with `BLS EndTest`, since `BLS` is intended to be used with unsigned integers.

Tracing a program can be a slow and tedious process. Fortunately, there is good news. The number of basic errors (i.e., blunders) made by the assembly language programmer is relatively small. After a little practice the programmer builds up a data bank of errors and can often rapidly detect and correct an error.

We have introduced the 68000, its instruction set, and its fundamental addressing modes. We can now begin to build on this knowledge and examine some of the 68000's more advanced instructions and addressing modes. Along the way we will look at the design and construction of assembly language programs.

Summary

- The 68000 has eight general purpose data registers, D0 to D7. The data registers behave identically, in the sense that whatever you can do to Di you can also do to Dj. Data registers are used to hold frequently accessed data and it's up to the programmer to decide which data is located in memory and which in registers.
- The data registers are all 32 bits wide. You can treat them as 8-bit registers by suffixing instructions with `.B`. For example, `ADD.B D0,D1` adds the eight least-significant bits of D0 to the least-significant bits of D1. Similarly, you can treat them as 16-bit registers by suffixing instructions with `.W`. When a data register takes part in an 8-bit or a 16-bit operation, only the low-order 8 or low-order 16 bits are affected by the operation.
- The 68000 has a condition code register, CCR, whose contents are updated after most (but not all) instructions. The CCR tells us whether the last operation yielded a zero result, a carry-out, a negative result, or an arithmetic overflow.
- The 68000 has eight 32-bit address registers, A0 to A7. These registers are used as pointer registers when accessing data structures. A0 to A6 are general-purpose address registers and A7 is a dedicated stack pointer used to implement subroutine calls.
- A minimum 68000 instruction set might be:

```
MOVE.B <ea>,Di        MOVE.B Di,<ea>        MOVE.B Di,Dj
ADD.B  <ea>,Di        ADD.B  Di,<ea>
SUB.B  <ea>,Di        SUB.B  Di,<ea>
CMP.B  <ea>,Di
BCC    label          BCS    label
BEQ    label          BNE    label
BSR    label          RTS
```

- The 68000 decides between two courses of action by means of conditional instructions of the form B$_{cc}$ `label`, where `cc` represents the condition tested. For example, the instruction `BEQ Loop` means *if the result of the last operation was zero, jump to the line labelled by "`Loop`", otherwise execute the next instruction in sequence.*

- A group of instructions, called a subroutine or procedure, can be called (i.e., executed) by means of the instruction `BSR Subroutine`, where "`Subroutine`" is the label of the subroutine called. A return from the subroutine to the instruction immediately following the call is made by means of an `RTS` instruction at the end of the subroutine.
- When the 68000 performs an operation it usually updates one or more bits in the CCR. Not all operations affect the bits of the CCR (`BSR` and `RTS` don't). The 68000 *blindly* updates its CCR. For example, it sets the N-bit (negative bit) of the CCR if the most-significant bit of the result is 1. The N-bit is set, even if the operation was not on a numeric value.
- The 68000 supports four logical operations: NOT, AND, OR, EOR. A NOT operation inverts all bits of the operand. An AND is used to mask bits (set them to zero). An OR is used to set bits (set them to 1). An EOR is used to toggle bits (change their state).
- The 68000's three fundamental addressing modes are: absolute, literal, and address register indirect. These are represented by: `MOVE.B Val6,D0`, `MOVE,B #Day,D0`, and `MOVE.B (A0),D0`, respectively.
- Absolute addressing is used to specify the actual location of an operand in memory.
- Literal or immediate addressing is used to specify a literal source operand (i.e., the actual data to be used in the operation). A literal operand is specified by prefixing it with a "#" symbol. For example, `MOVE.B #5,D0` puts the number 5 into data register D0.
- Address register indirect addressing is indicated by enclosing the address register in round brackets, e.g., `MOVE (A0),D0`. In this case the source operand address is in address register A0. This addressing mode is called *indirect* because the address is specified indirectly via an address register. Since the contents of an address register can be modified, it is possible to step through a table or list of data items in memory.
- A 68000 assembly language program is designed and tested in the following way (when using the Teesside cross-assembler/simulator):
  i. Construct a suitable algorithm.
  ii. Code it into 68000 assembly language.
  iii. Use any ASCII text editor to create a source file with the extension `.X68` (e.g., `MyFile.X68`).
  iv. Use the cross-assembler to generate an assembled listing file and an object binary file; for example, `X68K MyFile -l` creates the listing file `MyFile.lis` and the binary (code) file `MyFile.bin`.
  v. If there are any errors in the listing file, correct them in the source file, and reassemble it.

vi. Use the simulator to run the 68000 machine code; for example, `E68K MyFile`. The simulator's breakpoint and trace facilities permit you to monitor the execution of the program. If you detect errors, you should reassemble the source file.

Problems

1. What is a data register and how is it used? Why do you think that the 68000 has *eight* data registers? What is the optimum number of data registers?

2. What is the difference between a *data register* and a *memory location*?

3. What is the difference between a *data register* and an *address register*?

4. What is the function of the 68000's condition code register, and what are the functions of the individual bits of the CCR?

5. What does the expression `[M([A4])]` mean in plain English?

6. How are the symbols $, #, and % used in 68000 assembly language?

7. What is a sequence control instruction?

8. What is a subroutine?

9. What does *typing* (as in weak typing or strong typing) mean?

10. Is an assembly language weakly or strongly typed, and why?

11. In what way is the 68000's `MOVE` instruction its most versatile instruction?

12. How would I divide the contents of data register D3 by 16?

13. How would I multiply the contents of data register D6 by 32?

14. When you shift the contents of a register (or a memory location) left or right, what happens to the bit that falls off the end?

15. The operation `MOVE.B #25,Temp` loads the number 25 into memory location `Temp`. The operation

```
Temp DC.B 25
```

loads 25 into `Temp`. What is the difference between these two operations?

16. Write a program to load memory location 1000_{16} with 0, location 1001_{16} with 1, location 1002_{16} with 2, and so on up to location $10FF_{16}$.

17. Memory location 8000_{16} contains the bits ABCDEFGH and memory location 8002_{16} contains the bits PQRSTUVW. Write a 68000 program to implement the following function:

```
IF (A + B)C = 1  THEN Q := 1, R := 0

IF (DE + GH) = 1 THEN Q := 0; S := 1, V := 0
```

18. Consider the algebraic equation: $x_i = y_j + z_k - 2 + C$. If this equation were to be encoded into 68000 assembly language statements, what addressing modes would you use?

19. Write a program to calculate the number of days in a given month of the year. The month is in D0 (in the range 0 to 11) and the length of the month is to be returned in D1 (28, 30, or 31).

20. Describe the functions provided by the 68000 simulator that help you to debug a program you are developing.

CHAPTER 4

The 68000's Addressing Modes

In Chapter 3, we provided a gentle introduction to the 68000's instruction set and addressing modes. We are now going to look at two aspects of the 68000 in more detail, its registers and its addressing modes. It would be wonderful if all microprocessors had simple, regular instruction sets. Unfortunately, the requirements of the microprocessor designer, the economics of chip design, and the need to pack as many instructions as possible into a 16-bit format don't permit the construction of an elegant instruction set. We begin by describing the 68000's ability to operate with 8-bit, 16-bit, and 32-bit data values, and then look at its addressing modes. In this and the next chapter, we develop and refine our simple model of the 68000.

However, it is impossible to do justice to the topics of the 68000's instruction set and its addressing modes independently of each other. If we describe a powerful addressing mode, we cannot provide a realistic example of its applications without using some of the 68000's instructions we have not yet introduced. Equally, we cannot show how some of the sophisticated instructions are used without employing addressing modes we have not described. One way of overcoming the problem of dealing with topics in isolation is to break with the tradition of teaching a subject topic by topic in a linear fashion. Consequently, we will introduce new instructions and features of the 68000 whenever they are needed.

4.1 Variable Length Operands

We first look at how the 68000 supports *variable-length operations* by permitting instructions to act on 8-bit, 16-bit, or 32-bit operands. But before we do that, we demonstrate how the programmer would go about synthesizing 16-bit or larger integers using only the 8-bit instruction set and registers already described.

Most first-generation microprocessors were characterized as 8-bit devices. That is, they store data in 8-bit units; internal data registers are 8 bits wide; and their instructions operate on 8-bit values. There are also 4-bit microprocessors, but these highly-specialized devices are largely employed as simple controllers in electronic systems (e.g., a microwave oven or a washing machine). Therefore, we will confine all our discussions to microprocessors with 8-bit and larger wordlengths.

An 8-bit wordlength is ideal for applications involving ASCII-encoded text, since a 7-bit character fits nicely into a byte. The remaining bit can be used as a *tag* to distinguish, for example, between a printing character and a character used to control the format of the text (e.g., to select between bold or italic fonts). An 8-bit wordlength is less than optimum for arithmetic operations, as it permits a precision of only one in 256. Don't panic — 8-bit machines do not limit you to integers less than 256. You can always chain together consecutive 8-bit bytes to form a longer word. For example, if we wish to employ 16-bit numbers, we just use two bytes *side-by-side* to create a 16-bit number. Arithmetic on this 16-bit value is called *multiple precision* arithmetic. We will now demonstrate how you can perform operations on 24-bit numbers in an 8-bit world.

Multiple Precision Arithmetic

Suppose we have an 8-bit machine but need to use a 24-bit wordlength to handle unsigned integers in the range 0 to $2^{24} - 1$ (i.e., 0 to 16M). A 24-bit number, P, can be stored in memory as three consecutive 8-bit values at addresses P_L, P_M, and P_U. The subscripts, L, M, and U, follow the convention of calling the least-significant byte the *lower* byte, the intermediate byte the *middle* byte, and the most-significant byte the *upper* byte.

Let's look at a 24-bit addition expressed in high-level language in the form `R := P + Q`. The memory map of Figure 4.1 shows how the data is actually arranged in memory. P, Q, and R each take up three consecutive memory locations, starting at address N. We can also represent the same data elements *logically*, as if they were really 24-bit entities. In order to add P to Q we have to add P_L to Q_L, P_M to Q_M, and then P_U to Q_U.

```
MOVE.B   PL,DO         Add the lower-order bytes
ADD.B    QL,DO
MOVE.B   DO,RL
MOVE.B   PM,DO         Add the middle-order bytes
ADD.B    QM,DO
MOVE.B   DO,RM
MOVE.B   PU,DO         Add the upper-order bytes
ADD.B    QU,DO
MOVE.B   DO,RU
```

This fragment of code suffers from two problems. The first is that it is tedious and unwieldy (but that's the price you have to pay for multiple precision arithmetic). The second, and rather more serious, problem is that it doesn't always work. When the lower-order pair of bytes are added (i.e., Q_L to P_L), any carry-out from the most-significant bit must be added to the sum of the next pair of bytes. That is, the carry-out from bit 7 generated by the addition of P_L and Q_L must be added to bit 0 when P_M and Q_M are added.

Let's look at an example in decimal arithmetic. When we add the decimal numbers 123 + 236 we get 359 (i.e., 9 = 3 + 6, 5 = 2 + 3, 3 = 1 + 2). If we add 123 + 239, the least-significant pair of digits 3 + 9 give a total of 2 together with a carry-out digit to the next column. When we add the digits in the second column, we add 2 + 3 plus the carry-in from the column on the right to get 6. The same procedure is executed

Figure 4.1 Chaining bytes to create longer wordlengths

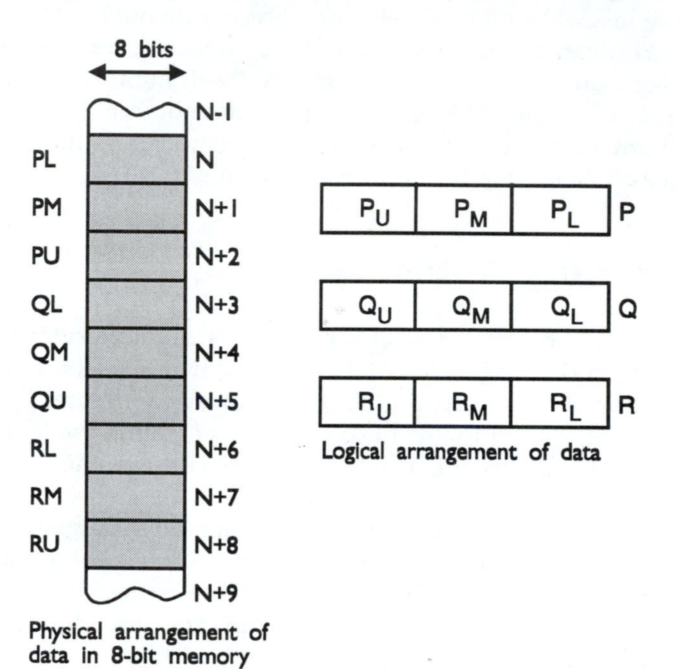

when performing binary arithmetic. The next step is to show how microprocessor designers have made it relatively easy to implement multiple precision arithmetic.

Using Special Instructions to Perform Multiple Precision Arithmetic

The designers of the 68000 have anticipated the needs of those who wish to perform multi-precision arithmetic by creating a special instruction, `ADDX` (add extended), that adds the contents of a source location to the contents of a destination location plus the X-bit of the condition code register, and deposits the result in the destination location. The instruction `ADDX.B D0,D1` is expressed in RTL terms as:

`[D1(0:7)] ← [D1(0:7)] + [D0(0:7)] + [X].`

Remember that the X-bit is a copy of the carry-bit, C-bit, generated during certain arithmetic and shifting operations. When two numbers are added together, the X-bit is set to the same value as the carry-out. By executing an `ADDX` instruction to add a pair of numbers, any previous carry-out is added to the total. You may wonder why the 68000 has both an X-bit and a carry bit. Basically, the X-bit is used only to perform extended length operations, whereas the C-bit is a carry-out of the most significant bit of the operand(s).

To demonstrate how the `ADDX` instruction enables us to perform multi-precision addition, we will implement the 24-bit addition described by Figure 4.1.

```
MOVE.B  PL,D0       Add the lower-order bytes
ADD.B   QL,D0       (using the ADD instruction)
MOVE.B  D0,RL
MOVE.B  PM,D0       Add the middle-order bytes
MOVE.B  QM,D1
ADDX.B  D0,D1       (using the ADDX instruction)
MOVE.B  D1,RM
MOVE.B  PU,D0       Add the upper-order bytes
MOVE.B  QU,D1
ADDX.B  D0,D1       (using the ADDX instruction)
MOVE.B  D1,RU
```

The first operation, ADD.B QL,D0, adds together the two lower-order 8-bit bytes (P_L to Q_L). Any carry-out from the most-significant bit position (i.e., into bit 8) is stored in the X-bit of the CCR. When the two middle-order 8-bit bytes are added together by ADDX.B D0,D1, the carry-out, recorded by the X-bit, is added to their sum. The same process is continued to form the sum of the high-order bytes. Although the 68000's MOVE instruction modifies the CCR, it does not affect the state of the X-bit. Consequently, the MOVE instructions between the ADD and the ADDX operation do not alter the state of the X-bit.

This fragment of code uses the instruction ADDX.B D0,D1, rather than ADDX.B QM,D0 as you might expect. In order for the 68000's designers to cram in as many instructions as possible, some infrequently used instructions lack the full complement of addressing modes. The ADDX instruction takes just two addressing modes. One is data-register-to-data-register with the assembly language form ADDX Di,Dj. The other has the assembly language form ADDX (Ai)+,(Aj)+ that we meet again later in this chapter. By the way, how can a programmer tell what addressing modes a given instruction can take? The answer is that programmers just have to learn what addressing modes can be used with what instructions. If this makes you depressed, remember that the 68000 has a more regular instruction set than many other microprocessors. One of the most common programming errors made by a beginner is the invention of non-existent addressing modes.

Bytes, Words, and Longwords

Chaining bytes together to form a longer wordlength is inefficient, and a much better way of extending precision is to increase the size of the processor's wordlength. Up to now we have largely regarded the 68000 as a microprocessor with an 8-bit wordlength. The actual 68000 supports a 32-bit wordlength, has 32-bit internal registers, and can perform 32-bit operations on data. The 68000's 32-bit wordlength means that it can represent and process unsigned integers with a range from 0 to $2^{32} - 1$ (i.e., 0 to 4 gigabytes), without resorting to extended precision. By the way, although the 68000 has a 32-bit architecture, it has a 16-bit interface to memory and is internally implemented as a 16-bit processor. The only effect of using a 16-bit bus is to reduce the performance (throughput) of the processor. The more powerful 68020 and 68030 processors have 32-bit architectures and full 32-bit interfaces to memory.

Let's stress again that the 68000 is a sophisticated processor that supports 8-bit, 16-bit, and 32-bit operations. Operations on 32-bit values are indicated by appending a `.L` suffix to the appropriate mnemonic, and operations on 16-bit operands are indicated by appending a `.W` suffix. Each 68000 instruction has an *intrinsic* or default wordlength. Most instructions are 16 bits by default. If you omit a suffix from an instruction, the suffix `.W` is usually assumed. That is, `ADD.W D0,D1` and `ADD D0,D1` have identical effects. However, we would advise you always to include the `.W` suffix to identify clearly the size of the operand. We can easily visualize the organization of the 68000's registers (Figure 4.2).

Figure 4.2 The byte, word, and longword

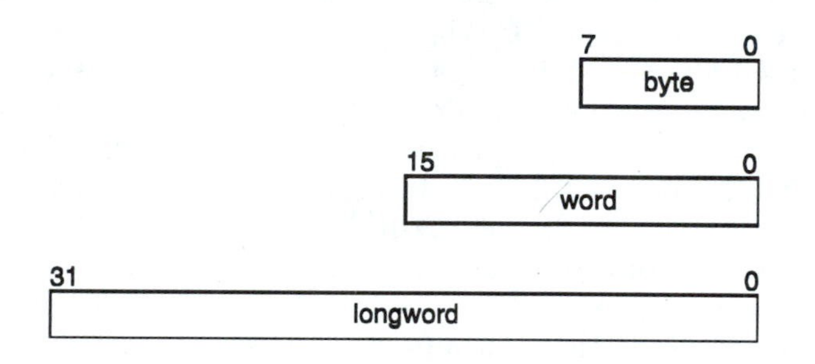

The following three examples demonstrate how the `MOVE` instruction can be applied to byte, word, and longword operands. Figure 4.3 illustrates the corresponding data flow.

Assembly language	RTL form	Action
`MOVE.B $1234,D3`	`[D3(0:7)] ← [M($1234)]`	The 8-bit contents of memory location $1234 are moved into register D3.
`MOVE.W $1234,D3`	`[D3(0:15)] ← [M($1234)]`	The 16-bit contents of memory locations $1234 and $1235 are moved into the lower-order word of D3.
`MOVE.L $1234,D3`	`[D3(0:31)] ← [M($1234)]`	The 16-bit contents of locations $1234 and $1235 are copied into the upper word of D3. The contents of memory locations $1236 and $1237 are copied into the lower word of D3.

Figure 4.3 Byte, word, and longword data transfers

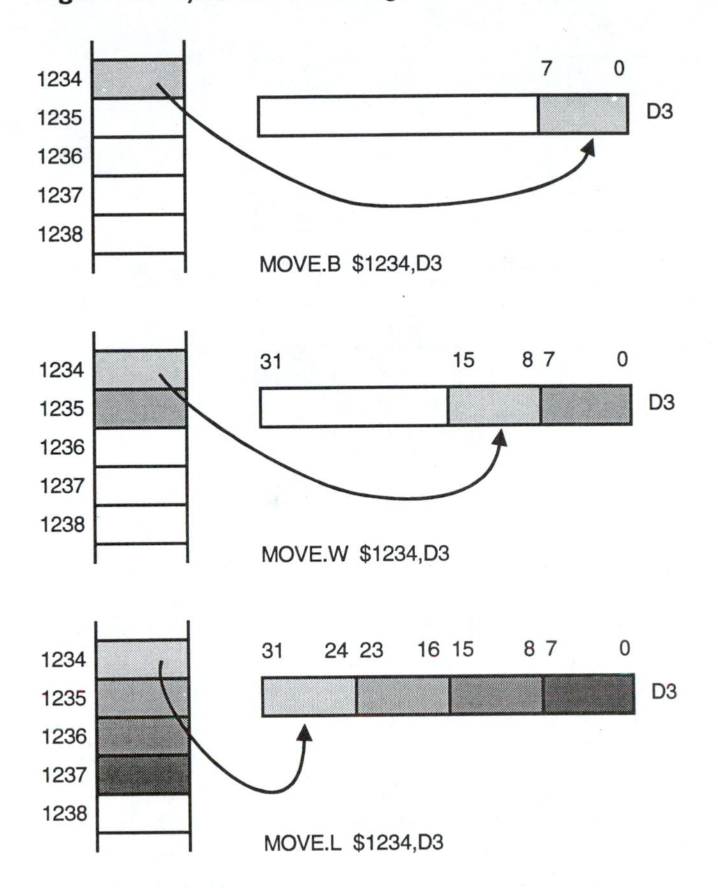

At this stage we need to say something about the way in which the 68000's memory is arranged. The 68000's memory is said to be *byte-addressed*. That is, memory locations have sequential byte addresses numbered 0,1,2,3,4... When the 68000 reads a byte, it gets it from a single location. When it reads a word (i.e., two bytes), it gets it from two *consecutive* byte locations. When it reads a longword (i.e., 4 bytes), it gets it from four consecutive locations. The 68000 hardware and its memory interface memory automatically take care of the generation of successive byte addresses when accessing an operand. In general, this is an operational detail and does not affect the assembly language programmer (i.e., the programmer does not have to worry about how the 68000 moves information between its registers and its memory). The only restriction placed on the programmer by the 68000's byte/word/longword accessing mechanism is that all word and long-word accesses must be to an *even* address. We will return to this topic later.

If the 68000 has 32-bit registers, what happens when a *byte* operation is performed? Figure 4.4 shows, by means of shading, what happens when a register takes part in an operation. A longword operation involves all bits 0 to 31 of the register; a word operation involves bits 0 to 15 of the register; and a byte operation

Figure 4.4 Dividing a longword into words and bytes

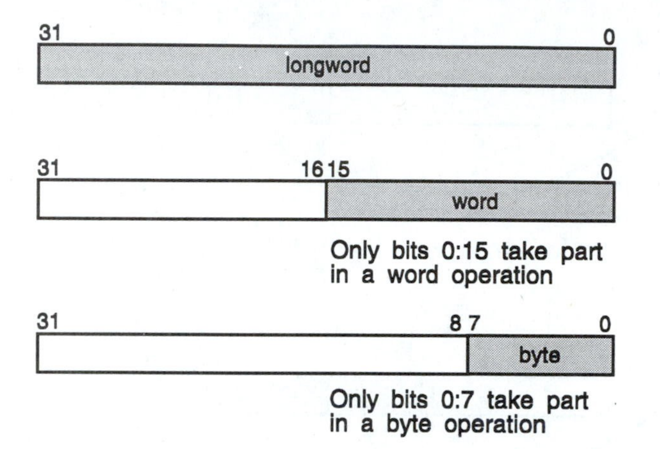

involves bits 0 to 7 of the register. That is, only the *low-order byte* participates in byte operations, and only the *low-order word* participates in word operations.

When a register (or a memory location) takes part in an operation, the bits of the 68000's condition code register, CCR, are updated accordingly. Where, then, does the *carry bit* come from when operations other than a longword are performed? The carry bit represents the carry-out from bits 31, 15, or 7, for longword, word, and byte operands, respectively. Figure 4.5 demonstrates 8-bit addition using the operation `ADD.B D0,D1`.

An add operation such as `ADD.B D0,D1` affects only the eight low-order bits of D1. The carry bit resulting from the addition of bit 7 of D0 to bit 7 of D1 is copied into the CCR's carry flag, and bits 8 through 31 of D1 remain unchanged. If `[D0] = $12345678` and `[D1] = $13579BDF`, the operation `ADD.B D0,D1` results in `[D1] = $13579B57` and the carry flag is set to 1. Had we performed a word operation with `ADD.W D0,D1`, the carry bit would have been set to the carry-out resulting from the addition of the two bit 15s.

The ability of the 68000 to support byte, word, and longword operands is a trap waiting for the unwary. Consider a program to input decimal digits in the range 0 to 9 (i.e., 00000000_2 to 00001001_2) and add them to a running total kept in D1. Suppose that we call a subroutine, `GetDigit`, to input a digit into D0(0:7). That is, a return from the subroutine is made with the digit in the lowest-order byte of D0. Since the running total might become larger than 255, we will store it in the lower-order word of D1, D1(0:15), to cater for a result in the range 0 to 65,535. We can now perform our addition:

```
BSR     GetDigit        Read a single digit into D0(0:7)
ADD.B   D0,D1           Total := Total + Digit
```

We hope you can see the problem raised by the above code. This code adds the low-order *byte* in D0 to the low-order *word* in D1, by means of a `.B` operation, and neglects any carry-out generated by the addition of the byte in register D0 to

Figure 4.5 Byte arithmetic applied to a longword register

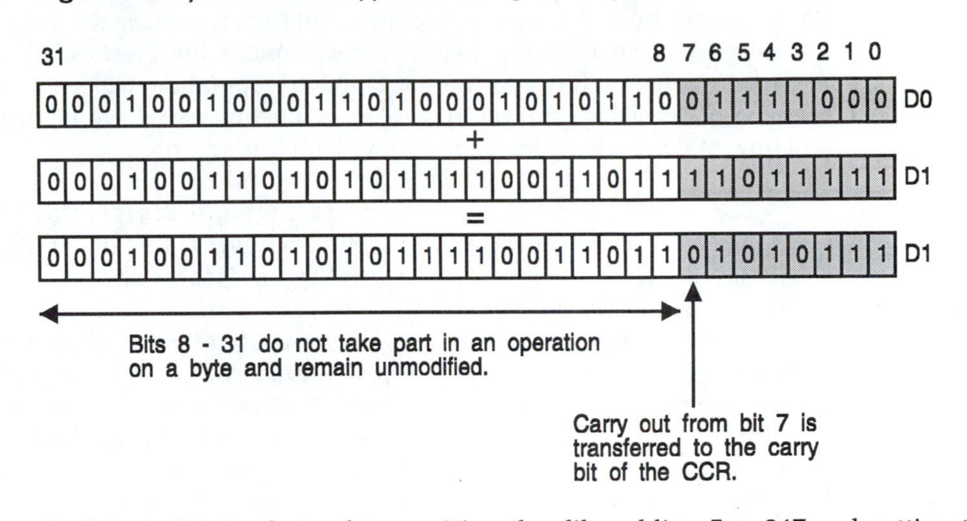

the byte in D1. In decimal terms, it's rather like adding 5 to 247 and getting the result 242, because we neglected the carry-out from the addition of the 5 and the 7. Figure 4.6 demonstrates the effect of an `ADD.B` operation.

We can easily solve the problem by performing a `.W` addition as follows.

```
BSR    GetDigit                Read a single digit into D0(0:7)
ADD.W  D0,D1                   Total := Total + Digit
```

If there's one thing you should learn from this book, it's *never trust no-one*! This solution is also incorrect. We have performed a `.W` addition to solve the problem caused by the carry bit. Unfortunately, we have now created a new problem by adding D0(0:15) to D1(0:15) (i.e., two 16-bit words are added together). The subroutine that inputs the digit puts it in bits D0(0:7), which is a *byte* location. What is the value of bits D0(8:15), the higher-order byte of the low-order

Figure 4.6 The effect of an `ADD.B` instruction on mixed byte/word data values

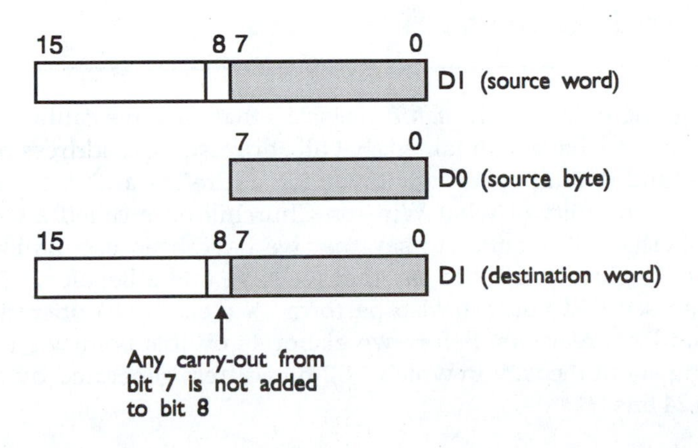

word in D0, after the subroutine has been called? We don't know. Unless we know exactly how the input subroutine, `GetDigit`, works, we have no way of knowing what bits D0(8:15) are after the subroutine has been called. If these bits are set to zero, performing the `ADD.W D0,D1` operation will have the desired effect. If these bits are not set to zero, an erroneous result will be produced. We will now make a third attempt at dealing with this addition.

```
BSR     GetDigit                   Read a single digit into D0(0:7)
AND.W   #%0000000011111111,D0      Mask out bits 8 to 15 of D0
ADD.W   D0,D1                      Total := Total + Number
```

At last we have solved all the problems caused by the addition of numbers of differing size. The byte input in D0(0:7) is *expanded* to a word by ANDing the 8 high-order bits of D0(0:15) with zero and ANDing the 8 low-order bits with 1. In practice, you would probably write the instruction as `AND.W #$00FF,D0` rather than `AND.W #%0000000011111111,D0`. If you knew that the subroutine `GetDigit` did not modify bits 8:15 of data register D0, you could use the following sequence:

```
CLR.L   D0                         Clear D0 before calling subroutine
BSR     GetDigit                   Read a single digit into D0(0:7)
ADD.W   D0,D1                      Total := Total + Number
```

You shouldn't think that the use of mixed-size operands in arithmetic is difficult. It isn't. But you have to be very careful when using two operands of different size. In general, you must perform an operation appropriate to the *larger* operand and *expand* the smaller operand to the same size.

4.2 Addresses and Address Registers

We are now going to look at the *address registers* that we have cautiously avoided so far. Up to now we have maintained that all addresses and address registers are 32-bit values, and that all operations involving addresses are `.L` operations. We have, in fact, been guilty of what Winston Churchill once called a *terminological inexactitude*. Perhaps it is better to say that we have been economical with the truth. In plain English, you could say that we have told a lie, albeit a white one. We shall soon see that you can also perform `.W` (i.e., 16-bit) operations on the contents of address registers. Before we elaborate on this point we are going to say something about the way in which a 32-bit address generated by the 68000 is truncated to 24 bits.

The 68000's Address Bus

The 68000 has 32-bit address registers and a 32-bit program counter. Consequently, the addresses used to specify the location of instructions and operands in memory are 32-bits wide, enabling the 68000 to address any one of 2^{32} byte locations. However, in order to squeeze the 68000 into a 64-pin package, address lines A_{24} to A_{31} within the chip are not connected to address pins, as Figure 4.7 demonstrates. Since the state of address bits A_{24} to A_{31} inside the chip has no effect on the address sent to memory, the 68000 can access only 2^{24} (i.e., 16M) bytes. For most practical purposes, 16M bytes is more than adequate. If you need to employ a larger memory, you have to use a 68020, etc., that has a full 32-bit address bus.

The effect of not connecting address lines A_{24} to A_{31} to address pins is that the state of these bits within the 68000 does not affect the operation of the 68000 system. For example, if you write a program to access memory locations $00123456 and $27123456, you will access the *same* physical location. The most-significant eight bits of the address generated by the 68000 do not affect the physical memory. Consequently, most assembly language programmers use six hexadecimal characters to express a 68000 address, rather than eight characters. However, unless you wish to write programs larger than 16M bytes, you will not be inconvenienced by the 68000's 24-address line restriction.

Of course, when the 68000 saves an address in memory (either as an operand or an instruction), the address is stored in memory as two words (i.e., 32 bits).

Figure 4.7 depicts the 68000's address bus as A_{00} to A_{23}. From the *programmer's* point of view the 68000 does indeed have 24 address lines, but from the *hardware designer's* point of view, the 68000 has only 23 address lines A_{01} to A_{23}. Address line A_{00} is implemented *indirectly* by means of two pins — upper data strobe and lower data strobe. Address lines A_{01} to A_{23} select a word at an even address, and the data strobes select a byte or the whole word. The lower data strobe selects the byte at the odd address, and the upper data strobe selects the byte at the even address. If the strobes are asserted together, the entire word is selected. For the purpose of this text, we shall continue to regard the 68000 as having a 24-bit address bus.

Figure 4.7 The 68000's 24-bit address bus

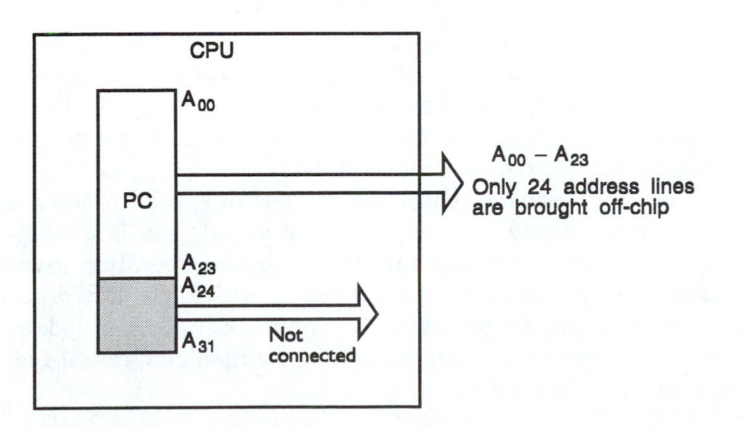

Like most processors, the 68000 regards memory locations as sequential, starting with 000...00 and ending with 111...11. What, then, is the location *before* 000...00 and *after* 111...11? Since numbers *wrap around*, the number before 000...00 is 111...11 and the number after 111...11 is 000...00. There is nothing mysterious about this fact. If you buy a car straight off the production line, its odometer will read 000000. If you drive it backward for a mile it will read 999999. What we have just said is normally of little practical consequence. However, a bug in a program might lead to addresses wrapping round zero in this fashion.

Special Instructions for Address Registers

The 68000's instruction set includes special instructions that operate on the *contents* of address registers. For example, the MOVE instruction doesn't allow an operand to be loaded into an *address* register. Although MOVE.L A1,D0 is legal, the inverse operation MOVE.L D1,A0 is not. The designers of the 68000 have made a determined effort to force the programmer to appreciate that address registers behave differently than data registers. If you wish to use an address register as a *destination* operand, you must write MOVEA rather than MOVE, ADDA rather than ADD, SUBA rather than SUB, and CMPA rather than CMP. These restrictions are consistent with the philosophy of segregating addresses and data, and attempt to make it difficult for the programmer to carelessly corrupt an address. We will now look at the difference between the ADD and ADDA instructions (the same rules apply to the other special instructions used to access address registers).

The ADDA instruction is almost identical to the ADD instruction and must be used whenever the destination of the result is an address register. ADDA must, of course, be used only with word and longword operands. Executing an ADDA instruction has no effect on the contents of the condition code register. For example, the effect of ADDA.L D3,A4 is to add the entire contents of data register D3 to address register A4 and deposit the result in A4, while leaving all the bits of the CCR unchanged.

In practice, it is not necessary to create special instructions to access the 68000's address registers. No fundamental law of computer science states that a processor must have distinct address and data registers. The designers of the 68000 have said that if you write ADD.L D3,A4 rather than ADDA.L D3,A4, the *official* assembler will flag an error, and you must edit your program to correct it. However, some assembler writers have ignored this convention and permit you to write either ADD.L D3,A4 or ADDA.L D3,A4. The assembler itself automatically chooses the appropriate operation code.

We must stress that the reason for the use of special mnemonics to access address registers is purely to emphasize that an address is a single entity and that the contents of the CCR are not affected by an operation involving an address register. The philosophy of segregating addresses and data arouses the same passions in some people as the introduction of seat-belt legislation once did in others. Some programmers have told me that they resent being forced to write programs in a *certain* way.

Word Operations on the Contents of Address Registers

When we said that the 68000's address registers support only longword, `.L`, operations, we were not entirely correct. A data register supports byte, word, and longword operations because the type of information it holds may represent many possible items (e.g., text, number, instruction). An address register holds a 32-bit address which is a *single entity*, and therefore word and byte operations are not relevant. This statement is true in the sense that the contents of an address register represent a 32-bit address, and that the address cannot be subdivided into independent fields or smaller units. However, the 68000 does, in fact, permit word, `.W`, operations on the contents of an address register.

Although you can apply a word operation to the contents of an address register, the effect is that of a *longword* operation. Suppose you decide to add 4 to the contents of address register A0. The instruction `ADDA.W #4,A0` will have the desired effect. The 68000 reads the 16-bit immediate word operand $0004 from the instruction and *extends* it to a longword $00000004. Then the 68000 adds this longword to the contents of A4. The advantage of permitting a word operation on an address register is simple — it requires sixteen less bits to encode `ADDA.W #4,A0` than to encode `ADDA.L #4,A0`. Note that byte operations are not permitted on address registers.

Now consider the following example. Suppose you write the instruction `ADDA.W #$FFF4,A0`. The purpose of this instruction is to add the immediate value of $FFF4_{16}$ to the contents of A0. Before the addition takes place the immediate value must be extended to 32 bits. You might think that the extension is from $FFF4_{16}$ to $0000FFF4_{16}$ — it is not. The 16-bit immediate value $FFF4_{16}$ is extended to the 32-bit value $FFFFFFF4_{16}$.

In order to understand why $FFF4_{16}$ is extended to $FFFFFFF4_{16}$, you have to appreciate one small, but vital, fact. All the 68000's addresses are considered to be *signed* values. Thus, when a 16-bit value is extended to a 32-bit value, it is *sign extended*. If the most-significant bit of the word is zero, bits 16 to 31 of the sign-extension are set to zero. If the most-significant bit of the word is one, bits 16 to 31 of the sign-extension are set to one. For example, the 16-bit signed value $FFF4_{16}$ represents the signed number -12_{10}. When extended to 32-bits it becomes $FFFFFFF4_{16}$, which is also -12_{10}.

Having stated that a 16-bit operand is sign-extended when it is used in an operation whose destination is an address register, we have only one thing left to explain. Why on earth should an address be a *signed value*? After all, you cannot have a negative address. In fact it makes good sense to implement negative addresses. Suppose that address register A0 contains the value 100020_{16} and you execute the instruction `ADDA.W #$FFF4,A0`. The immediate value $FFF4 is extended to 32 bits (i.e., $FFFFFFF4) and added to the contents of address register A0. Since the immediate value is negative (i.e., it represents -12_{10}), the final contents of address register A0 are 100014_{16}, which is 12_{10} less than the original value in A0. If you imagine that a positive address implies a movement in memory in the *forward* direction toward higher addresses, a negative address implies a movement in the *reverse* direction.

Figure 4.8 demonstrates the addition of positive and negative constants to the contents of an address register. Note that the '-3' would be stored as its 16-bit two's complement value 1111111111111101_2 (i.e., $FFFD_{16}$ in hexadecimal). If the contents of address register A0 point to memory location N, adding or subtracting a number from A0 allows us to point to a memory location below or above N, respectively (above and below are with respect to the convention that diagrams show memory with the low address at the top). We shall soon see that the 68000's ability to move an address pointer is one of its most powerful facilities.

Short and Long Absolute Addressing

Up to now, we have generally regarded absolute addresses as 32-bit values. In fact, the 68000 provides two types of absolute addressing: absolute *short* addressing and absolute *long* addressing. In absolute short addressing, the address of the operand is a 16-bit word following the instruction that is sign-extended to 32 bits before it is used to access the operand. Consequently, absolute short addresses in the range $0000 to $7FFF (i.e., 0000000000000000_2 to 0111111111111111_2) are sign-extended to $00000000 to $00007FFF, while absolute short addresses in the range $8000 to $FFFF (i.e., 1000000000000000_2 to 1111111111111111_2) are sign-extended to $FFFF8000 to $FFFFFFFF. Figure 4.9 shows that you can use absolute short addressing to access only the top and the bottom 32K bytes of memory space. Note that throughout this book, the *top* of memory is the region of *low* addresses.

Absolute long addressing requires two 16-bit words following an instruction to generate a 32-bit absolute address and allows the whole of memory to be accessed. Consider the operation:

```
Display EQU    $00806312
        .
        .
        .
        MOVE.B D0,Display
```

When `MOVE.B D0,Display` is encoded, the address of the operand `Display` is stored in memory as two consecutive words — 0080_{16} and 6312_{16}. Of course, when the 68000 actually executes the instruction, only 806316_{16} appears on the address bus, because address lines A_{24} to A_{31} inside the chip are not connected to pins.

You may wonder why the 68000 implements absolute long and short addressing. A 16-bit address can be read from memory faster than a 32-bit address. Program execution time is reduced by locating frequently used variables in the first (or last) 32K bytes of memory space. Remember that speed is one of the reasons for employing assembly language. You don't have to worry about long and short forms of addressing modes, as the assembler *automatically* selects the appropriate version. If you write `MOVE.B Temp6,D3`, the assembler selects absolute short addressing for `Temp6` if it can. The assembly language programmer should locate variables in the regions of memory accessible by short absolute addresses.

Figure 4.8 The effect of modifying an address register

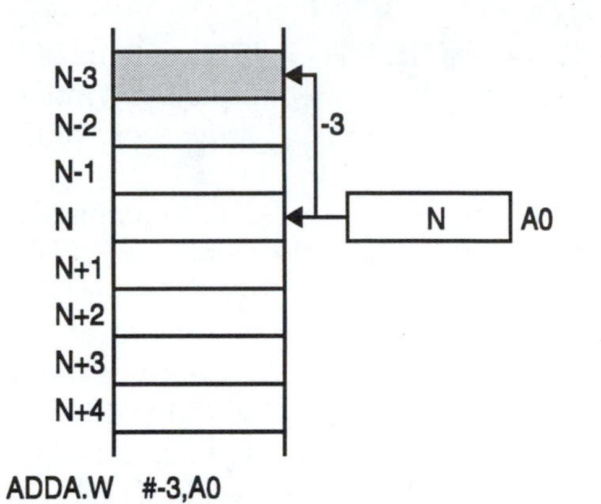

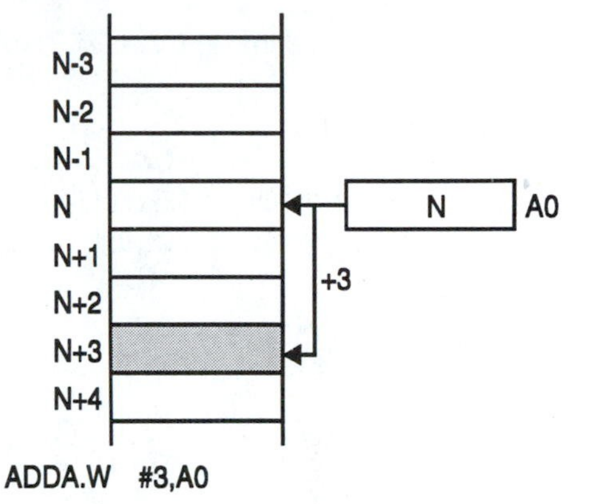

The address register A0 initially points at memory location N. After the instruction it points at N - 3.

The address register A0 initially points at memory location N. After the instruction it points at N + 3.

Consider the following three examples of absolute addressing. As elsewhere, addresses are provided in numeric form to make it easier to see what is happening. In practice, programmers would almost always use symbolic addresses.

Assembly language	RTL form	Action
`MOVE.L D3,$1234`	`[M($1234)] ← [D3(0:31)]`	The contents of register D3 are copied into memory location $1234. Because the 68000's memory is byte-organized, locations $1234 to $1237 each contain a byte of the operand.

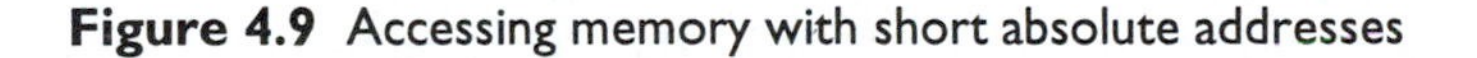

Figure 4.9 Accessing memory with short absolute addresses

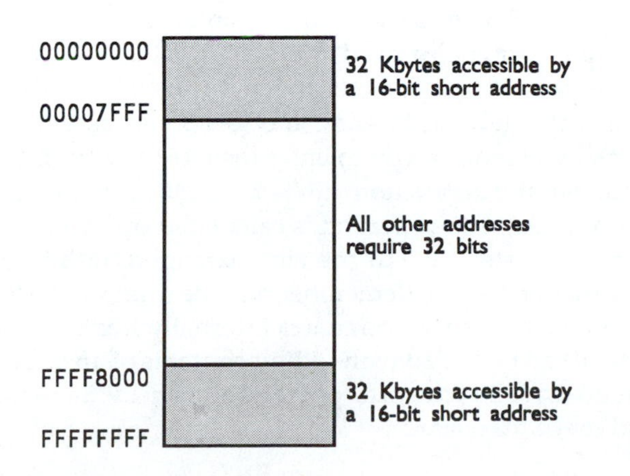

Assembly language	RTL form	Action
`MOVE.W $1234,D3`	`[D3(0:15)] ← [M($1234)]`	The contents of location $1234 are moved into the lower-order word of D3.
`PTM EQU $FFFFC120` `MOVE.L PTM,D2`	`[D2(0:15)]←[M($FFFFC120))]`	The contents of memory location $FFFFC120 are moved into register D2. Note that the 32-bit address $FFFFC120 is encoded as the short address $C120 and is automatically sign-extended to 32 bits ($FFFFC120) when the instruction is executed.

4.3 Address Register Indirect Addressing with Displacement

In the register indirect addressing mode, the effective address of an operand is given by the contents of an address register. For example, the effective address of the destination operand in the instruction `MOVE.B D6,(A6)` is given by the contents of address register A6. We can extend this addressing mode by specifying an *offset* or *displacement* (i.e., a constant) as part of the instruction. This addressing mode is called *address register indirect with displacement* and is written in the form `(d16,Ai)`, where `d16` represents a 16-bit signed displacement. The range over which operands can be accessed with respect to address register Ai is limited because the displacement is only 16 bits, rather than the 32 bits required to provide a comprehensive indexed addressing mode. That is, the displacement can specify a location +32,767 bytes ahead or -32,768 bytes back from the contents of an address register.

The 68000 assembler originally used the syntax `d16(Ai)` to indicate this addressing mode. However, it was later changed to the form `(d16,Ai)` to be consistent with the 68020's new addressing modes. Some assemblers like the Teesside cross-assembler still require you to enter the `d16(Ai)` syntax.

Consider the action of the instruction `MOVE.W (8,A0),D0` illustrated in Figure 4.10. The effective address of the operand is calculated by adding the contents of the specified address register, A0, to the sign-extended 16-bit displacement forming part of the instruction (i.e., 8). Remember that the term *sign-extended* means that the 16-bit two's complement displacement is internally transformed into a 32-bit signed value, so that it can be added to the 32-bit contents of an address register.

In RTL terms, the effective address of an operand is given by: `EA = d16 + [Ai]`. Two examples should make this clear.

Assembly language	RTL definition	Action
`MOVE.L (12,A4),D3`	`[D3] ← [M($0012 + [A4])]`	The contents of the memory location whose address is given by 12 plus the contents of register A4 are moved into register D3.
`MOVE.W (-$04,A1),D0`	`[D0] ← [M(-$04 + [A1])]`	The contents of the memory location whose address is given by the contents of register A1 minus 4 are moved into register D0. The offset -4, is encoded as the 16-bit value $FFFC.

In Figure 4.10 the instruction `MOVE.W (8,A0),D0` loads data register D0 with the contents of the memory location whose address is 8 bytes greater than the value in A0. In this example, we used a *numeric* displacement to describe this addressing mode, although the programmer would frequently employ a *symbolic* displacement; for example, `MOVE.W (Time,A0),D3`.

Address register indirect addressing is frequently used to support a table or a list of data items. Consider the following example.

```
Error    EQU    0              Equate these names to data items
Status   EQU    1
Input    EQU    2
Flag     EQU    3
         .
         .
         ORG    $2000          The data region is to begin here
DataArea EQU    *              We will call location 2000₁₆ "DataArea"
         DS.B   1024           Let's save some storage here
```

Figure 4.10 Using an offset to access an operand with respect to an address register

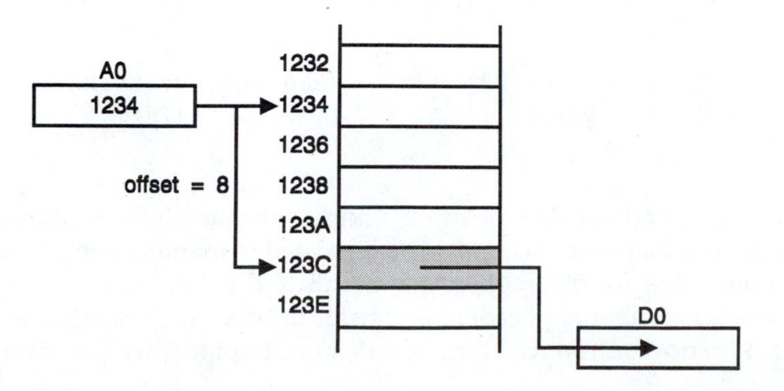

```
.
.
MOVEA.L #DataArea,A0   A0 points to area of memory containing data
.
MOVE.B  (Error,A0),D3  Read the error status (at offset 0)
CMPI.B  #12,(Flag,A0)  Is Flag = 12? (Flag is at offset 3)
.
```

Figure 4.11 provides a memory map for this problem in which the four data items are equated to the values 0 to 3, and address register A0 is loaded with the value 2000_{16}. The assembler directive, or *pseudoinstruction*, `ORG $2000` sets the assembler's location counter to 2000_{16}. Any program or data loaded into memory will therefore be loaded at address 2000_{16} onwards.

We would like to call location 2000_{16} `DataArea` but unfortunately the `ORG` assembler directive cannot take a label. We can, however, use an `EQU` directive of the form `DataArea EQU *` to equate `DataArea` to the current value of the location counter (i.e., 2000_{16}) which is represented by an asterisk. When the instruction `MOVEA.L #DataArea,A0` is executed, the 32-bit literal `DataArea` is copied into A0. Since `DataArea` was equated to the location counter and the value of the location counter is 2000_{16}, address register A0 is loaded with 2000_{16} and now points to the start of the data area.

The instruction `MOVE.B (Error,A0),D3` copies the byte pointed at by A0 plus the offset `Error` into D3. The offset `Error` is zero and the first item in the data area is accessed. The next instruction, `CMPI.B #12,(Flag,A0)`, specifies an operand at an offset of `Flag` (i.e., 3) from A0 and accesses the byte at address 2003_{16}. The instruction `CMPI` stands for *compare immediate* and permits you to compare the contents of a memory location with a literal.

We could have written the preceding fragment of code without resorting to address register indirect addressing, as the following code demonstrates.

```
        ORG     $2000            Start of the data area
Error   DS.B    1
Status  DS.B    1
Input   DS.B    1
Flag    DS.B    1
        .
        .
        MOVE.B  Error,D3         Read the error status
        CMPI.B  #12,Flag         Is Flag = 12?
        .
```

One of the advantages of using address register indirect addressing with displacement is that the data area can be relocated in memory simply by modifying the contents of the address register that points to the data area.

Consider another application of register indirect addressing to access a look-up table. When we introduced binary codes in Chapter 1, we described the ASCII

Figure 4.11 Example of the use of address offsets

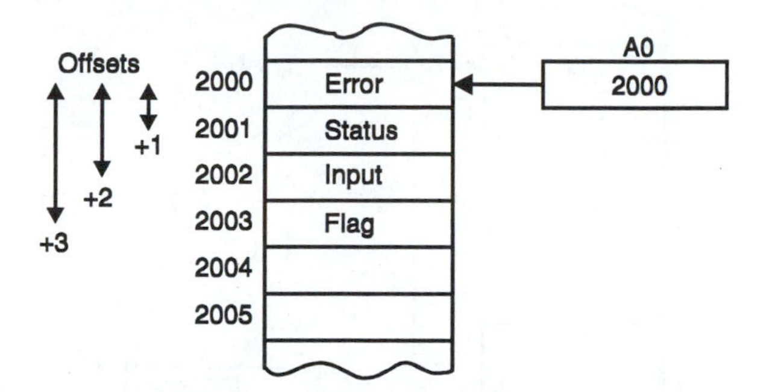

code used to represent alphanumeric characters. The assembly language programmer often needs to convert between the ASCII code for a symbol and the binary value it represents. For example, the symbol "5" is printed by sending the binary value 00110101 to a printer. We can use two techniques to convert a 4-bit hexadecimal digit (HexVal) into its ASCII character form (HexChar). One is to take an algorithmic approach:

```
HexChar := HexVal + $30;
IF HexChar > $39 THEN HexChar := HexChar + $07.
```

The other conversion technique is to resort to a simple *look-up table*. That is, the input value to be converted is used to look-up the output value in a table (a look-up table behaves rather like a dictionary). In the following example, the longword contents of data register D2 are to be printed as a string of eight ASCII-encoded hexadecimal characters using a look-up table to perform the translation. The look-up table, described in Figure 4.12, starts at address TRANS and has 16 entries corresponding to the ASCII codes 30_{16} to 39_{16} and 41_{16} to 46_{16} (i.e., the codes for the symbols 0 to 9 and A to F, respectively). In this example we assume that A0 contains the value $0C used to index into the table to access location TRANS+$0C. This location contains the hexadecimal value $43; that is, the ASCII representation of the character "C". In this example we are using the constant d16 in the effective address (d16,Ai) to indicate the *start* of the table and the contents of the address register as an *index* into the table. Most applications of address register indirect addressing with a constant offset use the address register as a pointer to the table and use the fixed constant d16 as an index into the table.

The following code implements the character translation. Note that the subroutine PRINT_CHAR prints the contents of D0(0:7) as an ASCII character.

```
*     D2.L contains the longword to be printed as 8 hexadecimal characters
*     D1.W is used as a counter to count the 8 characters to be printed
*     D0.B is used to send a character to the print subroutine
*     D3.L is used as a temporary register
*     TRANS is the address of the translation table
*
```

Figure 4.12 Example of application of address register indirect addressing with offset

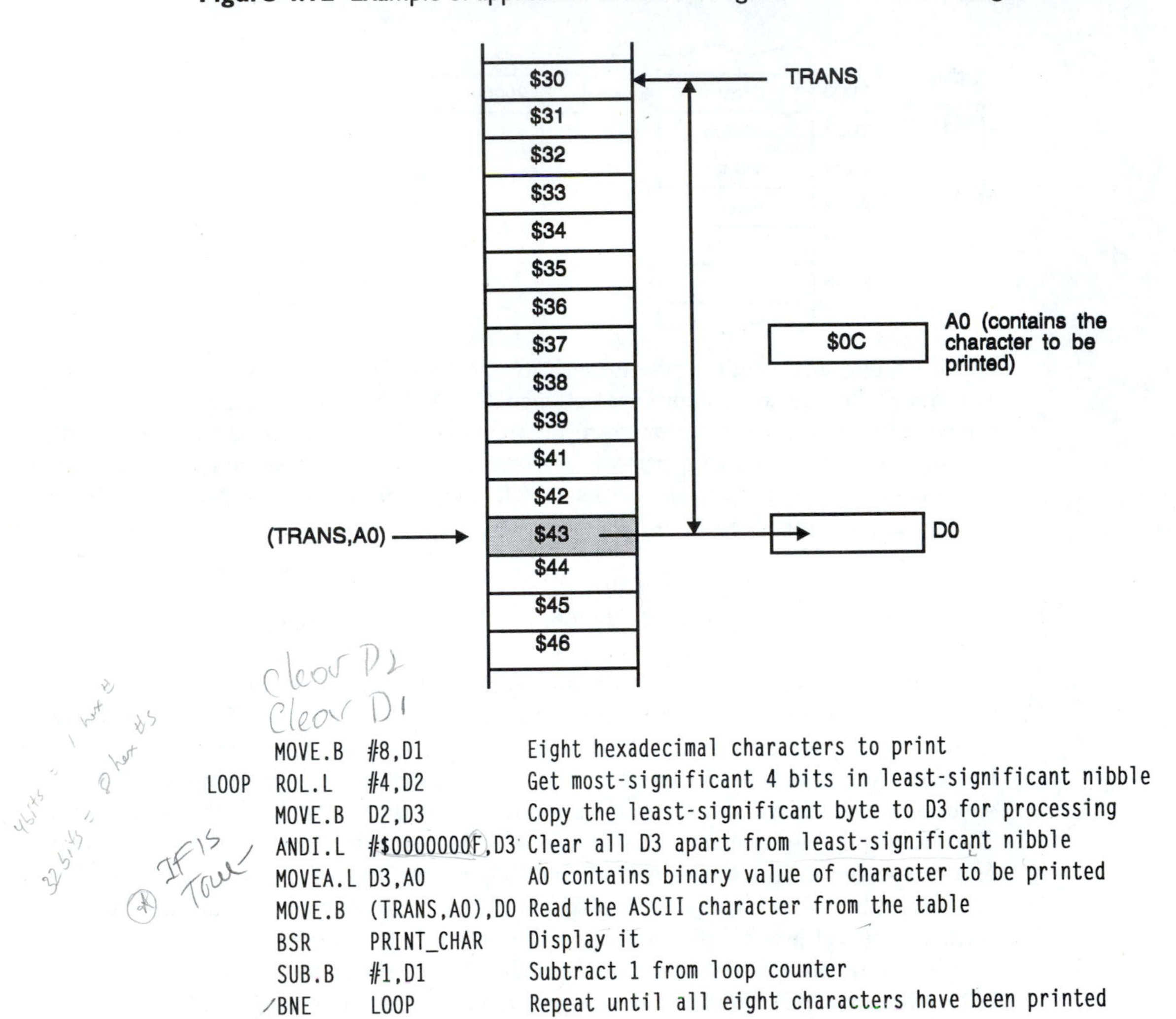


```
      MOVE.B  #8,D1          Eight hexadecimal characters to print
LOOP  ROL.L   #4,D2          Get most-significant 4 bits in least-significant nibble
      MOVE.B  D2,D3          Copy the least-significant byte to D3 for processing
      ANDI.L  #$0000000F,D3  Clear all D3 apart from least-significant nibble
      MOVEA.L D3,A0          A0 contains binary value of character to be printed
      MOVE.B  (TRANS,A0),D0  Read the ASCII character from the table
      BSR     PRINT_CHAR     Display it
      SUB.B   #1,D1          Subtract 1 from loop counter
      BNE     LOOP           Repeat until all eight characters have been printed
      .
TRANS DC.B    '0123456789ABCDEF' Translation table as ASCII string
```

When printing the string of eight characters in data register D2, we have to print first the most-significant character, because printers print from left-to-right. The `ROL.L #4,D2` instruction non-destructively shifts the most-significant nibble into the least-significant position. The least-significant byte of the string is copied to D3, masked with $0000000F to strip it to its four least-significant bits, and then copied to address register A0. Now it can be used as an offset into the character translation table starting at location `TRANS`. The appropriate ASCII character is read from the table and sent to the printer.

By the way, since the 68000's address register indirect addressing mode permits only a 16-bit offset, it follows that the table `TRANS` must lie in the region of memory from \$0000 to \$7FFF (or from \$FF8000 to \$FFFFFF). Later we shall see that the 68020 has an *indirect* addressing mode that makes it even easier to manipulate tables.

Before we look at an example of how address register indirect addressing is used in practice, we are going to introduce a very important instruction, LEA.

The Load Effective Address Instruction

The load effective address, LEA, instruction calculates the *effective address* of an operand and loads it into an address register. Although it looks quite innocent, the LEA instruction is one of the most powerful instructions provided by the 68000. I once read a book in which the author dismissed LEA as an unimportant minor instruction. A more incorrect statement would be hard to find. So, having given the LEA such a build-up, what does it do? Well, not a lot, really. An LEA instruction calculates an address and puts it in an address register. Since the LEA instruction always generates a 32-bit address, it is an intrinsically longword operation and therefore does not require a .L extension. Once the address of an operand has been calculated and deposited in an address register, the programmer can employ address register indirect addressing to access the operand. In other words the LEA instruction is a *facilitator* because it lets you do powerful things later. We will now look at what the LEA instruction does.

Assembly language	**RTL form**	**Action**
LEA \$0010FFFF,A5	[A5] ← \$0010FFFF	Load the address \$0010FFFF into address register A5.
LEA (A0),A5	[A5] ← [A0]	The contents of A0 are loaded into A5.
LEA (12,A0),A5	[A5] ← [A0] + 12	The contents of A0 plus 12 are loaded into A5.
LEA (12,A0,D4.L),A5	[A5] ← 12+[A0]+[D4]	The contents of A0 plus the contents of D4 plus 12 are loaded into A5.

Note that LEA \$0010FFFF,A5 deposits the 32-bit immediate value $0010FFFF_{16}$ in A5 because it is loading an effective address and not the contents of a memory location. This instruction has the same effect as MOVEA.L #\$0010FFFF,A5. You might reasonably wonder why the LEA instruction does not employ a '#' in front of a constant to denote that it is an immediate value. The LEA instruction takes as its source operand an *address* and deposits it in an address register. The longword value '\$0010FFFF' represents an address. By the way, there is a difference between the instructions MOVEA.W #address,Ai and LEA address,Ai. The

MOVEA.W instruction sign-extends a 16-bit value before it is loaded into an address register, while LEA takes a 32-bit operand which is not sign-extended.

The second example, LEA (A0),A5 copies the *contents* of A0 into A5 and is therefore identical to MOVEA.L A0,A5. Once again, you must appreciate that the source operand, (A0), is in round brackets because the effective address is given by the contents of A0.

In the third example, LEA (12,A0),A5, the effective address is the sum of the contents of address register A0 plus 12 and is deposited in A5. In the fourth example, LEA (12,A0,D4.L),A5, the address evaluated from the expression 12 + [A0] + [D4] is deposited in A5. We have not yet introduced this addressing mode (called *indexed* addressing) — all we need say here is that the effective address is computed from the sum of *two* pointer registers. If the instruction MOVEA (12,A0,D4.L),A5 had been used, the *contents* of that address would have been deposited in A5. Load effective address has been provided to avoid the time-consuming calculation of effective addresses by the CPU.

It is clearly more efficient to put the effective address into an address register by means of an LEA <ea>,An instruction and then use address register indirect addressing to access an operand, than to recalculate the address every time it is used. For example, if the operation ADD.W ($1C,A3,D2),D0 is to be repeated many times, it is better to execute an LEA ($1C,A3,D2),A5 instruction once and then to repeat ADD.W (A5),D0.

A realistic example of the application of the LEA instruction is provided later in this book when we consider the design of a command line interpreter. However, we will provide a short summary of the example here. A table in memory is pointed at by an address in address register A2. The length of each entry in the table is six bytes plus the value in the first element in the entry (Figure 4.13). We wish to calculate the address of the *next* entry in order to step through the table. We cannot simply add a fixed increment to the address register, since the length of each entry is not fixed. By executing MOVE.B (A2),D1, we can store the length of the current entry (less 6) in D1. Now, executing an LEA (6,A2,D1.W),A3 adds

Figure 4.13 Using LEA to access a table

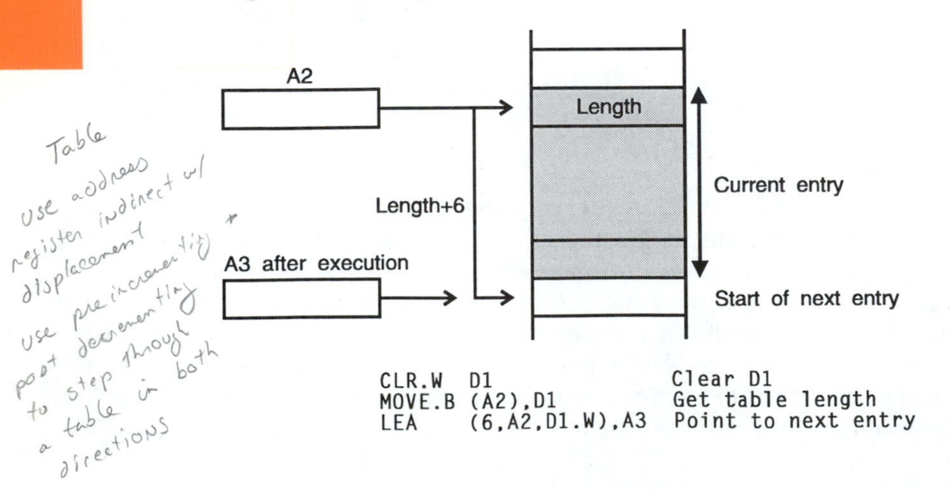

the contents of A2 to the contents of the lower-order word of D1 plus 6 and deposits the result in A3. That is, `[A3] ← [A2] + [D1(0:15)] + 6`. The contents of A3 are, of course, the address of the next entry in the table. That is, A3 now points to the next entry in the table. Note that `D1(8:15)` must be cleared by `CLR.W D1`, before carrying out the `LEA`.

Later in this chapter we shall discover that the `LEA` instruction has a very powerful application. It can be used to write programs that are called *position independent*. Such programs can be moved from one region of memory to another without having to modify the address of operands in memory.

4.4 An Application of Address Register Indirect Addressing

We're now going to examine an application of address register indirect addressing. This addressing mode is one of the most powerful tools available to the assembly language programmer. Essentially, it enables you to implement complex data structures and to access data from them quickly and efficiently. The expression *data structure* describes how information is organized and lies at the heart of programming. The terms used to describe data structures are frequently borrowed from everyday life (e.g., list, table, array, matrix, queue). Since an address register can be said to *point* to an item in a data structure, all we have to do to access another item in the structure is to *modify* the contents of the address register.

A good example of the use of address register indirect addressing is provided by the *linked list*. The simplest type of singly linked list is composed of a chain of elements, each of which is linked to its successor by an address (i.e., a pointer). Figure 4.14 illustrates a linked list in which each item is composed of two elements: a *head* and a *tail*. The head is an address that points to the next item in the linked list. The last element in the list has a null (i.e., zero) address. The tail of each item can be any type of data item. A typical linked list is a disk file. Each record that makes up the file contains a pointer to the next record. Note that we call the list *singly linked* because the links point in one direction only (i.e., the head points to the next item). In a *doubly linked* list, the links are *bidirectional* so that each head points to both the next and to the prior unit.

Figure 4.14 The linked list

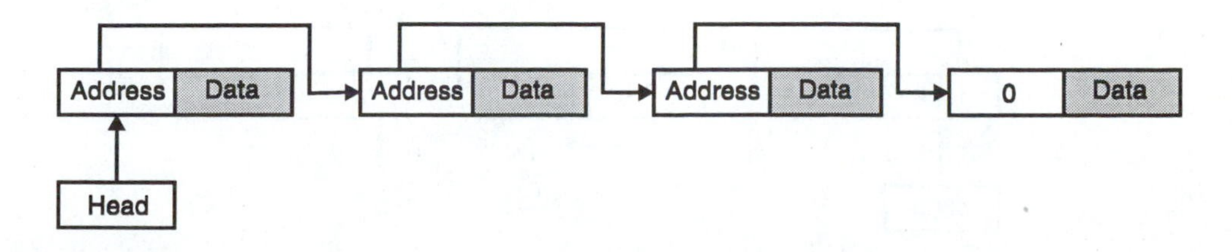

Suppose we wish to add a new element to the linked list by inserting it at the end. All we have to do is read the address field of the first element to locate the next (i.e., second) element. Then we read the address field of this element in order to move to the third element. The list can be traversed in this way until its end is reached. We know that the end of the list has been located when the address of the next element is zero. The following fragment of code inserts a new item into the list. Initially, the longword HEAD points to the first item in the list, and the longword NEW contains the address of the new item to be inserted.

We have introduced a new instruction, TST (i.e., test operand). The TST instruction reads the operand, compares it with zero, and sets the bits of the CCR accordingly. For example, TST D3 has a similar effect to CMP #0,D3 (there is a slight difference between these instructions as we shall see later).

Linked Lists

```
     LEA      HEAD,A0        A0 initially points to the start of the list
LOOP TST.L    (A0)           IF the address field pointed at by A0 is 0
     BEQ      EXIT              THEN exit
     MOVEA.L  (A0),A0           ELSE read the address of the next element
     BRA      LOOP           CONTINUE
EXIT LEA      NEW,A1         Pick up the address of the new element
     MOVE.L   A1,(A0)        Add new entry to end
     CLR.L    (A1)           Add new terminator
     MOVE.L   data,(4,A1)    Insert new data in list
```

1st 4 bytes stores address

The *work* is done by the MOVEA.L (A0),A0 instruction. Before executing this instruction, address register A0 holds the head of the *current* element and therefore points to the *next* element in the list. The source operand in MOVEA.L (A0),A0 is the longword pointed at by A0, which is a pointer to the next element in the chain. If you understand the role carried out by MOVEA.L (A0),A0, you will have few problems using address register indirect addressing. Figure 4.15 demonstrates a linked list before the insertion of a new element, and Figure 4.16 demonstrates the same list after the insertion of the element. The new element has the value $ABCD_{16}$ and is to be added at location 2004_{16}.

Figure 4.17 shows the memory map of the linked list both before and after the insertion of a new item.

The 68000, like any other modern processor, has a range of powerful facilities. However, without address register indirect addressing this processor would be severely limited and would not be able to handle complex data structures

Figure 4.15 Example of a linked list

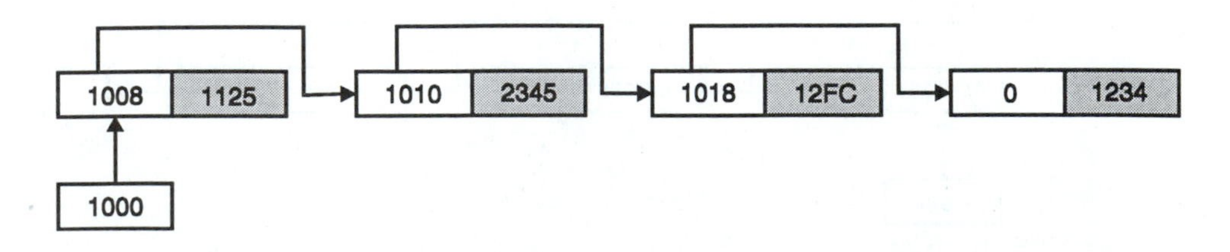

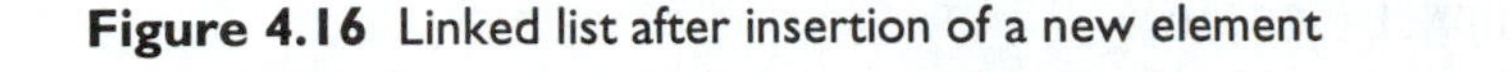
Figure 4.16 Linked list after insertion of a new element

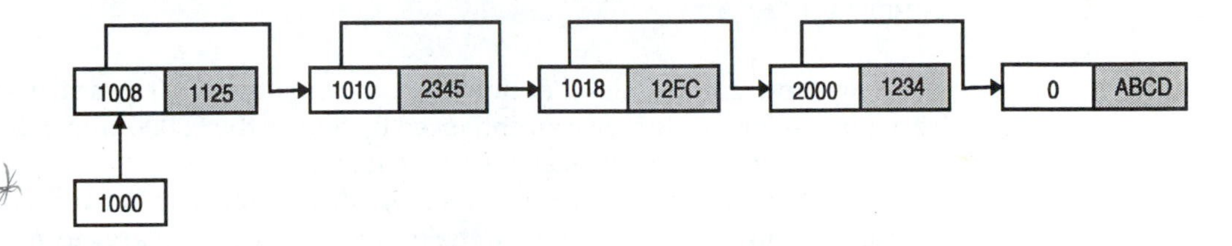

efficiently. When we introduce the 68000's more powerful successor, the 68020, we will discover that one of the most significant changes is the addition of even more indirect addressing modes to the 68000's existing addressing modes.

4.5 Address Register Indirect Addressing with Post-incrementing and Pre-decrementing

It makes sense, at least intuitively, to argue that an effective computer architecture should remove some of the burden from the programmer by converting a complex sequence of operations into a single instruction. You have already met this concept before — the single instruction `LEA (EA,A0,D3.L),A2` has the same effect as the sequence of three instructions:

Figure 4.17 Memory map of a linked list before and after inserting an item

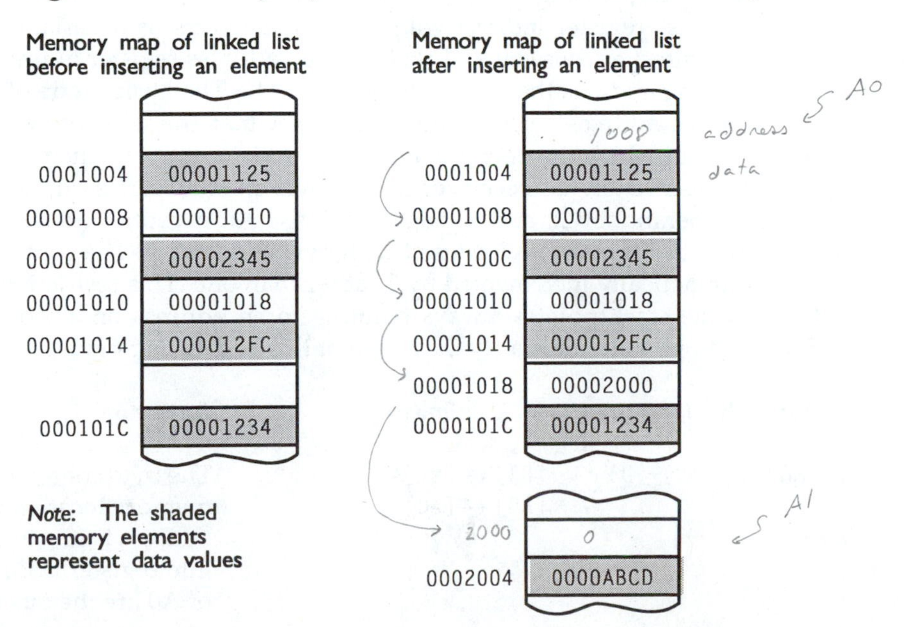

```
ADDA.L  D3,A0
ADDA.L  #12,A0
MOVEA.L A0,A2.
```

We are now going to look at how the power of the 68000's address register indirect addressing mode has been enhanced by letting the 68000 *automatically* alter the contents of the pointer register each time it is used.

Let's first look at a typical application of address register indirect addressing in accessing a data structure in which the individual elements are stored consecutively. For example, consider the following fragment of a program designed to fill a 16-element table (i.e., array) of bytes, called BUFFER, with zeros.

```
     MOVE.B  #16,D0      Set up a counter for 16 elements
     LEA     BUFFER,A0   A0 points to the first element of the array
LOOP CLR.B   (A0)        Clear the element pointed at by A0
     ADDA.L  #1,A0       Move the pointer to point to the next element
     SUB.B   #1,D0       Decrement the element counter
     BNE     LOOP        Repeat until the count is zero
```

There is nothing wrong with this fragment of code and it does what it is supposed to. However, note that when we use A0 as a pointer in an instruction like CLR.B (A0), we follow it by the instruction ADDA.L #1,A0 to bump up the pointer. As you can imagine, this second instruction frequently follows a memory access employing address register indirect addressing. The designers of the 68000 have provided two addressing modes that improve its efficiency: address register indirect with *post-incrementing* addressing, and address register indirect with *pre-decrementing* addressing.

Address register indirect with post-incrementing is a variation of address register indirect addressing and is also called address register indirect with auto-incrementing and has the assembly form (Ai)+. The basic mode of operation is the same as address register indirect, except that the contents of the address register from which the operand address is derived are incremented by 1, 2, or 4 after the instruction has been executed. A byte operand causes an increment by 1, a word operand by 2, and a longword by 4. An exception to this rule occurs when the stack pointer, A7, is used with byte addressing. The contents of A7 are then automatically incremented by 2 rather than one. This restriction is intended to keep the stack pointer always pointing to an address on a word boundary. Some examples should clarify the action of this addressing mode.

Assembly language	**RTL definition**	**Action**
MOVE.B (A0)+,D3	[D3] ← [M([A0])] [A0] ← [A0] + 1	The byte contents of the memory location whose address is in A0 are copied into register D3. The contents of A0 are then increased by 1.

`MOVE.L (A0)+,D3`	`[D3] ← [M([A0])]` `[A0] ← [A0] + 4`	The longword contents of the memory location whose address is in A0 are copied into register D3. The contents of address register A0 are then increased by 4.
`MOVE.W (A7)+,D4`	`[D4(0:15)] ← [M([A7])]` `[A7] ← [A7] + 2`	The 16-bit contents of the location whose address is in A7 are copied into the lower-order 16 bits of D4. The contents of address register A7 are then increased by 2.
`MOVE.B (A7)+,D4`	`[D4(0:7)] ← [M([A7])]` `[A7] ← [A7] + 2`	The 8-bit contents of the location whose address is in address register A7 are copied into the lower-order 8 bits of D4. The contents of address register A7 are then increased by two, rather than one, because A7 is the system stack pointer.

Figure 4.18 illustrates the effect of the 68000's auto-incrementing mechanism, and describes the state of the system before and after the execution of a `MOVE.B (A0)+,D0`. Before the instruction is executed, A0 points at location 1001. After execution, it points at location 1002. The increment is by one because the operand is a byte.

Consider again the fragment of a program designed to fill a 16-element array of byte with zeros. This time we will use the post-incrementing variation of address register indirect addressing.

Figure 4.18 The effect of pre-incrementing on an address register

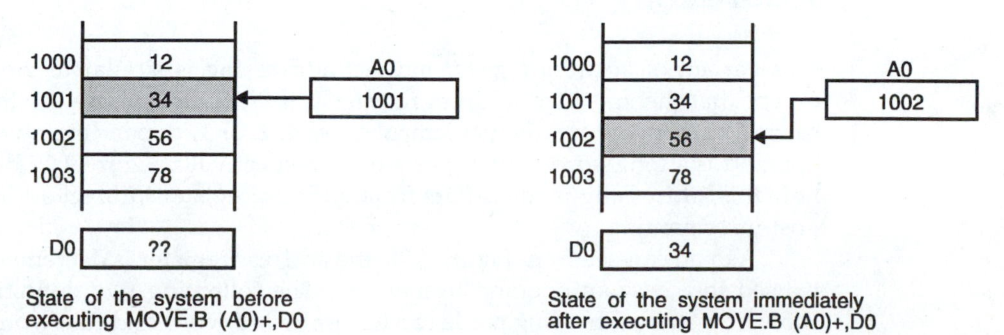

```
     MOVE.B  #16,D0        Set up a counter for 16 elements
     LEA     BUFFER,A0     A0 points to the first element of the array
LOOP CLR.B   (A0)+         Clear element and move the pointer to the next
     SUB.B   #1,D0         Decrement the element counter
     BNE     LOOP          Repeat until the count is zero
```

Let's look at a second example of this addressing mode. Suppose that a program employs two data tables, each N bytes long, and it is necessary to compare the contents of the tables, element by element, to determine whether they are identical. The following program will do this. The work is done by the CMPM (A0)+,(A1)+ instruction, that compares the contents of the element pointed at by A0 with the contents of the element pointed at by A1, and then increments both pointers. The mnemonic CMPM used in this program stands for *compare memory with memory*.

```
TABLE_1   EQU      $002000       Location of Table 1
TABLE_2   EQU      $003000       Location of Table 2
N         EQU      $30           48 elements in each table
          .
          .
          LEA      TABLE_1,A0    A0 points to the top of Table 1
          LEA      TABLE_2,A1    A1 points to the top of Table 2
          MOVE.B   #N,D0         D0 is the element counter
NEXT      CMPM.B   (A0)+,(A1)+   Compare a pair of elements
          BNE      FAIL          IF not the same THEN exit to FAIL
          SUB.B    #1,D0         ELSE decrement the element counter
          BNE      NEXT          REPEAT until all done
SUCCESS   .                      Deal with success (all matched)
          .
          .
FAIL      .                      Deal with fail (not all matched)
```

Address Register Indirect with Pre-decrementing Addressing

This variant of address register indirect addressing is similar to the one above, except that the specified address register is *decremented* before the instruction is carried out. As before, the decrement is by 4, 2, or 1, depending on whether the operand is a longword, a word, or a byte, respectively. Figure 4.19 demonstrates how this addressing mode differs from address register indirect addressing with post-incrementing.

As you can see from Figure 4.19, the address register is decremented *before* it is used to access an operand in memory. The following two definitions in RTL show how this addressing mode can be applied to word and longword operands.

Assembly language	RTL definition	Action
`MOVE.W -(A7),D4`	`[A7] ← [A7] - 2` `[D4(0:15)] ← [M([A7])]`	The contents of address register A7 are first decremented by 2. The contents of the memory location pointed at by A7 are moved into data register D4.
`MOVE.L -(A0),D3`	`[A0] ← [A0] - 4` `[D3] ← [M([A0])]`	The contents of address register A0 are first decremented by 4. The contents of the memory location pointed at by A0 are then moved into register D3.

Why does the 68000 support both *post*-incrementing and *pre*-decrementing, and why does one mode act on the contents of an address register *after* it is used to access an operand, and the other *before* it is used? One reason for implementing both incrementing and decrementing modes is that the programmer can step through a table in both directions (from top-down or from bottom-up). We explain why the 68000 implements *post*-incrementing and *pre*-decrementing when we demonstrate how these addressing modes are used to implement the stack.

We now demonstrate an example that employs both of these two addressing modes. The order of the 16 bytes in a region of memory called BUFFER is to be reversed. For example, the sequence 12,34,56,78,90,11,22,33,44,55,66,77,88,99,45,27 is to be reversed to give the sequence 27,45,99,88,77,66,55,44,33,22,11,90,78,56,34,12. There are many ways of reversing the numbers. One method is to swap the top element with the bottom element, and then the next element down from the top with the next element up from the bottom, and so on.

Figure 4.19 The effect of auto-decrementing on an address register

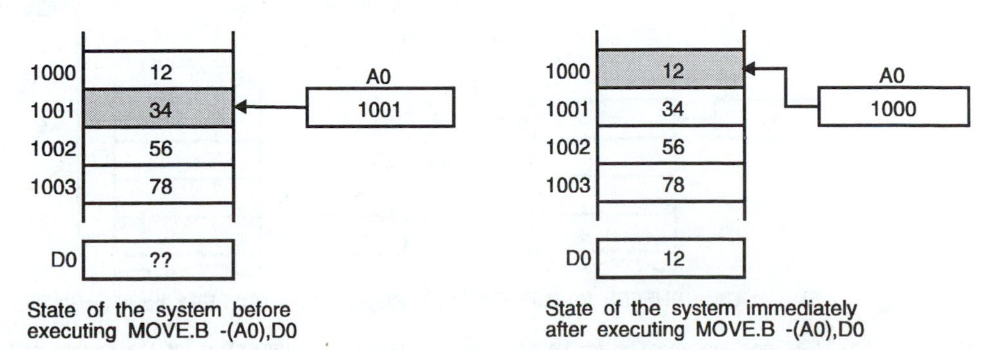

We will employ a very simple but rather crude technique to reverse the order of the numbers. The contents of BUFFER are copied into a second region of memory, BUFFER1, using the instruction MOVE.B (A0)+,(A1)+. Post-incrementing is used for both source and destination operands; see Figure 4.20. Having made a copy of the data, all we have to do is copy the contents of BUFFER1 back into BUFFER in the *reverse order*. To do this we set A0 pointing to the top of BUFFER and move down, and leave A1 pointing to the bottom of BUFFER1 and move up. This time the data transfer is carried out by MOVE.B -(A1),(A0)+.

```
      MOVE.B  #16,D0         Set up a counter for 16 elements
      LEA     BUFFER,A0      A0 points to first element of the source array
      LEA     BUFFER1,A1     A1 points to first element of the destination array
LOOP  MOVE.B  (A0)+,(A1)+    Move an element from its source to its destination
      SUB.B   #1,D0          Decrement the element counter
      BNE     LOOP           Repeat until the count = zero and all elements moved
*
*     At the end of this operation A0 points to one location beyond
*     (i.e., higher than) BUFFER and A1 points to one location beyond BUFFER1
*
      MOVE.B  #16,D0         Reset the counter for 16 elements
      LEA     BUFFER,A0      A0 points to first element of the destination array
LOOP1 MOVE.B  -(A1),(A0)+    Move an element from BUFFER1 to BUFFER
      SUB.B   #1,D0          Decrement the element counter
      BNE     LOOP1          Repeat until the count is zero
```

Figure 4.20 Using both post-incrementing and pre-decrementing

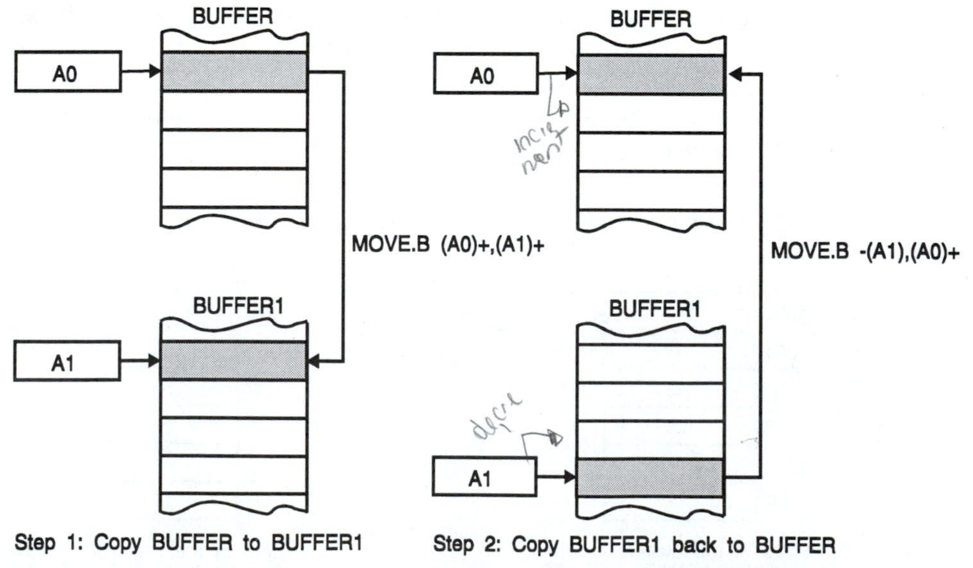
BUFFER
A0
MOVE.B (A0)+,(A1)+
BUFFER1
A1
Step 1: Copy BUFFER to BUFFER1
Use post-incrementing for both source and destination operands
BUFFER
A0
MOVE.B -(A1),(A0)+
BUFFER1
A1
Step 2: Copy BUFFER1 back to BUFFER
Use pre-decrementing for the source and post-incrementing for the destination operands

Post-incrementing and Pre-decrementing Addressing and the Stack

We have already introduced the stack as a mechanism used to implement subroutines. Whenever a subroutine is called, its *return address* is *pushed* on the top of the stack pointed at by address register A7. Whenever a return from a subroutine is made, the return address is *pulled* from the top of the stack.

It is possible to employ a stack as a general data structure and to push data onto it or to pull data off it as required. The 68000 supports eight stacks, as it has eight stack pointers, A0 to A7. Only A7 is used to support the subroutine return mechanism. The programmer is free to employ any of the other seven address registers as stack pointers. Consider the example of Figure 4.21 in which a stack is pointed at by address register A1. We first push two bytes onto this stack and then pull the bytes off the stack.

The stack grows *upward* in the direction of *low* addresses as items are added to the top of the stack (i.e., pushed). The stack pointer points to the top item on the stack and an item is pushed onto the stack in two stages. First, the stack pointer is decremented to point to the next free location above the old top-of-stack, TOS. Then the new item is stored at the location pointed at by the stack pointer.

Initially, in state A, the stack pointer, A1, points to the item at the top of the stack labelled TOS. In state B, the instruction `MOVE.B D0,-(A1)` pushes the low-order byte in data register D0 onto the stack. The stack pointer is automatically moved up (i.e., decremented) by the auto-decrementing addressing mode. Remember that throughout this text *low* addresses are at the *top* of a figure. In state C, the contents of D1 are pushed on the stack and the contents of A1 adjusted.

In state D, the instruction `MOVE.B (A1)+,D2` picks up the byte on the top of the stack, transfers it to register D2, and then moves the stack pointer down one byte to point to the new top of stack. Note that, since the stack pointer points at the top of the stack, the data to be pulled can be accessed without modifying the

Figure 4.21 Example of the use of an address register as a stack pointer

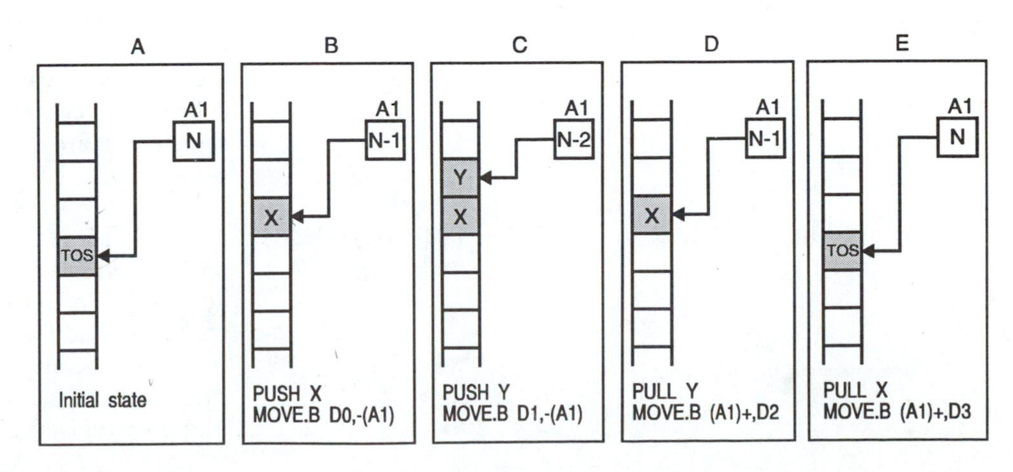

stack pointer. However, once it has been accessed the stack pointer must be moved down. Consequently, pulling an item off the stack requires the auto-decrementing addressing mode. It is for this reason that the 68000 implements pre-decrementing and post-incrementing addressing.

4.6 Address Register Indirect Addressing with Index

We have already introduced address register indirect addressing with a displacement that allows us to calculate the effective address `(d16,Ai)`. This addressing mode permits us to access an operand at a *fixed offset* (i.e., d16 bytes) from the location pointed at by an address register. It would be even better if we could specify a *variable offset* that can be changed as the program runs. Address register indirect addressing with *index* provides such a facility.

Address register indirect addressing with *index* forms the effective address of an operand by adding the contents of an address register to the contents of a *general register*, together with an 8-bit signed displacement. The general register may be an address register or a data register and is termed an *index* register. The assembly language form of this addressing mode is written as `(d8,An,Xn.W)` or `(d8,An,Xn.L)`. Note that the general register Xn may be An or Dn and that it may be a 16-bit value or a 32-bit value. If a 16-bit value is specified, the low-order 16 bits of the register are sign-extended to 32 bits before being added to the address register. In RTL form, the effective address of an operand is given by: `EA = d8 + [Ai] + [Xj]`. The following examples show how an indexed address is computed.

Assembly language	**RTL definition**	**Action**
`MOVE.L (A1,D0.L),D3`	[D3] ← [M([A1] + [D0])]	The contents of the memory location whose effective address is given by the contents of address register A1 plus the contents of data register D0 are moved into register D3.
`MOVE.L (3,A1,D0.L),D3`	[D3] ← [M(3 + [A1] + [D0])]	The contents of the memory location whose effective address is given by the contents of address register A1 plus the contents of D0 plus a constant, 3, are moved into data register D3.

`MOVE.L (3,A1,D0.W),D3`	`[D3] ←[M(3+[A1]+[D0(0:15)])]`	The contents of the memory location whose effective address is given by the contents of address register A1 plus the sign-extended contents of the low-word of D0 plus a constant, 3, are moved into register D3.

Figure 4.22 illustrates the effect of executing `MOVE.B (3,A1,D0.L),D3` for the case in which address register A1 points at memory location $1232 and data register D0 contains the integer 6. The effective address of the operand is given by the contents of address register A1 plus the contents of data register D0 plus 3 (i.e., $00001232 + $00000006 + $00000003 = $0000123B).

Indexed addressing is the most general and most complex type of addressing mode we have so far encountered. Some notes are needed to bring out its special features.

- The 8-bit displacement is a signed, two's complement value, offering a range of -128 to +127. An 8-bit displacement is provided simply because there are only 8 bits in the instruction code left for this purpose after the operation has been specified by the other bits.
- The contents of the 32-bit index register may be treated as a 32-bit longword or a 16-bit word. For example, `MOVE.L (12,A1,D0.L),D3` forms the effec-

Figure 4.22 An example of indexed addressing

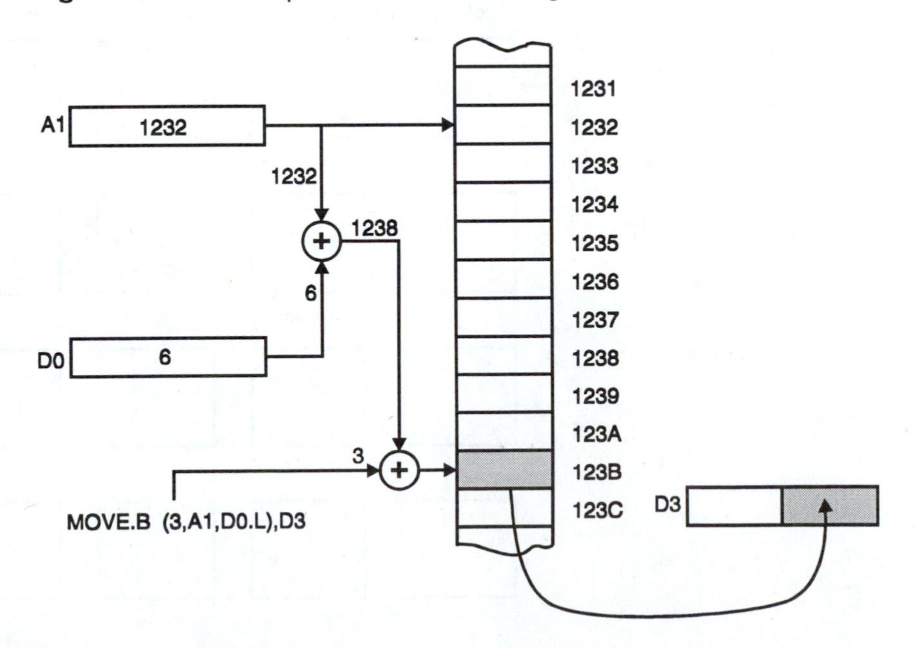

tive address of the source operand by adding the entire contents of D0 to A1 plus 12. The instruction `MOVE.L (12,A1,D0.W),D3` forms the effective address of the operand by adding the lower-order contents of D0 (i.e., bits 0 through 15, sign-extended to 32 bits) to the contents of A1 plus 12.

Using Address Register Indirect Addressing to Access Array Elements

Consider the four-column by four-row data structure, called an *array* or *matrix*, of Figure 4.23. In this figure we have drawn the array elements as boxes, because we are concerned only with how you locate a box and not with the size or the contents of the boxes. The box in which we are interested is shaded and lies in column 3 of row 2.

We can access the shaded box (i.e., the selected array element) by pointing to the first box in the array and then calculating where the desired box is with respect to the first box. In Figure 4.23 we have used address register A0 to point to the location of the first element in the array, D0, to indicate the row in which the box falls, and the column offset to indicate the box's column location.

The 68000's indexed addressing mode facilitates the handling of *two-dimensional tables* or *arrays* of the type shown by Figure 4.23. Since a computer's memory system is one-dimensional, an array must be stored in memory column by col-

Figure 4.23 Accessing an element in an array

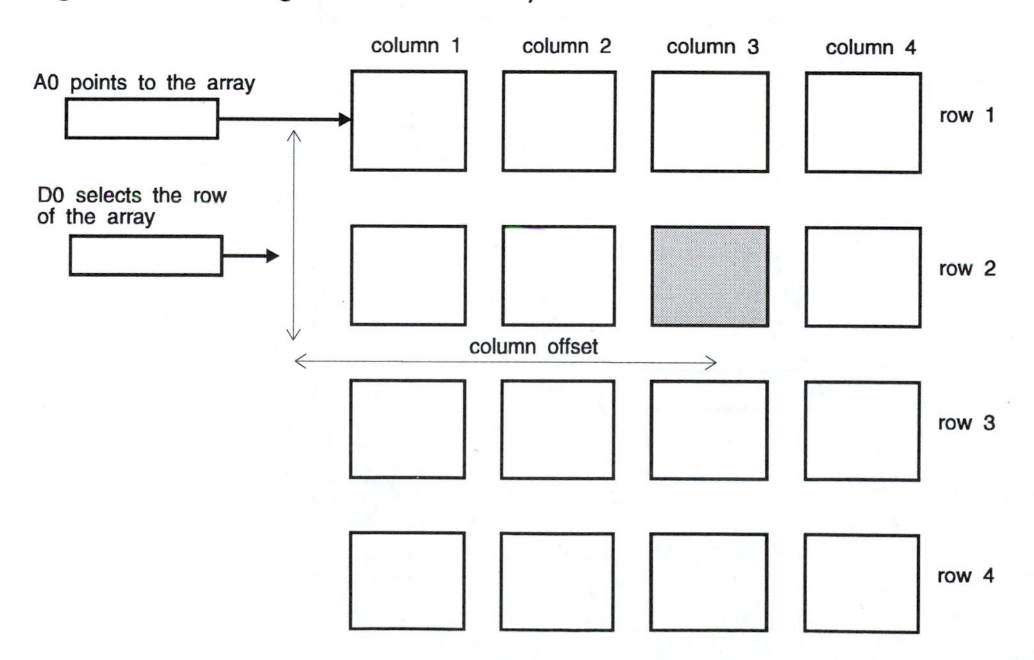

umn (or row by row). Figure 4.24 shows how an element in an array can be accessed using the 68000's address register indirect with index addressing mode.

Suppose we need to access the *seventh* item on a row in a table of byte-sized records. If the head of the table is pointed at by A0, and the location of the row in which the record falls (measured from the start of the table) is in data register D0, the operation `MOVE.B (6,A0,D0.L),D1` will access the required item. The offset is six rather than seven, because the offset for the first element is zero. By the way, the contents of register D0 are not the row number, but the total number of elements from the first row (i.e., rows x elements per row).

Consider another example of this addressing mode. The instruction `MOVE.B (3,A0,D0.W),D2` loads the contents of data register D2 with the contents of the memory location whose effective address is `(3,A0,D0.W)`. This address is given by the contents of A0 plus the contents of D0.W (sign-extended to 32 bits) plus the constant 3_{16}. Address register A0 points to the base of the data structure. The contents of D0 are added to A0, and then the required data item is located by adding in the offset (i.e., 3). We are now going to demonstrate how you might employ this versatile addressing mode.

As we have said, address register indirect addressing provides a particularly useful tool for the accessing of elements in multi-dimensional arrays or matrices. The m-row by n-column matrix A can be written in the form:

$$\begin{array}{cccccc} a_{1,1} & a_{1,2} & a_{1,3} & a_{1,4} & \cdots & a_{1,n} \\ a_{2,1} & a_{2,2} & a_{2,3} & a_{2,4} & \cdots & a_{2,n} \\ \cdot & & & & \cdots & \cdot \\ \cdot & & & & \cdots & \cdot \\ a_{m,1} & a_{m,2} & a_{m,3} & a_{m,4} & \cdots & a_{m,n} \end{array}$$

Figure 4.24 Illustration of address register indirect with indexing addressing mode

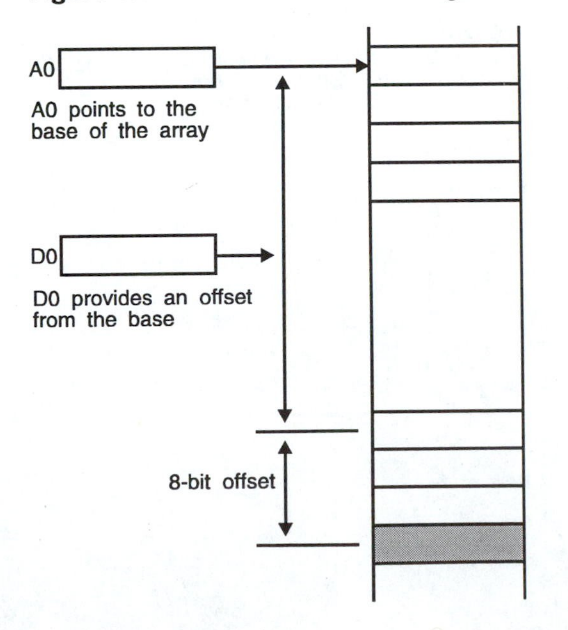
A0
A0 points to the base of the array
D0
D0 provides an offset from the base
8-bit offset

Since memory is essentially a one-dimensional array, the two-dimensional matrix must be mapped onto the memory array. We do this by storing a matrix as a series of rows (or columns), one after the other. If the matrix is stored as sequential rows, we speak of *row order*. If the matrix is stored in row order, its memory map might look like that of Figure 4.25.

Consider the location of element $a_{i,j}$ in the matrix A. Assume that the first element (i.e., $a_{1,1}$) is located in memory at address A. Row i starts at location A + (i - 1)n, since there are n elements in a row. Element $a_{i,j}$ has the address A + (i - 1)n + j - 1. The subscripts i and j appear in the expression as (i - 1) and (j - 1), respectively, because the array starts at element $a_{1,1}$ rather than $a_{0,0}$.

Let's look at a 3 x 3 matrix of byte elements:

$$\begin{matrix} a_{1,1} & a_{1,2} & a_{1,3} \\ a_{2,1} & a_{2,2} & a_{2,3} \\ a_{3,1} & a_{3,2} & a_{3,3} \end{matrix} = \begin{matrix} 4 & 2 & 7 \\ 1 & 3 & 7 \\ 4 & 3 & 1 \end{matrix}$$

If the matrix is composed of byte elements and stored at location 1000_{16}, its memory map is like that of Figure 4.26.

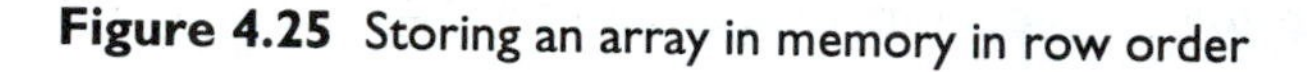

Figure 4.25 Storing an array in memory in row order

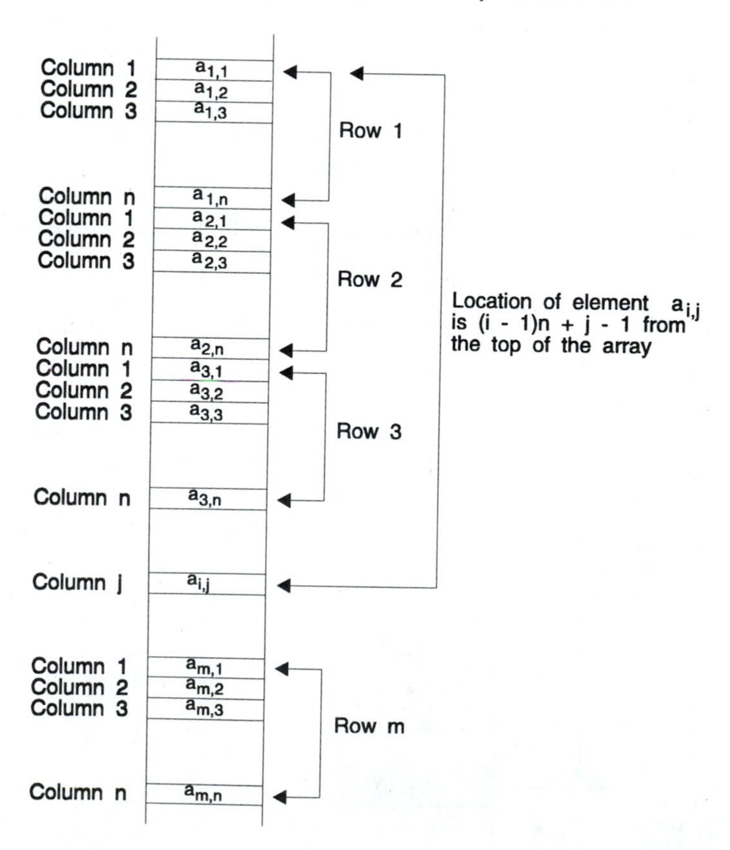

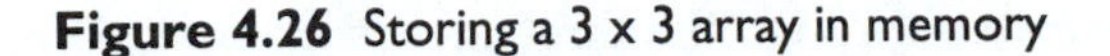

Figure 4.26 Storing a 3 x 3 array in memory

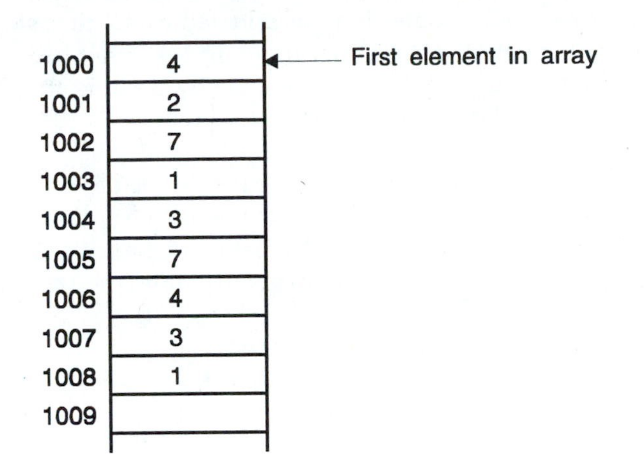

Consider the matrix sum C = A + B, where A, B, and C are m-row by n-column matrices and each matrix element is a byte value. The general element $c_{i,j}$ is defined as: $c_{i,j} = a_{i,j} + b_{i,j}$

If we use an address register to point to the start of a matrix, the element offset can be loaded into a data register and indexed addressing employed to access the desired element. The following fragment of code calculates the sum C := A + B.

```
m    EQU      <m>              Number of rows
n    EQU      <n>              Number of columns
A    EQU      <address>        Start address of matrix A
B    EQU      <address>        Start address of matrix B
C    EQU      <address>        Start address of matrix C
*
     LEA      A,A0             A0 points to the base of matrix A
     LEA      B,A1             A1 points to the base of matrix B
     LEA      C,A2             A2 points to the base of matrix C
     MOVE.W   #m,D0            D0 is the row counter
     CLR.W    D2               Clear the element offset
*
L2   MOVE.W   #n,D1            D1 is the column counter
L1   MOVE.B   (A0,D2.W),D6     Get element from A
     ADD.B    (A1,D2.W),D6     Add element from B
     MOVE.B   D6,(A2,D2.W)     Store sum in C
     ADD.W    #1,D2            Increment column element pointer
     SUB.W    #1,D1            Repeat until n columns added
     BNE      L1
     SUB.W    #1,D0            Repeat until m rows added
     BNE      L2
```

In this example, we step through the three arrays element by element. Two loops are used: an inner loop terminated by label L1 that steps through the elements of a given row, and an outer loop terminated by L2 that steps through the rows themselves. However, since all locations in the array are accessed sequentially, we could have simplified the above code by treating the three arrays as three one-dimensional tables as follows:

```
       LEA       A,A0          A0 points to base of matrix A
       LEA       B,A1          A1 points to base of matrix B
       LEA       C,A2          A2 points to base of matrix C
       MOVE.W    #m*n,D0       D0 is the element counter (m x n)
*
L1     MOVE.B    (A0)+,D6      Get element from A
       ADD.B     (A1)+,D6      Add element from B
       MOVE.B    D6,(A2)+      Store sum in C
       SUB.W     #1,D0         Repeat until m x n elements added
       BNE       L1
```

Consider a more general example of the manipulation of an array element. Suppose we have to calculate the effective address of element $a_{i,j}$ of m-row by n-column matrix A, where A0 points to the first element of the array, D0 contains i, and D1 contains j. We can use the following code:

```
LEA       A,A0          A0 points to base of array A
SUBQ.L    #1,D0         Rows are numbered from 1
MULU      #n,D0         D0 now contains the row offset
SUBQ.L    #1,D1         Columns are numbered from 1
ADD.L     D1,D0         D0 now contains column + row offset
ADDA.L    D0,A0         A0 now points to the i,j th element
```

Note that we have to subtract one from the row and column pointers, since the array subscripts are numbered from one rather than from zero. If the elements were larger than byte values (e.g., word, longword, or quadword = 4 words = 8 bytes), it would be necessary to *scale* the values in D0 and D1 accordingly (by 2, 4, or 16, respectively). By the way, we have chosen '.L' operations on data registers D0 and D1, since we later add D0.L to A0.

4.7 Program Counter Relative Addressing

Program counter relative addressing is very similar to register indirect addressing, except that the address of an operand is specified with respect to the contents of the *program counter* rather than with respect to the contents of an address register. Two forms of program counter relative addressing are implemented on the 68000:

program counter relative with displacement and program counter relative with index. The effective addresses generated by these modes are as follows.

Program counter with displacement: EA = [PC] + d16
Program counter with index: EA = [PC] + [Xn] + d8

It is important to stress that the only difference between the address register indirect and program counter relative addressing modes is that the former employs address register Ai as a base register and the latter the PC. The real difference between these modes lies in the way in which they are used by the programmer. The assembly language forms of these effective addresses are written (LABEL,PC) and (LABEL,PC,Xi), respectively. In this section we will consider only the (LABEL,PC) version of this addressing mode:

```
        MOVE.B   (TABLE,PC),D2
        .
        .
        .
TABLE   DC.B     Value1
        DC.B     Value2
```

The assembler uses the offset TABLE in the instruction MOVE.B (TABLE,PC),D2 to calculate the *difference* between the contents of the program counter and the address of the memory location TABLE (Figure 4.27). The result gives the 16-bit signed offset, d16, required by the instruction MOVE.B (TABLE,PC),D2. When the instruction is executed, the 16-bit offset is sign-extended to 32 bits and added to the contents of the PC to give the address of TABLE, and the operand at this address, Value1, is loaded into the lower-order byte of D2.

You should appreciate that the value of TABLE recorded in the instruction MOVE.B (TABLE,PC),D2 is not the actual value of TABLE (i.e., the absolute address of TABLE), but the distance between the current instruction and the location of TABLE (hence the term *relative addressing*).

The power of the program counter relative addressing mode is that it allows you to specify the address of an operand with respect to the program counter. If the program is moved (i.e., relocated) in memory, the address of the operand does not have to be recalculated. Therefore, program counter relative addressing enables the programmer to write *position independent code*, PIC, because the location of an operand is specified with respect to the instruction that accesses it. An advantage of this addressing mode is that the resulting position independent code can be placed in read-only memory and located anywhere in a processor's address space.

The 68000 permits only *source* operands to be specified by program counter relative addressing. Consequently, MOVE (LIST,PC),D2 is a legal operation, whereas MOVE D2,(LIST,PC) is illegal. It has been argued that program counter relative addressing should not be allowed to modify a source operand because this would make *self-modifying* code easy to write. This restriction on the use of

Figure 4.27 Example of program counter relative addressing

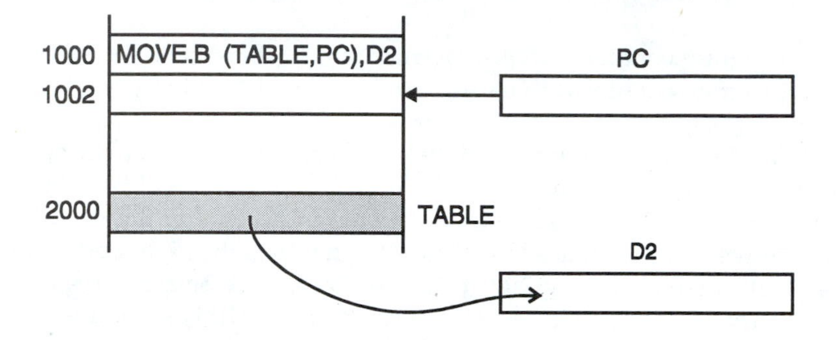

The operand TABLE is located at address $2000. The assembler calculates the difference between the current value of the PC ($1002) and the operand TABLE at $2000. This offset is $1FFE and is stored as part of the instruction.

PC relative addressing means that program counter relative addressing can be used only to *read* operands.

The LEA instruction we introduced earlier in this chapter can be used to generate destination operands with program counter relative addresses. As we have just said, program counter relative addressing permits the specification of source operands but not destination operands. Suppose we use LEA with program counter relative addressing to generate the *address* of a destination operand. This relative address is loaded into an address register and can now be used with register indirect addressing to achieve position independent code. The following example demonstrates how this is done.

Case 1

```
      MOVE.B (TABLE,PC),D0
        .
        .
        .
        .
TABLE   . . .
```

Case 2

```
      LEA     (TABLE,PC),A0
      MOVE.B  (A0),D0
        .
        .
        .
        .
TABLE   . . .
```

Both examples generate position independent code. In Case 1, program counter relative addressing is permitted because it is not used to modify a destination operand. In Case 2, program counter relative addressing is achieved by loading the relative address into A0.

We have now described all the 68000's addressing modes (the 68020 has even more). Before we conclude this chapter we will talk about *permitted addressing modes*.

4.8 Permitted Addressing Modes

The 68000 has a very *regular architecture* in the sense that it has no special-purpose data or address registers that can be used only in conjunction with certain instructions. However, it does not have *regular addressing* modes. Some instructions can be used with almost all the possible addressing modes while other instructions are limited to one addressing mode. This restriction on the use of addressing modes arises from the limited number of op-code/addressing mode combinations possible with a 16-bit instruction. The chip's designers have attempted to provide the most frequently used instructions with the greatest number of addressing modes. I am afraid that there is no simple way to learn which instruction can be used with what addressing modes. The appendix lists the legal addressing modes for each instruction. Figure 4.28 on page 168 provides a summary of the 68000's addressing modes according to instruction type.

Summary

- The contents of the 68000's data registers can be treated as 8-bit, 16-bit, or 32-bit units of data. Furthermore, you can allocate, say, one register to an 8-bit-wide operand, three registers to 16-bit-wide operands, and four registers to 32-bit-wide operands. You can have any combination of operand sizes, as long as the total is no more than eight.
- Operations on the contents of data registers or memory locations are specified by `.B`, `.W`, or `.L` suffixes for byte, word, and longword operations, respectively. In the absence of a suffix, the default size is normally a word.
- A longword operation affects all 32 bits of a data register. A word operation affects bits 0 to 15 of a data register and leaves bits 16 to 31 unaffected. A byte operation affects bits 0 to 7 of a data register and leaves bits 8 to 31 unaffected. You can, in fact, use the high-order bits of a register to store other data while the low-order bits are taking part in an operation. Using registers in this way is legal, although it might lead to confusion.
- A longword in memory location N occupies byte addresses N, N+1, N+2, and N+3. N has to be on a word boundary (i.e., N must be even). The most-significant byte is stored at the lowest address.
- The value of the CCR calculated after an operation is determined by the *size* of the operation. For example, if a `.B` operation is performed, the Z-bit is set if bits 0 to 7 are zero; the C-bit is set if a carry-out is generated from bit 7; and the N-bit is set if bit 7 is 1.

Figure 4.28 Summary of 68000 addressing modes

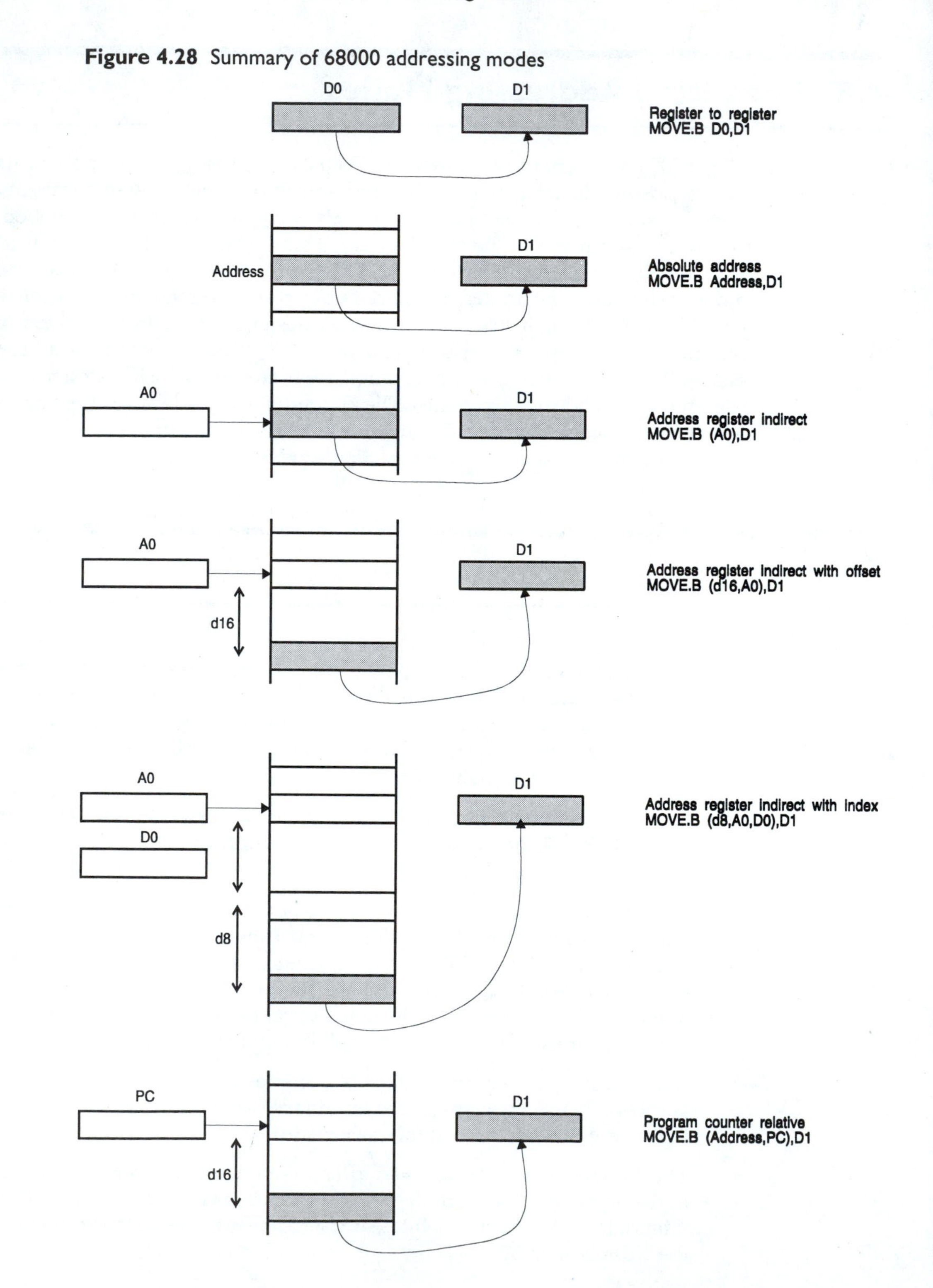

- Great care must be taken when working with operands of different size. You can't add an 8-bit value to a 16-bit value, directly. The 8-bit value must be converted to 16 bits and then a `.W` operation performed. One common error is to carelessly mix `.B`, `.W`, and `.L`, operations.
- Unless you have explicitly set bits 8 to 31 of a register to zero when working with byte values, never assume that these bits are zero. The same is true for bits 16 to 31 of a register when working with word values. You can clear bits 8 to 31 of D0 by means of `AND.L #$FF,D0`.
- The 68000's address registers each hold a 32-bit address.
- Although all addresses are 32 bits wide and all address bits take part in internal operations, only 24 of the address bits take part in the selection of a memory location. Address bits A_{24} to A_{31} have no effect on the selection of a location in external memory.
- An operation on an address register affects all 32 bits of the register. If a `.W` operation is performed, the 16-bit operand is sign-extended to 32 bits before it is used. That is, a `.W` operation like `ADDA.W #1,A0` yields a 32-bit result and affects the entire 32-bit address.
- If the destination operand is an address register, the contents of the CCR are not modified (unless the operation is a compare with address register, `CMPA`).
- If the destination operand is an address register, the mnemonics `MOVEA`, `ADDA`, `SUBA`, and `CMPA` are used to perform a move, addition, subtraction, and comparison, respectively.
- The 68000 has two forms of absolute addressing: long and short. A long absolute address is a 32-bit value that specifies the actual location of an operand. A long address can access any location in the 68000's address space. A short absolute address is stored as a 16-bit value that is sign-extended to 32 bits by the 68000 when it is used.
- A short absolute address permits the programmer to access efficiently data in the 32K-byte region $000000 to $007FFF at the top of memory, and the 32K-byte region $FF8000 to $FFFFFF at the bottom of memory. Note that you do not have to worry about selecting long or short absolute addresses. The assembler performs the appropriate selection automatically. For example, you just write the instruction `MOVE.W D3,Temp_6` and the assembler chooses an absolute short or long address, depending on the actual location of `Temp_6` in memory.
- The addressing mode `(d16,Ai)` is called address register indirect addressing with *displacement*. The displacement, `d16`, is a 16-bit signed constant that is added to the contents of address register Ai to give the actual address of the operand.
- The addressing mode `(d16,Ai)` was originally expressed in the form `d16(Ai)`. Older texts on the 68000 family and some assemblers still employ this syntax.

- The 16-bit offset is treated as a signed-value by the 68000 and is sign-extended to 32-bits before it is added to the specified address register. The range of locations that can be accessed by a 16-bit displacement is +32K or -32K bytes on either side of the specified address register.
- The displacement is invariably a symbolic value that is translated into a 16-bit constant by the assembler. For example, the programmer would write `MOVE (June,A3),D2` rather than `MOVE (6,A3),D2`.
- Address register indirect addressing with displacement is used to access data in a table. If Ai points to the top of a data table, the effective address `(ItemX,Ai)` points to `ItemX` in the table.
- The load effective address instruction, `LEA`, simply generates the address of an operand specified by its source address and loads this address into an address register. For example, `LEA (12,A2),A3` generates the effective source address `[A2]` + 12 and then deposits it in A3.
- The purpose of the `LEA` instruction is to generate an address for *later use* by instructions that employ address register indirect addressing.
- Address register indirect addressing permits the programmer to access complex data structures.
- In a linked list each data item contains two components: a data item and the address of the next item in the list.
- The linked list is scanned or traversed by reading the address field of an item and then using this address to locate the next item in the list.
- If address register A0 contains the address of the address field of the current item in the list, the instruction `MOVEA.L (A0),A0` reads the address field of the current item in memory and deposits it in A0. At the start of the instruction, A0 points to the *current* record, and at the end of the instruction, A0 points to the *next* record.
- Address register indirect addressing with post-incrementing is indicated to the assembler by `(Ai)+`. The 32-bit contents of address register Ai are automatically incremented *after* Ai has been used to access an operand.
- Address register indirect addressing with post-incrementing allows the programmer to step through the elements of an array (matrix or table) without having to explicitly modify the contents of the address register.
- The address register is incremented by 1 for byte operations, 2 for word operations, and 4 for longword operations.
- Address register indirect addressing with pre-decrementing is indicated by `-(Ai)` and causes address register Ai to be decremented by 1, 2, or 4 *before* it is used to access an operand.
- Address registers A0 to A6 can be used as general-purpose user stack pointers. Address register A7 is the system stack pointer used to store subroutine

return addresses. You may use A7 as a general-purpose stack pointer but you must be careful not to corrupt subroutine return addresses (by poor programming).

- Address register indirect addressing with pre-decrementing and post-incrementing have two legal forms, `-(Ai)` and `(Ai)+`, respectively. This addressing mode does not permit offsets and therefore the effective addresses `-(d16,Ai)` and `(d16,Ai)+` are illegal.
- Address register indirect with index addressing mode is the most versatile and powerful of 68000's addressing mode. The effective address of an operand is given by `(d8,Ai,Xi)`, where `d8` is an 8-bit signed constant providing an offset in the range -128 to +127. Xi is a general-purpose register (i.e., an address or a data register).
- The general-purpose register, Xi, may be a longword value (i.e., `Xi.L`) or a word value (i.e., `Xi.W`). If it is a word, it is first sign-extended to 32 bits before it is added to the address register.
- The 8-bit constant and the general-purpose register are both sign-extended to 32 bits before they are added to the contents of the specified address register.
- Address register indirect addressing with index makes it easy to access two-dimensional arrays.
- In program counter relative addressing, the effective address of an operand is specified by its location with respect to the current value in the program counter. In principle, there is no fundamental difference between address register indirect addressing and program counter relative addressing. However, these two addressing modes play different roles in assembly language programming.
- The two program counter relative addressing modes are written `(d16,PC)` and `(d8,PC,Xi)`. The offsets are normally the symbolic names of the variables accessed by these addressing modes. The assembler automatically calculates the value of the offset.
- Program counter relative addressing can be used only to read operands — it cannot be used to modify operands.
- The purpose of program counter relative addressing is to enable the programmer to write position independent code, PIC. Programs written using program counter relative addressing can be relocated anywhere in memory without recalculating the addresses of operands.
- The key to program counter relative addressing is the `LEA`, load effective address instruction. For example, `LEA (DATA2,PC),A0` calculates the effective address of `DATA2` with respect to the program counter and puts the address in A0. You can then access `DATA2` by means of address register indirect addressing — `MOVE.B (A0),D1`.

Problems

1. Computers have 8-bit, 16-bit, 24-bit, and 32-bit wordlengths. What is the *optimum* computer wordlength (and what do we mean by optimum)?

2. What is multiple precision arithmetic?

3. What is the difference between the 68000's ADD and ADDX instructions?

4. When the 68000 performs an operation on one byte of a 32-bit register, what happens to the other three bytes?

5. The 68000 updates its CCR after most of its operations. How does the size of the 68000's operation (byte, word, longword) affect the computation of the CCR's Z-, N-, C-, V-, and X-bits?

6. How would you set the middle two bytes of a data register to zero without modifying bits 0 to 7 and 24 to 31? We have not yet encountered the 68000's instructions that let you do this — try anticipating them.

7. Why does the following code work,

```
MOVE.B Val1,D0      Val2 := Val2 + Val1
MOVE.B Val2,D1
ADDX.B D0,D1
```

whereas the following code does not work?

```
ADD.B  Temp1,D0     Test := Test + Temp1
MOVE.B D0,Test
BVS    Error        Deal with error on overflow
```

8. What is wrong with the following sequence of instructions?

```
MOVE.B D0,$1234
MOVE.W D1,$1235
MOVE.L D2,$1237
```

9. The 68000 supports mixed-sized operations. As long as the programmer is working with constant-sized operands (e.g., all bytes or all words), few problems arise. However, when the programmer employs mixed operand sizes, great care must be taken. Why?

10. What is the difference between a 68000 address register and a 68000 data register?

11. Why does the 68000 have special versions of instructions (e.g., ADDA) exclusively for operations on address registers?

12. In what way do a .W operation on a data register and a .W operation on an address register differ?

13. What is the meaning of a *negative* address?

14. What is the difference between *short* and *long* absolute addresses? How does the programmer decide which one to use?

15. What is the effect of each of these instructions?

a. `CLR $1234`
b. `CLR #1234`
c. `CLR 1234`
d. `CLR #$1234`
e. `CLR (A0)`
f. `CLR (5,A0)`
g. `CLR (A0,5)`

16. What is the effect of the instruction `CLR.B (-2,A0)` and how would the offset, -2, be encoded (i.e., what binary pattern is used to represent -2)?

17. Given that [A0] = 00001238_{16}, describe the effect of the following instructions:

a. `CLR.B (A0)`
b. `CLR.W (4,A0)`
c. `CLR.W (-4,A0)`
d. `ADDA.L #10,A0`
e. `ADDA.L #$8000,A0`
f. `ADDA.W #$8000,A0`

18. What is the minimum number of addressing modes required by the 68000 to execute programs? If we said, "What is the *reasonable minimum* number of addressing modes required by the 68000 to execute programs," would it make any difference?

19. What does the load effective address instruction, LEA, do and why is it so important?

20. If [A0] = $12345678, [D0] = $10008124, Table = -12, List = $C0, what is the effect of the following instructions?

a. `LEA (A0),A1`
b. `LEA (Table,A0),A1`

c. `LEA (List,A0),A1`
d. `LEA (A0,D0),A1`
e. `LEA (A0,D0.L),A1`
f. `LEA (Table,A0,D0.L),A1`
g. `LEA (List,A0,D0),A1`
h. `LEA List,A1`
i. `LEA Table,A1`

21. By means of diagrams explain the meaning of a *singly-linked list* and a *doubly-linked list*?

22. A singly-linked list, whose first element is pointed at by A0, consists of elements whose head is a 32-bit address pointing to the next element in the list and a variable-length tail. The tail may be of any length greater than 4 bytes. The last element in the list points to the null address zero. Write a program to search the list for an element whose tail begins with the word in data register D0.

23. Why does the 68000 have a *post*-incrementing addressing mode and a *pre*-decrementing addressing mode? Would the provision of a *post*-incrementing addressing mode and a *pre*-decrementing addressing mode be equally suitable? What about a post-decrementing and a post-incrementing pair of addressing modes?

24. How does the 68000 choose an increment of 1, 2, or 4 when performing post-incrementing?

25. A string of ASCII-encoded characters is stored in consecutive memory locations, starting at location $1000, and is terminated by the carriage-return character, $0D. Write a subroutine to determine whether the string is a *palindrome* (i.e., it reads the same forwards and backwards — for example, hannah). If the string is a palindrome, load address register with zero. If it is not a palindrome, clear A0.

26. The 68000's address register indirect addressing mode with index is written in the assembly language form `(d,An,Xi)`. What are `d` and `Xi` and what are their *ranges*?

27. A 6-row by 4-column matrix is composed of 3-byte elements. The first element is located at $1000. Write a program to load the element whose address is i,j into data register D0.

28. What is the purpose of program counter relative addressing?

29. In the following fragment of code, the values of `July` recorded in the op-codes of the two instructions `MOVE.W July,D0` and `MOVE.W (July,PC),D0` are different. Why, and in what way do they differ?

Case 1

```
        MOVE.W July,D0
        .
        .
July    DS.W   1
```

Case 2

```
        MOVE.W (July,PC),D0
        .
        .
July    DS.W   1
```

CHAPTER 5

The 68000's Instruction Set

We are now going to look at some familiar 68000 instructions in more detail and introduce some new ones. We begin with the 68000's data movement instructions that move or *copy* data from one place to another. Then we look at its arithmetic instructions that enable you to implement fixed-point integer arithmetic. These are followed by the logical and bit operations that manipulate the individual bits of an operand. Finally, we look at the 68000's conditional instructions that permit us to implement high-level language constructs like the `CASE`, the `REPEAT...UNTIL`, and the `IF...THEN...ELSE` construct.

5.1 Data Movement and Arithmetic Operations

The 68000's instruction set includes 13 data movement operations that all copy information from one place to another. These instructions are: `MOVE`, `MOVEA`, `MOVE to CCR`, `MOVE to SR`, `MOVE from SR`, `MOVE USP`, `MOVEM`, `MOVEQ`, `MOVEP`, `LEA`, `PEA`, `EXG`, and `SWAP`. We encountered some of these when we introduced the 68000's addressing modes, but have included them here for the sake of completeness. Note that one of the above move instructions, `MOVE to SR`, modifies the status byte of the 68000's status register and may not be executed when the 68000 is operating in its *user mode*. We will briefly explain the purpose of the 68000's *user* and *supervisor* modes in this chapter, and then discuss them in detail when we introduce exception handling and interrupts in Chapter 10.

Data Movement Instructions

`MOVE` The move instruction copies an 8-, 16-, or 32-bit value from one memory location or register to another memory location or register. All the addressing modes discussed so far can be used to specify the source of the data or its destination, with three exceptions. Immediate addressing, address register direct addressing, and program counter relative addressing cannot be used to specify a destination (e.g., `MOVE D0,(Table,PC)` is illegal). The V- and C-bits of the CCR are cleared by a `MOVE`; the N- and Z-bits are updated according to the value of the destination operand; and the X-bit is unaffected. For example, the operation `MOVE.B #$80,D0` results in the XNZVC bits of the CCR being set to ?1000. The

"?" indicates that the state of the X-bit is unknown, as it is not modified by the MOVE operation.

MOVEA The MOVEA copies a source operand to an address register. This mnemonic has been implemented solely to make an explicit distinction between operations on addresses and on data. Like most other instructions operating on the contents of an address register, the MOVEA doesn't affect the state of the CCR. There is another difference between a MOVE and MOVEA. The MOVE instruction can take a byte, word, or longword extension, while MOVEA can take only .W and .L extensions. Moreover, a MOVEA.W instruction sign-extends the 16-bit operand to 32 bits and then loads it into the target address register. For example, the .W operation MOVEA.W #$8C00,A0 loads A0 with the longword value $FFFF8C00. The .W operation MOVEA.L #$8C00,A0 loads A0 with the longword value $00008C00.

MOVEQ The move quick instruction moves a 32-bit literal value in the range -128 to +127 to one of the eight data registers. The data to be moved is a byte that is *sign-extended* to 32 bits before it is copied to its destination. Therefore, although this instruction moves an 8-bit value, it yields a 32-bit result. For example, the operation MOVEQ #-3,D2 has the effect of loading the value $FFFFFFFD into data register D2. The purpose of the MOVEQ instruction is to enhance the performance of the 68000, because MOVEQ is encoded in less bits than a MOVE instruction. Since MOVEQ appears to be a byte operation and yet, due to sign-extension, behaves as a longword operation, you should be careful when using it. For example, if you write MOVEQ #4,D0 intending to load a *byte* into D0, you will also clear bits 8 to 31 of D0. The 68000 implements two other *quick* instructions, ADDQ and SUBQ. However, these support operands in the range 1 to 8 that are not sign-extended.

MOVE to CCR This instruction moves data to the condition code register and is classified as a word operation. A MOVE <ea>,CCR operation copies the *lower-order byte* of the operand at the specified effective address into the CCR. The higher-order byte of the operand is ignored. That is, the instruction is technically a word length operation but has the effect of a byte operation. MOVE to CCR allows you to preset the CCR. For example, MOVE #0,CCR clears all the CCR's bits. Similarly, MOVE #%00001,CCR sets the carry bit and clears all other bits in the CCR. Note that the 68000 does not implement the corresponding MOVE from CCR instruction. If you wish to read the CCR, you have to use a MOVE from SR instruction and then extract the low-order byte.

MOVE to SR, MOVE from SR The first of these instructions copies a word to the 68000's status register, and the second copies a word from the status register. Their assembly language forms are MOVE <ea>,SR and MOVE SR,<ea>, respectively. Both these instructions and the MOVE USP (i.e., move user stack pointer) to be described next are closely connected with the 68000's exception handling mechanism. You may wish to omit these instructions on a first reading.

Figure 5.1 describes the 68000's status register which contains two bytes: a *system byte* (or status byte) and a *user byte* (or condition code register). The status

Figure 5.1 The 68000's status register

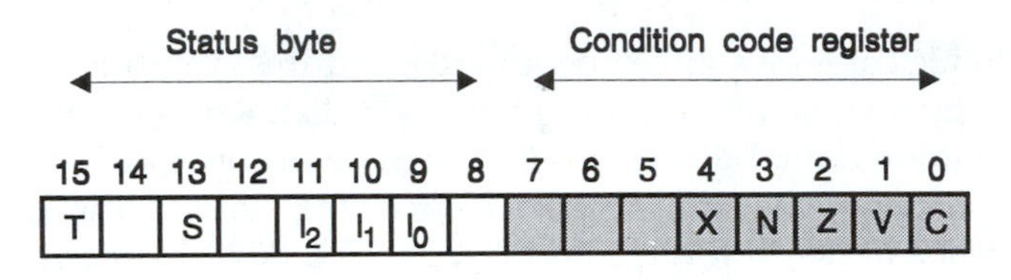

byte contains information that defines the operational status of the 68000. At this stage we are not concerned with the 68000's operating modes. All we need say is that it can run in a *user* mode (S = 0) and in a *supervisor* mode (S = 1). The operating system runs in the supervisor mode, and user programs (i.e., application programs) run in the less privileged user mode. The T-bit, or *trace-bit*, of the status byte can be set to force an operating system call after each instruction is executed. Setting the T-bit puts the 68000 into a *diagnostic mode* and permits a program to be traced (or monitored), instruction by instruction. We'll return to the trace bit when we describe the 68000's exception handling facilities.

The status byte's three interrupt mask bits, I_2, I_1, I_0, control the way in which the 68000 responds to interrupts. If the current value of the interrupt mask is *n*, the 68000 will respond only to interrupts with a priority greater than *n*.

Unlike the other instructions we have met, the `MOVE to SR` instruction is *privileged*. Instructions that modify the operating mode of the 68000 are said to be privileged and may not be executed when the 68000 is operating in its user mode (i.e., when S = 0). Only the operating system is able to modify the 68000's operating conditions (you wouldn't want to allow someone in a multi-user system to do anything that would bring the entire system to a halt). This topic is covered in more detail in Chapter 10 and is entirely irrelevant to the basic operation of the 68000. Privileged instructions can be executed only by operating system software and are not required by user (i.e., applications) programmers.

If the 68000 is running in the user mode and an attempt is made to modify the contents of the 68000's status byte, a *privilege exception* occurs and the operating system is called upon to deal with the problem.

The `MOVE from SR` instruction copies the 68000's status register to the specified effective address (e.g., `MOVE SR,D3`) which allows you to examine either the state of the CCR or the status byte. This instruction is not privileged because it does not affect the state of the status register. However, `MOVE from SR` is made privileged on the 68010, 68020, etc. We explain why in Chapter 10.

`MOVE USP` The 68000 has *nine* address registers: A0 to A6 plus two A7s. As you know, address register A7 is used as a stack pointer to keep track of subroutine return addresses. One A7 is associated with the *user* mode and one with the *supervisor* mode. These A7s are called USP (user stack pointer) and SSP (supervisor stack pointer), respectively, whenever it is necessary to distinguish between them. By employing two A7s, the user programmer cannot inadvertently modify the operating system's own stack pointer.

When the 68000 is operating in its supervisor mode, MOVE.L USP,An and MOVE.L An,USP, transfer the user SP to address register An and vice versa. The only permitted addressing mode is address register to address register.

These two instructions allow the operating system to manipulate the user stack. For example, the operating system can set up the user stack before a user task is run by means of the following code:

```
MOVEA.L #NEW_STACK,A0   Copy address of user's stack into A0
MOVE.L  A0,USP          Transfer it to the user stack pointer
```

When the 68000 is in the user mode, the SSP is entirely hidden from the user and cannot be accessed by the user. Once again, we should emphasize that the MOVE.L USP,An and MOVE.L An,USP instructions are used only by the designer of operating systems and related software.

MOVEM The move multiple register instruction offers the programmer a very efficient way of transferring a *group* of the 68000's registers to or from memory with a single instruction. Its assembly language form is:

```
     MOVEM <register list>,<ea>
or   MOVEM <ea>,<register list>.
```

MOVEM operates only on words or longwords. The effect of MOVEM is to transfer the contents of the *group* of registers specified by register list to consecutive memory locations or to restore them from consecutive memory locations. Programmers frequently use this instruction to save working registers on entering a subroutine and to retrieve them at the end of the subroutine. The expression *working registers* describes the registers that are modified or corrupted by the subroutine. The contents of the CCR are not affected by a MOVEM.

The *register list* is defined as: Ai-Aj/Dp-Dq, where the hyphen indicates a sequence of contiguous registers and the back slash separates lists of registers. For example, the register list A0-A4/D3-D7 specifies address registers A0 to A4 and data registers D3 to D7, inclusive. Similarly, A0/A3/A5-A6/D0-D2/D5 specifies address registers: A0, A3, A5, A6, and data registers D0, D1, D2, and D5. MOVEM.L D0-D7/A0-A6,-(A7) pushes all the data registers and address registers A0 to A6 onto the system stack (see Figure 5.2). MOVEM.L (A7)+,D0-D7/A0-A6 has the reverse effect of pulling the registers off the stack. Note that *word* operands transferred by MOVEM.W are sign-extended to 32 bits when loaded into either address or *data* registers. A typical application of the MOVEM instruction is:

```
ProcA MOVEM.L A0-A3/D0-D7,-(A7)  Dump all working registers on the stack
      .
      .                          Body of procedure
      .
      MOVEM.L (A7)+,A0-A3/D0-D7  Restore all registers to their old values
      RTS
```

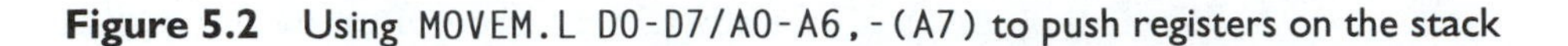

Figure 5.2 Using `MOVEM.L D0-D7/A0-A6,-(A7)` to push registers on the stack

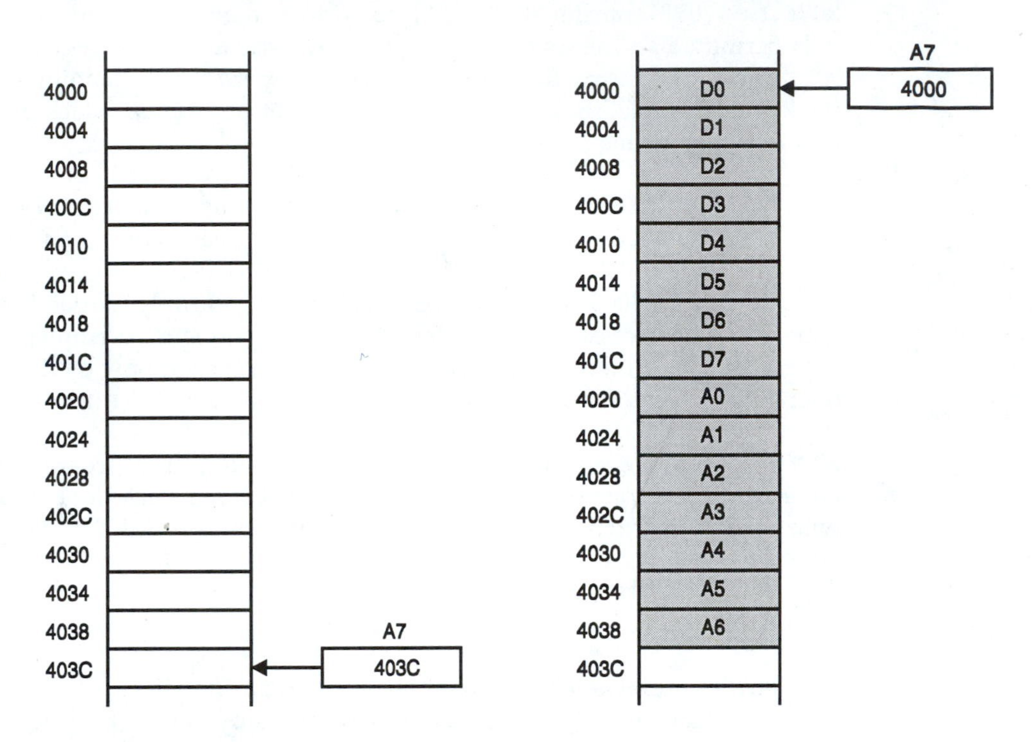

a. State of the stack before executing MOVEM.L D0-D7/A0-A6,-(A7)

b. State of the stack after executing MOVEM.L D0-D7/A0-A6,-(A7)

Although the next chapter covers assemblers in more detail, it's worth pointing out that the assembly language programmer doesn't have to spend a lot of time writing register lists. The following fragment of code from an assembler demonstrates how the assembler directive `REG` is used to *bind* a name to a register list. This name can then be used in conjunction with the `MOVEM` instruction.

```
Source file: REG.X68
Assembled on: 93-01-19 at: 22:20:12
          by: X68K PC-2.0 Copyright (c) University of Teesside 1989,93
Defaults: ORG $0/FORMAT/OPT A,BRL,CEX,CL,FRL,MC,MD,NOMEX,NOPCO

 1 00000400                    ORG      $400
 2          00007FFF  SAVE_ALL: REG      A0-A6/D0-D7          ;Define a "large" register list
 3          00003387  SAVE_FEW: REG      A0-A1/A4-A5/D0-D2/D7 ;Define a "small" register list
 4                    *
 5 00000400 48E7FFFE  PROCA:   MOVEM.L  SAVE_ALL,-(A7)       ;Save all registers
 6                    *
 7                    *
 8 00000404 4CDF7FFF           MOVEM.L  (A7)+,SAVE_ALL       ;Restore them before returning
```

```
 9 00000408 4E75                RTS
10                      *
11 0000040A 48E7E1CC    PROCB:  MOVEM.L  SAVE_FEW,-(A7)     ;Save a few registers
12                      *
13                      *
14 0000040E 4CDF3387            MOVEM.L  (A7)+,SAVE_FEW     ;Restore them before returning
15 00000412 4E75                RTS
16          00000400            END      $400

Lines: 16, Errors: 0, Warnings: 0.
```

MOVEP The move peripheral instruction copies words or longwords to or from an 8-bit memory-mapped peripheral. We cover input/output techniques in a later chapter and are including this instruction here to demonstrate the range of the 68000's move operations. Byte-oriented peripherals are connected to the 68000's data bus in such a way that *consecutive* bytes in the peripheral are mapped onto *successive odd* (or even) addresses in the 68000's memory space. Figure 5.3 shows a peripheral with four internal registers together with a section of the 68000's address map.

Figure 5.3 Byte and word address space

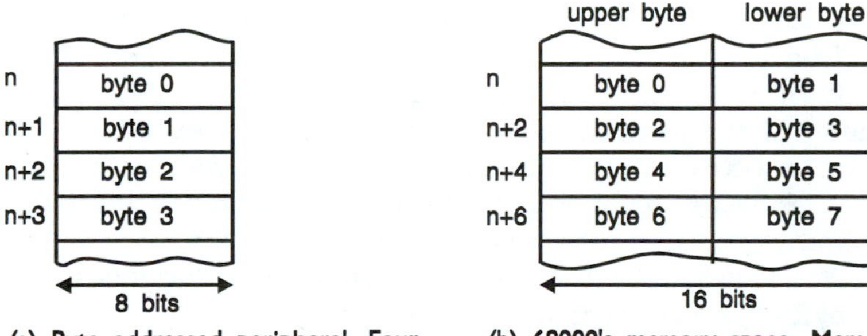

(a) Byte addressed peripheral. Four locations occupy consecutive bytes.

(b) 68000's memory space. Memory locations are 16 bits wide and occupy two bytes.

Since the peripheral is eight bits wide, it cannot be connected to all the 68000's 16 data bits. It must either be connected to bits d_0 to d_7 (the lower byte) or to bits d_8 to d_{15} (the upper byte). As you can see from Figure 5.3, the peripheral's four bytes take up *consecutive* locations in the peripheral. However, when the 8-bit peripheral is wired to the 68000's 16-bit address bus, the consecutive *bytes* of the peripheral are mapped onto consecutive *words* of the 68000's memory space. That is, each of the peripheral's locations is separated by *two* bytes rather than one.

Consider a device with four 8-bit registers connected to data bits d_0 to d_7 and memory-mapped at address \$080001, shown in Figure 5.4. The four 8-bit registers appear to the 68000 programmer as locations \$080001, \$080003, \$080005, and \$080007. Note that, in this diagram, locations \$080000, \$080002, \$080004, and \$080006 do not exist and cannot be accessed.

Figure 5.4 Example of a memory-mapped byte-wide peripheral

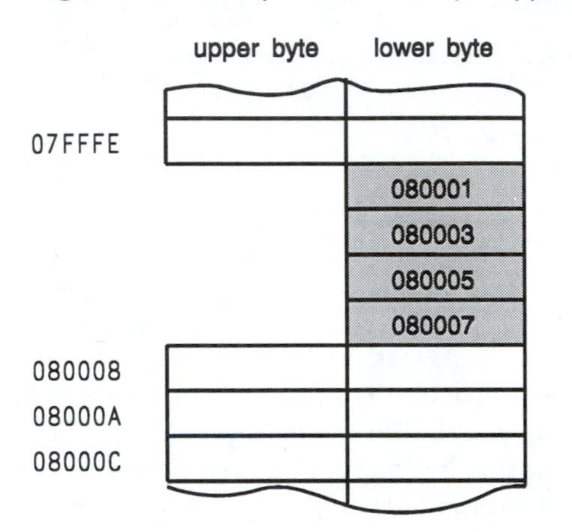

You can't more than one byte at a time to an 8-bit peripheral by means of a conventional MOVE.W or a MOVE.L instruction, because these instructions move a word or a longword to *consecutive bytes* in memory. MOVEP automatically moves a word or a longword between a data register and a byte-wide, memory-mapped peripheral. The contents of the data register are moved to consecutive even (or odd) byte addresses. For example, MOVEP.L D2,(0,A0) copies the four bytes in D2 to the addresses given by: [A0] + 0, [A0] + 2, [A0] + 4, [A0] + 6.

The assembler form of the MOVEP instruction is: MOVEP Dx,(d16,Ay) or MOVEP (d16,Ay),Dx. Only register indirect with displacement addressing is permitted with this instruction. Figure 5.5 demonstrates how the four bytes in D0 are copied to successive odd addresses in memory, starting with location 080001_{16} by means of the instruction MOVEP.L D0,(A0). Note that the most-significant byte in the data register is transferred to the lowest address.

Without the MOVEP instruction it would take the following code to move four bytes to a memory-mapped peripheral.

Figure 5.5 Using the MOVEP.L instruction to transfer peripheral data to memory

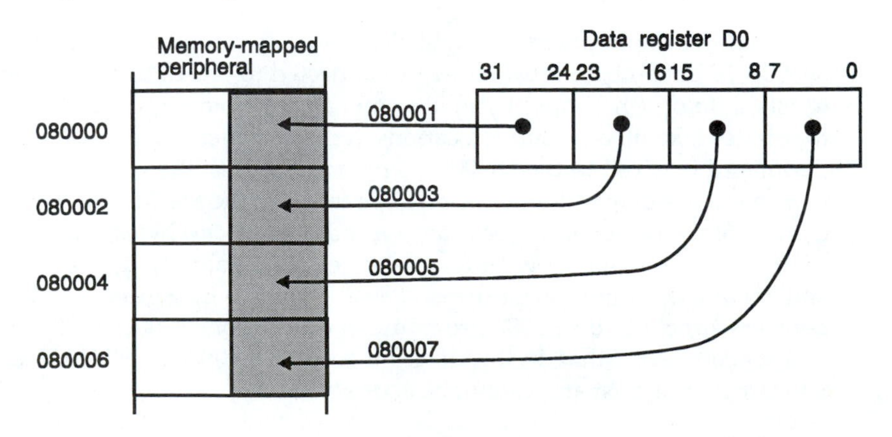

```
LEA      Peri,A0       A0 points to the memory-mapped peripheral
MOVE.B   D0,(6,A0)     Move the least-significant byte of D0 to the peripheral
ROR.L    #8,D0         Rotate D0 to get the next 8 bits
MOVE.B   D0,(4,A0)     Move the next byte, bits 8 to 15, to the peripheral
ROR.L    #8,D0
MOVE.B   D0,(2,A0)
ROR.L    #8,D0
MOVE.B   D0,(0,A0)
ROR.L    #8,D0         After four rotations D0 is back to its old value
```

Each time a byte is moved, it is copied to an address two bytes away from the previous address. The `ROR.L #8,D0` following each data movement rotates the longword in D0 eight bits to move the next byte into the least-significant position. By using the `MOVEP` instruction, the entire code can be reduced to:

```
LEA      Peri,A0     A0 points to the memory-mapped peripheral
MOVEP.L D0,(A0)      Move the longword in D0 to the peripheral
```

Notice how compact the code is. It is unlikely that you would use the `MOVEP` instruction for anything other than input/output operations, although it might come in useful if you ever need to interleave bytes. Suppose that D0 contains bytes D,C,B,A, and D1 contains bytes S,R,Q,P, and you wish to order the bytes D,S, C,R, B,Q, A,P in memory. You could write:

```
LEA      Mem,A0        A0 points to destination of data in memory
MOVEP.L D0,(A0)        Move DCBA in D0 to the alternate locations
MOVEP.L D1,(1,A0)      Now interleave D1
```

Figure 5.6 shows that if the contents of D0 are 0D,0C,0B,0A, and the contents of D1 are 11,22,33,44, and Mem = 1000_{16}, executing the above code results in the sequence of bytes 0D,11,0C,22,0B,33,0A,44.

`EXG` The *exchange* instruction exchanges the entire 32-bit contents of two registers. In RTL terms, `[Xi]` ← `[Xj]`; `[Xj]` ← `[Xi]`, where Xi and Xj represent any data or address registers. The data transfers are carried out simultaneously. Although `EXG` allows the contents of two data registers to be exchanged, an important application is in transferring a value calculated in a data register to an address register. The contents of the condition code register are not affected by an `EXG` instruction. You might use this instruction to copy an address register to a data register for processing, and then again to copy it back to the original address register. For example, we can use the following code to calculate 10*[A4]+6.

```
EXG      A4,D2       Get A4 in D2 without destroying old D2
MULU     #10,D2      Multiply address by 10
ADD.L    #6,D2       Add 6
EXG      A4,D2       Copy address back to A4 and restore D2
```

Figure 5.6 Using the MOVEP instruction to interleave bytes

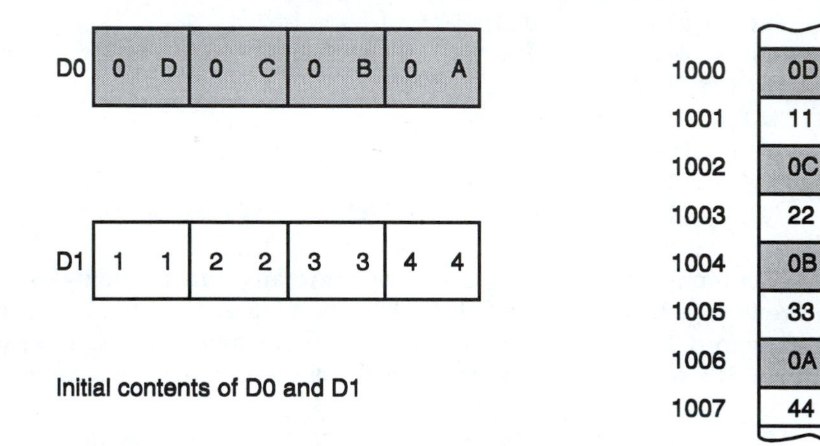

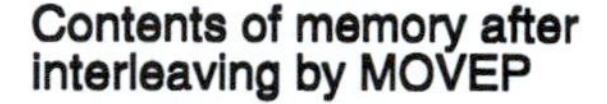

Another application of EXG might be in conjunction with subroutine calls. If a particular procedure operates on D1 and D2, but your data is in D4 and D7, you could execute:

```
EXG   D4,D1         Copy D4 to D1 and D7 to D2
EXG   D7,D2
BSR   Subroutine    Now call the subroutine
EXG   D4,D1
EXG   D7,D2         Transfer results to D4 and D7
```

When proofreading this manuscript I first thought that there was an error in the above code. I am accustomed to reading 68000 instructions expressed in the form `operation source,destination`. After the subroutine has been called, the instruction `EXG D4,D1` copies the parameter in data register D1 to D4 (and also copies D4 to D1). This instruction is not wrong, but I do find it misleading. The code below demonstrates how you could rewrite the code to make its meaning slightly clearer.

```
EXG    D4,D1         Copy D4 to D1 and D7 to D2
EXG    D7,D2
BSR    Subroutine    Now call the subroutine
EXG    D1,D4
EXG    D2,D7         Transfer results back to D4 and D7
```

As in the case of the MOVEP instruction, the provision of an EXG instruction is not necessary. It just helps the programmer to produce more efficient code.

SWAP The SWAP instruction has the assembler form SWAP Dn and exchanges the upper- and lower-order words of a data register. In RTL terms, SWAP Di is expressed as: [Di(16:31)] ← [Di(0:15)]; [Di(0:15)] ← [Di(16:31)]. This instruction has been provided because all operations on 16-bit words act only on bits 0 to 15 of a register. For example, if D0 contains the bytes PQRS, SWAP D0 results in RSPQ. By using a SWAP, the high-order word Di(16:31) can be moved to the low-order word and then operated on by a word-mode instruction. The SWAP instruction affects the CCR exactly like the MOVE.L instruction. For example, if D1 contains the 32-bit value $13ACF045, executing SWAP D1 results in the loading of D1 with $F04513AC (the CCR's N-bit is set to 1 and its Z-bit to 0).

A particularly common application of the SWAP instruction is in conjunction with the DIVU (divide) instruction. Although we have not yet covered division, all we need say at this stage is that the instruction DIVU #10,D0 divides the 32-bit contents of register D0 by ten to produce a quotient and a remainder. The 16-bit quotient is stored in the low-order word of D0 and the 16-bit remainder is stored in the high-order word of D0. For example, if D0 contains the binary equivalent of 1234_{10}, the action of DIVU #10,D0 is to leave the quotient 123_{10} in D0(0:15) and the remainder 4_{10} in D0(16:31). Since the divide instruction packs a quotient and a remainder into the *same* register, we often have to unpack them using the SWAP instruction. Figure 5.7 illustrates the execution of the following code.

```
DIVU    #10,D0      Divide D0 by 10
MOVE.W  D0,D1       Get the quotient in D1
.                   and process it
SWAP    D0          Move the remainder to D0(0:15)
MOVE.W  D0,D2       Copy the remainder to D2
.                   and process it
```

LEA We have already encountered the load effective address instruction that copies an effective address into an address register. This instruction simply calculates an effective address and loads it into an address register. The real purpose of an LEA <ea>,Ai instruction is to facilitate the later use of address register indirect addressing. Consider the following example.

Figure 5.7 Illustration of the SWAP instruction

Dividend — Initial value of D0

Remainder | Quotient — D0 after DIVU #10,D0

| Quotient — D1 after MOVE.W D0,D1

Quotient | Remainder — D0 after SWAP D0

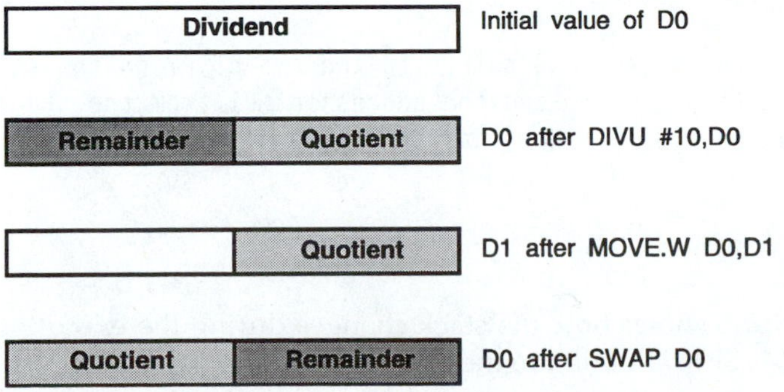

```
        LEA     (Table1,PC),A0   A0 points to start of Table1
        LEA     (Table2,PC),A1   A1 points to start of Table2
        MOVE.B  #12,D0           12 elements to add
Loop    MOVE.B  (A0)+,D1         FOR i = 0 to 11
        ADD.B   D1,(A1)+            Table2(i) := Table2(i) + Table1(i)
        SUB.B   #1,D0
        BNE     Loop             END_FOR
        .
        .
        .
Table1 DS.B   12
Table2 DS.B   12
```

In this fragment of code the LEA instruction sets up pointers to the two tables. The rest of the code then calculates Table2 := Table1 + Table2. Note that the LEA instruction permits us to use *position independent code.*

PEA The push effective address, PEA, instruction calculates an effective address and pushes it onto the stack pointed at by address register A7 (i.e., the stack pointer). The only difference between PEA and LEA is that LEA deposits an effective address in an address register, while PEA pushes it onto the stack. Thus, PEA <ea> is equivalent to LEA <ea>,Ai followed by MOVE.L Ai,-(SP), except that the PEA does not modify Ai.

PEA is used to push the address of an operand onto the stack prior to calling a subroutine. The subroutine can read this address from the stack, load it into an address register, and access the actual data by means of address register indirect addressing. That is, PEA is used to pass a parameter to a subroutine by its address. Consider the following example (we look at subroutines in more detail in a later chapter).

```
P1  DS.W     1           Parameter P1 (one word)
    .
    PEA      P1          Push the address of P1 on the stack
    BSR      ABC         Call subroutine 'ABC'
    LEA      (4,SP),SP   Increment stack pointer to remove address of P1
    .
    .
ABC LEA      (4,SP),A0   A0 points to address of P1 on the stack
    MOVEA.L  (A0),A0     Read the address of P1 from the stack
    MOVE.W   (A0),D0     Now read the value of P1
    .
    .
    RTS
```

Figure 5.8 shows how the stack changes during the execution of the above code. LEA (4,SP),A0 adds 4 to the stack pointer and copies the result into A0. It is

Figure 5.8 The stack and parameter passing

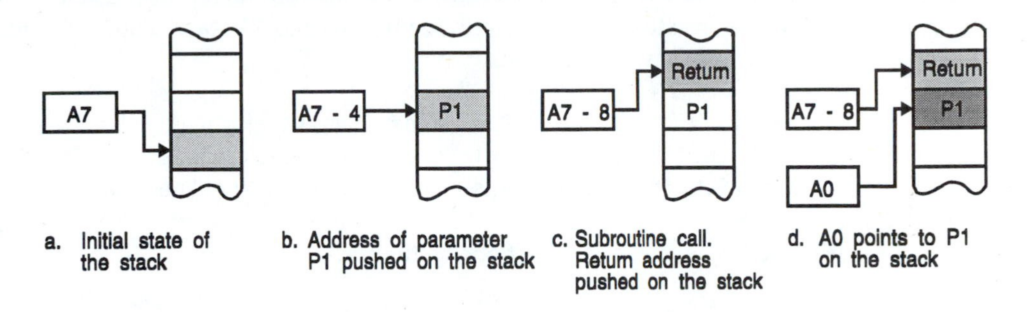

necessary to add 4 to the stack pointer because the address of P1 is *buried* on the stack under the return address. At this stage, A0 is pointing at the address of P1 on the stack. The instruction `MOVEA.L (A0),A0` reads the address of P1 from the stack and puts it in A0. Finally, the instruction `MOVE.W (A0),D0` uses the address of P1 to access P1 itself.

Now that we've covered all the 68000's data move instructions, we are going to look at the instructions that operate on numeric data.

Arithmetic Operations

The 68000 has a conventional set of arithmetic operations, all of which are *integer* operations. Floating-point operations are not supported by the 68000. Except for division, multiplication, and operations whose destination is an address register, arithmetic operations act on 8,- 16-, or 32-bit entities. The arithmetic group of instructions includes: `ADD`, `ADDA`, `ADDQ`, `ADDI`, `ADDX`, `SUB`, `SUBA`, `SUBQ`, `SUBI`, `SUBX`, `NEG`, `NEGX`, `CLR`, `CMP`, `DIVS`, `DIVU`, `MULS`, `MULU`. Note that the 68000 also has three special-purpose arithmetic operations designed to facilitate calculations in BCD (i.e., binary coded decimal): `ABCD` — add decimal with extend, `SBCD` —subtract decimal with extend, and `NBCD` — negate decimal with extend.

`ADD` The add instruction adds the contents of a source location to the contents of a destination location and deposits the result in the destination location. Either the source or destination must be a data register. Direct memory-to-memory additions are not permitted with this instruction. The `ADD` instruction cannot be used to modify the contents of an address register. The following code evaluates the Pascal-like expression `R := P + Q`.

```
MOVE.W  P,D0
ADD.W   Q,D0
MOVE.W  D0,R
```

By using the 68000's address register indirect addressing with either pre-decrementing or post-incrementing, we can perform arithmetic operations directly on the contents of the stack (this is true of subtraction as well as addition). The following code demonstrates how two words on the top of the stack, P and Q, are pulled off the stack and the result, R, pushed onto the stack. Figure 5.9 demonstrates the effect of these instructions.

```
MOVE.W  (SP)+,D0   Pull P off the stack
ADD.W   (SP)+,D0   Pull Q off the stack and add it to P
MOVE.W  D0,-(SP)   Push the result, R, on the stack
```

Figure 5.9 Arithmetic operations on the stack

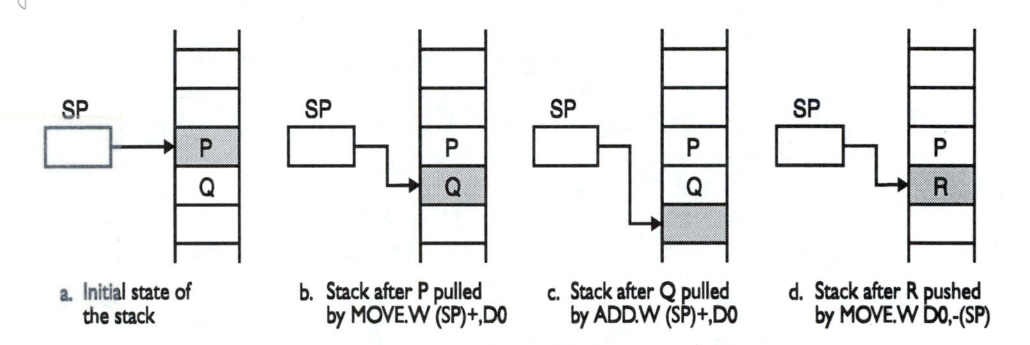
SP
P
Q
SP
P
Q
SP
P
Q
SP
P
R
a. Initial state of the stack
b. Stack after P pulled by MOVE.W (SP)+,D0
c. Stack after Q pulled by ADD.W (SP)+,D0
d. Stack after R pushed by MOVE.W D0,-(SP)

ADDA The ADDA instruction is necessary whenever the *destination* of the result is an *address register*. ADDA must, of course, be used only with word and longword operands. Note that ADDA.W sign-extends the operand before adding it to the contents of an address register. For example, ADDA.W #$8122,A0 sign-extends the 16-bit operand 8122_{16} to $FFFF8122_{16}$ before adding it to the 32-bit contents of A0. Executing an ADDA instruction has no effect on the contents of the condition code register. For example, ADDA.L D3,A4 adds the entire contents of D3 to A4 and deposits the result in A4, leaving the CCR unmodified.

ADDQ The add quick instruction, ADDQ, adds a literal (i.e., constant) value in the range 1 to 8 to the contents of a memory location or a register. ADDQ takes byte, word, and longword operands. ADDQ can also be applied to the contents of an address register — in which case the CCR is not modified and only .W and .L operands are permitted. The term *quick* is employed because the instruction format of ADDQ includes the 3-bit constant to be added to the destination operand. Therefore, an ADDQ #4,D1 is executed faster than the corresponding ADD #4,D1, because it has a 16-bit instruction rather than a 32-bit instruction.

Some assemblers automatically select the ADDQ version of an add instruction if the operand is a literal in the range 1 to 8. Other assemblers do not use an ADDQ instruction unless the programmer explicitly specifies it. I once assembled a program using two different assemblers and was surprised to find that they produced different binary code files. After checking the listing files, I discovered that one assembler automatically converted ADDs to ADDQs wherever possible and the other didn't.

Finally, we will remind you that there is an important difference between ADDQ (and SUBQ) and MOVEQ. The move quick instruction moves an 8-bit constant in the range -128 to +127 and this value is sign-extended to 32 bits.

ADDI The add immediate instruction, ADDI, adds a literal value of a byte, word, or longword to the contents of the destination operand and stores the result in the destination. The destination may be a memory location or a data register. Note that the instructions ADD #$1234,D4 and ADDI #$1234,D4 are equivalent. However, the ADDI and ADD # instructions are coded *differently*, because ADDI permits a literal to be added to the contents of a memory location, unlike an ADD # instruction. For example, ADDI.W #$1234,(A0) adds the constant $1234 to the contents of the memory location pointed at by A0.

ADDX The add extended instruction behaves exactly like an ADD instruction, except that the X-bit is also added to the source plus the destination operand. This instruction facilitates extended arithmetic (e.g., 64-bit arithmetic). Note that the Z-bit of the CCR is cleared if the result is non-zero and left unchanged otherwise. The only addressing modes permited for ADDX are data register to data register, and memory to memory — ADDX -(Ax),-(Ay).

Some assemblers permit only the use of the ADD # mnemonic and automatically provide the appropriate ADDI op-code if the destination is not a data register. Indeed, some assemblers permit only the mnemonic 'ADD' and then automatically choose the appropriate op-code for ADD, ADDI, ADDA or ADDQ, depending on the circumstances. For example, if the literal is in the range 1 to 8, the code for ADDQ is selected; otherwise, the code for ADD is selected. Typical instruction translations that might be made by a typical assembler are:

```
ADD #5,D3      becomes  ADDQ #5,D3
ADD #12,D3     becomes  ADD  #12,D3
ADD #12,T7     becomes  ADDI #12,T7
ADD D3,A4      becomes  ADDA D3,A4
```

CLR The clear instruction loads the contents of the specified data register or memory location with zero. It can operate on byte, word, and longword operands. The assembly language form of this instruction is CLR <ea> and its effect is essentially the same as MOVE #0,<ea>. The X-bit of the CCR is unchanged, the Z-bit set to 1, and the N- and V-bits are cleared. There is no explicit instruction to clear the contents of an address register. You can use SUB.L An,An or MOVEA.L #0,An to clear the contents of address register An.

DIVS, DIVU The 68000 implements two operations to carry out integer division with the assembly language forms DIVU <ea>,Dn and DIVS <ea>,Dn. DIVU performs unsigned division and DIVS operates on two's complement numbers. The 32-bit longword in data register Dn is divided by the 16-bit word at the effective address given in the instruction. The quotient is a 16-bit value and is deposited in the lower-order word of Dn. The remainder is stored in the upper-order word of

Dn. All source addressing modes are permitted except address register direct (i.e., you cannot divide by the contents of an address register).

For example, if `[D4] = $00005678`, the operation `DIVU #$1234,D4` divides the 32 bits in data register D4 by the 16-bit literal value 1234_{16} to yield the longword result `[D4] = $0DA80004`. This result is interpreted as a 16-bit quotient $0004 in the low-order word of D4, and a 16-bit remainder $0DA8 in D4(16:31). If you don't require the remainder and are going to use the quotient in a 32-bit operation, you must remember to clear bits 16 to 31 of the destination register after a division. That is:

```
DIVU  #$1234,D4        Divide D4 by 10
AND.L #$0000FFFF,D4    Delete the remainder in D4(16:31)
```

The X-bit of the CCR is not affected by a division operation and the C-bit is cleared. The N-bit is set if the quotient is negative and is cleared otherwise. However, if either overflow or divide by zero takes place, the state of the N-bit is undefined. The Z-bit is set by a zero quotient and cleared by a non-zero quotient. Like the N-bit, it too is undefined after an overflow or a divide-by-zero operation. The V-bit is set by an overflow during the division and cleared otherwise. If an attempt is made to divide by zero, the V-bit is undefined. Note also that an attempt to divide by zero results in a divide-by-zero-exception and a call to the operating system (see Chapter 10 for further details). If nothing else, the 68000's divide instructions demonstrate that you have to be very careful when interpreting the state of the CCR after the execution of an instruction.

Consider the following example of the `DIVU` instruction in the conversion of a binary value in the range 0 to 255 into three BCD characters. For example, the binary value 11001100_2 is converted to 0000 0010 0000 0100 (i.e., 204_{10}). The source is initially in the low-order byte of data register D0 and the result is deposited in the low-order word of D0. Note that the following code uses the instruction `LSL.W #4,D0`, that we have not yet introduced. This instruction is a *logical shift left* and has the effect of shifting the contents of the low-order word of D0 four places left. Zeros fill the vacated positions so that the value 1011000000110011 becomes 0000001100110000 after four shifts left.

```
CLR.L   D1            Clear all D1 as DIVU requires a 32-bit dividend
MOVE.B  D0,D1         Copy the source to D1
DIVU    #100,D1       Divide D1 by 100 to get 100s digit in D1(0:15)
MOVE.W  D1,D0         Save the 100s digit in D0 in least-significant position
SWAP    D1            Move remainder in D1(16:31) to low-order word
AND.L   #$0000FFFF,D1 Clear most-significant word of D1 before division
DIVU    #10,D1        Divide remainder by 10 to get 10s digit in D1(0:15)
LSL.W   #4,D0         Move 100s digit in result one place left
OR.W    D1,D0         Insert 10s digit into result
LSL.W   #4,D0         Move both 100s and 10s digits one place left
SWAP    D1            Move remainder in D1(16:31) to low-order word
OR.W    D1,D0         Insert ones digit in least-significant place
```

We will trace the execution of this code for [D0] = 11001100. The values of D0 and D1 in the comment field are the values after the execution of the operation.

```
CLR.L   D1          D0 = 000000CC  D1 = 00000000
MOVE.B  D0,D1       D0 = 000000CC  D1 = 000000CC
DIVU    #100,D1     D0 = 000000CC  D1 = 00040002
MOVE.W  D1,D0       D0 = 00000002  D1 = 00040002
SWAP    D1          D0 = 00000002  D1 = 00020004
AND.L   #$FFFF,D1   D0 = 00000002  D1 = 00000004
DIVU    #10,D1      D0 = 00000002  D1 = 00040000
LSL.W   #4,D0       D0 = 00000020  D1 = 00040000
OR.W    D1,D0       D0 = 00000020  D1 = 00040000
LSL.W   #4,D0       D0 = 00000200  D1 = 00040000
SWAP    D1          D0 = 00000200  D1 = 00000004
OR.W    D1,D0       D0 = 00000204  D1 = 00000004
```

MULS, MULU As with division, two multiplication instructions are available. MULS forms the product of two 16-bit signed two's complement integers, and MULU forms the product of two 16-bit unsigned integers. The assembly language forms are MULS <ea>,Dn and MULU <ea>,Dn. Multiplication is a 16-bit operation that multiplies the low-order 16-bit word in Dn by the 16-bit word at the effective address in the operand. The 32-bit longword product is deposited in Dn. The multiply instructions clear the V- and C-bits of the CCR and leave the X-bit unchanged. The Z-bit is set on a zero result and the N-bit on a negative result.

For example, if [D4] = $ABCD5678, the operation MULU.W #$1234,D4 results in $06260060, because the 68000 multiplies the 16-bit literal $1234 by the lower-order word of D4 (i.e., $5678) to give the 32-bit result $06260060. The operation MULU.W Dn,Dn squares the low-order word in Dn to give a 32-bit result.

SUB, SUBA, SUBQ, SUBI, SUBX These five arithmetic operations are the subtraction equivalents of ADD, ADDA, ADDQ, ADDI, and ADDX, respectively. Each instruction subtracts the source operand from the destination operand and places the result in the destination operand. For example, SUBI.B #$30,D0 is interpreted as [D0(0:7)] ← [D0(0:7)] - $30.

NEG The negate instruction subtracts the destination operand from zero and deposits the result at the destination address. This is a monadic operation and has the assembly language form NEG, <ea>. The operand address may be a memory location or a data register, but not an address register. This instruction simply forms the two's complement of an operand. The following subroutine, MODULUS, returns the modulus of the signed longword in D0.

```
MODULUS TST.L   D0      IF D0 > 0 THEN Exit
        BPL     Exit
        NEG.L   D0                 ELSE D0 := -D0
Exit    RTS
```

`NEGX` The negate with extend instruction forms the two's complement of an operand minus the X-bit. The `NEGX` instruction enables you to implement multiple precision arithmetic. Suppose you have a 64-bit value. If you negate the low-order 32-bits, a carry-out will be generated (unless the number was zero). When you negate the upper-order 32-bits, you have to subtract the number from zero *together with the carry-in from the low-order 32-bits.*

All bits of the CCR are modified by a `NEGX` instruction. As in the case of the `ADDX` and, `SUBX`, instructions, the Z-bit behaves differently to most other instructions. If the result of a `NEGX` is non-zero, the Z-bit is cleared. However, if the result is zero, the Z-bit is *unchanged* and is not set to 1 as you might expect. The programmer sets the Z-bit to 1 initially (i.e., before performing a `NEGX`). If the operation yields a zero result, Z remains 1. If the result is non-zero, Z is cleared to indicate this fact. Consequently, you can apply `NEGX` to a chain of bytes, words, or longwords during multiple precision arithmetic and test for zero over the entire result. Consider the use of `NEGX` to perform a 64-bit negation over two longwords.

```
MOVE.W #%00000100,CCR   Clear the CCR and set the Z-bit
NEG.L  Low              Negate the low-order longword
NEGX.L High             Now negate the high-order longword
```

`EXT` The sign-extend instruction has the assembly language form `EXT.W Dn` or `EXT.L Dn`. The `EXT.W` version sign-extends the low-order byte in Dn to 16 bits by copying bit Dn(7) to bits Dn(8:15). Similarly, the `EXT.L` version sign-extends the low-order word in Dn to 32 bits by copying bit Dn(15) to bits Dn(16:31). For example, if [D0] = $1234B021_{16}$ executing `EXT.W D0` results in [D0] = 12340021_{16}. Executing `EXT.L D0` results in [D0] = $FFFFB021_{16}$.

Consider an application from the world of digital signal processing. An analog to digital converter, ADC, periodically converts an input signal into a 16-bit signed-value. The processor divides each of these samples by 42 and transmits it to a digital to analog converter. Assume that address register A0 points at the memory-mapped ADC and that the DAC is located at the next word address.

```
MOVE.W   (A0),D0    Read a 16-bit sample from the ADC
EXT.L    D0         Extend it to 32 bits
DIVS     #42,D0     Divide it by 42
MOVE.W   D0,2(A0)   Write it to the DAC
```

The sign-extension is necessary because the `DIVS` instruction requires a 32-bit dividend. Note that the `MOVE.W D0,2(A0)` operation transfers the quotient to the DAC, leaving the remainder in the upper-order word of D0.

BCD Arithmetic

skip

Like all conventional computers, the 68000 family uses binary arithmetic and represents signed integers in two's complement form. Since pure *8421-weighted* binary arithmetic is not always convenient in a binary world, many processors support a limited form of BCD arithmetic. BCD arithmetic avoids the need to convert from decimal form to binary form before carrying out binary arithmetic, and then the need to convert from binary to decimal form after the arithmetic.

The 68000's instruction set includes three instructions that support BCD arithmetic: `ABCD`, `SBCD`, and `NBCD`. These three instructions are: add, subtract, and negate, respectively. Each of these instructions operates on a packed BCD byte (i.e., a BCD digit occupies a nibble and two BCD digits fit into a byte). As the 68000's BCD instructions are fairly specialized, they do not support a wide range of addressing modes. For example, `ABCD` and `SBCD` can be used only with the data register direct and address register indirect with pre-decrementing modes. That is, the only two valid addressing modes are:

`ABCD Di,Dj` and `ABCD -(Ai),-(Aj).`

The add BCD with extend, `ABCD`, instruction adds together two BCD digits packed into a byte, together with the X-bit from the CCR. The RTL definition of `ABCD` is `[destination]` ← `[destination]` + `[source]` + `[X]`. Similarly, the `SBCD` instruction performs the subtraction of the source operand together with the X-bit from the destination operand. The RTL definition of the `SBCD` instruction is `[destination]` ← `[destination] - [source] - [X]`.

The `NBCD <ea>` instruction subtracts the specified operand from zero together with the X-bit and forms the ten's complement of the operand if X = 0, or the nine's complement if X = 1.

Figure 5.10 illustrates the action of these three BCD instructions for the case in which X is initially zero. Figure 5.11 illustrates the effect of the same operations using the same data for the case in which X is initially one.

Each of these BCD instructions employs the X-bit of the CCR, because the 68000's BCD instructions are intended to be used in chained calculations (i.e., operations on a string of BCD digits). As each pair of digits is added (subtracted), the X-bit records the carry (borrow) and can be used in an operation on the next pair of digits. The BCD instructions operate on the Z-bit of the CCR in a special way (just like the `NEGX` instruction). The result of a BCD operation clears the Z-bit if the result is non-zero. If, however, the result is zero, the Z-bit is *unaffected*. The reason for this behavior is easy to understand if you consider the addition of, say, three pairs of BCD digits:

```
122056
248023
370079
```

As you can see, the addition of the pair of digits in the thousands and hundreds columns (i.e., 20 + 80) yields a zero result in these columns, even though the whole result (i.e., 370079) is not zero.

A simple example will suffice to demonstrate the application of BCD instructions. Assume that two strings of BCD digits are stored in memory. One

Figure 5.10 The effect of the 68000's BCD instructions (X = 0 initially)

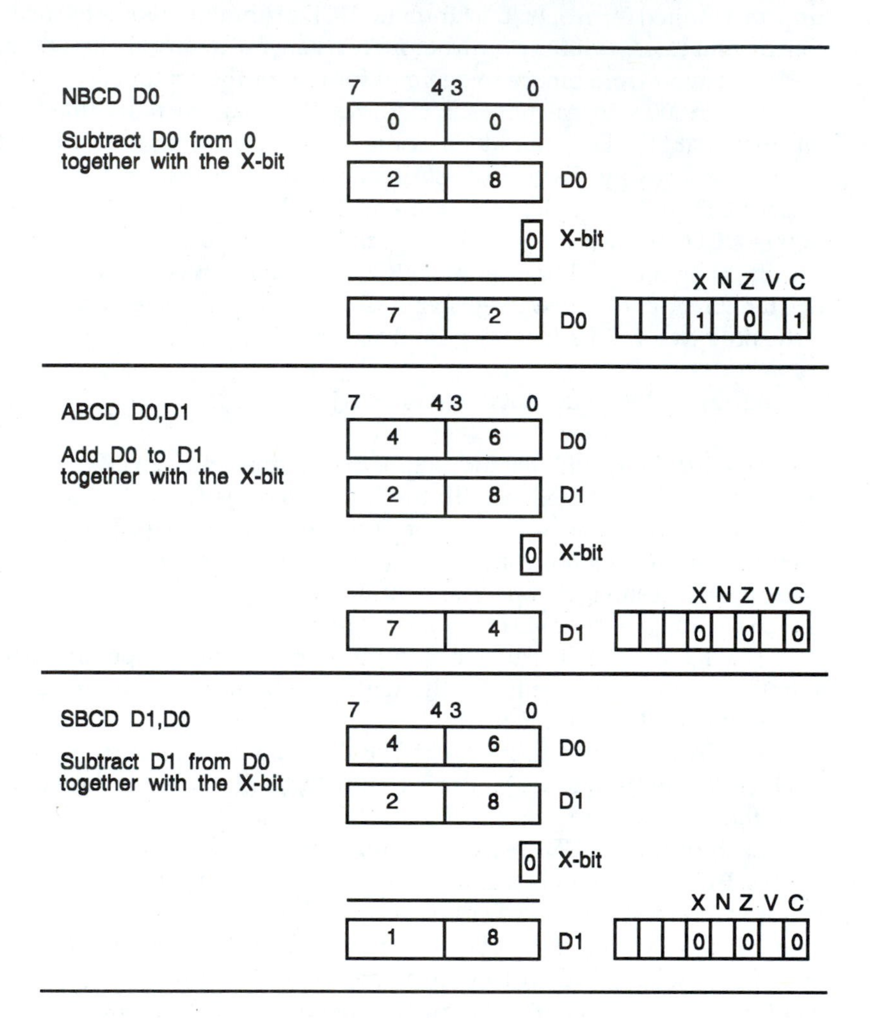

string starts at location `String1` and the other at location `String2`. When the strings of BCD digits are added together, the destination location is to be `String2` (the original `String2` is overwritten). Both strings are made up of 12 BCD digits, each of which requires six bytes (since the BCD digits are packed two to a byte). Note also that the strings must be stored so that their least-significant byte is at the *highest* address, because the auto-decrementing mode starts at a high address and moves toward lower addresses. For example, if `String1` is 123456123456 and `String2` is 001122334455, the result will be `String2` = 124578457911.

```
MOVE.W #5,D0           Six bytes in the string to be added
MOVE.W #%00100,CCR     Clear the X-bit of CCR and set Z-bit
LEA    String1+6,A0    A0 points at end of source string + 1
LEA    String2+6,A1    A1 points at end of destination string + 1
```

Figure 5.11 The effect of the 68000's BCD instructions (X = 1 initially)

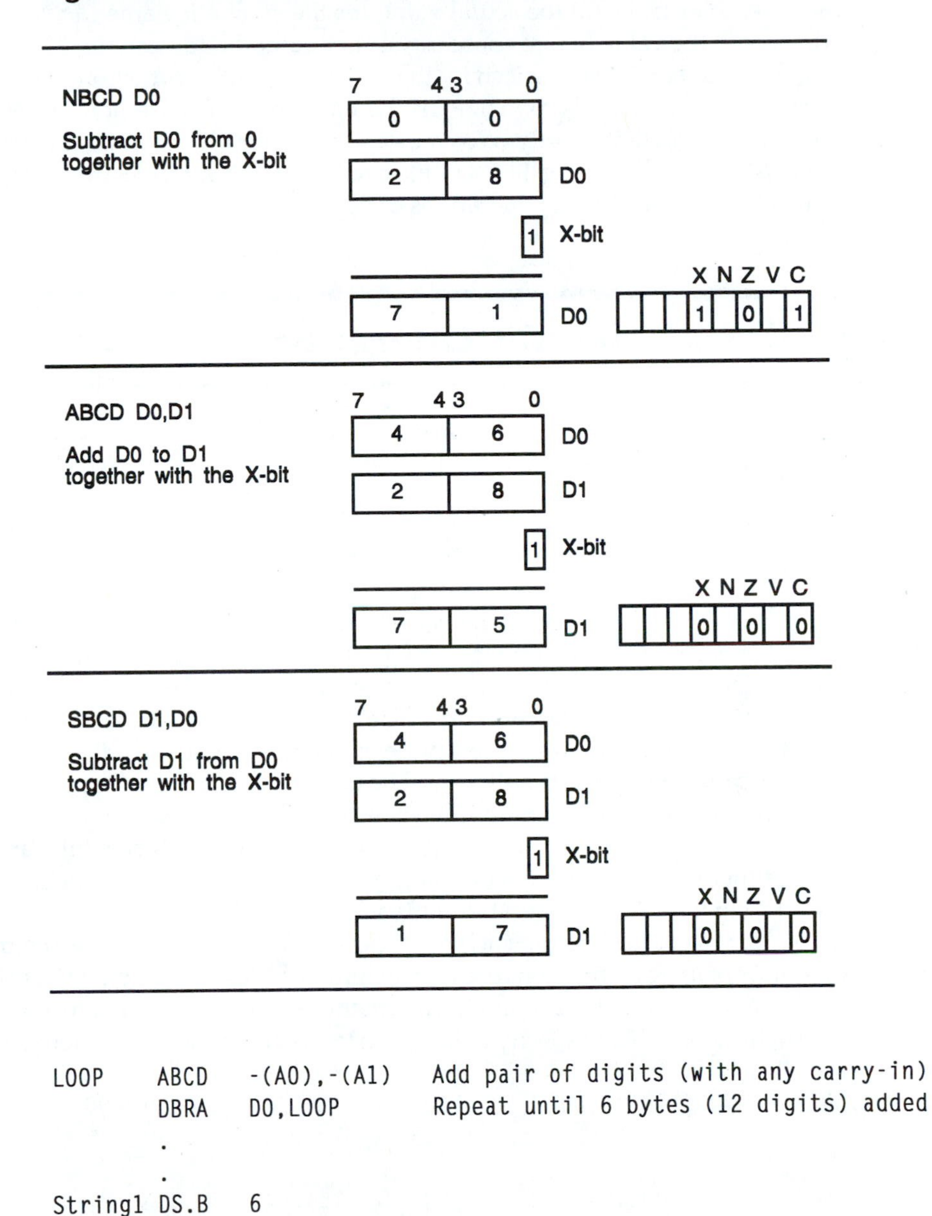


```
LOOP    ABCD   -(A0),-(A1)   Add pair of digits (with any carry-in)
        DBRA   D0,LOOP       Repeat until 6 bytes (12 digits) added
        .
        .
String1 DS.B   6
String2 DS.B   6
```

Each time the loop is executed, a pair of digits are added and any carry recorded by the X-bit. The X-bit is added to the running total on the next pass round the loop. When the addresses of the two strings are initially loaded into address registers, the values loaded are the addresses of the *ends of the strings plus one* (e.g., `LEA String1+6,A0`), since the auto-decrementing addressing will subtract one before it is first used. We can write the expression `String1+6` because the assembler looks up the actual value for the symbolic name `String1` and then adds six to the value before encoding as an operand address.

The loop is mechanized by the `DBRA D0,LOOP` instruction that decrements register D0 and returns to "`LOOP`" if the contents of D0 are not -1. We describe the `DBRA` instruction shortly. We cannot use the simple counting mechanism (i.e., `SUBQ #1,D0` followed by `BNE LOOP`), because `SUBQ` modifies the very X-bit on which the BCD arithmetic operations depend.

5.2 Logical, Shift, and Bit Operations

In the previous section we looked at operations on numeric data. Now we are going to look at operations that process the individual bits of a byte, word, or longword.

Logical Operations

The 68000 implements four Boolean operations: AND, OR, EOR, and NOT. All logical operations can be applied to longword, word, and byte operands. Additionally, logical operations can, with immediate addressing, be applied to the contents of the status register or the condition code register. Operations on the 68000's status register are carried out to alter the mode of operation of the 68000 and are *privileged* (i.e., they may be executed only when the 68000 is running in its supervisor mode).

In general, logical operations are used to modify one or more *fields* of an operand. An AND masks out bits (i.e., clears them), an OR sets bits, and an EOR toggles them (i.e., causes them to change state). The following instructions illustrate the effect of these logical operations. In each case an immediate operand is used. The low-order byte of D0 before each operation is 11110000_2. Note that the immediate operands of these instructions use the mnemonics `ANDI`, `ORI`, and `EORI`. As in the case of the `ADDI` and `SUBI` arithmetic instructions, the immediate forms of logical instructions are able to specify a data register as an operand or a memory location.

```
ANDI.B  #%10100110,D0        [D0] ← 10100110 . 11110000
                             [D0] ← 10100000

ORI.B   #%10100110,D0        [D0] ← 10100110 + 11110000
                             [D0] ← 11110110

EORI.B  #%10100110,D0        [D0] ← 10100110 ⊕ 11110000
                             [D0] ← 01010110
```

Logical operations affect the same bits of the CCR as `MOVE` instructions. That is, the X-bit is unmodified, the V- and C-bits are set to zero, and the N- and Z-bits are set according to the result. The assembly language forms of the logical operations supported by the 68000 are listed below. Logical operations can be applied to byte, word, or longword operands.

```
AND  <ea>,Dn
AND  Dn,<ea>
ANDI #<data>,<ea>
ANDI #<data>,CCR
ANDI #<data>,SR          (privileged)

EOR  Dn,<ea>
EORI #<data>,<ea>
EORI #<data>,CCR
EORI #<data>,SR          (privileged)

NOT  <ea>

OR   <ea>,Dn
OR   Dn,<ea>
ORI  #<data>,<ea>
ORI  #<data>,CCR
ORI  #<data>,SR          (privileged)
```

Remember that an operation labelled *privileged* can be executed only when the 68000 is operating in its supervisor mode. As you can see, the privileged instructions are those that operate on the supervisor byte of the 68000's status register. Note that logical instructions are not entirely symmetric. For example, the operation `EOR <ea>,Dn` is not permitted.

Consider an application of the `AND` operation. A subroutine that inputs an ASCII-encoded character might return a 7-bit code plus a parity bit in the most-significant bit position. Suppose your application is *case-insensitive* and doesn't distinguish between, say, the strings `Jan` and `JAN`. A glance at the ASCII table tells us that the difference between the upper- and lowercase versions of a letter is that bit 5 is clear in the uppercase character and set in the lowercase character. For example, "A" = 01000001 and "a" = 01100001. We can use the `AND` operation both to clear any parity bit (i.e., bit 7) and to convert lower- to uppercase.

```
BSR     GetChar           Read an ASCII-encoded character into D1
ANDI.B  #%01011111,D1     Strip the parity bit and clear bit 5
```

If, instead, we want to convert uppercase to lowercase, we can employ the `OR` instruction as follows:

```
BSR     GetChar           Read an ASCII-encoded character into D1
ANDI.B  #%01111111,D1     Strip the parity bit
ORI.B   #%00100000,D0     Set bit 5
```

Suppose we have a value in memory, PQRSTUVW, and we wish to clear bits P and Q, set bits R and S, and toggle bits T, U, V, and W. We could use the following sequence of logical operations.

```
ANDI.B  #$3F,PQRSTUVW     Clear bits P and Q
ORI.B   #$30,PQRSTUVW     Set bits R and S
EORI.B  #$0F,PQRSTUVW     Toggle bits T to W
```

Shift Operations

In a shift operation all the bits of the operand are moved one or more places left or right, subject to the variations described below. The 68000 is particularly well endowed with shift operations. Note that a circular shift is also called a *rotate.* Shift operations act on bytes, words and longwords in data registers, but only words in memory.

Shifts are categorized as *logical, arithmetic,* or *circular,* and are described in Figures 5.12 to 5.14. The symbol *C* denotes the carry-bit of the condition code register and *X* denotes its extend bit. In a *logical shift,* Figure 5.12, all the bits are moved left or right and zero enters at the input of the shifter. A logical shift left is indicated by `LSL` and a shift right by `LSR`. The bit shifted out of one end of the register is placed in the carry flag of the CCR and also the X flag.

Figure 5.13 describes the *arithmetic shift*. An arithmetic shift left, `ASL`, is almost identical to a logical shift left. An arithmetic shift right, `ASR`, causes the most-significant bit, the sign-bit, to be propagated right and, therefore, preserves the correct sign of a two's complement value. For example, if the bytes 00101010 = 42_{10} and 10101010 = -86_{10} are shifted one place right by the instruction `ASR`, the results of the arithmetic shift are 00010101 = 21_{10} and 11010101 = -43_{10}, respectively. Arithmetic shifts are intended to be used in conjunction with two's complement arithmetic. Consider the following code that multiplies the word contents of memory location PQR by ten.

Figure 5.12 Logical shift

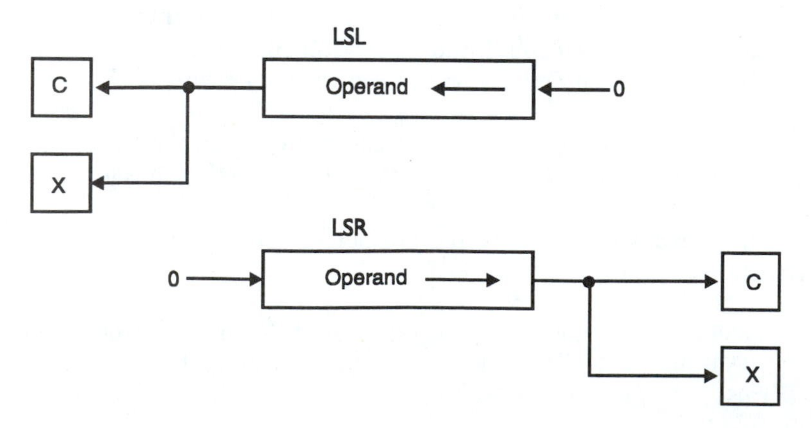

```
*          Multiply PQR by ten
           MOVE.W    PQR,D0         D0  := PQR
           ASL.W     PQR            PQR := PQR x 2
           ASL.W     PQR            PQR := PQR x 2  (i.e.,  4 PQR)
           ADD.W     D0,PQR         PQR := PQR + D0 (i.e.,  5 PQR)
           ASL.W     PQR            PQR := PQR x 2  (i.e., 10 PQR)
```

In a *circular shift*, Figure 5.14, the bit shifted out is moved to the position of the bit shifted in. The bits are shifted left by ROL, rotate left, and right by ROR, rotate right. No bit is lost during a circular shift. The bit copied from one end of the register to the other is also copied into the carry bit.

Arithmetic shift operations update all bits of the CCR. The N- and Z-bits are set or cleared as you would expect. The V-bit is set if the most-significant bit of the operand is changed at any time during the shift operation (i.e., if the number changes sign). Think about this — we'll return to it shortly. The C- and X-bits are set to the state of the last bit shifted out of the operand. If the shift count is zero (i.e., a shift length of zero that shifts no bits), C is cleared and X is unaffected.

Figure 5.13 Arithmetic shift

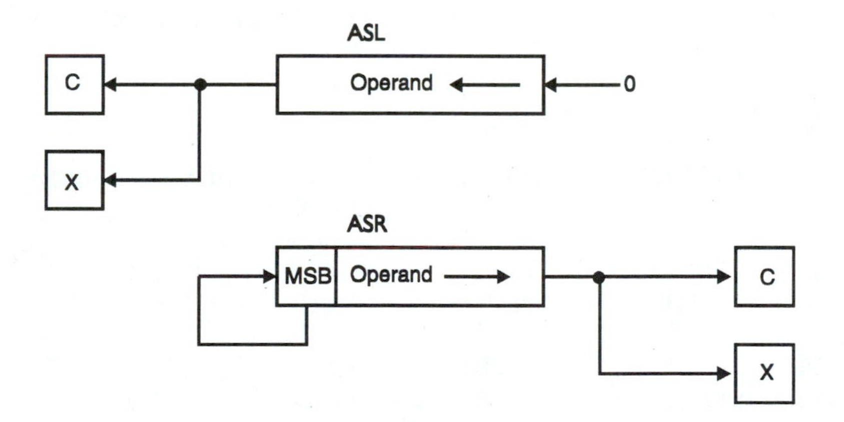

Figure 5.14 Rotate instruction (i.e., circular shift)

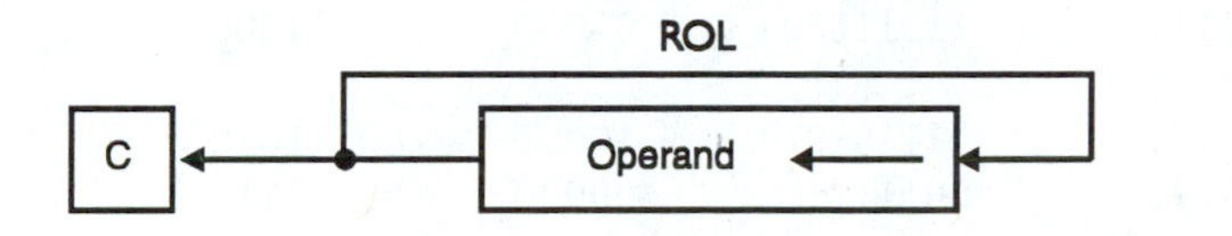

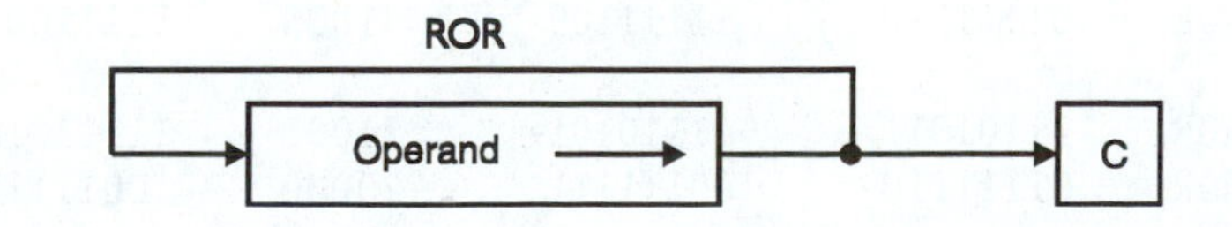

Logical shift operations behave in the same way, except that the overflow bit is set to zero. Rotate operations affect the CCR like logical shifts, except that the X-bit is not affected (i.e., it does not change state). Consider the examples in Table 5.1 in which an 8-bit byte is shifted by two places. The value of the data and the CCR after the first shift is *hypothetical*, as the programmer sees only the data and CCR at the end of the second shift.

We have not yet described the last of the 68000's shift instruction pairs, the *rotate through extend instructions*, ROXL and ROXR. These instructions behave rather like the ROL/ROR pair, except that the rotate includes the X-bit. That is, the shift takes place over 9 bits for a .B operation, over 17 bits for a .W operation, and over 33 bits for a .L operation. As you can see from Figure 5.15, the old value of the X-bit is shifted into the register (or memory location) and the bit shifted out is shifted into both the X-bit and the C-bit.

The rotate through extend instructions enable you to perform shifts over words longer than 32 bits. Suppose you have a 64-bit quadword in the register pair D1, D0 with the most-significant 32 bits in D1, and wish to perform a logical shift left over the entire 64-bit quadword.

```
*                        64-bit logical shift level over D1, D0
        LSL.L  #1,D0     Shift low-order longword one place left
        ROXL.L #1,D1     Shift high-order longword one place left
```

Table 5.1 The effect of shift operations

	Initial value	After first shift	CCR XNZVC	After second shift	CCR XNZVC
ASL	11101011	11010110	11001	10101100	11001
ASL	01111110	11111100	01010	11111000	11011
ASR	11101011	11110101	11001	11111010	11001
ASR	01111110	00111111	00000	00011111	10001
LSL	11101011	11010110	11001	10101100	11001
LSL	01111110	11111100	01000	11111000	11001
LSR	11101011	01110101	10001	00111010	10001
LSR	01111110	00111111	00000	00011111	10001
ROL	11101011	11010111	?1001	10101111	?1001
ROL	01111110	11111100	?1000	11111001	?1001
ROR	11101011	11110101	?1001	11111010	?1001
ROR	01111110	00111111	?0000	10011111	?1001

Executing LSL.L #1,D0 shifts the lower longword one place left and the bit shifted out of the left-hand end is copied into the X-bit. If we now execute ROXL.L #1,D1, the most-significant longword is shifted left and the bit shifted out of D0 is shifted into the least-significant bit of D1. Rotate through extend instructions are largely used to facilitate 64-bit arithmetic.

Figure 5.15 The rotate through extend instructions

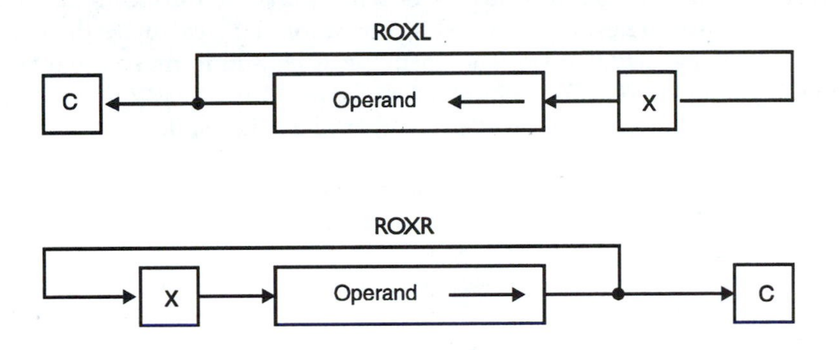

Assembly Language Form of Shift Operations

All eight shift instructions are expressed in assembly language form in one of three ways. We illustrate these three forms by means of the ASL (arithmetic shift left instruction).

Mode 1	`ASL Dx,Dy`	Shift data register Dy by Dx bits
Mode 2	`ASL #<data>,Dy`	Shift data register Dy by #data bits
Mode 3	`ASL <ea>`	Shift the contents of <ea> by one place

A shift instruction can be applied to a byte, word, or longword operand with the exception of mode 3 shifts that act only on words in memory. In mode 1, the *source* operand, Dx, specifies the number of places by which the destination operand, Dy, is to be shifted. Dy may be shifted by 1 to 32 bits. For example, if D0 contains 2, the operation LSR.B D0,D2 has the effect of shifting the low-order byte of register D2 right by two places.

In mode 2, the literal, #<data>, specifies the number of places by which Dy is to be shifted. This literal must be in the range 1 to 8 (take care — it's not 1 to 32 as you might think). For example, the operation LSR.W #4,D2 shifts the contents of D2.W four places right. In mode 3, the *word* memory location specified by the effective address, <ea>, is shifted one place. Many microprocessors permit only the *static* shifts of modes 2 and 3. Mode 1 shifts are called *dynamic* shifts, because the number of bits to be shifted is computed at run-time.

Note that an arithmetic shift left and a logical shift left operation are virtually identical. In each case, all the bits are shifted one place left. The bit shifted out enters the carry bit and extend bit of the CCR and a zero enters the vacated position (the least-significant bit). There is, however, one tiny difference between an ASL and an LSL. Since an arithmetic left shift multiplies a number by two, it is possible for the most-significant bit of the value being shifted to change sign and therefore generate an arithmetic overflow. The V-bit of the CCR is set if this event occurs during an ASL. For example, suppose we shift the 8-bit value 01111111_2 = $+127_{10}$ one place left arithmetically. The new value is $11111110_2 = -2_{10}$, the V-bit is set to indicate that arithmetic overflow has taken place. However, since logical operations are applied to strings of bits, a LSL instruction clears the V-bit.

As a simple example of a shift operation, let's calculate the even parity of bits 0 to 6, inclusive, in D0. The parity bit is chosen to make the total number of bits in the byte even. For example, if the byte is P1010111, P is made equal to 1 to give 11010111, which has an even number of 1's (i.e., six).

parity bits

(*) on test study very carefully

```
*     Calculate a parity bit for D0 over bits 0 to 6
*     Bit 7 of D0 is set/cleared to provide even parity
*     D0.B contains the initial data
*     D1 and D2 are used as working registers and are modified
*
      CLR.B   D1              Clear the bit counter
      ANDI.B  #%01111111,D0   Clear the parity bit initially
      MOVE.B  #7,D2           Seven bits to test
Next  ROR.B   #1,D0           Rotate D0 one place right
      BCC     Zero            IF the bit rotated out was 1
      ADDQ.B  #1,D1             THEN add one to the parity count
Zero  SUB.B   #1,D2           REPEAT UNTIL all 7 bits tested
      BNE     Next
      ROR.B   #1,D0           Do a final rotate to restore D0
      LSR.B   #1,D1           Move least-significant bit of parity count into C bit
      BCC     Exit            Exit on even number
      ORI.B   #%10000000,D0   Else set the parity bit
Exit  ...
```

Test even or odd #

This code takes D0 and rotates it right seven times. As each bit falls off one end and is moved into the other end of D0, the carry bit of the CCR is set or cleared, accordingly. If the carry bit is set, we increment the one's counter D1. After seven shifts, D1 contains the number of 1's in bits 0 to 6 of D0. A further rotate restores D0 to the state it was in originally, because rotating a byte eight times has no overall effect.

At this stage the ones counter, D1, contains a value in the range 0 to 7. If the value is odd, we need to set the most-significant bit of D0. By shifting D1 one place right, the least-significant bit is copied into the C-bit of the CCR. The C-bit is set to 1 if the one's count was odd, and is set to 0 otherwise. Finally, we can test the carry bit and then use an OR instruction to set the parity bit if C = 1. We haven't implemented the most efficient way of calculating a parity bit — see if you can improve on it. We shall soon discover that other instructions like the bit test, BSET, can be used to calculate the parity of a byte.

Bit Manipulation

The 68000 provides four instructions that act on a *single bit* of an operand, rather than on the entire operand. The selected bit is first tested to determine whether it is a one or a zero, and then it is operated on according to the actual instruction. In each of the bit manipulation instructions, the complement of the selected bit is moved to the Z-bit of the CCR and then the bit is left unchanged, set, cleared, or toggled (i.e., *complemented* or *inverted*). The N, V, C, and X bits of the CCR are not affected by bit operations. Bit manipulation instructions are applied to one bit of a byte if the operand is in memory, or to one bit of a longword if the operand is in a data register. Consequently, it is not necessary to apply an extension to `BTST`, since `BTST #1,D0` is a longword operation and `BTST #1,(A4)` is a byte operation. The four instructions in this group are: `BTST`, `BSET`, `BCLR`, and `BCHG`.

`BTST` The bit test instruction tests a bit of an operand. If the tested bit is zero, the Z-bit of the condition code register is set. A bit test does not affect the value of the operand under test in any way. The assembly language form is either `BTST #<data>,<ea>` or `BTST Dn,<ea>`. The location of the bit to be tested is specified as a *constant*, or as the contents of a data register. For example, `BTST #3,(A0)` tests bit 3 of the byte in memory pointed at by A0. `BTST D0,(A0)` tests the bit in the byte pointed at by A0 whose bit-number is in D0.

All four bit manipulation instructions have the same format. If the 68000's instruction set did not include this instruction, you would test bit four of D0 (i.e., `BTST.B #4,D0`) by executing:

```
AND.B #%00010000,D0    Isolate bit 4
```

The `AND` instruction updates the Z-bit of the CCR, but also destroys the contents of D0 (unlike the `BTST`).

`BSET` Bit test and set causes the Z-bit of the CCR to be set if the specified bit is zero, and then forces the specified bit of the operand to be set to one.

`BCLR` Bit test and clear works exactly like `BSET`, except that the specified bit is cleared (i.e., forced to zero) after it has been tested.

`BCHG` Bit test and change causes the value of the specified bit to be reflected in the Z-bit of the CCR and then toggles (i.e., inverts) the state of the specified bit.

The bit test group of instructions is used when dealing with peripherals (i.e., input/output devices such as serial interfaces and disk controllers). Most memory-mapped peripherals have registers containing control and status bits that have to be frequently tested, set, and cleared.

In order to apply these four instructions, two items of information are needed by the assembler: the effective address of the operand, and the position of the bit to be tested. All addressing modes are available to these four instructions for the destination operand, except immediate addressing, address register direct, and program counter relative addressing. The location of the bit to be tested is specified in one of two ways — an absolute (i.e., constant or *static*) form in the instruction, or the contents of a data register (i.e., *dynamically*). In the following example, BSET is used as an illustration, but any of the members of this group could have been chosen. The assembly language forms of these instructions are:

```
BTST  Dn,<ea>
BTST  #<data>,<ea>
```

If the destination address is a memory location, the source operand is treated as a *modulo-8* value. The modulo-8 value of N is the remainder when N is divided by 8. For example, if the source operand is 18, bit 2 is tested. If the destination address is a data register, the source operand is treated as a modulo-32 value. As an example of bit-manipulation, consider a subroutine to count the number of ones in a byte. On entry to the subroutine, D0(0:7) contains the byte to be tested, and on exit it contains the number of ones in the byte. No other registers must be modified by the subroutine. For example, if D0 = 11011001 before the subroutine is called, D0 = 00000101 after calling the subroutine.

```
*  D0 = input/output register (only D0.B is modified)
*  D1 = one's counter (not modified by the subroutine)
*  D2 = pointer to bit of D0 to be tested (not modified by the subroutine)
*
ONES_COUNT MOVEM.L  D1-D2,-(A7)    Save working registers on the stack
           CLR.B    D1             Clear the one's counter
           MOVEQ    #7,D2          D2 initially points at most-significant bit of D0.B
NEXT_BIT   BTST     D2,D0          Test the D2th bit of D0
           BEQ      LOOP_TEST      IF zero THEN nothing more to do
           ADDQ.B   #1,D1                   ELSE increment 1's count
LOOP_TEST  SUBQ.B   #1,D2          Decrement bit pointer
           BPL      NEXT_BIT       REPEAT UNTIL count negative
           MOVE.B   D1,D0          Transfer one's count to D0
           MOVEM.L  (A7)+,D1-D2    Restore working registers
           RTS                     Return
```

Now that we've looked at the arithmetic operations and instructions acting on individual bits, we examine instructions used to implement conditional behavior.

5.3 Program Control and the 68000

The computational power of all computers lies in their ability to choose between two or more courses of action on the basis of the available information. Without such powers of decision, the computer would be almost entirely worthless, as it could only execute a long list of instructions blindly. Consider the game of chess. When you make your move, you have to decide which of several possible legal moves to select. If you write a program to simulate a chess player, you require instructions that make a decision and then select between two or more possible courses of action.

Instructions that make decisions at the *assembly language level* are used to implement constructs like IF...THEN...ELSE and REPEAT...UNTIL at the *high-level language level*. Although we have already met some program control instructions (e.g., the conditional branch), we are now going to look at the family of instructions that enable the programmer to synthesize high-level control constructs.

Compare Instructions

A computer chooses between two courses of action by examining the state of one or more bits in its CCR and associating one action with one outcome of the test and another action with the other outcome. Although the CCR's bits are updated after certain instructions have been executed, two instructions can be used to explicitly update the CCR. These are the *bit test* that we have already met, and the compare. Some programmers regard the compare instruction as belonging to the *arithmetic group* because it performs a subtraction.

CMP, CMPA, CMPI, CMPM The compare instructions all subtract the contents of one register (or memory location) from another register (or memory location) and update the contents of the condition code register accordingly. The N-, Z-, V-, and C-bits of the CCR are all updated and the X-bit remains unaffected. The difference between the subtract operation (i.e., SUB Q,R) and the compare operation (i.e., CMP Q,R) is that a subtraction evaluates P = R - Q and retains P, while the compare operation merely evaluates R - Q and does not keep the result. All instructions in this group, except CMPA, take byte, word, or longword operands.

The basic compare instruction, CMP, compares the source operand with the destination operand. The destination operand must be a data register and the source operand may be specified by any of the 68000's addressing modes. For example:

```
CMP.L D0,D1         evaluates   [D1(0:31)] - [D0(0:31)]
CMP.B TEMP1,D3      evaluates   [D3(0:7)]  - [(TEMP1)]
CMP.L TEMP1,D3      evaluates   [D3(0:31)] - [(TEMP1)]
CMP.W (A3),D2       evaluates   [D2(0:15)] - [M([A3])]
```

Note that CMP <ea1>,<ea2> evaluates [<ea2>] - [<ea1>] so that the first operand is subtracted from the second. Some other microprocessors write their mnemonics with operands in the reverse order. CMP has three variations on its basic form: CMPA, CMPI, and CMPM. The assembly language form of the CMPA instruction is CMPA <ea>,Ai and is used to compare the operand at the specified effective address with the contents of address register Ai. As in the case of all instructions operating on an address register, CMPA operates only on word and longword operands. However, unlike other instructions that act on the contents of an address register, CMPA sets the N, Z, V, and C bits of the CCR.

CMPI, *compare immediate*, is used to execute a comparison with a literal and is written: CMPI #<data>,<ea>. Like the corresponding ADDI, the CMPI instruction allows a literal operation to be performed with the contents of a memory location. Consider two examples of CMPI operations.

```
CMPI.B  #$07,D3     evaluates [D3(0:7)] - 7
CMPI.W  #$07,TEMP   evaluates [M(TEMP)] - 7
```

CMPM, *compare memory with memory*, is one of the few instructions that permits a direct memory-to-memory operation. Only one addressing mode is allowed with CMPM, register indirect with auto incrementing — CMPM (Ai)+,(Aj)+. The CMPM instruction is used to compare the contents of two tables, element by element. The following example demonstrates how the CMPM instruction is used to compare two blocks of memory for equality.

```
BLOCK1    EQU      <address 1>
BLOCK2    EQU      <address 2>
SIZE      EQU      <number of words in block>
          LEA      BLOCK1,A0        A0 points to the first block
          LEA      BLOCK2,A1        A1 points to the second block
          MOVE.W   #SIZE,D0         D0 is the word counter
LOOP      CMPM.W   (A0)+,(A1)+      Compare a pair of words
          BNE      NOT_SAME         IF not same THEN exit
          SUBQ.W   #1,D0                         ELSE repeat until all done
          BNE      LOOP
ALL_SAME                            Exit here if they are the same
          .
          .
NOT_SAME                            Exit here if they are different
```

The compare instructions prepare the way for the operations that force the 68000 to perform one of two courses of action — the conditional branch instructions.

Branch Instructions

The 68000 provides the programmer with a toolkit containing three instructions for the implementation of all conditional control structures. These instructions are:

Bcc <label>	Branch to label on condition cc true
BRA <label>	Branch to label unconditionally
DBcc Dn,<label>	Test condition cc, decrement, and branch

The first two of these instructions are entirely conventional and are found on all microprocessors. The third one is more unusual and is not provided by most microprocessors. We did not include the JMP instruction in this list, although JMP <label> is functionally identical to BRA <label>. I always use the BRA instruction for unconditional branches and the JMP for computed branches. We will shortly return to this point.

Branch Conditionally

There are 14 versions of the Bcc d8 or Bcc d16 instruction, where cc stands for one of 14 logical conditions. If the specified condition, cc, is true, a branch is made to the instruction whose address is d8 or d16 locations onward from the start of the next instruction. That is, branch instructions take a relative address. The number of locations branched is specified by either an 8-bit value, d8, or a 16-bit value, d16. The displacement, d8 or d16, forms part of the instruction and is an 8- or 16-bit signed two's complement value, permitting a branch forward or backward from the location following the instruction Bcc d8 or Bcc d16. An 8-bit signed-offset allows a branch of +127 bytes forward or -128 bytes backward, and a 16-bit offset provides a range of +32,767 bytes forward and -32,768 bytes backward. Note that the branch is from the current value of the PC which is the location of the branch instruction *plus two*.

Some assemblers for the 68000 automatically select the *short* 8-bit displacement or the *long* 16-bit displacement, according to the distance to be branched. Otherwise, you have to write Bcc.S to force a short 8-bit branch. The extension, ".S", selects the 8-bit displacement. Remember that the numerical value of d8 or d16 is *automatically calculated* by the assembler, as the displacement is invariably in the form of a label rather than an address. The two examples below, Code 1 and Code 2, demonstrate that you can use a numerical branch or a symbolic branch address. Very few programmers would ever employ an expression like *+6 to indicate a branch six bytes on from the current value of the PC.

Code 1

```
        BRA *+6
        NOP
        NOP
Next    ...
```

Code 2

```
        BRA Next
        NOP
        NOP
Next    ...
```

Table 5.2 defines the 14 possible values of cc. After an arithmetic or logical operation is carried out (together with certain other operations), the values of the Z, N, C, V flags in the condition code register are updated accordingly. These flag bits are then used to determine whether the appropriate logical condition is true or false. For example, BCS LABEL causes the state of the carry-bit to be tested. If it is set, a branch is made to the point in the program called LABEL. Otherwise, the instruction immediately following BCS LABEL is executed.

We can divide branch instructions into two classes: those that branch on an *unsigned* condition and those that branch on a *signed* condition. For example, suppose that you execute the operation `CMP.B #$F1,D0`, where [D0] = 1. This comparison causes the evaluation of 00000001_2 - 11110001_2. Now, if we assume that the numbers represent *signed* values, the comparison is (+1) - (-15). However, if we assume that the numbers are *unsigned*, the comparison is 1 - 241. As you can see, the *interpretation* of the comparison depends on whether the arithmetic is signed or unsigned. This comparison results in Z = 1, N = 0, V = 0, and C = 1 (the carry is set because a borrow is generated).

Table 5.2 The 68000's conditional branch instructions

Mnemonic	Instruction	Branch on
BCC (BHS)	branch on carry clear	$\overline{C}$
BCS (BLO)	branch on carry set	C
BEQ	branch on equal	Z
BGE	branch on greater than or equal	$N.V + \overline{N}.\overline{V}$
BGT	branch on greater than	$N.V.\overline{Z} + \overline{N}.\overline{V}.\overline{Z}$
BHI	branch on higher than	$\overline{C}.\overline{Z}$
BLE	branch on less than or equal	$Z+\overline{N}.V+N.\overline{V}$
BLS	branch on lower than or same	C + Z
BLT	branch on less than	$N.\overline{V} + \overline{N}.V$
BMI	branch on minus (i.e., negative)	N
BNE	branch on not equal	$\overline{Z}$
BPL	branch on plus (i.e., positive)	$\overline{N}$
BVC	branch on overflow clear	$\overline{V}$
BVS	branch on overflow set	V

Branch instructions used in conjunction with *signed* arithmetic are: `BGE`, `BGT`, `BLE`, and `BLT`. Branch instructions used in conjunction with *unsigned* arithmetic are: `BHI`, `BCC`, `BLS`, and `BCS`.

Some assemblers let you write `BHS` (*branch on higher or same*) instead of `BCC`, and `BLO` (*branch on lower or same*) rather than `BCS`. For example, if [D0] = $C0 and we perform the operation `CMP.B #$25,D0`, the result is $C0_{16}$ - 25_{16} = $9B_{16}$, and Z = 0, N = 1, V = 0, C = 0. If we perform a `BHS`, the branch will be taken because $C0 is higher than $25 when using unsigned arithmetic. If we use a signed branch, `BGE`, the branch will not be taken, because $C0 is less than $25 (i.e., -64 is less than 37).

By the way, you should take care with the `BMI` (branch on minus) instruction. Since arithmetic overflow causes a sign change, the `BMI` instruction cannot reliably be used to branch on a negative result. If you don't believe me, consider:

```
MOVE.B  #$40,D0    Load D0 with 40₁₆ = 64₁₀
MOVE.B  #$60,D1    Load D1 with 60₁₆ = 96₁₀
ADD.B   D0,D1      [D1] := [D1] + [D0], 160₁₀ = 64₁₀ + 96₁₀ = A0₁₆
BMI     Minus
```

The most-significant bit of the result, $A0_{16}$, is 1 and the N-bit of the CCR is set. Consequently, the branch on minus will be taken. Many programmers employ BMI simply as a means of testing the most-significant bit of an operand.

Example of the Use of Conditional Instructions

As an example of the application of conditional branch instructions, consider the conversion of hexadecimal values to their ASCII-character equivalents. The relationship between ASCII-encoded characters and their binary or hexadecimal equivalents is provided in Table 5.3.

The algorithm for the conversion of a hexadecimal value into its ASCII-encoded equivalent can readily be derived from this table. In what follows, Hex_Val represents a single hexadecimal digit in the range 0 to 15 (i.e., 0 to F_{16}), and Char_Code the ASCII-encoded character equivalent. Thus, if Hex_Val = $A, the corresponding value of Char_Code is $41. By inspecting Table 5.3, we can derive a relationship between Char_Code and Hex_Val.

```
Char_Code := Hex_Val + $30
IF Hex_Val > $39 THEN Char_Code := Char_Code + $07.
```

We can now write down the algorithm in 68000 assembly language.

```
*     Convert a 4-bit hexadecimal value into its ASCII-encoded
*     character equivalent
*
      MOVE.B  Hex_Val,D0      Get Hex_Val value to be converted into D0
      ADDI.B  #$30,D0         Add $30 to it
      CMPI.B  #$39,D0         Test for hex values in the range $0A to $0F
      BLS.S   EXIT            IF not in range $0A to $0F THEN exit
      ADDQ.B  #$07,D0                                     ELSE add 7
*
EXIT  MOVE.B  D0,Char_Code    Save result in Char_code
```

Unconditional Branches

The 68000's two unconditional branch/jump instructions are BRA and JMP. The unconditional branch *always* instruction, BRA, forces a branch to the instruction whose address is indicated by the label following the BRA mnemonic. An 8-bit or 16-bit signed offset follows the op-code for BRA, providing a maximum branching range of 32K bytes on either side of the current instruction. The unconditional branch is equivalent to the GOTO instruction in high-level languages. The 68000 also has a jump instruction, JMP, that is functionally equivalent to the branch, BRA. The only difference is that a BRA instruction uses only relative addressing, whereas a JMP instruction may employ any of the following addressing modes.

```
JMP (An)
JMP (d16,An)
JMP (d8,An,Xi)
JMP Absolute_address
JMP (d16,PC)
JMP (d8,PC,Xi)
```

As an example of the application of an unconditional branch, consider the implementation of the CASE statement found in many high-level languages. The CASE statement permits you to select one of a number of outcomes depending on the value of a given variable. A typical Pascal CASE statement might be:

```
CASE Day OF
             Sat:      Hours := 4;
             Sun:      Hours := 2;
             Mon:      Hours := 5;
             Fri:      Hours := 2;
             Weekday: Hours := 8
   END;
```

In the following program, the variable TEST contains the integer used to determine which of three courses of action (labelled ACT1, ACT2, ACT3) is to be carried out. If TEST contains a value greater than 2, suitable code is provided to deal with the problem (labelled EXCEPTION).

Table 5.3 ASCII to hexadecimal conversion

ASCII-character	Hexadecimal code
0	30
1	31
2	32
3	33
4	34
5	35
6	36
7	37
8	38
9	39
A	41
B	42
C	43
D	44
E	45
F	46

The ASCII string "12AC" is equivalent to the binary sequence

00110001 00110010 01000001 01000011

```
CASE        MOVE.B TEST,D0      Put the value of TEST in D0
            BEQ    ACT1         If zero then carry out ACT1
            SUBQ.B #1,D0        Decrement TEST
            BEQ    ACT2         If zero then carry out ACT2
            SUBQ.B #1,D0        Decrement TEST
            BEQ    ACT3         If zero then carry out ACT3
EXCEPTION   ...                 Else deal with the exception

            BRA    EXIT         Leave CASE

ACT1        ...                 Execute action 1
            BRA    EXIT         Leave CASE

ACT2        ...                 Execute action 2
            BRA    EXIT         Leave CASE

ACT3        ...                 Execute action 3
            ...
            ...
                                (Fall through to exit)
EXIT        ...                 Single exit point for CASE
```

The above method of implementing a CASE statement is not unique, and would not be used if there were many more possible values of TEST. A better method is to use a JMP with a computed address such as JMP (d8,A0,D3), where D3 contains a value that is a function of TEST. Consider the following example of the computed jump.

```
        CLR.L   D0            We will need to clear bits 8 to 31 of D0
        LEA     JUMPTAB,A0    A0 points to the start of the jump table
        MOVE.B  TEST,D0
        ASL.L   #2,D0         Multiply TEST by 4 (each address is a longword)
        MOVEA.L (A0,D0),A0    Read the address of the action from the table
        JMP     (A0)          Perform indirect jump to appropriate action
        .
        .
JUMPTAB DC.L    Action1       Thread through to the appropriate action
        DC.L    Action2
        DC.L    Action3
        .
        .
Action1 .                     Code for Action1
        .
Action2 .                     Code for Action2
        .
Action3 .                     Code for Action3
```

TEST contains an integer in the range 0 to n-1, and an action is associated with each of these n values. The value of TEST is multiplied by 4 by ASL.L #2,D0 because the address for each action is a longword. We use a shift operation for multiplication because it is faster than a MULU #4,D0. The offset in D0 is added to A0 to produce a pointer to the address of the desired action. This address is read from the table of addresses and loaded into A0 by the MOVEA.L (A0,D0),A0 instruction. Finally, the indirect jump, JMP (A0), forces a jump to the code for the action to be taken. Although the computed jump provides a multiway branch facility, it also makes it more difficult to debug a program. You can trace through the program only by means of a simulator or emulator because the address of the next operation after the computed jump is not known until runtime.

Test Condition, Decrement, and Branch

The 68000's *test condition, decrement, and branch* instruction, DBcc, is not found in 8-bit microprocessors and provides a powerful way of implementing loop mechanisms. As in the case of the Bcc instruction, there are 14 possible computed values of the condition, cc, plus the two static (i.e., constant) values, cc = T (true) and cc = F (false). When cc = T (i.e., DBT = decrement and branch on true), the tested condition is always true, and when cc = F (i.e., DBF = decrement and branch on false), the tested condition is always false.

The DBcc instruction makes it easy to execute a loop a given number of times and has the assembly language form DBcc Dn,<label>, where Dn is a data register, and <label> is a label used to specify a branch address. The label is assembled to a 16-bit signed displacement which permits a range of 32K bytes.

When the 68000 encounters a DBcc instruction, it first carries out the test defined by the cc field. If the result of the test is true, the branch is not taken and the next instruction in sequence is executed (i.e., the loop is exited). Note that this has the opposite effect to a Bcc instruction — a Bcc instruction *takes the branch* on condition cc true, whereas a DBcc instruction *exits the branch* on condition cc true.

If the specified condition, cc, is not true, the low-order 16-bits of register Dn are decremented by 1. If the resulting contents of Dn are equal to -1, the next instruction in sequence is executed (i.e., the loop is exited). Otherwise, a branch to <label> is made. The DBcc instruction can be defined in RTL as:

```
IF cc TRUE THEN EXIT
            ELSE
                BEGIN
                [Dn] := [Dn] - 1
                IF [Dn] = -1 THEN EXIT
                               ELSE [PC]  ← label
                END_IF
                END
END_IF
EXIT:
```

One of the reviewers of my draft manuscript, Bill Neumann at ASU, suggested that this pseudocode be rewritten. We are providing Bill's version below to demonstrate how you can express the action of DBcc in two different ways.

```
IF cc FALSE THEN
            BEGIN
            [Dn] := [Dn] - 1
            IF [Dn] ≠ -1 THEN [PC] ← label
            END_IF
            END
END_IF
EXIT:
```

Unlike the Bcc instruction, DBcc allows the condition F (i.e., false) to be specified by cc. For example, DBF Dn,<label> always causes Dn to be decremented and a branch made to <label> until the contents of Dn are -1. Many assemblers, permit the use of the mnemonic DBRA instead of DBF.

DBcc is used to *mechanize* the loop. Suppose we wish to calculate the sum of 1+2+3+4+...+199+200. This operation involves adding together 200 numbers (i.e., executing a loop 200 times). Consider the following fragment of code.

```
      MOVE.W  #199,D0      Load the loop counter D0 with 199
      CLR.W   D1           Clear the total
      MOVEQ   #1,D2        Preset the increment
NEXT  ADD.W   D2,D1        Add current increment to the total
      ADDQ.W  #1,D2        Bump up the increment
      DBF     D0,NEXT      Decrement D0 and branch if not -1
```

Register D0 is pre-loaded with 199 and the DO-loop entered. The loop simply adds an increment in D2 to the total in D1. Each pass through the loop adds one to the increment. On the first pass through the loop, the DBF D0,NEXT instruction causes D0 to be decremented by 1 to yield 198. A branch is then made to NEXT, the start of the body of the loop. When D0 contains 0, the next execution of DBF D0,NEXT yields -1, and the loop is terminated. Note that the loop is repeated [Dn] + 1 times (i.e., 199+1=200 times). In order to execute the loop *200* times we had to load the counter with *199*. By the way, the above program can be made more efficient — we leave it to you to improve it.

Interestingly, DBcc Dn,<label> works only with 16-bit values in Dn. That is, loops greater than 65,536 cannot be implemented directly by this instruction. This restriction has been made to speed up the operation of the DBcc, because a 32-bit test, decrement, and branch instruction would take longer than the 16-bit version. The 68000 is internally organized as a 16-bit machine and two operations have to be carried out to implement a 32-bit operation.

The DBcc instruction is designed to support applications in which one of two conditions may terminate a loop. One condition is the loop count in the specified data register, and the other is the condition specified by the test.

Consider the following application of a DBcc instruction. Data is received by a program and processed as a block of up to 256 words. If the word $FFFF occurs in the input stream, the processing is terminated. The data to be input is stored in a memory location INPUT by some external device. Another memory location, READY, has its least-significant bit cleared to zero if there is no data in INPUT waiting to be read. If the least-significant bit of READY is true, data can be read from INPUT. The act of reading from INPUT automatically clears the least-significant bit of READY. This behavior corresponds closely to real input mechanisms.

```
*    WordCount := 255
*    REPEAT
*         Read Word
*         {Process it}
*         WordCount := WordCount - 1
*    UNTIL WordCount = -1 OR Word = $FFFF
*
     MOVE.W  #255,D1       Setup D1 as a counter - max block size 256
WAIT BTST    #0,READY      Test bit 0 of READY
     BEQ     WAIT          Repeat test until not zero
     MOVE.W  INPUT,D0      Get input and move it to D0
     .                     )
     .                     Process input
     .                     )
     CMPI.W  #$FFFF,D0     Test input for terminator
     DBEQ    D1,WAIT       Continue for 256 cycles or until terminator found
```

Note that it is very easy to make a mistake with the DBcc instruction. The lower-order word of the data register specified by the DBcc is decremented as explained earlier. Therefore, this register must be set up by a *word* operation. Sometimes, it is easy to think that a .B operation is sufficient if the loop count is less than 255.

5.4 Miscellaneous Instructions

At this point we briefly introduce some of the 68000's instructions that do not fall neatly into any of the groups described earlier.

Set Byte Conditionally

This is a somewhat unusual instruction and is not found in most other microprocessors. The assembly language form is Scc <ea>, where cc is one of the 14 logical tests in Table 5.2, an <ea> is an effective address. For example, SEQ means set operand on zero condition code, and SPL means set operand on positive condition code. When Scc is encountered by the 68000, it evaluates the condition specified by cc, and, if true, sets all eight bits of the byte specified by <ea>. If the condition is false, all eight bits at the effective address are cleared. After Scc <ea> has been executed, the contents of <ea> are, therefore, either $00 or $FF. Note that the size of this instruction is intrinsically a byte.

The best way of looking at Scc is to regard it as doing the groundwork for a deferred test. Suppose we carry out an operation and it is necessary to note whether the result was positive, for later processing. We will perform the operation first without the help of an Scc instruction and then with it.

```
         BSR      GET_DATA    Get input to be tested in D0
         CLR.B    FLAG        Clear FLAG
         TST.L    D0          Test result
         BMI.S    NEXT        If negative then exit with FLAG = 0
         MOVE.B   #$FF,FLAG   Else set all the bits of FLAG
NEXT     ...                  Continue
```

In this example, the longword in D0 returned by GET_DATA is to be tested. The operation TST.L D0 (i.e., test the contents of D0) is equivalent to CMP.L #0,D0. If the result is negative, 0 is stored in FLAG, otherwise $FF is stored. This operation requires four instructions. By using SPL (set on positive) we can simplify it to:

```
         BSR      GET_DATA    Get input to be tested in D0
         TST.L    D0          Test result
         SPL      FLAG        If negative then FLAG = 0 else FLAG = $FF
```

The use of SPL saves two instructions and also requires less time to run than the version of the program using a BMI instruction.

Check Register against Bounds

The check register against bounds instruction has the assembly language form CHK <ea>,Dn and checks the lower-order word of data register Dn against two bounds. One bound is zero and the other is the bound located at the specified effective address. If Dn(0:15) < 0 or if Dn(0:15) > [ea], then a call to the operating system is made (this call is known as an exception). Otherwise, the next instruction in sequence is executed. Operating system calls and exceptions are covered when we deal with interrupts and exceptions in Chapter 10.

The CHK instruction is intended to help prevent the programmer accessing an array element outside the bounds of the array. The following example demonstrates how you might use a CHK instruction.

```
*        Array Q is an m-row by n-column matrix with m x n elements
*        Each element of the array is a byte
```

```
*       Calculate the location of the element in Q's ith row & jth column
*       The row location is in D0.W and column location is in D1.W
*
        MULU  #n,D0      Calculate the row offset
        ADD.W D0,D1      Locate the element's position from element 0,0
        CHK   #m*n-1,D1  Check that element is in the array
        ....             Process element if OK (otherwise call OS)
```

In this example the position of the byte-sized element with respect to the start of the array must be from 0 to (n x m - 1). The CHK instruction checks for an element within this range and calls the operating system if it is outside the range. Note that you cannot use the CHK instruction unless you can also write (or at least use) the appropriate CHK exception handler that runs under the control of the operating system. We wrote the check instruction in the form CHK #m*n-1,D1 because the assembler will evaluate (m x n - 1) when the instruction is assembled. Some programmers might have written:

```
UpperBnd EQU   m*n-1
          .
          .
          .
          CHK   #UpperBnd,D1
```

In general, the user programmer does not use this instruction directly. The CHK instruction is of greater interest to the compiler writer who may wish to provide run-time checking of array bounds.

When we look at the 68020 in Chapter 12 we will discover that the CHK instruction is greatly enhanced and permits both address and data register to be checked, byte, word, and longword values to be checked, and both variable lower and upper bounds to be specified.

No Operation

The 68000's *no operation* instruction has no effect on the CPU other than to advance the program counter to the next instruction. A NOP wastes time and memory space. *Time* because it must be read from memory, interpreted, and executed — *space* because it takes up a word of memory space.

Programmers sometimes use a NOP instruction to generate a defined time delay, because a NOP requires four clock cycles. Consider the following code in which two events must be separated by an additional 2.5μs over the normal delay incurred by the MOVE.B D1,Output.

```
          MOVE.B  Input,D0      Read data from the input
          NOP
          NOP
```

```
        NOP                     Make delay of 5 NOPs = 5 x 4 = 20
        NOP                     clock cycles (2500ns at 8 MHz clock)
        NOP
        MOVE.B  D1,Output       Send data to the output
```

Some programmers use NOPs to patch a program. They insert several NOPs when writing it. If the program has bugs, they replace the NOPs by a jump to the code which fixes the bug. The following example shows how you might do this.

Case 1 (code before the fix)

```
*       Insert two NOPs here
        NOP
        NOP
*       Continue normally
```

Case 2 (code after the fix)

```
*       Now patch the fix in
        JSR  Fix
*
*       Return to here and continue
        .
Fix     {Fix the problem here}
        .
        RTS   Return
```

The code on the left, Case 1, includes two NOPs. A NOP instruction takes 16 bits and therefore these instructions take up four bytes of memory space. In the code on the right, Case 2, the NOPs have been replaced by the instruction JSR Fix which also takes four bytes (for a 16-bit address). The subroutine Fix contains the code that deals with the problem.

We present this example as a demonstration of a possible application of the no operation instruction. No respectable programmer should write code in the above way (i.e., hack it and then leave room for fixes). Of course, you know of such programmers — but this is something you would never ever do yourself. Isn't it?

You can also use the NOP instruction as a *marker* in a program. You might wish to search a region of machine code for a certain point. The machine code for the NOP instruction is $4E71_{16}$. I sometimes use two NOPs in sequence to create a unique marker $4E714E71_{16}$. Two NOPs avoid the situation in which the end of one instruction is 4E and the start of the next is 71 — giving a false marker.

Although we deal with the 68030 in Chapter 12, it is worth noting that this processor uses the NOP instruction in a special way. The 68030 overlaps instruction execution under certain circumstances. By placing a NOP between two instructions, the processor is forced to complete the instruction preceding the NOP before starting the one after it. This use of the NOP to perform bus synchronization is rather esoteric and most programmers will never need to use it.

Reset

The RESET instruction is a privileged instruction that, when executed, forces the 68000's RESET* output pin active-low for 124 clock periods. Asserting RESET* resets any device connected to this pin, but has no effect on the 68000 itself. There are very few circumstances when you would use a RESET instruction.

Return from Exception

The return from exception instruction, RTE, terminates an exception or interrupt handling routine in the same way that an RTS terminates a subroutine call. The RTS pulls a return address off the stack and loads it into the program counter. The RTE instruction pulls a word off the stack and loads it into the status register, then pulls the return address off the stack. The RTE instruction is privileged and we shall meet it again when we cover interrupts and exceptions in Chapter 10.

Stop

As its mnemonic suggests, the STOP instruction stops the 68000. Its assembly language form is STOP #n, where n is a 16-bit word. When a STOP is encountered, the literal value n is loaded into the 68000's status register and the processor ceases to execute further instructions. Normal processing continues only when a trace, interrupt, or reset exception occurs. STOP is, of course, a privileged instruction. The average 68000 programmer might never employ a STOP instruction, as its use is so specialized (i.e., stop processing and wait for an interrupt). You can use it to terminate assembly language programs running under the Teesside 68000 simulator. A STOP #$2700 stops the simulation, clears the CCR, sets the supervisor mode, clears the trace bit, and sets the interrupt mask to level 7.

Test and Set

The test and set instruction has the assembly language form TAS <ea> and tests the byte specified by the effective address. If the byte is zero or negative, the Z- and N-flags of the CCR are set accordingly. The V- and C-flags are always cleared, and the X-flag is unaffected. Bit 7 (the most-significant bit) of the operand is then set to one. That is, the instruction executes the operation:

```
IF [M(ea)] = 0 THEN Z  ←  1
IF [M(ea)] < 0 THEN N  ←  1
[M(ea){7}]  ←  1.
```

You might be tempted to think that you could replace a TAS instruction by a test instruction followed by a set instruction (i.e., an OR instruction). The following code appears to do what a TAS does.

```
TST.B <ea>
ORI.B #$80,<ea>.
```

Although the TAS performs no useful *computation*, it facilitates the synchronization of processors in a *multiprocessor system*. Most computers have a single processor. However, it is possible to design a *multiprocessor* in which a number of processors cooperate to speed up the execution of certain tasks.

Consider a multiprocessor system with two processors A and B. Suppose processor A wishes to access some memory common to both processors. Processor A cannot access this memory until processor B has given it up. The system's designer can define a special bit in the shared memory, called a *semaphore*. If the memory is in use, the semaphore is set; otherwise, it is clear. When processor A wishes to access the shared memory, it reads the state of the semaphore. If the semaphore is set, the memory is in use and processor A must try again later.

If A finds the semaphore clear, A sets it to claim the memory. Processor A now has sole right to access the shared memory. End of story. Consider the situation in which both processors A and B access the shared memory at almost the same time. Processor A might read the semaphore first and find that it is clear. Because of the nature of shared memory (i.e., its hardware design), it is possible for processor B to read the semaphore *before* A has set it. You could therefore have a situation in which both processors A and B attempt to claim the memory.

The TAS instruction executes a special *read-modify-write cycle*, in which the operand is first read to carry out the test and then written to in order to set bit 7. The TAS instruction is said to be *indivisible* because the read-modify-write cycle is always executed to completion and cannot be interrupted by another processor requesting the bus between the test and set phases of the instruction. Unless you are designing multiprocessors, you are unlikely to use the TAS instruction.

Trap on Overflow

If the overflow bit, V, of the CCR is set, executing a TRAPV instruction causes the TRAPV exception to be raised and a call to the operating system made. If $V = 0$, a TRAPV instruction has no effect other than to advance the PC to the start of the next instruction. The following code demonstrates how TRAPV is used.

```
ADD.L D0,D4         Perform the addition
TRAPV               Call the operating system on overflow
```

You can use a TRAPV instruction code to call an operating system function if arithmetic overflow takes place. This is another instruction whose use is fairly specialized. Like the TAS and CHK instructions, TRAPV is most likely to be used by the compiler writer or the operating systems designer.

Illegal

The 68000 has a *legal* illegal instruction. Honestly. Whenever the 68000 reads an op-code during a fetch cycle and finds that it does not correspond to one of its legal instructions, the 68000 generates an illegal instruction exception and calls the operating system. This approach makes good sense because it helps the 68000 to recover if ever a jump is made to a region of data.

Some *currently illegal* bit patterns might be allocated to new instructions in later members of the 68000 family. However, the bit pattern $4AFC_{16}$ will always be treated by any member of the 68000 family as an illegal instruction. You can use this instruction to force a call to the illegal instruction exception handler.

We have introduced the 68000's instruction set and will not encounter new instructions until we cover the 68020 microprocessor. In the following chapters we concentrate on the design of programs and the programming of a microprocessor system.

Summary

- The `MOVEM` instruction can be used to push a group of registers onto the system stack at the start of a subroutine, and then to restore the registers at the end of the subroutine. For example, you can push a group of registers on the stack by means of the instruction `MOVEM.L A2-A4/D1-D5,-(A7)` and then retrieve them by means of `MOVEM.L (A7)+,A2-A4/D1-D5`. Note that `MOVEM` is used only in conjunction with word and longword operands.

- The `MOVEQ`, move *quick*, instruction loads a value in the range -128 to +127 into a data register. The operation is intrinsically 32 bits because the data is *sign-extended* as it is moved. The only reason for implementing this instruction is that it is fast.

- The `MOVEP` instruction is highly specialized and is used to transfer data between a byte-wide memory-mapped peripheral and a data register. Since the peripheral is only 8 bits wide and the 68000's data bus is 16 bits wide, successive locations in the peripheral are separated by two bytes. This instruction therefore moves two or four bytes between a register and successive odd (or even) addresses in memory.

- The `EXG` instruction exchanges the 32-bit contents of two registers. For example, `EXG D3,A6` copies the contents of D3 into A6, and the contents of A6 into D3. The contents of the CCR are unmodified.

- The `SWAP` instruction swaps the upper- and lower-order words of a data register. `SWAP` is used largely to access the high-order word of a register (because all the 68000's word operations act on the lower-order word of a register). This instruction updates the N- and Z-bits of the CCR.

- The `PEA` instruction calculates an effective address (just like `LEA`) and then pushes it on the stack. This instruction is used to pass parameters to a subroutine by their address.

- The ADDQ instruction adds a literal (i.e., constant or immediate value) in the range 1 to 8 to the contents of a memory location or an address register. Operands may be byte, word, or longword values.
- The ADDI instruction adds an immediate value to a destination operand. The destination may be a memory location or a data register.
- The DIVU and DIVS instructions perform unsigned and signed division, respectively. They both divide the 32-bit value in the destination data register by the 16-bit value at the source effective address to give a 16-bit quotient and a 16-bit result. The remainder is placed in the most-significant word of the destination and the quotient is placed in the least-significant word.
- The MULU and MULS instructions multiply two unsigned or signed 16-bit values, respectively. The product is a 32-bit value and is loaded into the destination data register.
- The 68000 supports four logical operations: AND, OR, NOT, EOR. An AND operation clears (masks) bits, an OR operation sets them, a NOT operation inverts bits, and EOR toggles (inverts) them.
- The NEG operation negates a numeric value (i.e., it calculates its two's complement by subtracting it from zero).
- The NEGX instruction subtracts the operand from zero together with the X-bit of the CCR. Its purpose is to permit chained operations (i.e., multi-precision operations) because it includes the carry, recorded in the X-bit, generated by the previous operation.
- The EXT instruction sign-extends a signed byte to a word (EXT.W) or a signed word to a longword (EXT.L).
- Three instructions support binary coded decimal arithmetic, ABCD (add), SBCD (subtract), and NBCD (negate). These instructions are designed to be used in chained BCD arithmetic. They take very limited addressing modes (i.e., ABCD Di,Dj and ABCD -(Ai),-(Aj), only). Like the NEGX instruction, each of these instructions includes the X-bit of the CCR in the calculation. Note that they are all byte operations and act on two packed BCD digits.
- The ABCD, SBCD, NBCD, ADDX, SUBX, NEGX instructions all treat the Z-bit of the CCR in a special way. If the operation results in a non-zero result, the Z-bit is cleared to reflect this fact. However, if the operation yields a zero result, the Z-bit is not set to one. The programmer must set the Z-bit to one before using these instructions. This behavior permits you to operate on multilength data because the Z-bit is not set to 1 when part of a number is zero.
- The 68000 supports logical, arithmetic, and circular shifts. The mnemonics are LSL, LSR (*logical shift*), ASL, ASR (*arithmetic shift*), ROL, ROR (*rotate*), and ROXL, ROXR (rotate through the X-bit). Logical shifts shift unsigned values and arithmetic shifts shift signed values. The rotate operations are *non-destructive* because the bit shifted out at one end is shifted into the other

end. Take care with the bits of the CCR as the outcome depends on the nature of the shift. The number of places to be shifted can be specified by a literal (e.g., `ASL.B #4,D0`) or by the contents of a data register (e.g., `ROR.L D3,D0` rotates D0 right by the number of places in D3). Note that if you use a literal, you can shift only 1 to 8 places. You can also shift the contents of a memory location, but by one position only.

- The 68000 supports four operations that test a single bit — `BTST`, `BSET`, `BCLR`, and `BCHG`. These have two forms: `BTST #n,<ea>` which tests the nth bit of the operand at the effective address, and `BTST Dn,<ea>` which tests the bit of the effective address whose number is in Dn.
- A `BTST` operation simply tests a bit and sets the Z-bit of the CCR. The `BSET` also forces the tested bit to 1 after the test, the `BCLR` tests the bit and then clears it, and the `BCHG` tests the bit and then toggles it.
- The `CMP` instruction compares the source operand with the destination operand by subtracting the source from the destination. The N, Z, V, and C bits of the CCR are updated. The `CMP` instruction has `CMPA`, `CMPI`, and `CMPM` versions. The *compare memory with memory* version has the assembly language form `CMPM (Ai)+,(Aj)+` and is used to compare two regions of memory on a byte-by-byte basis.
- The 68000 has 14 conditional branch instructions of the form `Bcc label` that force a branch to the line labelled by `label` if the tested condition is true. All conditions are determined by the state of the bits in the CCR.
- When choosing a branch condition, take care to select the appropriate one. Some operate with signed values and some with unsigned values.
- The `Bcc` instruction permits a branching range of +129 bytes or -126 bytes (8-bit offset) or +32K bytes or -32K bytes (16-bit offset). The addressing mode is program counter relative, and the offset is calculated automatically by the assembler. If you wish to select a short branch with a d8 offset, you should use the `.S` suffix and write `Bcc.S`.
- The `JMP <ea>` instruction forces a jump to the effective address. It is frequently used in the form `JMP (Ai,Dj)`, where register Dj contains a dynamic offset. This form makes it easy to construct jump tables (used to implement `CASE` statements).
- The test condition, decrement, and branch instruction is used to simplify the design of loop constructs. Its assembly language form is `DBcc Dn,label`. If condition cc is true, the loop is exited. If cc is false, the contents of Dn are decremented by 1 and the loop repeated. Looping stops when the contents of Dn are decremented to -1. Note that the value in Dn is a 16-bit value. Note also that `DBcc Dn,label` is equivalent to *repeat [Dn]+1 times or until condition cc is true.*

Problems

1. How many data move instructions does the 68000 have, and how many does it actually need?

2. Why is the 68000's MOVE instruction its most versatile instruction?

3. The MOVEQ instruction is potentially dangerous. Why?

4. What does the MOVEM instruction do and how is it used?

5. Which of the following instructions modify the state of the CCR: MOVEA, MOVEM, MOVEQ, EXG, SWAP?

6. What is the difference between a SWAP and an EXG instruction?

7. What is the difference between an LEA and a PEA instruction?

8. Why are the two "quick" instructions ADDQ and MOVEQ inconsistent?

9. Explain the operation of the MOVEP instruction and indicate why the 68000 implements this instruction.

10. What is the difference between these instructions?

    ADDA.W #$8000,A0 and
    ADDA.L #$8000,A0?

11. If [D0] = $10000003 and [D1] = $00001234, what is the effect of the operation DIVU D0,D1?

12. Write a routine to multiply the 32-bit unsigned contents of D0 by the 32-bit unsigned contents of D1 and put the 64-bit result in the register pair D6,D7.

13. What is special about the ADDX, NEGX, and SUBX group of instructions? How and why are these instructions used?

14. What is the difference between the 68000's CCR and status byte? How can the programmer access each of them?

15. Under what circumstances would the systems programmer use the operation MOVE USP,A2?

16. Write a program to multiply the 4-digit BCD number pointed at by A0 by the 4-digit BCD number pointed at by A1. The result is to be pointed at by A0.

17. Instructions like `ADDX` are able to clear the Z-bit of the CCR on a non-zero result but cannot set it on a zero result. Why?

18. Describe how each of the 68000's shift instructions affects the bits of the CCR.

19. Consider the two operations Case 1 and Case 2.

Case 1	**Case 2**
`ASL.W #1,D0` `ASL.W #1,D0`	`ASL.W #2,D0`

The actions of both Case 1 and Case 2 appear to perform the same operation. Why (and under what circumstances) do they yield different results?

20. Assume that D0 contains the value \$A1B2C374. For each of the operations below, evaluate the value in D0 and the state of the CCR. Assume that initially X = 1.

a.	`ASL.B #1,D0`	b.	`ASL.L #4,D0`	c.	`ASR.L #4,D0`
d.	`LSL.B #2,D0`	e.	`LSR.L #4,D0`	f.	`ASR.L #8,D0`
g.	`ROL.B #1,D0`	h.	`ROR.L #4,D0`	i.	`ROXL.L #4,D0`
j.	`ASL.W #3,D0`	k.	`ASR.W #3,D0`	l.	`ROXR.L #6,D0`

21. Explain how the divide instructions affect the CCR.

22. What is the difference between a static shift and a dynamic shift? Give examples of each. If the 68000 did not permit dynamic shifts, how would you implement them in terms of static shift operations?

23. Assume that the date is stored in D0(23:0) in packed BCD format as: YYMMDD. Write a routine to calculate the time, in seconds, elapsed since the year 1900.

24. Write a program to *transpose* m-row by m-column matrix A. A matrix is transposed by mapping element $a_{i,j}$ onto element $a_{j,i}$ for all values of i and j.

25. Suppose you wanted to clear all data and address registers except A7. How would you do this rapidly?

26. Why does the 68000 include the `MOVEA` instruction when the `LEA` instruction can do the same job?

27. If the 68000 lacked an ADDX or a SUBX instruction, how would you simulate it using other instructions?

28. What are the relative advantages of the BRA and JMP instructions?

29. What is the difference between signed and unsigned branch instructions?

30. Why is the BMI instruction potentially dangerous?

31. Write a subroutine to find the location of the most-significant 1 in a 32-bit data register. For example, if the 32-bit contents of data D0 register are 00000100101111111010101010101010, the subroutine would return the value 5 (because the right-most one is in bit position 5 measured from the left-hand end).

CHAPTER 6

Introduction to Assembly Language Programming

At first sight it might seem rather strange to introduce a chapter on assembly language programming at such a late stage in this book. However, up to this point we have largely looked at simple examples of assembly language programs, because we have been more concerned with the basics — the 68000's instruction set and its addressing modes. Now we are going to spend a little time thinking about how we go about *designing* assembly language programs. An engineer armed only with a knowledge of a microprocessor's assembly language would probably design rather poor programs. We begin with some of the ideas that lie at the heart of program design such as *top-down design* and *structured programming*. The final part of this chapter is devoted to *pseudocode* that provides a practical way of designing programs.

6.1 The Design Process

Nearly 20 years ago a friend described the newly emerged microprocessor as *the last bastion of the amateur*. He was referring to the effect that the introduction of the microprocessor had on some programmers. Computer technology (both hardware and software) had come a long way by the mid-1970s and there was little room in the industry for a sloppy or non-professional approach to the design of systems. The introduction of the microprocessor soon changed that situation. Its very low cost made it available to anyone who wanted to investigate its properties. Consequently, people with little or no formal training or experience were using and programming microcomputers. Moreover, the tiny memories available in the 1970s and the lack of development tools forced many to write programs in assembly language and to develop bad habits such as spaghetti (i.e., unstructured) programming. Some were even forced to write programs in hexadecimal machine-code form.

Today, the status quo of the 1970s has been resumed, and a professional approach to software design for microprocessors is the rule rather than the exception. It is no longer acceptable to throw together assembly language programs. This change in attitude is not due to new fashions in programming or to sheer snobbishness. It's because of the poor results that follow from an approach to programming in which the programmer attempts to solve a problem by immediately attempting to code it. We can identify five important concepts in the design of software for microprocessor systems. These are:

- Top-down design
- Modularity
- Structured programming
- Testability
- Recoverability.

All these topics are relevant to the design of both assembly language and high-level language programs. Indeed, we would argue that these topics are equally relevant to the design of hardware.

Top-down Design

Top-down design or *programming by stepwise refinement* offers a method of handling large and complex programs by decomposing them into more tractable components. Top-down design is an iterative process that seeks to separate the *goals* of program design from their *means of achievement*. In other words, you first decide *what* you want to do, and later you think about *how* to do it. A task is initially expressed in terms of a number of subtasks. Each of these subtasks is, in turn, broken down into further subtasks. Figure 6.1 illustrates this point. At the top of Figure 6.1 (the highest level of abstraction) is only a statement of the problem to be solved. The solution may be broken down into two separate ac-

Figure 6.1 Top-down decomposition of a task

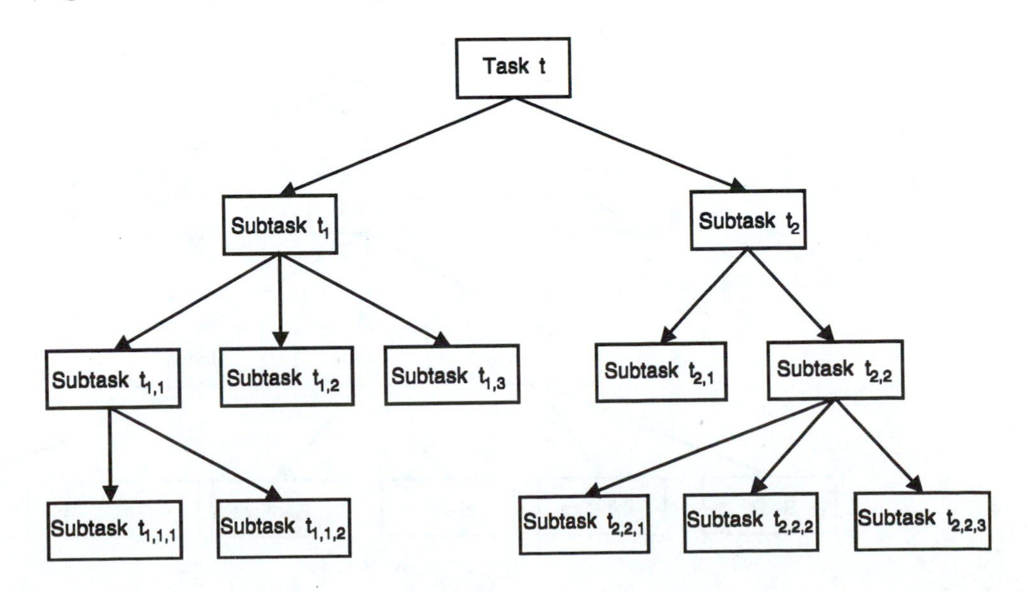

tions, t_1 and t_2. At this level no details about the implementation of these subtasks need be considered. In turn, subtask t_1 may be divided into three further subtasks, $t_{1,1}$, $t_{1,2}$, and $t_{1,3}$. This process is repeated until each subtask has been fully elaborated and is expressed in terms of the most primitive actions available to the designer. Generally speaking, the lowest level of a task is, according to Shooman, *"... small enough to grasp mentally and to code at one sitting in a straightforward and uncomplicated manner."*

Figure 6.2 shows how this approach can be applied to the design of a disk operating system. Here the problem has, initially, been split up by a functional decomposition in which each of the subtasks carries out a well defined function. At the highest level of task definition, the operating system is split into three basic subtasks: a command line interpreter, a task scheduler, and a disk file manager (DFM). Note that in Figure 6.2 some of the subtasks appear several times. For example, the create, delete, and update subtasks each require the same lower-level subtask (i.e., seek).

Let's consider the disk file manager (DFM in Figure 6.2) that can itself be broken down into a number of operations on the data structures stored on the disk, and so on. The top-down approach groups together relevant operations and their associated data at specific levels, and attempts to stop irrelevant detail obscuring the action at a particular level. Thus, the operating system's file utilities provided by the DFM perform only the operations of create, delete, and update on a file. At this level, the operations carried out by the DFM on files are not concerned with the organization and structure of the files. In everyday terms it's like booking a flight — you pay for a journey from one airport to another and do not have to worry about lower-level tasks like refuelling the aircraft.

Figure 6.2 Example of top-down decomposition

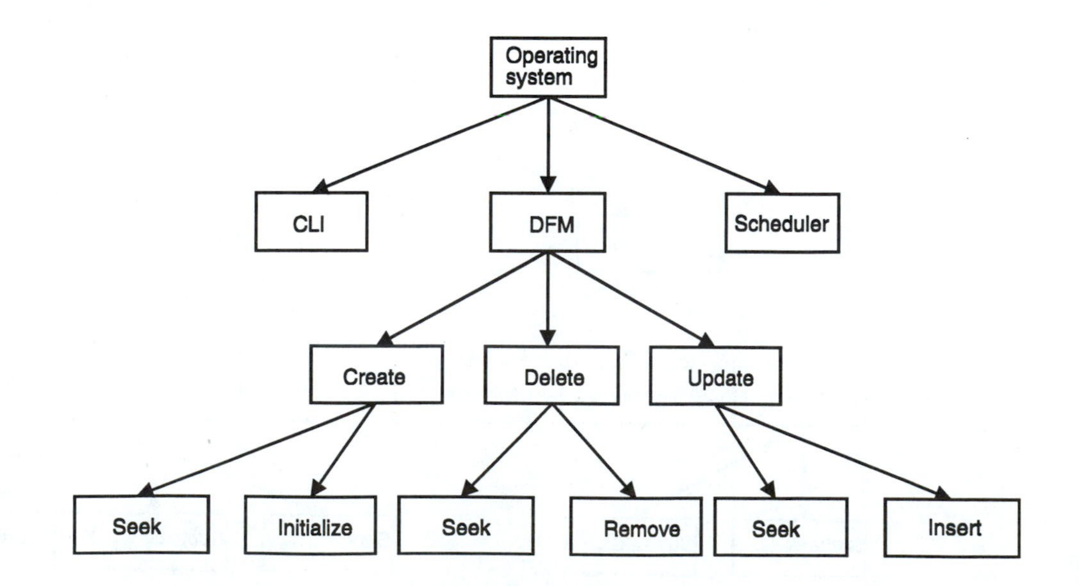

The DFM is concerned only with operations on individual records and relegates any operations involving the disk drive itself to lower-level software. For example, the module called *Create* calls a subtask *Seek* when it needs to locate a particular record. That is, the module "Create" does not have to know anything about the physical nature of the disk drive or how data is stored.

An advantage of the top-down approach is that each subtask can be validated independently of its own subtasks. A program no longer has to be tested as a single entity, and its individual components, the subtasks, can be tested independently just like their hardware equivalents.

One popular approach to program construction is called *top-down design and bottom-up coding*. You start by decomposing the problem into levels of abstraction (top-down design), and then implement the system by coding the lowest levels first (bottom-up coding). For example, suppose you are designing a word processor. After initially performing a conventional top-down design, you might first write the code to deal with the basic input/output of characters, then the formatting of the input/output, and so on.

System Specification

Before a system can be designed, it must first be *specified*. The specification of a system provides a statement of the *goals* that it must achieve. By setting these goals, the specifier of the system gives the designer a method of objectively assessing any system he or she creates. Sometimes the specification can be related to an international standard. For example, you might say that a personal computer has a *serial interface*. You might say that it has a *serial interface conforming to the EIA RS232D standard*. In the latter case, a copy of the specification can be obtained by the buyer to verify the operation of the system.

A tightly specified system is generally more reliable than a loosely specified system, because a tight specification covers all possible eventualities (i.e., operating and input conditions). A loosely specified system may fail if certain input conditions have not been anticipated. For example, a command in a machine code debugger may be designed to display the contents of memory locations between `Address_1` and `Address_2`. A poorly designed program would start by displaying the contents of `Address_1`, `Address_1` + 1, and so on until `Address_2` is located. A tightly specified system would first check that `Address_1` was actually less than `Address_2`.

In addition to design goals it can be helpful to define *non-goals*. In other words, as well as defining what a system should do, we should also state what the system does not attempt to achieve. This approach does not mean that obvious non-goals need be stated. There's little point in saying that a personal computer is not intended to fry eggs. A non-goal provides the system with explicit limitations that might otherwise be unclear. For example, an engineer may state that security is a non-goal of an interface designed to store data on a floppy disk. This non-goal means that the data on the disk is not encrypted and can be read by any other interface without undergoing a decryption process. However, it does not mean that the source data cannot be suitably encrypted by the user *before* it is presented to the interface for recording on the disk.

Modular Design

One of the most significant features of modern electronic systems is their *modularity*. A complex circuit is invariably decomposed into several less complex subsystems, called modules. The advantage of such an approach is that the modules can be designed and tested independently of the parent system. A module made by one manufacturer can be replaced by one from another manufacturer, as long as the two modules have the same interface to the rest of the system and are functionally equivalent.

In the world of software, modularity is an attempt to treat software like hardware by creating software elements, called *modules*. The principal requirement of a software module is that it is concerned only with a single, logically coherent task. For example, the software used to convert an input x into an output $sin(x)$ can be considered as a module — it takes an input, x, and returns a result, the sine of x. It performs a single function and is *logically coherent* because it carries out only those actions necessary to calculate $sin(x)$. Figure 6.3 illustrates the concept of a module.

Figure 6.3 The module

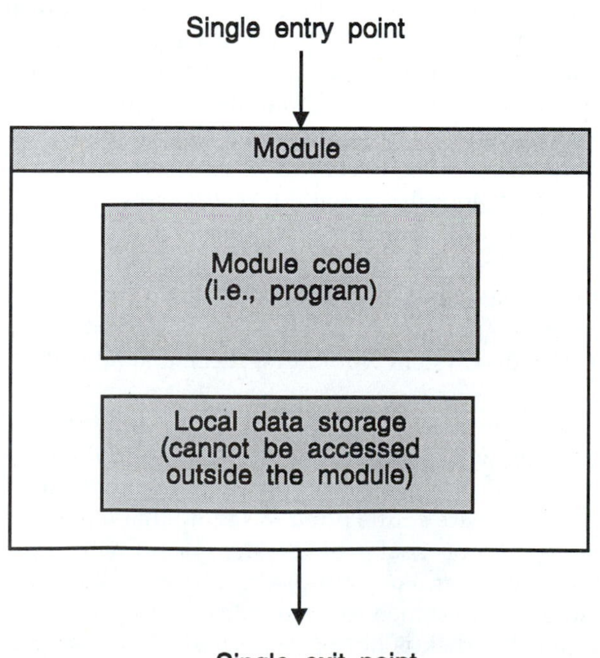

A module is not just an arbitrary segment of a larger program. An old story is told about the dark ages of programming when a programmer was asked to introduce modularity into his programs. He took a ruler and drew a line after every 75 statements. Those were his modules.

A software module is analogous to a hardware element because it has a number of inputs and a number of outputs and can be *plugged into a system*. Incoming information is processed by the module to yield one or more outputs. The internal operation of the module is both irrelevant to and hidden from its user, just as the transport properties of electrons in a semiconductor are irrelevant to the programmer of a microprocessor. The advantages of software modules are broadly the same as those of hardware modules. Modules can be tested and verified independently of the parent system, and they can be supplied by manufacturers who know little or nothing about the parent system.

The disk file monitor introduced earlier can be regarded as such a module. It is entered at one point with details of the action to be carried out passed as a parameter to the module. A return is made from the module to its calling point with parameters passed back as appropriate. It is no more reasonable to enter a module at some other point than it is to drill a hole in a floppy disk controller and to attach a wire directly to the silicon chip. Indeed, the whole point of modular design is to make the design, production, and testing of hardware and software almost identical. For example, the module designed to calculate *sin(x)* takes an input, x, and generates an output together with any relevant status information. Passing information to the inner working of the module or getting information from within the module is quite meaningless to higher-level modules.

Module Coupling and Module Strength

A module is sometimes described in terms of two properties: *coupling* and *strength*. Coupling, as its name suggests, indicates how information is shared between modules. Tightly coupled modules share common data areas. This situation is regarded as undesirable, because this makes it harder to isolate the action of one module from the action of others.

A module exhibiting loose coupling has data that is entirely independent of other modules. Any data it accesses is strictly *private* to the module and does not interact with other modules. This makes it easier to debug systems, because the programmer knows that data associated with a module is not modified by other modules. Therefore, errors can be speedily localized.

The strength of a module is a measure of its *modularity*. In the story we told about the programmer who divided a program into units of 75 lines, the resulting modularity was very weak because the modules were entirely arbitrary and functions were distributed between modules in a random fashion.

A strong module is one that performs a single task. For example, the *sin(x)* module discussed above is strong because it does nothing more than calculate the sine of an input x. A weak module performs more than one task. For example, a module dealing with input from the keyboard, a disk drive, and a serial interface is weak. It is a module in the sense that it performs logically related functions (i.e., input), but it is weak because several functions have been grouped together.

The strength of a module is important, because strong modules are relatively easy to test as they perform only one function. Equally, you can readily replace a strong module from one supplier with a functionally equivalent module from another supplier. A weak module carries out several tasks, making it more difficult to test or to replace.

Testability

In an ideal world where all stories have a happy ending and toast invariably falls buttered-side up, design for testability would not be needed. Unfortunately, real-world components sometimes fail, circuits are incorrectly assembled, and the tracks on printed circuit boards break or get shorted together. Therefore, although a microcomputer has been well designed, when finished it may sit miserably on the bench refusing to do anything.

Manufacturers of microcomputers must test their products. The amount of effort put into testing depends on the application for which the computer is intended and the economics of the situation. A manufacturer churning out ultra-low-cost microcomputers may perform minimal testing and offer a free replacement policy, as it is cheaper to replace (say) 1% of the computers than to test all of them fully. The designer of an automatic landing system for a large jet has to deliver a reliable product. In this case, simply letting the consumers do their own testing in the field might have unfortunate consequences.

Design for testability is as important for software components as for hardware elements. The designers of electronic hardware approach testability by providing test points at which the value of a signal can be examined as it flows through a system. Because almost all digital systems operate in a *closed-loop* mode, the test engineer must break the loop in order to separate cause and effect. A closed-loop system is one in which the output is fed back to the input. Typically, breaking the loop involves either single-stepping the processor, or injecting known information into, say, the processor's address pins and observing the effect on the rest of the system. The same tests can be applied to software design. Test points become the interfaces between the modules. Open-loop testing is carried out by breaking the links between software modules, just as is done with hardware. Generally, test points in software are implemented by *breakpoints*. These breakpoints are software markers that print out a snapshot of the state of the processor whenever they are encountered during the execution of a program.

Two philosophies of program testing exist: *top-down* and *bottom-up*. Bottom-up testing involves testing the lowest-level components of a system first. For example, the DFM of Figure 6.2 might be tested by first verifying the operation of software primitives such as read/write a sector of a disk. When the lowest-level components have been tested, the next level up is examined. In this case, it might be the *insert module* of the DFM that uses the read/write sector primitives that have already been tested.

Bottom-up testing is complete when the highest level of the system has been tested. This is an attractive philosophy because it is relatively easy to implement. A module is tested by constructing the software equivalent of the *fixture* or *jig* used to test hardware in industry. The software fixture provides a testing environment for the module that cannot stand on its own. For example, a module designed to read sectors would require software to pass the necessary track and sector addresses to it, examine the returned data, and display the error status at the end of an operation. Note that high-level modules may call low-level modules during the test, safe in the knowledge that the low-level modules have already been tested.

In top-down testing, the highest levels are tested first. This does not require much in the way of a fixture, as the high-level modules are tested by giving them the task that the application software is to perform. This philosophy is in line with the top-down design approach, and has the advantage that major design errors may be spotted in an early phase. Because high-level modules call lower-level modules, top-down testing cannot rely on untested lower-level modules. In order to deal with this, lower-level modules are replaced by *stubs*. These stubs are dummies that simulate the modules they replace. For example, in testing the DFM, a *read sector* operation must be replaced by a stub that obtains data from an array rather than from the disk drive.

In recent years the term *white box testing* has become fashionable. A *black box* represents the unknown. That is, we cannot say anything about the contents of the black box or how it operates. All we can do is observe its inputs and outputs, but not its *internal state*. Its converse, the white box, represents the known, or at least the partially known. The implication of this statement is that a black box must be tested exhaustively with all possible inputs. If we know something about the contents of the black box (i.e., how it operates and how it is put together), we can devise tests that exercise its various functions. We call such a system a white box.

For example, a 16-bit by 16-bit multiplier can be treated as a black box by applying to it all possible combinations of inputs (i.e., 2^{32} different values). If we know, however, that each of the two inputs is constrained to lie within certain ranges due to some property of the system using the multiplier, white box testing can be performed by testing only those inputs that can occur in practice.

Recoverability

Recoverability, or exception handling, is the ability of a system to cope with erroneous data and to recover from certain classes of error. For example, software sometimes encounters faulty data and you have to filter it out just as you would with noise in a hardware system.

You have to decide what action to take if a software module fails to achieve the intended results because of faulty software or hardware elsewhere in the system. For example, what does a DFM do if one of the procedures it calls fails to execute an operation such as reading a track? Clearly, permitting the system to hang-up in an infinite polling-loop is as bad as letting a hardware device hang-up because it has not received an electrical handshake from a peripheral. Designers should always consider the possible forms that recovery may take. Several attempts can be made to read the track to distinguish between a soft and a hard error. If the error persists, a graceful recovery may be attempted. A user who has spent two hours editing a text file would not be very happy if the system collapsed due to a minor error. A better approach would be to save as much of the text as possible and then report the error.

Exception handling is, to some extent, a controversial subject, as a poorly designed or ill-conceived error recovery mechanism may be far worse than nothing at all. The 68000 provides several exception-handling mechanisms, some at a software level and some at a hardware level. Now that we have introduced some of the background topics, the next step is to describe *structured programming* that makes it possible to design programs easily. Well, mostly.

6.2 Structured Programming

Structured programming offers a semiformal method of writing programs and avoids ad hoc methods of program design (i.e., hacking). The purpose of structured programming is threefold: it improves programmer productivity, it makes the resulting programs easier to read, and it yields more reliable programs. Essentially, structured programming techniques start from the axiom that all programs can be constructed from three fundamental components: the *sequence*, a generalized *looping* mechanism, and a *decision* mechanism. The rise of structured programming is largely attributed to the overenthusiastic use of the `GOTO` (i.e., `JMP` or `BRA`) by programmers in the 1960s. A program with many `GOTO`s provides a messy flow of control, making it very difficult to understand or to debug. Programs liberally peppered with `GOTO`s are sometimes called *spaghetti* programs (because the flow of control is so complicated, tangled, and convoluted).

The *sequence* consists of a linear list of actions that are executed in order, one by one. If the sequence is P1, P2, P3, action P1 is executed first, then P2, and then P3. The actions represented by P1, and so on, may be single operations or processes. The *process* is similar to the module described above and has only one entry point and one exit point. Indeed, it is this very *process* that is expanded into subtasks during top-down design.

The looping mechanism permits a sequence to be carried out a number of times. In many high-level languages, the looping mechanism takes the form `DO WHILE` or `REPEAT UNTIL`. The decision mechanism, which often surfaces as the `IF THEN ELSE` construct, allows one of two courses of action to be chosen, depending on the value of a test variable. By combining these three elements, any program can be constructed without using the `GOTO` statement.

Today the pendulum has swung a little way back, and a few programmers consider the total banishment of the `GOTO` to be a little unwise. In small doses, it can be used to good effect to produce a more elegant program than would otherwise result from sticking rigidly to the philosophy of structured programming.

Because of the importance of the decision and the loop mechanisms in structured programming, we examine their form in high-level language and show how they can be implemented in assembly language.

The Conditional Structure

The ability of a computer to make decisions can be called *conditional behavior*. As well as showing how the 68000 implements conditional behavior, it is worthwhile demonstrating how such behavior appears in high-level languages. Consider two entities, L and S. Entity L is a logical expression yielding a single logical value that may be true or false. Entity S is a statement that causes some action to be carried out. In what follows, the term *conditional behavior* is called *control action*, the most primitive form of which is expressed as:

```
IF L THEN S
```

The logical expression L is evaluated and, if true, action S is carried out. If L is false, S is not carried out, and the next action following this control action is executed. Figure 6.4 illustrates this basic IF L THEN S construct. Consider, for example, the conditional expression `IF INPUT_1 > INPUT_2 THEN OUTPUT := 4`. If, say, INPUT_1 = 5 and INPUT_2 = 3, the logical expression is true and OUTPUT is made equal to 4. This conditional expression is depicted in C as:

```
if(expression) statement
```

Figure 6.4 The IF...THEN construct

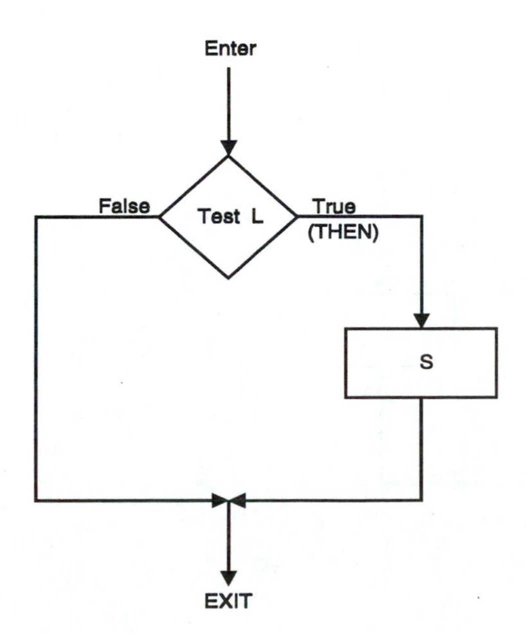

We can express the `IF...THEN` control action in a more useful form as:

```
IF L THEN S1 ELSE S2
```

Here, S1 and S2 are alternative statements. If L is true, then S1 is carried out; otherwise, S2 is carried out. There are no circumstances where neither actions S1 nor S2 or both action S1 and S2, may be carried out. Figure 6.5 illustrates this construct in diagrammatic form. This conditional expression is expressed in C as:

```
if(expression) statement1 else statement2
```

Now consider the expression:

```
IF INPUT_1 > INPUT_2 THEN OUTPUT := 4 ELSE OUTPUT := 7.
```

In this case, if INPUT_1 is greater than INPUT_2, OUTPUT is made equal to 4. Otherwise, it is made equal to 7.

The CASE Construct

The IF L THEN S1 ELSE S2 control action can be extended to a more general form in which one of several possible statements, S1, S2, ... Sn, is executed. As the logical expression L yields only a two-valued result, an expression generating an

Figure 6.5 The IF...THEN...ELSE construct

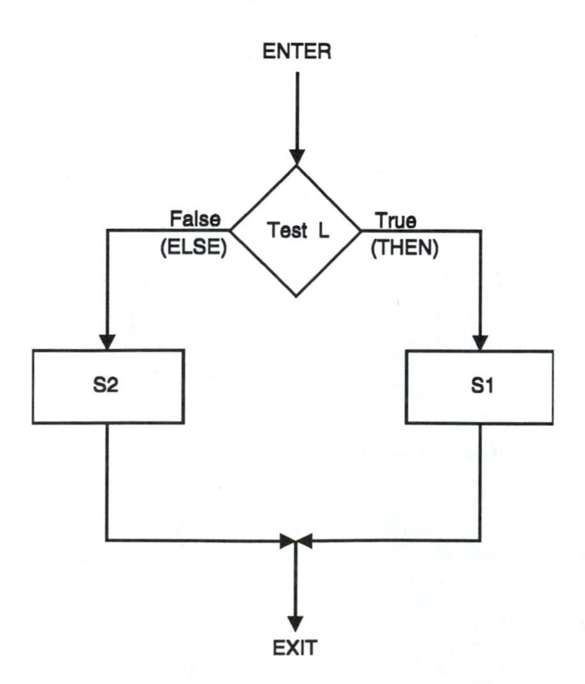

integer value must be used to make the choice between S1, S2, ... Sn. The multiple-choice control action is called a CASE construct in Pascal, a SELECT construct in some versions of BASIC, and a switch construct in C.

Consider the following example in C in which a student's exam result is divided by 20 to produce an integer mark with the value 0, 1, 2, 3, or 4. The switch construct assigns a letter or grade to each of the five results. After each case has been processed, the break forces an explicit termination of the switch construct.

```
Grade(mark)
    int mark;
    {
        char G;
        switch (mark/20)
            {
                case 0: G = 'E'; break;
                case 1: G = 'D'; break;
                case 2: G = 'C'; break;
                case 3: G = 'B'; break;
                case 4: G = 'A'; break;
            }
        return (G);
    }
```

For our current purposes we introduce the CASE construct, which is written:

```
CASE I OF
        I1:  S1
        I2:  S2
        .    .
        .    .
        .    .
        In:  Sn
END_CASE
```

The integer expression, I, is evaluated and, if it is equal to Ii, statement Si is executed. All statements Sj, where j ≠ i, are ignored. If I does not yield a value in the range I1 to In, an error may be flagged by the operating system. In some high-level languages, an exception is raised which provides an alternative course of action called an exception. Figure 6.6 illustrates the CASE statement.

Looping Mechanisms

In addition to the IF and the CASE statements, all high-level languages include a looping mechanism that permits the repeated execution of a statement S. Four major variants of the looping mechanism are as follows.

Figure 6.6 The CASE construct

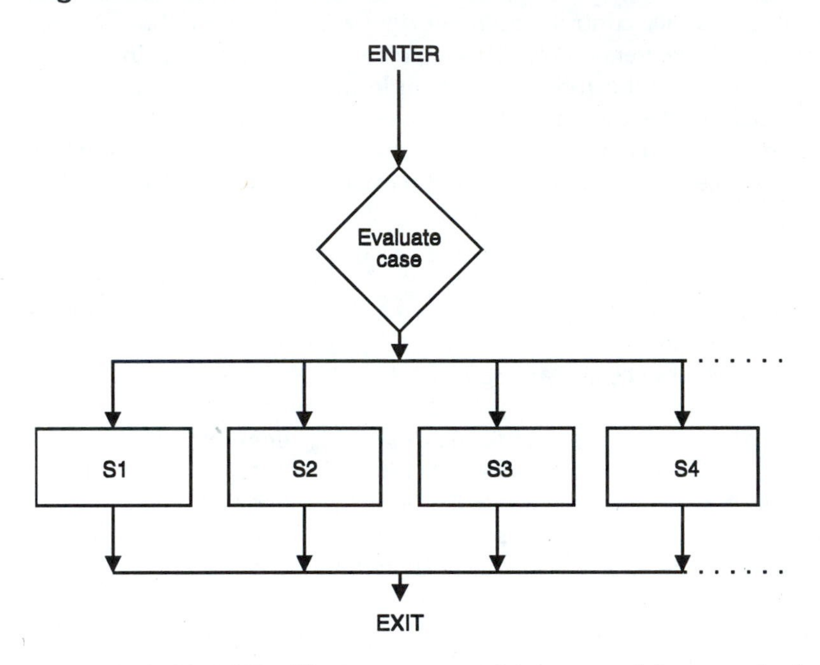

- DO S FOREVER Here, statement S is repeated forever. As forever is an awfully long time, S must contain some way of abandoning or exiting the loop. Typically, an IF L THEN EXIT mechanism is used to leave the loop — EXIT is a label that identifies a statement outside the loop. Strictly speaking, the IF THEN EXIT conditional behavior does not conform to the philosophy of structured programming. However, life without it can be very difficult.

- FOR I := N1 TO N2 DO S In this case, the control variable I is given the successive integer values N1, N1 + 1, ..., N2, and statement S is executed once for each of the N2 - N1 + 1 values of I. The FOR loop is the most conventional looping mechanism and is found in most high-level languages. Some programmers avoid this construct, as it leads to apparent ambiguity if N1 = N2 or if N1 > N2. In practice, each high-level language defines exactly what does happen when N1 > N2. Unfortunately, they don't all do the same thing.

- WHILE L DO S Whenever a WHILE construct is executed, the logical expression L is first evaluated, and, if it yields a true value, the statement S is carried out. Then the WHILE construct is repeated. If, however, L is false, statement S is not carried out and the WHILE construct is not repeated. Figure 6.7 illustrates the action of a WHILE construct.

- REPEAT S UNTIL L Statement S is first carried out, then the logical expression L evaluated. If logical expression L is true, the next statement following the REPEAT...UNTIL L is carried out. If L is false, statement S is repeated. The difference between the actions of REPEAT...UNTIL and WHILE...DO is

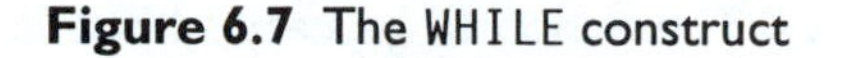

Figure 6.7 The WHILE construct

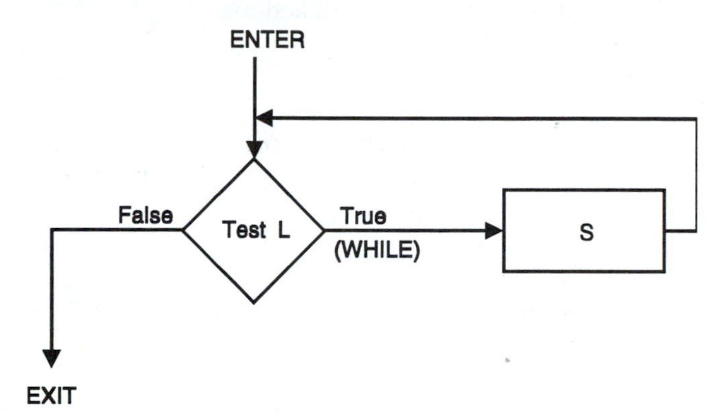

that the former causes S to be executed at least once and continues until L is true, while the latter tests L first and then executes S only if L is true. The REPEAT...UNTIL tests the logical expression *after* executing S, whereas the WHILE...DO tests the logical expression *before* executing S. Furthermore, REPEAT...UNTIL terminates when L is *true*, but WHILE...DO terminates when L is *false*. Figure 6.8 illustrates the REPEAT...UNTIL construct.

Implementing Conditional Expressions in Assembly Language

We can readily translate all the above high-level constructs into assembly language form. In the following examples, the logical value to be tested is in register

Figure 6.8 The REPEAT...UNTIL construct

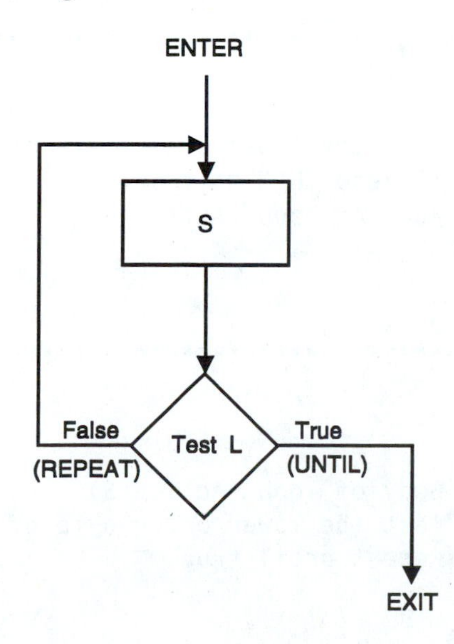

D0 and is either 1 (true) or 0 (false). The action to be carried out is represented simply by S, and consists of any sequence of actions.

```
*       IF L THEN S
*
        TST.B   D0                 Test the lower-order byte of D0
        BEQ     EXIT               IF zero THEN exit ELSE continue
        S                          Action S
EXIT

************************************************************
*       IF L THEN S1 ELSE S2
*
        TST.B   D0                 Test the lower-order byte of D0
        BEQ     ELSE               IF zero execute S2 (the ELSE part)
        S1                         IF not zero THEN S1
        BRA     EXIT               Skip past ELSE part
ELSE    S2                         Action S2 (ELSE part)
EXIT

************************************************************
*       FOR I = N1 TO N2
*
        MOVE.B  #N1,D0             D0 is the loop counter
NEXT    S                          Action S (body of loop)
        ADDQ.B  #1,D0              Increment loop counter
        CMP.B   #N2+1,D0           Test for end of loop
        BNE     NEXT               REPEAT UNTIL counter = N2 + 1
EXIT
************************************************************
*       WHILE L DO S
*
REPEAT  TST.B   D0                 Test the lower-order byte of D0
        BEQ     EXIT               IF zero THEN quit loop
        S                          Body of loop (Action S)
        BRA     REPEAT             REPEAT
EXIT

************************************************************

*       REPEAT S UNTIL L
*
NEXT    S                          Body of loop (Action S)
        TST.B   D0                 Test the lower-order byte of D0
        BEQ     NEXT               Repeat until true
EXIT
```

```
*********************************************************
*       FOR I = N DOWNTO -1    {using the DBRA instruction}
*
        MOVE.W   #N,D1         D1 is the loop counter
NEXT    S                      Action S (body of loop)
        DBRA     D1,NEXT       Decrement D1 and loop if not -1
```

The actual implementation of any control construct is a matter of taste. However, you should appreciate the following points. First, you just cannot avoid using the BRA or JMP instructions to skip over instructions you do not wish to execute. You should therefore design your control constructs with the maximum clarity. Second, one of the most common errors is to use the wrong test (e.g., using a signed comparison rather than an unsigned comparison).

6.3 Pseudocode (PDL)

Having spent some time looking at the 68000's *grammar* (its instruction set and addressing modes), we are now going to present a technique for the production of assembly language programs. Advances in computer hardware have been matched by the development of software. The very first computers were programmed in machine code. The art of programming later progressed to assembly language, to early high-level languages like FORTRAN and then to block structured high-level languages such as Pascal or Modula 2. While a high-level language (HLL) compiled into machine code by an automatic compiler may not execute as rapidly as an equivalent program written by hand in assembly language, few would argue against the proposition that the high-level language greatly improves *programmer productivity* and *program reliability* over assembly language programming.

The history of computer science, like everything else, involves two steps forward and one step backward. First-generation microprocessor-based machines traditionally had small random access memories and poor secondary storage facilities. Consequently, relatively primitive languages like BASIC evolved, and these were often interpreted rather than compiled. An interpreted BASIC program is very much slower than either a compiled BASIC or a pure assembly language program. The rise of interpreted BASIC forced those who required low execution times to take one step (or was it a giant leap?) backward and return to assembly language programming. Indeed, some applications such as animated graphics or speech processing make assembly language programming a viable option even today.

One of the reasons for the unpopularity of assembly language (at least in professional circles) is the difficulty of writing, debugging, documenting, and maintaining such programs. As we have said, the productivity of a programmer writing in, say, Pascal is almost certainly far greater than that of one writing in assembly language.

Even though the early 1980s saw a return (by some) to assembly language programming, there is now no excuse for developing assembly language programs in a haphazard or ad hoc form. Today's programmers have a great many tools that were not available to their predecessors to help them write programs. One software tool employed by some assembly language programmers is called *pseudocode* or a *program design language* (PDL). A PDL offers a way of writing assembly language programs using both top-down design techniques and structured constructs. Unlike real high-level languages, the PDL is a personalized pseudo-HLL. That is, programmers may design their own PDLs. This does not mean that a PDL is sloppy. It means that the conventions adopted by one programmer may not be those adopted by another, but any given PDL must be self-consistent.

Characteristics of a PDL

A PDL is nothing more than a convenient method of writing down an algorithm before it is coded into assembly language form. For example, a flowchart could be considered to be one type of PDL. Clearly, it is easier to code a program in PDL form than to try coding a problem into assembly language without going through any intermediate steps. The features of a PDL are summed up as follows:

- PDL represents a practical compromise between a high-level language and an assembly language. It lacks the *complexity* of the former and the *obscurity* of the latter.
- A PDL is to facilitate the production of reliable code in circumstances where a high-level language is not available or not appropriate. PDL provides an intermediate step when writing assembly language programs.
- A PDL shares some of the features of HLLs, but rejects their overall complexity. For example, it supports good programming techniques including top-down design, modularity, and structured programming. Similarly, it may support primitive data structures.
- A PDL provides a shorthand notation for the description of algorithms and allows the use of plain English words to substitute for entire expressions. This feature gives it its strong top-down design facilities.
- A PDL is extensible — its syntax can be extended to deal with the specific task to which it is applied.

Pseudocode Design Example 1

The best way to introduce a PDL is by means of an example. We shall therefore present a simple problem in words and then develop a solution in PDL. Finally, we use the PDL to write a suitable 68000 assembly language program.

A sequence of ASCII characters is stored at memory location $600 onward (each character occupies one byte). Such a sequence of characters is called a *string*. A second string of equal length is stored at memory location $700 onward. Each string ends with the character $0D (i.e., the ASCII value for a *carriage return*). Our first problem is to write a program to determine whether these two strings are identical. If they are identical, the value $FF is to be placed in data register D0. If they are not identical, the value $00 is placed in D0.

Solution

We begin by drawing a memory map for the problem, Figure 6.9. As an ancient Chinese philosopher once said, "A picture is worth a thousand lines of code."

Pseudocode We can compare the two strings by taking the first pair of characters and comparing them. If they are the same, we continue and compare the next pair, and so on. If we find that one of the characters is $0D, the process is complete and the strings are the same. If, at any time, the two characters of a pair do not match, we can abandon the search, as the strings are different.

Although you would probably immediately write detailed pseudocode for this simple algorithm, we'll take the top-down approach. The first-level pseudocode is given on page 244. Note that the variable `Match` is set to the boolean value `false` initially to indicate that the strings are not equal. The characters are matched a pair at a time and, if they do not match, an exit is made (with `Match` already set to `false`). If the loop is terminated when a $0D is found, `Match` is set to `true`.

Figure 6.9 Memory map for a string-matching problem

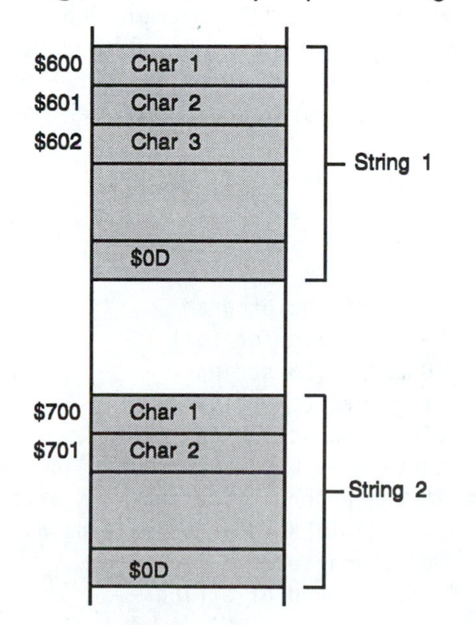

```
Match := false
REPEAT
   Read a pair of characters
   IF they do not match then EXIT
UNTIL a character = $0D
Match := true
EXIT
```

At this stage we have indicated *what* we want to do, but not *how* to do it. We can fill in the gaps in the second-level pseudocode.

```
Match := false
Set Pointer1 to point to String1
Set Pointer2 to point to String2
REPEAT
     Read Character pointed at by Pointer1
     Compare with Character pointed at by Pointer2
     IF they do not match THEN EXIT
     Pointer1 := Pointer1 + 1
     Pointer2 := Pointer2 + 1
UNTIL Character = $0D
Match := true
EXIT
```

As you can see, the pseudocode is a mixture of plain English plus a few Pascal-like constructs. For example, the symbol ":=" indicates an assignment, and we can make liberal use of control constructs like REPEAT...UNTIL. At this stage the pseudocode is detailed enough to write a 68000 program. Before we begin, we should note that we are performing operations involving 8-bit characters. Therefore, operations on the strings will use the .B suffix. Of course, operations on *pointers* (i.e., the contents of address registers) are longword operations.

```
*          D0                    Error flag
*          A0                    Pointer to string 1
*          A1                    Pointer to string 2
*
           ORG       $400        Start of the program
           MOVE.B    #$00,D0     Set the flag for fail
           MOVEA.L   #$600,A0    A0 points to string1
           MOVEA.L   #$700,A1    A1 points to string2
REPEAT     MOVE.B    (A0),D1     Get a character from string1
           CMP.B     (A1),D1     Compare it with a character in string2
           BNE       Exit        IF characters different THEN exit
           ADDA.L    #1,A0       IF same THEN point to next pair
           ADDA.L    #1,A1        of characters.
           CMP.B     #$0D,D1     Test for end of strings
```

could also use cmpm (A0)+, (A1)+

Carriage Return

```
          BNE     REPEAT     IF not end THEN compare next pair
          MOVE.B  #$FF,D0               ELSE set flag for success
EXIT      .                  Exit point
          .
          ORG     $600
Pointer1  DS.B    <length of string1>
          ORG     $700
Pointer2  DS.B    <length of string2>
```

Once you've written a program, it's reasonable to look at the program again to determine whether it can be improved. By *improved,* we mean made more *compact,* or more *efficient,* or even more *elegant*. We will have another go at encoding the problem.

```
Car_Ret   EQU     $0D
          ORG     $400              Start of program
          CLR.B   D0                Set the flag for fail
          LEA     Pointer1,A0A0 points to string 1
          LEA     Pointer2,A1A1 points to string 2
REPEAT    MOVE.B  (A0),D1           Get character from string1
          CMP.B   (A1),D1           Compare it with string2
          BNE     Exit              IF different THEN EXIT
          LEA     (1,A0),A0         IF same THEN point to next
          LEA     (1,A1),A1         pair of characters.
          CMP.B   #Car_Ret,D1Test for end of strings
          BNE     REPEAT            IF not end THEN compare next pair
          MOVE.B  #$FF,D0                      ELSE set D0 for success
EXIT      .                         Exit
          .
          ORG     $600
Pointer1  DS.B    <length of string1>
          ORG     $700
Pointer2  DS.B    <length of string2>
```

Notes

- We have used *symbolic names* rather than actual values. The program is clearer if we use `Pointer1` than if we use `$600`.
- Most 68000 programmers would probably not use the instruction `MOVEA.L #Pointer1,A0`. A better instruction that does the same thing is `LEA Pointer1,A0`.
- The load effective address instruction, `LEA (1,A0),A0`, means *add 1 to the contents of address register A0 and then load the result into A0.* It has the same effect as an `ADD.L #1,A0` instruction.

- The EQU assembler directive *equates* a name to a value, so that we can later write the name. It is easier to appreciate that 'CMP.B #Car_ret,D0' means compare the contents of data register D0 with the ASCII equivalent of a carriage return than CMP.B #$0D,D0.

We can improve the program still further.

```
CarRet    EQU     $0D
          ORG     $400                 Start of program
          CLR.B   D0                   Set the match flag to fail
          LEA     (Pointer1,PC),A0     A0 points to String 1
          LEA     (Pointer2,PC),A1     A1 points to String 2
REPEAT    MOVE.B  (A0)+,D1             Get character from String1 and increment pointer1
          CMP.B   (A1)+,D1             Compare it with character from String2 and
*                                      increment Pointer2
          BNE     Exit                 IF different THEN exit
          CMP.B   #Car_Ret,D1          Test for end of strings
          BNE     REPEAT               IF not end THEN compare next pair
          SUBQ.B  #1,D0                         ELSE set D0 for success
EXIT      ...                          Exit
          ORG     $600
Pointer1  DS.B    <length of string1>
          ORG     $700
Pointer2  DS.B    <length of string2>
```

Notes

- The operation MOVE.B (A0)+,D1 copies the byte pointed at by address register A0 into data register D1, and then increments the contents of A0 by one. By employing address register indirect with auto-incrementing, we avoid having to explicitly increment the pointer registers.
- We set the flag (i.e., D0) to $FF after a successful match by subtracting one from D0 with a SUBQ #1,D0 instruction, since we know that the contents of D0 are initially 0. That is, the value of 0 - 1 is FF_{16}.
- We have used program counter relative addressing when loading the two pointer registers. This makes the code relocatable because it does not use fixed addresses.

6.4 Examples of Pseudocode Design

We now look at two further examples of the design of simple programs. The first describes an input/output buffer and the second emulates a simple four-function calculator.

A Circular Buffer

Many computers cannot process data as they receive it. Instead, the data is placed in a store until it is ready for use by the computer. This store is traditionally called a *buffer* and is analogous to a doctor's waiting room. The number of people in the waiting room increases as new patients enter, and decreases as patients are treated by the doctor. Consider a 68000-based system with a software module that is employed by both input and output routines, and whose function is to buffer data. When it's called by the input routine, it adds a character to the buffer. When it's called by the output routine, it removes a character from the buffer. Below are the operational parameters of the subroutine.

- Register D0 is to be used for character input and output. The character is an 8-bit value and occupies the lowest-order byte of D0.
- Register D1 contains the code 0, 1, or 2 on entering the subroutine.

  Code 0 means clear the buffer and reset all pointers.

  Code 1 means place the character in D0 into the buffer.

  Code 2 means remove a character from the buffer and place it in D0.

  We assume that a higher-level module ensures that only one of 0, 1, or 2 is passed to the module (i.e., invalid operation codes cannot occur).
- The location of the first entry in the buffer is at $010000 and the buffer size is 1024 bytes. Pointers and storage may be placed after the end of the buffer.
- If the buffer is full, the addition of a new character overwrites the oldest character in the buffer. In this case, bit 31 of D0 is set to indicate overflow and cleared otherwise.
- If the buffer is empty, removing a new character results in the contents of the lower byte of D0 being set to zero and its most-significant bit set.
- Apart from D0, no other registers are to be modified by calling the subroutine.

The first step in solving the problem is to construct a diagram, Figure 6.10, to help us visualize the buffer.

Figure 6.10 shows the memory map corresponding to buffer. We can draw this diagram at an early stage, as it is relatively straightforward. It is clear that a region of 1024 bytes ($400) must be reserved for the buffer together with at least two 32-bit pointers. `IN_POINTER` points to the location of the next free position into which a new character is to be placed, and `OUT_POINTER` points to the location of the next character to be removed from the buffer. At the right-hand side of the diagram is the logical arrangement of the circular buffer. This arrangement provides the programmer with a better mental image of how the process is to operate.

The first level of abstraction in PDL is to determine the overall action the module is to perform. This can be written as:

Figure 6.10 The circular buffer

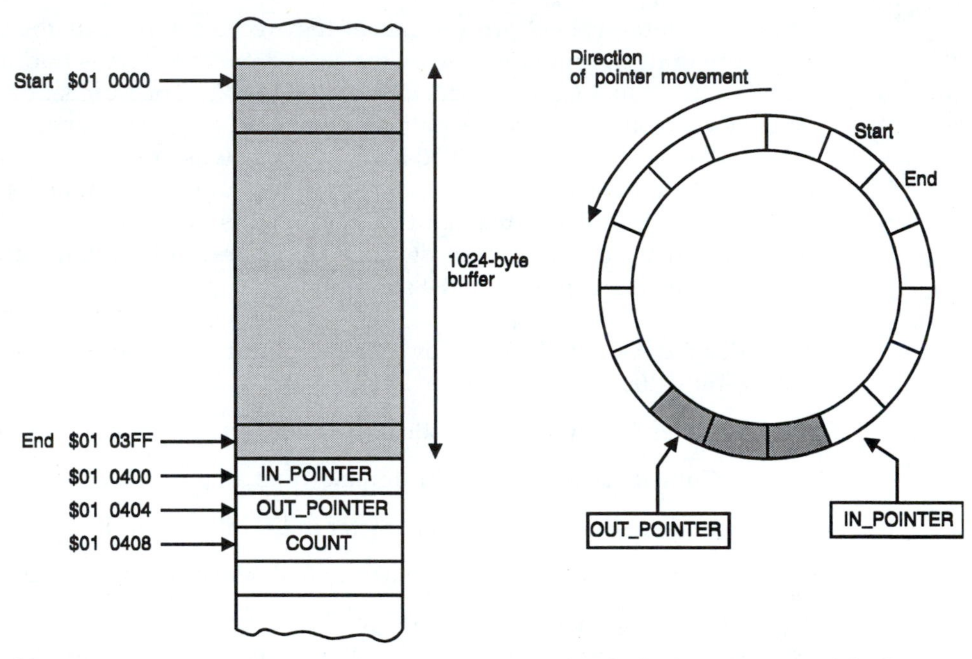


```
Module Circular_buffer
       Save working registers
       CASE OF:
                 D1 = 0:   Initialize system
                 D1 = 1:   Input a character
                 D1 = 2:   Output a character
       END_CASE
       Restore working registers
End Circular_buffer
```

Once again, note that the above PDL description is written in almost plain English. Any programmer should be able to follow another programmer's PDL. Try doing that in LISP or FORTH. At this, the highest level of abstraction, we have provided no indication of how any action is to be carried out and the only control structure is the selection of one of three possible functions. The next step is to elaborate on some of these actions.

```
Module Circular_buffer
       Save working registers
       IF [D1] = 0 THEN Initialize END_IF
       IF [D1] = 1 THEN Input_character END_IF
       IF [D1] = 2 THEN Output_character END_IF
```

```
        Restore working registers
End Circular_buffer

Initialize
          Count := 0
          In_pointer := Start
          Out_pointer := Start
End Initialize

Input_character
               Store new character
               Deal with any overflow
End Input_character
Output_character
               IF buffer NOT empty THEN Get_character_from_buffer
                                   ELSE return null character
               END_IF
End Output_character
```

The PDL is now fairly detailed. Both the module selection and the initialization routines are complete. We still have to work on the input and output routines because of the difficulty in dealing with the effects of *overflow* and *underflow* in a circular buffer.

Looking at the circular buffer of Figure 6.10, it seems reasonable to determine the state of the buffer by means of a variable, COUNT, that indicates the number of characters in the buffer. If COUNT is greater than zero and less than its maximum value, a new character can be added or one removed without any difficulty. If COUNT is zero, the buffer is *empty* and we can *add* a character but not remove one. If COUNT is equal to its maximum value and therefore the buffer is *full*, each new character must *overwrite* the oldest character as specified by the program requirements. This last step is tricky because the next character to be output (the oldest character in the buffer) is overwritten by the latest character. Therefore, the next character to be output will now be the oldest surviving character, and the pointer to the output must be moved to reflect this.

Sometimes it is helpful to draw a simplified picture of the system to enable you to walk through the design. Figure 6.11 on page 251 shows a buffer with four locations. Initially, in state (a), the buffer is empty and both pointers point to the same location. At state (b), a character is entered, the counter incremented, and the input pointer moved to the next free position. States (c) to (e) show the buffer successively filling up to its maximum count of 4. If another character is now input, as in state (f), the oldest character in the buffer is overwritten.

It is not necessary to rewrite the entire module in PDL. We will concentrate on the input and output routines and then begin assembly language coding. Since the logical buffer is circular while the physical buffer is not, we must wrap the physical buffer round. That is, when the last location in the physical buffer is filled, we must move back to the start of the buffer.

```
Input_character
   Store new character at In_pointer
   In_pointer := In_pointer + 1
   IF In_pointer > End THEN In_pointer := Start END_IF
   IF Count < Max THEN Count := Count + 1
                  ELSE
                  BEGIN
                  Set overflow flag
                  Out_pointer := Out_pointer + 1
                  IF Out_pointer > End THEN Out_pointer := Start END_IF
                  END
   END_IF
End Input_character
Output_character
   IF Count = 0 THEN return null and set underflow flag
                ELSE
                BEGIN
                  Count := Count - 1
                  Get character pointed at by Out_pointer
                  Out_pointer := Out_pointer + 1
                  IF Out_pointer > End THEN Out_pointer := Start END_IF
                END
   END_IF
End Output_character
```

The program design language has now done its job and we can translate the routines into the appropriate assembly language.

```
CIRC    EQU     *                     This module implements a circular buffer
        MOVEM.L A0-A1,-(SP)           Save working registers
        BCLR.L  #31,D0                Clear bit 31 of D0 (no error)
        CMPI.B  #0,D1                 Test for initialize request
        BNE.S   CIRC1                 IF not 0 THEN next test
        BSR.S   INITIAL               IF 0 THEN perform initialize
        BRA.S   CIRC3                  and exit
CIRC1   CMPI.B  #1,D1                 Test for input request
        BNE.S   CIRC2                 IF not input THEN must be output
        BSR.S   INPUT                 IF 1 THEN INPUT
        BRA.S   CIRC3                  and exit
CIRC2   BSR.S   OUTPUT                By default OUTPUT
CIRC3   MOVEM.L (SP)+,A0-A1           Restore working registers
        RTS                           End CIRCULAR
*
INITIAL EQU *   This module sets up the circular buffer
        CLR.W   COUNT                 Initialize counter and pointers
        MOVE.L  #BUFFER,IN_POINTER    Set up In_pointer
```

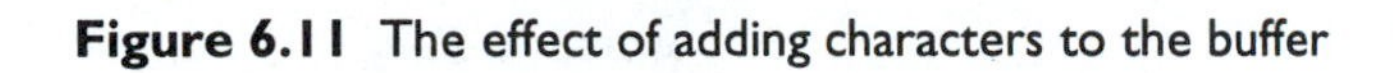

Figure 6.11 The effect of adding characters to the buffer

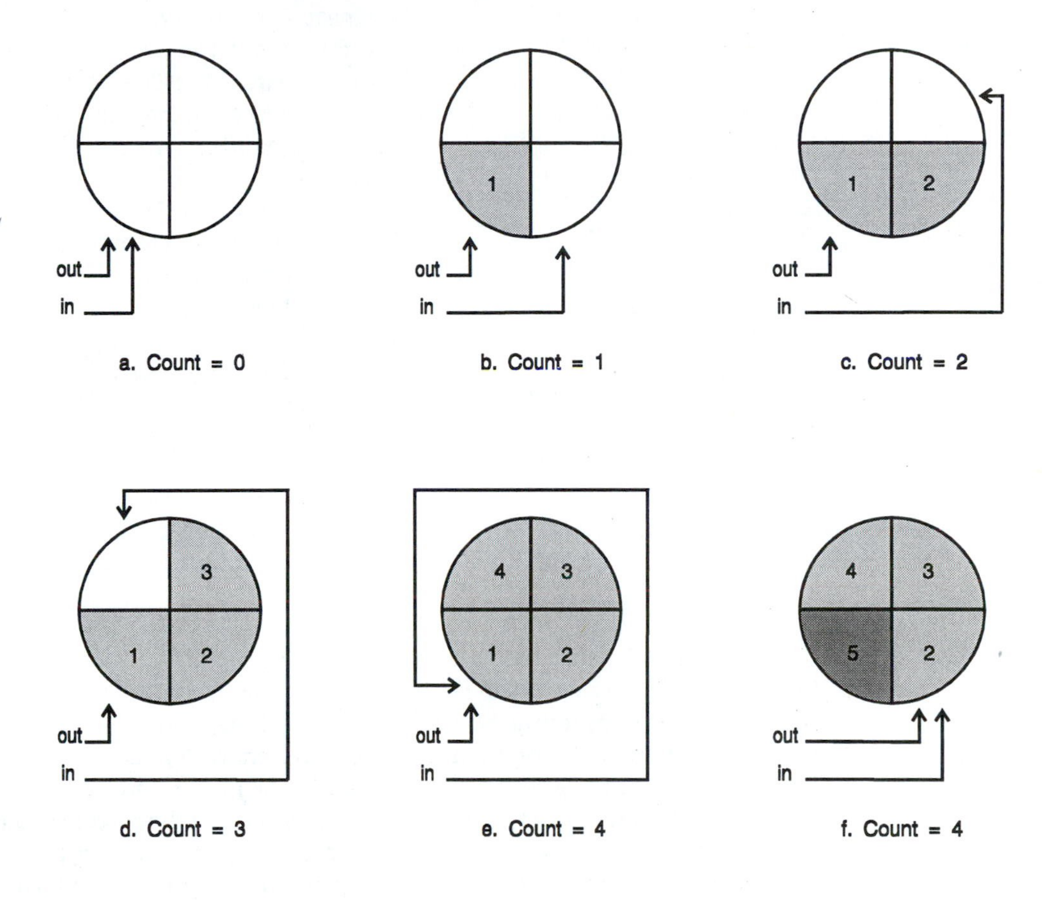


```
        MOVE.L  #BUFFER,OUT_POINTER Set up Out_pointer
        RTS
*
INPUT   EQU *   This module stores a character in the buffer
        MOVEA.L IN_POINTER,A0       Get pointer to input
        MOVE.B  D0,(A0)+            Store char in buffer, update pointer
        CMPA.L  #END+1,A0           Test for wrap-round
        BNE.S   INPUT1              IF not end THEN skip reposition
        MOVEA.L #START,A0           Reposition input pointer
INPUT1  MOVE.L  A0,IN_POINTER       Save updated pointer
        CMPI.W  #MAX,COUNT          Is buffer full?
        BEQ.S   INPUT2              IF full THEN deal with overflow
        ADDQ.B  #1,COUNT                    ELSE increment character count
        RTS                          and return
```

```
INPUT2   BSET.L   #31,D0              Set overflow flag
         MOVEA.L  OUT_POINTER,A0      Get output pointer
         LEA      1(A0),A0            Increment Out_pointer
         CMPA.L   #END+1,A0           Test for wrap-round
         BNE.S    INPUT3              IF not wrap-round THEN skip fix
         MOVEA.L  #START,A0           ELSE wrap-round Out_pointer
INPUT3   MOVE.L   A0,OUT_POINTER      Update Out_pointer in memory
         RTS                            and return
OUTPUT   TST.W    COUNT               Examine state of buffer
         BNE.S    OUTPUT1             IF buffer not empty output char
         CLR.B    D0                  ELSE return null output
         BSET.L   #31,D0               set underflow flag
         RTS                           and exit
OUTPUT1  SUBI.W   #1,COUNT            Decrement COUNT for removal
         MOVEA.L  OUT_POINTER,A0      Point to next char to be output
         MOVE.B   (A0)+,D0            Get character and update pointer
         CMPA.L   #END+1,A0           Test for wrap-round
         BNE.S    OUTPUT2             IF not wrap-round THEN exit
         MOVEA.L  #START,A0           ELSE wrap-round Out_pointer
OUTPUT2  MOVE.L   A0,OUT_POINTER      Restore Out_pointer in memory
         RTS
```

Now that we've designed and coded the buffer, the next step is to test it. The following code is the assembled circular buffer program with the necessary *driver* routines. The program inputs two characters at a time, and implements an 8-byte buffer. The first character of a pair is the control character (i.e., 0 = initialize, 1 = input, and 2 = output). For example, to initialize the buffer you type `0X`, where `X` is any character. If you type `1Y`, the character `Y` is stored in the next free place in the buffer. If you type `2Z`, the next character to be output is displayed. After each operation, the contents of the buffer are printed and the current value of the variable `count` displayed.

```
Source file: BUFFER.X68
Assembled on: 93-01-20 at: 21:46:00
        by: X68K PC-2.0 Copyright (c) University of Teesside 1989,93
Defaults: ORG $0/FORMAT/OPT A,BRL,CEX,CL,FRL,MC,MD,NOMEX,NOPCO

  1 00000400                          ORG     $400
  2 00000400 4DF900001012             LEA     HEADER,A6       ;Print heading
  3 00000406 6100015E                 BSR     PRINT
  4 0000040A 61000104      NEXT:      BSR     GETCHAR         ;REPEAT
  5 0000040E 04010030                 SUB.B   #$30,D1         ;Get number (0, 1, or 2)
  6 00000412 1E01                     MOVE.B  D1,D7           ;Save it
  7 00000414 610000FA                 BSR     GETCHAR         ;Get character for input routine
  8 00000418 1001                     MOVE.B  D1,D0
  9 0000041A 1207                     MOVE.B  D7,D1           ;Restore number
```

```
10 0000041C 6100002A               BSR      CIRC            ;Call the buffer routine
11 00000420 0C010002               CMP.B    #2,D1           ;IF we did output THEN print the character
12 00000424 66000018               BNE      CONT
13 00000428 1C00                   MOVE.B   D0,D6           ;Save character from output
14 0000042A 610000F4               BSR      NEWLINE
15 0000042E 4DF900001023           LEA      OUT,A6          ;Print "output"
16 00000434 61000130               BSR      PRINT
17 00000438 1206                   MOVE.B   D6,D1
18 0000043A 610000DC               BSR      PUTCHAR         ;Now print the character
19 0000043E 610000E0     CONT:     BSR      NEWLINE
20 00000442 610000E8               BSR      DISPLAY         ;Display the buffer contents
21 00000446 60C2                   BRA      NEXT            ;FOREVER
22                       *
23          00000448     CIRC:     EQU      *               ;This module implements a circular buffer
24 00000448 48E700C0               MOVEM.L  A0-A1,-(SP)     ;Save working registers
25 0000044C 0880001F               BCLR     #31,D0          ;Clear bit 31 of D0 (no error)
26 00000450 0C010000               CMPI.B   #0,D1           ;Test for initialize request
27 00000454 6604                   BNE.S    CIRC1           ;IF not 0 THEN next test
28 00000456 6114                   BSR.S    INITIAL         ;IF 0 THEN perform initialize
29 00000458 600C                   BRA.S    CIRC3           ;and exit
30 0000045A 0C010001     CIRC1:    CMPI.B   #1,D1           ;Test for input request
31 0000045E 6604                   BNE.S    CIRC2           ;IF not input THEN must be output
32 00000460 6126                   BSR.S    INPUT           ;IF 1 THEN INPUT
33 00000462 6002                   BRA.S    CIRC3           ;and exit
34 00000464 6174         CIRC2:    BSR.S    OUTPUT          ;By default OUTPUT
35 00000466 4CDF0300     CIRC3:    MOVEM.L  (SP)+,A0-A1     ;Restore working registers
36 0000046A 4E75                   RTS                      ;End CIRCULAR
37                       *
38          0000046C     INITIAL:  EQU      *               ;This module sets up the circular buffer
39 0000046C 427900001008           CLR.W    COUNT           ;Initialize counter and pointers
40 00000472 23FC00001000           MOVE.L   #BUFFER,IN_POINT ;Set up In_pointer
            0000100A
41 0000047C 23FC00001000           MOVE.L   #BUFFER,OUT_POINT ;Set up Out_pointer
            0000100E
42 00000486 4E75                   RTS
43                       *
44          00000488     INPUT:    EQU      *               ;This module stores a character in the buffer
45 00000488 20790000100A           MOVEA.L  IN_POINT,A0     ;Get pointer to input
46 0000048E 10C0                   MOVE.B   D0,(A0)+        ;Store char in buffer, update pointer
47 00000490 B1FC00001008           CMPA.L   #LAST+1,A0      ;Test for wrap-round
48 00000496 6606                   BNE.S    INPUT1          ;IF not end THEN skip reposition
49 00000498 207C00001000           MOVEA.L  #START,A0       ;Reposition input pointer
50 0000049E 23C80000100A INPUT1:   MOVE.L   A0,IN_POINT     ;Save updated pointer
51 000004A4 0C7900080000           CMPI.W   #MAX,COUNT      ;Is buffer full?
            1008
52 000004AC 6708                   BEQ.S    INPUT2          ;IF full THEN deal with overflow
```

```
53 000004AE 527900001008            ADDQ.W   #1,COUNT        ;ELSE increment character count
54 000004B4 4E75                    RTS                      ;and return
55 000004B6 08C0001F     INPUT2:    BSET     #31,D0          ;Set overflow flag
56 000004BA 20790000100E            MOVEA.L  OUT_POINT,A0    ;Get output pointer
57 000004C0 41E80001                LEA      1(A0),A0        ;Increment Out_pointer
58 000004C4 B1FC00001008            CMPA.L   #LAST+1,A0      ;Test for wrap-round
59 000004CA 6606                    BNE.S    INPUT3          ;IF not wrap-round THEN skip fix
60 000004CC 207C00001000            MOVEA.L  #START,A0       ;ELSE wrap-round Out_pointer
61 000004D2 23C80000100E INPUT3:    MOVE.L   A0,OUT_POINT    ;Update Out_pointer in memory
62 000004D8 4E75                    RTS                      ;and return
63                       *
64 000004DA 4A7900001008 OUTPUT:    TST.W    COUNT           ;Examine state of buffer
65 000004E0 6608                    BNE.S    OUTPUT1         ;IF buffer not empty output char
66 000004E2 4200                    CLR.B    D0              ;ELSE return null output
67 000004E4 08C0001F                BSET     #31,D0          ;set underflow flag
68 000004E8 4E75                    RTS                      ;and exit
69 000004EA 047900010000 OUTPUT1:   SUBI.W   #1,COUNT        ;Decrement COUNT for removal
            1008
70 000004F2 20790000100E            MOVEA.L  OUT_POINT,A0    ;Point to next char to be output
71 000004F8 1018                    MOVE.B   (A0)+,D0        ;Get character and update pointer
72 000004FA B1FC00001008            CMPA.L   #LAST+1,A0      ;Test for wrap-round
73 00000500 6606                    BNE.S    OUTPUT2         ;IF not wrap-round THEN exit
74 00000502 207C00001000            MOVEA.L  #START,A0       ;ELSE wrap-round Out_pointer
75 00000508 23C80000100E OUTPUT2:   MOVE.L   A0,OUT_POINT    ;Restore Out_pointer in memory
76 0000050E 4E75                    RTS
77                       *
78 00000510 103C0005     GETCHAR:   MOVE.B   #5,D0           ;Read an ASCII character into D1
79 00000514 4E4F                    TRAP     #15
80 00000516 4E75                    RTS
81 00000518 103C0006     PUTCHAR:   MOVE.B   #6,D0           ;Print the ASCII character in D1
82 0000051C 4E4F                    TRAP     #15
83 0000051E 4E75                    RTS
84 00000520 123C000D     NEWLINE:   MOVE.B   #$0D,D1         ;New Line
85 00000524 61F2                    BSR      PUTCHAR
86 00000526 123C000A                MOVE.B   #$0A,D1
87 0000052A 60EC                    BRA      PUTCHAR
88 0000052C 48E7FFFE     DISPLAY:   MOVEM.L  A0-A6/D0-D7,-(A7) ;Display the buffer contents
89 00000530 61EE                    BSR      NEWLINE
90 00000532 43F900001000            LEA      BUFFER,A1
91 00000538 3E3C0007                MOVE.W   #7,D7
92 0000053C 1219         DIS1:      MOVE.B   (A1)+,D1
93 0000053E 61D8                    BSR      PUTCHAR         ;Print a character
94 00000540 123C0020                MOVE.B   #$20,D1         ;and a space
95 00000544 51CFFFF6                DBRA     D7,DIS1
96 00000548 4DF900001037            LEA      COUNTER,A6      ;Print header before current count
97 0000054E 61000016                BSR      PRINT
```

```
 98 00000552 323900001008          MOVE.W    COUNT,D1         ;Display count
 99 00000558 06010030              ADD.B     #$30,D1
100 0000055C 61BA                  BSR       PUTCHAR
101 0000055E 61C0                  BSR       NEWLINE
102 00000560 4CDF7FFF              MOVEM.L   (A7)+,A0-A6/D0-D7
103 00000564 4E75                  RTS
104                       *
105 00000566 48E7FFFC     PRINT:   MOVEM.L   A0-A5/D0-D7,-(A7) ;Print the string pointed at by A6
106 0000056A 121E         PRINT1:  MOVE.B    (A6)+,D1
107 0000056C 6700000A              BEQ       PRINT2
108 00000570 103C0006              MOVE.B    #6,D0
109 00000574 4E4F                  TRAP      #15
110 00000576 60F2                  BRA       PRINT1
111 00000578 4CDF3FFF     PRINT2:  MOVEM.L   (A7)+,A0-A5/D0-D7
112 0000057C 4E75                  RTS
113                       *
114 00001000                       ORG       $1000
115          00001000     START:   EQU       *
116          00001007     LAST:    EQU       *+7
117          00000008     MAX:     EQU       8
118 00001000 00000008     BUFFER:  DS.B      8
119 00001008 00000002     COUNT:   DS.W      1
120 0000100A 00000004     IN_POINT: DS.L     1
121 0000100E 00000004     OUT_POINT: DS.L     1
122 00001012 43697263756C HEADER:  DC.B      'Circular Buffer ',0
             617220427566
             6665722000
123 00001023 436861726163 OUT:     DC.B      'Character output = ',0
             746572206F75
             747075742030
             2000
124 00001037 2020436F756E COUNTER: DC.B      '  Count = ',0
             74203D2000
125          00000400              END       $400

Lines: 125, Errors: 0, Warnings: 0.
```

A Four-function Calculator

For our next example we design a primitive four-function calculator. This calculator performs operations of the form X@Y=, where X and Y are both positive integer decimal numbers and @ represents one of the four operators: +, -, /, or *. The calculator must be able to input positive integers in the range 0 to +999. That is, it should be able to operate on numbers of variable length (e.g., 123+345, 12+123, 1+34, 234+23, etc.). The division operation need provide only an integer quotient (e.g., 15/2 = 7). The only two external procedures available for use by the calculator are: `GetChar` and `PutChar`. `GetChar` receives an ASCII-encoded character from the keyboard and puts it into data register D1.B, and `PutChar` displays the ASCII-encoded character in D1.B on the display device.

Remember that the purpose of this example is to provide a vehicle to take us through all the stages of problem solving, from the initial algorithm to the final assembly language code. You should appreciate that the solution we present is neither the only solution nor the optimum solution. In particular, we do not present a complete solution, since some loose ends remain untied. Finally, there is a flaw in the program. We will discuss the program's limitations shortly.

The Pseudocode

The first step is to express the problem at its highest level of abstraction in pseudocode.

```
Module Calculator
        Get input
        Perform calculation
        Print result
End Calculator
```

The calculator requires three fundamental operations: input the string to be evaluated, process the input, and output the result. The next step is to think about the problem a little before expanding the level 1 pseudocode. We have to ask ourselves: 'what are we really trying to do?' and 'are there any problems?'.

The input required by the program is a single string of ASCII-encoded characters in the form `X@Y=`. For example, `123+84=` is a valid input. The string consists of four components: number, operator, number, =. We initially have to decide between two ways of processing the input. The first is to input the entire string, store it in memory, and then process it. The second is to process (i.e., evaluate) the string as it is entered character-by-character. If we input the whole string before analyzing and processing it, we can easily edit it. For example, if we type `123+12` and then realize that the operator is wrong, we can type three backspaces and retype the string as `123-12`. Treating the input as a string means that no processing is performed until the '=' terminator has been entered. This approach requires memory storage to hold the input string until it is evaluated.

On the other hand, processing the input character-by-character as it is entered is very easy, since you can immediately act on the input without first storing it. Of course, processing the input character-by-character means that input errors cannot readily be corrected. For example, if you type 123 instead of 132, it's too late to do anything about it. In this current example, we are going to process the input as it is entered to keep the solution as simple as possible.

The specification of the problem was incomplete, as it has nothing to say about errors in the data. Any computer program that receives input must take into account the possibility of erroneous data. In this case we should look for invalid numbers and operators (e.g., strings of the form `12Z+23=` or `123!345=`).

A particular difficulty posed by this problem is the nature of the input. The number of digits in each operand is a variable in the range one to three. Consequently, the end of a sequence of digits can be identified only indirectly when a non-valid digit is input. For example, the first operand in the expression 5+6= is 5, while the first operand in the expression 15+123= is 15. In each case, the non-numeric character, +, terminates the string of digits that make up the operand.

Having thought about the input, the next step is to consider the way in which it is processed. At this stage only one decision need be considered: whether to perform the operations in BCD arithmetic or in binary arithmetic. Binary arithmetic is easier to perform on a 68000 than BCD arithmetic, but it does mean that the decimal input must be first converted into binary form prior to processing, and the result of the calculation must be converted to decimal form before it is displayed. We will use binary arithmetic to avoid writing decimal multiplication and division routines.

Finally, we have to think about the output. Since the result is to be displayed as a *decimal* integer, the internal binary representation must be converted to decimal form before it can be displayed. Consideration must also be given to the handling of invalid or out-of-range results. Most calculators simply display an E for an erroneous result. The program presented below does not deal with an erroneous result — that is left as an example for the reader.

These considerations can be taken into account when we write down the second-level pseudocode for the calculator. In the following example we expand the pseudocode to include comments. Any text of the form `{this is a comment}` is to be regarded as a comment. Since procedures need parameters, we indicate these in the conventional way. For example, the procedure `GetNumber(Number, LastChar)` is called and returns the number entered from the keyboard and the last character entered.

```
Module Calculator
       GetNumber(Number, LastChar)
       Operand1 := Number {Save first number}
       Operator := LastChar
       GetNumber(Number, LastChar)
       {Perform calculation}
       CASE LastChar OF:
                        +: Perform addition
                        -: Perform subtraction
                        *: Perform multiplication
                        /: Perform division
       END_CASE
       {Print result}
       Convert binary result to decimal form
       Print result
End Calculator
```

```
GetNumber(Number, LastChar)
        Number := 0
        REPEAT
              GetChar(Char)
              IF Char = valid digit THEN
                                  Number := Number*10 + (Char - $30)
              END_IF
        UNTIL Char ≠ valid digit
        LastChar := Char {i.e., the non-digit entered last}
End GetNumber
```

Converting a string of decimal characters into a binary number is easy. As each ASCII-encoded character is received and validated, 30_{16} is subtracted to convert the 8-bit ASCII character into a 4-bit value in the range 0 to 9. This value is added to ten times the running total (i.e., the variable `Number` in the above pseudocode). At this stage we have described the sequence of actions necessary to implement the calculator, but have not looked at their detailed implementation.

In the third-level pseudocode, we have to describe each action in sufficient detail to make its assembly language encoding fairly effortless. Most of the detail is concerned with converting between decimal and binary values, and vice versa. A binary result is converted into decimal form by dividing the result by 10 to obtain a quotient and a remainder. The remainder is in the range 0 to 9 and is converted to an ASCII digit and printed. The quotient is then divided by ten and the process continued. For example, if the binary value is 256 (i.e., 100000000_2), dividing it by ten yields 25 (i.e., 11001_2) and a remainder 6 (i.e., 0110_2).

However, you cannot print the successive digits in the order they are calculated (otherwise, the result 1345 would be displayed as 5431). The order of the digits to be printed must be reversed. We can do this by pushing them onto a stack or by shifting them along a register.

```
Module Calculator
        GetNumber(Number, LastChar)
        Operand1 := Number {Save first number}
        Operator := LastChar
        GetNumber(Number, LastChar)
        {Perform calculation}
        CASE LastChar OF
                        +: Perform addition
                        -: Perform subtraction
                        *: Perform multiplication
                        /: Perform division
        {The arithmetic yields a binary value "Result"}
        END_CASE
        {Print result}
        ResultString := 0 {Used to accumulate the result}
        FOR I = 1 to 8 {eight 4-bit digits in a longword}
```

```
            Digit := Result mod 10 {find the remainder}
            Add Digit to ResultString
            Result := Result/10
            Shift ResultString 4 places left
            ShiftString := ShiftString + Digit
        END_FOR
        FOR I = 1 to 8
            Digit := ShiftString
            Mask Digit to 4 bits
            Digit := Digit+$30 {convert to ASCII in range 0 to 9}
            Print Digit
            Shift ShiftString 4 places right
        END_FOR
End Module

GetNumber
        Number := 0
        REPEAT
           GetChar(Char)
           IF Char = valid digit THEN
                              Number := Number*10 + (Char - $30)
           END_IF
        UNTIL Char ≠ valid digit
        LastChar := Char {i.e., the non-digit entered last}
End GetNumber
```

We need to say a few words concerning the way in which the result is printed. Since a longword can hold eight 4-bit BCD digits, we perform eight cycles of binary to BCD conversion. If you divide an integer by ten, the remainder is the least-significant digit. As each BCD digit is obtained, it is shifted into a string (`ShiftString`). Note that this accumulates the digits *backwards* — the numerical value 1234 would be stored as 4321. A second `FOR` loop takes each of the BCD digits from the string and converts them to ASCII form before printing them. Since this prints the least-significant digit first, the digits are printed in the correct sequence. We can now translate the level-three pseudocode into 68000 assembly language.

To simplify testing, the calculator program is written as a subroutine and called by a simple repetitive test program. After each call, a new line and carriage return is printed. If the last character entered is a '!', the program is terminated.

```
*       This section of the program provides a framework within
*       which we can test the module "calc".
*
        ORG     $400            Start of the program
        LEA     $1000,A7        Set up the stack pointer
Repeat  BSR     Calc            Call the calculator
        MOVE.B  #$0D,D1         Move to a new line
```

```
        BSR     PutChar
        MOVE.B  #$0A,D1
        BSR     PutChar
        CMP.B   #'!',D5             Test for last character = "!"
        BNE     Repeat              REPEAT UNTIL  "!" found
        STOP    #$2700
*
*   Calculator register usage:
*   D0: Used by input/output routines GetChar/PutChar
*   D1: Used by input/output routines GetChar/PutChar
*   D2: Holds number from routine GetNum
*   D3: Holds the first operand
*   D4: Holds the second operand
*   D5: Holds the operator (+, -, *, /)
*   D6: Used as a counter in output routines
*   D7: Used to accumulate digits during the output
*
Calc    BSR     GetNum              Get the first number and terminator
        MOVE.L  D2,D3               Save the number in D3
        MOVE.B  D4,D5               Save the operator in D5
        BSR     GetNum              Get the second number in D2
        CMP.B   #'+',D5             IF operator = plus THEN add operands
        BEQ     Plus
        CMP.B   #'-',D5             IF operator = minus THEN subtract operands
        BEQ     Minus
        CMP.B   #'*',D5             IF operator = multiply THEN multiply operands
        BEQ     Mult
        CMP.B   #'/',D5             IF operator = divide THEN divide operands
        BEQ     Div
        BRA     EXIT                Exit on non-valid operator
*
*                                   CASE Operator OF:
Plus    ADD.L   D2,D3                   Add:      Result := operand1+operand2
        BRA     Output
Minus   SUB.L   D2,D3                   Subtract: Result := operand1-operand2
        BRA     Output
Mult    MULU    D2,D3                   Multiply: Result := operand1*operand2
        BRA     Output
Div     DIVU    D2,D3                   Divide:   Result := operand1/operand2
        AND.L   #$0000FFFF,D3                     (delete the remainder)
        BRA     Output              END_CASE
*
Output  CLR.L   D7                  Clear output (i.e., ShiftString := 0)
        MOVE.W  #7,D6               FOR I = 1 TO 8
Output1 LSL.L   #4,D7                 Shift ShiftString left four places
        DIVU    #10,D3                Result/10 to get remainder in D3(16:31)
```

```
         SWAP    D3                   Get remainder in D3(0:15)
         MOVE.B  D3,D1                Copy remainder to D1 (BCD in range 0 - 9)
         SWAP    D3                   Restore D3 (i.e., quotient to D3(0:15))
         AND.L   #$0000FFFF,D3        Mask out the remainder
         ADD.B   D1,D7                Copy the digit to ShiftString
         DBRA    D6,Output1         END_FOR
*
         MOVE.W  #7,D6              FOR I = 1 to 8
Output2  CLR.B   D1                   Clear the 'character holder'
         MOVE.B  D7,D1                Move 2 BCD chars from ShiftString to D1
         AND.B   #$0F,D1              Mask to D1 to a single character
         ADD.B   #$30,D1              Convert BCD value to ASCII form
         BSR     PutChar              Print a character
         LSR.L   #4,D7                Shift ShiftString four places right
         DBRA    D6,Output2         END_FOR
         RTS
*
EXIT     RTS
*
* Get a number in D2 and an operator in D4
*
GetNum   CLR.L   D2                 Clear Number
GetNum1  BSR     GetChar            REPEAT Get an ASCII-encoded character
         CMP.B   #$30,D1              IF Char < $30 OR Char > $30
         BMI     NotNum                  THEN EXIT
         CMP.B   #$39,D1
         BGT     NotNum
         AND.L   #$000000FF,D1        Mask character to 8 bits
         SUB.B   #$30,D1              Convert ASCII to 4-bit binary
         MULU    #10,D2               Number := Number + 10 * Digit
         ADD.L   D1,D2
         BRA     GetNum1              Continue
NotNum   MOVE.B  D1,D4              Save last character (terminator) in D4
         RTS
*
GetChar  MOVE.B  #5,D0              Get an ASCII character in D1.B
         TRAP    #15
         RTS
PutChar  MOVE.B  #6,D0              Print the ASCII character in D1.B
         TRAP    #15
         RTS
*
         END     $400
```

The solution to the calculator problem is limited in three ways. First, the input is not validated (i.e., it does not deal with errors like 123Z46). Second, leading zeros are printed (e.g., the input string 2+2= results in the output 00000004). Finally, it does not handle negative numbers.

We can now use the cross-assembler to assemble the above program to produce suitable 68000 code. This stage detects any errors of syntax.

```
Source file: CALC.X68
Assembled on: 91-03-16 at: 18:37:40
        by: X68K PC-1.9 Copyright (c) University of Teesside 1989,92
Defaults: ORG $0/FORMAT/OPT A,BRL,CEX,CL,FRL,MC,MD,NOMEX,NOPCO

  1 00000400                   ORG     $400          ;Start of the program
  2 00000400 4FF81000          LEA     $1000,A7      ;Set up the stack pointer
  3 00000404 6100001C  REPEAT: BSR     CALC          ;Call the calculator
  4 00000408 123C000D          MOVE.B  #$0D,D1       ;Move to a new line
  5 0000040C 610000D2          BSR     PUTCHAR
  6 00000410 123C000A          MOVE.B  #$0A,D1
  7 00000414 610000CA          BSR     PUTCHAR
  8 00000418 0C050021          CMP.B   #'!',D5       ;Test for last character = "!"
  9 0000041C 66E6              BNE     REPEAT        ;REPEAT UNTIL  "!" found
 10 0000041E 4E722700          STOP    #$2700
 11                    *
 12                    * Register usage:
 13                    *  D0: Used by input/output routines GetChar/PutChar
 14                    *  D1: Used by input/output routines GetChar/PutChar
 15                    *  D2: Holds number from routine GetNum
 16                    *  D3: Holds last non-BCD character from GetNum (and result)
 17                    *  D4: Holds the second operand
 18                    *  D5: Holds the operator (+, -, *, /)
 19                    *  D6: Used as a counter in output routines
 20                    *  D7: ShiftString (used to accumulate digits during output)
 21                    *
 22 00000422 61000088  CALC:   BSR     GETNUM        ;Get first number
 23 00000426 2602              MOVE.L  D2,D3         ;Save it in D3
 24 00000428 1A04              MOVE.B  D4,D5         ;Save the operator in D5
 25 0000042A 61000080          BSR     GETNUM        ;Get second number in D2
 26 0000042E 0C05002B          CMP.B   #'+',D5       ;IF operator = plus THEN add operands
 27 00000432 6700001E          BEQ     PLUS
 28 00000436 0C05002D          CMP.B   #'-',D5       ;IF operator = minus THEN subtract operands
 29 0000043A 6700001C          BEQ     MINUS
 30 0000043E 0C05002A          CMP.B   #'*',D5       ;IF operator = * THEN multiply operands
 31 00000442 6700001A          BEQ     MULT
 32 00000446 0C05002F          CMP.B   #'/',D5       ;IF operator = divide THEN divide operands
 33 0000044A 67000018          BEQ     DIV
 34 0000044E 6000005A          BRA     EXIT          ;Exit on non-valid operator
 35                    *
 36                    *                        CASE of:
```

```
37 00000452 D682          PLUS:     ADD.L   D2,D3            ;Add: Result := operand1+operand2
38 00000454 6000001A                BRA     OUTPUT
39 00000458 9682          MINUS:    SUB.L   D2,D3            ;Subtract: Result := operand1-operand2
40 0000045A 60000014                BRA     OUTPUT
41 0000045E C6C2          MULT:     MULU    D2,D3            ;Multiply: Result := operand1*operand2
42 00000460 6000000E                BRA     OUTPUT
43 00000464 86C2          DIV:      DIVU    D2,D3            ;Divide: Result := operand1/operand2
44 00000466 02830000FFFF            AND.L   #$0000FFFF,D3    ;(delete the remainder)
45 0000046C 60000002                BRA     OUTPUT           ;END_CASE
46                        *
47 00000470 4287          OUTPUT:   CLR.L   D7               ;Clear output (i.e., ShiftString := 0)
48 00000472 3C3C0007                MOVE.W  #7,D6            ;FOR I = 1 TO 8
49 00000476 E98F          OUTPUT1:  LSL.L   #4,D7            ;Shift ShiftString left four places
50 00000478 86FC000A                DIVU    #10,D3           ;Result/10 to get remainder in D3(16:31)
51 0000047C 4843                    SWAP    D3               ;Get remainder in D3(0:15)
52 0000047E 1203                    MOVE.B  D3,D1            ;Copy remainder to D1 (BCD 0 - 9)
53 00000480 4843                    SWAP    D3               ;Restore D3 (i.e., quotient to D3(0:15))
54 00000482 02830000FFFF            AND.L   #$0000FFFF,D3    ;Mask out the remainder
55 00000488 DE01                    ADD.B   D1,D7            ;Copy the digit to ShiftString
56 0000048A 51CEFFEA                DBRA    D6,OUTPUT1       ;END_FOR
57                        *
58 0000048E 3C3C0007                MOVE.W  #7,D6            ;FOR I = 1 to 8
59 00000492 4201          OUTPUT2:  CLR.B   D1               ;Clear the 'character holder'
60 00000494 1207                    MOVE.B  D7,D1            ;Move 2 BCD chars from ShiftString to D1
61 00000496 0201000F                AND.B   #$0F,D1          ;Mask to D1 to a single character
62 0000049A 06010030                ADD.B   #$30,D1          ;Convert BCD value to ASCII form
63 0000049E 61000040                BSR     PUTCHAR          ;Print a character
64 000004A2 E88F                    LSR.L   #4,D7            ;Shift ShiftString four places right
65 000004A4 51CEFFEC                DBRA    D6,OUTPUT2       ;END_FOR
66 000004A8 4E75                    RTS
67                        *
68 000004AA 4E75          EXIT:     RTS

69                        *
70                        * Get a number in D2 and an operator in D4
71                        *
72 000004AC 4282          GETNUM:   CLR.L   D2               ;Clear Number
73 000004AE 61000028      GETNUM1:  BSR     GETCHAR          ;REPEAT Get an ASCII-encoded character
74 000004B2 0C010030                CMP.B   #$30,D1          ;IF Char < $30 OR Char > $30
75 000004B6 6B00001C                BMI     NOTNUM           ;THEN EXIT
76 000004BA 0C010039                CMP.B   #$39,D1
77 000004BE 6E000014                BGT     NOTNUM
78 000004C2 0281000000FF            AND.L   #$000000FF,D1    ;Mask character to 8 bits
79 000004C8 04010030                SUB.B   #$30,D1          ;Convert ASCII to 4-bit binary
80 000004CC C4FC000A                MULU    #10,D2           ;Number := Number + 10 * Digit
81 000004D0 D481                    ADD.L   D1,D2
```

```
82 000004D2 60DA              BRA     GETNUM1      ;Continue
83 000004D4 1801     NOTNUM:  MOVE.B  D1,D4        ;Save last character (terminator) in D4
84 000004D6 4E75              RTS
85                   *
86 000004D8 103C0005 GETCHAR: MOVE.B  #5,D0        ;Get an ASCII character in D1.B
87 000004DC 4E4F              TRAP    #15
88 000004DE 4E75              RTS
89                   *
90 000004E0 103C0006 PUTCHAR: MOVE.B  #6,D0        ;Print the ASCII character in D1.B
91 000004E4 4E4F              TRAP    #15
92 000004E6 4E75              RTS
93                   *
94          00000400          END     $400

Lines: 94, Errors: 0, Warnings: 0.
```

6.5 Pseudocode Design: Using PDL to Design a Command Line Interpreter (CLI)

A more extended example of the application of a PDL to the construction of assembly language programs is provided by the design of a *command line interpreter* (CLI). A CLI is a program that collects a line of text from a processor's keyboard, parses it to extract a command, and then executes the command. The CLI is found in many text editors, word processors, monitors, and even interpreted languages.

The particular CLI to be designed here is intended to form part of a microprocessor system's monitor. Such a monitor is part of many low-cost, single-board, educational computers and permits you to enter an assembly language program, preset memory locations, and then to execute the program. Frequently, a monitor provides debugging aids, allowing you to follow the execution of your program by displaying intermediate results on a display terminal. The specifications of the CLI are given below.

- Text is entered into a 64-character (i.e., 64-byte) buffer, and is terminated by a carriage return. The longest string that may be input is 63 characters plus a carriage return. Note that the carriage return is to be stored in the buffer.
- In addition to the carriage return terminator, the CLI is to recognize the back space and the abort (ASCII control A) characters. A back space deletes the last character received. If the line buffer is already empty, inputting one or more back spaces has no effect. An abort character forces an immediate exit from the CLI routine with the carry-bit of the CCR set to a logical one. Otherwise, a return from the CLI is made only after a carriage return has

been received. In this case, the state of the carry-bit of the CCR is determined by the interpreter part of the routine.

- Should more than 63 alphanumeric characters be received, representing a buffer overflow condition, the message *Buffer full — reenter command* is displayed on a new line, a further new line command issued, and the input procedure initiated as if the routine were being invoked for the first time.
- Once a command line has been successfully input, it is interpreted. The first word in the input buffer is matched against the entries in a command table. If a successful match is found, the appropriate jump (transfer) address of the command is loaded into address register A0, the carry-bit of the CCR cleared, and a return from the CLI made. If a match cannot be found, a return from the CLI is executed with the carry-bit of the CCR set. In this case, the contents of address register A0 are undefined.
- The CLI may assume the existence of a subroutine `GETCHAR` that, when called, returns with an 8-bit ASCII character in the lower byte of data register D0. It is assumed that `GETCHAR` takes care of any character echo. The subroutine `NEWLINE` moves the cursor to the left-hand column of the next line on the display. The subroutine `PSTRING` displays the text string composed of ASCII characters occupying consecutive byte locations and terminated by a null (zero byte) pointed at by address register A4. These three subroutines have no effect on any register not mentioned above.
- No register other than A0 and the CCR must be affected by the CLI following a return from the CLI.

The structure of the command table is defined in Figure 6.12, and Figure 6.13 provides an example of a typical data structure. Each entry has four fields:

a. a byte that indicates the length of the command name in the table

b. a byte that indicates the minimum number of characters which must be matched — for example, MEM, MEMO, MEMOR, or MEMORY may be typed to invoke the MEMORY command

c. the command string itself, whose length is given by entry (a)

d. the 4-byte transfer address.

Now that we've specified the problem, the next step is to translate it into a sufficiently elaborated PDL. The lowest level of the PDL should make the translation into 68000 assembly language possible without complex steps. That is, no single PDL statement should require translating into a stream of assembly language requiring embedded branch instructions.

PDL Level 1

The lowest level of PDL is rather trivial and is included here for the sake of completeness and to illustrate the process of top-down design.

Figure 6.12 The command table

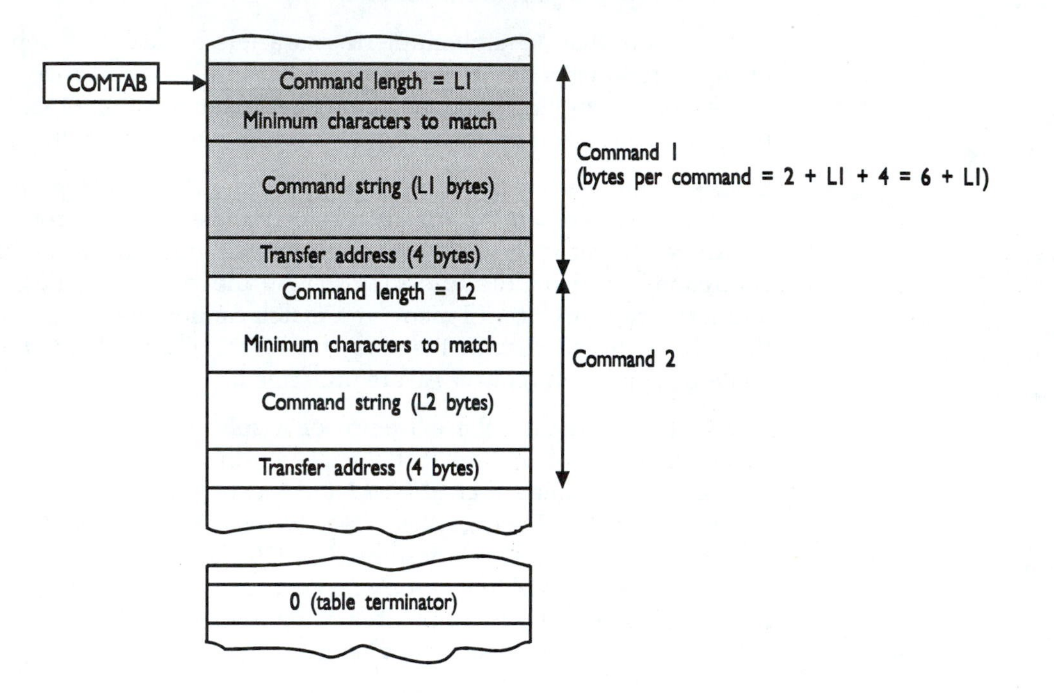


```
Module  CLI
        Initialize variables
        Assemble the input line
        Interpret the line of text
        Exit
End CLI
```

PDL Level 2

At Level 2 we cannot fully elaborate any part of the Level 1 design, as we do not, for example, yet know what variables must be initialized. We do know that registers must not be corrupted by this program, so we can make a note of this. Similarly, we can begin to consider the type of actions to be carried out by the major steps in Level 1 above.

```
Module  CLI
        Initialize
                    Save working registers
                    Initialize any variables
                    Clear buffer
```

```
        Assemble
                    REPEAT
                              Get character
                              Deal with special cases
                              Store character
                    UNTIL     character = carriage return
        Interpret
                    Match command with command table
                    IF successful THEN Extract transfer address
                                  ELSE Set error indicator
                    END_IF
        Exit
                    Restore working registers
End CLI
```

Figure 6.13 Example of a command table

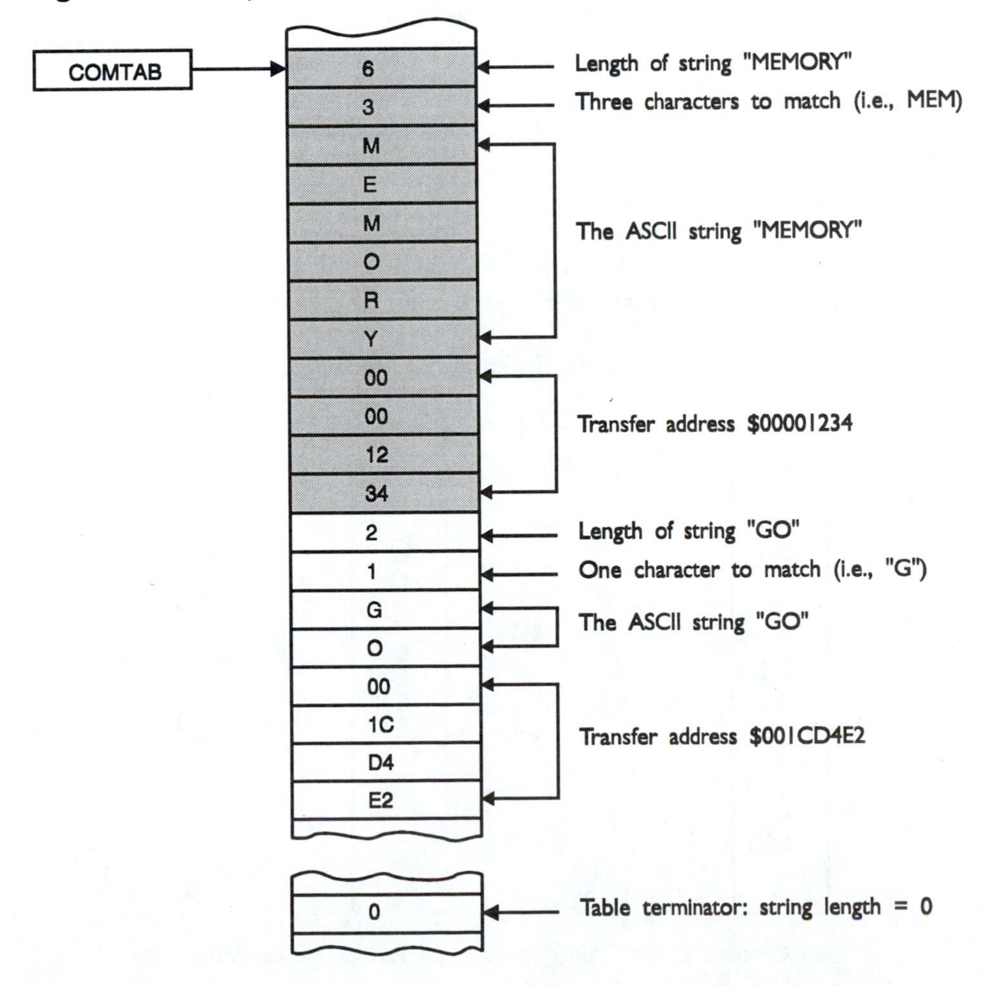

PDL Level 3

At this stage, we can begin to consider the details of the actions defined at Level 2. Perhaps another picture will help. Figure 6.14 shows the state of the system after a command `MEMORY 1234 1240` has been entered into the line buffer. The software is to compare the first string in the input buffer (i.e., `MEMORY`) with the entries in the command table. In this case, the command `MEMORY` is in the table and the software should return the address 00008764.

If we are lucky, this is the highest level of PDL necessary; otherwise, another iteration will be necessary. A little common sense tells us that this routine requires few variables other than pointers for the line buffer and for the table with which the match is made. However, there are enough variables to require some form of type declaration block as an aid to understanding the program.

As this example is more complex than the previous one in the sense that it requires a number of constants and variables, we can extend the PDL to include a declaration block. This block lists variables and constants, indicates the size of data items, and presets values as appropriate. An element of an array is indicated by square brackets — the ith element of the input buffer is indicated by `Buffer[I]`.

```
Initialize
             Array of bytes: Comtab
             Array of bytes: Buffer
             Longword:       Buffer_pointer
             Longword:       Table_pointer
             Longword:       Next_entry
             Longword:       Transfer_address
```

Figure 6.14 Comparing the line buffer with entries in the command table

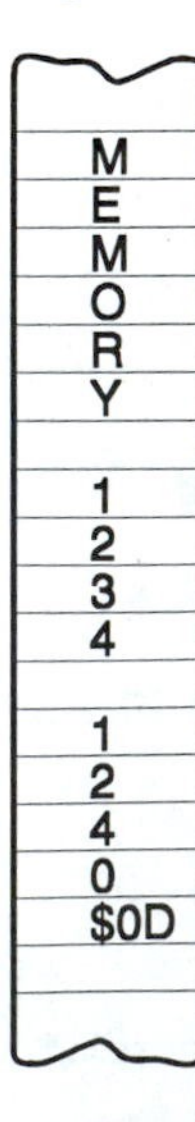

(a) Contents of the input buffer

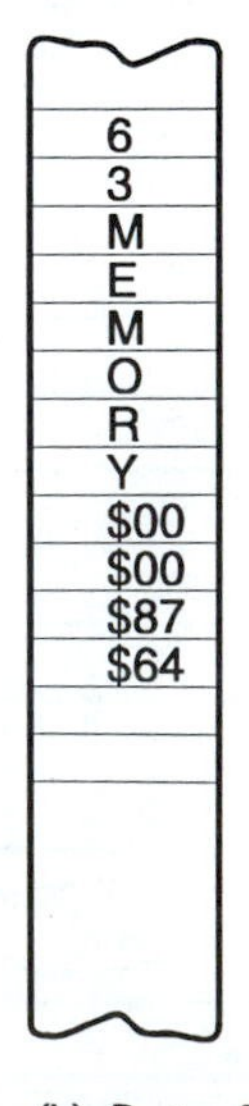

(b) Part of the command table

```
            Byte:           String_length
            Byte:           Chars_to_match
            Byte:           Buffer_length = 64
            Equate:         Buffer_start = Buffer
            Equate:         Buffer_end:=Buffer_start+Buffer_length-1
            Equate:         Table_address = Comtab
            Equate:         Back_space = $08
            Equate:         Abort = $01
            Equate:         Carriage_return = $0D
            Save working registers on system stack
            Clear_buffer
End Initialize

Clear_buffer
               FOR I := 0 TO Buffer_length - 1
                   Buffer[I] := 0
               END_FOR
End Clear_buffer

Assemble
         REPEAT
            IF Buffer_pointer = Buffer_end + 1 {Deal with buffer full}
               THEN BEGIN
                    Newline
                    Display error message
                    Newline
                    Buffer_pointer := Buffer_start
                    Clear_buffer
                    END
            END_IF
            Get character
            IF character = Abort THEN Set Error_flag
               Exit {leave the interpreter}
            END_IF
            IF character = Back_space
               THEN BEGIN
                    IF Buffer_pointer > Buffer_start
                       THEN Buffer_pointer := Buffer_pointer - 1
                    END_IF
                    END
               ELSE BEGIN
                    Store character
                    Buffer_pointer := Buffer_pointer + 1
                    END
            END_IF
         UNTIL character = carriage_return
End Assemble
```

```
Interpret
            Table_pointer := Table_address
            REPEAT
                Buffer_pointer := Buffer
                String_length := Comtab[Table_pointer]
                IF String_length = 0 THEN
                                          Set Fail
                                          Exit
                END_IF
                Chars_to_match := Comtab[Table_pointer + 1]
                Next_entry := Table_pointer + String_length + 6
                Table_pointer := Table_pointer + 2
                Match
                IF Success THEN
                           BEGIN
                           Transfer_address := Comtab[Next_entry-4]
                           Exit
                           END
                           ELSE Table_pointer := Next_entry
                END_IF
            END_REPEAT
End Interpret

Match
           REPEAT
              IF Buffer[Buffer_pointer] ≠ Comtab[Table_pointer]
                 THEN BEGIN
                      Set Fail
                      Exit Match
                      END
                 ELSE BEGIN
                      Table_pointer := Table_pointer + 1
                      Buffer_pointer := Buffer_pointer + 1
                      END
              END_IF
           Chars_to_match := Chars_to_match - 1
           UNTIL Chars_to_match = 0
           Set Success
End Match

Exit
           Restore registers from system stack
           Return
End Exit
```

Having reached this level of detail, we can readily translate the PDL into 68000 assembly language. As no fixed storage is necessary, apart from the line buffer, we can use the 68000's registers as the pointers and temporary storage.

```
BUFFER_LENGTH   EQU   64
BACK_SPACE      EQU   $08
ABORT           EQU   $01
CARRIAGE_RETURN EQU   $0D
BUFFER_START    EQU   $008000
BUFFER_END      EQU   BUFFER_START+BUFFER_LENGTH-1
COMTAB          EQU   $009000
*
* Registers A0  = Pointer to transfer address
*           A1  = Buffer_pointer
*           A2  = Table_pointer
*           A3  = Next_entry
*           A4  = Pointer to error message
*           D0  = Contains character from console
*           D1  = String_length
*           D2  = Chars_to_match
*
**********************************************************************
*
* INITIALIZE saves the working registers and clears the line buffer
* Note that EXIT restores the registers
*
            ORG     $001000               Program origin
INITIALIZE  MOVEM.L A1-A4/D0-D2,-(SP)     Save working registers
            BSR.S   CLEAR                 Clear the buffer
*                                         Fall through to ASSEMBLE
*
**********************************************************************
*
*  ASSEMBLE reads a line of text and stores it in the line buffer
*  A back space deletes the previous character and an abort forces
*  an exit from the CLI with the C flag set
*
ASSEMBLE    CMPA.L  #BUFFER_END+1,A1      Test for end of line buffer
            BNE.S   ASSEMBLE1             IF not full THEN continue
            BSR     NEWLINE               ELSE display error
            LEA     MESSAGE,A4            Point to message
            BSR     PSTRING               Print it
            BSR     NEWLINE
            BSR.S   CLEAR                 Clear buffer
ASSEMBLE1   BSR     GETCHAR               Input an ASCII character
            CMPI.B  #ABORT,D0             Test for abort
```

```
            BNE.S    ASSEMBLE2           IF not abort THEN skip exit
            ORI.W    #$01,CCR            ELSE set error flag
            BRA.S    EXIT                and leave CLI
ASSEMBLE2   CMPI.B   #BACK_SPACE,D0      Test for back space
            BNE.S    ASSEMBLE3           IF not back space THEN skip
            CMPA.L   #BUFFER_START,A1    Test for buffer empty
            BEQ.S    ASSEMBLE1           IF empty THEN get new char
            LEA      (-1,A1),A1          ELSE move back pointer
            BRA.S    ASSEMBLE1           and get new character
ASSEMBLE3   MOVE.B   D0,(A1)+            Store the input
            CMPI.B   #CARRIAGE_RETURN,D0 Test for terminator
            BNE.S    ASSEMBLE            IF not terminator repeat
*                                         ELSE fall through to interpret
***********************************************************************
*
*  INTERPRET takes the first word in the line buffer and tries
*  to match it with an entry in COMTAB
*
INTERPRET   LEA      COMTAB,A2           A2 points to command table
INTERPRET1  LEA      BUFFER_START,A1     A1 points to line buffer
            CLR.L    D1                  Clear all 32 bits of D1
            MOVE.B   (A2),D1             Get string length
            BNE.S    INTERPRET2          IF not zero THEN continue
            ORI.W    #$01,CCR            ELSE set error flag
            BRA.S    EXIT                and return from CLI
INTERPRET2  MOVE.B   (1,A2),D2           Get chars to match from table
            LEA      (6,A2,D1),A3        Get address of next entry
            LEA      (2,A2),A2           Point to string in COMTAB
            BSR.S    MATCH               Match it with line buffer
            BCC.S    INTERPRET3          IF C clear THEN success
            LEA      A3,A2               ELSE point to next entry
            BRA      INTERPRET1          Try again
INTERPRET3  LEA      (-4,A3),A0          Get transfer address pointer
            MOVEA.L (A0),A0              Get transfer address in COMTAB
*                                        Fall through to exit
***********************************************************************
*  EXIT is the only way out of CLI. Restore registers and exit with
*  C clear for success and C set for failure. If C = 0, the
*  transfer address is in A0
*
EXIT        MOVEM.L (SP)+,A1-A4/D0-D2   Restore registers
            RTS                          Return
*
CLEAR       MOVE.W   #BUFFER_LENGTH-1,D0 Clear the input buffer
            LEA      BUFFER_START,A1
CLEAR1      CLR.B    (A1)+
```

```
            DBF     CLEAR1,D0
            LEA     BUFFER_START,A1      Reset A1 to start of buffer
            RTS
*
**********************************************************************
*
*  MATCH compares two strings for equality. The number of characters
*  to match is in D2
*  A1 = Buffer_pointer, A2 = Table_pointer
*

MATCH       CMPM.B  (A1)+,(A2)+          Compare two characters
            BNE.S   MATCH1               IF not equal THEN return
            SUBQ.B  #1,D2                ELSE decrement match count
            BNE.S   MATCH                IF not zero test next pair
            ANDI.W  #$FE,CCR             ELSE success, clear C flag
            RTS                          and return
MATCH1      ORI.W   #$01,CCR             Set fail flag
            RTS
```

Summary

- *Top-down* design provides a method of constructing programs. You start by defining the major steps or operations that comprise the program. Then you expand or elaborate these actions.
- Top-down design enables you to separate *what* a program does from *how* it does it.
- Before you begin to design a program you should specify what exactly it is to do. If you omit this step, the program may not perform as you expected (especially when presented with erroneous or unusual data).
- *Modular design* goes hand-in-hand with *top-down* design. By turning actions into modules with a single input and a single exit point, you can design and test a program's functional parts independently of each other. Just as importantly, well designed modules can help prevent undesired interactions between the various elements of a program (e.g., the accidental use of a register for two purposes at different parts of a program).
- Programmers should always consider how their programs (and the modules of a program) are to be tested.
- Even *well designed* programs fail. You should attempt to anticipate possible failure modes and design the program to help recovery from them.

- *Structured programming* refers to a set of tools that makes it easier to construct programs. A structured program is composed of a sequence of actions that are executed one-by-one. These actions can be represented by *processes* or *modules*. Decisions are made by `IF...THEN...ELSE`, `DO...WHILE`, `REPEAT...UNTIL` and `FOR` loops. Perhaps the most important aspect of structured programming is the avoidance of spaghetti programming (i.e., unstructured jumps or gotos that make it impossible for a human to follow the flow of a program).
- Pseudocode offers the programmer an intermediate step between an algorithm and a low-level language.

Problems

1. Top-down programming is the *recommended* way of designing programs. Is top-down design conceptually obvious? Is it used in everyday activities? Are your vacations planned in a top-down fashion?

2. What are the advantages of top-down programming? Are there any disadvantages?

3. What is *spaghetti programming* and why is it so unpopular?

4. What constitutes a *good* program?

5. Describe the basic structures from which all programs can be constructed.

6. How do the concepts of *module coupling* and *module strength* affect the programmer?

7. How would you apply the concept of *testability* to the four-function calculator we described earlier in this chapter?

8. What is the difference between *black box* and *white box* testing?

9. Why do top-down design and structured programming make it apparently harder to implement recoverability?

10. If software doesn't rust, why must it be maintained?

11. In Conway's Game of Life, the universe consists of an array of two-dimensional cells (like a chess board). Each cell may be occupied or unoccupied, and each has eight immediate neighbors. The state of the system changes each T seconds, when the occupancy of each cell is recalculated. If a cell is currently occupied, it becomes empty in the next cycle if 0, 1, or 4 or more of

its immediate neighbors are occupied. If a cell is unoccupied, but exactly three of its neighbors are occupied, the cell becomes occupied in the next cycle. Write an assembly program to simulate Conway's life for a 64 x 64 universe. Take care with the edges of the universe.

12. Write a program to take a 6-digit BCD-encoded unsigned number in D0 and convert it into 32-bit IEEE floating-point format.

CHAPTER 7

Subroutines and Parameters

We introduced the subroutine in Chapter 3 and made extensive use of it in Chapter 4. In this chapter we focus our attention on two important aspects of subroutines. The first is the way in which a subroutine communicates with the program that called it. The second is the way in which a subroutine employs memory for the temporary storage of its data. Before covering these topics, we review the basic principles of subroutines.

7.1 Basic Principles of Subroutines

A subroutine is a piece of code that can be *called* from some point in a program to carry out a certain task. Although we used the expression *piece of code* to describe a subroutine, it is not just any bunch of consecutive instructions. A subroutine is a *coherent* sequence of instructions that carries out a well defined function. One of the reasons for employing subroutines is that a particular subroutine can be used many times in a program without the need to rewrite the same sequence of instructions over and over again. Programmers also employ subroutines to make a program more readable, as we shall soon see.

Let's look at a simple subroutine called `GetChar` whose function is to read a character from the keyboard and put it into data register D0. In this example, we have provided the assembler listing file that gives the addresses (i.e., locations) of the instructions in order to facilitate our explanation of the subroutine call and the return mechanisms.

Address Data   Label Instruction   Comment field

```
000FFA  41F900004000          LEA     TABLE,A0  A0 points to destination of data
001000  61000206      NextChr BSR     GetChar   REPEAT Read a character
001004  10C0                  MOVE.B  D0,(A0)+         Store it in memory
001006  0C00000D              CMP.B   #$0D,D0   UNTIL  Character = carriage return
00100A  66F4                  BNE     NextChr
                              .
                              .
                              .
```

```
001102  61000104              BSR     GetChar   Read character and quit if "Q"
001106  0C000051              CMP.B   #'Q',D0
00110A  67000EF4              BEQ     QUIT
                              .
                              .
001208  123900008000 GetChar  MOVE.B  ACIAC,D1  Read ACIA status
00120E  08010001              BTST    #1,D1
001212  66F4                  BNE     GetChar   UNTIL ACIA ready
001214  103900008002          MOVE.B  ACIAD,D0  Get character into D0
00121A  4E75                  RTS               Return
```

In this example we call the subroutine GetChar *twice*. The first time is from address 001000_{16} in the loop that stores successive characters in a table. The second time is from address 001102_{16} when we read a single character and test it to see whether it is a letter Q.

The way in which the branch or jump to the subroutine is executed is very simple: the 68000's program counter is loaded with the starting address of the subroutine. When the assembler encounters the line NextChr BSR GetChar at address 001000_{16}, the symbolic location GetChar is translated into the *difference* between the address of the BSR instruction plus two and the address of the actual subroutine. This difference is the offset used by the BSR instruction and is computed as $1208_{16} - 1002_{16} = 206_{16}$. The instruction is, effectively, BSR $206. Note that the value of the PC used in the offset calculation is the address of the BSR *plus 2*, because the PC is automatically incremented by 2 as soon as the BSR is read. The following code repeats the above fragment of a program with the branch marked.

```
000FFA  41F900004000          LEA      TABLE,A0
001000  61000206     NextChr  BSR      GetChar    ─┐
001004  10C0                  MOVE.B   D0,(A0)+    │
001006  0C00000D              CMP.B    #$0D,D0     │   Carriage Return
00100A  66F4                  BNE      NextChr     │
                              .                    │1208-(1000+2)=206
001102  61000104              BSR      Getchar     │
001106  0C000051              CMP.B    #'Q',D0     │
00110A  67000EF4              BEQ      QUIT        │
                              .                    │
001208  123900008000 GetChar  MOVE.B   ACIAC,D0   ─┘
```

When the program is executed, the 68000 reads the offset from the instruction at location 001000_{16} (i.e., 206_{16}), and adds it to the contents of the program counter (i.e., 1002_{16}) to get 1208_{16}, which is loaded into the program counter to force a jump to the subroutine.

When the assembler next encounters the BSR GetChar instruction at address 001102_{16}, the computed offset is different (i.e., $1208_{16} - 1104_{16}$) and the relative branch offset becomes $104. By the way, the second subroutine call uses the long form of the BSR instruction which takes a 16-bit signed offset. We could have written `BSR.S GetChar` which would have taken an 8-bit offset and shortened the code by one word.

A more interesting aspect of subroutines is the way in which a return is made automatically to the appropriate calling point. Whenever a subroutine call is made, the return address (i.e., the location of the next instruction after the call), is pushed onto the stack pointed at by A7. Figure 7.1 demonstrates the action of the first subroutine call in which the return address 1004_{16} is pushed onto the stack — assumed that the stack pointer is initially pointing at address 4008_{16}.

Figure 7.1 Pushing the subroutine return address on the stack

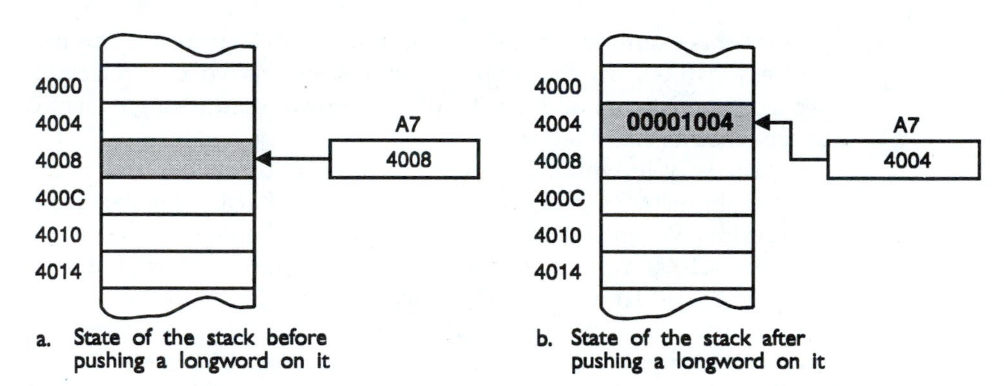

We can define the effect of the instruction `BSR d8`, where d8 is an 8-bit offset, in RTL as:

```
[A7]         ← [A7] - 4     Pre-decrement the stack pointer
[M([A7])]    ← [PC]         Push the program counter on the stack
[PC]         ← [PC] + d8    Load the program counter with the target address
```

When the subroutine has been executed, an `RTS` instruction pulls the longword off the top of the stack and deposits it in the program counter to effect the return. The action of the `RTS` instruction in RTL is:

```
[PC] ← [M([A7])]          Pull return address off the stack
[A7] ← [A7] + 4           Post-increment the stack pointer
```

You can perform a return from subroutine without the use of an `RTS` instruction by means of an indirect `JMP` instruction as follows.

```
MOVEA.L (A7)+,A0    A0 now holds the return address
JMP     (A0)        Jump to the return address
```

The `MOVEA.L (A7)+,A0` instruction reads the contents of memory pointed at by the stack pointer and deposits it in A0. Since the stack pointer points at the return address, the return address is loaded into A0. The post-incrementing addressing mode steps the stack pointer past the return address and cleans up the stack. The second instruction, `JMP (A0)`, forces a jump to the location whose address is in address register A0. As this is the subroutine return address, a return from subroutine is executed.

Nested Subroutines

If subroutines were called one at a time from a main program and a return made before the next subroutine was called, we would not need to use the stack to store the return address. Any register or fixed location could be used to save the appropriate return address. In practice, a stack is needed to manage subroutine return addresses because subroutines may be *nested*. That is, a main program may call a subroutine and the subroutine may itself call a subroutine, and so on. Figure 7.2 illustrates nested subroutines, in which subroutines B, C, and D are completely enclosed by the subroutine that calls them. For example, from subroutine C you can return only to subroutine B and not to subroutine A or to the main program. We can represent the nesting in Figure 7.2 by the following code:

Figure 7.2 Nested subroutines

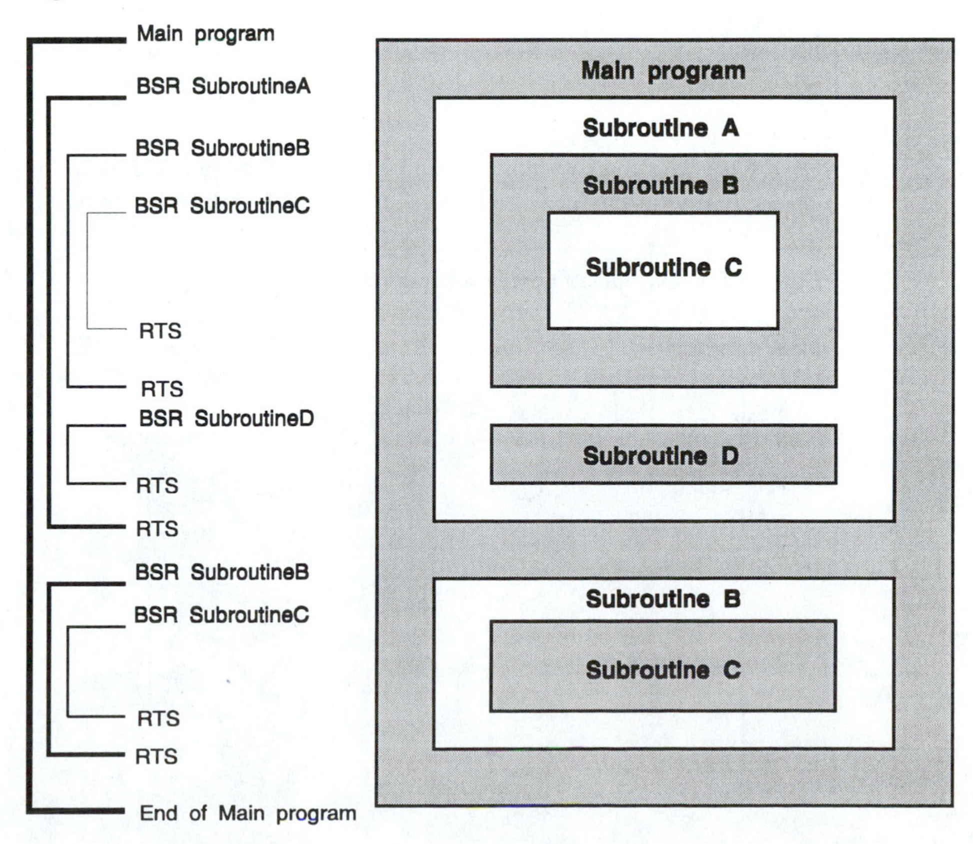

```
MainProgram  .
             BSR SubroutineA
             .
             BSR SubroutineB
             .
             END
SubroutineA  .
             BSR SubroutineB
             .
             BSR SubroutineD
             .
             .
             RTS
SubroutineB  .
             BSR SubroutineC
             .
             .
             RTS
SubroutineC  .
             .
             RTS
SubroutineD  .
             .
             RTS
```

Consider the two nested subroutines in Figure 7.3. The main program first calls Sub1 and then Sub1 calls Sub2, as the following code demonstrates.

Figure 7.3 Example of nested subroutines

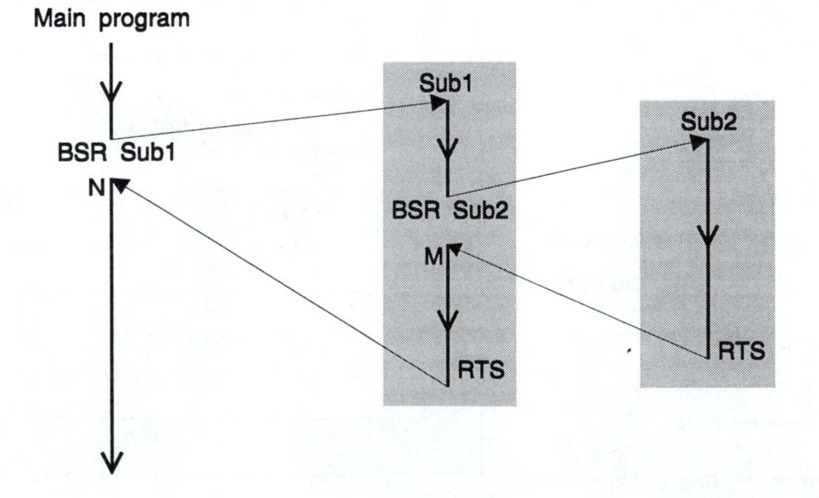

```
MainProgram
                  .
                  BSR  Sub1
                  .
                  END
Sub1              .               (subroutine 1)
                  .
                  .
                  BSR  Sub2
                  .
                  .
                  RTS             (return point from subroutine 1)
Sub2              .               (subroutine 2)
                  .
                  RTS             (return point from subroutine 2)
```

When subroutine Sub1 is called, the return address N is pushed on to the top of the stack — Figure 7.4a. When Sub2 is called from Sub1, return address M is pushed on the stack — Figure 7.4b. Return address, M, is now the new top of the stack. When the second subroutine has been executed to completion, the RTS instruction is executed and a return made to the address at the top of the stack. This return address is M, which is the point in Sub1 immediately after the point at which Sub2 was called. The state of the stack at this point is given in Figure 7.4c. Sub1 is now completed and the RTS at its end loads the address at the top of the stack (i.e., N) into the PC to force a return to the main program. Figure 7.4d illustrates this state, which is the state of the stack before Sub1 was first called.

Subroutines can be nested to any depth by using the stack to save subroutine return addresses, as long as the stack doesn't run out of space for return addresses. The average depth of subroutine nesting is about five. However, systems employing *recursive techniques* may nest subroutines to a much greater depth. A recursive subroutine is a subroutine that calls itself.

Figure 7.4 State of the stack during the execution of the code of Figure 7.3

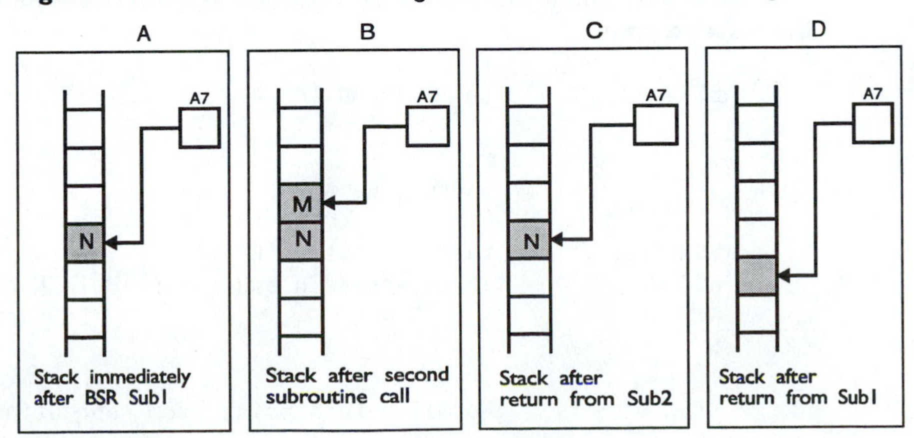

The term *nesting* implies that one subroutine is entirely enclosed within another, as Figures 7.2 and 7.3 illustrate. In Figure 7.3 a return from Sub2 is made to Sub1. Suppose we write the code in such a way as to return directly from Sub2 to the main program should certain errors arise during the execution of Sub2. Such an operation is possible, as the following code demonstrates.

```
Sub2  .
      .
      BEQ  Exit        Detect a problem here
      .
      .
      RTS              Normal return point to Sub1
Exit  LEA  (4,A7),A7   Move past the normal return point
      RTS              and return directly to main program
```

Sometimes we might think that the detection of an error (or any other event) in a subroutine is best dealt with by returning directly to a higher-level subroutine or to the main program. As you can see from the above code, all you have to do is to skip past the intermediate return addresses on the stack that takes you to the next level of nested subroutine. If you wish to skip two levels of subroutine, you would replace the LEA (4,A7),A7 instruction by LEA (8,A7),A7.

Forcing an exit from a subroutine in this way is regarded as very bad practice indeed. The resulting code is difficult to read; the fundamental rules of structured programming are violated, and the likelihood of introducing a serious bug in the program is increased. We have pointed out that it is possible to skip one or more levels when returning from a subroutine simply to illustrate how the stack functions. The fact that an assembly language *permits* you to do something doesn't make it either right or desirable. You can even force a jump out of a subroutine to any point in the program by means of an unconditional branch or jump (e.g., BRA Escape or JMP PANIC). Doing this would be a nightmare because the state of the stack would be difficult to determine. The number of return addresses left on the stack would depend on the point from which you jumped. You can always employ the following technique to return from a nested SubB to the main program.

```
      BSR  SubA        Call subroutine A
      .
      .
SubA  .                Subroutine A
      .
      BSR SubB         Call subroutine B
      BCC OK           IF no error in subroutine B THEN continue
      RTS                                             ELSE return
OK    .
      .
      RTS              Normal return point from subroutine A
```

```
SubB  .                   Subroutine B
      .
      .
      ORI  #%00001,CCR Escape from here - Set C-bit for error
      BRA  Rtrn           Now exit
      .
      .
Rtrn  RTS                 Normal return
```

In this example the main program calls subroutine A which in turn calls subroutine B. If an error occurs in subroutine B, a return is made to subroutine A. However, subroutine A checks the state of the C-bit immediately after the return and forces a return to the main program if it is set.

Subroutines and Readability

We have said that subroutines are useful because they remove the need to rewrite chunks of code (like an input routine) that are repeated many times in a program. I sometimes use subroutines to improve the readability of a program. Consider the following example.

```
* Calculator program
*
        BSR Initialize   Set up the constants
Again   BSR GetNum1      Read the first number
        BSR GetOpr       Read the operator: one of +,-,*,/
        BSR GetNum2      Read the second number
        BSR Calc         Perform the calculation
        BSR PrintRes     Display the result
        .
        .
GetNum1 BSR GetDigit     Get a digit from the keyboard
        BSR Valid        Test for a valid digit in range 0 to 9
        BCS DigErr       IF not valid digit THEN deal with error
        BSR SaveDigit                       ELSE save it
        .
        RTS
```

This fragment of code demonstrates how you might employ subroutines to make your program highly readable and to comply with the ideas of *top-down design*. The first part of the program consists of a series of subroutines that are called one by one to carry out the function of the program. In this case we might not use, say, the subroutine Calc more than once in the program. The only reason for constructing the subroutine called Calc is to enhance the readability of the program. For example, if the subroutine Calc were replaced by the appropriate in-line code, anyone reading the program would find it much harder to follow. Even the subroutine GetNum1 called by the main program is little more than another sequence of subroutine calls. Now that we have covered the basics, we are going to look at how information is passed to and from a subroutine.

7.2 Parameter Passing

We employ a subroutine to carry out a specific action. For example, in a high-level language a subroutine SIN(X) might be called to calculate the sine of X. A subroutine operates on data that we can call its *input* to produce an *output*. Therefore, a subroutine must transfer information between itself and the program calling it. The only exception occurs when a subroutine is called to trigger an event. For example, a subroutine can be designed to start the motor of a disk drive or to sound an alarm. Simply calling the appropriate subroutine causes the pre-determined action to take place. No explicit data is transferred between the subroutine and the program calling it. In this chapter we deal with two interrelated topics: the way in which information is actually transferred to and from a subroutine, and the way in which the subroutine allocates memory for its own use.

Consider first the application of subroutines to inputting or outputting data. Obtaining a character from a keyboard, or transferring one to a CRT terminal, requires the execution of a number of instructions and is inherently device-dependent (i.e., it involves the control of specific hardware like a keyboard controller). As even a fundamentally simple operation like reading the value of a key might require tens of machine-level operations, input and output transactions are frequently implemented by calling the appropriate input or output subroutine.

Suppose a program calls the subroutine Put_Char to display a single character on a CRT terminal. The character is printed by calling the subroutine with BSR Put_Char. The entire process of locating the subroutine and returning from it takes place automatically, thanks to the assembler that calculates the relative address, and to the processor's stack mechanism that looks after the return address. In this example, the program has to transfer just one item of information to the subroutine, namely the ASCII code of the character to be displayed.

When only a single character is passed to the subroutine, one of the eight data registers serves as a handy vehicle to transfer the information from the calling program to the subroutine. For example, if the character to be displayed is in a table pointed at by A0 and register D0 is used to carry the character to the subroutine, we can write the following code.

```
MOVE.B  (A0),D0      Pick up the character to be printed
BSR     Put_Char      and print it
```

Programmers frequently transfer data between a program and a subroutine by means of one or more registers, particularly when the quantity of data to be transferred is either very small or the processor is well endowed with registers. Transferring data via a register permits both *position independent code* and *re-entrancy*. In Chapter 4 we introduced position independent code, PIC, that does not use absolute addresses to refer to operands. You can move a block of PIC from one place in memory to another without recalculating the address of operands. If data is transferred to a subroutine in a data register, position independence is guaranteed because no absolute memory location is involved in the transfer of data. Since, in our example, the character to be output is transferred in D0, we can relocate the program anywhere in memory because D0 does not have an address in memory.

The following fragment of code demonstrates the passing of a parameter via a memory location (in this example it is HoldChr).

```
HoldChr DS.B    1               Reserve a location for byte to print
        .
        .
        MOVE.B (A0),HoldChr  Move the character to be printed to
        BSR    Put_Chr        HoldChr and print it
        .
        .
Put_Chr MOVE.B HoldChr,D0     Read the character from HoldChr
        .
        RTS
```

If the program is ever relocated in a different part of memory, it must be recompiled, because the source operand used by MOVE.B HoldChr,D0 will have a new value. However, before we continue with parameter passing, we have to take a short diversion and look at the re-entrant subroutine.

The only disadvantage in passing information to and from subroutines via registers is that it reduces the number of registers available for use by the programmer. Moreover, the quantity of information that can be transferred is limited by the number of registers. In the case of the 68000, it is theoretically possible to transfer up to 15 long words of data in this way. This total is made up of eight data registers and seven address registers. Address register A7, the stack pointer, cannot itself be used to transfer data.

Re-entrant Subroutines

Re-entrancy is such an important concept that we have a short time-out to introduce it. Suppose you are reading a book and a friend asks to borrow it before you've finished it. You lend the book to your friend who returns it after reading it. You then continue reading from where you left off. This is an everyday example of re-entrancy. All that is needed to make the process re-entrant is for you to remember where you had reached before you lent the book, and for your friend not to rip out pages (i.e., the data should not be corrupted).

A subroutine is re-entrant if a program can begin executing it; if another program can take control of the subroutine and run it *before* the first program has completed it; and if it can be returned to the first program without problems. This sequence is entirely analogous to the example of the book. A necessary requirement for re-entrancy is that the program that takes over the subroutine must leave it in the state in which the program found it.

Why is re-entrancy a desirable feature in a computer? More importantly, why should a program wish to borrow a subroutine that is being used by another program? In a *multitasking* system, several programs or *tasks* run *simultaneously*. To be more precise, the operating system switches between user tasks so rapidly that to an outsider (e.g., a human operator) they appear to be running simultaneously. Imagine a situation in which task A is running and using subroutine X to perform a certain function. Suppose the operating system switches tasks while A is still in the middle of subroutine X. Task B also uses subroutine X. When B has finished with subroutine X, it must leave the subroutine in exactly the same condition it found X. Otherwise, the multitasking system simply would not work.

We're not going to deal with multitasking and interrupt processing further at this stage, because they are covered in Chapter 10. What we need to discuss now is the way in which a subroutine can be re-used by another program. Let's return to the example of the output subroutine in which data is transferred to the subroutine in a data register, D0. Re-entrancy is possible as long as the re-use of the subroutine saves the contents of the register employed to transfer the data *before* it is re-used. If this sounds confusing, think about Goldilocks. If she had put the porridge on the back-burner, made herself a new helping, eaten it, washed the dishes, and then restored the original porridge to the front of the stove before the three bears got home, she would have saved herself a lot of trouble.

Consider the situation of Figure 7.5. Initially, task A is running subroutine X and the operating system forces a task switch. In order to preserve task A's *working environment*, all its registers must be saved in a safe place. Before the processor is given to task B, all task B's registers are restored from their safe place. When the operating system intervenes again to switch back to task A, B's registers are saved and A's registers are restored. Task A, which was running subroutine X at the time a switch occurred, is in the same state as before the switch.

It is important to appreciate that task switching should be *invisible* to a subroutine and should be able to occur at any point in the code. Consider the following code in which a task switch takes place at the indicated point (i.e., immediately after the execution of the instruction `BTST.B #0,D1`).

Figure 7.5 Switching in a multitasking environment

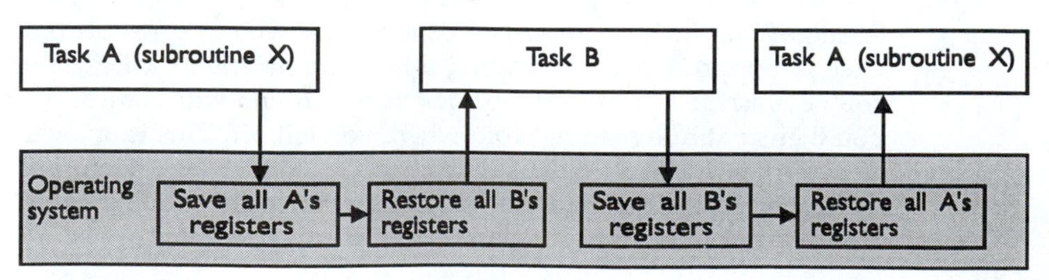

```
         MOVE.B   (A0),D0      Pick up the character to be printed
         BSR      Put_Char     and print it
         .
         .
Put_Char MOVE.W   D1,-(A7)     Push D1 on the stack
         MOVE.B   ACIAC,D1     Read status
         BTST.B   #0,D1
         .
         .
         MOVE.W   (A7)+,D1     Restore D1
         RTS
```

Task switching takes place here. The contents of all registers are saved and the subroutine can be re-used by the interrupting program. When the task is to be run again, all its registers must be restored.

As you can see from the above code, registers A0 and D1 are in use at the time the task is switched. The contents of these registers must be safely saved and retrieved before the task is run again

If data is saved in an *absolute* address location, it will be corrupted by the re-use of the subroutine. For example, suppose we store a loop counter in memory location $1234. At some point during the execution of the program, the contents of $1234 indicate how many times a loop has been executed. If an interrupt occurs and another program runs the same subroutine, the new program will also store its loop count in memory location $1234. Consequently, the re-use of the subroutine corrupts any of its data held in fixed or *static* memory locations.

Mechanisms for Parameter Passing

At this point, we are going to talk about how parameters are passed. The two basic ways of passing parameters are transfer by *value* and transfer by *reference*. In the former, the *actual* parameter is transferred, while in the latter, the *address* of the parameter is passed between program and subroutine. Figure 7.6 demonstrates these two concepts.

This distinction between *value* and *reference* is important because it affects the way in which parameters are handled. When passed by value, the subroutine receives a copy of the parameter. Therefore, if the parameter is modified by the subroutine, the *new value* does not affect the *old value* of the parameter elsewhere in the program. In other words, passing a parameter by value causes the parameter to be cloned and the clone to be used by the subroutine.

Figure 7.6 Passing a parameter by value and by reference

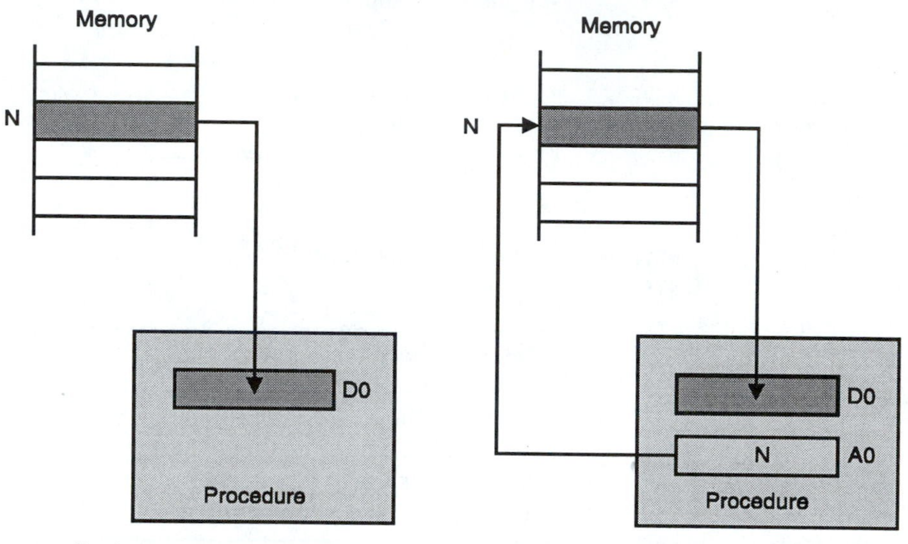

(a) Passing a parameter by value. The data in the memory location is copied to the subroutine.

(b) Passing a parameter by reference. The address of the parameter is passed to the subroutine.

When a parameter is passed by address (i.e., by reference), the subroutine receives a *pointer* to the parameter. In this case, there is only one copy of the parameter, and the subroutine is able to access this unique value because it knows the address of the parameter. If the subroutine modifies the parameter, it is modified globally and not only within the subroutine.

The actual mechanism by which information is passed to a subroutine generally falls into one of three categories: a register, a memory location, or the stack. We have already seen how a register is used to transfer an actual value. A region of memory can be treated as a mailbox and used by both calling program and subroutine, with one placing data in the mailbox and the other emptying it. However, it is the stack mechanism that offers the most convenient method of transferring information between a subroutine and its calling program.

Passing Parameters by Reference (i.e., Address)

Let's look at an example of how you might pass parameters to a subroutine by *reference*. Suppose a subroutine is written to search a region of memory containing a text string for the first occurrence of a particular sequence of characters called a substring. The sequence we are looking for (i.e., the substring) is stored as a string in another region of memory. In this example, the subroutine requires four pieces of information: the starting and ending addresses of both the region of text to be searched and the substring to be used in the matching process. Figure 7.7 provides a memory map for this problem.

The information required by the subroutine is the four addresses, 00 1000_{16}, 00 $100B_{16}$, 00 1100_{16}, and 00 1103_{16} that indicate the starting and ending points of the two strings. Note that in this example we are passing the parameters by reference, because the subroutine receives their addresses. We are not passing the actual parameters (i.e., the text string and the substring) themselves. After the subroutine has been executed, it returns the value 00 1007_{16}. While it is possible to transfer all these addresses via four address registers, an alternative technique is to assemble the four parameters into a data block in memory and then pass the address of this block. Figure 7.8 shows how the block is arranged.

The only information required by the subroutine is the address 00 2000_{16}, which points to the first item in the block of parameters stored in memory. The following fragment of code shows how the subroutine is called and how the subroutine deals with the information passed to it. In this example, address register A0 is used to pass the address of the parameter block to the subroutine, and A3, A4, A5, and A6 are used by the subroutine to point to the beginning and end of both the text and the string to be matched. Note that, in practice, we would replace the four `MOVE.L (A0)+,Ai` instructions in the subroutine `MATCH` by a single `MOVEM.L (A0)+,A3-A6` instruction.

```
         ORG      $2000          Set up the parameter block
StrtStrg DS.L     1              Pointer to start of string to be matched
EndStrg  DS.L     1              Pointer to its end
StrtSub  DS.L     1              Pointer to start of substring to be matched
```

Figure 7.7 Memory map of a string-matching problem

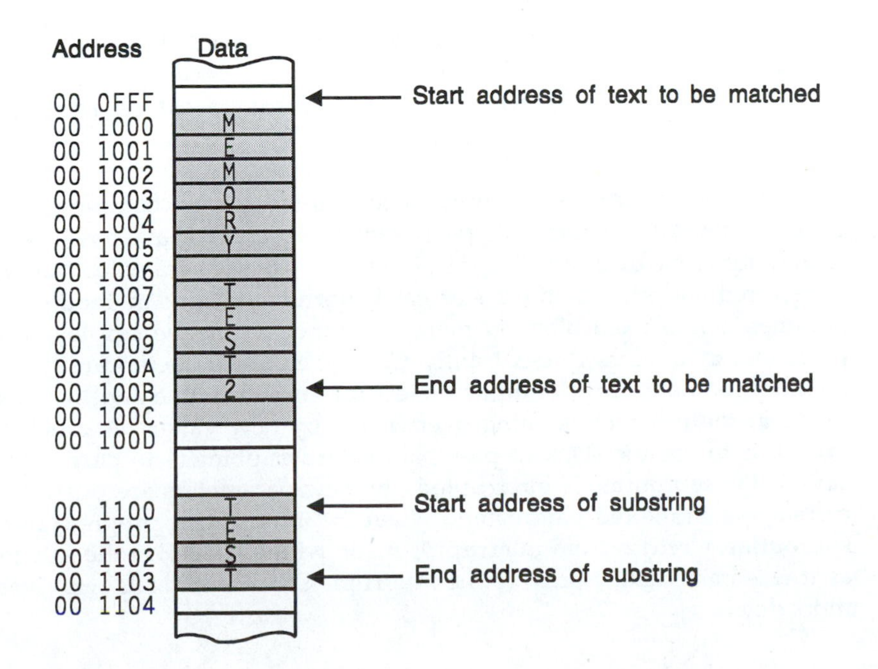

Figure 7.8 The parameter block for Figure 7.7

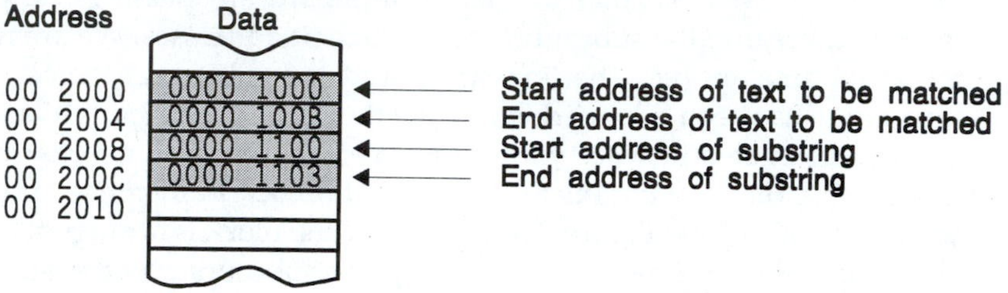

```
EndSub  DS.L      1                Pointer to its end
        .
        .
        .
        LEA       $002000,A0       Set up address of parameter block in A0
        BSR       MATCH            Call string matching subroutine
        .
        .
        .
MATCH   MOVEM.L A3-A6,-(A7)        Save address registers on stack
        MOVEA.L (A0)+,A3           Put start address of text in A3
        MOVEA.L (A0)+,A4           Put end address of text in A4
        MOVEA.L (A0)+,A5           Put start address of string in A5
        MOVEA.L (A0)+,A6           Put end address of string in A6
        .
        .                          Body of the subroutine
        .
        MOVEM.L (A7)+,A3-A6        Restore address registers from stack
        RTS
```

The four addresses required by the subroutine are obtained by using address register A0 as a pointer to the parameter block. We have passed a parameter to the subroutine by its address (i.e., the address of the parameter block) and have transferred this address in register A0. Unfortunately, this method of storing the parameters in a fixed block in memory cannot always be employed, as such a program cannot be used re-entrantly. Clearly, if the parameters are stored in static memory locations, any attempt to interrupt the subroutine and to re-use it must result in the parameters being overwritten by new values. A much better approach is to use the stack to pass parameters or pointers to parameters. In this case, if the subroutine is interrupted, the new parameters are pushed higher on the stack than the old parameters. When the interrupting program has used the subroutine, a return from interrupt is made with the stack in the same condition as it was immediately prior to the interrupt. Chapter 10 deals with interrupts in more detail.

The Stack and Parameter Passing

The stack is not only useful for storing subroutine return addresses in such a way that subroutines may call further subroutines; it can also be used to transfer information to and from a subroutine. All that need be done is to push the parameters (or their addresses) on the stack before calling the subroutine. Since we are transferring parameters by reference, we will exploit the PEA instruction. The program fragment below shows how this is done for our string-matching algorithm. In this example, we transfer all four parameters by *reference*.

```
         PEA       TEXT_START      Push start and end addresses
         PEA       TEXT_END        of string to search
         PEA       STRING_START    Push start and end addresses
         PEA       STRING_END      of substring used in search
         BSR       STRING_MATCH    Call subroutine for matching
         LEA       (16,A7),A7      Adjust stack pointer
         .
         .
         .
MATCH    LEA       (4,A7),A0       Put pointer to parameters in A0
         MOVEM.L   (A0)+,A3-A6     Get parameters off stack
         .
         .                         Body of subroutine
         .
         RTS
```

The instruction PEA, push effective address, is used to push the address of the four operands onto the stack. We could have used MOVE.L #TEXT_START,(A7) to do the same job, but the PEA instruction is much neater. Figure 7.9a shows the initial state of the stack, and Figure 7.9b shows the stack immediately after the four addresses have been pushed. Figure 7.9c shows the state of the stack on subroutine entry, with the return address on top of the parameters.

The first instruction in the subroutine, LEA (4,A7),A0, loads address register A0 with the starting address of the last parameter pushed on the stack. We add 4 to the stack pointer because the return address is at the top of the stack. After the subroutine has been called, the instruction MOVEM.L (A0)+,A3-A6 pulls the four addresses off the stack, and deposits them in address registers A3 to A6 for use as required. Note that these parameters are left on the stack after a return from subroutine is executed (Figure 7.9d).

In the calling program, the instruction LEA (16,A7),A7 is executed after a return from subroutine has been made. This replaces the contents of the stack pointer with the contents of the stack pointer plus 16. Consequently, the stack pointer is restored to the position it was in before the four 32-bit parameters were pushed on the stack (Figure 7.9a).

Since the subroutine makes use of address registers A3 to A6 to point to the four parameters, any information in these registers is corrupted by the string matching subroutine. Most programmers would employ a MOVEM instruction to save these registers on the stack before they are used by the subroutine, and then restore them immediately before executing a return from subroutine. The following code demonstrates this point.

```
MATCH    MOVEM.L  A0/A3-A6,-(A7)    Save working registers
         LEA      (24,A7),A0        Load A0 with pointer to parameters
         MOVEM.L  (A0)+,A3-A6       Get parameters off stack
         .
         .                          Body of subroutine
         .
         MOVEM.L  (A7)+,A0/A3-A6    Restore working registers
         RTS
```

Note that we use an offset of 24 in LEA (24,A7),A0 because the parameters are buried under the return address and the five registers saved by the MOVEM.L A0/A3-A6,-(A7). Passing parameters on the stack facilitates position independent code and re-entrant programming. If a subroutine is interrupted, the stack builds upward and information currently on the stack is not overwritten.

7.3 The Stack and Local Variables

In addition to the parameters passed between a subroutine and the calling program, a subroutine sometimes needs a certain amount of *local workspace* for its

Figure 7.9 State of the stack during parameter passing

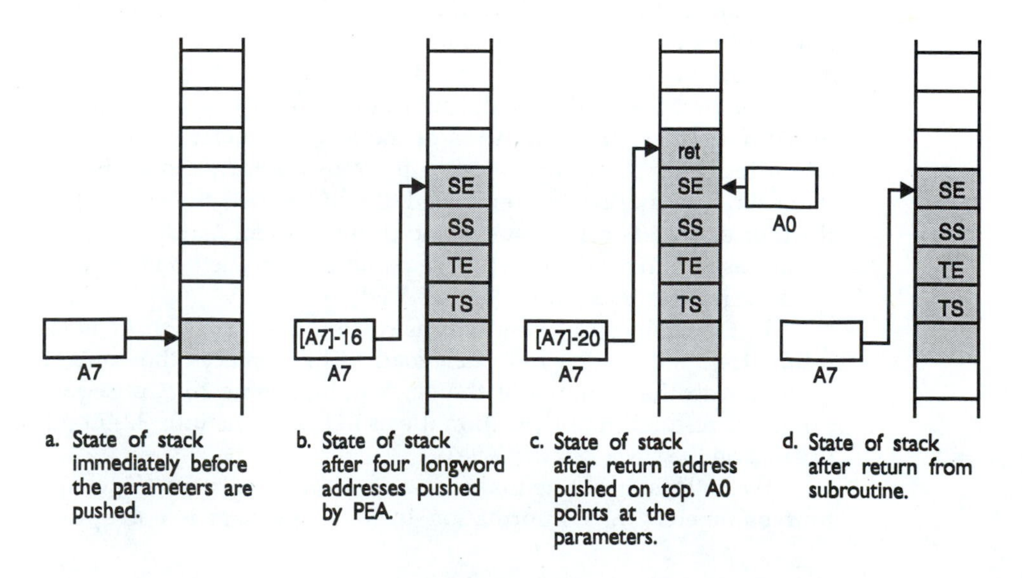
A7
[A7]-16
A7
SE
SS
TE
TS
[A7]-20
A7
ret
SE
SS
TE
TS
A0
A7
SE
SS
TE
TS
a. State of stack immediately before the parameters are pushed.
b. State of stack after four longword addresses pushed by PEA.
c. State of stack after return address pushed on top. A0 points at the parameters.
d. State of stack after return from subroutine.

temporary variables. Here, the word *local* means that the workspace is private to the subroutine, and is never accessed by the calling program or by other subroutines that the current subroutine might call. You can sometimes allocate a region of the system's memory space to a subroutine that requires a work area at the time the program is written. You simply reserve fixed locations for the subroutine's variables in a process called static allocation. This approach is satisfactory for subroutines that are not going to be used *re-entrantly* or *recursively*. A subroutine is used recursively if it calls itself.

If a subroutine is to be made re-entrant or used recursively, its local variables must be bound up not only with the subroutine itself, but with the *actual occasion of its use*. In other words, each time the subroutine is called, a new workspace must be assigned to it. For example, suppose a subroutine is being used by task A. Workspace is allocated for use by the subroutine's variables. Suppose now that a task switch takes place while task A is still executing the subroutine and that the subroutine is used by task B. Clearly, task B must be allocated new workspace for its own variables, if it is not to corrupt task A's variables. Once again, the stack provides a convenient mechanism for implementing the dynamic allocation of workspace. This type of storage allocation is called *dynamic* because it is allocated to variables when they are created and then de-allocated when the variables are no longer required.

Two items closely associated with dynamic storage techniques for subroutines are the *stack frame* (SF) and the *frame pointer* (FP). The stack frame is a region of temporary storage at the top of the current stack. The frame pointer, which is normally an address register, points to the bottom of the stack frame. Figure 7.10a illustrates the state of the stack after a subroutine call, and Figure 7.10b illustrates the stack frame that has been created on top of the subroutine's return address.

Figure 7.10b also shows how a stack-frame is created merely by moving the stack pointer up by *d* locations at the start of a subroutine. Remember that the 68000's stack grows toward the *low* end of memory, and therefore the stack pointer is *decremented*. It is perhaps unfortunate that we talk of the stack *growing* as the address in the stack pointer gets *smaller*.

Figure 7.10 The stack frame

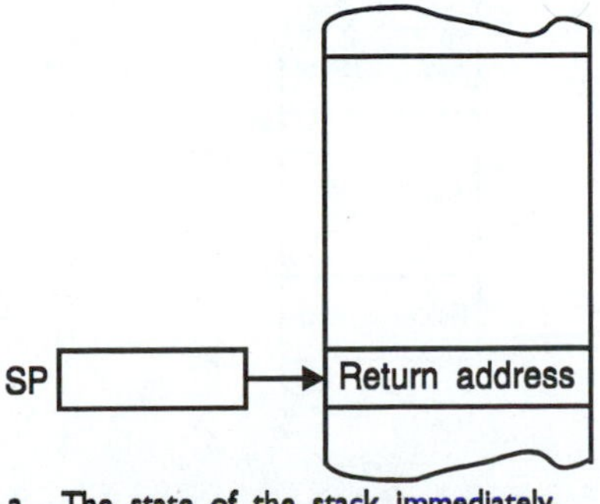

a. The state of the stack immediately after a subroutine call.

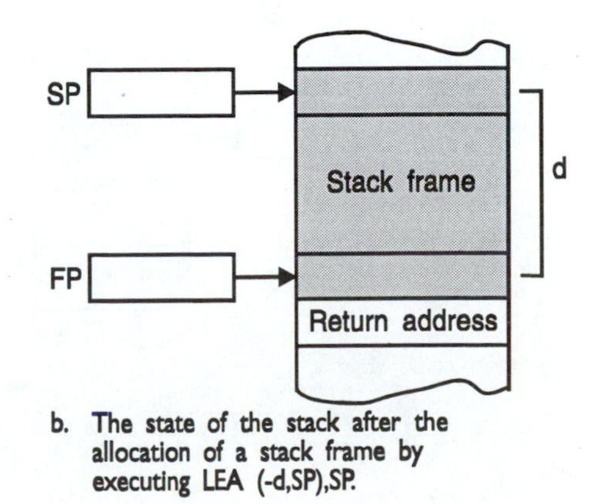

b. The state of the stack after the allocation of a stack frame by executing LEA (-d,SP),SP.

Reserving 100 words of memory is achieved by executing the operation LEA (-200,A7),A7. The offset is *200*, as all 68000 offsets are *byte* values. Once the stack frame has been created, local variables can be accessed by any addressing mode that uses A7 as a pointer. However, most programmers access variables via the frame pointer. For example, we could write the code:

```
AnySub  LEA    (-4,A7),A6      Set up A6 as the frame pointer
        LEA    (-200,A7),A7    Create the stack frame
        .
        .                      The subroutine proper
        .
        LEA    (200,A7),A7     Collapse the stack frame
        RTS                    and return from subroutine
```

Before a return from subroutine is made, the stack frame must be collapsed by an LEA (200,A7),A7 instruction. We shall soon see that the 68000 has two special instructions, LINK and UNLK, that automatically manage the stack frame.

Figure 7.11 shows how stack frames can be built on top of each other. In this example, subroutine A has a stack frame at the top of the stack while the subroutine is being executed (Figure 7.11a). Suppose that subroutine A calls subroutine B, and that subroutine B also builds a stack frame on top of the stack. This situation is illustrated by Figure 7.11b. Subroutine B accesses data in its stack frame and any data that A was using is safe in its own stack frame.

Figure 7.11 Growth of stack frames

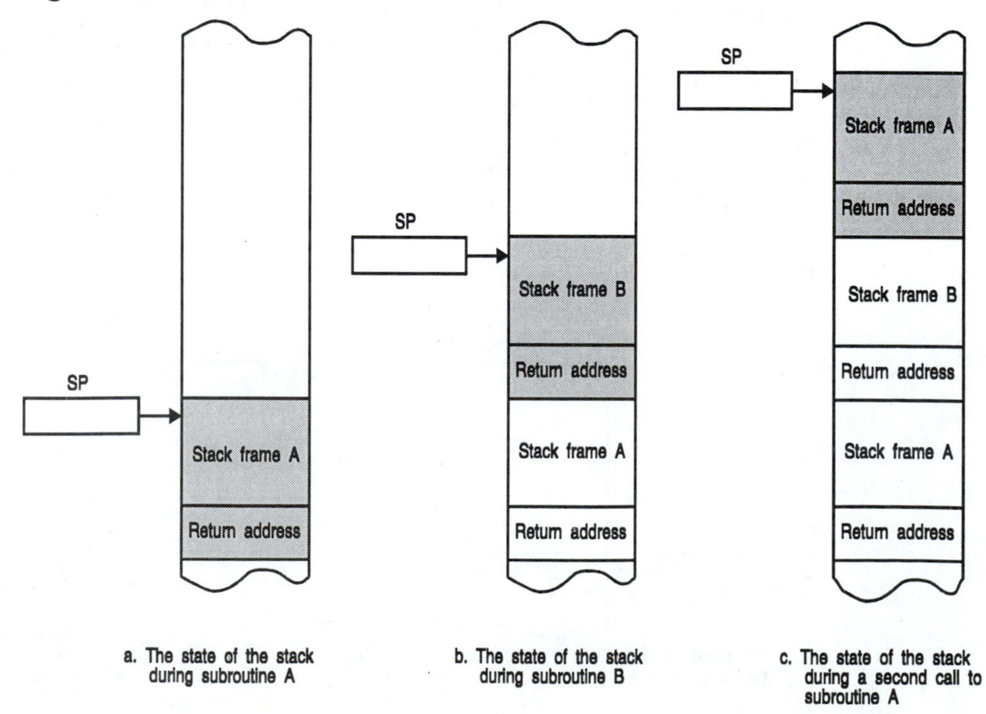

Suppose that subroutine B calls subroutine A and a new stack frame is built on subroutine A's stack (Figure 7.11c). Subroutine A now has *two* stack frames, one for each of its calls. Since the scratchpad data used during each of A's runs is located in *separate* stack frames, these two runs do not interfere with each other. When the second use of subroutine A is completed, the stack frame is collapsed and the situation is the same as Figure 7.11b. When subroutine B has been completed, its stack frame is collapsed, and we are once more at the state of Figure 7.11a.

The Link and Unlink Instructions

The 68000 mechanizes the simple scheme of Figure 7.11 by a complementary pair of instructions, `LINK` and `UNLK` (link and unlink). These are relatively complex in terms of their detailed implementation, but are conceptually simple. The great advantage of `LINK` and `UNLK` is their ability to let the 68000 manage the stack automatically and to make the memory allocation scheme entirely re-entrant.

If the stack frame storage mechanism is to support re-entrant programming, a new stack frame must be reserved each time a subroutine is called. Since each successive stack frame is of a (possibly) variable size, its length must be preserved. The 68000 uses an address register for this purpose. In the following description of `LINK` and `UNLK`, a 16-bit signed constant, *d*, represents the size of the stack frame. At the start of a subroutine `LINK` creates the stack frame belonging to this subroutine. The following code achieves the desired effect:

```
Sub1    LINK    A1,#-64         Allocate 64 bytes (16 long words) of storage
                                in this stack frame - use A1 as frame pointer
        .
        .
        .                       Body of the subroutine
```

The minus sign in the `LINK` instruction is needed because the stack grows toward low memory and the stack pointer must be decremented. We can define the action executed by `LINK A1,#-64` in RTL terms.

```
LINK: [SP]       ← [SP] - 4    Decrement the stack pointer by 4
      [M([SP])]  ← [A1]        Push the contents of address register A1
      [A1]       ← [SP]        Save stack pointer in A1
      [SP]       ← [SP] - 64   Move stack pointer up by 64 locations
```

Clearly, a `LINK` instruction does more than simply move the stack pointer to the top of the stack frame. Figure 7.12 shows the stages of the execution of a `LINK A1,#-64`. The first operation performed by the `LINK` is to save the old contents of A1 by pushing them on the stack (Figure 7.12b). Pushing the old value of the frame pointer on the stack makes sense because the previous stack pointer can later be restored when the current stack frame is thrown away. Similarly, the old value of the stack pointer is preserved in A1 (Figure 7.12c). In other words, the `LINK` destroys no information and, therefore, it is possible to undo the effect of a

Figure 7.12 Executing a LINK instruction

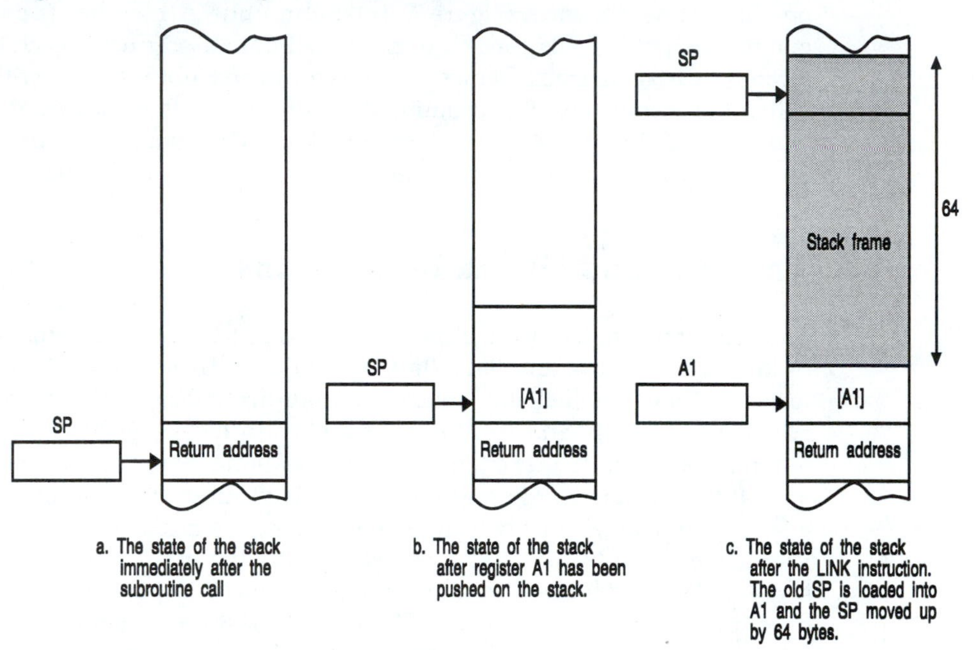
SP
Return address
SP
[A1]
Return address
SP
A1
Stack frame
64
[A1]
Return address
a. The state of the stack immediately after the subroutine call
b. The state of the stack after register A1 has been pushed on the stack.
c. The state of the stack after the LINK instruction. The old SP is loaded into A1 and the SP moved up by 64 bytes.

LINK at some later time. Note that in Figure 7.12c the stack pointer is moved to the top of the stack frame, and the region between this point and the saved value of A1 is available to the subroutine. Address register A1 points at the base of the stack frame and can be used to reference data on the stack frame.

The next step is to demonstrate how the work of the LINK instruction is undone at the end of the subroutine. The following code shows how you would use an UNLK instruction before returning from a subroutine.

```
Sub1  LINK   A1,#-64     Allocate 64 bytes (16 long words) of storage
      .                  in this stack frame - use A1 as frame pointer
      .
      .                  Body of the subroutine
      .
      UNLK   A1          De-allocate Subroutine 1's stack frame
      RTS                Return to calling point.
```

The RTL definition of UNLK A1 is:

```
UNLK:   [SP]  ← [A1]
        [A1]  ← [M([SP])]
        [SP]  ← [SP] + 4
```

The stack pointer is first loaded with the contents of address register A1. Remember that A1 contains the value of the stack pointer just before the stack frame was created. By doing this, the stack frame collapses. The next step is to pop the top item off the stack and place it in A1. This has two effects. It returns both the stack and the contents of A1 to the points they were in before LINK was executed. Following the UNLK, a return from subroutine can be made.

Subroutines often employ several data and address registers. Since this re-use of registers by another subroutine will overwrite and corrupt their contents, it is necessary to save any registers that will be used by the new subroutine and then restore them. Many programmers employ a MOVEM instruction to save registers on the stack at the start of a subroutine. Consider the following fragment of code.

```
Sub1 LINK     A1,#-64                 Allocate 64 bytes of storage in this
*                                     stack frame - use A1 as frame pointer
     MOVEM.L D0-D7/A3-A6,-(SP)        Save working registers on the stack
     .
     .                                Body of the subroutine
     .
     MOVEM.L (SP)+,D0-D7/A3-A6        Restore working registers
     UNLK     A1                      De-allocate the subroutine stack frame
     RTS                              Return to calling point.
```

Note that we wrote *SP* rather than *A7* — they are interchangeable. The state of the stack before the subroutine call, after the call and the LINK operation, and after a MOVEM instruction is described in Figure 7.13. In this example, the working registers, D0 to D7 and A3 to A6, are saved on the stack. The stack frame's temporary storage allocation is 64 bytes and address register A1 is used by LINK. Note that we first create the stack frame by the LINK instruction and then push the working registers by means of the MOVEM. That is, the sequence is:

```
LINK     A1,#-64              Create the stack frame
MOVEM.L D0-D7/A3-A6,-(SP)     Save the registers
.
.
MOVEM.L (SP)+,D0-D7/A3-A6     Restore registers
UNLK     A1                   Delete the stack frame
```

Alternatively, we could have chosen the sequence:

```
MOVEM.L D0-D7/A3-A6,-(SP)     Save the registers
LINK     A1,#-64              Create the stack frame
.
.
UNLK     A1                   Delete the stack frame
MOVEM.L (SP)+,D0-D7/A3-A6     Restore registers
```

In this case, the working registers are stacked *before* creating the stack frame. Although both techniques are equally valid, the recommended approach is to

Figure 7.13 Using a MOVEM to save registers

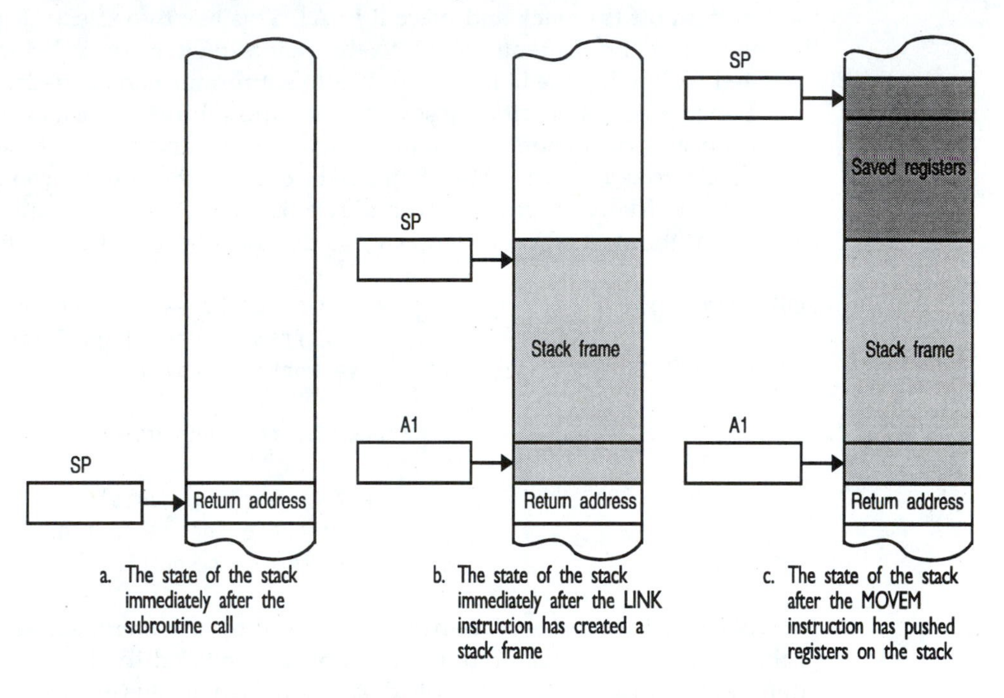

a. The state of the stack immediately after the subroutine call

b. The state of the stack immediately after the LINK instruction has created a stack frame

c. The state of the stack after the MOVEM instruction has pushed registers on the stack

create the stack frame *before* saving the registers. The only reason for adopting this sequence is that it sometimes makes it easier to manipulate the stack frame.

In Figure 7.12 the `LINK` instruction sets address register A1, the frame pointer, to point to the base of the stack frame. Any temporary variables created by the subroutine can be accessed by means of the frame pointer. For example, `MOVE.W D3,(-2,A1)`, stores the low-order word in data register D3 in the stack frame immediately on top of the saved value of A1.

Example of the Use of Local Variables

Consider the following example of a subroutine that uses the `LINK/UNLK` pair. A subroutine, `CALC`, uses three parameters, P, Q, and R. Parameters P and Q are called by *value*, and parameter R by *reference*. This subroutine calculates the value of $(P^2+Q^2)/(P^2-Q^2)$. Although it is not always necessary to save working registers on the stack, we do it in the following example because it is good practice. Having to remember that calling a certain subroutine corrupts the values of D3 and A4 is not as good as ensuring that D3 and A4 are not corrupted by the subroutine. We use the `LINK` instruction to create a stack frame (which can be used by the subroutine as temporary storage).

```
*                              D0 contains value of P
*                              D1 contains value of Q
```

```
MAIN MOVE.W  D0,-(SP)       Push the value of P on the stack
     MOVE.W  D1,-(SP)       Push the value of Q on the stack
     PEA     R              Push the reference to R on the stack
     BSR     CALC           Call the subroutine
     LEA     (8,SP),SP      Clean up the stack by removing P, Q, R
     .
     .
     STOP    #$2700         End of program
     .
     .
CALC LINK    A0,#-14        Establish 3 longword + 1 word frame for the stack
     MOVEM.L D6/A6,-(SP)    Save working registers on the stack
     MOVE.W  (14,A0),D6     Get value of P from stack
     MULU    D6,D6          Calculate P^2
     MOVE.L  D6,(-4,A0)     Save P^2 on the stack frame
     MOVE.W  (12,A0),D6     Now get Q from the stack
     MULU    D6,D6          Calculate Q^2
     MOVE.L  D6,(-8,A0)     Store Q^2 on the stack frame
     MOVE.L  (-4,A0),D6     Get P^2
     ADD.L   (-8,A0),D6     Add Q^2
     MOVE.L  D6,(-12,A0)    Store P^2+Q^2 on the stack frame
     MOVE.L  (-4,A0),D6     Get P^2
     SUB.L   (-8,A0),D6     Subtract Q^2
     MOVE.W  D6,(-14,A0)    Store P^2-Q^2 on the stack frame (as a word)
     MOVE.L  (-12,A0),D6    Copy P^2+Q^2 from stack frame to D6
     DIVU    (-14,A0),D6    Calculate (P^2+Q^2)/(P^2-Q^2) Note 16-bit arithmetic
     LEA     (8,A0),A6      Get pointer to address of R
     MOVEA.L (A6),A6        Get actual address of R
     MOVE.W  D6,(A6)        Modify R in the calling routine
     MOVEM.L (SP)+,D6/A6    Restore working registers
     UNLK    A0             Collapse the stack frame
     RTS
```

This code is rather cumbersome because we have not optimized the usage of registers. That is, we did not really have to use a stack frame, because the 68000 has sufficient registers to hold all temporary variables. However, a compiler for a high-level language might produce an assembly language output similar to the above code. Figure 7.14 provides a composite illustration of how the stack grows during the execution of this code.

In order to clarify the process we have just described, we will walk through the code and present a snapshot of the stack as the instructions are executed. Before the subroutine is called, the two MOVE.W <register>,-(SP) instructions push the *values* of the parameters P and Q onto the stack. The following instruction, PEA R, pushes the *address* of R onto the stack. P and Q each take up a word, and R takes up a longword because the address of R is pushed rather than its value. Figure 7.15 illustrates the stack at this stage.

Figure 7.14 State of the stack during the execution of the CALC subroutine

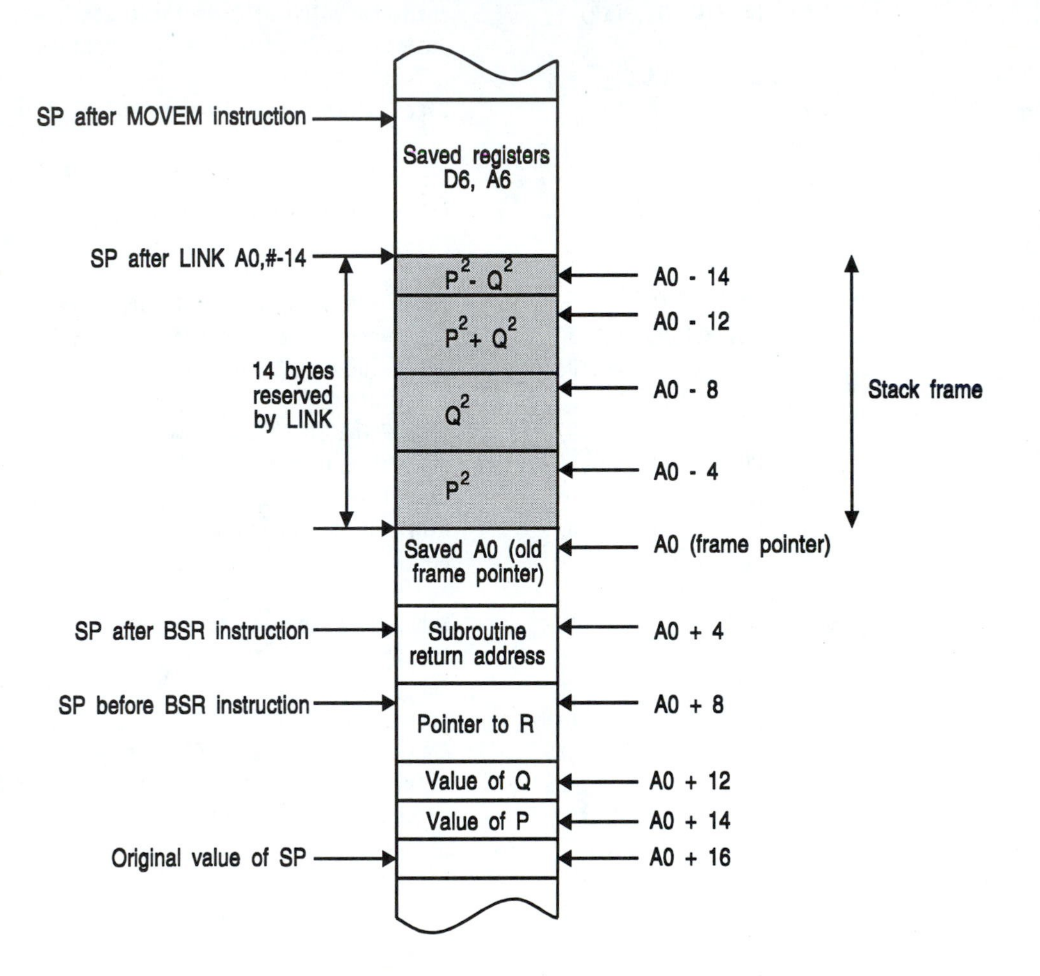

Figure 7.15 State of the stack after pushing the parameters

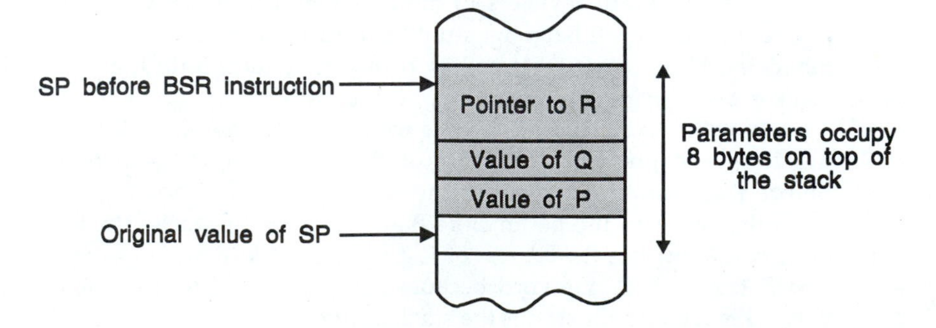

After calling the subroutine, the `LINK A0,#-14` instruction pushes the old value of A0 (the frame pointer) on the stack and moves the stack pointer up by 14 bytes (Figure 7.16). A0 is loaded with the value of the stack pointer immediately before it was moved up 14 bytes. That is, A0 is the current frame pointer and is now pointing at the base of the subroutine's stack frame. We can access all stack values by reference to A0 — we could also use the SP as a reference. It is better to use A0 (i.e., the FP) rather than the stack pointer to access items in the stack frame, since A0 will be constant for the duration of this procedure, while the stack pointer may be modified. Note that the stack frame is 14 bytes because it holds three longwords (P^2, Q^2, P^2+Q^2) and a word (P^2-Q^2). We save P^2-Q^2 as a word, because the 68000 requires that a divisor be a word value.

Figure 7.17 illustrates the next stage in the process in which the two working registers, D6 and A6, are saved on the stack above the stack frame. This operation takes up two longwords on the stack. You should appreciate that we save the registers *after* creating a stack frame to simplify the pointer arithmetic. If the saved registers were below the stack frame, the offsets of the parameters on the stack (i.e., P, Q, and R) would depend on the number of registers saved. By putting the registers on top of the stack frame, we do not need to worry about the space taken by the saved registers.

At this stage, Figure 7.17, the work of the subroutine can be carried out and parameters accessed from the stack. Note that we get P and Q as values from the

Figure 7.16 State of the stack after the subroutine call and `LINK A0,#-14`

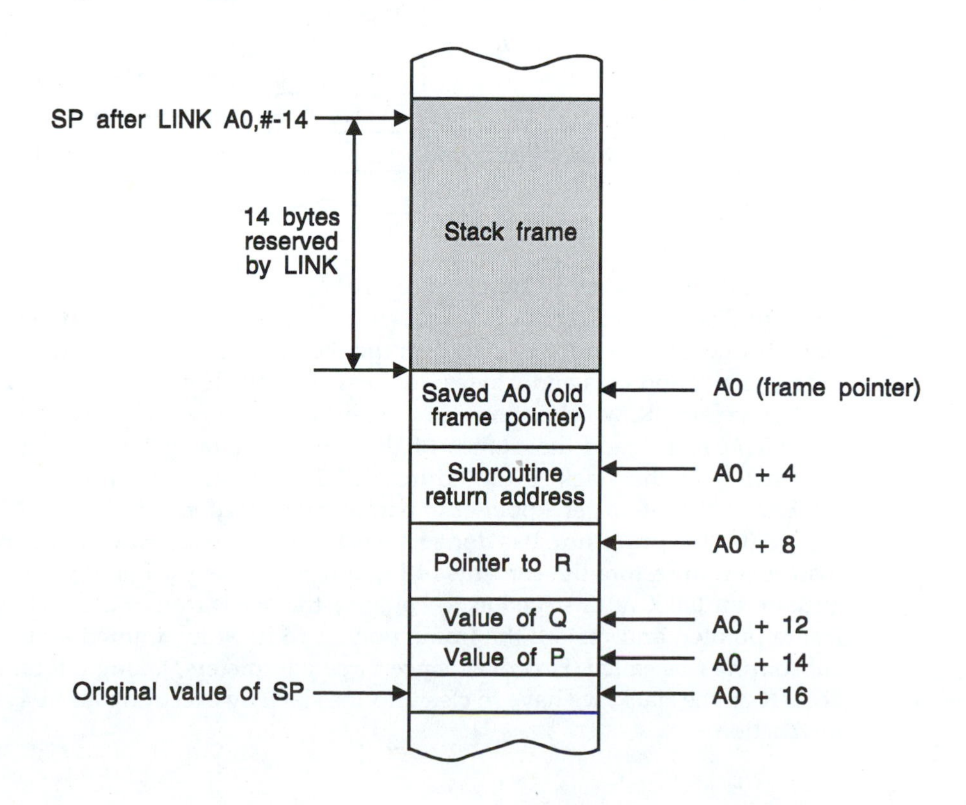

Figure 7.17 State of the stack after saving the working registers

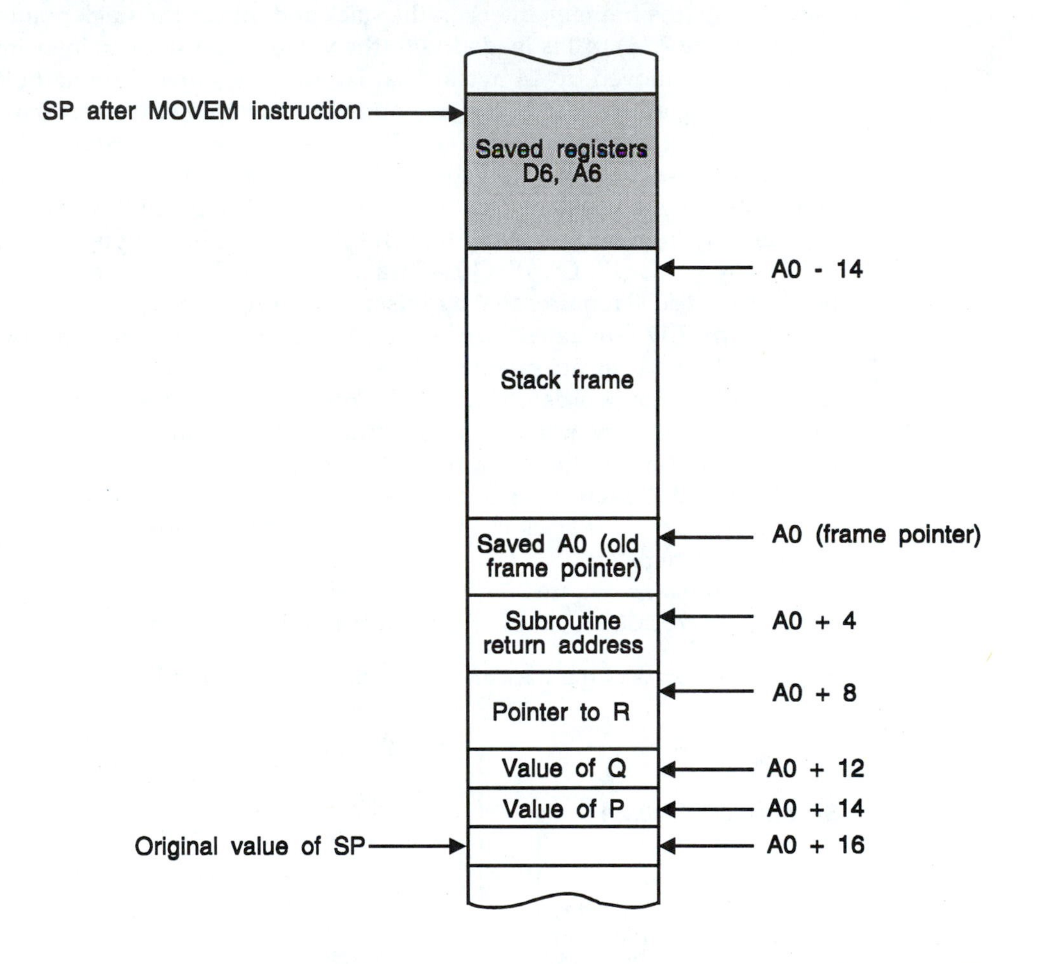

stack, and then access R indirectly by reading its address from the stack. The value of P is accessed by `MOVE.W (14,A0),D6` because A0 points to the base of the stack frame and we have to step past the old A0 (4 bytes), the return address (4 bytes), and R and Q (4 and 2 bytes) to access P itself. The total offset is 14 bytes.

Accessing R, which is passed by reference, is more complex. The instruction `LEA (8,A0),A6` loads the *address* of the parameter on the stack. However, the parameter on the stack is the address of R and we must use the instruction `MOVEA.L (A6),A6` to get a pointer to parameter R itself in A6.

After the procedure has done its work, we first use a `MOVEM.L (SP)+,D6/A6` instruction to restore the contents of D6 and A6 to their original values. We then execute an `UNLK A0` instruction to collapse the stack frame, discard the current frame pointer, and restore the frame pointer to its value immediately before the subroutine was called. However, since three parameters, taking a total of 8 bytes, are left on the stack, we have to clean up the stack by executing an `LEA (8,SP),SP` instruction.

Since the `LINK` and `UNLK` instructions are very important to the programmer who is going to write sophisticated 68000 programs, we will use the 68000 simulator to go through the example. Before we can do that, we need to build the subroutine into a test fixture.

We have set up an initial value for the stack pointer, the variables P and Q, the contents of A0, A6, and D6, and have reserved a memory location for R (i.e., 2000_{16}). The contents of registers A0, A6, and D6 have been set to the values \$A0A0A0A0, \$A6A6A6A6, and \$D6D6D6D6, respectively, so that they will stand out when we look at the stack. The initial values of P and Q are 7 and 6, respectively.

The following output is from the Teesside cross-assembler. Remember that the cross-assembler accepts only the original 68000 assembly language form of address register indirect addressing. For example, `MOVE -12(A0),D2` must be entered rather than `MOVE (-12,A0),D2`.

```
Source file: LINK1.X68
Assembled on: 93-02-02 at: 19:59:19
          by: X68K PC-2.0 Copyright (c) University of Teesside 1989,93
Defaults: ORG $0/FORMAT/OPT A,BRL,CEX,CL,FRL,MC,MD,NOMEX,NOPCO

 1 00000400                       ORG      $400
 2 00000400 2E7C00001026          MOVEA.L  #$1026,SP     ;Set up the stack pointer
 3 00000406 203C00000007          MOVE.L   #7,D0         ;Set up P
 4 0000040C 223C00000006          MOVE.L   #6,D1         ;Set up Q
 5 00000412 4DF9A6A6A6A6          LEA      $A6A6A6A6,A6  ;Set up dummy A6
 6 00000418 41F9A0A0A0A0          LEA      $A0A0A0A0,A0  ;Set up dummy A0
 7 0000041E 2C3CD6D6D6D6          MOVE.L   #$D6D6D6D6,D6 ;Set up dummy D6
 8                          *
 9 00000424 3F00                  MOVE.W   D0,-(SP)      ;Push value of P on stack
10 00000426 3F01                  MOVE.W   D1,-(SP)      ;Push value of Q on stack
11 00000428 487900002000          PEA      R             ;Push reference to R on stack
12 0000042E 6100000A              BSR      CALC          ;Call the subroutine
13 00000432 4FEF0008              LEA      8(SP),SP      ;Clean up stack by removing P,Q,R
14 00000436 4E722700              STOP     #$2700
15                          *
16 0000043A 4E50FFF2        CALC: LINK     A0,#-14       ;Stack frame=3 longwords + word
17 0000043E 48E70202              MOVEM.L  D6/A6,-(SP)   ;Save working registers on the stack
18 00000442 3C28000E              MOVE.W   14(A0),D6     ;Get value of P from stack
19 00000446 CCC6                  MULU     D6,D6         ;Calculate P^2
20 00000448 2146FFFC              MOVE.L   D6,-4(A0)     ;Save P^2 on the stack frame
21 0000044C 3C28000C              MOVE.W   12(A0),D6     ;Now get Q from the stack
22 00000450 CCC6                  MULU     D6,D6         ;Calculate Q^2
23 00000452 2146FFF8              MOVE.L   D6,-8(A0)     ;Store Q^2 on the stack frame
24 00000456 2C28FFFC              MOVE.L   -4(A0),D6     ;Get P^2
25 0000045A DCA8FFF8              ADD.L    -8(A0),D6     ;Add Q^2
26 0000045E 2146FFF4              MOVE.L   D6,-12(A0)    ;Store P^2+Q^2 on the stack frame
```

```
27 00000462 2C28FFFC          MOVE.L   -4(A0),D6       ;Get P²
28 00000466 9CA8FFF8          SUB.L    -8(A0),D6       ;Subtract Q²
29 0000046A 3146FFF2          MOVE.W   D6,-14(A0)      ;Store P²-Q² on the stack frame
30 0000046E 2C28FFF4          MOVE.L   -12(A0),D6      ;Read P²+Q² from the stack frame
31 00000472 8CE8FFF2          DIVU     -14(A0),D6      ;Calculate (P²+Q²)/(P²-Q²)
32 00000476 4DE80008          LEA      8(A0),A6        ;Get pointer to address of R
33 0000047A 2C56              MOVEA.L  (A6),A6         ;Get actual address of R
34 0000047C 3C86              MOVE.W   D6,(A6)         ;Modify R in the calling routine
35 0000047E 4CDF4040          MOVEM.L  (SP)+,D6/A6     ;Restore working registers
36 00000482 4E58              UNLK     A0              ;Collapse the stack frame
37 00000484 4E75              RTS
38                     *
39 00002000                   ORG      $2000
40 00002000 00000004   R:     DS.L     1
41          00000400          END      $400

Lines: 41, Errors: 0, Warnings: 0.
```

We will load the program and run it under the trace mode. Every so often, we will examine the state of the stack by using the simulator's MD (memory display function). Remember that the stack grows toward lower addresses.

```
OK, e68k LINKDEMO
[E68k 4.0 Copyright (c) Teesside Polytechnic 1989]

Address space 0 to ^10485759 (10240kbytes), MAXKB is ^4.
Loading binary file "LINK.BIN".
Start address: 000400, Low: 00000400, High: 00000485

>DF

PC=000400 SR=2000 SS=00A00000 US=00000000         X=0
A0=00000000 A1=00000000 A2=00000000 A3=00000000 N=0
A4=00000000 A5=00000000 A6=00000000 A7=00A00000 Z=0
D0=00000000 D1=00000000 D2=00000000 D3=00000000 V=0
D4=00000000 D5=00000000 D6=00000000 D7=00000000 C=0
———>MOVEA.L  #4134,SP

>TR

PC=000406 SR=2000 SS=00001026 US=00000000         X=0
A0=00000000 A1=00000000 A2=00000000 A3=00000000 N=0
A4=00000000 A5=00000000 A6=00000000 A7=00001026 Z=0
D0=00000000 D1=00000000 D2=00000000 D3=00000000 V=0
D4=00000000 D5=00000000 D6=00000000 D7=00000000 C=0
———>MOVE.L   #7,D0
```

```
Trace>

PC=00040C SR=2000 SS=00001026 US=00000000          X=0
A0=00000000 A1=00000000 A2=00000000 A3=00000000 N=0
A4=00000000 A5=00000000 A6=00000000 A7=00001026 Z=0
D0=00000007 D1=00000000 D2=00000000 D3=00000000 V=0
D4=00000000 D5=00000000 D6=00000000 D7=00000000 C=0
———>MOVE.L    #6,D1

Trace>

PC=000412 SR=2000 SS=00001026 US=00000000          X=0
A0=00000000 A1=00000000 A2=00000000 A3=00000000 N=0
A4=00000000 A5=00000000 A6=00000000 A7=00001026 Z=0
D0=00000007 D1=00000006 D2=00000000 D3=00000000 V=0
D4=00000000 D5=00000000 D6=00000000 D7=00000000 C=0
———>LEA       $A6A6A6A6,A6

Trace>

PC=000418 SR=2000 SS=00001026 US=00000000          X=0
A0=00000000 A1=00000000 A2=00000000 A3=00000000 N=0
A4=00000000 A5=00000000 A6=A6A6A6A6 A7=00001026 Z=0
D0=00000007 D1=00000006 D2=00000000 D3=00000000 V=0
D4=00000000 D5=00000000 D6=00000000 D7=00000000 C=0
———>LEA       $A0A0A0A0,A0

Trace>

PC=00041E SR=2000 SS=00001026 US=00000000          X=0
A0=A0A0A0A0 A1=00000000 A2=00000000 A3=00000000 N=0
A4=00000000 A5=00000000 A6=A6A6A6A6 A7=00001026 Z=0
D0=00000007 D1=00000006 D2=00000000 D3=00000000 V=0
D4=00000000 D5=00000000 D6=00000000 D7=00000000 C=0
———>MOVE.L    #-690563370,D6

Trace>

PC=000424 SR=2008 SS=00001026 US=00000000          X=0
A0=A0A0A0A0 A1=00000000 A2=00000000 A3=00000000 N=1
A4=00000000 A5=00000000 A6=A6A6A6A6 A7=00001026 Z=0
D0=00000007 D1=00000006 D2=00000000 D3=00000000 V=0
D4=00000000 D5=00000000 D6=D6D6D6D6 D7=00000000 C=0
———>MOVE.W    D0,-(SP)

Trace>
```

```
PC=000426 SR=2000 SS=00001024 US=00000000           X=0
A0=A0A0A0A0 A1=00000000 A2=00000000 A3=00000000 N=0
A4=00000000 A5=00000000 A6=A6A6A6A6 A7=00001024 Z=0
D0=00000007 D1=00000006 D2=00000000 D3=00000000 V=0
D4=00000000 D5=00000000 D6=D6D6D6D6 D7=00000000 C=0
———>MOVE.W   D1,-(SP)

Trace>

PC=000428 SR=2000 SS=00001022 US=00000000           X=0
A0=A0A0A0A0 A1=00000000 A2=00000000 A3=00000000 N=0
A4=00000000 A5=00000000 A6=A6A6A6A6 A7=00001022 Z=0
D0=00000007 D1=00000006 D2=00000000 D3=00000000 V=0
D4=00000000 D5=00000000 D6=D6D6D6D6 D7=00000000 C=0
———>PEA      $00002000

Trace>

PC=00042E SR=2000 SS=0000101E US=00000000           X=0
A0=A0A0A0A0 A1=00000000 A2=00000000 A3=00000000 N=0
A4=00000000 A5=00000000 A6=A6A6A6A6 A7=0000101E Z=0
D0=00000007 D1=00000006 D2=00000000 D3=00000000 V=0
D4=00000000 D5=00000000 D6=D6D6D6D6 D7=00000000 C=0
———>BSR.L    $0000043A
```

At this stage we have set up the initial values of A0, A6, D0, D1, and D6. Note that the initial value of the stack pointer, A7, was 1026_{16}, and is now $101E_{16}$ because we have pushed eight bytes on the stack. We will now start executing the program proper using the values we have set up.

```
>MD 1016

001016  00 00 00 00 00 00 00 00 00 00 20 00 00 06 00 07
001026  00 00 00 00 00 00 00 00 00 00 00 00 00 00 00 00.
```

As you can see, the stack contains the values 00002000_{16}, 0006_{16}, and 0007_{16}. We will continue tracing.

```
>TR

PC=00043A SR=2000 SS=0000101A US=00000000           X=0
A0=A0A0A0A0 A1=00000000 A2=00000000 A3=00000000 N=0
A4=00000000 A5=00000000 A6=A6A6A6A6 A7=0000101A Z=0
D0=00000007 D1=00000006 D2=00000000 D3=00000000 V=0
D4=00000000 D5=00000000 D6=D6D6D6D6 D7=00000000 C=0
———>LINK     A0,#-14
```

```
Trace>

PC=00043E SR=2000 SS=00001008 US=00000000          X=0
A0=00001016 A1=00000000 A2=00000000 A3=00000000 N=0
A4=00000000 A5=00000000 A6=A6A6A6A6 A7=00001008 Z=0
A4=00000000 A5=00000000 A6=A6A6A6A6 A7=00001008 Z=0
D0=00000007 D1=00000006 D2=00000000 D3=00000000 V=0
D4=00000000 D5=00000000 D6=D6D6D6D6 D7=00000000 C=0
---------->MOVEM.L  A6/D6,-(SP)

Trace>

PC=000442 SR=2000 SS=00001000 US=00000000          X=0
A0=00001016 A1=00000000 A2=00000000 A3=00000000 N=0
A4=00000000 A5=00000000 A6=A6A6A6A6 A7=00001000 Z=0
D0=00000007 D1=00000006 D2=00000000 D3=00000000 V=0
D4=00000000 D5=00000000 D6=D6D6D6D6 D7=00000000 C=0
---------->MOVE.W   14(A0),D6
```

We have just pushed registers A6 and D6 on the stack and are about to execute the instruction `MOVE.W (14,A0),D6`. Let's have another look at the current state of the stack.

```
>MD 1000

001000  D6 D6 D6 D6 A6 A6 A6 A6 00 00 00 00 00 00 00 00
001010  00 00 00 00 00 00 A0 A0 A0 A0 00 00 04 32 00 00
001020  20 00 00 06 00 07 00 00 00 00 00 00 00 00 00 00.
```

Starting at the top of the stack and moving down we have: the contents of registers D6 and A6 saved by the `MOVEM.L` instruction, 14 bytes of zero comprising the stack frame itself (the simulator initializes memory to zero), the original value of A0 (the frame pointer), the subroutine return address (00000432_{16}), and the eight bytes pushed on the stack before calling the subroutine. We will continue to trace the program.

```
>TR

PC=000446 SR=2000 SS=00001000 US=00000000          X=0
A0=00001016 A1=00000000 A2=00000000 A3=00000000 N=0
A4=00000000 A5=00000000 A6=A6A6A6A6 A7=00001000 Z=0
D0=00000007 D1=00000006 D2=00000000 D3=00000000 V=0
D4=00000000 D5=00000000 D6=D6D60007 D7=00000000 C=0
——————>MULU     D6,D6

Trace>
```

```
PC=000448 SR=2008 SS=00001000 US=00000000          X=0
A0=00001016 A1=00000000 A2=00000000 A3=00000000 N=1
A4=00000000 A5=00000000 A6=A6A6A6A6 A7=00001000 Z=0
D0=00000007 D1=00000006 D2=00000000 D3=00000000 V=0
D4=00000000 D5=00000000 D6=00000031 D7=00000000 C=0
——>MOVE.L   D6,-4(A0)
```

The instruction `MULU D6,D6` squares the contents of D6 (i.e., P = 7) to provide a result 31_{16} (this is $49_{10} = 7^2$).

```
Trace>
PC=00044C SR=2000 SS=00001000 US=00000000          X=0
A0=00001016 A1=00000000 A2=00000000 A3=00000000 N=0
A4=00000000 A5=00000000 A6=A6A6A6A6 A7=00001000 Z=0
D0=00000007 D1=00000006 D2=00000000 D3=00000000 V=0
D4=00000000 D5=00000000 D6=00000031 D7=00000000 C=0
——>MOVE.W   12(A0),D6

Trace>

PC=000450 SR=2000 SS=00001000 US=00000000          X=0
A0=00001016 A1=00000000 A2=00000000 A3=00000000 N=0
A4=00000000 A5=00000000 A6=A6A6A6A6 A7=00001000 Z=0
D0=00000007 D1=00000006 D2=00000000 D3=00000000 V=0
D4=00000000 D5=00000000 D6=00000006 D7=00000000 C=0
——>MULU     D6,D6

Trace>

PC=000452 SR=2000 SS=00001000 US=00000000          X=0
A0=00001016 A1=00000000 A2=00000000 A3=00000000 N=0
A4=00000000 A5=00000000 A6=A6A6A6A6 A7=00001000 Z=0
D0=00000007 D1=00000006 D2=00000000 D3=00000000 V=0
D4=00000000 D5=00000000 D6=00000024 D7=00000000 C=0
——>MOVE.L   D6,-8(A0)

Trace>

PC=000456 SR=2000 SS=00001000 US=00000000          X=0
A0=00001016 A1=00000000 A2=00000000 A3=00000000 N=0
A4=00000000 A5=00000000 A6=A6A6A6A6 A7=00001000 Z=0
D0=00000007 D1=00000006 D2=00000000 D3=00000000 V=0
D4=00000000 D5=00000000 D6=00000024 D7=00000000 C=0
——>MOVE.L   -4(A0),D6

Trace>
```

```
PC=00045A SR=2000 SS=00001000 US=00000000          X=0
A0=00001016 A1=00000000 A2=00000000 A3=00000000 N=0
A4=00000000 A5=00000000 A6=A6A6A6A6 A7=00001000 Z=0
D0=00000007 D1=00000006 D2=00000000 D3=00000000 V=0
D4=00000000 D5=00000000 D6=00000031 D7=00000000 C=0
------->ADD.L    -8(A0),D6

Trace>
PC=00045E SR=2000 SS=00001000 US=00000000          X=0
A0=00001016 A1=00000000 A2=00000000 A3=00000000 N=0
A4=00000000 A5=00000000 A6=A6A6A6A6 A7=00001000 Z=0
D0=00000007 D1=00000006 D2=00000000 D3=00000000 V=0
D4=00000000 D5=00000000 D6=00000055 D7=00000000 C=0
------->MOVE.L   D6,-12(A0)
```

We have now added $P^2 + Q^2 = 49 + 36 = 85 = 55_{16}$. As you can see, this value is currently stored in data register D6.

```
Trace>

PC=000462 SR=2000 SS=00001000 US=00000000          X=0
A0=00001016 A1=00000000 A2=00000000 A3=00000000 N=0
A4=00000000 A5=00000000 A6=A6A6A6A6 A7=00001000 Z=0
D0=00000007 D1=00000006 D2=00000000 D3=00000000 V=0
D4=00000000 D5=00000000 D6=00000055 D7=00000000 C=0
------->MOVE.L   -4(A0),D6

Trace>

PC=000466 SR=2000 SS=00001000 US=00000000          X=0
A0=00001016 A1=00000000 A2=00000000 A3=00000000 N=0
A4=00000000 A5=00000000 A6=A6A6A6A6 A7=00001000 Z=0
D0=00000007 D1=00000006 D2=00000000 D3=00000000 V=0
D4=00000000 D5=00000000 D6=00000031 D7=00000000 C=0
------->SUB.L    -8(A0),D6

Trace>

PC=00046A SR=2000 SS=00001000 US=00000000          X=0
A0=00001016 A1=00000000 A2=00000000 A3=00000000 N=0
A4=00000000 A5=00000000 A6=A6A6A6A6 A7=00001000 Z=0
D0=00000007 D1=00000006 D2=00000000 D3=00000000 V=0
D4=00000000 D5=00000000 D6=0000000D D7=00000000 C=0
------->MOVE.W   D6,-14(A0)
```

At this point, the contents of D6 are $P^2 - Q^2 = 49 - 36 = 13 = D_{16}$.

```
Trace>
PC=00046E SR=2000 SS=00001000 US=00000000          X=0
A0=00001016 A1=00000000 A2=00000000 A3=00000000 N=0
A4=00000000 A5=00000000 A6=A6A6A6A6 A7=00001000 Z=0
D0=00000007 D1=00000006 D2=00000000 D3=00000000 V=0
D4=00000000 D5=00000000 D6=0000000D D7=00000000 C=0
------>MOVE.L   -12(A0),D6

Trace>
PC=000472 SR=2000 SS=00001000 US=00000000          X=0
A0=00001016 A1=00000000 A2=00000000 A3=00000000 N=0
A4=00000000 A5=00000000 A6=A6A6A6A6 A7=00001000 Z=0
D0=00000007 D1=00000006 D2=00000000 D3=00000000 V=0
D4=00000000 D5=00000000 D6=00000055 D7=00000000 C=0
------>DIVU     -14(A0),D6

Trace>
PC=000476 SR=2000 SS=00001000 US=00000000          X=0
A0=00001016 A1=00000000 A2=00000000 A3=00000000 N=0
A4=00000000 A5=00000000 A6=A6A6A6A6 A7=00001000 Z=0
D0=00000007 D1=00000006 D2=00000000 D3=00000000 V=0
D4=00000000 D5=00000000 D6=00070006 D7=00000000 C=0
------>LEA      8(A0),A6
```

We have now performed the division of $(P^2 + Q^2)/(P^2 - Q^2) = 85/13$, which is 6 remainder 7. This is the value currently in D6. Before continuing, we will have another look at the stack.

```
>MD 1000

001000  D6 D6 D6 D6 A6 A6 A6 A6 00 0D 00 00 00 55 00 00
001010  00 24 00 00 00 31 A0 A0 A0 A0 00 00 04 32 00 00
001020  20 00 00 06 00 07 00 00 00 00 00 00 00 00 00 00
001030  00 00 00 00 00 00 00 00 00 00 00 00 00 00 00 00.
```

We continue tracing. The next two instructions fetch the address of parameter R and then store the result (i.e., 6) at this location.

```
>tr

PC=00047A SR=2000 SS=00001000 US=00000000          X=0
A0=00001016 A1=00000000 A2=00000000 A3=00000000 N=0
A4=00000000 A5=00000000 A6=0000101E A7=00001000 Z=0
D0=00000007 D1=00000006 D2=00000000 D3=00000000 V=0
D4=00000000 D5=00000000 D6=00070006 D7=00000000 C=0
------>MOVEA.L  (A6),A6
```

```
Trace>
PC=00047C SR=2000 SS=00001000 US=00000000          X=0
A0=00001016 A1=00000000 A2=00000000 A3=00000000 N=0
A4=00000000 A5=00000000 A6=00002000 A7=00001000 Z=0
D0=00000007 D1=00000006 D2=00000000 D3=00000000 V=0
D4=00000000 D5=00000000 D6=00070006 D7=00000000 C=0
------>MOVE.W   D6,(A6)
```

Now that the work's all done, we can pack up and go home. The saved registers are first retrieved and the stack frame collapsed.

```
Trace>

PC=00047E SR=2000 SS=00001000 US=00000000          X=0
A0=00001016 A1=00000000 A2=00000000 A3=00000000 N=0
A4=00000000 A5=00000000 A6=00002000 A7=00001000 Z=0
D0=00000007 D1=00000006 D2=00000000 D3=00000000 V=0
D4=00000000 D5=00000000 D6=00070006 D7=00000000 C=0
------>MOVEM.L  (SP)+,A6/D6

Trace>

PC=000482 SR=2000 SS=00001008 US=00000000          X=0
A0=00001016 A1=00000000 A2=00000000 A3=00000000 N=0
A4=00000000 A5=00000000 A6=A6A6A6A6 A7=00001008 Z=0
D0=00000007 D1=00000006 D2=00000000 D3=00000000 V=0
D4=00000000 D5=00000000 D6=D6D6D6D6 D7=00000000 C=0
------>UNLK     A0

Trace>

PC=000484 SR=2000 SS=0000101A US=00000000          X=0
A0=A0A0A0A0 A1=00000000 A2=00000000 A3=00000000 N=0
A4=00000000 A5=00000000 A6=A6A6A6A6 A7=0000101A Z=0
D0=00000007 D1=00000006 D2=00000000 D3=00000000 V=0
D4=00000000 D5=00000000 D6=D6D6D6D6 D7=00000000 C=0
------>RTS

Trace>

PC=000432 SR=2000 SS=0000101E US=00000000          X=0
A0=A0A0A0A0 A1=00000000 A2=00000000 A3=00000000 N=0
A4=00000000 A5=00000000 A6=A6A6A6A6 A7=0000101E Z=0
D0=00000007 D1=00000006 D2=00000000 D3=00000000 V=0
D4=00000000 D5=00000000 D6=D6D6D6D6 D7=00000000 C=0
------>LEA      8(SP),SP
```

```
Trace>

PC=000436 SR=2000 SS=00001026 US=00000000          X=0
A0=A0A0A0A0 A1=00000000 A2=00000000 A3=00000000 N=0
A4=00000000 A5=00000000 A6=A6A6A6A6 A7=00001026 Z=0
D0=00000007 D1=00000006 D2=00000000 D3=00000000 V=0
D4=00000000 D5=00000000 D6=D6D6D6D6 D7=00000000 C=0
------->STOP      #9984
```

The final step is to examine the contents of memory location 2000_{16}. It should contain the result (i.e., 6).

```
>MD 2000

002000  00 06 00 00 00 00 00 00 00 00 00 00 00 00 00 00.
```

It does, it does. This example should demonstrate three things. The first is how the two instructions `LINK` and `UNLK` work. The second is how the stack can be used to pass parameters to a procedure both by reference and by value. The third is the inherent difficulty in writing assembly language programs. For example, in order to reference parameter P, we had to execute `MOVE.W (14,A0)`, which meant that we had to have a clear picture of the location of P with respect to the frame pointer, A0. The chance of making a mistake in dealing with the stack in programs like this is very great indeed. Such complex stack manipulation is best left to compilers.

7.4 Recursion

There are not a lot of highly original jokes associated with computer science. One of the few is 'Question: What's the dictionary definition of recursion?' 'Answer: See recursion.' An object is said to be defined *recursively* if it is defined in terms of itself. A procedure is said to be recursive if it calls itself.

At first sight, recursion sounds odd, if not ridiculous. Many programmers do not encounter recursion early in a computer science course. They first meet *iteration*, which might be said to be the *opposite* of recursion. The textbook example most commonly used to introduce recursion is the factorial. Factorial n is written n! and defined as:

$$n! = n \times (n-1) \times (n-2) \times (n-3) \times \ldots \times 3 \times 2 \times 1.$$

For example, 6! = 6x5x4x3x2x1 = 720. We can immediately write down an iterative procedure to evaluate n!.

```
Factorial_N = 1
I := N
REPEAT
    Factorial_N := Factorial * I
    I := I - 1
UNTIL I = 0
```

This procedure can readily be translated into 68000 code as:

```
*                               D0 contains initial value of N
*                               D1 contains the 16-bit result
*
        MOVE.W   #1,D1          Set initial value of the result to 1
Loop    MULU     D0,D1          REPEAT N_Factorial:= N_Factorial * I
        SUBQ.W   #1,D0                 I := I - 1
        BNE      Loop           UNTIL I = 0
        RTS
```

This code provides a correct result, provided that it is less than 65,535. We now look at the recursive definition of factorial n. That is:

n! = n x (n-1)!

This definition is recursive because n! is defined in terms of another factorial value; i.e., (n-1)!. Using the same definition we can evaluate (n-1)! from (n-1) x (n-2)!, and so on. When we reach 1!, we substitute the value 1! = 1, and then evaluate the expression for n!. The pseudocode construct of a recursive procedure to evaluate n! is:

```
Module Factorial(N)
                  IF N = 1 THEN Factorial(N) := 1
                           ELSE Factorial(N) := N*Factorial(N-1)
                  END_IF
End Factorial
```

The procedure `Factorial(N)` calls itself. It is, therefore, a recursive procedure. Let's walk through this procedure for n = 4.

1. The procedure is called initially with n = 4. The `IF` part yields a false value because N is not 1, and the `ELSE` part begins to calculate 4*`Factorial(3)`.
2. The `ELSE` part calls `Factorial(3)` recursively. The `IF` part of the procedure yields a false value and the `ELSE` part begins to calculate the expression 3*`Factorial(2)`.
3. The `ELSE` part calls `Factorial(2)`. Again, the `IF` part yields a false value and the `ELSE` part begins to calculate 2*`Factorial(1)`.

4. The ELSE part calls Factorial(1). Now, the IF part yields a true value and a return from Factorial(1) with the value 1 is made.

5. A return is made from Factorial(1) to Factorial(2). The expression 2*Factorial(1) is calculated, which is 2*1 = 2.

6. The return from Factorial(2) to Factorial(3) results in the evaluation of 3*Factorial(2), which is 3*2 = 6.

7. Finally, the return from Factorial(3) to Factorial(4) results in 4*6 = 24.

As you can see from this walkthrough, the arithmetic is performed after the procedure has been called recursively n times. In order to implement recursion, successive return addresses (and any workspace) are stored on the stack.

We can now write a recursive factorial procedure in assembly language.

```
*         Factorial(N)
*         On entry D0 holds N and on exit D0 holds N!
*
Factor    MOVE.W  D0,-(A7)      Push N on the stack
          SUBQ.W  #1,D0         N := N - 1
          BEQ     ExitFact      IF N = 1 THEN Factorial(N) := 1
          BSR     Factor                 ELSE
          MULU    (A7)+,D0               Factorial(N) := N*Factorial(N-1)
          RTS
ExitFact  MOVE.W  (A7)+,D0      Clean up the stack
          RTS                   and return
```

This recursive procedure can be tested by running it with N = 4. The following output is from the cross-assembler. Note that we have included a section that sets up the stack and calls the recursive procedure.

will be on exam

```
Source file: FACT.X68
Assembled on: 92-11-27 at: 16:11:38
          by: X68K PC-1.9 Copyright (c) University of Teesside 1989,92
Defaults: ORG $0/FORMAT/OPT A,BRL,CEX,CL,FRL,MC,MD,NOMEX,NOPCO

  1                             *        Factorial(N)
  2                             *        On entry D0 holds N and on exit D0 holds N!
  3                             *
  4  00000400                            ORG      $400
  5  00000400 4FF81000                   LEA      $1000,A7      ;Set up the stack pointer
  6  00000404 203C00000004               MOVE.L   #4,D0
  7  0000040A 61000006                   BSR      FACTOR        ;Evaluate 4!
  8  0000040E 4E722700                   STOP     #$2700
  9                             *
 10  00000412 3F00              FACTOR:  MOVE.W   D0,-(A7)      ;Push N on the stack
 11  00000414 5340                       SUBQ.W   #1,D0         ;N := N - 1
```

```
12  00000416 67000008             BEQ       EXITFACT     ;IF N = 0 THEN Factorial_N := 1
13  0000041A 61F6                 BSR       FACTOR       ;ELSE
14  0000041C C0DF                 MULU      (A7)+,D0     ;Factorial_N:=N*Factorial_N(N-1)
15  0000041E 4E75                 RTS
16  00000420 301F       EXITFACT: MOVE.W    (A7)+,D0     ;Clean up the stack
17  00000422 4E75                 RTS                    ;and return
18                      *
19           00000400             END       $400

Lines: 19, Errors: 0, Warnings: 0.
```

When the procedure is first called, register D0.W, which contains 4, is pushed on the stack and then decremented. The procedure is called again if the contents of D0 are not zero. This sequence is repeated and the stack looks like Figure 7.18. Note that we have drawn the stack in an unusual fashion, with alternate words and longwords. The longwords are the successive return addresses, and the words are the values pushed on the stack by the procedure.

We will use the Teesside simulator to walk through the program itself. After loading the program, we use the DF, display formatted register command, and then the trace command.

```
df

PC=000400 SR=2000 SS=00A00000 US=00000000          X=0
A0=00000000 A1=00000000 A2=00000000 A3=00000000 N=0
A4=00000000 A5=00000000 A6=00000000 A7=00A00000 Z=0
D0=00000000 D1=00000000 D2=00000000 D3=00000000 V=0
D4=00000000 D5=00000000 D6=00000000 D7=00000000 C=0
———>LEA      $1000,SP
```

Figure 7.18 Growth of the stack during the calculation of 4!

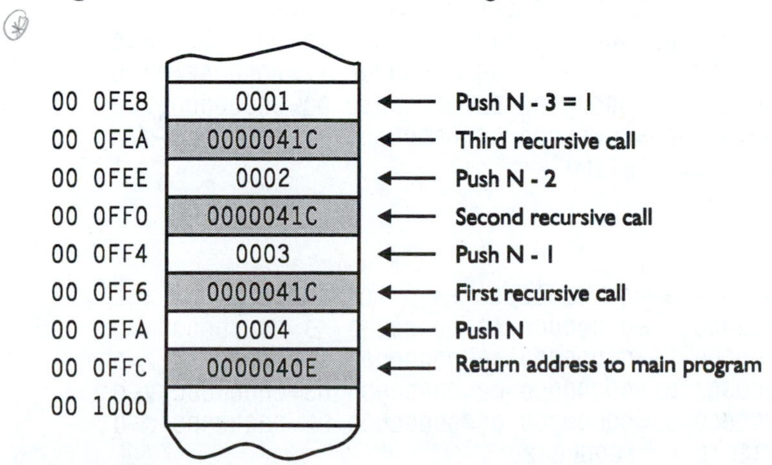

```
>tr

PC=000404 SR=2000 SS=00001000 US=00000000          X=0
A0=00000000 A1=00000000 A2=00000000 A3=00000000 N=0
A4=00000000 A5=00000000 A6=00000000 A7=00001000 Z=0
D0=00000000 D1=00000000 D2=00000000 D3=00000000 V=0
D4=00000000 D5=00000000 D6=00000000 D7=00000000 C=0
———>MOVE.L    #4,D0

Trace>

PC=00040A SR=2000 SS=00001000 US=00000000          X=0
A0=00000000 A1=00000000 A2=00000000 A3=00000000 N=0
A4=00000000 A5=00000000 A6=00000000 A7=00001000 Z=0
D0=00000004 D1=00000000 D2=00000000 D3=00000000 V=0
D4=00000000 D5=00000000 D6=00000000 D7=00000000 C=0
———>BSR.L     $00000412
```

Here's where we call the procedure FACTOR for the first time.

```
Trace>

PC=000412 SR=2000 SS=00000FFC US=00000000          X=0
A0=00000000 A1=00000000 A2=00000000 A3=00000000 N=0
A4=00000000 A5=00000000 A6=00000000 A7=00000FFC Z=0
D0=00000004 D1=00000000 D2=00000000 D3=00000000 V=0
D4=00000000 D5=00000000 D6=00000000 D7=00000000 C=0
———>MOVE.W    D0,-(SP)

Trace>

PC=000414 SR=2000 SS=00000FFA US=00000000          X=0
A0=00000000 A1=00000000 A2=00000000 A3=00000000 N=0
A4=00000000 A5=00000000 A6=00000000 A7=00000FFA Z=0
D0=00000004 D1=00000000 D2=00000000 D3=00000000 V=0
D4=00000000 D5=00000000 D6=00000000 D7=00000000 C=0
———>SUBQ.W    #1,D0

Trace>

PC=000416 SR=2000 SS=00000FFA US=00000000          X=0
A0=00000000 A1=00000000 A2=00000000 A3=00000000 N=0
A4=00000000 A5=00000000 A6=00000000 A7=00000FFA Z=0
D0=00000003 D1=00000000 D2=00000000 D3=00000000 V=0
D4=00000000 D5=00000000 D6=00000000 D7=00000000 C=0
———>BEQ.L     $00000420
```

```
Trace>

PC=00041A SR=2000 SS=00000FFA US=00000000          X=0
A0=00000000 A1=00000000 A2=00000000 A3=00000000 N=0
A4=00000000 A5=00000000 A6=00000000 A7=00000FFA Z=0
D0=00000003 D1=00000000 D2=00000000 D3=00000000 V=0
D4=00000000 D5=00000000 D6=00000000 D7=00000000 C=0
———>BSR.S    $00000412
```

And here's our second call — this time it's recursive.

```
Trace>

PC=000412 SR=2000 SS=00000FF6 US=00000000          X=0
A0=00000000 A1=00000000 A2=00000000 A3=00000000 N=0
A4=00000000 A5=00000000 A6=00000000 A7=00000FF6 Z=0
D0=00000003 D1=00000000 D2=00000000 D3=00000000 V=0
D4=00000000 D5=00000000 D6=00000000 D7=00000000 C=0
———>MOVE.W   D0,-(SP)

Trace>

PC=000414 SR=2000 SS=00000FF4 US=00000000          X=0
A0=00000000 A1=00000000 A2=00000000 A3=00000000 N=0
A4=00000000 A5=00000000 A6=00000000 A7=00000FF4 Z=0
D0=00000003 D1=00000000 D2=00000000 D3=00000000 V=0
D4=00000000 D5=00000000 D6=00000000 D7=00000000 C=0
———>SUBQ.W   #1,D0

Trace>

PC=000416 SR=2000 SS=00000FF4 US=00000000          X=0
A0=00000000 A1=00000000 A2=00000000 A3=00000000 N=0
A4=00000000 A5=00000000 A6=00000000 A7=00000FF4 Z=0
D0=00000002 D1=00000000 D2=00000000 D3=00000000 V=0
D4=00000000 D5=00000000 D6=00000000 D7=00000000 C=0
———>BEQ.L    $00000420

Trace>

PC=00041A SR=2000 SS=00000FF4 US=00000000          X=0
A0=00000000 A1=00000000 A2=00000000 A3=00000000 N=0
A4=00000000 A5=00000000 A6=00000000 A7=00000FF4 Z=0
D0=00000002 D1=00000000 D2=00000000 D3=00000000 V=0
D4=00000000 D5=00000000 D6=00000000 D7=00000000 C=0
———>BSR.S    $00000412
```

This is the third call.

```
Trace>

PC=000412 SR=2000 SS=00000FF0 US=00000000          X=0
A0=00000000 A1=00000000 A2=00000000 A3=00000000 N=0
A4=00000000 A5=00000000 A6=00000000 A7=00000FF0 Z=0
D0=00000002 D1=00000000 D2=00000000 D3=00000000 V=0
D4=00000000 D5=00000000 D6=00000000 D7=00000000 C=0
———>MOVE.W   D0,-(SP)

Trace>

PC=000414 SR=2000 SS=00000FEE US=00000000          X=0
A0=00000000 A1=00000000 A2=00000000 A3=00000000 N=0
A4=00000000 A5=00000000 A6=00000000 A7=00000FEE Z=0
D0=00000002 D1=00000000 D2=00000000 D3=00000000 V=0
D4=00000000 D5=00000000 D6=00000000 D7=00000000 C=0
———>SUBQ.W   #1,D0

Trace>

PC=000416 SR=2000 SS=00000FEE US=00000000          X=0
A0=00000000 A1=00000000 A2=00000000 A3=00000000 N=0
A4=00000000 A5=00000000 A6=00000000 A7=00000FEE Z=0
D0=00000001 D1=00000000 D2=00000000 D3=00000000 V=0
D4=00000000 D5=00000000 D6=00000000 D7=00000000 C=0
———>BEQ.L    $00000420

Trace>
PC=00041A SR=2000 SS=00000FEE US=00000000          X=0
A0=00000000 A1=00000000 A2=00000000 A3=00000000 N=0
A4=00000000 A5=00000000 A6=00000000 A7=00000FEE Z=0
D0=00000001 D1=00000000 D2=00000000 D3=00000000 V=0
D4=00000000 D5=00000000 D6=00000000 D7=00000000 C=0
———>BSR.S    $00000412
```

This is the fourth and final call to the recursive procedure.

```
Trace>
PC=000412 SR=2000 SS=00000FEA US=00000000          X=0
A0=00000000 A1=00000000 A2=00000000 A3=00000000 N=0
A4=00000000 A5=00000000 A6=00000000 A7=00000FEA Z=0
D0=00000001 D1=00000000 D2=00000000 D3=00000000 V=0
D4=00000000 D5=00000000 D6=00000000 D7=00000000 C=0
———>MOVE.W   D0,-(SP)
```

```
Trace>

PC=000414 SR=2000 SS=00000FE8 US=00000000          X=0
A0=00000000 A1=00000000 A2=00000000 A3=00000000 N=0
A4=00000000 A5=00000000 A6=00000000 A7=00000FE8 Z=0
D0=00000001 D1=00000000 D2=00000000 D3=00000000 V=0
D4=00000000 D5=00000000 D6=00000000 D7=00000000 C=0
——>SUBQ.W   #1,D0

Trace>

PC=000416 SR=2004 SS=00000FE8 US=00000000          X=0
A0=00000000 A1=00000000 A2=00000000 A3=00000000 N=0
A4=00000000 A5=00000000 A6=00000000 A7=00000FE8 Z=1
D0=00000000 D1=00000000 D2=00000000 D3=00000000 V=0
D4=00000000 D5=00000000 D6=00000000 D7=00000000 C=0
——>BEQ.L    $00000420
```

Let's have a look at the contents of the stack by using the memory display function.

```
>MD FE8

000FE8  00 01 00 00 04 1C 00 02 00 00 04 1C 00 03 00 00
000FF8  04 1C 00 04 00 00 04 0E 00 00 00 00 00 00 00 00.
```

This memory map is the same as that of Figure 7.18, and shows the alternating sequence of data values and return addresses.

```
>TR

PC=000420 SR=2004 SS=00000FE8 US=00000000          X=0
A0=00000000 A1=00000000 A2=00000000 A3=00000000 N=0
A4=00000000 A5=00000000 A6=00000000 A7=00000FE8 Z=1
D0=00000000 D1=00000000 D2=00000000 D3=00000000 V=0
D4=00000000 D5=00000000 D6=00000000 D7=00000000 C=0
——>MOVE.W   (SP)+,D0
Trace>


PC=000422 SR=2000 SS=00000FEA US=00000000          X=0
A0=00000000 A1=00000000 A2=00000000 A3=00000000 N=0
A4=00000000 A5=00000000 A6=00000000 A7=00000FEA Z=0
D0=00000001 D1=00000000 D2=00000000 D3=00000000 V=0
D4=00000000 D5=00000000 D6=00000000 D7=00000000 C=0
——>RTS
```

Now we begin the returns and unwind the stack.

```
Trace>
PC=00041C SR=2000 SS=00000FEE US=00000000          X=0
A0=00000000 A1=00000000 A2=00000000 A3=00000000 N=0
A4=00000000 A5=00000000 A6=00000000 A7=00000FEE Z=0
D0=00000001 D1=00000000 D2=00000000 D3=00000000 V=0
D4=00000000 D5=00000000 D6=00000000 D7=00000000 C=0
———>MULU      (SP)+,D0
```

We pop an element off the stack and perform a multiplication.

```
>TR

PC=00041E SR=2000 SS=00000FF0 US=00000000          X=0
A0=00000000 A1=00000000 A2=00000000 A3=00000000 N=0
A4=00000000 A5=00000000 A6=00000000 A7=00000FF0 Z=0
D0=00000002 D1=00000000 D2=00000000 D3=00000000 V=0
D4=00000000 D5=00000000 D6=00000000 D7=00000000 C=0
———>RTS

Trace>
PC=00041C SR=2000 SS=00000FF4 US=00000000          X=0
A0=00000000 A1=00000000 A2=00000000 A3=00000000 N=0
A4=00000000 A5=00000000 A6=00000000 A7=00000FF4 Z=0
D0=00000002 D1=00000000 D2=00000000 D3=00000000 V=0
D4=00000000 D5=00000000 D6=00000000 D7=00000000 C=0
———>MULU      (SP)+,D0
```

Here, we perform the next multiplication.

```
>TR
PC=00041E SR=2000 SS=00000FF6 US=00000000          X=0
A0=00000000 A1=00000000 A2=00000000 A3=00000000 N=0
A4=00000000 A5=00000000 A6=00000000 A7=00000FF6 Z=0
D0=00000006 D1=00000000 D2=00000000 D3=00000000 V=0
D4=00000000 D5=00000000 D6=00000000 D7=00000000 C=0
———>RTS

Trace>
PC=00041C SR=2000 SS=00000FFA US=00000000          X=0
A0=00000000 A1=00000000 A2=00000000 A3=00000000 N=0
A4=00000000 A5=00000000 A6=00000000 A7=00000FFA Z=0
D0=00000006 D1=00000000 D2=00000000 D3=00000000 V=0
D4=00000000 D5=00000000 D6=00000000 D7=00000000 C=0
———>MULU      (SP)+,D0
```

This is the last multiplication. After this, the contents of data register D0 should be 4!, i.e., 24 or $18.

```
Trace>

PC=00041E SR=2000 SS=00000FFC US=00000000          X=0
A0=00000000 A1=00000000 A2=00000000 A3=00000000 N=0
A4=00000000 A5=00000000 A6=00000000 A7=00000FFC Z=0
D0=00000018 D1=00000000 D2=00000000 D3=00000000 V=0
D4=00000000 D5=00000000 D6=00000000 D7=00000000 C=0
------->RTS

Trace>

PC=00040E SR=2000 SS=00001000 US=00000000          X=0
A0=00000000 A1=00000000 A2=00000000 A3=00000000 N=0
A4=00000000 A5=00000000 A6=00000000 A7=00001000 Z=0
D0=00000018 D1=00000000 D2=00000000 D3=00000000 V=0
D4=00000000 D5=00000000 D6=00000000 D7=00000000 C=0
------->STOP     #9984
```

This text does not use recursive techniques elsewhere. We have introduced recursion here because it makes use of the 68000's stack mechanism. Moreover, if the recursive procedure requires local workspace, the 68000's LINK and UNLK instructions can be used to create a stack frame for each recursive call.

Recursive algorithms are found in many branches of computer science, particularly data structures. Many algorithms that are expressed iteratively can also be expressed in a recursive form. For example, consider the algorithm to detect whether a string is a palindrome or not (i.e., the string reads the same left to right as right to left, e.g., hannah). The iterative approach leads to an algorithm like:

```
Module Palindrome
        Set Found := true
        Set LeftPointer to leftmost character
        Set RightPointer to rightmost character
        REPEAT
                IF Char[LeftPointer] ≠ Char[RightPointer]
                    THEN
                    Found := false
                    EXIT
                END_IF
                LeftPointer := LeftPointer + 1
                RightPointer := RightPointer - 1
        UNTIL LeftPointer = RightPointer OR LeftPointer+1 = RightPointer
End Palindrome
```

We can also use a recursive algorithm to detect a palindrome. A string is a palindrome if the outermost two characters are equal and the remainder of the string is a palindrome.

```
Initialize Palindrome
        Left := pointer to start of string
        Right := pointer to end of string
Module Palindrome
        IF (Left < Right)
           THEN RETURN([M(LEFT)]=[M(RIGHT)]) AND Palindrome(Left+1,Right-1)
           ELSE RETURN(TRUE)
        END_IF
End Palindrome
```

The recursive algorithm says that a word is a palindrome if its outermost characters are equal and the remaining characters are a palindrome. One of my students, Roger Allen, wrote the following 68000 recursive procedure to check whether a string is a palindrome. We have added a function call and a mechanism to display the result (using the Teesside 68000 simulator).

```
Source file: PALINDRM.X68
Assembled on: 93-02-02 at: 21:03:08
        by: X68K PC-2.0 Copyright (c) University of Teesside 1989,93
Defaults: ORG $0/FORMAT/OPT A,BRL,CEX,CL,FRL,MC,MD,NOMEX,NOPCO
  1 00000400                        ORG     $400
  2 00000400 4FF81000               LEA     $1000,A7        ;Set up the stack
  3 00000404 487900002000           PEA     PAL_STRT        ;Push the start address of the string
  4 0000040A 487900000464           PEA     PAL_END         ;Push the end address of the string + 1
  5 00000410 61000024               BSR     PALINDRM        ;Test for a palindrome
  6 00000414 4FEF0008               LEA     8(A7),A7        ;Clean up the stack by removing addresses
  7 00000418 43F900000464           LEA     NOT_PAL,A1      ;Let's say it wasn't a palindrome
  8 0000041E 4A40                   TST.W   D0              ;If it wasn't go and say so
  9 00000420 66000008               BNE     P_STRING
 10 00000424 43F90000047D           LEA     IS_PAL,A1       ;If it was then say so
 11 0000042A 1219         P_STRING: MOVE.B  (A1)+,D1        ;Pick up the size of the message
 12 0000042C 103C0000               MOVE.B  #0,D0           ;D0 = 0 to print a string
 13 00000430 4E4F                   TRAP    #15             ;Call the O/S
 14 00000432 4E722700               STOP    #$2700
 15                       *
 16 00000436 4240         PALINDRM: CLR.W   D0              ;Clear D0 to assume palindrome
 17 00000438 48E740C0               MOVEM.L D1/A0-A1,-(A7)  ;Save working registers
 18 0000043C 4CEF03000010           MOVEM.L 16(A7),A0-A1    ;Load the start/end address in A0, A1
 19 00000442 B3C8                   CMPA.L  A0,A1           ;Test for "middle" or palindrome
 20 00000444 6C18                   BGE.S   PAL_EXIT        ;If at the middle, we can exit
 21 00000446 1219                   MOVE.B  (A1)+,D1        ;Otherwise, pick up a character
 22 00000448 41E8FFFF               LEA     -1(A0),A0       ;and move the other pointer back
```

```
23 0000044C 48E700C0                MOVEM.L  A0-A1,-(A7)          ;Save the two pointers
24 00000450 61E4                    BSR      PALINDRM             ;and see if what's left is a palindrome
25 00000452 4FEF0008                LEA      8(A7),A7             ;Remove the parameters after the subroutine
26 00000456 B218                    CMP.B    (A0)+,D1             ;Compare left character with right character
27 00000458 6704                    BEQ.S    PAL_EXIT             ;If same then all is well
28 0000045A 303CFFFF                MOVE.W   #-1,D0               ;otherwise set not_palindrome flag
29 0000045E 4CDF0302       PAL_EXIT: MOVEM.L (A7)+,D1/A0-A1       ;Restore registers before leaving
30 00000462 4E75                    RTS
31                         *
32          00000464       PAL_END: EQU      *
33 00000464 180A0D497420 NOT_PAL:  DC.B     24,$0A,$0D,'It is not a palindrome'
            6973206E6F74
            20612070616C
            696E64726F6D
            65
34 0000047D 140A0D497420 IS_PAL:   DC.B     20,$0A,$0D,'It is a palindrome'
            697320612070
            616C696E6472
            6F6D65
35 00002000                         ORG      $2000                ;Let's put the palindrome somewhere memorable
36 00002000 41424241       PAL_STRT: DC.B    'ABBA'               ;A test palindrome
37          00000400                END      $400

Lines: 37, Errors: 0, Warnings: 0.
```

Note that the Teesside 68000 simulator's TRAP #15 instruction is used to output a string. The parameter 0 in data register D0 tells the trap handler to output the string pointed at by address register A1 whose length is given by the contents of data register D1.

Summary

- A subroutine is called by means of either a BSR or a JSR instruction. A return from the subroutine to the instruction immediately after the calling point (i.e., after the BSR or JSR) is made by means of an RTS instruction.
- You should leave a subroutine at a single point (i.e., at the RTS instruction). You should never use a BRA, JMP, or Bcc instruction to jump out of a subroutine into another subroutine or the main program.
- The 68000's stack pointed at by A7 automatically handles subroutine return addresses. Subroutines may be nested; that is, one subroutine may call another subroutine. You must not corrupt the stack pointed at by A7; otherwise, the subroutine return mechanism will fail and the system will crash.

- The BSR instruction employs a relative address. Fortunately, the address is calculated automatically by the assembler. If you know that the subroutine is situated within 128 bytes of the calling point, you can use the short form of the branch — BSR.S. If you use a JSR instruction, the address is *absolute* and the code is not position-independent.
- Although subroutines are primarily used to avoid rewriting bits of frequently used code, they can also be used to facilitate top-down design and to enhance program readability.
- Subroutines usually operate on data. Data can be passed to and from a subroutine by putting it in registers, memory, or on the stack. The stack is the preferred transfer mechanism.
- Passing data to and from a subroutine via the stack aids the writing of re-entrant subroutines.
- Data may be passed to a subroutine by means of its value or by its address. When it is passed by value, a copy of the parameter is transferred. When it is passed by address (or by reference), the address of the parameter is passed to the subroutine. You can use the push effective address instruction, PEA, to push an address on the stack before calling a subroutine.
- A subroutine frequently requires local variables. These variables are intermediate values required by the subroutine. Local values are not accessed from outside the subroutine and are discarded after the subroutine is exited. Local variables are normally created on the top of the stack.
- The 68000 includes a LINK and UNLK instruction pair that create a stack frame for local variables and collapse the stack frame, respectively.

Problems

1. If the BSR and RTS instructions did not exist, what two instruction sequences would you use to synthesize these instructions?

2. Why is the BSR instruction generally preferred to the JSR?

3. When does a JSR instruction prove invaluable?

4. What is the difference between passing a parameter to a subroutine by value and passing it by reference?

5. What is a *local variable* and why should it be local?

6. Explain how the 68000's LINK and UNLK instructions help to implement a stack frame.

7. Will the following code work?

```
        BSR Sub1            Call subroutine 1
        .
Sub1    BSR Sub2            Call subroutine 2
        .
        .
        RTS                 Return from subroutine 1
Sub2    .
        .
        LEA (4,A7),A7       Step past the return address
        RTS                 Return directly to the main program
```

8. Why does the `LINK` instruction invariably take a *negative* literal as a parameter?

9. Write a recursive subroutine to calculate the nth value in the Fibonacci sequence: 1, 1, 3, 5, 8, 13. The first two terms in the sequence are 1, 1, and the value of any other term is given by the sum of the two immediately preceding terms.

CHAPTER 8

Assemblers

It might seem a little strange to include a chapter on assemblers so late in a book on assembly language for the 68000 family. So far, we have not devoted a lot of time to assemblers as such for four good reasons. First, it takes some time for the programmer to become familiar with an assembler, just like any other item of computer software. Second, although there is only one 68000 architecture, there are many commercially available assemblers and cross-assemblers. Each of these programs has its own assembler syntax and facilities. Third, the fundamental purpose of this book is to teach the 68000's architecture and its assembly language. Going into fine detail about specific assemblers too early in the text would get in the way of this objective. The fourth reason is that there are assemblers and assemblers! Some of today's assemblers do rather more than translate a series of mnemonics into machine code. In this chapter we are going to concentrate on the features offered by these advanced assemblers.

8.1 Overview of the Assembler

We begin this chapter with a revision of the basic principles of an assembler. An assembler takes a source file, invariably in ASCII-encoded text, and uses it to create at least two new files. One is a *listing* file that contains the assembled source program complete with the addresses of instructions and data plus any error and warning messages. The other is a *binary* or *object* file that contains the binary code produced by the assembler. Some assemblers also produce a *symbol table* that provides a reference to every symbolic name used by the programmer.

The simplest assembler of all is the type provided by the monitor program of many single-board computers (e.g., Motorola's TUTOR). Such an assembler is able only to translate mnemonics into machine code. The programmer enters mnemonics and *absolute addresses,* as the assembler cannot deal with assembler directives and symbolic names.

One of the few practical facilities provided by this form of primitive assembler is the ability to translate an absolute branch address into a relative branch address. For example, you might type `BEQ $081234`, where 81234_{16} is the *actual* branch target address, and the assembler will compute the appropriate *offset* required when the CPU generates an effective address at run-time. Branching back to a line you have already entered is easy, but branching forward is virtually impossible. You cannot branch to a location if you do not know its address. One solution is to branch to a dummy location, finish the program, and then re-enter the branch instruction with the actual target address.

A *two-pass* assembler is so-called because the source file is read twice. On the first pass the assembler analyzes the syntax and detects errors. Mnemonics are translated into machine code and all symbolic names are entered into a symbol table. This symbol table is used by the assembler to convert all names and labels to actual (i.e., numeric) values.

Backward references are translated from symbolic to actual values on the first pass. Forward references cannot be resolved until the second pass. Consider the following example.

```
Space   EQU     $20
        .
Next    MOVE.B $8000,D1
        BEQ     Next
        MOVE.B $8002,D0
        CMP.B   #Space,D0
        BEQ     EndLine
        .
        .
EndLine .
```

In this fragment of code, the instructions `BEQ Next` and `CMP.B #Space,D0` can both be translated into machine code on the first pass, as the assembler already knows the actual values of the symbolic names. However, the instruction `BEQ EndLine` cannot be translated into machine code on the first pass, because the assembler does not yet know the location to which `EndLine` refers.

Some assemblers operate as we have described by doing as much translation as possible in pass one, and then fill in the gaps (i.e., forward references) in pass two. Other assemblers simply perform syntax checking and compile a symbol table in pass one. They then perform the actual translation during pass two. This latter type of assembler is more suited to the 68000.

The 68000 contains some instructions that cannot be translated into machine code until the operand is known. Consider the instruction `MOVE.B #2000,Temp5`. On the first pass the assembler knows the instruction (`MOVE.B`) and the source operand (`#2000`). It does not yet know the value of the destination operand (`Temp5`) — assuming that this is a forward reference. However, the 68000 cannot encode this instruction; it does not yet know whether the operand has a 16-bit address or a 32-bit address. Remember that an operand in the first 32K bytes of memory space (or the last 32K bytes) can be encoded as a 16-bit short absolute address. Not only is it impossible to calculate the appropriate code for the instruction, it is also impossible to generate the address of the next instruction. A longword operand takes four bytes while a word operand takes two bytes. It is, in fact, possible for the assembler to assume a worst-case value in such instances, and always default to a longword if the size of an operand is unknown.

You can create a symbol table when assembling a file. The following example was produced by the Teesside cross-assembler with the option OPT CRE that has the effect of producing a symbol table at the end of the assembled listing.

```
Source file: CALC.X68
Assembled on: 93-04-22 at: 18:50:58
          by: X68K PC-2.1 Copyright (c) University of Teesside 1989,93
Defaults: ORG $0/FORMAT/OPT A,BRL,CEX,CL,FRL,MC,MD,NOMEX,NOPCO

    1                                 OPT     CRE
    2  00000400                       ORG     $400          ;Start of the program
    3  00000400 4FF81000              LEA     $1000,A7      ;Set up the stack pointer
    4  00000404 6100001C    REPEAT:   BSR     CALC          ;Call the calculator
    5  00000408 123C000D              MOVE.B  #$0D,D1       ;Move to a new line
    6  0000040C 610000D2              BSR     PUTCHAR
    7  00000410 123C000A              MOVE.B  #$0A,D1
    8  00000414 610000CA              BSR     PUTCHAR
    9  00000418 0C050021              CMP.B   #'!',D5       ;Test for last character = "!"
   10  0000041C 66E6                  BNE     REPEAT        ;REPEAT UNTIL  "!" found
   11  0000041E 4E722700              STOP    #$2700
   12                       *
   13                       * Register usage:
   14                       *   D0: Used by input/output routines GetChar/PutChar
   15                       *   D1: Used by input/output routines GetChar/PutChar
   16                       *   D2: Holds number from routine GetNum
   17                       *   D3: Holds last non-BCD character from GetNum (and result)
   18                       *   D4: Holds the second operand
   19                       *   D5: Holds the operator (+, -, *, /)
   20                       *   D6: Used as a counter in output routines
   21                       *   D7: ShiftString (used to accumulate digits during output)
   22                       *
   23  00000422 61000088    CALC:     BSR     GETNUM        ;Get first number
   24  00000426 2602                  MOVE.L  D2,D3         ;Save it in D3
   25  00000428 1A04                  MOVE.B  D4,D5         ;Save the operator in D5
   26  0000042A 61000080              BSR     GETNUM        ;Get second number in D2
   27  0000042E 0C05002B              CMP.B   #'+',D5       ;IF operator = plus THEN add operands
   28  00000432 6700001E              BEQ     PLUS
   29  00000436 0C05002D              CMP.B   #'-',D5       ;IF operator = minus THEN subtract operands
   30  0000043A 6700001C              BEQ     MINUS
   31  0000043E 0C05002A              CMP.B   #'*',D5       ;IF operator = * THEN multiply operands
   32  00000442 6700001A              BEQ     MULT
   33  00000446 0C05002F              CMP.B   #'/',D5       ;IF operator = divide THEN divide operands
   34  0000044A 67000018              BEQ     DIV
   35  0000044E 6000005A              BRA     EXIT          ;Exit on non-valid operator
   36                       *
   37                       *                       CASE of:
```

```
38 00000452 D682         PLUS:     ADD.L   D2,D3          ;Add: Result := operand1+operand2
39 00000454 6000001A               BRA     OUTPUT
40 00000458 9682         MINUS:    SUB.L   D2,D3          ;Subtract: Result := operand1-operand2
41 0000045A 60000014               BRA     OUTPUT
42 0000045E C6C2         MULT:     MULU    D2,D3          ;Multiply: Result := operand1*operand2
43 00000460 6000000E               BRA     OUTPUT
44 00000464 86C2         DIV:      DIVU    D2,D3          ;Divide: Result := operand1/operand2
45 00000466 02830000FFFF           AND.L   #$0000FFFF,D3  ;(delete the remainder)
46 0000046C 60000002               BRA     OUTPUT         ;ENDCASE
47                       *
48 00000470 4287         OUTPUT:   CLR.L   D7             ;Clear output (i.e., ShiftString)
49 00000472 3C3C0007               MOVE.W  #7,D6          ;FOR I = 1 TO 8
50 00000476 E98F         OUTPUT1:  LSL.L   #4,D7          ;Shift ShiftString left four places
51 00000478 86FC000A               DIVU    #10,D3         ;Result/10 to get remainder in D3(16:31)
52 0000047C 4843                   SWAP    D3             ;Get remainder in D3(0:15)
53 0000047E 1203                   MOVE.B  D3,D1          ;Copy remainder to D1 (BCD in range 0 - 9)
54 00000480 4843                   SWAP    D3             ;Restore D3 (i.e., quotient to D3(0:15))
55 00000482 02830000FFFF           AND.L   #$0000FFFF,D3  ;Mask out the remainder
56 00000488 DE01                   ADD.B   D1,D7          ;Copy the digit to ShiftString
57 0000048A 51CEFFEA               DBRA    D6,OUTPUT1     ;ENDFOR
58                       *
59 0000048E 3C3C0007               MOVE.W  #7,D6          ;FOR I = 1 to 8
60 00000492 4201         OUTPUT2:  CLR.B   D1             ;Clear the 'character holder'
61 00000494 1207                   MOVE.B  D7,D1          ;Move 2 BCD chars from Shift String to D1
62 00000496 0201000F               AND.B   #$0F,D1        ;Mask to D1 to a single character
63 0000049A 06010030               ADD.B   #$30,D1        ;Convert BCD value to ASCII form
64 0000049E 61000040               BSR     PUTCHAR        ;Print a character
65 000004A2 E88F                   LSR.L   #4,D7          ;Shift ShiftString four places right
66 000004A4 51CEFFEC               DBRA    D6,OUTPUT2     ;ENDFOR
67 000004A8 4E75                   RTS
68                       *
69 000004AA 4E75         EXIT:     RTS
70                       *
71                       * Get a number in D2 and an operator in D4
72                       *
73 000004AC 4282         GETNUM:   CLR.L   D2             ;Clear Number
74 000004AE 61000028     GETNUM1:  BSR     GETCHAR        ;REPEAT Get an ASCII-encoded character
75 000004B2 0C010030               CMP.B   #$30,D1        ;IF Char < $30 OR Char > $30
76 000004B6 6B00001C               BMI     NOTNUM         ;THEN EXIT
77 000004BA 0C010039               CMP.B   #$39,D1
78 000004BE 6E000014               BGT     NOTNUM
79 000004C2 0281000000FF           AND.L   #$000000FF,D1  ;Mask character to 8 bits
80 000004C8 04010030               SUB.B   #$30,D1        ;Convert ASCII to 4-bit binary
81 000004CC C4FC000A               MULU    #10,D2         ;Number := Number + 10 * Digit
82 000004D0 D481                   ADD.L   D1,D2
83 000004D2 60DA                   BRA     GETNUM1        ;Continue
```

```
 84 000004D4 1801      NOTNUM:  MOVE.B   D1,D4            ;Save last character (terminator) in D4
 85 000004D6 4E75               RTS
 86                    *
 87 000004D8 103C0005  GETCHAR: MOVE.B   #5,D0            ;Get an ASCII character in D1.B
 88 000004DC 4E4F               TRAP     #15
 89 000004DE 4E75               RTS
 90                    *
 91 000004E0 103C0006  PUTCHAR: MOVE.B   #6,D0            ;Print the ASCII character in D1.B
 92 000004E4 4E4F               TRAP     #15
 93 000004E6 4E75               RTS
 94                    *
 95          00000400           END      $400

Lines: 95, Errors: 0, Warnings: 0.


SYMBOL  TABLE INFORMATION

Symbol-name   Type   Value     Decl  Cross reference line numbers
CALC          LABEL  00000422    23    4.
DIV           LABEL  00000464    44   34.
EXIT          LABEL  000004AA    69   35.
GETCHAR       LABEL  000004D8    87   74.
GETNUM        LABEL  000004AC    73   23,  26.
GETNUM1       LABEL  000004AE    74   83.
MINUS         LABEL  00000458    40   30.
MULT          LABEL  0000045E    42   32.
NOTNUM        LABEL  000004D4    84   76,  78.
OUTPUT        LABEL  00000470    48   39,  41,  43,  46.
OUTPUT1       LABEL  00000476    50   57.
OUTPUT2       LABEL  00000492    60   66.
PLUS          LABEL  00000452    38   28.
PUTCHAR       LABEL  000004E0    91    6,   8,  64.
REPEAT        LABEL  00000404     4   10.
```

Further Assembler Directives

Up to now we have limited ourselves to the absolute minimum assembler directives (i.e., ORG, EQU, DC, DS, END). Here we shall introduce some of the facilities offered by typical assemblers.

In this section we describe the 68000's assembler in greater detail. I'm afraid that there's no way to break this to you gently — I am going to confuse you by *redefining* a notation that we've already used. Motorola's literature describing the 68000 assembler uses the notation [parameter] to indicate an element in a parameter that is optional (i.e., it may be omitted). This notation is commonly used to describe computer languages — it's just a little unfortunate that we've already employed it in RTL to indicate the contents of a memory location or register.

SECTION The `SECTION` assembler directive is used to create a *relocatable program section.* Essentially, `SECTION` performs a similar function to the `ORG` assembler directive by resetting the location counter prior to assembling a chunk (i.e., section) of code. The format of this assembler directive is:

```
[<name>] SECTION  <number>.
```

The optional field, `[<name>]`, allows you to name the section. The `<number>` field indicates a named common area within the section and is a mandatory parameter. The section number must be in the range 0 to 15.

68000 assembly language programs can be written as a number of relocatable modules that are later *linked* together. Each of these modules may contain code, data, and references to code and data defined in other modules. The code contained in these modules can be divided into 16 sections.

OFFSET The `OFFSET` directive is used to define a table of offsets. A programmer might use these offsets (or displacements) in conjunction with the address register indirect addressing with displacement mode — `(d16,Ai)`. The format of the `OFFSET` directive is:

```
    OFFSET <expression>
```

where `<expression>` provides the value (i.e., address) at which the table is to begin. The expression must provide an *absolute* value and cannot, therefore, contain any forward, undefined, or external references. Note that this directive must be terminated by an `ORG` or `SECTION` directive. Consider the following fragment of code.

```
*   These values are the offsets into the array of vowels. Each
*   element consists of the vowel's name and the vowel-count
*
       OFFSET  0
A      DS.W    1
E      DS.W    1
I      DS.W    1
O      DS.W    1
U      DS.W    1
*
       SECTION 8
       .
       .
       .
       MOVE.W  (U,A0),D0   Read the entry for "U"
```

In this example, the instruction MOVE.W (U,A0),D0 employs the offset (or displacement) 8 for the symbolic value 'U'. Why 8 rather than 4? Because each of the values is a *word* quantity in the table of offsets, and 'U' is at offset 8 if we regard 'A' as being at offset 0. The OFFSET directive could be replaced by a sequence of EQU directives.

```
A       EQU     0
E       EQU     2
I       EQU     4
O       EQU     6
U       EQU     8
```

As you can see, the EQU directive gives programmers more work to do by forcing them to count the offsets in the table.

REG The REG directive allows the programmer to define a *register list*. The format of this directive is

```
<label>  REG  <register list>.
```

You use the REG directive to save time and effort. Indeed, it is no more than a special version of the EQU assembler directive designed for the register lists employed by the MOVEM instruction. Suppose you are writing a program with a large number of subroutines. Some of these subroutines save all working registers (i.e., A0 to A6 and D0 to D7), while others save only A6 and D4 and D5. To save the effort involved in writing the full form of the MOVEM instruction each time, we can write:

```
SaveAll REG      A0-A6/D0-D7
SaveFew REG      A6/D4-D5
         .
Calc     MOVEM.L SaveAll,-(SP)    Dump all registers on stack
         .
         .
         .
         MOVEM.L (SP)+,SaveAll    Restore all registers
         RTS
         .
         .
         .
OutPut   MOVEM.L SaveFew,-(SP)    Dump working registers
         .
         .
         .
         MOVEM.L (SP)+,SaveFew    Restore working registers
         RTS
```

The following listing was produced by this fragment of code.

```
Source file: REG.X68
Assembled on: 93-01-19 at: 22:20:12
        by: X68K PC-2.0 Copyright (c) University of Teesside 1989,93
Defaults: ORG $0/FORMAT/OPT A,BRL,CEX,CL,FRL,MC,MD,NOMEX,NOPCO

 1 00000400                          ORG      $400
 2          00007FFF    SAVE_ALL: REG      A0-A6/D0-D7          ;Define a "large" register list
 3          00003387    SAVE_FEW: REG      A0-A1/A4-A5/D0-D2/D7 ;Define a "small" register list
 4                      *
 5 00000400 48E7FFFE    PROCA:    MOVEM.L  SAVE_ALL,-(A7)       ;Save all registers
 6                      *
 7                      *
 8 00000404 4CDF7FFF              MOVEM.L  (A7)+,SAVE_ALL       ;Restore them before returning
 9 00000408 4E75                  RTS
10                      *
11 0000040A 48E7E1CC    PROCB:    MOVEM.L  SAVE_FEW,-(A7)       ;Save a few registers
12                      *
13                      *
14 0000040E 4CDF3387              MOVEM.L  (A7)+,SAVE_FEW       ;Restore them before returning
15 00000412 4E75                  RTS
16          00000400              END      $400

Lines: 16, Errors: 0, Warnings: 0.
```

SET The SET directive is very similar to the EQU and has an identical effect. For example, the following two assembler directives do exactly the same thing.

```
Comma    EQU    $2C
Term     SET    $2C
```

Although the effect of EQU and SET are identical (i.e., they bind a symbolic name to a numeric value), an important difference exists in the way in which they are used. An EQU directive is written in stone and once a name has been linked to a value, that value cannot be changed in the source program. However, the SET directive can be used several times to change the value associated with a symbolic name. For example, in the above code, the symbolic name Comma is equated to its ASCII value and Comma will always be bound to $2C_{16}$. The symbolic name Term (short for *terminator*) is set to the value $2C_{16}$ (the same as the comma). Initially in the program, we might be using a comma as a terminator and we might write CMP.B #Term,D0. Later in the program, we might be using a carriage return as a terminator. We can simply reassign the symbolic name Term to a new value by writing the line:

```
Term     SET    $0D.
```

DCB The DCB, define constant block, assembler directive is related to the DS and DC directives. Define constant block means: *fill a block of memory space with a given value*. It is really just like a repeated DC directive. Its format is:

```
[<label>] DCB[.<size>] <length>,<value>.
```

The <label> field is an optional label that may be used to refer to the start of the constant block. The optional <size> field defines the size of the constant (i.e., .B, .W or .L). The <length> field defines the length of the block (i.e., the number of entries), and the <value> field defines the value of the constant to be loaded. The following three lines are examples of legal DCB statements.

```
        DCB     12,0
Data    DCB.B   100,$FF
List    DCB.L   $A0,$12345678
```

For example, the assembler directive

```
Table DCB.B   64,$0D
```

fills a 64-byte block of memory with the value $0D_{16}$. This directive is used to preset memory to a known value. You can, for example, clear 1024 bytes of memory simply by using a DCB.L $256,0 directive.

8.2 Assemblers and Modularity

The very nature of a textbook like this (and the course it's designed to accompany) means that it is impractical to present the type of real problems that the programmer will encounter in practice. Consider, for example, a programmer who is asked to design a monitor and controller for an internal combustion engine. The programmer may have to write the program in assembly language because speed of execution is vital and because it has to directly control the engine. Such a program is very much larger than the type of illustrative fragments we have provided in this text. Indeed, it may be so large that several programmers have to work on it in parallel. Consequently, you cannot write this program as a single module — it must be divided into units that can be developed independently of each other and then put together. Modern assemblers support large projects of this nature.

Consider the system of Figure 8.1 that has three large subroutines A, B, and C, each of which is written by separate programmers. It is not necessary, of course, to employ three programmers. A single programmer could design all modules. What is important is that the program is subdivided into separate units that can be designed and tested independently of each other. Let's look at a fragment of a program and ask ourselves what we need to do to modularize it.

```
MASKWORD EQU     $0000FFFF        Mask used to clear high-order word
MASKBYTE EQU     $0F              Mask used to clear high-order nibble
ACIAC    EQU     $800001          ACIA control byte
ACIAD    EQU     ACIAC+2          ACIA data port
         ORG     $400             Start of the program
         LEA     $1000,A7         Set up the stack pointer
Repeat   BSR     Calc             Call the calculator
         MOVE.B  #$0D,D1          Move to a new line
         BSR     PutChar
         MOVE.B  #$0A,D1
         BSR     PutChar
         CMP.B   #'!',D5          Test for last character = "!"
         BNE     Repeat           REPEAT UNTIL  "!" found
         STOP    #$2700
*
Calc     BSR     GetNum           Get first number
         MOVE.L  D2,D3            Save it in D3
         MOVE.B  D4,D5            Save the operator in D5
         BSR     GetNum           Get second number in D2
         CMP.B   #'+',D5          IF operator = "+" THEN add operands
         BEQ     Plus
         CMP.B   #'-',D5          IF operator = "-" THEN subtract operands
         BEQ     Minus
         CMP.B   #'*',D5          IF operator = "*" THEN multiply operands
         BEQ     Mult
         CMP.B   #'/',D5          IF operator = "/" THEN divide operands
         BEQ     Div
```

Figure 8.1 A program composed of three separate modules

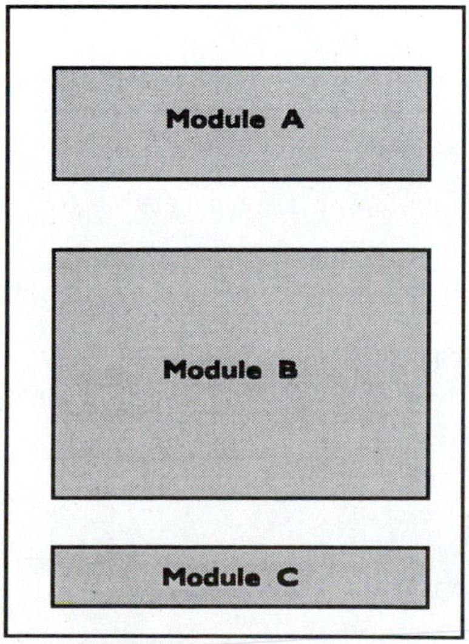

Module A

Module B

Module C

```
         BRA      EXIT                 Exit on non-valid operator
*                                      CASE of:
Plus     ADD.L    D2,D3                 Add: Result := operand1+operand2
         BRA      Output
Minus    SUB.L    D2,D3                 Subtract: Result := operand1-operand2
         BRA      Output
Mult     MULU     D2,D3                 Multiply: Result := operand1*operand2
         BRA      Output
Div      DIVU     D2,D3                 Divide: Result := operand1/operand2
         AND.L    #$0000FFFF,D3              (delete the remainder)
         BRA      Output               END_CASE
*
Output   CLR.L    D7                 Clear output (i.e., ShiftString)
         MOVE.W   #7,D6              FOR I = 1 TO 8
Output1  LSL.L    #4,D7                Shift ShiftString left four places
         DIVU     #10,D3               Result/10 to get remainder in D3(16:31)
         SWAP     D3                   Get remainder in D3(0:15)
         MOVE.B   D3,D1                Copy remainder to D1 (BCD in range 0 - 9)
         SWAP     D3                   Restore D3 (i.e., quotient to D3(0:15))
         AND.L    #MASKWORD,D3         Mask out the remainder
         ADD.B    D1,D7                Copy the digit to ShiftString
         DBRA     D6,Output1         END_FOR
         MOVE.W   #7,D6              FOR I = 1 to 8
Output2  CLR.B    D1                   Clear the 'character holder'
         MOVE.B   D7,D1                Move 2 BCD chars from ShiftString to D1
         AND.B    #MASKBYTE,D1         Mask to D1 to a single character
         ADD.B    #$30,D1              Convert BCD value to ASCII form
         BSR      PutChar              Print a character
         LSR.L    #4,D7                Shift ShiftString four places right
         DBRA     D6,Output2         END_FOR
EXIT     RTS
*
PutChar  BTST     #1,ACIAC           REPEAT
         BNE      PutChar              Read ACIAC status
         MOVE.B   D1,ACIAD           UNTIL ACIA ready
         RTS                         Print character and return
*
* Get a number in D2 and an operator in D4
*
GetNum   CLR.L    D2                 Clear Number
GetNum1  BSR      GetChar            REPEAT Get an ASCII-encoded character
         CMP.B    #$30,D1              IF Char < $30 OR Char > $30
         BMI      NotNum                  THEN EXIT
         CMP.B    #$39,D1
         BGT      NotNum
         AND.L    #$000000FF,D1        Mask character to 8 bits
```

```
         SUB.B    #$30,D1            Convert ASCII to 4-bit binary
         MULU     #10,D2             Number := Number + 10 * Digit
         ADD.L    D1,D2
         BRA      GetNum1           Continue
NotNum   MOVE.B   D1,D4             Save last character (terminator) in D4
         RTS
GetChar  MOVE.B   #5,D0             Get an ASCII character in D1.B
         TRAP     #15
         RTS
PutChar  MOVE.B   #6,D0             Print the ASCII character in D1.B
         TRAP     #15
         RTS
         END      $400
```

We have placed part of this program in a block to indicate that we wish to convert it into a module. What problems does this cause? If we were to take this block out of the program and then try to assemble the remaining program, we would generate four errors. The four lines in the main program that reference this block (i.e., `BRA Output`) cannot be assembled, because the branch target address, `Output`, is missing.

Consider now the module `Output`. This module too cannot be assembled independently. In this case the symbolic names `MASKWORD` and `MASKBYTE` that have been equated to numeric values in the main program would trigger errors, as they are undefined within the module. Similarly, the symbolic addresses `ACIAC` and `ACIAD` are also undefined in the module. If programs are to be written as *independent modules,* two things are necessary. The first is a mechanism to tell a module that some of its resources are currently unavailable, but will be provided later. The second is a framework that will take the independently compiled modules and link them together, supplying all the missing addresses and data values. Many assemblers now supply all the ingredients for *separate assembly.*

Three new assembler directives, `IDNT`, `XDEF`, and `XREF` are used to write modules. The `IDNT` (i.e., identify) directive identifies and labels a relocatable object module. Every module requires an `IDNT` directive to identify it so that it can be handled by the program that puts together (i.e., links) the individual modules. The format of the `IDNT` directive is:

```
<module name> IDNT <version>,<revision> [<comment>].
```

Note that `IDNT` must have a name to distinguish this module from other modules. The version and revision fields are used to supply the version and revision number of the module — an important consideration when a large-scale project is being developed and software is frequently being corrected and updated. As ever, the comment field is optional. The following line might be used to identify our output module.

```
NUMDISP  IDNT    1,2                Display a number module
```

The XDEF and XREF directives tell the assembler about external symbols (i.e., symbols that are defined or used in another module). The XDEF (external symbol definition) has the format XDEF <symbol>[,<symbol>]..., where <symbol> represents a symbol used in another module. That is, XDEF tells the linker which symbols are to be exported to other modules. In our example we might write:

```
CALC     IDNT    1,1           The main routine of the calculator program
         XDEF    MASKWORD,MASKBYTE,ACIAC,ACIAD
*
MASKWORD EQU     $0000FFFF     Mask used to clear high-order word
MASKBYTE EQU     $0F           Mask used to clear high-order nibble
ACIAC    EQU     $800001       ACIA control byte
ACIAD    EQU     ACIAC+2       ACIA data port
         ORG     $400          Start of the program
         LEA     $1000,A7      Set up the stack pointer
Repeat   BSR     Calc          Call the calculator
         MOVE.B  #$0D,D1       Move to a new line
         BSR     PutChar
         .
         .
         .
```

The line XDEF MASKWORD,MASKBYTE,ACIAC,ACIAD tells the module (called CALC by the IDNT directive) that the four symbols MASKWORD, MASKBYTE, ACIAC, ACIAD are to be exported to other modules. These symbols are equates, subroutine addresses, and references to memory locations.

The XREF (external reference) assembler directive performs the inverse function to the XDEF. This assembler directive lists the symbols referred to in the current module that are actually defined in other modules. XREF has the format XREF <symbol>[,<symbol>]...,.

When there are several modules in a system, it is necessary to tell the assembler from which module the symbols are to be imported. In such cases the syntax of the XREF is XREF <section:symbol>[,<section:symbol>]..., where section indicates the module from which the symbols are to be imported. In terms of the previous example we might write:

```
        XREF    MASKWORD,MASKBYTE,ACIAC,ACIAD
Output  CLR.L   D7              Clear output (i.e., ShiftString)
        MOVE.W  #7,D6           FOR I = 1 TO 8
Output1 LSL.L   #4,D7             Shift ShiftString left four places
        DIVU    #10,D3            Result/10 to get remainder in D3(16:31)
        SWAP    D3                Get remainder in D3(0:15)
        MOVE.B  D3,D1             Copy remainder to D1 (BCD in range 0 - 9)
        SWAP    D3                Restore D3 (i.e., quotient to D3(0:15))
        AND.L   #MASKWORD,D3      Mask out the remainder
        ADD.B   D1,D7             Copy the digit to ShiftString
```

```
         DBRA    D6,Output1        END_FOR
         MOVE.W  #7,D6             FOR I = 1 to 8
Output2  CLR.B   D1                  Clear the 'character holder'
         MOVE.B  D7,D1               Move 2 BCD chars from ShiftString to D1
         AND.B   #MASKBYTE,D1        Mask D1 to a single character
         ADD.B   #$30,D1             Convert BCD value to ASCII form
         BSR     PutChar             Print a character
         LSR.L   #4,D7               Shift ShiftString four places right
         DBRA    D6,Output2        END_FOR
EXIT     RTS
PutChar  BTST    #1,ACIAC          REPEAT
         BNE     PutChar             Read ACIAC status
         MOVE.B  D1,ACIAD          UNTIL ACIA ready
         RTS                       Print character and return
```

Figure 8.2 represents the relationship between the modules diagrammatically.

8.3 The Macro Assembler

Many assemblers include a *macro* facility. The term macro is a word that stands for an entire group of instructions. Using macros is a two-stage process. First, you define a macro by enclosing a sequence of instructions between the words MACRO (*define macro*) and ENDM (*end macro definition*), and then you call the macro when you wish to execute these instructions. Note that the macro definition command

Figure 8.2 The use of XDEF and XREF directives

```
CALC      IDNT  1,1
          XDEF  MASKWORD,MASKBYTE
          XDEF  ACIAC,ACIAD
          .
          .
          .
```

```
NUMDISP   IDNT  1,2
          XREF  MASKWORD,MASKBYTE
          XREF  ACIAC,ACIAD
          .
          .
          .
```

must have a label in order for the macro to be given a name. For example, we could take a group of instructions and turn them into a macro as follows.

```
AddMul    MACRO                   Macro definition of 'AddMul'
          ADD.B     #7,D0           D0 := D0 + 7
          AND.W     #$00FF,D0       Mask D0 to a word
          MULU      #12,D0          D0 := D0 x 12
          ENDM                    End macro definition of AddMul
```

This is an example of a macro with the name `AddMul`. We can make use of the instructions enclosed by `MACRO...ENDM` in a program just by writing the name of the macro, `AddMul`. Note that we do not have to call the macro by means of a `BSR`, `JSR`, or `CALL` command. Consider the following example.

```
          MOVE.B    Val2,D0       Get Val2
          AddMul                  Call a macro to process it
          .
          .
          MOVE.B    Val3,D0       Get Val3
          AddMul                  Call a macro to process it
          .
          .
```

What then is the difference between a *macro* and a *subroutine*? They both permit a group of instructions to be defined as a named entity and then called up when they are needed. A subroutine is called by a `BSR` or `JSR` instruction, and a macro simply by using its name.

A subroutine is a piece of code that is assembled and, when called, a branch is made to its entry. On its completion the `RTS` (return from subroutine) instruction forces a jump to the instruction immediately following the branch. Each subroutine appears once only in the binary code (i.e., in the machine code) but may be called many times from different locations within the program.

A macro is simply a form of *shorthand*. When a group of instructions is defined by a macro, the group is placed wherever the macro occurs in the source program. No call or return mechanism is associated with macros. In the example we provided, the macro `AddMul` is replaced twice by the *actual code* specified by the macro when the program is assembled. If you attempted to find a trace of the macro in the assembled machine code, it would be impossible.

Once again let me say that a macro is just a form of shorthand used to avoid repeatedly writing a group of instructions. Therefore, a macro does not have to be the type of coherent unity that you might expect from a subroutine. In the above example, the code we have chosen is coherent, and we could have used a subroutine to do the same job. If we had used a subroutine, the resulting machine code would be shorter than the macro version (because there would be only one subroutine, an `RTS` instruction and a `BSR AddMul` for each subroutine call). If there are n macro calls, the code of the macro is repeated n times, each one replacing the macro. The macro version of the program runs faster than the subroutine version because there is no call and return overhead associated with the macro.

When a program using macros is assembled, the macros are not replaced by the instructions in the macro definitions in the *listing* file — unless the programmer includes the assembler directive `OPT MEX` in the program. This assembler directive means *turn on macro expansion.* The default mode of assemblers is to list macros but not their expansion.

Macros and Parameters

The macro we just described operates on the contents of data register D0. Suppose that somewhere else in the program we must also perform the same sequence of operations on D1. Does this mean that we have to write two almost identical macros? The answer to this question is no, since parameters can be passed to macros. These parameters are used when the macro is converted into a sequence of assembly language instructions. Consider the following example.

```
AddMul  MACRO                   Macro definition of AddMul
        ADD.B    #7,\1          Reg := Reg + 7
        AND.W    #$00FF,\1      Mask Reg to a byte of a word
        MULU     #12,\1         Reg := Reg x 12
        ENDM                    End macro definition of AddMul
```

The register name 'D0' has been replaced by the symbol '\1'. All this symbol means is *please replace me by parameter number one when the macro is called.* We can make use of this macro by writing:

```
        MOVE.B   Val2,D0        Get Val2
        AddMul   D0             Process it
        .
        .
        MOVE.B   Val3,D1        Get Val3
        AddMul   D1             Process it
```

When the macro is expanded the first time, it is replaced by its code, and the symbol `\1` is replaced by `D0` wherever it occurs. When the macro is expanded a second time, the symbol `\1` is replaced by `D1`.

If a macro has several parameters, the parameters used in the macro definition are written `\1, \2, \3, \4`, and so on. The number refers to the *order* of the parameters in the macro call. For example, a macro call might be:

```
    Clear D3 D4 A1 A3,
```

with the result that when the macro call is expanded, parameter `\1` is replaced by `D3`, parameter `\2` is replaced by `D4`, and so on.

Suppose you write a large program that heavily uses subroutines and these subroutines employ local storage and also save registers. You might create two macros (`Call` and `Return`) to help you.

```
Call     MACRO
         MOVEM.L  \1,-(SP)              Save working registers
         LINK     \2,#\3                Create a stack frame
         ENDM

Return   UNLK     \1                    Collapse the stack frame
         MOVEM.L  (SP)+,\2              Retrieve saved registers
         ENDM
```

These macros demonstrate that we can write the code to perform the desired action, but omit the list of the actual registers and constants until we make the appropriate macro call. Note that a macro parameter number refers only to the *order* in which the parameter is used within the macro. In the `Call` macro, parameter \1 provides the list of registers to be saved, while in the `Return` macro, parameter \1 is the address register used by the `UNLK`. These macros can be used in the following way.

```
Check Call   D0-D4/A0-A3,A4,-6  Save working registers and create stack frame
        .
        .                       Body of the subroutine
        .
      Return A4,D0-D4/A0-A3     Collapse the frame and restore registers
```

We have taken the macro call definition and the macro calls and converted them into a dummy program. The following listing provides the output from the assembler. Note that the directive `OPT MEX` forces the macro calls to be expanded in a listing. We will return to this option and discuss it in more detail later.

```
Source file: T3.X68
Assembled on: 92-11-29 at: 15:13:54
          by: X68K PC-1.9 Copyright (c) University of Teesside 1989,92
Defaults: ORG $0/FORMAT/OPT A,BRL,CEX,CL,FRL,MC,MD,NOMEX,NOPCO

 1                  ( 1)         OPT      MEX
 2 00000400         ( 2)         ORG      $400
 3                  ( 3) CALL:   MACRO
 4                  ( 4)         MOVEM.L \1,-(SP)          ;Save working registers
 5                  ( 5)         LINK     \2,#\3           ;Create a stack frame
 6                  ( 6)         ENDM
 7                  ( 7) *
 8                  ( 8) RETURN: MACRO
 9                  ( 9)         UNLK     \1               ;Collapse the stack frame
```

```
10                    (10)         MOVEM.L (SP)+,\2          ;Retrieve saved registers
11                    (11)         ENDM
12                    (12) *
13 00000400           (13) CHECK:  CALL    D0-D4/A0-A3,A4,-6 ;Save working regs and create frame
14 00000400 48E7F8F0  (13)         MOVEM.L D0-D4/A0-A3,-(SP) ;Save working registers
15 00000404 4E54FFFA  (13)         LINK    A4,#-6            ;Create a stack frame
16                    (14) *       .
17                    (15) *       .                          Body of the subroutine
18                    (16) *       .
19 00000408           (17)         RETURN  A4,D0-D4/A0-A3    ;Collapse frame and restore regs
20 00000408 4E5C      (17)         UNLK    A4                ;Collapse the stack frame
21 0000040A 4CDF0F1F  (17)         MOVEM.L (SP)+,D0-D4/A0-A3 ;Retrieve saved registers
22                    (18) *
23          00000400  (19)         END     $400
Lines: 23, Errors: 0, Warnings: 0.
```

Macros and Labels

Programmers often wish to employ labels in macros just as they do in subroutines. Since each macro is replaced by its code when it is called, labels cannot, apparently, be used in the definition of a macro. Suppose a macro has a label `NextVal` and is called twice in a program. The label `NextVal` will appear *twice* in the expanded assembly language code and therefore an assembly error will be flagged, because using the same label twice is illegal.

Fortunately, macro assemblers do provide a mechanism to avoid the multiple definition of labels. The programmer uses a label within the definition of a macro and the assembler automatically *renames* this label each time the macro is expanded. Labels are written in the form `<string>\@` or `\@<string>`, where `<string>` is a user-defined string of up to four characters. A valid macro label might be `Next\@`. When the macro assembler encounters the label `Next\@`, it translates the label into the form `Next.nnn`, where `.nnn` represents a three-digit decimal number automatically chosen by the assembler. This `.nnn` extension ensures the uniqueness of labels in consecutive macro calls. We will now look at a simple example of the use of labels. Consider the following code.

```
        OPT     MEX
        ORG     $400
SUM     MACRO                   \1 = Start, \2 = Stop, \3 = Sum
        CLR.W   \3              Total := 0
        ADDQ.W  #1,\2           Stop := Stop + 1
SUM1\@ ADD.W   \1,\3           FOR i = Start to Stop
        ADD.W   #1,\1                 Total := Total + i
        CMP.W   \1,\2
        BNE     Sum1\@          UNTIL Start = Stop
        ENDM
```

```
MOVE.W   #1,D1
MOVE.W   #10,D2
SUM      D1,D2,D3
MOVE.W   #5,D3
MOVE.W   #10,D4
SUM      D3,D4,D5
STOP     #$2700
END      $400
```

This program adds the sequence of numbers, i, i+1, i+2, ..., n, twice. The first time the macro is called is for the range 1 to 10, and the second time for the range 5 to 10. In this example, we use the label SUM1\@. When the code is assembled, we get:

```
Source file: T4.X68
Assembled on: 92-11-29 at: 16:04:57
        by: X68K PC-1.9 Copyright (c) University of Teesside 1989,92
Defaults: ORG $0/FORMAT/OPT A,BRL,CEX,CL,FRL,MC,MD,NOMEX,NOPCO

 1                     ( 1)           OPT      MEX
 2 00000400            ( 2)           ORG      $400
 3                     ( 3) SUM:      MACRO              ;\1 = Start, \2 = Stop, \3 = Sum
 4                     ( 4)           CLR.W    \3        ;Total := 0
 5                     ( 5)           ADDQ.W   #1,\2     ;Stop := Stop + 1
 6                     ( 6) SUM1\@:    ADD.W    \1,\3     ;FOR i = Start to Stop
 7                     ( 7)           ADD.W    #1,\1     ;Total := Total + i
 8                     ( 8)           CMP.W    \1,\2
 9                     ( 9)           BNE      SUM1\@    ;UNTIL Start = Stop
10                     (10)           ENDM
11                     (11) *
12 00000400 323C0001   (12)           MOVE.W   #1,D1
13 00000404 343C000A   (13)           MOVE.W   #10,D2
14 00000408            (14)           SUM      D1,D2,D3
15 00000408 4243       (14)           CLR.W    D3        ;Total := 0
16 0000040A 5242       (14)           ADDQ.W   #1,D2     ;Stop := Stop + 1
17 0000040C D641       (14) SUM1.000: ADD.W    D1,D3     ;FOR i = Start to Stop
18 0000040E 5241       (14)           ADD.W    #1,D1     ;Total := Total + i
19 00000410 B441       (14)           CMP.W    D1,D2
20 00000412 66F8       (14)           BNE      SUM1.000  ;UNTIL Start = Stop
21 00000414 363C0005   (15)           MOVE.W   #5,D3
22 00000418 383C000A   (16)           MOVE.W   #10,D4
23 0000041C            (17)           SUM      D3,D4,D5
24 0000041C 4245       (17)           CLR.W    D5        ;Total := 0
25 0000041E 5244       (17)           ADDQ.W   #1,D4     ;Stop := Stop + 1
26 00000420 DA43       (17) SUM1.001: ADD.W    D3,D5     ;FOR i = Start to Stop
27 00000422 5243       (17)           ADD.W    #1,D3     ;Total := Total + i
```

```
28 00000424 B843        (17)        CMP.W    D3,D4
29 00000426 66F8        (17)        BNE      SUM1.001     ;UNTIL Start = Stop
30 00000428 4E722700    (18)        STOP     #$2700
31          00000400    (19)        END      $400

Lines: 31, Errors: 0, Warnings: 0.
```

You can see that in the macro expansion, the label SUM1\@ is first translated into SUM1.000 and then into SUM1.001 by the assembler.

Example of the Use of Macros

We now look at the design of a program that makes use of a macro. The program is intended to calculate the date of Easter Sunday given the year. We have taken the algorithm from Structured Programming in Assembly Language for the IBM PC by William G. Runnion. The program uses a rather complex algorithm to calculate the date in March or April on which Easter Sunday falls. This algorithm makes repeated use of the mathematical function *x mod y* (verbalized as x modulo y) that yields the remainder when x is divided by y. For example, the value of 50 mod 6 is 2, because 50 = 6 x 8 + 2.

We could have turned the mod function into a subroutine, but instead chose to use a macro. The macro takes two parameters, \1 and \2. The first parameter is the source of the number whose modulo we wish to calculate, and the second parameter is the actual modulo number. For example, if we wish to calculate the value of the contents of D3 modulo 5, we would invoke the parameter with mod D3,5.

```
*       Program to calculate the date of Easter Sunday for a given year
*
*       A := Year mod 19
*       X := (19A+24)mod 30
*       B := 2(Year mod 4)
*       C := 4(Year mod 7)
*       Y := (6X+5+B+C)mod 7
*       DATE := Y+X+22
*       IF Date > 31 THEN
*                   Date := Date - 31
*                   IF Date = 26 THEN Date := 19
*                   END_IF
*                   Month := April
*               ELSE
*               Month := March
*       END_IF
*
        OPT     MEX             Expand macros
```

```
         ORG     $400
Start    BSR     GetYear          Year in D1
         CMP.W   #0,D1            Exit if Year = 0
         BEQ     End_Calc         Else perform a calculation
         BSR     CalcDate
         BSR     PrtDate
         BRA     Start
End_Calc STOP    #$2700
*
GetYear  LEA     Query,A1         Ask for the year
         BSR     P_String
         MOVE.B  #4,D0            Input year into D1
         TRAP    #15
         RTS
*
PrtDate  LEA     Answer,A1        Display the answer
         BSR     P_String
         CMP.B   #1,D6            IF D6=1 THEN Month = April
         BEQ     Print_Ap
         LEA     March,A1         IF not April THEN Month = March
         BSR     P_String
         BRA     Date             Print the date (skip April)
Print_Ap LEA     April,A1         Print April
         BSR     P_String
Date     MOVE.B  #3,D0            Print the date
         MOVE.W  D5,D1
         TRAP    #15
         LEA     NewLine,A1
         BRA     P_String
*
*        MACRO to calculate x mod y
*
MOD      MACRO   Calculate x mod y
         DIVU    #\2,\1           Calculate quotient and remainder
         SWAP    \1               Get remainder in lower word of \1
         AND.L   #$0000FFFF,\1    Remove upper-order word
         ENDM
*
CalcDate AND.L   #$0000FFFF,D1    Make sure Year in D1 is 16 bits
         MOVE.L  D1,D2            Copy Year to D2
         MOD     D2,19            A := Year mod 19
         MULU    #19,D2           X := (19A + 24) mod 30
         ADD.W   #24,D2
         MOD     D2,30
         MOVE.L  D1,D3            Copy Year to D3
         MOD     D3,4             B := 2(Year mod 4)
```

```
        MULU    #2,D3
        MOVE.L  D1,D4           Copy Year to D4
        MOD     D4,7            C := 4(Year mod 7)
        MULU    #4,D4
        MOVE.W  D2,D5           Copy X to D5
        MULU    #6,D5           Y := (6X + 5 + B + C) mod 7
        ADD.W   #5,D5
        ADD.W   D3,D5
        ADD.W   D4,D5
        MOD     D5,7
        ADD.W   D2,D5           Date := Y + X + 22
        ADD.W   #22,D5
        CMP.W   #31,D5          IF Date > 31
        BLE     Mar                     THEN
        SUB.W   #31,D5                  Date := Date - 31
        CMP.W   26,D5                   IF Date := 26
        BNE     Apr                             THEN Date := 19
        MOVE.W  #19,D5
        BRA     Apr
Mar     MOVE.W  #0,D6           Set D6 := 0 for March
        RTS
Apr     MOVE.W  #1,D6           Set D6 := 1 for April
        RTS
*
*       Print the string pointed at by A1 and terminated by 0
P_StringMOVE.B  (A1)+,D1        Read a character
        BEQ     P_Exit          IF null THEN Exit
        MOVE.B  #6,D0                   ELSE Print the character
        BRA     P_String        REPEAT
P_Exit  RTS                     Exit
*
Query   DC.B    'Enter year  ',0
Answer  DC.B    'Date of Easter = ',0
March   DC.B    'March ',0
April   DC.B    'April ',0
NewLine DC.B    $0A,$0D,0
        END     Start
```

The program was assembled with the 68000 cross-assembler to provide the following source listing. The option MEX (i.e., macro expansion) was employed to force the expansion of the macros.

```
1                     ( 1) *Program to calculate the date of Easter Sunday for a given year
2                     ( 2) *
3                     ( 3) *      A := Year mod 19
4                     ( 4) *      X := (19A+24)mod 30
```

```
 5                          (  5) *        B := 2(Year mod 4)
 6                          (  6) *        C := 4(Year mod 7)
 7                          (  7) *        Y := (6X+5+B+C)mod 7
 8                          (  8) *        DATE := Y+X+22
 9                          (  9) *        IF Date > 31 THEN
10                          ( 10) *                            Date := Date - 31
11                          ( 11) *                            IF Date = 26 THEN Date := 19
12                          ( 12) *                            END_IF
13                          ( 13) *                            Month := April
14                          ( 14) *                    ELSE
15                          ( 15) *                            Month := March
16                          ( 16) *        END_IF
17                          ( 17) *
18                          ( 18)          OPT     MEX          ;Expand macros
19 00000400                 ( 19)          ORG     $400
20 00000400 61000018        ( 20) START:   BSR     GETYEAR      ;Year in D1
21 00000404 0C410000        ( 21)          CMP.W   #0,D1        ;Exit if Year = 0
22 00000408 6700000C        ( 22)          BEQ     END_CALC     ;Else perform a calculation
23 0000040C 6100005A        ( 23)          BSR     CALCDATE
24 00000410 6100001A        ( 24)          BSR     PRTDATE
25 00000414 60EA            ( 25)          BRA     START
26 00000416 4E722700        ( 26) END_CALC: STOP   #$2700
27                          ( 27) *
28 0000041A 43F90000050A ( 28) GETYEAR:  LEA     QUERY,A1     ;Ask for the year
29 00000420 610000D8        ( 29)          BSR     P_STRING
30 00000424 103C0004        ( 30)          MOVE.B  #4,D0        ;Input year into D1
31 00000428 4E4F            ( 31)          TRAP    #15
32 0000042A 4E75            ( 32)          RTS
33                          ( 33) *
34 0000042C 43F900000517 ( 34) PRTDATE:  LEA     ANSWER,A1    ;Display the answer
35 00000432 610000C6        ( 35)          BSR     P_STRING
36 00000436 0C060001        ( 36)          CMP.B   #1,D6        ;IF D6=1 THEN Month = April
37 0000043A 67000010        ( 37)          BEQ     PRINT_AP
38 0000043E 43F900000529 ( 38)          LEA     MARCH,A1     ;IF not April THEN Month=March
39 00000444 610000B4        ( 39)          BSR     P_STRING
40 00000448 6000000C        ( 40)          BRA     DATE         ;Print the date (skip April)
41 0000044C 43F900000530 ( 41) PRINT_AP: LEA     APRIL,A1     ;Print April
42 00000452 610000A6        ( 42)          BSR     P_STRING
43 00000456 103C0003        ( 43) DATE:    MOVE.B  #3,D0        ;Print the date
44 0000045A 3205            ( 44)          MOVE.W  D5,D1
45 0000045C 4E4F            ( 45)          TRAP    #15
46 0000045E 43F900000537 ( 46)          LEA     NEWLINE,A1
47 00000464 60000094        ( 47)          BRA     P_STRING
48                          ( 48) *
49                          ( 49) *                MACRO to calculate x mod y
50                          ( 50) *
```

```
51                        ( 51) MOD:      MACRO                 ;Calculate x mod y
52                        ( 52)           DIVU      #\2,\1      ;Calculate quotient and remainder
53                        ( 53)           SWAP      \1          ;Get remainder in lower word of \1
54                        ( 54)           AND.L     #$0000FFFF,\1
55                        ( 55)           ENDM
56                        ( 56) *
57 00000468 02810000FFFF ( 57) CALCDATE: AND.L     #$0000FFFF,D1 ;Make sure Year in D1 is 16 bits
58 0000046E 2401         ( 58)           MOVE.L    D1,D2       ;Copy Year to D2
59 00000470              ( 59)           MOD       D2,19       ;A := Year mod 19
60 00000470 84FC0013     ( 59)           DIVU      #19,D2      ;Calculate quotient and remainder
61 00000474 4842         ( 59)           SWAP      D2          ;Get remainder in lower word of D2
62 00000476 02820000FFFF ( 59)           AND.L     #$0000FFFF,D2
63 0000047C C4FC0013     ( 60)           MULU      #19,D2      ;X := (19A + 24) mod 30
64 00000480 06420018     ( 61)           ADD.W     #24,D2
65 00000484              ( 62)           MOD       D2,30
66 00000484 84FC001E     ( 62)           DIVU      #30,D2      ;Calculate quotient and remainder
67 00000488 4842         ( 62)           SWAP      D2          ;Get remainder in lower word of D2
68 0000048A 02820000FFFF ( 62)           AND.L     #$0000FFFF,D2
69 00000490 2601         ( 63)           MOVE.L    D1,D3       ;Copy Year to D3
70 00000492              ( 64)           MOD       D3,4        ;B := 2(Year mod 4)
71 00000492 86FC0004     ( 64)           DIVU      #4,D3       ;Calculate quotient and remainder
72 00000496 4843         ( 64)           SWAP      D3          ;Get remainder in lower word of D3
73 00000498 02830000FFFF ( 64)           AND.L     #$0000FFFF,D3
74 0000049E C6FC0002     ( 65)           MULU      #2,D3
75 000004A2 2801         ( 66)           MOVE.L    D1,D4       ;Copy Year to D4
76 000004A4              ( 67)           MOD       D4,7        ;C := 4(Year mod 7)
77 000004A4 88FC0007     ( 67)           DIVU      #7,D4       ;Calculate quotient and remainder
78 000004A8 4844         ( 67)           SWAP      D4          ;Get remainder in lower word of D4
79 000004AA 02840000FFFF ( 67)           AND.L     #$0000FFFF,D4
80 000004B0 C8FC0004     ( 68)           MULU      #4,D4
81 000004B4 3A02         ( 69)           MOVE.W    D2,D5       ;Copy X to D5
82 000004B6 CAFC0006     ( 70)           MULU      #6,D5       ;Y := (6X + 5 + B + C) mod 7
83 000004BA 5A45         ( 71)           ADD.W     #5,D5
84 000004BC DA43         ( 72)           ADD.W     D3,D5
85 000004BE DA44         ( 73)           ADD.W     D4,D5
86 000004C0              ( 74)           MOD       D5,7
87 000004C0 8AFC0007     ( 74)           DIVU      #7,D5       ;Calculate quotient and remainder
88 000004C4 4845         ( 74)           SWAP      D5          ;Get remainder in lower word of D5
89 000004C6 02850000FFFF ( 74)           AND.L     #$0000FFFF,D5
90 000004CC DA42         ( 75)           ADD.W     D2,D5       ;Date := Y + X + 22
91 000004CE 06450016     ( 76)           ADD.W     #22,D5
92 000004D2 0C45001F     ( 77)           CMP.W     #31,D5      ;IF Date > 31
93 000004D6 6F000016     ( 78)           BLE       MAR
94 000004DA 0445001F     ( 79)           SUB.W     #31,D5      ;Date := Date - 31
95 000004DE 0C45001A     ( 80)           CMP.W     #26,D5      ;IF Date := 26
96                        ( 81)
```

```
 97 000004E2 66000010     ( 82)           BNE      APR           ;THEN Date := 19
 98 000004E6 3A3C0013     ( 83)           MOVE.W   #19,D5
 99 000004EA 60000008     ( 84)           BRA      APR
100 000004EE 3C3C0000     ( 85) MAR:      MOVE.W   #0,D6         ;Set D6 := 0 for March
101 000004F2 4E75         ( 86)           RTS
102 000004F4 3C3C0001     ( 87) APR:      MOVE.W   #1,D6         ;Set D6 := 1 for April
103 000004F8 4E75         ( 88)           RTS
104                       ( 89) *
105                       ( 90) *      Print the string pointed at by A1 and terminated by 0
106 000004FA 1219         ( 91) P_STRING: MOVE.B   (A1)+,D1      ;Read a character
107 000004FC 6700000A     ( 92)           BEQ      P_EXIT        ;IF null THEN Exit
108 00000500 103C0006     ( 93)           MOVE.B   #6,D0         ;ELSE Print the character
109 00000504 4E4F         ( 94)           TRAP     #15
110 00000506 60F2         ( 95)           BRA      P_STRING      ;REPEAT
111 00000508 4E75         ( 96) P_EXIT:   RTS                    ;Exit
112                       ( 97) *
113 0000050A 456E74657220 ( 98) QUERY:    DC.B     'Enter year ',0
             796561722020
             00
114 00000517 44617465206F ( 99) ANSWER:   DC.B     'Date of Easter = ',0
             662045617374
             6572203D2000
115 00000529 4D6172636820 (100) MARCH:    DC.B     'March ',0
             00
116 00000530 417072696C20 (101) APRIL:    DC.B     'April ',0
             00
117 00000537 0A0D00       (102) NEWLINE:  DC.B     $0A,$0D,0
118          00000400     (103)           END      START

Lines: 118, Errors: 0, Warnings: 0.
```

8.4 Conditional Assembly and Structured Control

Some macro assemblers offer a facility called *conditional assembly* that allows statements to be assembled only if specified conditions are met. Since macros are translated into assembly language statements by the assembler at assemble time, the process of conditional macro assembly takes place when the program is assembled. In other words, conditional assembly does not take place at *run time*. The purpose of conditional macros is to permit the programmer to produce macros that can be modified according to the circumstances. You might write, for example, input/output routines that can be used to support both hard disks and floppy disks. Conditional assembly permits the routines to be tailored to the application at the time they are assembled.

We will describe conditional assembly by example. Several conditional constructs are defined by a conditional assembler. A conditional expression can be written as follows:

```
IFxx    <string1>,<string2>
        <statement>
ENDC
```

The condition xx is either C or NC. The mnemonic IFC means *if compare* or *if equal*, and the mnemonic IFNC means *if not compare* or *if not equal.* Both IFC and IFNC are applied to the two strings. For example, the expression IFC 'Date','\2' yields the value true if parameter 2 is equal to the string 'Date'.

The term <statement> stands for the body of the conditional expression and consists of any sequence of assembly language instructions. The statement is assembled if the conditional part yields a true value and is ignored otherwise. The term ENDC terminates the conditional statement.

Suppose we wish to create a macro that squares the contents of D0, but is sometimes required to scale the contents of D0 by ten before squaring.

```
Square MACRO                    Define a macro called 'Square'
       IFC    'scale','\1'      If parameter 1 is 'scale' then scale D0
       DIVU   #10,D0            Divide D0 by 10
       AND.L  #$0000FFFF,D0     Strip off the remainder
       ENDC                     End of conditions
       MULU   D0,D0             Perform the squaring
       ENDM                     End of macro definition
```

Note that the line IFC 'scale','\1' compares the string 'scale' with the string that is imported as parameter 1. When the program is written, the macro can be invoked by, say, Square No_scale if we do not wish to perform the scaling or by Square scale if we do.

```
1                      ( 1)
2                      ( 2)         OPT     MEX            ;Tell assembler to expand macros
3                      ( 3) SQUARE: MACRO                  ;Define a macro called Square
4                      ( 4)         IFC     'scale','\1'   ;IF called with parameter 'scale'
5                      ( 5)         DIVU    #10,D0         ;THEN divide D0 by 10
6                      ( 6)         AND.L   #$0000FFFF,D0  ;and strip the remainder
7                      ( 7)         ENDC                   ;End condition
8                      ( 8)         MULU    D0,D0          ;Square D0
9                      ( 9)         ENDM
10                     (10)
11 00000400            (11)         ORG     $400
12                     (12) *
13 00000400 303C0064   (13) START:  MOVE.W  #100,D0        ;Body of the program - load D0 with 100
14                     (14) *
```

```
15 00000404          (15)        SQUARE  No_scale              ;Square D0
16                   (15)        IFC     'scale','No_scale' ;IF called with parameter 'scale'
17                   (15)        DIVU    #10,D0                ;THEN divide D0 by 10
18                   (15)        AND.L   #$0000FFFF,D0         ;and strip the remainder
19                   (15)        ENDC                          ;End condition
20 00000404 COCO     (15)        MULU    D0,D0                 ;Square D0
21                   (16) *
22          00000400 (17)        END     START
```

In this example we have written a little dummy code round the macro and have then invoked it with the command `SQUARE No_scale`. The above listing provides the code of the macro because we selected the option `MEX`, (i.e., macro expansion). However, the conditional part of the macro (lines 17 and 18) is not assembled into 68000 assembly language instructions. The only part of the macro assembled is the instruction `MULU D0,D0` that falls outside the conditional structure. Now suppose we repeat the same exercise but invoke the macro with the statement `SQUARE Scale`.

```
1                        ( 1)
2                        ( 2)          OPT     MEX              ;Tell assembler to expand macros
3                        ( 3) SQUARE:  MACRO                    ;Define a macro called Square
4                        ( 4)          IFC     'scale','\1'     ;IF called with parameter 'scale'
5                        ( 5)          DIVU    #10,D0           ;THEN divide D0 by 10
6                        ( 6)          AND.L   #$0000FFFF,D0    ;and strip the remainder
7                        ( 7)          ENDC                     ;End condition
8                        ( 8)          MULU    D0,D0            ;Square D0
9                        ( 9)          ENDM
10                       (10)
11 00000400              (11)          ORG     $400
12                       (12) *
13 00000400 303C0064     (13) START:   MOVE.W  #100,D0          ;Body of the program - load D0 with 100
14                       (14) *
15 00000404              (15)          SQUARE  scale            ;Scale D0 by 10 and square it
16                       (15)          IFC     'scale','scale   ;IF called with parameter 'scale'
17 00000404 80FC000A     (15)          DIVU    #10,D0           ;THEN divide D0 by 10
18 00000408 02800000FFFF (15)          AND.L   #$0000FFFF,D0    ;and strip the remainder
19                       (15)          ENDC                     ;End condition
20 0000040E COCO         (15)          MULU    D0,D0            ;Square D0
21                       (16) *
22          00000400     (17)          END     START
```

In this case the condition tested is true because the string `'scale'` supplied as a parameter in the macro invocation matches the string in the instruction (line 15). As the test is true, the instructions on lines 17 and 18 are assembled.

The `IFxx` construct can take two forms, the first of which we have already seen. An alternative form is:

```
IFxx <expression>
<statement>
ENDC
```

where the condition xx is one of the following: EQ (equal to), NE (not equal to), LT (less than), LE (less than or equal to), GT (greater than), or GE (greater than or equal to). The expression is compared with the value zero and the specified statement assembled if the logical condition is true. The above example could be repeated using the following construct:

```
Square MACRO                  Define a macro called 'Square'
       IFLT  \1               If parameter 1 < 0  then perform scaling
       DIVU  #10,D0           Divide D0 by 10
       AND.L #$0000FFFF,D0    Strip off the remainder
       ENDC                   End of conditions
       MULU  D0,D0            Perform the squaring
       ENDM                   End of macro definition
```

In order to invoke the macro and perform scaling, we must call it with any negative parameter; e.g., Square -1.

Structured Control Statements

68000 assemblers may also include *structured control statements* that provide the assembler with pseudo-high-level language facilities. When we discussed program design and pseudocode, we translated our own high-level constructs into assembly language the hard way — by hand. A structured assembler automates this process. It is important to appreciate that there is a world of difference between *conditional macros* and *structured statements*. Conditional macros are evaluated at assemble time, and the specified condition is tested to determine whether or not a section of code should be assembled. Structured control statements are translated into machine code by the assembler and are executed at *run-time* (i.e., when the program is actually being executed). Structured macros enable the programmer to write basic high-level language constructs that are translated automatically into assembly language by the assembler.

The Motorola structured assembler reserves the following keywords (i.e., they may not be used by the programmer as symbolic names).

```
ELSE        ENDW        REPEAT
ENDF        FOR         UNTIL
ENDI        IF          WHILE
```

In addition to these reserved keywords, the structured assembler also recognizes the following keywords.

```
AND          DOWNTO TO
BY           OR
DO           THEN
```

These are *nonreserved* keywords because they may be used elsewhere by the programmer. That is, there is nothing to stop you calling a variable DOWNTO — apart from common sense.

Let's begin by looking at the FOR statement. The format syntax of the FOR statement is:

```
FOR[.<size>]   <op1> = <op2> TO <op3> [BY <op4>] DO[.<extent>]
               <statement>
ENDF
```

You can be forgiven for thinking that this statement looks impenetrable. All we need to remember is that square brackets, [ ], enclose optional parameters and angle brackets, < >, enclose values that must be supplied by the programmer.

The first parameter [.<size>] is optional and defines the size of the four operands <op1> to <op4>. This parameter may take the values .B, .W, or .L for byte, word, and longword operands, respectively. The four operands may be literals or the contents of a data register and are defined as follows:

<op1> Operand 1 is a user-defined operand whose memory location (or register) holds the FOR-counter.

<op2> Operand 2 provides the initial value of the FOR-counter.

<op3> Operand 3 provides the terminating value for the FOR-counter.

<op4> Operand 4 is optional and provides the increment by which the FOR-counter is stepped each time round the loop. The default value is one.

The optional <.extent> field applied to the DO defines whether the branch used to implement the loop is a short branch (i.e., .S) or a long branch (i.e., .L). The expression <statement> is the code that forms the body of the FOR loop and may be one or more lines of code. A possible FOR construct might therefore read:

```
FOR.L D0 = D1 TO D6 BY #5 DO.S
<statement>
ENDF
```

Consider the fragment of code to perform the addition of the first ten natural integers.

```
CLR.W   D0                       Clear the total
FOR.W   D1 = #1 TO #10 DO
ADD.W   D1,D0                    Add the loop count to the total
ENDF
```

In this example, operands 2 and 3 are literal values. This code was assembled using Motorola's M68MASM cross-assembler to provide the following output. Note that the Teesside cross-assembler does not support structured control statements. Lines that begin with a "+" symbol in the comment field are lines supplied by the assembler during the expansion of the control statement.

```
Motorola 68000 Family Assembler      (1.0 ) Wed Feb 03 16:58:33 1993

abs. rel.   LC   obj. code   source line
---- ----  ---- ---------   -----------
   1    1  0400               |          ORG     $400
   2    2  0400 4240          |          CLR.W   D0
   3    3  0402               |          FOR.W   D1 = #1 TO #10 DO
   4   1s  0402 323C 0001     +          MOVE.W  #1,D1
   5   2s  0406 6000 0004     +          BRA.W    .3
   6   3s  040A               +.1:
   7   4s  040A 5241          +          ADDQ.W  #1,D1
   8   5s  040C               +.3:
   9   6s  040C 0C41 000A     +          CMPI.W  #10,D1
  10   7s  0410 6E00 0006     +          BGT.W    .2
  11    4  0414 D041          |          ADD.W   D1,D0
  12    5  0416               |          ENDF
  13   1s  0416 60F2          +          BRA.W    .1
  14   2s  0418               +.2:
  15    6  0418               |          END
  15 lines assembled
```

We now present the syntax of the other structured statements. In each case we also provide a typical example of the syntax.

- The IF statement

```
IF[.<size>]     <expression> THEN[.<extent>]
          <statement>
ENDI
```

The IF statement behaves in a similar way to the conventional IF construct of high-level languages. The <expression> is one of the following:

```
<CC>        <CS>
<EQ>        <NE>
<GE>        <GT>
<HI>        <LE>
<LS>        <LT>
<MI>        <PL>
<VC>        <VS>
```

All these 14 expressions must be enclosed in angle brackets as shown, and refer to the same logical conditions as the 68000's Bcc instructions. That is, the CCR is tested by the selected condition. Consider the code:

```
ADD.L    D0,D1        Add mantissas
IF <VS> THEN          On overflow normalize the number
   ROXR.L #1,D1       Shift the mantissa right
   ADDQ.W #1,D2       and increment the exponent
ENDI
```

In this case, the condition tested (i.e., arithmetic overflow) results from the longword addition. We shall soon see that the condition tested can also be expressed as the relationship between two values (rather than by the state of the CCR). The following code was produced by the M68MASM assembler.

```
Motorola 68000 Family Assembler             (1.0 ) Wed Feb 03 19:48:47 1993

abs. rel.   LC   obj. code   source line
—— ——    — ———   ———-

  1   1  0400                |         ORG      $400
  2   2  0400 D280           |         ADD.L    D0,D1     Add mantissas
  3   3  0402                |         IF <VS> THEN       Normalize the number on overflow
  4  1s  0402 6900 0006      +         BVS.W    .1
  5   4  0406 E291           |         ROXR.L   #1,D1     Shift the mantissa right
  6   5  0408 5242           |         ADDQ.W   #1,D2     and increment the exponent
  7   6  040A                |         ENDI
  8  1s  040A                +.1:
  9   7  040A                |         END
  9 lines assembled
```

As in the case of the FOR loop, the optional term [.<extent>] controls the generation of long- or short-branch addresses and may be .S (short) or .L (long). An example of this construct might be:

```
IF.L ... THEN
.
.
ENDI
```

- The IF...THEN...ELSE statement

This construct is a simple expansion of the plain vanilla IF construct and is expressed in the form:

```
IF[.<size>] <expression> THEN[.<extent>]
                                <statement>
```

```
                ELSE[.<extent>]
                   <statement>
ENDI
```

A typical example of this construct might be:

```
IF.W D3 <GT> D4 THEN.S
             MOVE.B D3,D7
           ELSE.S
             MOVE.B Temp3,D7
             ADD.B  #5,D7
ENDI
```

Note that the expression in this construct is represented by `D3 <GT> D4`. The `IF`, `REPEAT`, and `WHILE` statements all define the `<expression>` part of the statement as either `<cc>` or `<operand1> <cc> <operand2>`. The value of `<cc>` is one of the 14 conditions described above. The following code shows how M68MASM assembles the `IF THEN ELSE` statement.

```
Motorola 68000 Family Assembler        (1.0 ) Wed Feb 03 17:39:48 1993
abs. rel.   LC   obj. code   source line
---- ----  ----  ---------   -----------
   1  1   0400               |       ORG  $400
   2  2   0000               |       IF.W D3 <GT> D4 THEN.S
   3  1s  0400 B843          +         CMP.W    D3,D4
   4  2s  0402 6C04          +         BGE.S    .1
   5  3   0404 1E03          |                           MOVE.B D3,D7
   6  4   0406               |                         ELSE.S
   7  1s  0406 6008          +         BRA.S    .2
   8  2s  0408               +.1:
   9  5   0408 1E39 0000     |                           MOVE.B Temp3,D7
   9      040C 0410          |
  10  6   040E 5A07          |                           ADD.B  #5,D7
  11  7   0410               |       ENDI
  12  1s  0410               +.2:
  13  8   0410               |Temp3 DS.B 1
  14  9   0411               |       END
  14 lines assembled
```

Another example of this construct might be:

```
IF.L #$1234 <LT> D6 THEN.S
                   CLR.L D6
                 ELSE
                   CLR.L D7
ENDI
```

- The REPEAT statement

```
REPEAT
                <statement>
UNTIL[.<size>]  <expression>
```

Example syntax:

```
CLR.W    D0
CLR.W    D1
REPEAT
  ADD.W  #1,D0
  ADD.W  D0,D1
UNTIL D0 <EQ> #10
```

The M68MASM assembler produces the following object code.

```
Motorola 68000 Family Assembler   (1.0 ) Wed Feb 03 17:03:10 1993

abs. rel.   LC   obj. code   source line
---- ----   ---- ---------   -----------
   1    1   0400              |       ORG   $400
   2    2   0400 4240         |       CLR.W D0
   3    3   0402 4241         |       CLR.W D1
   4    4   0404              |       REPEAT
   5   1s   0404             +.1:
   6    5   0404 5240         |           ADD.W #1,D0
   7    6   0406 D240         |           ADD.W D0,D1
   8    7   0408              |       UNTIL D0 <EQ> #10
   9   1s   0408 0C80 0000    +       CMPI.L  #10,D0
   9        040C 000A         +
  10   2s   040E 67F4         +       BEQ.W   .1
  11    8   0410              |       END
  11 lines assembled
```

- The WHILE statement

```
WHILE[.<size>]  <expression> DO[.<extent>]
                <statement>
ENDW
```

Example syntax:

```
CLR.W D0
CLR.W D1
WHILE.W D0 <LE> #9 DO
```

```
        ADD.W   #1,D0
        ADD.W   D0,D1
    ENDW
```

The M68MASM assembler produces the following output.

```
Motorola 68000 Family Assembler   (1.0 ) Wed Feb 03 17:09:08 1993

abs. rel.   LC   obj. code   source line
---- ---   ---- ---------

   1   1   0400                |        ORG    $400
   2   2   0400 4240           |        CLR.W D0
   3   3   0402 4241           |        CLR.W D1
   4   4   0404                |        WHILE.W D0 <LE> #9 D0
   5   1s  0404               +.1:
   6   2s  0404 0C40 0009     +         CMPI.W  #9,D0
   7   3s  0408 6E00 0008     +         BGT.W   .2
   8   5   040C 5240           |            ADD.W #1,D0
   9   6   040E D240           |            ADD.W D0,D1
  10   7   0410                |        ENDW
  11   1s  0410 60F2          +         BRA.W   .1
  12   2s  0412               +.2:
  13   8   0412                |        END
  13 lines assembled
```

Summary

- Most assemblers are two-pass assemblers. On the first pass they check the program's syntax and create a symbol table for all symbolic names and references. On the second pass they translate symbolic names into actual values.
- The SECTION n directive defines a relocatable program section (n = 0 to 15 and is user-definable).
- The OFFSET directive is used to create a table of offsets.
- The REG directive equates a register list to a symbolic name. For example, AllRegs REG D0-D7/A0-A6.
- The SET directive equates a symbolic name to a value and is similar to an EQU (except that SET can be used again to *redefine* the value of a symbol).
- The DCB directive loads a block of constants into memory. For example, the directive DCB.W 20,$AA is equivalent to 20 lines of DC.W $AA.

- A program can be written as a set of independent modules and the modules assembled independently. The IDNT assembler directive labels a module. The XDEF assembler directive lists the symbols that are defined externally to the module. That is, it lists the names whose values are defined in other modules. The XREF assembler directive tells the system that the following symbolic names are required by other modules. In other words, XDEF *imports* symbols and XREF *exports* them.

- A macro is a unit of code that is given a name by the programmer. Whenever the name is used in the program, it is replaced by the code of the macro. The macro is really a form of shorthand.

- A macro is defined by placing the unit of code between the assembler directives MACRO and ENDM.

- A macro can take parameters. When the macro is written, the first parameter is written as \1 and the second parameter as \2, and so on. When the programmer uses or invokes the macro, the parameters are replaced by actual values to be used in the expansion of the macro.

- You can use labels when writing macros. A label is written in the form <string>\@ and the symbol @ is automatically replaced by a new numeric value each time the macro is used in a program (to avoid using the same label twice or more).

- Conditional expressions can be embedded within macros. This permits the design of complex macros that can be tailored to suit a range of applications. For example, by supplying a certain parameter, the macro might be translated into one code sequence, and by supplying another parameter, the macro might be translated into another sequence.

- Some assemblers support structured control statements. These are templates that make it very easy to map high-level language control structures onto an assembler language.

- Structured statements can use the 14 conditions used by the 68000's B_{cc} instructions (i.e., <CC>, <CS>, <EQ>, <GE>, <GT>, <HI>, <LE>, <LS>, <LT>, <MI>, <NE>, <PL>, <VC>, and <VS>). Consider the statement

```
CLR.W  D0
REPEAT
       ADD.W #1,D0
       SUB.W D4,D3
UNTIL <LT>
```

  This construct seems strange because no expression is evaluated by the line UNTIL <LT>. The previous line, SUB.W D4,D3, subtracts D4 from D3 and sets the 68000's condition codes accordingly. When the UNTIL <LT> is encountered, the loop will be terminated if the previous operation resulted in a value less than zero.

Problems

1. Why is it harder for an assembler to resolve *forward* references than *backward* references?

2. Why does the 68000 assembler define a SET assembler directive as well as an EQU? When do you think that a SET directive might be used?

3. What does the DCB assembler directive do, and how might you use it?

4. What is the advantage of taking a modular approach to the design of assembly language programs, and what problems does it create?

5. How does the 68000 assembler deal with the problems posed by modular programming?

6. What is the difference between a *macro* and a *subroutine*?

7. What are the relative advantages of the macro and the subroutine, and when is each used?

8. *How* are parameters passed to a macro? *When* are parameters passed to a macro? *Why* are parameters passed to a macro?

9. What is the purpose of a conditional macro?

10. The conditional macro is evaluated at assemble-time, but the conditional statement is evaluated at run-time. What is the significance of *assemble-time* and *run-time* in this statement?

11. Examine the code in Chapter 11 and re-write it using the more advanced features of an assembler.

CHAPTER 9

Peripheral Programming

One of the most important reasons for teaching assembly language is that it is frequently well suited to applications involving input/output programming. Although the average assembly language programmer will never write anything as large as a compiler or an editor, he or she may well be asked to write *device drivers* that interface such programs to real input/output devices. In this chapter we are going to look at the basic principles underlying input/output and demonstrate how some practical I/O devices are programmed.

We also introduce the 68000's interrupt handling mechanism and show how it can be used to support input/output operations. However, we will not delve deeply into all aspects of interrupt handling, as the next chapter is devoted to the more general class of interrupts called *exceptions*.

Before we begin, we should make a distinction between two key aspects of input/output. The first is the *strategy* by which information is moved between the processor and external systems. The three strategies we describe in this chapter are programmed (or *polled*) I/O, interrupt-driven I/O, and direct memory access (DMA). The second aspect of I/O is the nature of the *hardware* employed to perform the I/O. In this chapter, we concentrate on typical parallel and serial I/O ports and show how the assembly language programmer accesses them.

9.1 Input/Output Strategies

Suppose that we have to transfer data from a *processor* to an *external system*. This data may be moved to a visual display terminal or hard disk drive, or even used to control the temperature of a chemical reaction in a vessel. How do we go about performing the data transfer? The simplest strategy is called *programmed data transfer* and involves nothing more than writing a suitable program to transfer data between the processor's memory and a peripheral.

The first question we should ask is, 'What special facilities does the 68000 have to enable it to perform I/O transactions?'. The answer to this sensible question is remarkably simple. The 68000 has no special I/O facilities whatsoever. Design engineers could have given the 68000 a special data bus to connect its internal buses to external peripherals. It could have been endowed with special instructions that transfer data between the CPU's registers and a peripheral (e.g.,

OUTPUT D3,DISK). The 68000 lacks both special purpose hardware and special purpose software devoted to input/output. Fortunately, any microprocessor able to access memory can also perform I/O transactions by using the *existing* CPU-memory information paths.

Figure 9.1 illustrates the structure of a typical arrangement of processor, memory, and input/output port. The I/O port is treated exactly like a memory location. Some I/O ports have more than one internal location and behave like a block of memory locations. The hardware systems designer interfaces the I/O port to the processor like any other memory block, and then the software writer can treat the I/O port as a memory location. This arrangement is called *memory-mapped I/O*. In Figure 9.1 the I/O port is memory-mapped at addresses $008000 to $008003. We will now look at the simplest of I/O strategies, programmed I/O.

Programmed I/O

All you have to do to transfer data between the CPU and an I/O port by means of *programmed I/O* is to employ any data movement instruction. Suppose we have a block of 256 bytes of data starting at address location $2000 and we wish to send them, one by one, to an output port mapped at memory location $8001. The following code will do what we require.

Figure 9.1 Processor, memory, and I/O

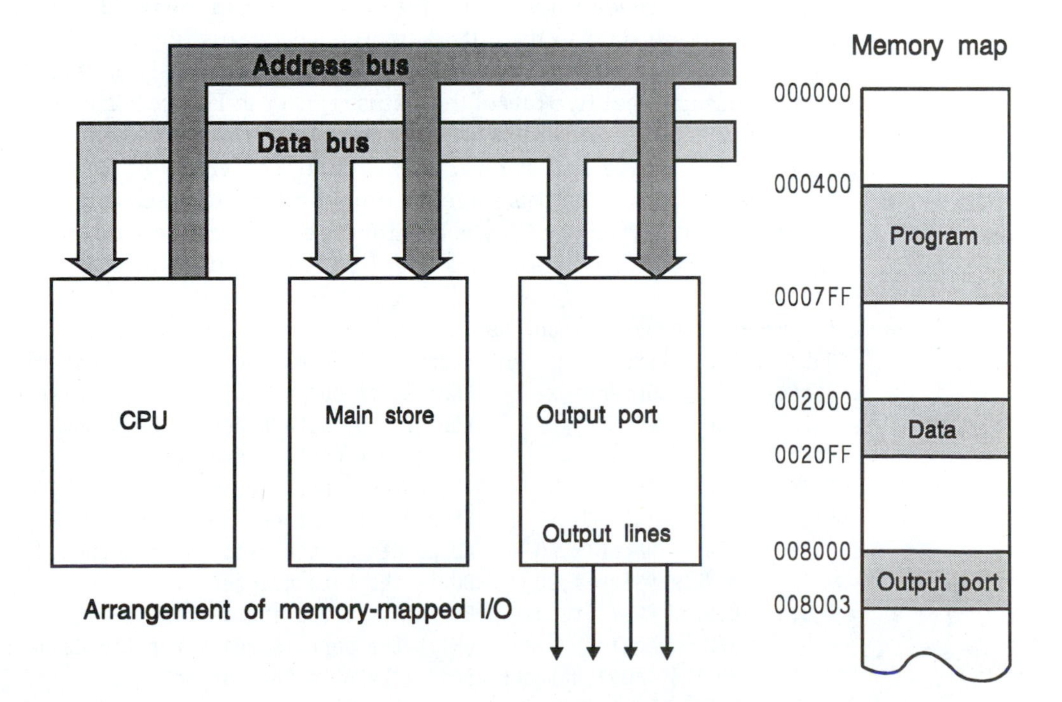

```
*       Programmed output routine
*
OutPort EQU     $8001               Address of output port
DataBlk EQU     $2000               Location of data block to be output
Size    EQU     256                 Size of data block
*
        LEA     DataBlk,A0          A0 points to the data to be output
        MOVE.W #Size-1,D0           D0 is the byte counter
Again   MOVE.B (A0)+,OutPort        REPEAT Send a byte to the output port
        DBRA    D0,Again            UNTIL all bytes have been transmitted
```

The above program transmits 256 bytes to the display device. Unfortunately, it suffers from one tiny, almost insignificant, problem. It doesn't work. There is no programming error in the code — the problem is caused by purely *operational* factors. A byte of data is transmitted to the output port each time the loop is executed. The `MOVE.B (A0)+,OutPort` instruction takes 20 clock cycles and a `DBRA` instruction takes 10 cycles when the branch is taken. Each time round the loop takes 30 clock cycles or 3.75 μs (assuming a 68000 operating with an 8 MHz clock that corresponds to a clock cycle of 125 ns). It is very unlikely that the display unit could accept data at this rate, and therefore most of the data will simply be lost. What we need is some method of *flow control*.

Virtually all real I/O ports include at least two registers. Figure 9.2 illustrates a hypothetical output port with two memory-mapped locations. One location is an *output data* register and the other a *status* register. The data register is used to transmit data to the outside world. The status register tells the processor things it needs to know about the status of the peripheral connected to the port. We will assume that two bits of the status register in Figure 9.2 are defined: bit 7 is a ready bit, RDY, and bit 0 is an error bit, ERR. The ready bit is set if the port is able to send a byte and is clear if it cannot. The error bit is set if an error in transmission occurs. At this stage we do not intend to discuss how these bits are set or cleared. All we need say is that they are set or cleared *automatically* by logic within the port. We will now look at our polled output routine again.

```
*   Programmed output routine
OutPort EQU     $8001               Address of output port data register
Status  EQU     OutPort+2           Address of output port status register
DataBlk EQU     $2000               Location of data block to be output
ERR     EQU     0                   The output port's error bit
RDY     EQU     7                   The output port's ready bit
Size    EQU     256                 Size of data block
        LEA     DataBlk,A0          A0 points to the data to be output
        MOVE.W #Size-1,D0           D0 is the byte counter
Again   BTST    #RDY,Status         REPEAT Test the status bit
        BEQ     Again               UNTIL the port is ready for the data
        MOVE.B (A0)+,OutPort        Send a byte to the output port
        DBRA    D0,Again            Repeat until all bytes have been transmitted
```

Figure 9.2 Structure of a memory-mapped I/O port

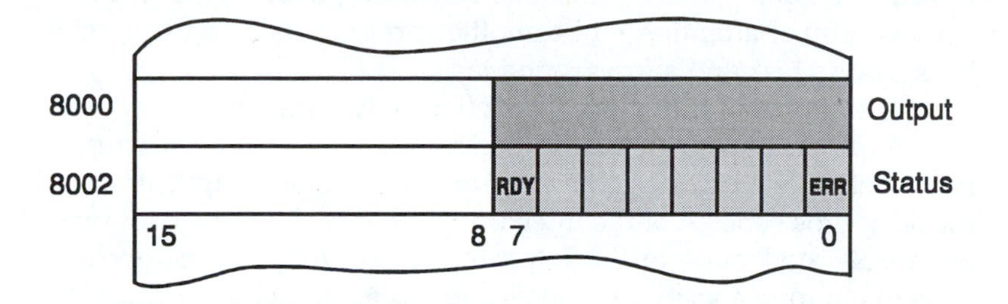

The port status register's ready bit, RDY, is tested by `BTST #RDY,Status`. If RDY is zero (i.e., the output port is not ready to send data), the CCR's Z-bit is set and the following instruction, `BEQ Again`, forces a branch back to the instruction that tests the ready bit. That is, the ready bit is continually read until it is set, indicating that a data transfer may be made. When the RDY bit is set by the port's internal logic, the branch is not taken and the output transfer takes place. By the way, in this example we don't use the error bit in the status word. If we wished, we could modify the routine to perform an error test before performing the data transfer in the following way.

```
Again   BTST    #RDY,Status     REPEAT Test the status bit
        BEQ     Again           UNTIL the port is ready for the data
        BTST    #ERR,Status     Check for error
        BNE     Error           If input in error then deal with it
        MOVE.B  (A0)+,OutPort   Send a byte to the output port
        DBRA    D0,Again        Repeat until all bytes have been transmitted
```

The loop that tests the ready bit until it is set is called a *polling loop*. Programmed I/O involving a polling loop is very easy to implement and requires minimal hardware and software overheads. Unfortunately, it is *inefficient*.

It takes a brave person to define *efficiency* in the context of a computer. For our present purposes we will define efficiency as the ratio of what a processor actually achieves to what it could achieve. We appreciate that this definition is unsatisfactory, because it attempts to distinguish between *useful* operations and useless operations. However, we want to stress that executing a polling loop over and over again gets no useful work done. If the processor were not polling the output port, it could be executing a different task. A more efficient alternative to polled I/O is *interrupt-driven I/O,* in which the processor is able to perform a useful task until the I/O port is actually ready to transfer data.

Interrupt-driven I/O

Sophisticated applications of computers cannot afford to let the CPU waste its time executing a polling loop, as we have just described. Such computers have two or more tasks to perform and can always find something useful to do. For example, a queue of programs may be waiting to be run, or some peripheral may need continual attention while another program is being run, or there may be a background task and a foreground task.

A technique for dealing more effectively with input/output transactions has been implemented on all microprocessors and is called an *interrupt handling mechanism*. Before we can describe *interrupt-driven input/output* in detail, we need to introduce the concept of the interrupt. In the world of the computer, the interrupt is entirely analogous to the interrupt in everyday life. Suppose I'm teaching a class of computer science students and I'm in the action of drawing a diagram on the board. Suppose also that a student cannot read my writing. The student might call out and *interrupt* my drawing. I deal with the student's problem and then return to my original activity. Interrupts in both the computer and human worlds have three components: a mechanism whereby an external device can *request attention* or service, a mechanism that permits the processor to *respond* to the interrupt, and a mechanism that enables the processor to *return* to the point it had reached immediately before the interrupt.

An interrupt request line, IRQ, is connected between the peripheral and the CPU. The IRQ line is an input to the CPU. Whenever the peripheral is ready to take part in an input/output operation, it *asserts* the IRQ line and invites the CPU to deal with the transaction. We use the term *assert* to indicate that a signal is put in the logical state that causes its "named action" to take place. The CPU is free to carry out other tasks between interrupt requests from the peripheral.

Since we need to describe the 68000's interrupt handling mechanism in order to explain its role in I/O operations, we will provide the simplest possible description of the 68000's interrupt mechanism here. Figure 9.3 illustrates the principal components of the 68000's interrupt handling mechanism. A peripheral uses one of the seven interrupt request lines, IRQ1* to IRQ7*, to request the 68000's attention. Remember that the asterisk after IRQ1, etc., indicates that an interrupt is requested by an *electrically low signal*. If the peripheral asserts interrupt request line IRQi*, we say that an interrupt request at level i has been made. The 68000's interrupt request inputs are *prioritized* so that an interrupt at level j takes precedence over an interrupt at level i, if $j > i$.

The three *interrupt mask* bits in the 68000's status register, I2, I1, I0, can be set to any value in the range 0 to 7 (usually by means of a `MOVE #data,SR` instruction). If the current value of the interrupt mask is p, then any interrupt request at a level equal to or less than p will not be serviced. Only interrupts at a level $p+1$ or greater will be serviced. The interrupt mask permits the 68000 to ignore interrupts it does not wish to service for the time being. For example, the 68000 may be transferring a file to the hard disk. During this operation it might not wish to respond to any interrupt request from a slow device like a keyboard.

When the 68000 responds to an interrupt, it executes an *interrupt acknowledge* or IACK cycle. The IACK cycle takes place at the hardware level and is invisible to the programmer. The device that requested the interrupt detects the IACK cycle and provides the 68000 with an *interrupt vector number*. In effect, the interruptor is saying, 'It was I — here is my name.' The 68000 uses this number to locate the

Figure 9.3 68000's interrupt handling mechanism

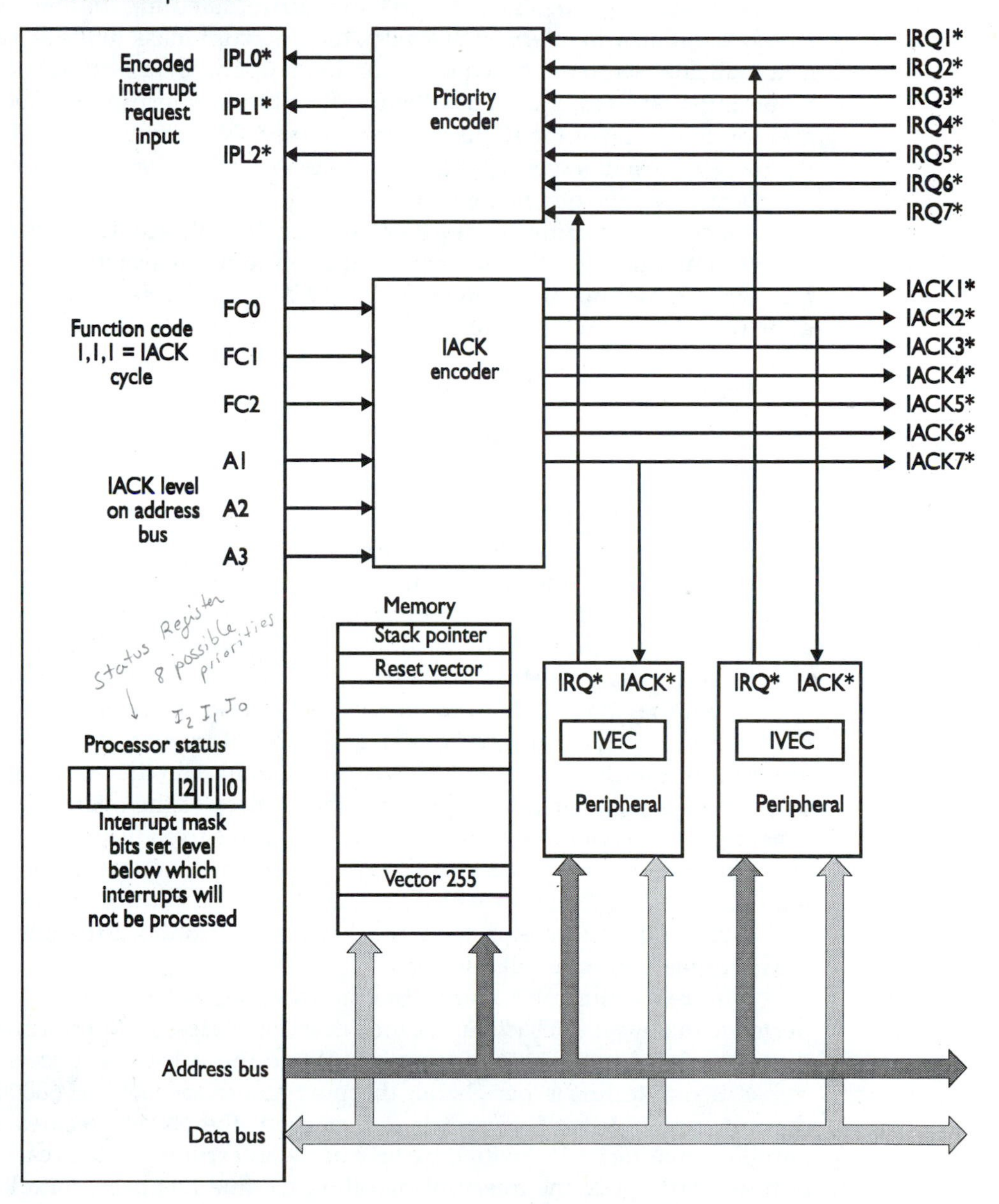

address of the interrupt handling routine that will service the needs of the device that requested the interrupt. The appropriate interrupt vector is located in a 256-longword table in the region of memory from $000000 to $0003FF.

The 68000 pushes the current value of the program counter onto the stack together with a copy of the status register at the time of the interrupt. This information is required to restore the processor to its pre-interrupt state after the interrupt has been processed. Finally, the program counter is loaded with the interrupt vector and a jump made to the interrupt handling routine. By the way, if a device initiates an interrupt at level p, the interrupt mask in the status register is automatically set to p. Consequently, an interrupt at level p cannot be interrupted by an interrupt with a level less than $p+1$, until the interrupt has been processed and the old value of the status register restored. Note that, the 68000 always sets the S-bit of its status register and switches to its supervisor state before processing an interrupt or any other exception.

Once the interrupt request has been dealt with, the last instruction in the interrupt handling routine, RTE (return from exception), has the effect of returning the 68000 to the state it was in immediately before the interrupt. We can write an interrupt handler in the form:

```
IntHandl  ...
          ...      Code of interrupt handler
          ...
          RTE      Final instruction returns to calling point
```

For our current purposes, we can think of the RTE instruction as being virtually the same as the RTS instruction used to terminate a subroutine. The only real difference between a *subroutine* and an *interrupt request* is that the programmer puts a subroutine in a program by writing BSR Subroutine, while an external device jams an interrupt request into the instruction flow whenever the device feels like it.

From the above brief introduction, we can see that dealing with interrupts requires a marriage of hardware and software. Moreover, the 68000's interrupts are usually handled by the operating system running in the *supervisor* state rather than by user programs running under the control of the operating system. It is considered good practice to insulate user programs from the actual peripherals in any computer system. A user program may say to the operating system, 'Please, may I transfer a file to the disk?'. It cannot, however, control the actual hardware that carries out the transfer. Now that we have introduced the interrupt we can return to the topic of input/output.

Figure 9.4 illustrates a port that employs the 68000's interrupt mechanism to perform an output transaction. Assume that the processor is performing a task not directly related to the output operation. When the output port becomes ready to transmit a byte to the peripheral, the port asserts one of the 68000's interrupt request inputs. If the interrupt is not masked, the 68000 executes an interrupt handling routine to deal with the cause of the interrupt (i.e., the ready state of the output port). Once the interrupt handling routine has been completed, an RTE instruction allows the processor to return to the point it had reached before the interrupt.

The advantage of interrupt-driven I/O is that it does not tie up the processor in a futile polling loop, and a data transfer is made only when the port is ready for it. Consider a simple example of the interrupt-driven output using the same arrangement we introduced when describing polled I/O.

Figure 9.4 I/O port using interrupts

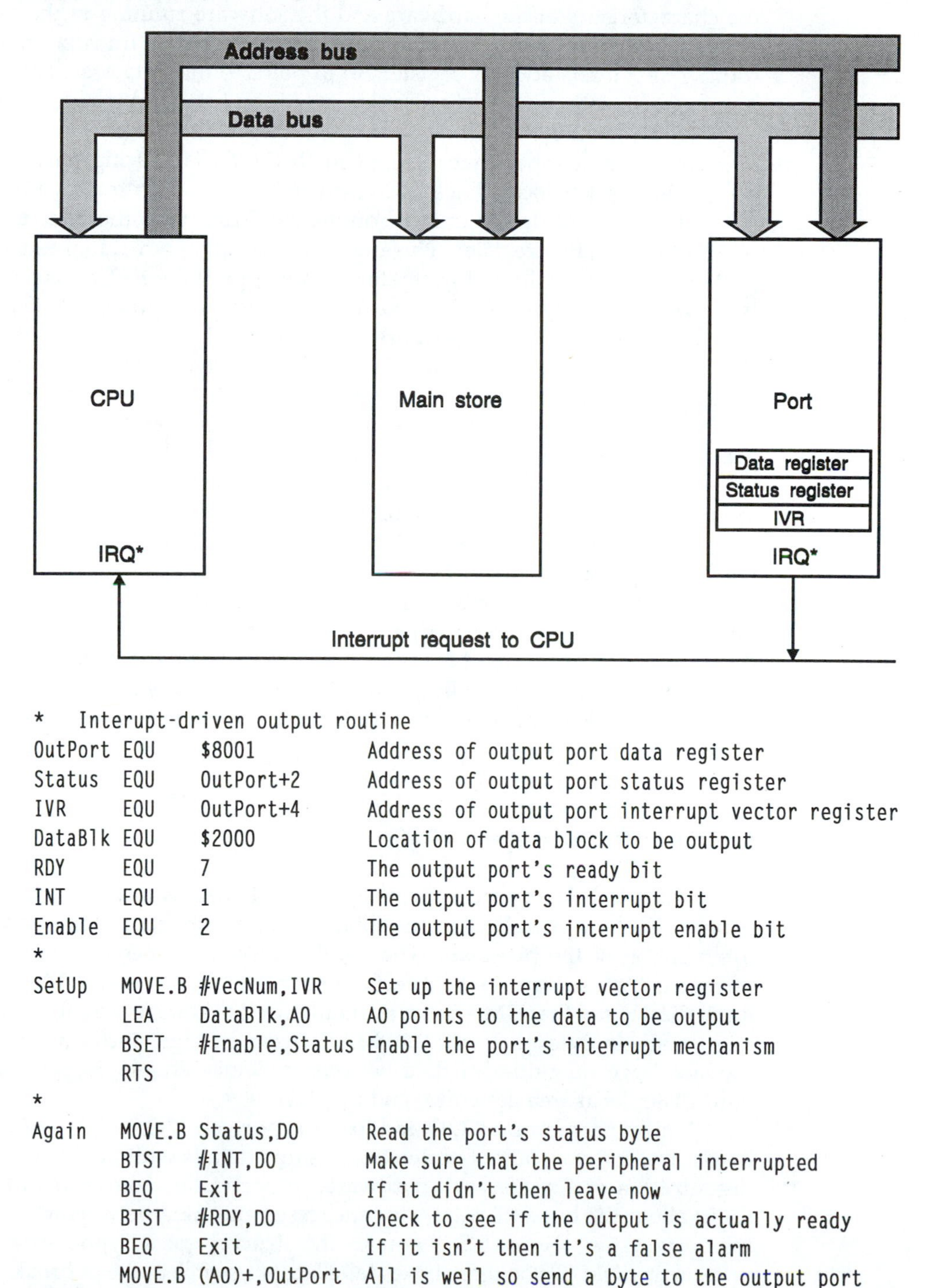


```
*   Interupt-driven output routine
OutPort EQU    $8001           Address of output port data register
Status  EQU    OutPort+2       Address of output port status register
IVR     EQU    OutPort+4       Address of output port interrupt vector register
DataBlk EQU    $2000           Location of data block to be output
RDY     EQU    7               The output port's ready bit
INT     EQU    1               The output port's interrupt bit
Enable  EQU    2               The output port's interrupt enable bit
*
SetUp   MOVE.B #VecNum,IVR     Set up the interrupt vector register
        LEA    DataBlk,A0      A0 points to the data to be output
        BSET   #Enable,Status  Enable the port's interrupt mechanism
        RTS
*
Again   MOVE.B Status,D0       Read the port's status byte
        BTST   #INT,D0         Make sure that the peripheral interrupted
        BEQ    Exit            If it didn't then leave now
        BTST   #RDY,D0         Check to see if the output is actually ready
        BEQ    Exit            If it isn't then it's a false alarm
        MOVE.B (A0)+,OutPort   All is well so send a byte to the output port
Exit    RTE                    Return from exception
```

This example is illustrative, as the precise details of a real system depend on the characteristics of the hardware and the software running in the background. Note that we have to employ some code to set up the data transfer. In this case the routine `SetUp` sets address register A0 to point to the data area, and enables the output port's interrupt mechanism by setting an interrupt enable bit in its status register. This routine also loads a vector number into the port's interrupt vector register — the number that is supplied to the 68000 during an IACK cycle. A pointer to `Again` is located at location `4xVecNum` in the exception vecor table.

The interrupt-driven output routine itself first tests the interrupt bit of the output port's status register. This may seem strange, but it makes sure that this particular device really did generate the interrupt. If the INT bit is not set, an exit from the routine is made. The next step is to test the ready bit, RDY. Again, you might feel that this is not necessary. Most practical I/O ports can perform many different functions, and it is necessary to make sure that the interrupt was generated by the function you wish to perform. If both these conditions are met, the output transfer is made, and the return from exception instruction causes the processor to return to normal processing.

As you can see, interrupt-driven I/O requires at least two steps. First, you set up the transaction by configuring the peripheral, and then you execute an interrupt routine each time an I/O operation is requested. Note that the processor's interrupt mechanism does not itself perform the I/O transaction. The interrupt is employed by the peripheral to inform the processor that an I/O operation can take place. The actual input or output data transfer is performed by an instruction of the form `MOVE <port>,<destination>` or `MOVE <destination>,<port>`.

We are now going to discuss the third I/O strategy available to the systems designer. This is direct memory access, which does not employ the processor to transfer data.

Direct Memory Access

The principle behind direct memory access, DMA, is exactly what you would expect. The peripheral transfers data *directly* to or from memory without the intervention of the processor. That is, the processor does not transfer data but relegates it to a servant — the direct memory access controller, DMAC. We should immediately say that DMA is not normally used to transfer small amounts of data (e.g., a single byte or a word). DMA is used in high performance systems to transfer large quantities of data between memory and high-speed devices like hard disks, local area networks, and displays.

Unlike programmed I/O and interrupt-driven I/O, DMA requires complex hardware, and you cannot understand its fine details without a knowledge of the organization of a processor's hardware. In particular, you need to know something about the buses that transfer addresses and data between the CPU and its memory system. Figure 9.5 illustrates the structure of a hypothetical microcomputer using DMA. The three boxes labelled *Bus switch* make or break connections to the system bus. In *normal operation* the CPU has control of the system's address and data bus. That is, it provides the address of a memory location and then reads or writes to the memory. In order to write a block of data to a peripheral, the processor must perform the following sequence of operations.

```
Point to source of data
REPEAT
      Read a byte of data
      Transfer it to the peripheral
      Point to next data location
UNTIL All data has been transferred.
```

This pseudocode demonstrates that the processor is performing a sequence of simple operations over and over again. Since a von Neumann machine usually requires two accesses per instruction (one to read the instruction and one to access the operand), it is not an optimum architecture for an application like transferring large blocks of data. In a system with DMA, the processor goes to sleep during the data transfer. A special purpose digital machine, the DMA controller or DMAC takes over control of the processor's buses and directly transfers data between the memory and the peripheral.

Figure 9.5 I/O using DMA

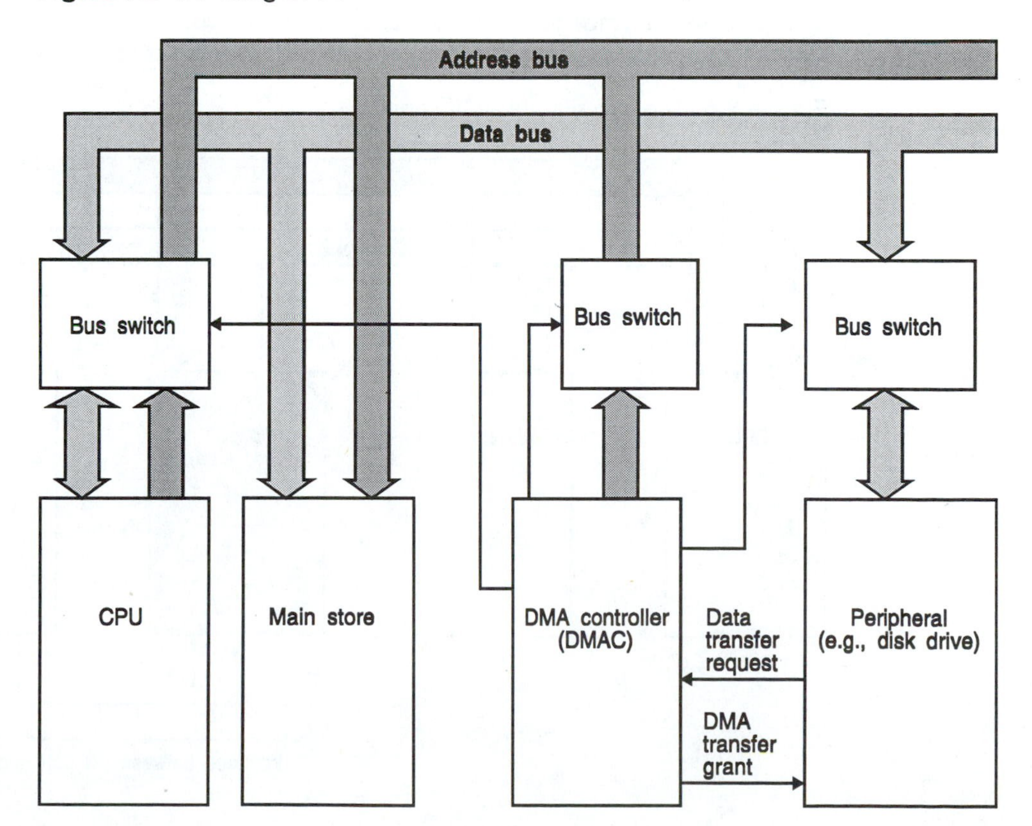

Let's look at an example. Suppose the processor wishes to transfer 10K bytes from memory to a hard disk. The processor first executes a program to set up the DMA controller, which is itself a memory-mapped peripheral. The DMA controller is told how much data is to be transferred, whether the data is to be transferred to or from memory, and where it is to go to in memory (or where it is to come from). Once the DMA controller has been given its orders, it is enabled (usually by setting an enable bit in one of its registers), and it goes about its task. Since the DMA controller is organized to perform high-speed data transfers, it does not have to read a program from memory. All the operations that it carries out are *hard-wired* (i.e., built into it). The DMA controller is able to cycle through addresses and transfer data to or from memory automatically.

After the DMA transfer has taken place, the DMA controller wakes up the processor which takes control of the address and data buses again. In fact, when a DMA operation has taken place, the DMA controller might interrupt the processor to tell it that the data has now been moved. We will not look at DMA controllers in further detail, as they belong in texts on microprocessor interfacing.

Having looked at the three *strategies* by which data is moved into and out of a microprocessor system, the next step is to consider the *hardware* that actually carries out the task. All modern microprocessor systems (see Figure 9.6) are constructed around a processor chip, memory components, and interface chips.

An interface chip has two *sides*. One side is called the *CPU side* and is designed to interface to the processor's bus. Once correctly interfaced, the CPU side of the interface chip looks like a set of memory-mapped registers to the

Figure 9.6 The two sides of an I/O port

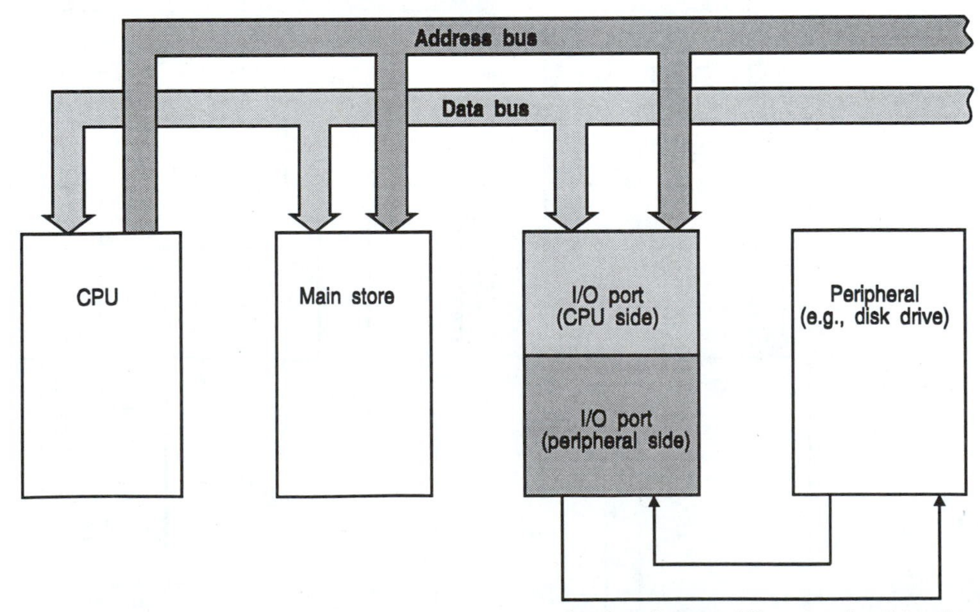

programmer. The other side of the interface chip is the *peripheral side* and is composed of all the data paths and signals required to support the peripheral. A floppy disk controller interface might have lines to transfer data between the controller and the floppy disk drive, plus lines that detect the index hole, control reading, and writing operations, and switch the drive motor on and off. A clock/calendar chip might have very few pins — two for a quartz crystal that keeps the time as it does in a digital watch, plus a pin to supply power from a battery when the computer is switched off.

9.2 The 68230 Parallel Interface/Timer

We are now going to look at the way in which a sophisticated parallel interface is implemented by the 68230 PI/T (parallel interface/timer). The 68230 is a general-purpose peripheral, with the primary function of an 8- or a 16-bit parallel interface between a computer and an external system, and a secondary function as a *programmable timer*. A parallel interface is able to transfer data to and from the processor a byte at a time, as opposed to a serial interface that transfers data a bit at a time. Before we look at the 68230 itself, we have to introduce some important concepts related to input/output techniques.

I/O Fundamentals

The 68230 PI/T furnishes the systems designer with two attributes that are fundamental to all but the most primitive I/O ports. These are *handshaking* and *buffering* facilities. Handshaking permits data transfers to be *interlocked* with an external activity (e.g., a disk drive), so that data is moved at a rate in keeping with the peripheral's capacity. *Interlocked* means that the next action cannot go ahead until the current action has been completed. Note that an I/O transfer using handshaking is also called a *closed-loop* data transfer.

Buffering is a facility that permits an *overlap* in the transfer of data between the CPU and the PI/T, and between the PI/T and its associated peripheral. For example, the PI/T may be obtaining the next byte of data from a disk controller, while the CPU is reading the last byte from the PI/T. Buffering requires temporary storage. In everyday terms, a doctor's waiting room is a buffer because it allows patients to enter and wait while the doctor is busy treating the current patient.

Input Handshaking and the PI/T

We are now going to look at how the PI/T reads data from a peripheral and sends it to the host CPU. Figure 9.7 illustrates the sequence of events taking place when

Figure 9.7 Closed-loop data transfer and the input handshake

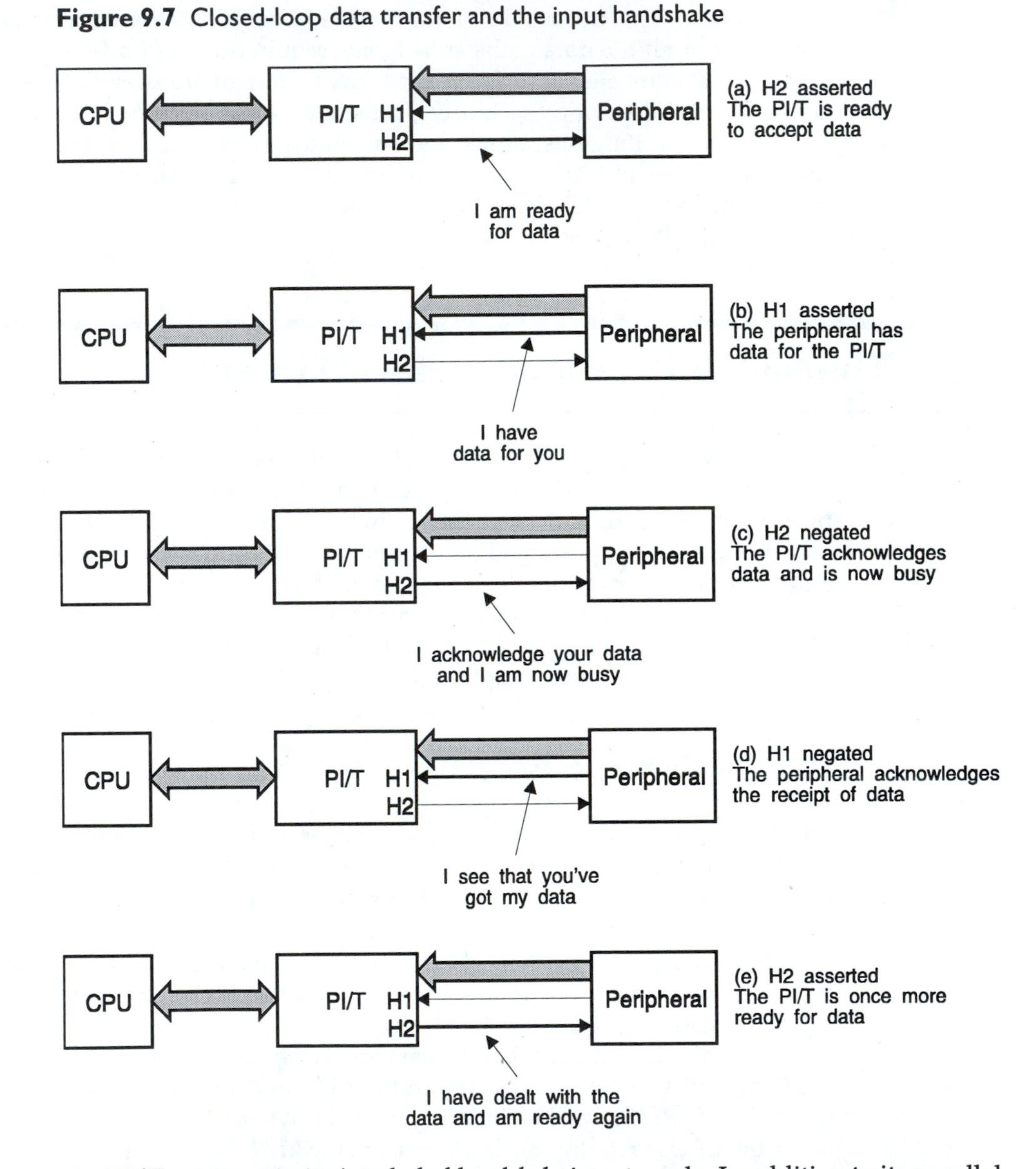

the PI/T operates in its *interlocked handshake* input mode. In addition to its parallel data interface, the PI/T has two *data transfer control lines*: an edge-sensitive input H1 and an output H2. The term *edge-sensitive* means that a state-of-change is detected by the H1 input, rather than a steady signal. That is, a 0 to 1 transition or a 1 to 0 transition triggers an operation. The 68230 permits the active state of H1 and H2 to be programmed by the user. That is, you can select whether you want a change to a high state or to a low state to indicate an event. The PI/T's parallel data paths in Figure 9.7 are shaded. However, at the moment we are concerned more with the operation of the handshake mechanism.

Like most peripherals the PI/T has internal status bits or *flags* that indicate the status of the device. For example, flag H1S is set by an active transition on handshake input H1. The PI/T's status flags can be read by the CPU at any time. We now describe how the PI/T executes a single-buffered input operation. The 68230 can operate in several different modes and these will be described later.

At point (a) in Figure 9.7, control output line H2 from the PI/T is in its asserted state, indicating to the peripheral that the PI/T is ready to receive data. We assume that the peripheral has a control input that can detect the state of H2 and initiate a data transfer when H2 is asserted. At state (b), the peripheral forces an active transition on the PI/T's H1 input, informing the PI/T that data is now available on its data input bus. Asserting H1 sets a status bit within the PI/T and generates an interrupt request if the PI/T is so programmed.

At state (c), the PI/T's H2 output is negated, informing the peripheral that the data has been accepted. Equally, the PI/T is no longer in a position to receive further data. At state (d), H1 is negated by the peripheral. This informs the PI/T that the peripheral has acknowledged the data transfer. At state (e), the PI/T asserts H2 to indicate that it is once more ready to receive data from the peripheral. At this stage, the system is in the same condition as state (a), and a new cycle may commence.

As you can see, the entire process involves a handshake (or dialogue) between the PI/T and the peripheral. This handshake is interlocked because each step in the dialogue must be completed before the next step is triggered.

The PI/T's data input is *double-buffered*, as Figure 9.8 illustrates. Double-buffering means that the PI/T is able to receive a new input while storing the previous input. Data can be transferred at almost the maximum rate at which the CPU can read the PI/T, without information being lost. Had the PI/T been supplied with many more buffers (making it a FIFO), instantaneous data rates of several times the host processor's transfer rate could have been supported.

The timing diagram of two successive interlocked handshake input transfers is given in Figure 9.9. The lines with arrows in Figure 9.9 are intended to emphasize *cause and effect*. Two cycles are shown because the PI/T is double-buffered. The active-low H2 output from the PI/T indicates whether the PI/T is able to

Figure 9.8 The PI/T's double-buffered input

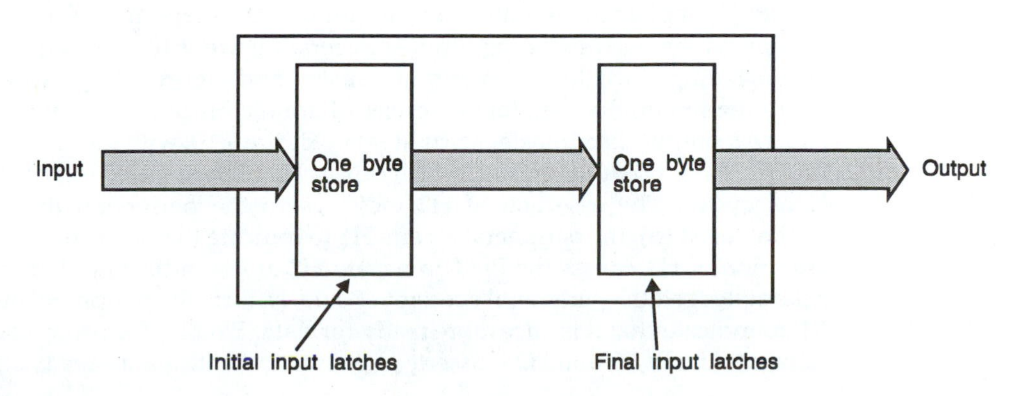

Figure 9.9 Two consecutive input cycles using double-buffered input

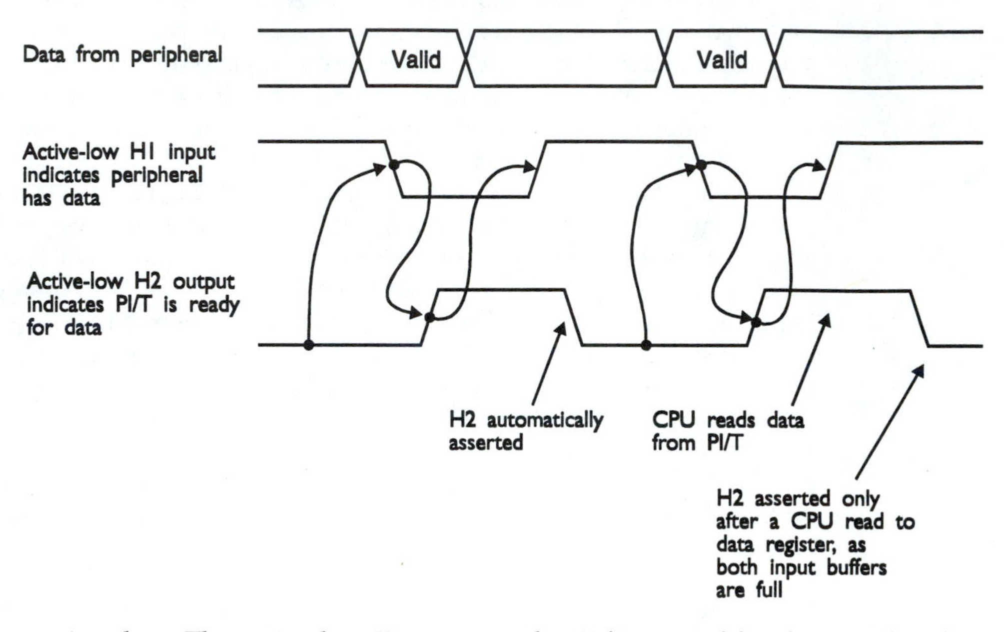

receive data. The active-low H1 input to the PI/T is used by the peripheral to indicate that it has data for the PI/T to read.

In the first cycle, H2 is negated after H1 has been asserted, and H2 is re-asserted automatically after approximately four clock cycles. H2 is asserted because the input has been transferred from PI/T's *initial input latches* to its *final input latches*, and the initial input latches are once more free to accept data (see Figure 9.8). However, on the second input cycle, H2 remains inactive-high (i.e., it is not self-clearing), because both input buffers are full. The H2 output re-asserts itself only when the CPU reads from the PI/T and empties the final input latches.

Output Handshaking

The PI/T implements double-buffered output transfers in very much the same fashion as the corresponding input transfers. Figure 9.10 shows the sequence of events taking place during an output transfer, and Figure 9.11 provides the corresponding timing diagram for two cycles of double-buffered output.

An output data transfer starts at (a) in Figure 9.10, with the CPU loading data into the PI/T's output register, which causes H2 to be asserted after a delay of two clock cycles. The assertion of H2 indicates to the peripheral that the data is available. At (b), the peripheral asserts H1 to indicate that it has read the data. The assertion of H1 causes the PI/T to negate H2 at (c), indicating that the PI/T has acknowledged the peripheral's receipt of data. In turn, the peripheral negates H1 at (d) to indicate that it is once more ready for data. Finally, the processor loads new data into the PI/T, and H2 is asserted again to indicate a data-ready state — (e).

Figure 9.10 Closed-loop data transfer and the output handshake

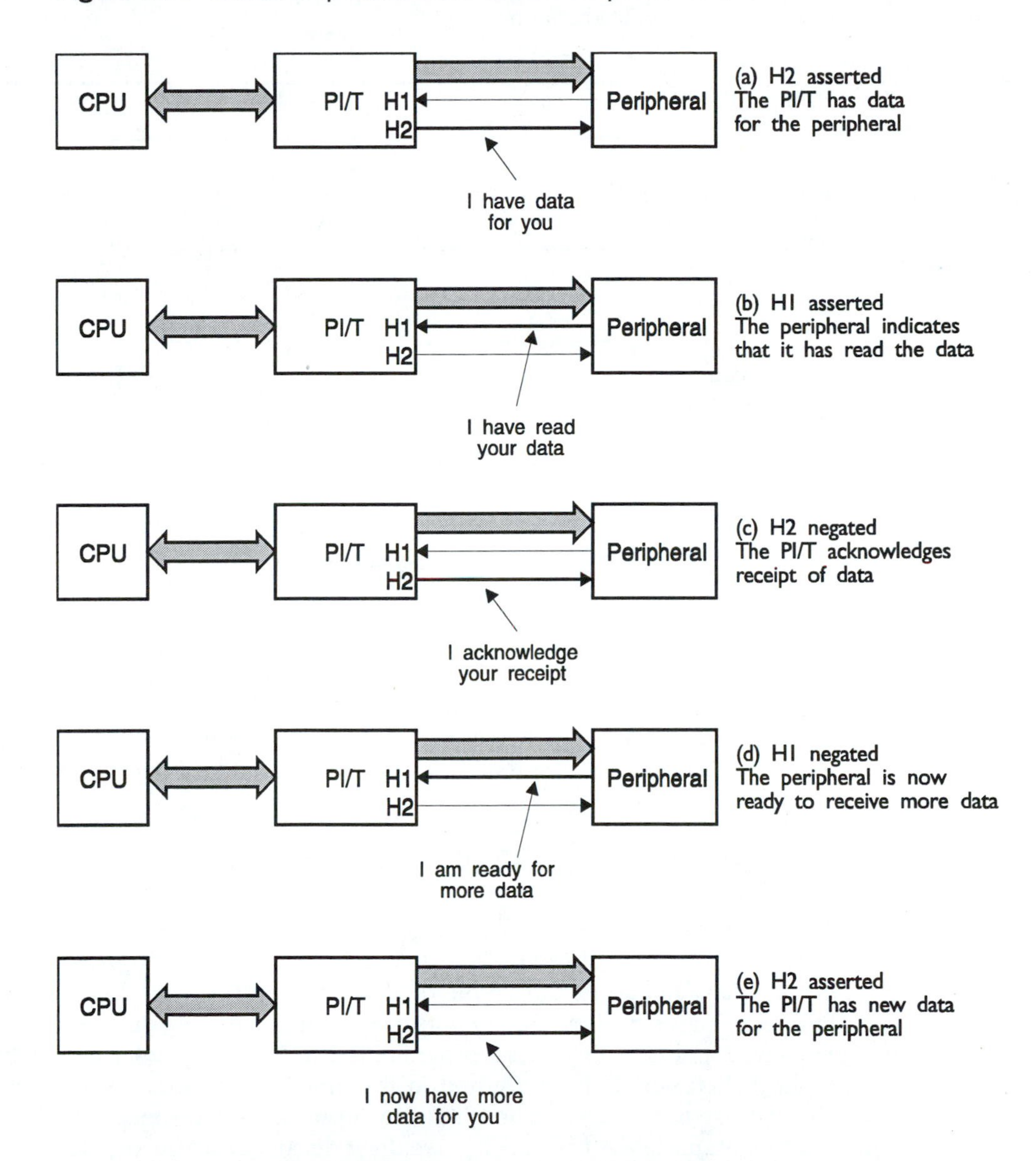
CPU
PI/T H1
H2
Peripheral
(a) H2 asserted
The PI/T has data
for the peripheral
I have data
for you
(b) H1 asserted
The peripheral indicates
that it has read the data
I have read
your data
(c) H2 negated
The PI/T acknowledges
receipt of data
I acknowledge
your receipt
(d) H1 negated
The peripheral is now
ready to receive more data
I am ready for
more data
(e) H2 asserted
The PI/T has new data
for the peripheral
I now have more
data for you

The timing diagram of Figure 9.11 also illustrates the effect of double-buffering on an output data transfer. Initially, both the PI/T's output buffers are empty. When data is first loaded into the PI/T by the CPU, the data is transferred to the chip's output terminals and H2 asserted. At this point, one of the two output buffers is full. The buffer connected to the CPU is called the *initial O/P buffer*, and that connected to the output pins is called the *final output buffer*.

When the next write to the PI/T's data register is made, the data is not immediately transferred to the output buffer, and therefore the PI/T is in a busy

Figure 9.11 Two consecutive cycles of output using interlocked handshaking and double-buffering

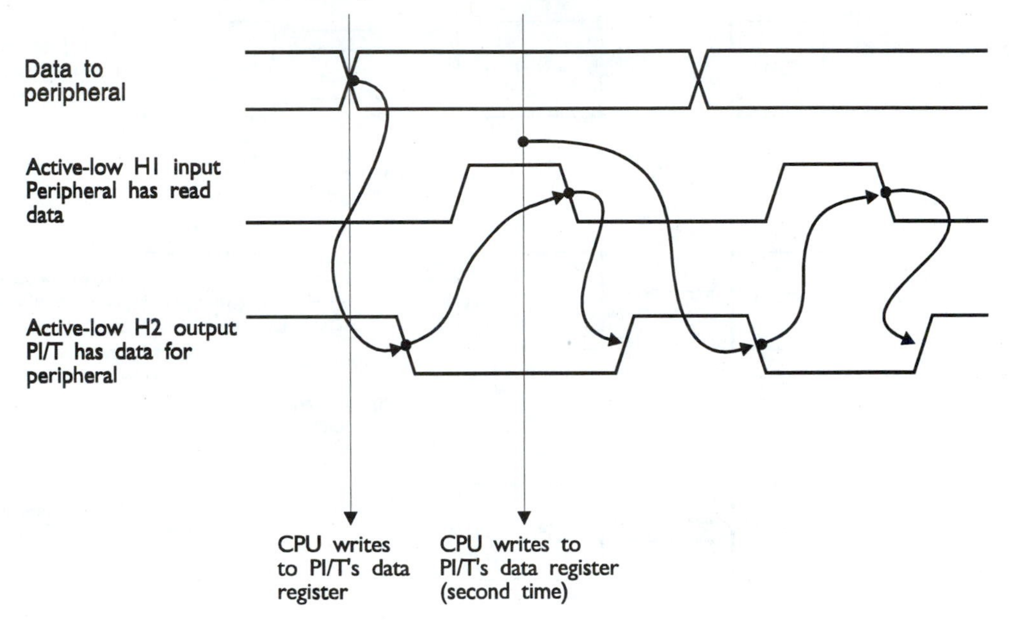

state and cannot accept new data. Only when H1 is asserted by the peripheral does the PI/T transfer its latest data to the output register. Now the PI/T may once more accept data from the CPU. As in the case of input transfers, the CPU knows when the PI/T is ready for data by examining the state of the H1 flag bit, H1S.

Structure of the 68230 PI/T

The 68230 PI/T has a 48-pin package that interfaces to the 68000 via eight data pins. A key feature of the PI/T is its two independent 8-bit, programmable I/O ports (A and B) and its third 8-bit, dual-function port C. You can program port C to operate as a simple 8-bit I/O port without handshaking and double-buffering. However, port C can also be programmed to act as an interface to the PI/T's timer. When port C is configured in the timer mode, some of its pins perform special system functions. The PI/T also supports DMA operation in conjunction with a suitable DMA controller — we don't discuss DMA in any detail in this text.

Figure 9.12 provides a block diagram of the internal structure of the PI/T. The two 8-bit ports A and B and their handshake lines (H1, H2 for port A, and H3, H4 for port B) are entirely *application-dependent* and can be used as required. Since port C is a dual-function port, some of its pins in Figure 9.12 have two labels.

The PI/T's timer takes up three pins of port C (PC2 = T_{IN}, PC3 = T_{OUT}, PC7 = TIACK*). The PI/T supports a basic 24-bit timer that has an optional (i.e., programmable) 5-bit input-prescaler that can divide the clock pulses by 32. The counter counts either clock pulses from the 68000 or pulses at the PC2/T_{IN} pin.

Figure 9.12 Structure of the PI/T

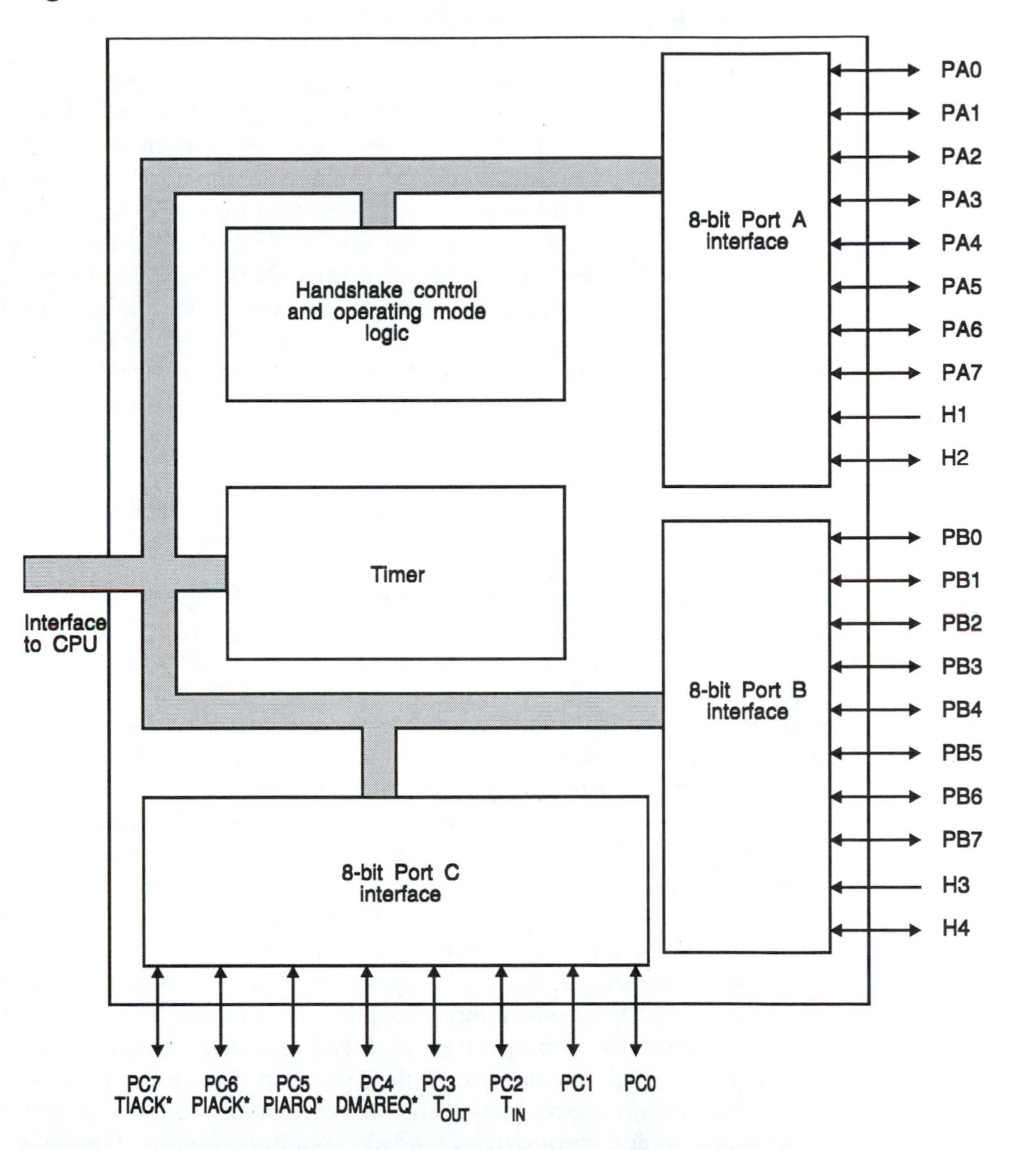

The timer output, PC3/T_{OUT}, generates the single or periodic pulses required by external equipment, or it can be connected to one of the CPU's interrupt request inputs to generate timed interrupts. The PC7/TIACK* input must be derived from the 68000's IACK logic and acknowledges an interrupt generated by an active transition on PC3/T_{OUT}.

Three other system functions are provided by port C. Pin PC5/PIRQ* is a composite interrupt request output for parallel ports A and B. Pin PC6/PIACK* is the corresponding interrupt acknowledge input to the PI/T. PC4/DMAREQ* is a DMA request output from the PI/T, and may be used in conjunction with an external DMA controller to request a DMA operation between the peripheral connected to port A or B and the system memory. We do not cover the PI/T's DMA functions further in this text.

The PI/T is a relatively sophisticated peripheral whose internal operation is controlled by 23 memory-mapped registers. Five address inputs to the PI/T, RS1 to RS5, permit one of 32 possible internal locations to be accessed. A particular register is selected by addressing the region of memory allocated to the PI/T. Table 9.1 defines the addresses and the names of the PI/T's 23 internal registers. Out of the 32 possible addresses on RS1 to RS5, nine values are not used and, when accessed, these null registers return the value $00. Although Table 9.1 appears complex, the registers can be divided into functional groups: PI/T control and status registers, port A/B data and data direction registers, and counter registers. The functions controlled by these registers are dealt with later. Registers labelled *R/W* can be either written to or read from, and registers labelled *R only* may only be read from.

Operating Modes of the PI/T

At first sight, the PI/T appears to be an exceedingly complex chip, because it supports so many different *modes of operation*. Each of the PI/T's operating modes is really very straightforward. The A and B ports of the PI/T can be configured to operate in four fundamental modes (modes 0, 1, 2, and 3), together with three submodes of mode 0 and two submodes of mode 1. These modes select the type of buffering (none/single/double), and whether the A and B ports operate as independent 8-bit ports or as a combined 16-bit port.

Figure 9.13 provides an overview of the PI/T's operating modes. The chosen operating mode of the PI/T is determined by the state of bits 7 and 6 of its general control register (PGCR). That is, both port A and port B must operate in the same mode. However, the individual port control registers may be programmed to permit independent submodes of ports A and B.

Before looking at the PI/T's operating modes in detail, we need to mention the options that they select. First, the 8-bit ports A and B can be used as separate ports, or as a single 16-bit port. Second, you can select the type of buffering on the ports (none, single, or double). Third, you can choose the type of handshaking procedure governing the data transfer. You can also determine whether the PI/T operates in an interrupt-driven mode or in a polled mode. Finally, you can select whether the asserted state (i.e., true state) of an input is electrically high or low.

Mode 0 Operation

Figure 9.14 shows the PI/T's mode 0 operation and its submodes 00, 01, and 1X. These submodes are so-called because they are chosen by loading bits 7 and 6 of the relevant port control register (PACR or PBCR) with the appropriate control code. The submode 1X means that bit 7 is set to 1 and bit 6 may be either 1 or 0. In each

Table 9.1 The 68230's internal register set

Register select bits RS5	RS4	RS3	RS2	RS1	Offset	Mnemonic	Description	Type
0	0	0	0	0	$00	PGCR	Port general control register	R/W
0	0	0	0	1	$02	PSRR	Port service request register	R/W
0	0	0	1	0	$04	PADDR	Port A data direction register	R/W
0	0	0	1	1	$06	PBDDR	Port B data direction register	R/W
0	0	1	0	0	$08	PCDDR	Port C data direction register	R/W
0	0	1	0	1	$0A	PIVR	Port interrupt vector register	R/W
0	0	1	1	0	$0C	PACR	Port A control register	R/W
0	0	1	1	1	$0E	PBCR	Port B control register	R/W
0	1	0	0	0	$10	PADR	Port A data register	R/W
0	1	0	0	1	$12	PBDR	Port B data register	R/W
0	1	0	1	0	$14	PAAR	Port A alternate register	R only
0	1	0	1	1	$16	PBAR	Port B alternate register	R only
0	1	1	0	0	$18	PCDR	Port C data register	R/W
0	1	1	0	1	$1A	PSR	Port status register	R/W
0	1	1	1	0	$1C		Null	
0	1	1	1	1	$1E		Null	
1	0	0	0	0	$20	TCR	Timer control register	R/W
1	0	0	0	1	$22	TIVR	Timer interrupt vector register	R/W
1	0	0	1	0	$24		Null	
1	0	0	1	1	$26	CPRH	Counter preload register high	R/W
1	0	1	0	0	$28	CPRM	Counter preload register middle	R/W
1	0	1	0	1	$2A	CPRL	Counter preload register low	R/W
1	0	1	1	0	$2C		Null	
1	0	1	1	1	$2E	CNTRH	Counter register high	R only
1	1	0	0	0	$30	CNTRM	Counter register middle	R only
1	1	0	0	1	$32	CNTRL	Counter register low	R only
1	1	0	1	0	$34	TSR	Timer status register	R/W
1	1	0	1	1	$36		Null	
1	1	1	0	0	$38		Null	
1	1	1	0	1	$3A		Null	
1	1	1	1	0	$3C		Null	
1	1	1	1	1	$3E		Null	

case, the bits of the PADDR, PGCR, and PACR registers that determine the relevant mode are given in Figure 9.14. Only port A is shown because port B behaves exactly the same as port A (except that H3 acts like H1 and H4 acts like H2).

The PI/T's port A and port B data registers are remarkably versatile. Not only can you configure a port as an input port or an output port, you can configure an *individual bit* of a port as an input or an output. In modes 0 and 1, a *data direction register* (DDR) is associated with each port. PADDR controls port A and PBDDR independently controls port B. Each bit of the data direction register

Figure 9.13 The PI/T's modes of operation

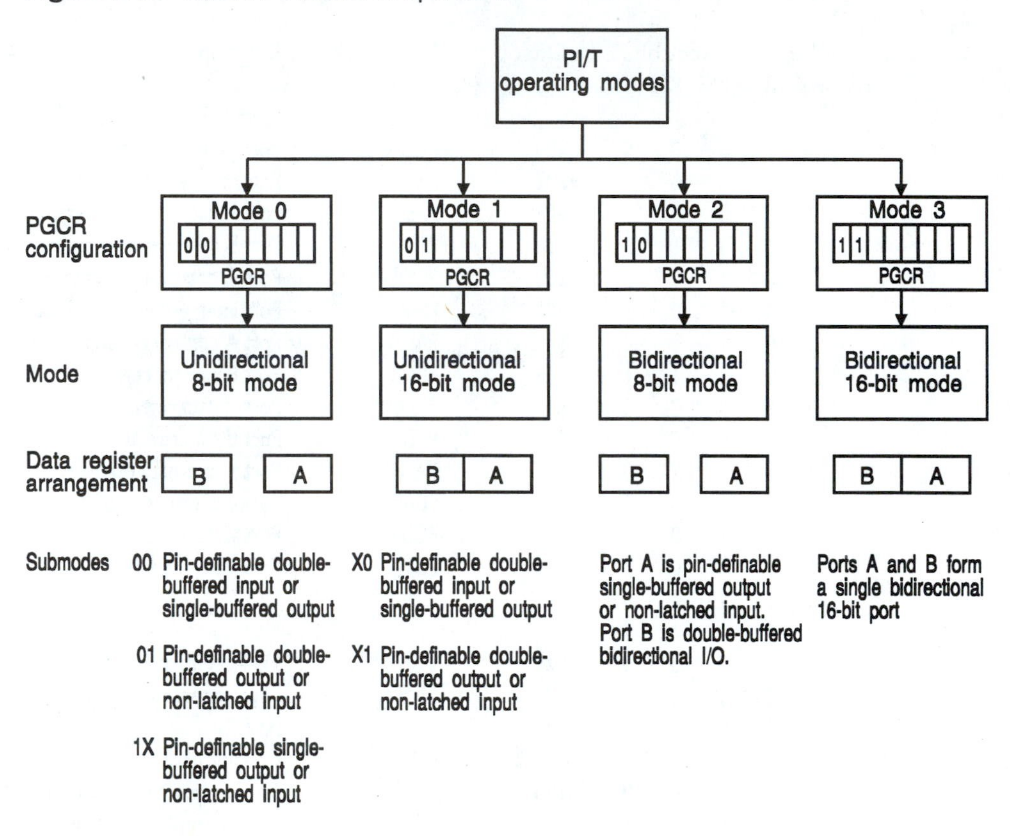

determines whether the corresponding bit of the port is an input or an output. A logical zero in bit *i* of a DDR defines the corresponding bit of the port as an *input*. A logical one defines the bit as an *output*. The contents of both DDRs are set to zero after a reset. For example, if PADDR is loaded with 00011111_2, bits PA7 to PA5 are configured as *inputs* and bits PA4 to PA0 are configured as *outputs*. This operating mode is called *unidirectional*, because it is changed only by resetting the PI/T or by reconfiguring the DDRs.

It can be seen from Figure 9.14 that the submodes differ in terms of the buffering they permit on their inputs and outputs. The direction of data transfer that permits double-buffering is known as the *primary data direction* of the port. Data transfers in the primary direction are controlled by handshake pins H1, H2 for port A, and H3, H4 for port B.

Mode 0, Submode 00

The PI/T provides *double-buffered input* in the primary direction in submode 00. Output from the PI/T is *single-buffered*; see Figure 9.14. This mode is therefore used largely to receive data from devices such as analog to digital converters. Data is

Figure 9.14 The PI/T in mode 0 (unidirectional 8-bit mode)

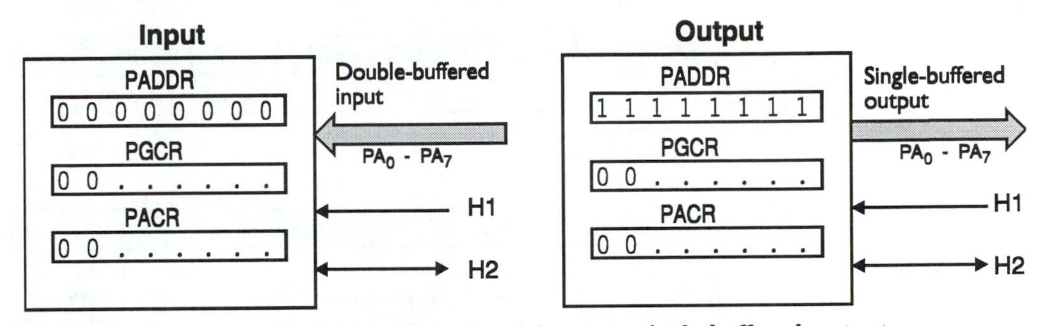

Submode 00 Pin-definable double-buffered input or single-buffered output

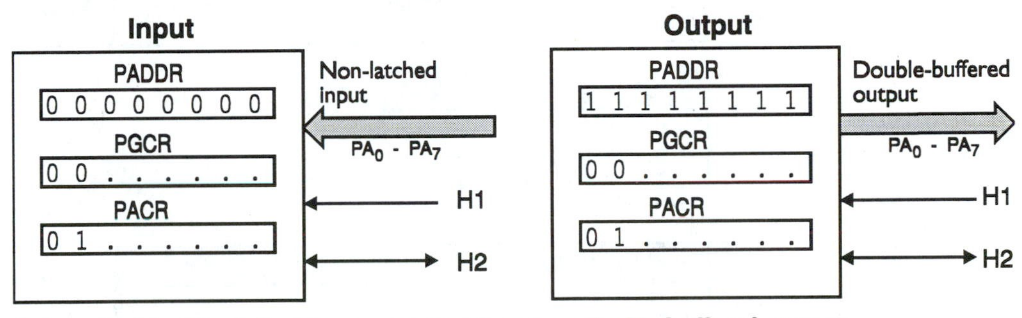

Submode 01 Pin-definable non-latched input or double-buffered output

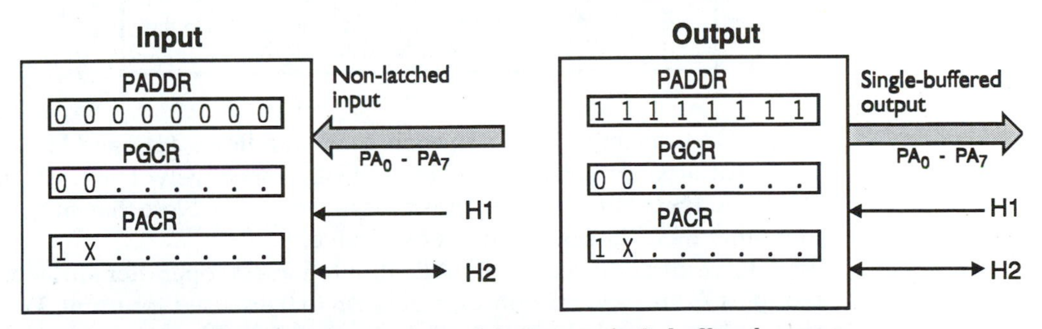

Submode 1X Pin-definable non-latched input or single-buffered output

latched into the PI/T's input register by the asserted edge of control input H1, and H2 behaves according to its programmed function defined in Table 9.2. Up to now, we have discussed only the interlocked handshake mode offered by H1 and H2. It can be seen from Table 9.2 that bits PACR5 to PACR3 of the port A control register may be used to define H2 as a simple output (i.e., an output at a logical 0 or logical 1 level), an interlocked handshake output, or a *pulsed* handshake output. In the latter case, the H2 output is asserted as in the interlocked mode of Figures 9.7 and 9.9, but is negated automatically after approximately four clock cycles. You can also

Table 9.2 Port A control register (PACR) in mode 0, submode 00

Bit	PACR7	PACR6	PACR5	PACR4	PACR3	PACR2	PACR1	PACR0
Function	0	0	H2 control			H2 int	H1 control	

Submode 00 (PACR7, PACR6)

PACR5	PACR4	PACR3	H2 control	H2S
0	X	X	Edge-sensitive input	Set on asserted edge
1	0	0	Output - negated	Always clear
1	0	1	Output - asserted	Always clear
1	1	0	Output - interlocked handshake	Always clear
1	1	1	Output - pulsed handshake	Always clear

PACR2	H2 interrupt enable
0	H2 interrupt disabled
1	H2 interrupt enabled

PACR1	PACR0	H1 control
0	X	H1 interrupt and DMA request disabled
1	0	H1 interrupt and DMA request enabled
X	X	H1S status set if data is available in the double-buffered input path

define H2 as an input pin if you wish. In what follows, H1S and H2S are the status bits associated with control lines H1 and H2, respectively. These status bits are bits PSR0 and PSR1 of the PI/T's port status register, PSR. Note that an X in Table 9.2 (or any other table) indicates a don't care value.

Consider an example of mode 0, submode 00 operation, in which bits 7 and 6 of port A are used for output, and bits 5 to 0 are used for input. We will assume that pulsed-handshaking is used in the primary direction (i.e., input) and that interrupts are disabled. Note that the port general control register is set up *twice*. The first is to select mode 00 and the second time is to enable the handshaking mechanism (we will discuss the PGCR in more detail later). The subroutine `SetUp` configures the PI/T, and the subroutines `Send` and `Receive` are called to perform output and input, respectively.

```
PIT     EQU    $008001              Base address of PI/T
PGCR    EQU    $00                  Offset of port general control register
PACR    EQU    $0C                  Offset of port A control register
PADDR   EQU    $04                  Offset of port A data direction register
```

```
PADR    EQU    $10                        Offset of port A data register
SetUp   LEA    PIT,A0                     A0 points to the PI/T
        MOVE.B #%00000000,(PGCR,A0)  Set up PGCR for mode 0 operation
        MOVE.B #%00111000,(PACR,A0)  Set up PACR for submode 00 operation
*                                         with pulsed H2 output handshake
        MOVE.B #%11000000,(PADDR,A0) Set up bits 7,6 as outputs, bits 5 to 0 as inputs
        MOVE.B #%00010011,(PGCR,A0)  Enable H12 and make H1,H2 active-high
*                                         note that H12 is set to enable H1 and H2
        RTS
Send    MOVE.B D0,(PADR,A0)               Write to Port A to output bits 7 and 6
        RTS
Receive MOVE.B (PADR,A0),D0               Read from port A to input bits 5 to 0
        RTS
```

Mode 0, Submode 01

In submode 01, the primary data direction is from the PI/T and *double-buffered output* is provided. You would use this mode to send data to an output device such as a printer or a digital to analog converter. Input in this mode is non-latched. That is, the input read by the CPU reflects the state of the input pin at the moment it is read. Table 9.3 shows the structure of the port A control register in submode 01. This table is almost exactly the same as Table 9.2, with the exception of the submode control fields and H1 status control bit, PACR0. When PACR0 = 0, the H1 status bit is set if either port A initial or final output latches can accept data and is clear otherwise. When PACR0 = 1, the H1 status bit is set if both port A output latches are empty and is clear otherwise. In other words, we can program H1S to indicate the state *fully empty* or the state *half empty*.

Consider a simple example of mode 0, submode 01 operation, in which all eight bits of port A are used for output. This time we will assume that interlocked handshaking is used and that interrupts are disabled.

```
PIT     EQU    $008001                    Base address of PI/T
PGCR    EQU    $00                        Offset of port general control register
PACR    EQU    $0C                        Offset of port A control register
PADDR   EQU    $04                        Offset of port A data direction register
PADR    EQU    $10                        Offset of port A data register
SetUp   LEA    PIT,A0                     Point to PIT
        MOVE.B #%00000000,(PGCR,A0)  Set up PGCR for mode 0 operation
        MOVE.B #%01110000,(PACR,A0)  Set up PACR for submode 01 operation
*                                         and H2 interlocked handshake output
        MOVE.B #%11111111,(PADDR,A0) Set up all bits as outputs
        MOVE.B #%00010011,(PGCR,A0)  Enable H12 and make H1,H2 active-high
        RTS
Send    MOVE.B D0,(PADR,A0)               Write to Port A to output bits 7 and 6
        RTS
```

Mode 0, Submode 1X

In mode 0, submode 1X (see Figure 9.14), simple bit I/O is available in both directions. Double-buffered I/O cannot be used in either direction. Data read from a pin programmed as an input is the instantaneous non-latched signal at that pin. Data written to an output is single-buffered. This mode provides a simple general-purpose bit input/output facility in which the various bits of port A (or port B) may be used individually to perform input and output as required.

H1 is an edge-sensitive input only, and plays no part in any handshaking procedure. The H2 pin may be programmed as an *edge-sensitive input* that sets status bit H2S when asserted. As in the case of the other submodes described previously, H2 can be programmed as an output and set or cleared under program control. Table 9.4 defines the options available in this submode.

Mode 1 Operation

In mode 1, the two 8-bit ports are combined to act as a single 16-bit port. This port is still a *unidirectional* port in the sense that the primary direction of data transfer

Table 9.3 Port A control register (PACR) in mode 0, submode 01

Bit	PACR7	PACR6	PACR5	PACR4	PACR3	PACR2	PACR1	PACR0
Function	0	1	H2 control			H2 int	H1 control	

Submode 01 (PACR7–PACR6)

PACR5	PACR4	PACR3	H2 control	H2S
0	X	X	Edge-sensitive input	Set on asserted edge
1	0	0	Output - negated	Always clear
1	0	1	Output - asserted	Always clear
1	1	0	Output - interlocked handshake	Always clear
1	1	1	Output - pulsed handshake	Always clear

PACR2	H2 interrupt enable
0	H2 interrupt disabled
1	H2 interrupt enabled

PACR1	PACR0	H1 control
0	0	H1 interrupt and DMA request disabled
1	1	H1 interrupt and DMA request enabled
X	0	H1S status set if initial or final output latches can accept data
X	1	H1S status set if both initial and final output latches empty

is associated with double-buffering and handshake control, and port A and B data direction registers define whether the individual bits of the 16-bit port are to act as inputs or outputs. Figure 9.15 illustrates the possible configurations of the PI/T in mode 1 operation. Mode 1 has two submodes. In submode X0 the input is double-buffered, and in submode X1 the output is double-buffered.

A combined port raises two problems — what do we do about the two pairs of handshake signals (H1, H2 and H3, H4), and what about the two port control registers (PACR and PBCR)? In mode 1, port B supplies the handshake signals and the control register (PBCR). The port A control register is used in conjunction with H1 and H2 to provide the 16-bit port with additional facilities. Table 9.5 defines the effect of the PACR on the mode 1 operation of the PI/T. In mode 1, the port A control register treats H1 as an edge-sensitive input, and H2 as an edge-sensitive input or an output that may be asserted or negated under programmer control.

Mode 1, Submode X0

In mode 1, submode X0, double-buffered inputs or single-buffered outputs of up to 16 bits are possible. Because the PI/T has only an 8-bit interface to the 68000, a 16-bit word must be transferred to the CPU as two bytes. Port A should be read *before* port B. For compatibility with the `MOVEP` instruction, port A should contain the most-significant byte of data. The operation `MOVEP.W (PADR,A0),D0` reads

Table 9.4 Port A control register (PACR) in mode 0, submode 1X

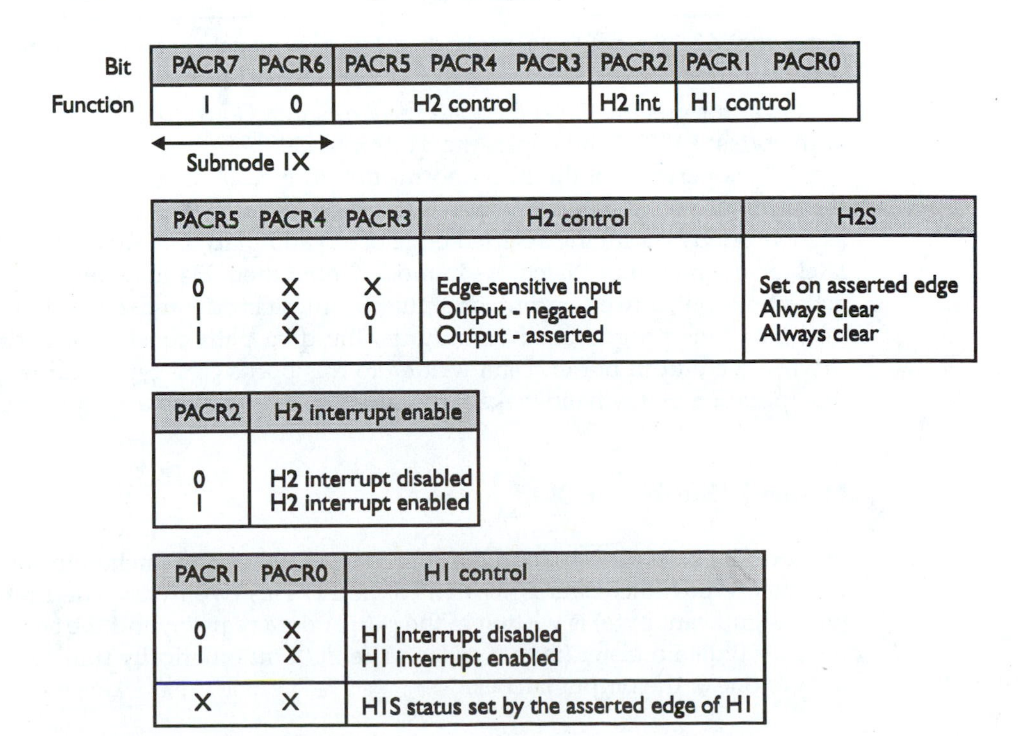

Bit	PACR7	PACR6	PACR5	PACR4	PACR3	PACR2	PACR1	PACR0
Function	1	0	H2 control			H2 int	H1 control	

Submode 1X (PACR7, PACR6)

PACR5	PACR4	PACR3	H2 control	H2S
0	X	X	Edge-sensitive input	Set on asserted edge
1	X	0	Output - negated	Always clear
1	X	1	Output - asserted	Always clear

PACR2	H2 interrupt enable
0	H2 interrupt disabled
1	H2 interrupt enabled

PACR1	PACR0	H1 control
0	X	H1 interrupt disabled
1	X	H1 interrupt enabled
X	X	H1S status set by the asserted edge of H1

Figure 9.15 The PI/T in mode 1 (unidirectional 16-bit mode)

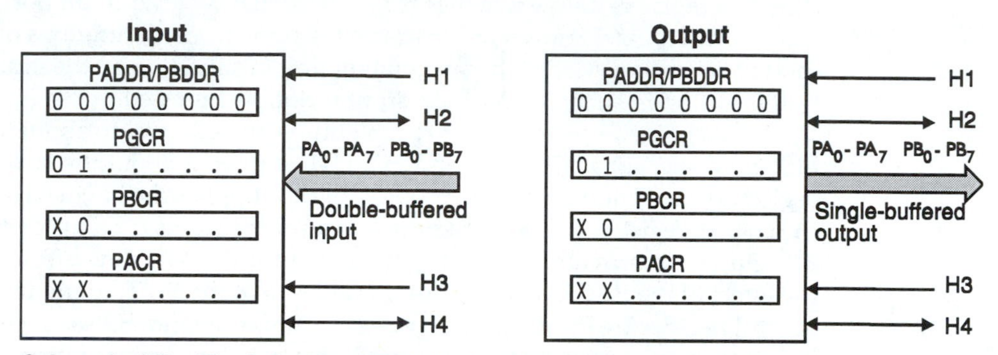

Submode X0 Pin-definable double-buffered input or single-buffered output

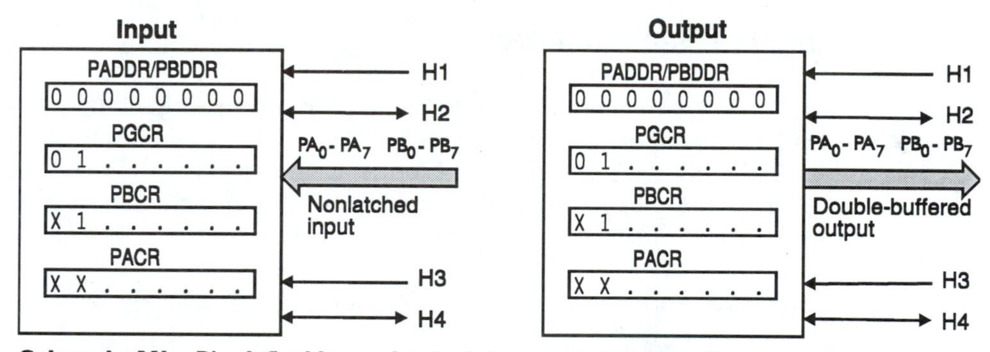

Submode X1 Pin-definable nonlatched input or double-buffered output

port A at the offset PADR and copies the 8 bits into D0(8:15), and then reads port B at the offset PADR+2 and copies the 8 bits into D0(0:7).

The operation of the 16-bit port is determined by the port B control register, the structure of which is given in Table 9.6. The signal at each input is latched asynchronously with the asserted edge of H3 and placed in either the initial input latch or the final input latch. As in mode 0 operation, H4 may be programmed to act as an input, a fixed output, or a pulsed/interlocked handshake signal.

For pins programmed as outputs, the data path consists of a single latch driving the output buffer. Data written to this port's data register does not affect the operation of any handshake pin, status bit, or any other aspect of the PI/T.

Mode 1, Submode X1

In mode 1, submode X1, double-buffered outputs, or non-latched inputs, of up to 16 bits are possible. Data is written to the PI/T as two bytes. The first byte (the most-significant byte) is written to the port A data register, and the second byte to the port B data register (in that order). The PI/T automatically transfers the 16-bit data to one of the output latches.

Table 9.5 The format of port A control register during a mode 1 operation

Bit	PACR7	PACR6	PACR5	PACR4	PACR3	PACR2	PACR1	PACR0
Function	0	0	H2 control			H2 int	H1 control	

Submode XX (don't care) (PACR7, PACR6)

PACR5	PACR4	PACR3	H2 control	H2S
0	X	X	Edge-sensitive input	Set on asserted edge
1	X	0	Output - negated	Always clear
1	X	1	Output - asserted	Always clear

PACR2	H2 interrupt enable
0	H2 interrupt disabled
1	H2 interrupt enabled

PACR1	PACR0	H1 control
0	X	H1 interrupt disabled
1	X	H1 interrupt enabled
X	X	H1S set by asserted edge of H1

The port A control register and associated handshake signals (H1 and H2) behave exactly as in mode 1, submode X0, defined by Tables 9.4 and 9.5.

In a similar fashion, the port B control register behaves rather like the same register in mode 1, submode X0. The only differences are in bits PBCR7, PBCR6, which are set to 0,1 to select this mode, and in bit PBCR0. When PBCR0 is zero, the H3S status bit is set when either the initial or final output latch of ports A and B can accept new data. Otherwise, it is clear. When PBCR0 is one, the H2S status bit is set when both the initial and final output latches of ports A and B are empty, and is clear otherwise.

Mode 2

Mode 2 offers *bidirectional* double-buffered I/O operations and is illustrated in Figure 9.16. Port A is used for bit I/O transfers with no associated handshake pins and provides unlatched input or single-buffered output. Individual pins of port A can be programmed as inputs or outputs by the setting or clearing bits in the port A data direction register.

Port B is the *work horse* in mode 2 and acts as a bi-directional, 8-bit, double-buffered I/O port. The handshake pins are all associated with port B and oper-

Table 9.6 Format of the port B control register during a mode 1, submode X0 operation

Bit	PBCR7	PBCR6	PBCR5	PBCR4	PBCR3	PBCR2	PBCR1	PBCR0
Function	0	0	H4 control			H4 int	H3 control	

Submode X0 (PBCR7, PBCR6)

PBCR5	PBCR4	PBCR3	H4 control	H4S
0	X	X	Edge-sensitive input	Set on asserted edge
1	0	0	Output - negated	Always clear
1	0	1	Output - asserted	Always clear
1	1	0	Output - interlocked handshake	Always clear
1	1	1	Output - pulsed handshake	Always clear

PBCR2	H4 interrupt enable
0	H4 interrupt disabled
1	H4 interrupt enabled

PACR1	PACR0	H3 control
0	X	H3 interrupt and DMA request disabled
1	X	H3 interrupt and DMA request enabled
X	0	H3S set if data present in double-buffered input path

ate in two pairs. H1, H2 control *output* transfers and H3, H4 control *input* transfers. The instantaneous direction of the data is determined by the H1 handshake pin. That is, the external device determines the direction of data transfer. The port B data direction register has no effect in this mode, because only byte I/O is permitted. Similarly, port A and B submode fields do not affect PI/T operation in mode 2.

The output buffers of port B are controlled by the level of the signal at the H1 input. When H1 is negated, the port B output buffers are enabled and its pins drive the output bus. Note that all eight buffers are enabled, because individual pins cannot be programmed as inputs or outputs in modes 2 and 3. Generally, H1 is negated by a peripheral in response to the assertion of H2, which indicates that new output data is present in the double-buffered latches. Following acceptance of the data, the peripheral asserts H1, disabling the port B output buffers. H1 also acts as an edge-sensitive input.

Double-buffered input transfers Data at the port B input pins is latched on the asserted edge of H3 and deposited in one of the input latches. The corresponding H3 status bit, H3S, is set whenever input data, that has not been read by the host computer, is present in the input latches. As in all other modes, H4 is programmable. In modes 2 and 3, H4 can be programmed to perform two functions.

1. H4 may be an output pin in the interlocked handshake mode. It is asserted when the port is ready to accept new data. It is negated asynchronously following the asserted edge of the H3 input. As soon as one of the input latches becomes ready to receive data, H4 is re-asserted. Once both input latches are full, H4 remains negated until data is read by the host processor.

2. H4 may be configured as an output pin in the pulsed input handshake mode. It is asserted when the input port is ready to receive new data exactly as above. It is automatically cleared (negated) approximately four clock cycles later. Should a subsequent active transition on H3 take place while H4 is asserted, H4 is negated asynchronously. That is, once the active-edge of H4 has been detected by the peripheral, new data may be loaded into the double-buffered input latches.

Double-buffered output transfers Data is written into one of the PI/T's port B output latches by the host processor. The peripheral connected to port B accepts data by asserting H1. The H1 status bit may be programmed to be set when *either* one or both output buffers is empty, or only when *both* output buffers are empty. H2 may be programmed to act in one of two modes.

1. H2 may be an output pin in the interlocked output handshake mode, and is asserted whenever the output latches are ready to accept new data from the host processor. H2 is negated asynchronously following the asserted edge of H1. As soon as one or both output latches become available, H2 is re-asserted. When both output latches are full, H2 remains asserted until at least one latch is emptied.

2. H2 may be an output pin in the pulsed output handshake protocol. It is asserted whenever the output latches are ready to accept new data, and

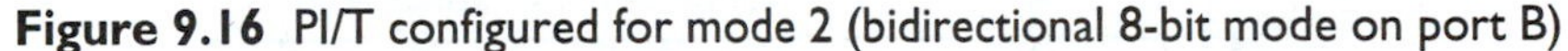
Figure 9.16 PI/T configured for mode 2 (bidirectional 8-bit mode on port B)

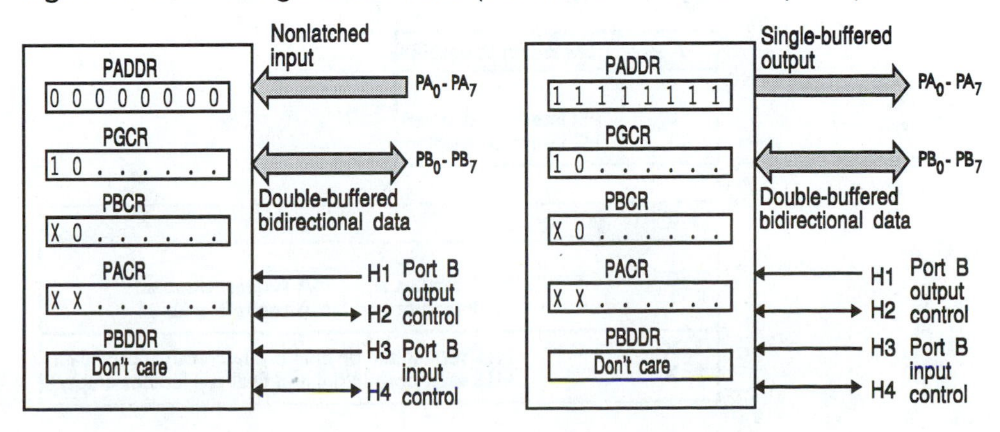

automatically negated approximately four clock cycles later. If the asserted edge of H1 is detected while H2 is asserted, H2 is negated asynchronously.

The programming of port control registers A and B in mode 2 is illustrated in Tables 9.7 and 9.8. Note that, in this mode, many bits are don't care functions.

Mode 3

Mode 3 operation is an extension of mode 2, and is illustrated in Figure 9.17. In mode 3, both ports A and B are dedicated to 16-bit double-buffered input/output transfers. As in mode 2, H1 and H2 control output transfers, and H3 and H4 control input transfers. The function of the port control registers (PACR, PBCR) is exactly the same as in mode 2, and therefore Tables 9.7 and 9.8 also apply to mode 3.

Mode 3 provides a relatively convenient way of transferring data between the PI/T and 16-bit peripherals at high speed. We use the word *relatively* because the PI/T interfaces to its host processor by an 8-bit data bus, allowing 16-bit peripheral-side data transfers but denying 16-bit CPU-side data transfers. However, by using the 68000's `MOVEP` (move peripheral instruction), 16-bit data transfers can be performed with only one instruction.

Table 9.7 Format of the port A control register during a mode 2 operation

Bit	PACR7	PACR6	PACR5	PACR4	PACR3	PACR2	PACR1	PACR0
Function	X	X	H2 control			H2 int	H1 control	

PACR7, PACR6: Submode XX (don't care)

PACR5	PACR4	PACR3	H2 control
X	X	0	Output - interlocked handshake
X	X	1	Output - pulsed handshake

PACR2	H2 interrupt enable
0	H2 interrupt disabled
1	H2 interrupt enabled

PACR1	PACR0	H1 control
0	X	H1 interrupt and DMA request disabled
1	0	H1 interrupt and DMA request enabled
X	0	H1S set if initial or final o/p latch can accept data
X	1	H1S set if both initial and final o/p latches empty

Figure 9.17 PI/T configured for mode 3 (bidirectional 16-bit mode)

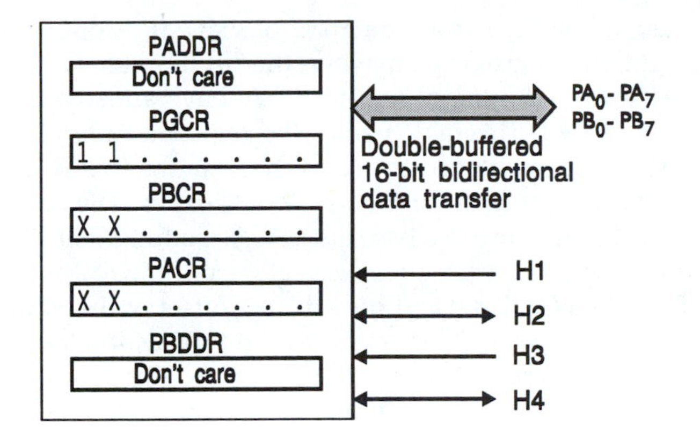

Table 9.8 Format of the port B control register during a mode 2 operation

Bit	PBCR7	PBCR6	PBCR5	PBCR4	PBCR3	PBCR2	PBCR1	PBCR0
Function	0	0	H4 control			H4 int	H3 control	

PBCR7, PBCR6: Submode XX (don't care)

PBCR5	PBCR4	PBCR3	H4 control
X	X	0	Output - interlocked handshake
X	X	1	Output - pulsed handshake

PBCR2	H4 interrupt enable
0	H4 interrupt disabled
1	H4 interrupt enabled

PBCR1	PBCR0	H1 control
0	X	H3 interrupt and DMA request disabled
1	0	H3 interrupt and DMA request enabled
X	X	H3S status set if data present in double-buffered input path

Registers of the PI/T

So far, we've looked at the PI/T from the point of view of its operational modes, and discussed its internal registers only when the need arose. Here, we provide further details on the PI/T's internal registers. We have already introduced the PI/T's registers in Table 9.1. This table names the registers and defines the register select lines, RS1 to RS5, employed to access them. In the following discussion, registers are also specified in terms of their *offset* addresses. The offset address is specified with respect to the device's base address, assuming that RS1 to RS5 are connected to address lines A_{01} to A_{05}, respectively, from the 68000. For example, the address offset of the PGCR is 0 and the address offset of the TSR, *timer status register*, is $34. In general, the PI/T is wired to the 68000's low-order data byte, D_{00} to D_{07}, and therefore mapped onto an odd address. Since the PI/T is a byte-wide device mapped onto a word-wide data bus, its registers are separated by two bytes.

Port General Control Register **(PGCR, offset = $00)**

Table 9.9 illustrates the structure of the PGCR, which is a global register. Bits 7 and 6 of the PGCR select the PI/T's operating mode. The H12 and H34 enable/disable fields, bits 5 and 4 of the PGCR, enable the handshake pairs H3, H4 and H1, H2. You have to set these control bits before you can make use of the control inputs and outputs. Doing this avoids spurious operation of the handshake lines before the PI/T has been fully configured.

Bits 3 to 0 determine the *sense* of the four handshake lines. That is, you can program the control lines to be active-high or active-low. All bits of the PGCR are cleared to zero after a reset operation.

Consider the effect of loading the value 00010011_2 into the port general control register. Table 9.9 tells us that we can rewrite this binary string as 00 01 0011 and interpret it as: mode 0, enable handshaking pair H12 and disable pair H34, set the asserted levels of H1, H2 to active-high, and set the assertion levels of H3, H4 to active-low.

Port Service Request Register **(PSRR, offset = $02)**

The port service request register is also a global register and determines the circumstances under which the PI/T may request service from the host processor (i.e., generate an interrupt or perform DMA). Table 9.10 gives the format of the PSRR, which is split into three logical fields: a service request field, SVCRQ, that determines whether the PI/T generates an interrupt or a DMA request when H1 or H3 is asserted; an operation select field that determines whether two of the dual-function pins belong to port C or perform special-purpose functions; and an interrupt-priority control field. A DMA request is signified by an active-low pulse on the DMAREQ* (direct memory access request) pin for three clock cycles. We do not cover DMA in detail in this text, but have included information on the PI/T's DMA operation for the sake of completeness.

Table 9.9 Format of the port general control register (PGCR)

Bit	PGCR7	PGCR6	PGCR5	PGCR4	PGCR3	PGCR2	PGCR1	PGCR0
Function	Port mode control		H34 enable	H12 enable	H4 sense	H3 sense	H2 sense	H1 sense

PGCR7	PGCR6	Port mode control
0	0	Mode 0
0	1	Mode 1
1	0	Mode 2
1	1	Mode 3

PGCR5	H3, H4 control
0	H34 disable
1	H34 enable

PGCR4	H1, H2 control
0	H12 disable
1	H12 enable

Table 9.10 Format of the port service request register (PSRR)

Bit	PSRR7	PSRR6	PSRR5	PSRR4	PSRR3	PSRR2	PSRR1	PSRR0
Function	X	SVCRQ (DMA control)		Interrupt control		Port interrupt priority		

PSRR4	PSRR3	Interrupt pin function	
0	0	PC5/PIRQ* = PC5 PC6/PIACK* = PC6	No interrupt support No interrupt support
0	1	PC5/PIRQ* = PIRQ* PC6/PIACK* = PC6	Autovectored interrupts supported Autovectored interrupts supported
1	0	PC5/PIRQ* = PC5 PC6/PIACK* = PIACK*	
1	1	PC5/PIRQ* = PIRQ* PC6/PIACK* = PIACK*	Vectored interrupts supported Vectored interrupts supported

Port interrupt priority PSRR2	PSRR1	PSRR0	Order of interrupt priority Highest ⟷ Lowest			
0	0	0	H1S	H2S	H3S	H4S
0	0	1	H2S	H1S	H3S	H4S
0	1	0	H1S	H2S	H4S	H3S
0	1	1	H2S	H1S	H4S	H3S
1	0	0	H3S	H4S	H1S	H2S
1	0	1	H3S	H4S	H2S	H1S
1	1	0	H4S	H3S	H1S	H2S
1	1	1	H4S	H3S	H2S	H1S

If bits PSRR6, PSRR5 of the PSRR are set to 0, 0, the PI/T does not support DMA and port C pin PC4 is a port C function. For example, Table 9.10 tells us that if we load the value $0\ 00\ 11\ 000_2$ into the PSRR, it will be interpreted as:

- Pin 5 of port C, PC5, is used to request an interrupt and PC6 is used to indicate an interrupt acknowledge. Pin 5 of port C is called PIRQ* — PI/T interrupt request. Pin 6 is called PIACK* — PI/T interrupt acknowledge.
- The order of the interrupt priority is: H1S, H2S, H3S, H4S, with H4S having the lowest priority.

Port Data Direction Registers **(PADDR, offset = $04)**
(PBDDR, offset = $06)
(PCDDR, offset = $08)

The port data direction registers select the direction and buffering characteristics of each of the appropriate port pins. A logical one in a PDDR bit makes the corresponding port I/O pin act as an output, while a logical zero makes the pin act as an input. All data direction registers are cleared to zero after a reset. The port C PDDR behaves in the same way as the other two PDDRs, and determines whether each dual-function chosen for port C operation is an input or an output pin. For example, loading the value 11000001_2 into PBDDR configures pins PB_7, PB_6, and PB_0 as outputs, and pins PB_5 to PB_1 as inputs.

Port Interrupt Vector Register **(PIVR, offset = $0A)**

We now describe the PI/T's interrupt facilities. Some readers might wish to read Chapter 10 before they tackle the PI/T's interrupt interface. When the parallel port section of the PI/T executes a vectored interrupt, it supplies the 68000 with one of four possible interrupt vector numbers. Each of these vector numbers is associated with a specific interrupt cause. When the 68000 reads the interrupt vector number during an IACK cycle, it multiplies the vector number by four to get the address of the appropriate exception handling routine.

The port interrupt vector register contains the upper-order six bits of the four port interrupt vectors. The contents of this register may be read in one of two ways: by an ordinary read cycle to the PIVR at offset address $0A, or by a port interrupt acknowledge bus cycle.

A programmed read of the PIVR during a normal read cycle presents no problems and the data returned is the contents of the PIVR. However, when the PIVR is read during an *interrupt acknowledge cycle*, the two least-significant bits are determined by the source of the interrupt, as illustrated in Table 9.11. For example, if the PIVR is loaded with 01101100_2 ($6C) during the PI/T's initialization phase, an interrupt initiated by H2 will yield an interrupt vector number of 01101101_2 ($6D). Similarly, an interrupt initiated by H4 will yield an interrupt vector number of 01101111_2 ($6F). This arrangement has been implemented to avoid the need for four separate vector number registers.

Table 9.11 Relationship between PIVR0, PIVR1, and interrupt source

Bit	PIVR7	PIVR6	PIVR5	PIVR4	PIVR3	PIVR2	PIVR1	PIVR0
Function	Interrupt vector number							
	User defined value						Selected automatically	

Interrupt source	PIVR1	PIVR0
H1	0	0
H2	0	1
H3	1	0
H4	1	1

When the PI/T is reset by hardware (e.g., during the system power-up), the port interrupt vector register is loaded with $0F. This is the *uninitialized vector* number, and forces interrupts to call a special routine to deal with this situation (see Chapter 10).

Port Control Registers **(PACR, offset = $0C)**
(PBCR, offset = $0E)

These two port control registers determine the submode operation of ports A and B, and control the operation of the handshake lines. We have already dealt with the programming of these two registers when we described the PI/T's operating modes.

Port Data Registers **(PADR, offset = $10)**
(PBDR, offset = $12)
(PCDR, offset = $18)

Port A and port B data registers are *holding registers* between the CPU-side of the PI/T and its input/output pins and internal buffer registers. These registers may be written to or read from at any time and are not affected by a reset of the PI/T.

The port C data register, PCDR (offset = $18), is a holding register for moving data to and from port C or its alternate-function pins. The exact nature of an information transfer depends on the type of cycle being executed (read or write), and on the way in which port C is configured. Table 9.12 shows how the PCDR is affected by read/write accesses. Pins configured as port C functions offer single-buffered output or non-latched input.

Note that it is possible to read *directly* the state of a dual-function pin even when it is used for non-port C functions. It is, of course, possible to generate non-port C functions *manually* by switching back to port C mode and writing to the PCDR. This register is readable and writable at all times, and is not affected by a hardware reset of the PI/T.

Table 9.12 Accessing the port C data register

Operation	Port C function		Alternate function	
	PCDDR = 0 (input)	PCDDR = 1 (output)	PCDDR = 0 (input)	PCDDR = 1 (output)
Read PCDR	Read pin	Read output register	Read pin	Read output register
Write PCDR	Output register buffer disabled	Output register buffer disabled	Write to output register	Write to output register

Port Alternate Registers **(PAAR, offset = $14)**
(PBAR, offset = $16)

Port A and port B alternate registers provide another way of reading the state of port A and B pins. Both PAAR and PBAR are read-only registers, and their contents reflect the actual *instantaneous* logic levels at the I/O pins. Writing to the PAAR or the PBAR registers has no other effect on the PI/T than to cause a dummy write cycle. These registers bypass the selected operating modes of ports A and B. That is, if you wish to know the state of the port A or port B pins, irrespective of the PI/T's operating mode, you just read from either PAAR or PBAR.

Port Status Register **(PSR, offset = $1A)**

The port status register is a global register that reflects the activity of the handshake pins. The format of the PSR is given in Table 9.13. Bits PSR7 to PSR4 show the *instantaneous* level at the respective handshake pin, and are independent of handshake pin sense bits in the PGCR. Bits PSR3 to PSR0 are the handshake status bits, HS1 to HS4, and are set or cleared as specified by the appropriate operating mode. Each of these bits is set when the appropriate handshake line is asserted. For example, if H1 is configured as active-low by clearing bit 0 of the PGCR, an electrically low level at the H1 pin will set PSR0 to 1 to indicate that status bit HS1 has been set.

Table 9.13 Format of the port status register, PSR

Bit	PSR7	PSR6	PSR5	PSR4	PSR3	PSR2	PSR1	PSR0
Function	H4 level	H3 level	H2 level	H1 level	H4S	H3S	H2S	H1S

Bits PSR7–PSR4: Bits set or cleared by instantaneous level at handshake pin

Bits PSR3–PSR0: Bits set by assertion of handshake pins as programmed

The state of bits PSR7 to PSR4 depends on the actual level of the signal at the control pins and, therefore, these bits are set or cleared automatically by the PI/T. Bits PSR3 to PSR0 are automatically set by the PI/T during a handshaking operation. However, these bits must be cleared manually by the programmer. Each bit is cleared by writing a one into it (not a zero, as you might expect). For example, you can clear all four bits by: `MOVE.B #$0F,(PSR,A0)`.

The Timer Port of the PI/T

The 68230 contains a single timer that interfaces to the host processor through the PI/T's same CPU-side pins as the parallel interface. It interfaces to external systems (or to the 68000's interrupt structure) through the alternate function pins of port C. Typical functions performed by the 68230 (or many other timers) are: the generation of square waves of programmable frequencies, the generation of single pulses of programmable duration, the production of single or periodic interrupts, and the measurement of frequency or elapsed time.

A timer is a simple device consisting of a counter that is clocked, typically, downward toward zero. By selecting the clock rate and the initial contents of the counter, it is possible to generate a specific delay between starting the counter and the moment it reaches zero. If the counter is reloaded every time it reaches zero, we have a method of implementing repetitive action.

Figure 9.18 shows the structure of the PI/T's timer together with its associated registers. The timer contains a 24-bit synchronous down-counter (CNTR) that is loaded from a 24-bit counter preload register (CPR). Since the PI/T has an 8-bit interface, each of these two 24-bit registers is implemented as three 8-bit registers (see Figure 9.18 on page 400). You can use the `MOVEP` instruction to preload the PI/T's counter preload register as follows.

```
MOVE.L  #$00123456,D0       Set up D0 with the count
MOVEP.L D0,(CPR,A0)         and dump it in the CPR.
```

Note that the address of the counter preload register has an offset of CPR = $24. If you look at Table 9.1 on page 381 that lists the PI/T's internal registers, you will find no register with this offset. The PI/T's counter and counter preload registers are each 24 bits (i.e., *three* bytes). However, the `MOVEP.L` instruction transfers *four* bytes between a register and *four* memory-mapped locations. Consequently, the PI/T has dummy locations at offset $24 for the CPR and at offset $2C for the CNTR. When you use a `MOVEP.L` instruction, the most-significant byte is transferred to or from the dummy register and the remaining three bytes are transferred to the correct registers. Finally, this arrangement provides the PI/T with an upgrade path to 32 bits.

The counter is clocked either by the 68000's own clock or by an external input applied to pin T_{IN}. The 68000's clock may, optionally, be prescaled by 32.

As the counter clocks downward, it eventually reaches zero and sets the *zero-detect status bit* (ZDS) of the timer status register (TSR). This event can be

Figure 9.18 Timer function of the 68230 PI/T

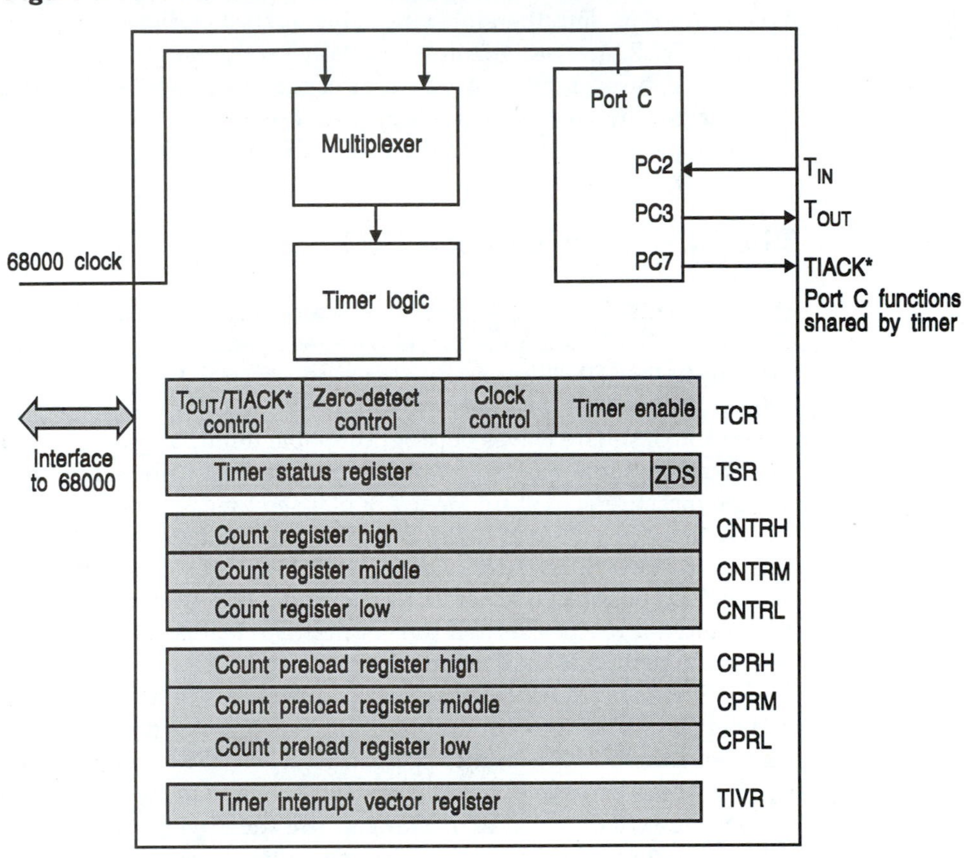

used to assert the T_{OUT} output from the timer. If T_{OUT} is connected to one of the processor's interrupt lines, an interrupt may be generated. Note that the ZDS bit is *set automatically* by the PI/T and is *cleared manually* by writing a one into it. That is, you can clear the ZDS bit by a `MOVE.B #1,(TSR,A0)` instruction.

The operating mode of the timer is determined by the *timer control register* (TCR), whose format is given in Table 9.14. The TCR controls:

1. The choice between the port C option and the timer option on three dual-function pins.
2. Whether the counter is loaded from the counter preload register or *rolls over* when zero detect is reached. When a counter rolls over, it is loaded with its maximum value following the count zero.
3. The source of the clock input (i.e., the 68000's own clock or an external user-supplied clock).
4. Whether the clock source is prescaled (i.e., divided by 32).
5. Whether the timer is enabled or disabled.

Table 9.14 Format of the timer control register

TCR7	TCR6	TCR5	T_{OUT}/TIACK* control
0	0	X	PC3/T_{OUT} and PC7/TIACK* are port C functions.
0	1	X	PC3/T_{OUT} is a timer function. In the run state T_{OUT} provides a square wave which is toggled on each zero-detect. The T_{OUT} pin is high in the halt state. PC7/TACK* is a port C function.
1	0	0	PC3/T_{OUT} is a timer function. In the run or halt state T_{OUT} is used as a timer request output. Timer interrupt is disabled. PC7/TIACK* is a port C function. Since interrupt request is disabled, the PI/T produces no response to an asserted TIACK*.
1	0	1	PC3/T_{OUT} is a timer function and is used as a timer interrupt request output. The timer interrupt is enabled and T_{OUT} is low whenever the ZDS bit is set to 1. PC7/TIACK* is a timer function, TIACK*, and acknowledges interrupts generated by the timer. Autovectored interrupts supported.
1	1	0	PC3/T_{OUT} is a timer function. In the run or halt state it is used as a timer interrupt request output. The timer interrupt is disabled and the pin always three-stated. PC7/TIACK* is a port C function.
1	1	1	PC3/T_{OUT} is a timer function and is used as a timer interrupt request output. The timer interrupt is enabled and T_{OUT} is low whenever the ZDS bit is set. PC7/TIACK* is a port C function. Autovectored interrupts are supported.

TCR4	Zero-detect control
0	The counter is loaded from the counter preload register on the first clock of the 24-bit counter after zero-detect; counting is then resumed.
1	The counter rolls over on zero-detect and then continues counting.

TCR2	TCR1	Clock Control
0	0	PC2/T_{IN} is a port C input function. Counter clock is CLK/32. The timer enable bit determines whether the timer is in the run of halt state.
0	1	PC2/T_{IN} is a timer input. Counter clock is CLK/32. The prescaler is decremented on the falling edge of the CLK input and the 24-bit counter is decremented when the counter rolls over from $00 to $1F. The timer is in the run state when the enable bit is one and the T_{IN} pin is high.
1	0	PC2/T_{IN} is a timer input prescaled by 32. The 24-bit counter is decremented when the prescaler rolls over from $00 to $1F. The timer enable bit selects the run or halt state.
1	1	PC2/T_{IN} is a timer input and prescaling is not used. The timer enable bit selects the run or halt state.

Table 9.14 (Cont.)

TCR0	Timer enable bit
0	Timer enabled
1	Timer disabled

The 68230 has an independent *timer interrupt vector register* (TIVR) that supplies an 8-bit interrupt vector number whenever the timer interrupt acknowledge pin, TIACK*, is asserted. The TIVR is automatically loaded with the uninitialized interrupt vector, $0F, following a hardware reset.

Timer States

The timer is always in one of two states: *running* or *halted*. The timer's state is determined by loading the appropriate value into the timer control register (Table 9.14). The characteristics of the two states are as follows:

1. *Halt state*
   a. The contents of the counter are stable (i.e., do not change), and can be reliably and repeatedly read from the count registers. The counter cannot reliably be read when the timer is running.
   b. The contents of the clock prescaler are forced to $1F.
   c. The ZDS bit is forced to zero, regardless of the contents of the 24-bit counter.
2. *Run state*
   a. The counter is clocked from the source programmed in the timer control register.
   b. The counter is not *reliably* readable.
   c. The prescaler is allowed to decrement if it is so programmed.
   d. The ZDS status bit is set when the counter makes a $000001 to $000000 transition.

Timer Applications

The timer section of the PI/T is not as complex as a first reading of Table 9.14 would suggest. We are going to illustrate the operation of the timer by describing some of its applications.

Real-Time Clock A real-time clock, RTC, generates an interrupt at periodic intervals. This interrupt may be used by the operating system to switch between several tasks in a multitasking system (see Chapter 10), or to update a record of the time-of-day. In this configuration, the T_{OUT} pin from the PI/T is connected to one of the host processor's interrupt request inputs, and the PI/T's TIACK* input used as an interrupt acknowledge input to the timer. The T_{IN} pin may be used as a clock input or the system clock may be selected. The format of the TCR needed to select the real-time clock mode is given in Table 9.15.

The host processor first loads the counter preload registers with a 24-bit value and then configures the TCR, as Table 9.15 indicates. The timer enable bit of the TCR may be set at any time counting is to begin — it need not be set during the timer initialization phase.

When the counter counts down from $000001 to $000000, the ZDS status bit is set and the T_{OUT} pin asserted to generate an interrupt request. At the next clock input to the 24-bit counter, the counter is loaded with the contents of the *counter preload register*. The host processor should clear the ZDS status bit to remove the source of the interrupt. The operation of the timer in this mode can be illustrated in pseudocode and by the timing diagram of Figure 9.19.

```
WHILE Timer_enable_bit = 1
          FOR I = Counter_preload_value DOWN_TO 0
                  Clear ZDS_bit
                  Negate TOUT
          END_FOR
       Set ZDS_bit
       Assert TOUT
END_WHILE
```

The following fragment of code demonstrates how you might set up a real-time clock.

```
PIT   EQU     $008001               Base address of PI/T
TCR   EQU     $20                   Offset of the timer control register
TIVR  EQU     $22                   Offset of the timer interrupt vector register
CPR   EQU     $24                   Offset of the counter preload register
TSR   EQU     $34                   Offset of the timer status register
Time  EQU     4*TimeVec             Location of the PI/T's interrupt handler
*                                   Note - TimeVec = value of interrupt number
*
SetUp LEA     PIT,A0
      MOVE.B  #TimeVec,(TIVR,A0)    Load the interrupt vector
      MOVE.L  #$00FFFFFF,D0         Set up the maximum count
      MOVEP.L D0,(CPR,A0)           and load it into the counter preload register
      MOVE.B  #%10100001,(TCR,A0)   Set up TCR
      RTS
```

Figure 9.19 Timing diagram of the PI/T in real-time clock mode

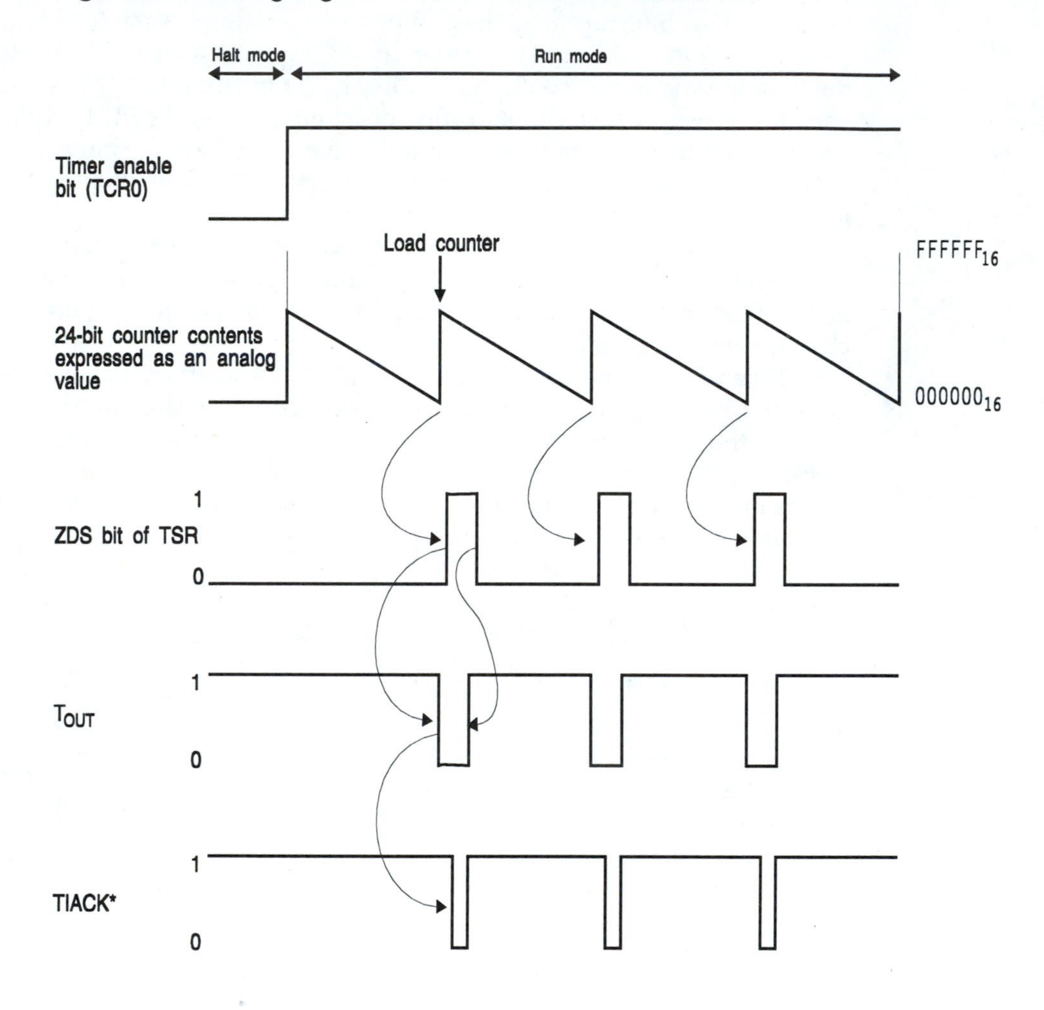

Table 9.15 Format of the TCR in the real-time clock mode

Bit	TCR7	TCR6	TCR5	TCR4	TCR3	TCR2	TCR1	TCR0
Value	1	X	1	0	0	0	0	1
Function	T_{OUT}/TIACK* control			ZD control		Clock control		Timer enable
	PC3/T_{OUT} = T_{OUT} Timer interrupt enabled. T_{OUT} set low when ZDS bit set to 1.			Counter reload on zero detect		Counter clock = CLK/32		Select run state

```
Time  MOVE.B  #1,(TSR,A0)          Clear the ZDS bit in the timer status register
        .                          Deal with the real-time interrupt
        .
      RTE                          Return from the interrupt
```

Square Wave Generator In this mode, the timer produces a square wave at its T_{OUT} terminal and interrupts are not generated. The format of the timer control register in the square wave mode is similar to that of the real-time mode — the only major difference is that bit 7 of the TCR is clear. A glance at Table 9.14 on page 401 reveals that TCR7 controls T_{OUT}. When TCR7 is clear, the signal at the T_{OUT} pin is toggled each time the counter counts down to zero and the ZDS bit is set.

Figure 9.20 provides a timing diagram for the timer in the square wave mode. The period is determined by the clock rate and the value loaded into the counter preload register. Note that the timer clock may be obtained from T_{IN} or from the 68000's own clock (which may be prescaled by 32).

Interrupt After Timeout In this mode, the timer generates an interrupt after a programmed period of time has elapsed. As in the case of the real-time clock, T_{OUT} is connected to the appropriate interrupt request line of the host processor, and TIACK* may be used as an interrupt acknowledge input. The source of timing may be derived from the system clock or from T_{IN}.

Table 9.16 gives the format of the TCR appropriate to this mode, and the other operating modes discussed here. This configuration is similar to the real-time clock

Figure 9.20 Timing diagram of the PI/T in the square wave generator mode

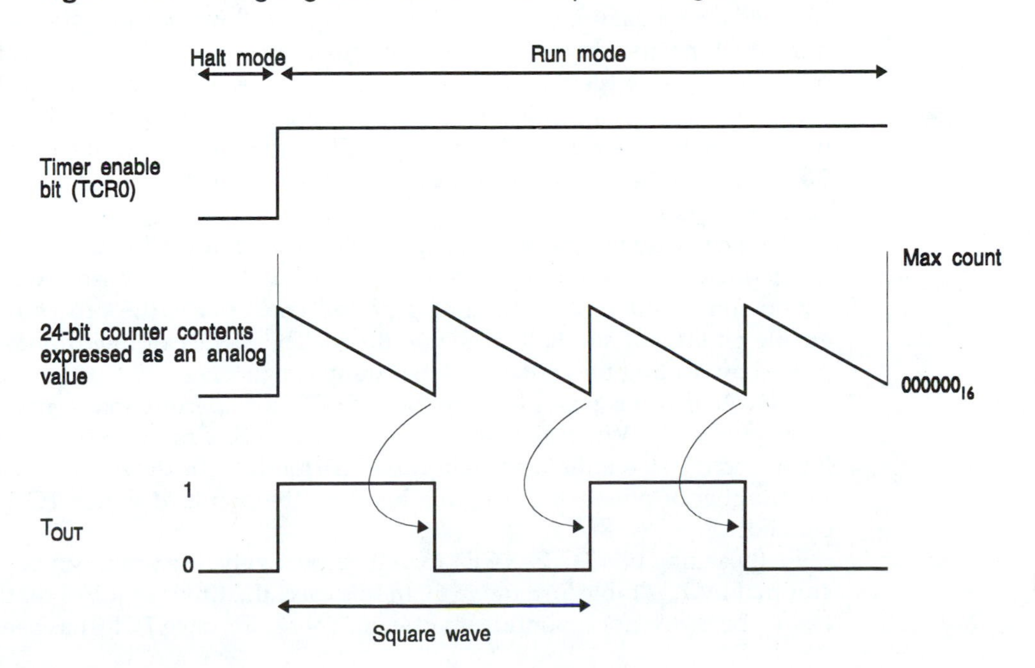

Table 9.16 Format of the TCR in various operating modes of the PI/T

	Mode	TCR7	TCR6	TCR5	TCR4	TCR3	TCR2	TCR1	TCR0
	1	1	X	1	0	0	00 or	1X	1
	2	0	1	X	0	0	00 or	1X	1
	3	1	X	1	1	0	00 or	1X	1
	4	0	0	X	1	0	0	0	1
	5	0	0	X	1	0	0	X	1
	6	1	X	1	1	0	0	1	1
Function		T_{OUT}/TIACK* control			ZD control		Clock control		Timer enable

Mode 1 = Real-time clock
Mode 2 = Square wave generator
Mode 3 = Interrupt after timeout
Mode 4 = Elapsed time measurement
Mode 5 = Pulse counter
Mode 6 = Period measurement

mode, except that the zero detect bit of the TCR is set. Consequently, when the counter reaches zero, it rolls over to its maximum value rather than being loaded from the counter load registers. Figure 9.21 illustrates this process. Once the interrupt has been serviced, the host processor can halt the timer and, if necessary, read the contents of the counter. At this point, the number in the counter gives an indication of the time elapsed between the interrupt request and its servicing.

Elapsed time measurement This configuration allows the host processor to determine the time that elapses between the triggering of the 68230 timer and its halting by clearing its enable bit. Table 9.16 (mode 4) gives the format of the TCR in this mode. The processor initializes the timer by loading its counter preload register with $FF FFFF (all ones) and setting its TCR, and then enables the timer by setting TCR0. The value $FF FFFF is selected because it provides the longest possible counting period.

Once TCR0 has been set, the prescaler counts down toward zero and decrements the counter each time it rolls over from $00 to $1F. When the event whose action signals the end of the timing period takes place, the processor clears the enable bit (TCR0) and halts the countdown. The processor determines the timing period by reading the contents of the counter registers.

Note that it is possible to program the TCR to permit an external clock to be connected to T_{IN}. When an external clock or pulse generator is connected to the T_{IN} input and the TCR initialized, as in Table 9.16 mode 5, the timer can be used to count the number of pulses at the T_{IN} pin between the points at which TCR0 is set and cleared.

By setting bits TCR2, TCR1 to 0, 1, respectively, the timer can be started and stopped by T_{IN} (Table 9.16 mode 6). In this case, the timer requires that both TCR0 *and* T_{IN} be high before counting may begin. Therefore, once TCR0 has been set under software control, counting begins only when T_{IN} goes high, and stops when T_{IN} returns low. If T_{IN} is controlled by external circuitry, the processor can determine the period between the positive and negative transition of T_{IN}.

Now that we've looked at the parallel interface, we are going to describe the serial interface used to link computers and peripherals.

Figure 9.21 Timing diagram of the PI/T in the interrupt after time-out generator mode

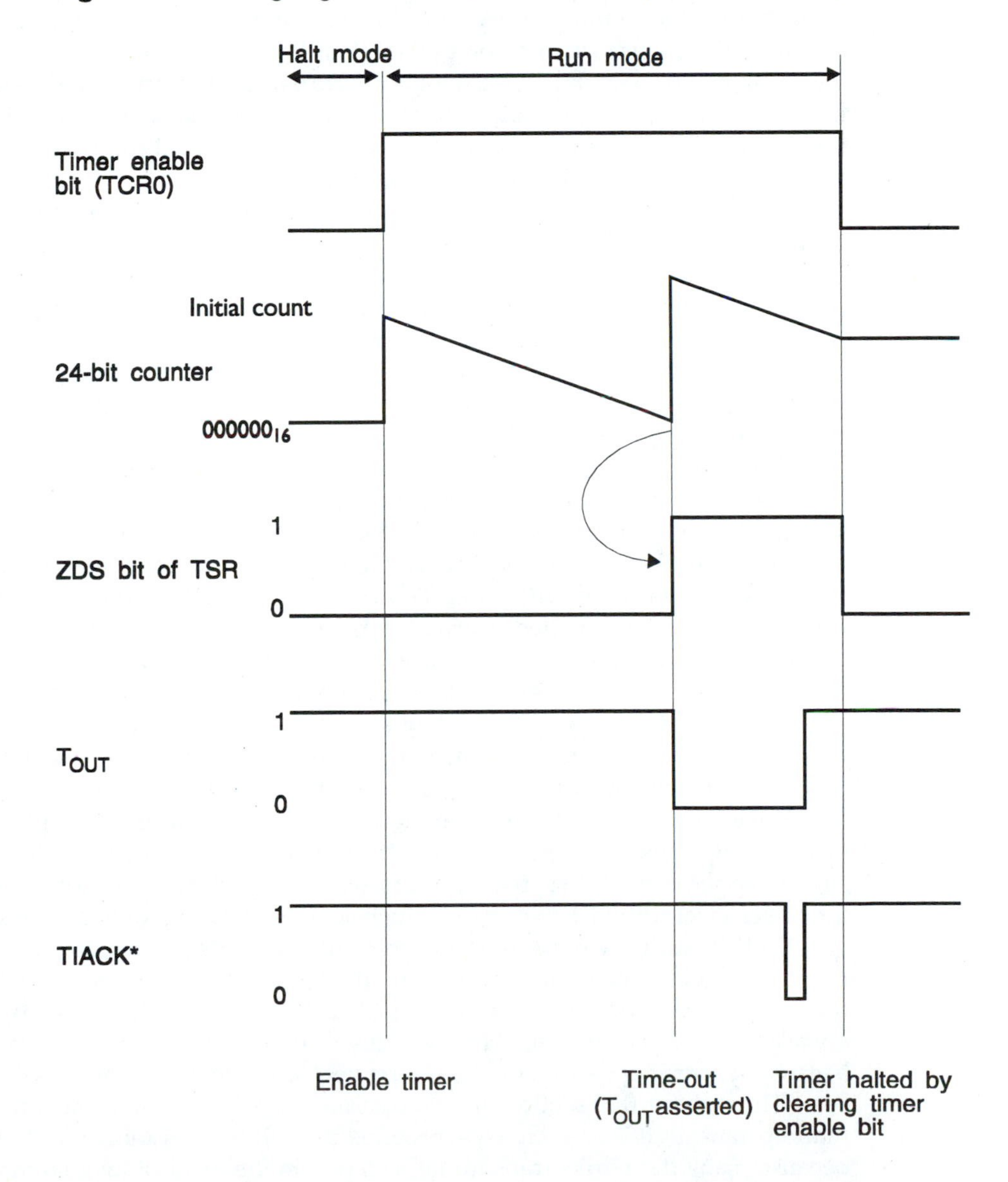

9.3 The Serial Interface

The vast majority of general-purpose microcomputers, except some entirely self-contained portable models, use a serial interface to communicate with remote peripherals such as CRT terminals. The serial interface, that moves information from point-to-point one bit at a time, is generally preferred to the parallel interface, that is able to move a group of bits simultaneously. This preference is not due to the high performance of a serial data link, but to its low cost, simplicity, and ease of use. We first describe how information is transmitted serially, and then examine a first-generation parallel-to-serial and serial-to-parallel chip that forms the interface between a microprocessor and a serial data link. Finally, we look at a more modern high-performance serial interface.

Figure 9.22 illustrates the basic serial data link between a computer and a CRT terminal. A CRT terminal requires a two-way data link, because information from the keyboard is transmitted to the computer and information from the computer is transmitted to the screen. The heart of the data link is the box labelled *serial interface*, that translates data between the form in which it is stored within the computer and the form in which it is transmitted over the data link. This function is often performed by a single device called an *asynchronous communications interface adaptor* — ACIA.

The *line drivers* in Figure 9.22 translate the voltage levels processed by the ACIA into a suitable form for sending over the transmission path. The transmission path itself is often just a twisted pair of conductors, which accounts for its very low cost. Some systems employ more esoteric transmission paths, such as fiber optics or infrared (IR) links. The connection between the line drivers and transmission path is labelled *plug and socket* in Figure 9.22 to emphasize that such mundane things as plugs become very important if interchangeability is required. International specifications cover this and other aspects of the data link.

The two items enclosed in clouds at the computer end of the data link in Figure 9.22 represent the software components of the data link. The lower cloud contains the software that directly controls the serial interface itself — the *device driver*. This software performs operations such as transmitting a single character, or receiving a character and checking it for certain types of error. On top of this sits the application-level software, that uses the primitive operations executed by the lower-level software to carry out such actions as listing a file on the screen.

In this section we examine the chip that converts information between the parallel form in which it is processed by the computer and the form in which it is transmitted over the data link. A serial data link operates in one of two modes: *asynchronous* or *synchronous*. Here we describe only the asynchronous data link because synchronous serial data links are best left to texts on networks.

Throughout this section, the word *character* refers to the basic unit of information transmitted over an asynchronous data link. This term has been chosen because many data links transmit information in the form of text, so that the unit of information corresponds to a printed character.

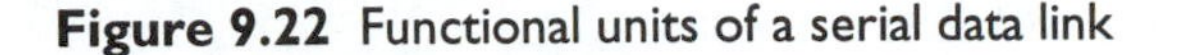

Figure 9.22 Functional units of a serial data link

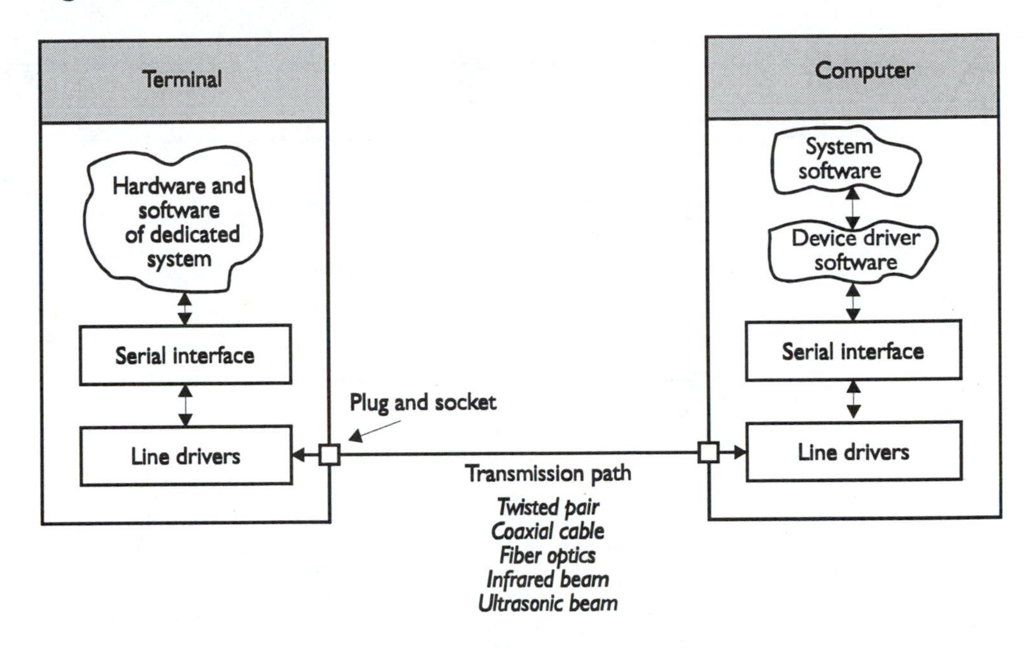

Asynchronous Serial Data Transmission

By far the most popular serial interface between a computer and its CRT terminal is the asynchronous serial interface. It is so-called because the transmitted data and the received data are not synchronized over any extended period, and therefore no special means of synchronizing the clocks at the transmitter and receiver is necessary. In fact, the asynchronous serial data link is a very old form of data transmission system, with its origin in the era of the teleprinter.

Serial data transmission systems have been around for a long time and are found in the telephone (human speech), Morse code, semaphore, and even the smoke signals once used by Native Americans. The fundamental problem encountered by all serial transmission systems is how to split the incoming datastream into individual units (i.e., bits) and how to group these units into words or *frames*. For example, in Morse code the dots and dashes of a letter are separated by an inter-symbol space, whereas the individual letters are separated by an inter-character space three times the duration of an inter-symbol space.

First, we examine how the data stream is divided into individual bits and the bits grouped into characters in an asynchronous serial data link. The key to the operation of this type of link is both simple and ingenious. Figure 9.23 gives the format of data transmitted over such a link.

An asynchronous serial data link is said to be *character oriented*, because information is transmitted in the form of groups of bits called characters. These characters are invariably units comprising seven or eight bits of *information* plus between two and four control bits. Asynchronous data is normally transmitted in

Figure 9.23 Format of asynchronous serial data

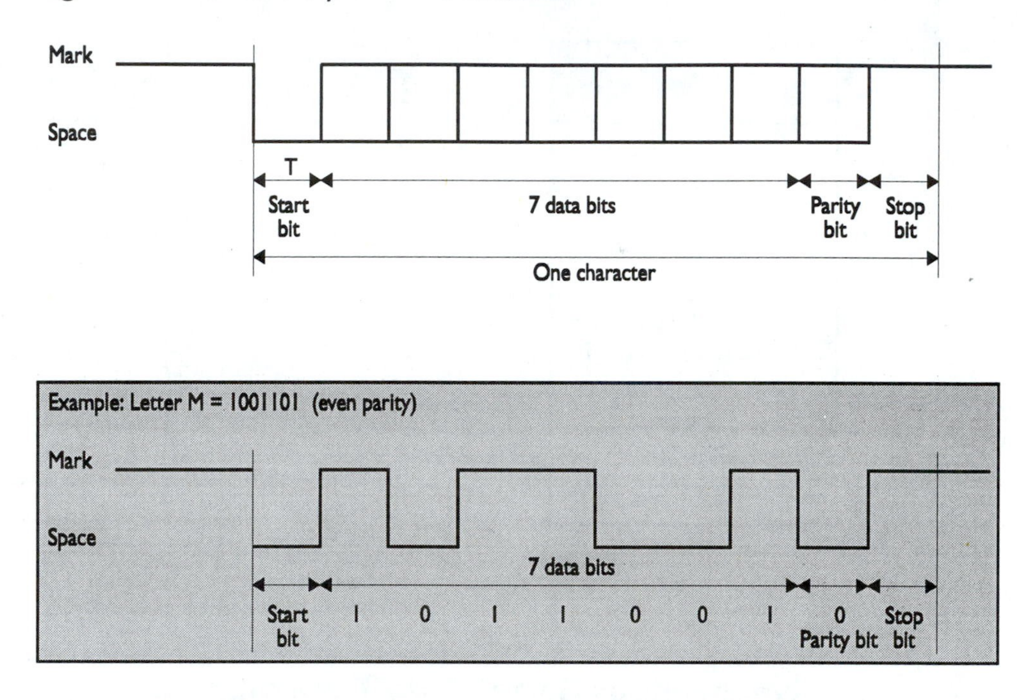

the form of ASCII-encoded characters. Initially, when no information is being transmitted, the line is in an idle state. Traditionally, the idle state is referred to as the *mark level*, which, by convention, corresponds to a logical one level.

When the transmitter wishes to send data, it first places the line in a *space* level (i.e., the complement of a mark) for one element period. This element is called the *start bit* and has a duration of T seconds. The transmitter then sends the character, one bit at a time, by placing each successive bit on the line for a duration of T seconds, until all bits have been transmitted. After this has been done, a single parity bit is calculated by the transmitter and sent after the data bits. Finally, the transmitter sends a *stop bit* at a mark level (i.e., the same level as the idle state) for one or two bit periods. Now the transmitter may send another character whenever it wishes. The only purpose of the stop bit is to provide a rest period for the receiver between consecutive characters.

As the data wordlength may be 7 or 8 bits with odd, even, or no parity bit, plus one or two stop bits, there are a total of 12 possible formats for serial data transmission. And this is before we consider that there are about seven commonly used values of T, the element duration. Consequently, connecting one serial link with another may be problematic because so many options are available.

At the receiving end of an asynchronous serial data link, the receiver continually monitors the line looking for a start bit. Once the start bit has been detected, the receiver waits until the end of the start bit and then samples the next N bits at their centers, using a clock generated locally by the receiver. As each incoming bit is sampled, it is used to construct a new character. When the received character has been assembled, its parity is calculated and compared with the received parity bit following the character. If they are not equal, a parity error flag is set to indicate a transmission error.

An obvious disadvantage of asynchronous data transmission is the need for a start, parity, and stop bit for each transmitted character. If 7-bit characters are used, the overall efficiency is only 7/(7+3)x100 = 70%. A less obvious disadvantage is due to the character-oriented nature of the data link. Whenever the data link connects a CRT terminal to a computer, few problems arise, as the terminal is itself character-oriented. However, if the data link is being used to, say, dump binary data to a magnetic tape, we run into a difficulty. If the data is arranged as 8-bit bytes with all 256 possible values corresponding to valid data elements, it is difficult (but not impossible) to embed control characters (e.g., tape start or stop) within the data stream, because the same character must be used both as pure data (i.e., part of the message) and for control purposes. If 7-bit characters are used, pure binary data cannot be transmitted in the form of one character per byte. Two characters are needed to record each byte, which is clearly inefficient.

The 6850 Asynchronous Communications Interface Adaptor (ACIA)

One of the first general-purpose interface chips was the *asynchronous communications interface adaptor*, or ACIA. This device relieves the system software of all the basic tasks involved in converting data between serial and parallel forms. That is, the ACIA contains almost all the logic necessary to provide an asynchronous data link between a computer and an external system.

Motorola's first ACIA was the 6850 illustrated in Figure 9.24. We describe this particular ACIA because it is much easier to understand than some of the newer serial interfaces, and is still widely found in microcomputers. Once you understand how the 6850 ACIA operates, you can read the data sheet of any other ACIA. After we have covered the 6850, we will briefly look at a more modern and more sophisticated device, the DUART.

Like any other digital device, the 6850 has a hardware model, a software model, and a functional model. We look at the hardware model first. From the designer's point of view, the 6850's hardware can be subdivided into three sections: the CPU side, the transmitter side, and the receiver side. Figure 9.24 illustrates the functional parts of the 6850 and its internal registers.

The CPU Side

The ACIA is a byte-oriented device and can be interfaced to either the 68000's lower-order byte or to its upper-order byte. The ACIA has a single register select line, RS, that determines the internal location (i.e., register) addressed by the processor. The 6850 has an interrupt request output, IRQ*, that can be connected

Figure 9.24 The 6850 ACIA

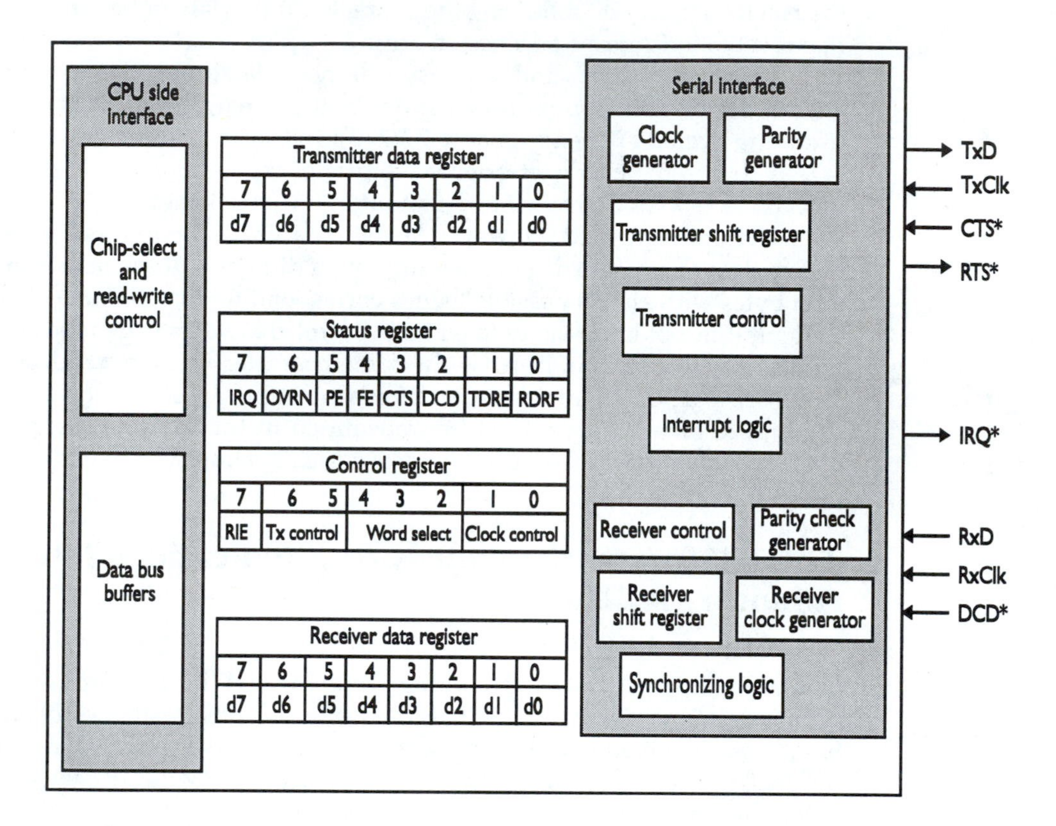

to any of the 68000's seven levels of interrupt request input. As the 6850 does not support vectored interrupts, autovectored interrupts must be used in the way to be described in Chapter 10.

Unusually, the 6850 does not have a reset input (i.e., a pin that can be used to clear its internal registers). Because there were not enough pins to support all the ACIA's functions, the manufacturer felt that a hardware reset was the most disposable of the functions. Some sections of the ACIA are reset automatically by an internal power-on-reset circuit. Afterward, a secondary reset can be performed by software, as we shall describe later.

The Receiver and Transmitter Sides of the ACIA

One of the great advantages of peripherals like the 6850 ACIA is that they isolate the CPU from the outside world both physically and logically. The *physical isolation* means that the engineer who is connecting a peripheral device to a microprocessor system does not have to worry about the electrical and timing requirements of the CPU itself. That is, all the engineer needs to understand about the ACIA is the nature of its transmitter- and receiver-side interfaces. Similarly, the peripheral performs a *logical isolation* by hiding the details of information transfer across it. For example, you can transmit a character from an ACIA by means of the instruction `MOVE.B D0,ACIA_DATA`, where register D0 contains the character to be transmitted and `ACIA_DATA` is the address of the data register in the ACIA. All the actions necessary to serialize the data and append start, parity, and stop bits are carried out automatically (i.e., invisibly) by the ACIA.

Here, only the essential details of the ACIA's transmitter and receiver sides are presented, because the way in which they function is described more fully later. The peripheral-side interface of the 6850 is divided into two entirely separate groups — the receiver group that forms the interface between the ACIA and a source of incoming data, and the transmitter group that forms the interface between the ACIA and the destination for outgoing data. *Incoming* and *outgoing* are used with respect to the ACIA. The nature of these signals is strongly affected by one particular role of the ACIA, that of an interface between a computer and the public switched telephone network, PSTN, via a *modem*.

Receiver side Incoming data to the ACIA is handled by three pins, RxD, RxCLK, DCD*. The RxD, receiver data input, pin receives serial data from the transmission path to which the ACIA is connected. A receiver clock must be provided at the RxCLK input pin. This clock may be either one-, sixteen-, or sixty-four times the rate at which bits are received at the data input terminal. Many modern ACIAs include on-chip receiver and transmitter clocks, relieving the system designer of the necessity of providing an additional external oscillator.

The third and last component of the receiver group is an active-low DCD*, data carrier detect, input. This input is intended for use in conjunction with a modem and, when low, indicates to the ACIA that the incoming data is valid. When inactive-high, DCD* indicates that the incoming data might be erroneous. This situation may arise if the level (i.e., signal strength) of the data received at the end of a telephone line drops below a pre-determined value, or if the connection itself is broken. The negation of DCD* is used to disable the ACIA's receiver.

Transmitter side The transmitter side of the ACIA comprises four pins: TxCLK, TxD, RTS*, and CTS*. The transmitter clock input, TxCLK, provides a timing signal from which the ACIA derives the timing of the transmitted signal elements. In most applications of the ACIA, the transmitter and receiver clocks are connected together and a common oscillator is used for both transmitter and receiver sides of the ACIA. Serial data is transmitted from the TxD, transmit data, pin of the ACIA to the serial transmission path.

An active-low request-to-send, RTS*, output indicates that the ACIA is ready to transmit information. This output is set or cleared under software control, and can be used to switch on any equipment needed to transmit the serial data over the data link.

An active-low clear-to-send, CTS*, input indicates to the transmitter side of ACIA that the external equipment used to transmit the serial data is ready. When negated, this input inhibits the transmission of data by the ACIA. CTS* is a modem signal indicating that the transmitter carrier is present and that transmission may go ahead.

Operation of the 6850 ACIA

The software model of the 6850 has four user-accessible registers, as defined in Table 9.17. These are: a transmit data register, TDR, a receive data register, RDR, a system control register, CR, and a system status register, SR. As there are four registers, but the ACIA has only a single register select input, RS, a way must be found to distinguish between registers. The ACIA uses its read/write input to do this. Two registers are read-only (i.e., RDR, SR), and two are write-only (TDR, CR). This is a perfectly logical, indeed an elegant, thing to do. But I don't like it. I am perfectly happy to accept read-only registers, but I am suspicious of the write-only variety because it is impossible to verify the contents of a write-only register. Suppose you have a program with a bug that executes an unintended write to a write-only register. You cannot detect the change by reading back the contents of the register. Moreover, certain 68000 instructions would play havoc with such an arrangement. For example, an `ORI #%01100000,ACIAC` instruction would read the contents of the ACIA's status register, perform a logical OR, and then write the result back to its control register — a meaningless operation.

Table 9.17 also gives the address of each register, assuming that the base address of the ACIA is $00 E001 and that it is connected to the low-order byte of the 68000's data bus. The purpose of this exercise is two-fold. It shows that the address of the lower-order byte is odd, and that the pairs of read-only and write-only registers are separated by two (i.e., $00 E001 and $00 E003).

Control Register

Because the ACIA is a versatile device that can be operated in any of several different modes, the control register permits the programmer to define its opera-

Table 9.17 Register selection scheme of the 6850

Address	RS	$R/\overline{W}$	Register type	Register function
00 E001	0	0	write only	control register (CR)
00 E001	0	1	read only	status register (SR)
00 E003	1	0	write only	transmit data register (TDR)
00 E003	1	1	read only	receive register (RDR)

RS = register select input
$R/\overline{W}$ = 0 = write
$R/\overline{W}$ = 1 = read

tional characteristics. This operation can even be performed dynamically at run-time, if the need ever arises. However, in almost all applications the ACIA is normally configured once only. Table 9.18 shows how the eight bits of the control register are grouped into four logical fields.

Table 9.18 Structure of the ACIA's control register

Bit	CR7	CR6 CR5	CR3 CR2 CR1	CR1 CR0
Function	receiver interrupt enable	transmitter control	word select	counter division

CR1	CR0	Division ratio
0	0	1
0	1	16
1	0	64
1	1	master reset

CR4	CR3	CR2	Word select			
			Data word length	Parity	Stop bits	Total bits
0	0	0	7	even	2	11
0	0	1	7	odd	2	11
0	1	0	7	even	1	10
0	1	1	7	odd	1	10
1	0	0	8	none	2	11
1	0	1	8	none	1	10
1	1	0	8	even	1	11
1	1	1	8	odd	1	11

CR6	CR5	Transmitter control	
		RTS*	Transmitter interrupt
0	0	low	disabled
0	1	low	enabled
1	0	high	disabled
1	1	low	disabled and break

CR7	Receiver interrupt enable
0	receiver may not interrupt
1	receiver may interrupt

Bits CR0 and CR1 determine the ratio between the transmitted or received bit rates and the transmitter and receiver clocks, respectively. The clocks operate at 1, 16, or 64 times the data rate. Most applications of the 6850 employ a receiver/transmitter clock at 16 times the data rate with CR1 = 0 and CR0 = 1. For example, if you wish to receive data at 1200 baud, you must provide a receiver clock input of 16 x 1200 = 19,200 Hz at the receiver clock input pin.

Setting CR1 = CR2 = 1 is a special case and performs a software reset of the ACIA. This reset clears all internal status bits, with the exception of the CTS and DCD bits of the status register. A software reset to the 6850 is invariably carried out during the initialization phase of the host processor's reset procedures.

The *word select* field, bits CR2, CR3, CR4, determines the format of the received or transmitted characters. The eight possible data formats are given in Table 9.18 on the previous page. These bits select also the type of parity (if any) and the number of stop bits. A common data format for the transmission of information between a processor and a CRT terminal is: start bit + 7 data bits + even parity + 1 stop bit. The corresponding value of CR4, CR3, CR2 is 0,1,0.

The *transmitter control* field, CR5 and CR6, selects the state of the active-low request-to-send (RTS*) output, and determines whether the transmitter section of the ACIA may generate an interrupt by asserting its IRQ* output. In most systems RTS* is set active-low whenever the ACIA is transmitting, because RTS* is used to activate equipment connected to the ACIA.

If the transmitter interrupt is enabled, an interrupt is generated by the transmitter whenever the transmit data register (TDR) is empty, signifying the need for new data from the CPU. When the ACIA's clear-to-send input, CTS*, is inactive-high, the TDR empty flag of the status register is held low, inhibiting any transmitter interrupt. If the ACIA is operated in a polled-data mode, interrupts are not necessary.

Setting both CR6 and CR5 to a logical one simultaneously creates a special case. When both these bits are high, a *break* is transmitted by the transmitter data output pin. A break is a condition in which the transmitter output is held continuously at the active level (i.e., space or TTL logical zero). This condition may be employed to force an interrupt at a distant receiver, because the asynchronous serial format precludes the existence of a space level for longer than about ten bit periods. The term *break* originates from the old current-loop data transmission system when a break was affected by disrupting (i.e., breaking) the flow of current round a loop.

The *receiver interrupt enable* field consists of one bit, CR7, that enables the generation of interrupts by the receiver when it is set (CR7 = 1) and disables receiver interrupts when it is clear (CR7 = 0). The receiver asserts its IRQ* output, assuming CR7 = 1, when the receiver data register full (RDRF) bit of the status register is set, indicating the presence of a new data character ready for the CPU to read. Two other circumstances also force a receiver interrupt. An overrun (covered later) sets the RDRF bit and generates an interrupt. Finally, a receiver interrupt can also be generated by a high-to-low transition at the active-low data-carrier-detect (DCD*) input, signifying a loss of the carrier from a modem. Note that CR7 is a composite interrupt enable bit, and enables all the three forms of receiver interrupt described above. It is impossible to enable either an interrupt caused by the RDR being empty or an interrupt caused by a positive transition at the DCD* pin alone.

Status Register

The eight bits of the read-only status register are depicted in Table 9.19 and serve to indicate the status of both the transmitter and receiver portions of the ACIA at any instant.

SR0 — Receiver data register full (RDRF) When set, this bit indicates that the receiver data register (RDR) is full, and a new word has been received. If the receiver interrupt is enabled by CR7 = 1, a logical one in SR0 also sets the interrupt status bit SR7 (i.e., IRQ). The RDRF bit is cleared either by reading the data in the receiver data register or by carrying out a software reset on the control register. Whenever the data-carrier-detect (DCD*) input is inactive-high, the RDRF bit remains clamped at a logical zero, indicating the absence of any valid input.

SR1 — Transmitter data register empty (TDRE) This is the transmitter counterpart of the RDRF bit. A logical one in SR1 indicates that the contents of the transmit data register (TDR) have been sent to the transmitter and that the register is now ready for new data from the processor. This bit is cleared either by loading the transmit data register or by performing a software reset. If the transmitter interrupt is enabled, a logical one in bit SR1 also sets bit SR7 (i.e., IRQ), of the status word. Note again that SR7 is a *composite* interrupt bit, because it is also set by an interrupt originating from the receiver side of the ACIA. If the clear-to-send input (CTS*) is inactive-high, the TDRE bit is held low, indicating that the terminal equipment is not ready for data.

SR2 — Data carrier detect (DCD) This status bit, associated with the receiver side of the ACIA, is employed when the ACIA is connected to the telephone network via a modem. Whenever the active-low DCD* input to the ACIA is inactive-high, SR2 is set. A high level on the DCD* line generally signifies that the incoming serial data is faulty. This also has the effect of clearing the SR0 (i.e., RDRF) bit, as possible erroneous input should not be interpreted as valid data.

When the DCD* input makes a low-to-high transition, not only is SR2 set, but the composite interrupt request bit, SR7, is also set if the receiver interrupt is

Table 9.19 Format of the status register

Bit	SR7	SR6	SR5	SR4	SR3	SR2	SR1	SR0
Function	IRQ	PE	OVRN	FE	CTS	DCD	TDRE	RDRF

enabled. Note that SR2 remains set even if the DCD* input later returns active-low. This is done to trap any occurrence of DCD* high, even if it goes high only briefly. To clear SR2, the CPU must read the contents of the status register and then the contents of the data register.

SR3 — Clear-to-send (CTS) The CTS bit directly reflects the status of the CTS* input on the ACIA's transmitter side. An active-low level on the CTS* input indicates that the transmitting device (i.e., the modem) is ready to receive serial data from the ACIA. If the CTS* input and, therefore, the CTS status bit are high, the transmit data register empty bit, SR1, is inhibited (clamped at a logical zero), and no data may be transmitted by the ACIA. Unlike the DCD status bit, the logical value of the CTS status bit is determined only by the CTS* input and is not affected by any software operation on the ACIA.

SR4 — Framing error (FE) A framing error is detected by the absence of a stop bit and indicates a synchronization (i.e., timing) error, a faulty transmission, or a break condition. The framing error status bit, SR4, is set whenever the ACIA determines that a received character is incorrectly framed by a start bit and a stop bit. The framing error status bit is automatically cleared or set during the receiver data transfer time, and is present throughout the time that the associated character is available.

SR5 — Receiver overrun (OVRN) The receiver overrun status bit is set when a character is received by the ACIA but is not read by the CPU before a subsequent character is received, overwriting the last character, which is now lost. Consequently, the receiver overrun bit indicates that one or more characters in the data stream have been lost. This status bit is set at the midpoint of the last bit of the second character received in succession without a read of the RDR having occurred. Synchronization of the incoming data is not affected by an overrun error. The error is due to the CPU not having read a character, rather than by any fault in the transmission and reception process. The overrun bit is cleared after reading data from the RDR or by a software reset.

SR6 — Parity error (PE) The parity error status bit is set whenever the received parity bit in the current character does not match the parity bit of the character generated locally in the ACIA from the received data bits. Odd or even parity may be selected by writing the appropriate code into bits CR2, CR3, and CR4 of the control register. If no parity is selected, then both the ACIA's transmitter parity generator and receiver parity checker are disabled. Once a parity error has been detected and the parity error status bit set, it remains set as long as the erroneous data remains in the receiver register.

SR7 — Interrupt request (IRQ) The interrupt request status bit is a composite active-high interrupt request flag, and is set whenever the ACIA wishes to interrupt the CPU, for whatever reason. The IRQ bit is set active-high by any of the following events:

1. Receiver data register full (SR0 set) and receiver interrupt enabled.
2. Transmitter data register empty (SR1 set) and transmitter interrupt enabled.
3. Data-carrier-detect status bit (SR2) set and receiver interrupt enabled.

Whenever SR7 is active-high, the IRQ* output from the ACIA is pulled low. The IRQ bit is cleared by a read from the RDR, by a write to the TDR, or by a software master reset.

Using the ACIA

The most daunting thing about many microprocessor interface chips is their sheer complexity. Often this complexity is more imaginary than real, because such peripherals are usually operated in only one of the many different modes that are software selectable. This is particularly true of the 6850 ACIA. Figure 9.25 shows how the 6850 is operated in a minimal mode. Only its serial data input, RxD, and output, TxD, are connected to an external system. The request to send output, RTS*, is left unconnected, and clear to send, CTS*, and data carrier detect, DCD*, inputs are both strapped to ground (i.e., an electrically low level) at the ACIA.

Figure 9.25 Using the ACIA in its minimal mode

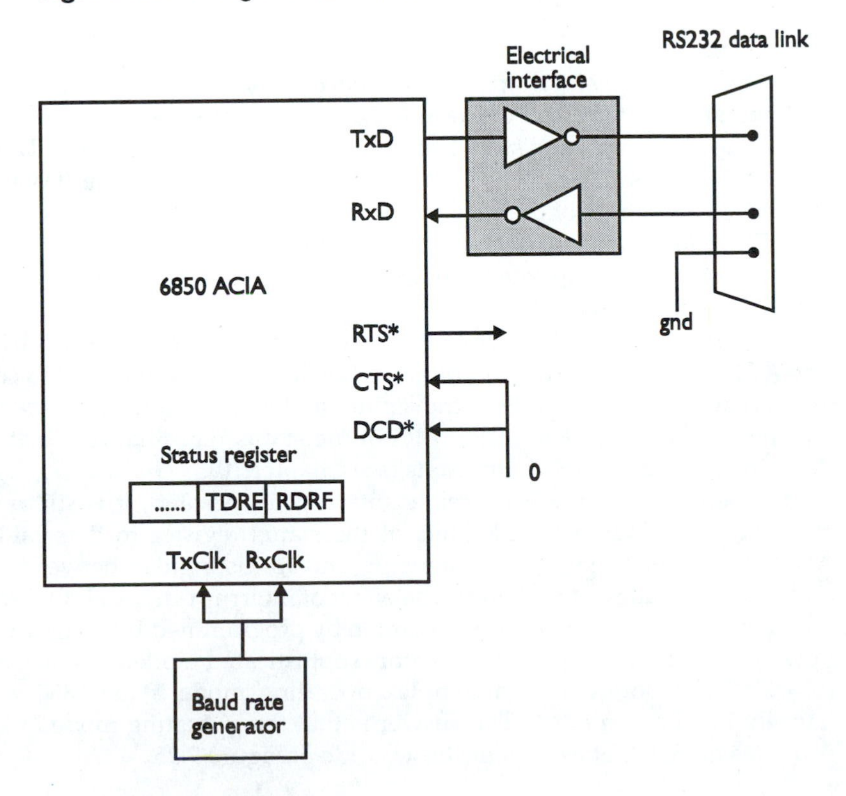

In a minimal, non-interrupt mode, bits 2 to 7 of the status register can be ignored. Of course, this throws away the error-detecting facilities of the ACIA. The software necessary to drive the ACIA in this minimal mode consists of three subroutines: an initialization, an input, and an output routine.

```
ACIAC      EQU     $00E001          Address of control/status registers
ACIAD      EQU     ACIAC+2          Address of the Tx/Rx data registers
RDRF       EQU     0                Receiver data register full
TDRE       EQU     1                Transmitter data register empty

INITIALIZE MOVE.B  #%00000011,ACIAC Reset the ACIA
           MOVE.B  #%00011001,ACIAC Set up control word - disable
           RTS                      interrupts, 8 data bits, even parity

INPUT      BTST    #RDRF,ACIAC      REPEAT
           BEQ     INPUT                 Poll RDRF bit
           MOVE.B  ACIAD,D0         UNTIL ACIA has data ready
           RTS                      Copy input to D0 and return

OUTPUT     BTST    #TDRE,ACIAC      REPEAT
           BEQ     OUTPUT                Poll TDRE bit
           MOVE.B  D0,ACIAD         UNTIL ACIA is ready for new data
           RTS
```

The `INITIALIZE` routine is called once before either input or output is carried out, and has the effect of executing a software reset on the ACIA followed by setting up its control register. The control word `%00011001` (see Table 9.18 on page 415) defines an 8-bit word with even parity and a clock rate (TxCLK, RxCLK) 16 times the data rate of the transmitted and received data.

The `INPUT` and `OUTPUT` routines are both entirely straightforward. Each routine tests the appropriate status bit and then reads data from or writes data to the ACIA's data register.

It is also possible to operate the ACIA in a minimal interrupt-driven mode. The ACIA's composite IRQ* output is connected to one of the 68000's seven levels of interrupt request input. Both transmitter and receiver interrupts are enabled by writing 1,0,1 into bits CR7, CR6, CR5 of the status register. Note that this ACIA does not support vectored interrupts (see Chapter 10).

When a transmitter or receiver interrupt is initiated, it is still necessary to examine the RDRF and TDRE bits of the status register to determine that the ACIA did indeed request the interrupt, and to distinguish between transmitter and receiver requests for service. The effect of interrupt-driven I/O is to eliminate the time-wasting polling routines required by programmed I/O. Figure 9.26 shows how the ACIA can be operated in a more sophisticated mode. You may be tempted to ask, "Why bother with a complex operating mode if the 6850 works quite happily in a basic mode?" The answer is that the operating mode in Figure 9.26 provides more facilities than the basic mode of Figure 9.25.

Figure 9.26 General-purpose interface using the 6850 ACIA

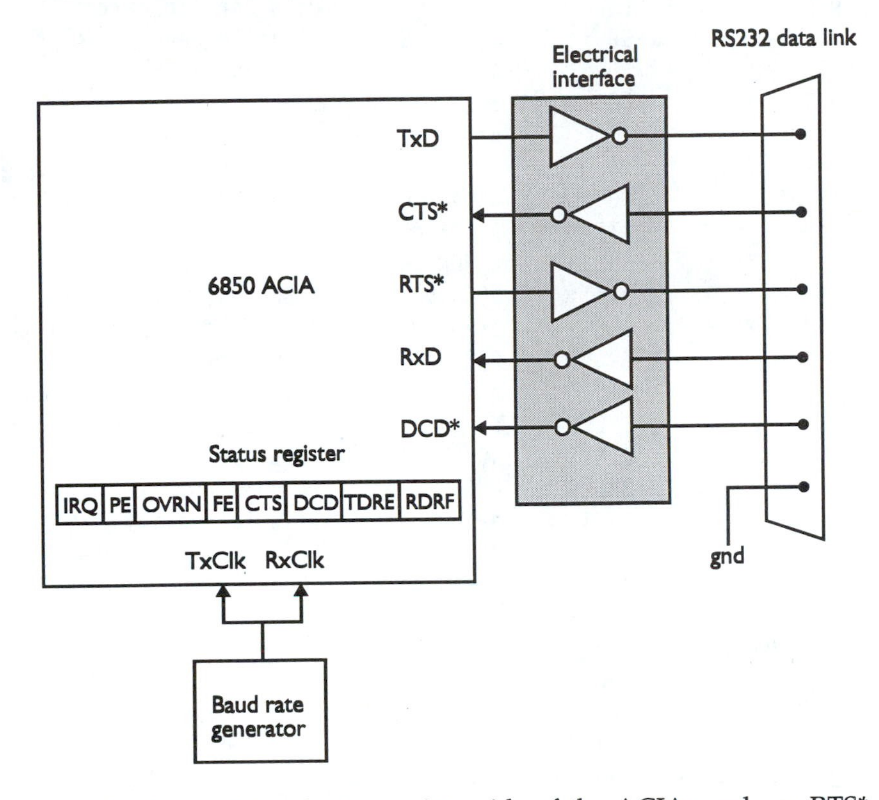

In Figure 9.26, the transmitter side of the ACIA sends an RTS* signal and receives a CTS* signal from the remote terminal equipment. Now the ACIA is able to say, "I am going to transmit data to you" by asserting RTS*, and "I know you are ready to receive my data" by detecting CTS* asserted. In the cut-down mode of Figure 9.25 on page 419, the ACIA simply sends data and hopes for the best.

Similarly, the ACIA's receiver uses the data carrier detect (DCD*) input to signal to the host computer that the receiver is in a position to receive data. If DCD* is negated, the terminal equipment is unable to send data to the ACIA.

The software necessary to receive data when operating the 6850 in its more sophisticated mode is considerably more complex than that of the previous example. It is not possible to provide a full input routine here, as such a routine would include recovery procedures from the errors detected by the 6850 ACIA. These procedures are, of course, dependent on the nature of the system and the protocol used to move data between a transmitter and receiver. However, the following fragment of an input routine gives some idea of how the 6850's status register is used.

```
ACIAC        EQU     <ACIA address>
ACIAD        EQU     ACIAC+2
RDRF         EQU     0              Receiver_data_register_full
```

```
TDRE            EQU      1              Transmitter_data_register_empty
DCD             EQU      2              Data_carrier_detect
CTS             EQU      3              Clear_to_send
FE              EQU      4              Framing_error
OVRN            EQU      5              Overrun
PE              EQU      6              Parity_error
*
INPUT           MOVE.B   ACIAC,DO       Get status from ACIA
                BTST     #RDRF,DO       Test for received character
                BNE.S    ERROR_CHECK    IF char received THEN test SR
                BTST     #DCD,DO        ELSE test for loss of signal
                BEQ      INPUT          REPEAT loop while CTS clear
                BRA.S    DCD_ERROR      ELSE deal with loss of signal
ERROR_CHECK     BTST     #FE,DO         Test for framing error
                BNE.S    FE_ERROR       IF framing error, deal with it
                BTST     #OVRN,DO       Test for overrun
                BNE.S    OVRN_ERROR     IF overrun, deal with it
                BTST     #PE,DO         Test for parity error
                BNE.S    PE_ERROR       IF parity error deal with it
                MOVE.B   ACIAD,DO       Load the input into DO
                BRA.S    EXIT
DCD_ERROR       Deal with loss of signal
                BRA.S    EXIT
FE_ERROR        Deal with framing error
                BRA.S    EXIT
OVRN_ERROR      Deal with overrun error
                BRA.S    EXIT
PE_ERROR        Deal with parity error
EXIT            RTS
```

9.4 The 68681 DUART

The 6850 ACIA is a first-generation interface device designed in the 1970s to work with the 8-bit 6800 microprocessor and is now rather long in the tooth (although it is still found in some 68000-based systems). Today's designers would rather implement an asynchronous serial interface with a more modern component like the 68681 DUART (i.e., dual universal asynchronous receiver/transmitter). The 68681 (from now on, we will just call it 'DUART') performs the same basic functions as a pair of 6850s plus a baud-rate generator. Designers prefer to use the DUART for the following reasons:

1. The DUART provides two *independent* asynchronous serial channels and therefore replaces two 6850 ACIAs.

2. The DUART has a full 68000 asynchronous bus interface, which means that it supports asynchronous data transfers and can supply a vector number during an interrupt acknowledge cycle.

3. The DUART has an on-chip programmable baud-rate generator, that saves both the cost and board space of a separate baud-rate generator. Moreover, the DUART's baud-rate generator can be programmed simply by loading an appropriate value into a clock select register. This feature makes it very easy to connect a system with a DUART to a communications system with an unknown baud rate. Communications systems based on the 6850 have to change their baud rate by altering links on the board, making it tedious to modify the baud rate. The DUART can receive and transmit at different baud rates (as can the 6850).

4. The DUART has a quadruple-buffered input so that up to four characters can be received in a burst before the host processor has to read the input stream. The host computer has to read each character from a 6850 as it is received (otherwise, an overrun will occur and characters will be lost). Similarly, the DUART has a double-buffered output, permitting one character to be transmitted while another is being loaded by the CPU.

5. The DUART has 14 I/O pins (six input, eight output) that can be used as modem-control pins, clock input and outputs, or as general-purpose input/output pins.

6. The 6850 has just one operating mode. The DUART can support several modes (e.g., a self-test loop-back mode).

Figure 9.27 illustrates the internal organization of a 68681 DUART.

Since the 68681 is so much better than the 6850, why then have we not used it to replace the 6850 in this chapter? The answer to this question is very simple. The 6850 is still widely used and is very much easier to understand than the more versatile DUART. However, because the DUART has become a standard in many 68000-based systems, we cannot neglect it. One way of approaching the DUART is to ignore its sophisticated functions and to treat it as an advanced ACIA. We will do this and describe how the DUART can used in a 68000 without going into fine detail. In short, we will treat the DUART as a black box.

The DUART's Registers

The DUART has 16 addressable registers, as illustrated in Table 9.20. Some registers are read-only, some write-only, and some read/write. For our current purposes, we concentrate on the DUART's five control registers that must be configured before it can be used as a transmitter or a receiver (the 6850 has just a single control register). Some registers are *global* and affect the operation of both the DUART's serial channels, while others are *local* to channel A or to channel B. In what follows, we use channel A registers to illustrate the DUART's operation.

Figure 9.27 Internal organization of a 68681 DUART

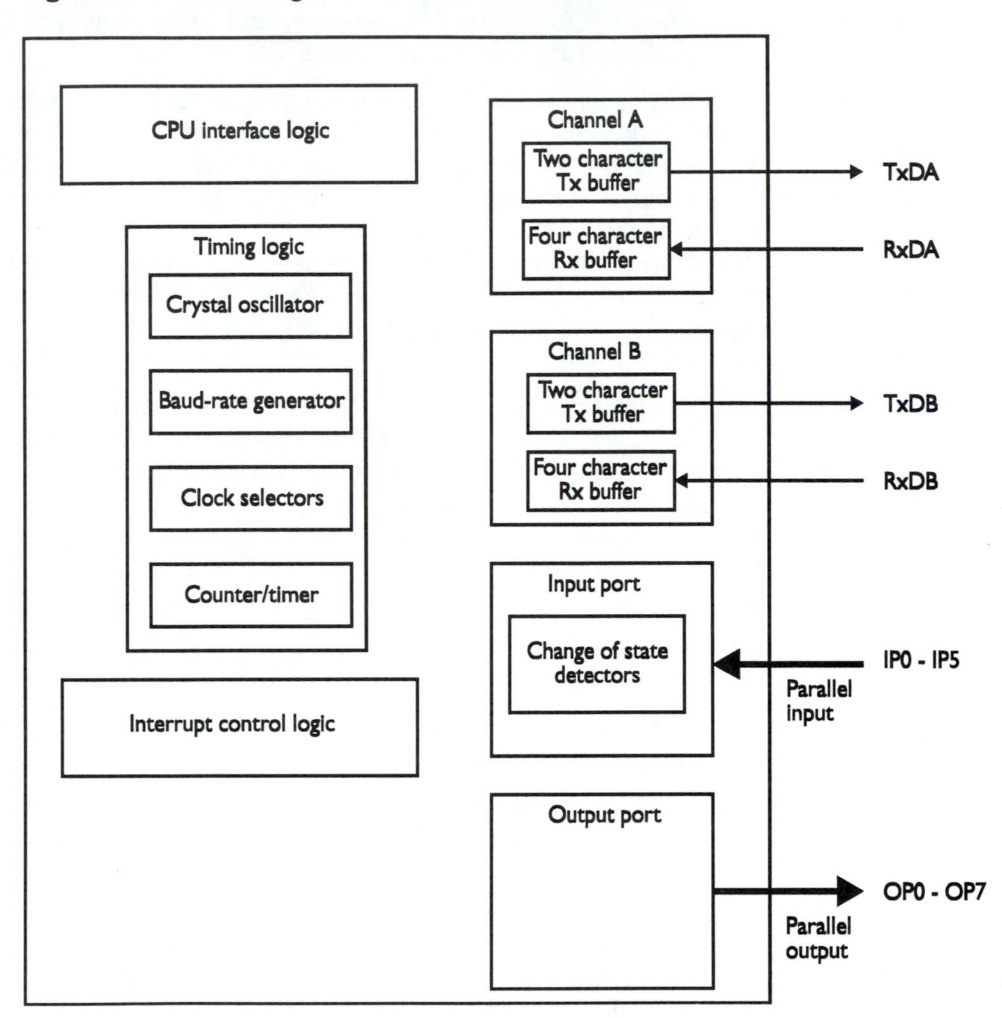

The five control registers are: MR1A (master register 1), MR2A (master register 2), CSRA (clock select register), CRA (command register), and ACR (auxiliary control register). Note that MR1A and MR2A share the *same* address. After a reset, control register MR1A is selected at the base address of the DUART. When MR1A is loaded with data by the host processor, MR2A is automatically selected at the same address. You can access MR1A again only by resetting the DUART or by executing a special *select MR1A* command. That is, the first time you write to the PI/T's base address following a reset, you access register MR1A. Further writes to the same address access register MR2A.

Table 9.21 provides a simplified extract from the DUART's data sheet that describes the five control registers. Modes of no interest to us here, such as the

Table 9.20 The DUART's registers

RS4	RS3	RS2	RS1	Read ($R/\overline{W}$ = 1)		Write ($R/\overline{W}$ = 0)	
0	0	0	0	Mode register A	MR1A,MR2A	Mode register A	MR1A,MR2A
0	0	0	1	Status register A	SRA	Clock select register A	CSRA
0	0	1	0			Command register A	CRA
0	0	1	1	Receiver buffer A	RBA	Transmitter buffer A	TBA
0	1	0	0	Input port change register	IPCR	Auxiliary control register	ACR
0	1	0	1	Interrupt status register	ISR	Interrupt mask register	IMR
0	1	1	0	Current MSB of counter	CUR	Counter/timer upper byte	CTUR
0	1	1	1	Current LSB of counter	CUL	Counter/timer lower byte	CTUL
1	0	0	0	Mode register B	MR1B, MR2B	Mode register B	MR1B, MR2B
1	0	0	1	Status register B	SRB	Clock select register B	CSRB
1	0	1	0			Command register B	CRB
1	0	1	1	Receiver buffer B	RBB	Transmitter buffer B	TBB
1	1	0	0	Interrupt vector register	IVR	Interrupt vector register	IVR
1	1	0	1	Input port (unlatched)		Output port configuration	OPCR
1	1	1	0	Start counter command		Output port bit set	
1	1	1	1	Stop counter command		Output port bit clear	

DUART's parallel I/O capabilities, have not been included in Table 9.21. The following notes provide sufficient details about the DUART's registers to enable you to use it in its basic operating mode.

The auxiliary control register, ACR, selects the DUART's clock source (internal or external), its baud rate set (there are two sets), and controls certain parallel input pins. Setting bit ACR7 to 0 selects baud rate set 1, and setting ACR7 to 1 selects set 2. For our purposes, ACR can be loaded with $80 to select baud rate set 2 and then forgotten about.

The clock-select register(s), CSRA and CSRB, permit the programmer to select the DUART's baud rate (CSRA selects the channel A baud rate and CSRB the channel B baud rate). Table 9.21 demonstrates that it is possible to select independent baud rates for transmission and reception. The values below can be loaded into the appropriate clock-select register to select the following popular baud rates for both transmission and reception, assuming that ACR7 is set to 1.

Value loaded in CSRA	**Baud rate (Tx and Rx)**
44_{16}	300
55_{16}	600
66_{16}	1,200
88_{16}	2,400
99_{16}	4,800
BB_{16}	9,600
CC_{16}	19,200

Each baud-rate value loaded into a clock-select register consists of two 4-bit values (bits 0 to 3 select the transmitter baud rate, and bits 4 to 7 select the re-

(continued on page 428)

Table 9.21 Structure of the DUART's registers

Clock-select register A (CSRA)	
Receiver clock	Transmitter clock
7 6 5 4 Baud rate 0 1 0 0 300 0 1 0 1 600 0 1 1 0 1200 1 0 0 0 2400 1 0 0 1 4800 1 0 1 1 9600 1 1 0 0 19200	3 2 1 0 Baud rate 0 1 0 0 300 0 1 0 1 600 0 1 1 0 1200 1 0 0 0 2400 1 0 0 1 4800 1 0 1 1 9600 1 1 0 0 19200

The clock select register allows you to specify the transmitter and receiver baud rates. There are two sets of baud-rate parameters — the set given here is set 2 which is selected when bit 7 of the ACR (auxiliary control register) is set to 1.

Channel A mode select register 1 (MR1A) and channel B mode select register 1 (MR1B)					
Rx RTS control 7	Rx IRQ select 6	Error mode 5	Parity mode 4 3	Parity type 2	Bits-per-character 1 0
0 = Disabled 1 = Enabled	0 = RxRDY 1 = FFULL	0 = Char 1 = Block	0 0 = With parity 1 0 = No parity	With parity 0 = even 1 = odd	1 0 = 7 1 1 = 8

The channel mode select register 1 defines the basic operating parameters of the DUART. Bit 6 determines whether an interrupt is generated by the receiver having at least one character, or the receiver having a full buffer. Bit 5 determines whether the error status refers to the most recent character received or to all characters in the receiver buffer.

Channel A command register (CRA) and channel B command register (CRB)			
Not used 7	Miscellaneous commands 6 5 4	Transmitter commands 3 2	Receiver commands 1 0
	0 0 0 No command 0 0 1 Reset MR pointer to MR1 0 1 0 Reset receiver 0 1 1 Reset transmitter 1 0 0 Reset error status	0 0 No action, stays in present mode 0 1 Transmitter enabled 1 0 Transmitter disabled	0 0 No action, stays in present mode 0 1 Receiver enabled 1 0 Receiver disabled

Channel A status register (SRA) and channel B status register (SRB)							
Received break 7	Framing error 6	Parity error 5	Overrun error 4	TxEMT 3	TxRDY 2	FFULL 1	RxRDY 0
0 = No 1 = Yes	0 = No 1 = Yes	0 = No 1 = Yes	0 = No 1 = Yes	0 = No 1 = Yes	0 = No 1 = Yes	0 = No 1 = Yes	0 = No 1 = Yes

The status register provides status information about the DUART's receiver and transmitter. Bits 3 and 2 define the status of the transmitter buffer. Bit 3 is set if the buffer is empty, and bit 2 is set if there is at least one position free. Bits 1 and 0 define the status of the receiver buffer. Bit 1 is set if the buffer is full and bit 0 is set if there is at least one character in the buffer.

Auxiliary control register (ACR)					
Baud rate select 7	Counter/timer mode and source 6 5 4	Delta IP3 IRQ 3	Delta IP2 IRQ 2	Delta IP1 IRQ 1	Delta IP0 IRQ 0
1	0 0 0 Counter mode, external clock on IP2 0 1 1 Counter mode, internal clock 1 0 0 Timer mode, external clock on IP2 1 1 0 Timer mode, internal clock	0 = Disabled 1 = Enabled			

The auxiliary control register performs miscellaneous functions. Bit 7 determines which of two sets of of baud rates are to be used by the CSR. Bits 6, 5, and 4 select the counter/timer operating mode and the source of the clock. Bits 3 to 0 are so-called delta bits. These permit the appropriate input port bit to generate an interrupt when it changes state.

Channel A mode register 2 (MR2A) and channel B mode register 2 (MR2B)			
Channel mode 7 6	Tx RTS control 5	CTS enable transmitter 4	Stop bit length 3 2 1 0
0 0 = Normal 0 1 = Automatic echo 1 0 = Local loopback 1 1 = Remote loopback	0 = disabled 1 = enabled	0 = disabled 1 = enabled	0 1 1 1 1 bit 1 1 1 1 2 bits

The channel mode register 2 determines the DUART's operating mode (normally bits 7,6 are set to 0,0). Bits 5 and 4 enable/disable the DUART's RTS and CTS functions used to implement flow control. Bits 3 to 0 select the length of the stop bit at the end of each character. Most systems use one bit (0,1,1,1).

ceiver baud rate). For example, the instruction `MOVE.B #$B8,CSRA` selects a receive rate of 9,600 baud and a transmit rate of 2,400 baud.

The channel mode control registers define the operating mode of the DUART (MR1A, MR2A for channel A, and MR1B, MR2B for channel B). Table 9.21 provides a simplified account of these bits. To operate the DUART in its normal, 8-bit character mode, with no parity, one stop bit, and no modem control functions activated, MR1A is loaded with 13_{16} and MR2A with 07_{16}. Remember that these registers share the same address, and that MR2A is selected automatically after MR1A has been loaded. That is:

```
MR1A  EQU     DUART_BASE
MR2A  EQU     MR1A          MR1A and MR2A have same address
      MOVE.B  #$13,MR1A     Load MR1A - no parity, 8 bits
      MOVE.B  #$07,MR2A     Now load MR2A - normal mode, 1 stop bit
```

The *command registers* (CRA and CRB) permit you to enable and disable a channel's receiver or transmitter, and to issue certain commands to the DUART. The command CRA(6:4) = 001 resets the master register pointer to MR1A. You can load CRA with $0A_{16}$ to disable both channels during its setting-up phase, and then load it with 05_{16} to enable its transmitter and receiver ports once its other registers have been set up.

The DUART's *status registers* (SRA and SRB) are very similar to their 6850 counterpart. The DUART's major innovations are SRA7 that detects when a break has been received, SRA3 (TxEMT) that indicates when the transmitter buffer is empty (i.e., there are no characters in the DUART's buffer waiting to be transmitted), and SRA1 (FFULL) that indicates when the receiver buffer is full (there are four received characters waiting to be read). You can, of course, forget about these new bits and operate the DUART exactly like the ACIA just by using the TxRDY and RxRDY bits of its status register.

The difference between the status-bits FFULL and RxRDY is that the FFULL flag is applied to the *whole* receiver buffer, while RxRDY tells us that there is at least one free place in the receiver buffer. Similarly, TxEMT tells us that there is no character in the transmitter buffer and that the buffer is completely empty, while TxRDY tells us that the DUART is ready for another character.

The DUART has sophisticated interrupt control and handling facilities (Figure 9.28). The interrupt vector register, IVR, provides a vector number when the DUART generates an interrupt and receives an IACK response from the 68000. If the IVR hasn't been loaded by the programmer since the last time the DUART was reset, the DUART supplies an uninitialized vector number during an IACK cycle.

The DUART has two interrupt control registers with identical formats: ISR is an interrupt status register whose bits are set when *interrupt generating* activities take place. IMR is an interrupt mask register whose bits are set by the programmer to enable an interrupt, or cleared to mask the interrupt. For example, ISR0 is set if TxRDYA is asserted to indicate that the channel A transmitter is ready for a character. If IMR0 is set to 1, the DUART will generate an interrupt when channel A is ready to transmit a character.

Figure 9.28 The DUART's interrupt control registers

Interrupt vector register IVR

Interrupt vector number bits D7 - D0

Interrupt status register ISR

Input port change	Delta break B	RxRDYB/ FFULLB	Counter/ timer	TxRDYB	Delta break A	RxRDYA/ FFULLA	TxRDYA

Interrupt mask register IMR

Input port change	Delta break B	RxRDYB/ FFULLB	Counter/ timer	TxRDYB	Delta break A	RxRDYA/ FFULLA	TxRDYA

We said earlier that the DUART has multi-purpose I/O pins that can be used as simple I/O pins, or used to perform special functions. Input pins IP0 to IP5 are configured by bits in the CSRA, CSRB, and ACR registers. These pins can be programmed to provide inputs for the DUART's timer/counter, its baud-rate generator, and its clear to send modem control. When MR2A4 = 1, pin IP0 acts as a channel A active-low clear-to-send input, CTS*. Similarly, IP1 can be configured as channel B's CTS* input by setting MR2B4 to 1. If the DUART is programmed to use its clear-to-send input, data is not transmitted by the DUART whenever CTS* is high. That is, the remote receiver can negate CTS* to stop the DUART sending further data. Figure 9.29 demonstrates how CTS* is used in conjunction with RTS*. The following code demonstrates how the DUART is configured to perform flow control.

```
*       Set MR2A4 = 1 and MR2A5 = 1 to configure output OP0 as a RTS output
*
        MOVE.B #$83,MR1A  Set up MR1A - enable Rx RTS, 8-bit chars, no parity
        MOVE.B #$27,MR2A  Automatically select MR2A - enable CTS, 1 stop bit
*
*       Note that RTS* must initially be asserted manually - after that
*       RTS* is asserted automatically whenever the receiver is ready to
*       receive more data. Note also that the contents of the DUART's
*       output port register are inverted before they are fed to the output
*       pins. That is, to assert RTS* low, it is necessary to load a one
*       into the appropriate bit of the OPR.
*
        MOVE.B #$01,OPR   Set output register bit 0 to assert RTS*
```

Figure 9.29 Performing flow control with CTS* and RTS*

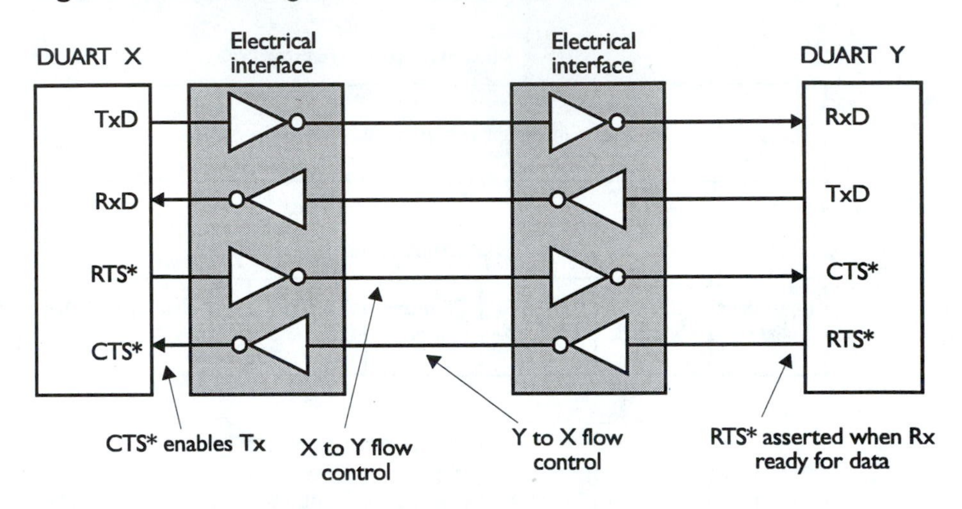

The 8-bit output port is controlled by an output port configuration register, OPCR, and certain bits of the ACR, MR1A, MR2A, MR1B, and MR2B registers. Output bits can be programmed as: simple outputs cleared and set under programmer control, timer and clock outputs, and status outputs. Some of the output functions that can be selected are:

Pin	Function	Action
OP0	RxRTSA*	Asserted if channel A Rx is able to receive a character
OP1	TxRTSA*	Negated if channel A Tx has nothing to transmit
OP2	RxRTSB*	Asserted if channel B Rx is able to receive a character
OP3	TxRTSB*	Negated if channel B Tx has nothing to transmit
OP4	RxRDYA	Asserted if channel A Rx has received a character
OP5	RxRDYB	Asserted if channel B Rx has received a character
OP6	TxRDYA	Asserted if channel A Tx ready for data
OP7	TxRDYB	Asserted if channel B Tx ready for data

Note the difference between the RxRTS* and TxRTS* functions. RxRTS* is used by a *receiver* to indicate to the remote transmitter that it (the receiver) is able to accept data. RxRTS* is connected to the transmitter's CTS* input to perform flow control (Figure 9.29). The TxRTS* function is used to indicate to a modem that the DUART has further data to transmit.

Programming the DUART

Once the DUART has been configured, it can be used to transmit and receive characters exactly like the 6850. The following fragment of code provides basic initialization, receive, and transmit routines for the DUART.

```
*           DUART equates
MR1A        EQU      1            Mode register 1
MR2A        EQU      1            Mode register 2 (same address as MR2A)
SRA         EQU      3            Status register
CSRA        EQU      3            Clock select register
CRA         EQU      5            Command register
RBA         EQU      7            Receiver buffer register (i.e., serial data in)
TBA         EQU      7            Transmitter buffer register (i.e., data out)
IPCR        EQU      9            Input port change register
ACR         EQU      9            Auxiliary control register
ISR         EQU     11            Interrupt status register
IMR         EQU     11            Interrupt mask register
IVR         EQU     25            Interrupt vector register
*
*           Initialize the DUART
*
INITIAL     LEA      DUART,A0     A0 points at DUART base address
*
*           Note the following three instructions are not necessary
*           after a hardware reset to the DUART. They are included to
*           show how the DUART is reset.
*
            MOVE.B  #$30,(CRA,A0)  Reset Port A transmitter
            MOVE.B  #$20,(CRA,A0)  Reset Port A receiver
            MOVE.B  #$10,(CRA,A0)  Reset Port A MR (mode register) pointer
*
*           Select baud rate, data format, and operating modes by
*           setting up the ACR, MR1, and MR2 registers.
*
            MOVE.B  #$80,(CACR,A0) Select baud rate set 2
            MOVE.B  #$BB,(CSRA,A0) Set both Rx and Tx speeds to 9600 baud
            MOVE.B  #$93,(MR1A,A0) Set Port A to 8-bit characters, no parity,
*                                  enable RxRTS output
            MOVE.B  #$37,(MR2A,A0) Select normal operating mode, enable
*                                  TxRTS, TxCTS, one stop bit
            MOVE.B  #$05,(CRA,A0)  Enable Port A transmitter and receiver
            RTS
*
*           Input a single character from Port A (polled mode) into D2
*
GET_CHAR    MOVEM.L D1/A0,-(SP)    Save working registers
            LEA     DUART,A0       A0 points to DUART base address
Input_poll  MOVE.B  (SRA,A0),D1    Read the Port A status register
            BTST    #RxRDY,D1      Test receiver ready status
            BEQ     Input_poll     UNTIL character received
            MOVE.B  (RBA,A0),D2    Read the character received by Port A
```

```
          MOVEM.L (SP)+,D1/A0    Restore working registers
          RTS
*
*         Transmit a single character in D0 from Port A (polled mode)
*
PUT_CHAR  MOVEM.L D1/A0,-(SP)    Save working registers
          LEA     DUART,A0       A0 points to DUART base address
Out_poll  MOVE.B  (SRA,A0),D1    Read Port A status register
          BTST    #TxRDY,D1      Test transmitter ready status
          BEQ     Out_poll       UNTIL transmitter ready
          MOVE.B  D0,(TBA,A0)    Transmit the character from Port A
          MOVEM.L (SP)+,D1/A0    Restore working registers
          RTS
*
```

In spite of the DUART's complexity, you can see that it may be operated in a simple, non-interrupt-driven, character-by-character input/output mode, exactly like the ACIA, once its registers have been set up. On at least one occasion, I have tested software written for a 6850-based system on a board with a DUART by making the following modifications to the 6850's I/O routines.

6850 I/O | **DUART I/O**

```
6850 I/O                         DUART I/O

SETUP  LEA     ACIA,A0           SETUP  LEA     DUART,A0
       MOVE.B  #$03,(A0)                MOVE.B  #$13,(A0)
       MOVE.B  #$15,(A0)                MOVE.B  #$07,(A0)
       RTS                              MOVE.B  #$BB,(2,A0)
                                        MOVE.B  #$05,(4,A0)
                                        RTS

       LEA     ACIA,A0                  LEA     DUART,A0
INPUT  BTST    #0,0(A0)          INPUT  BTST    #0,(2,A0)
       BNE     INPUT                    BEQ     INPUT
       MOVE.B  (2,A0),D0                MOVE.B  (6,A0),D0
       RTS                              RTS

OUTPUT BTST    #1,(0,A0)         OUTPUT BTST    #2,(2,A0)
       BNE     OUTPUT                   BEQ     OUTPUT
       MOVE.B  D0,(2,A0)                MOVE.B  D0,(6,A0)
```

Summary

- The input/output strategies (programmed, interrupt-driven, and DMA) define how we go about performing I/O. Irrespective of the actual I/O strategy, all 68000 systems employ memory-mapped I/O in which I/O ports are located within the processor's memory space.
- In a programmed data transfer, the processor simply reads from or writes to the appropriate memory-mapped I/O port whenever it wishes to perform I/O. Associated with programmed I/O is the *polling loop*, in which an I/O port is continually tested until it is ready to perform a data transfer.
- In interrupt-driven I/O, the peripheral interrupts the processor when it wishes to transfer data. This I/O strategy permits the processor to continue with another task until the port is ready. The I/O port interrupts the processor and the processor jumps to the interrupt handling routine. After the interrupt has been serviced, the processor returns to its pre-interrupt task. The interrupt shares many of the characteristics of the subroutine.
- The 68230 parallel interface/timer has two general-purpose 8-bit ports A and B plus a secondary 8-bit port C. Port C may act as a simple I/O port, or its pins may be devoted to special system functions. The PI/T can also act as a timer and generate pulses of programmed width or measure elapsed time.
- The PI/T has two control lines associated with port A (H1 and H2) and two control lines associated with port B (H3 and H4). Control lines H1 and H3 are inputs, and H2 and H4 may be programmed as inputs or outputs.
- The PI/T's A and B ports can be programmed to operate individually as independent 8-bit ports or together as a single 16-bit port. Furthermore, you can program individual bits of a port as input bits or output bits (i.e., the same port can perform both input and output operations at the bit level).
- Ports A and B offer double-buffered input and output. Double-buffered input means that the processor can read the PI/T's input port while the PI/T is simultaneously reading new data. Similarly, double-buffered output means that the processor can send new data to the PI/T while the PI/T is currently sending data to a peripheral.
- The PI/T's A and B ports can perform handshaking with the peripheral. That is, they can signal the peripheral that they are ready to transfer data, then transfer the data, and finally tell the peripheral that they have completed the transfer. This mechanism automatically performs flowcontrol, and permits the processor to operate at the speed dictated by the peripheral.
- Not only is the operation of the PI/T's control lines programmable, but also their *sense*. That is, you can program the control lines to be active-high or active-low, depending on the requirements of the application.

- The PI/T permits bidirectional I/O in which the direction of data transfer is controlled not by the PI/T but by the external peripheral. The peripheral signals the direction of data transfer by means of the PI/T's control pins.
- The PI/T's zero-detect status bit, ZDS, is set automatically by the PI/T when the down-counter, CNTR, reaches zero. However, the ZDS bit is cleared manually by writing a one into it. That is, you can clear the ZDS bit by `MOVE.B #1,(TSR,A0)`.
- The 6850 ACIA transmits and receives asynchronous serial data. In order to use an ACIA, you have to set up its control register to define the format of the serial data. Its status register indicates whether the ACIA is ready to transmit a character or has received a character. The status register also indicates receiver errors — parity, overrun, and framing.
- The 68681 DUART is a modern serial interface designed for use in 68000 systems. The DUART provides two serial channels and supports vectored interrupts. Unlike the 6850, the DUART includes a programmable, on-chip, baud-rate generator.
- The DUART operates in two modes: block and character. That is, it can operate on a character-by-character basis or on a small block of received/transmitted characters.

Problems

1. The following generic code is used as part of an interrupt-driven output routine to transmit data to a peripheral. This code contains a *semantic* error preventing it working with typical real-world peripherals. What is the error?

```
Again  BTST   #INT,D0          Make sure that the peripheral interrupted
       BEQ    Exit             If it didn't then leave now
       BTST   #RDY,D0          Check to see if the output is actually ready
       BEQ    Exit             If it isn't then it's a false alarm
       MOVE.B (A0)+,OutPort    All is well so send a byte to the output port
Exit   RTE                     Return
```

2. Explain the meaning of the following terms:

    a. Non-buffered input (or output)
    b. Single-buffered input (or output)
    c. Double-buffered input (or output)
    d. Open-loop data transfer
    e. Closed-loop data transfer
    f. Handshaking
    g. FIFO

3. What are the advantages and disadvantages of memory-mapped input/output?

4. Why does the 68230 PI/T have both a port A data register *and* a port A alternate register?

5. What is the effect of loading the following data values into the stated registers of a PI/T?

   a. $00 into PADDR (assume mode 0)
   b. $F0 into PADDR (assume mode 0)
   c. $FA into PADDR (assuming mode 0)
   d. $5A into PADDR (assume mode 3)
   e. $00 into PGCR
   f. $F0 into PGCR
   g. $18 into PSRR
   h. $6E into PACR (assume $1F in PGCR)

6. Define the following errors associated with asynchronous serial transmission systems, and state how each might occur in practice:

   a. Framing error
   b. Receiver overrun error
   c. Parity error

7. The control register of the 6850 ACIA is loaded with the value $B5. Define the operating characteristics of the ACIA.

8. The status register of the 6850 is found to contain $43. How is this value interpreted?

9. Write a procedure that permits the 6850 to operate in an interrupt-driven input mode. Whenever the ACIA receives a character, it generates a level 5 autovectored interrupt. Each new character is to be placed in an input buffer (together with its error status). The 68000 processor can access this buffer at any time to remove received characters (if any). State any assumptions you need to make about this problem.

10. Write a similar procedure to that of question 9 for the 68681 DUART.

11. The DUART has a programmable baud rate generator that is set by loading the appropriate value in the clock select register. This feature makes it possible to adapt to an *unknown* data rate. Write a subroutine that receives a string of carriage returns from a system (at an unknown speed) and adjusts the baud rate to match the incoming data. When the unknown baud rate has been determined, the DUART returns the ASCII string "Ready".

CHAPTER 10

68000 Interrupts and Exceptions

In this chapter we examine two closely related topics: *interrupts* and *exceptions*. You are already familiar with the interrupt, which is a specific instance of the more general exception. As its name suggests, an exception is an event that alters the normal execution of a program. We are going to describe how the 68000 implements exceptions in general and interrupts in particular. Exception handling is the one topic that most intimately combines the software and hardware aspects of a microprocessor.

An interrupt is a message to the CPU from an external device seeking attention. Such devices are frequently I/O peripherals like the serial and parallel interfaces described in Chapter 9. We demonstrate how the 68000 family implements its versatile interrupt handling mechanism. Other exceptions to be described range from bus errors caused when the processor fails to complete a memory access, to software errors caused by an attempt to execute an illegal instruction.

Since any discussion of exceptions covers several distinct but interrelated topics, a short overview of exception handling is provided to set the scene before we look at exceptions in detail.

10.1 Overview of 68000 Family Exception Handling Facilities

It is more difficult to write about the 68000's exception handling facilities than any of the 68000's other attributes. The difficulty stems from the way in which all aspects of exception handling are interrelated — you cannot describe one aspect without referring to the others. Moreover, exceptions bridge three important elements of a microcomputer: the hardware, the software, and the operating system.

Microprocessors belonging to the 68000 family operate in one of two states: *user* or *supervisor*. A 68000 runs its operating system in the supervisor state and its user or applications programs in the user state. One of the main differences between these two states is that each has its own stack pointer (the SSP in the supervisor mode and the USP in the user mode). Consequently, if a user program corrupts its stack, it does not cause the processor to crash, since the operating system's stack pointer is protected from the user. Of course, the actual applications program may crash.

Interrupts and exceptions are really *calls* to the *operating system,* and share many of the characteristics of the subroutine. These calls may be made explicitly by the programmer because he or she wishes to employ an operating system facility (e.g., output), or they may be made automatically by the 68000 in response to certain types of software or hardware error (e.g., an attempt to execute a non-valid instruction code or an attempt to access an address at which no memory exists). Some writers divide exceptions into two classes: *internal exceptions* are those generated by the execution of instructions, and *external exceptions* are those caused by actions taken by hardware outside the 68000 chip.

Each type of exception has its own *exception handler* that is responsible for dealing with the recovery from the event that caused the exception. Exception handlers are normally part of the operating system and are written by the systems programmer (rather than the applications programmer). Unlike the subroutine, an exception does not require an explicit address, since the sequence of actions taking place when an exception is triggered is predetermined by the designers of the microprocessor. Let's explain what we mean by this statement. If you call a subroutine to perform an input operation, you might write `BSR INPUT`, but if you call an exception handler to deal with arithmetic overflow, you might write `TRAPV` (i.e., trap on overflow). As you can see, the `TRAPV` instruction requires no address. The `TRAPV` instruction causes the 68000 to look for the address of the interrupt handler in a special table in memory called an *exception vector table.*

The *systems programmer* must put the address of the `TRAPV` exception handler at the appropriate place in the exception vector table. Similarly, systems programmers have to write the actual exception handlers themselves. We emphasize the term systems programmer because exception handlers are the province of the operating system designer. However, if you are designing a single board computer for use in an embedded system, there might be no distinction between user and supervisor states. Note that the term *exception processing* is used to describe the 68000's response to an exception, and *exception handling* is used to describe what the *user* does in response to an exception.

When a user program is running and it encounters an exception, the 68000 is forced into its supervisor state and the operating system deals with the exception. Of course, a supervisor state program remains in the supervisor state when it encounters an exception.

One important application of exceptions is in implementing input/output operations. For example, the user programmer may force a software exception, called a `TRAP`, when he or she wishes to perform I/O. By ensuring that all input and output transactions are carried out by the operating system, we make it possible to control the way in which users access I/O devices. More importantly, programmers are forced to perform I/O in a consistent fashion and are not able to access devices directly.

As we have already stated, once an exception is accepted by the 68000, it is dealt with by the appropriate exception handler routine. After the exception has been processed, a return from exception instruction, `RTE` (rather like an `RTS` instruction), restores the processor to the state it was in before the exception. As in the case of subroutines, exceptions can be nested to an arbitrary level. However, exceptions are prioritized and an exception with a high priority may not be interrupted by one with a lower priority.

10.2 Interrupts and Exceptions

An interrupt is clearly an *asynchronous event*, because the processor cannot know at which instant a peripheral such as a keyboard will generate an interrupt. In other words, the interrupt-generating activity (i.e., keyboard) bears no particular timing relationship to the activity the computer is carrying out between interrupts. When an interrupt occurs, the computer first decides whether to deal with it (i.e., to service it), or whether to ignore it for the time being. If the computer is doing something that must be completed, it ignores interrupts. The time between the CPU receiving an interrupt request and the time at which it responds is called the *interrupt latency*. Should the computer decide to respond to the interrupt, it must carry out the following sequence of actions:

- Complete its current instruction. All instructions are *indivisible*, which means that they must be executed to completion before the 68000 responds to an interrupt.
- The contents of the program counter must be saved in a safe place, in order to allow the program to continue from the point at which it was interrupted. All members of the 68000 family save the program counter on the stack so that interrupts can, themselves, be interrupted without losing their return addresses.
- The state of the processor (i.e., the 68000's status word) is saved on the stack as well as the PC. Clearly, it would be unwise to allow the interrupt service routine to modify, say, the value of the carry flag, so that an interrupt occurring *before* a BCC instruction would affect the operation of the BCC *after* the interrupt had been serviced. In general, the servicing of an interrupt should have no effect whatsoever on the execution of the interrupted program. This statement applies more strongly to hardware interrupts, rather than to exceptions whose origin lies in the software.
- A jump is then made to the location of the interrupt handling routine, which is executed like any other program. After this routine has been executed, a return from interrupt is made, the program counter restored, and the system status word returned to its pre-interrupt value.

The interrupt is transparent to the interrupted program, since the processor is returned to the state it was in immediately before the interrupt took place. Essentially, an interrupt does not modify the CCR. Before we examine the way in which the 68000 deals with interrupts, it is worthwhile mentioning some of the key concepts used in any discussion of interrupts and exceptions.

Nonmaskable Interrupts

An interrupt request is so-called because it is a request, and therefore carries the implication that it may be denied or *deferred*. Whenever an interrupt request is deferred, it is said to be *masked*. Sometimes it is necessary for the computer to respond to an interrupt no matter what it is doing. Most microprocessors have a special interrupt request input, called a *nonmaskable* interrupt request, NMI. This interrupt cannot be deferred and must always be serviced. A nonmaskable interrupt is normally reserved for events such as a loss of power. In this case, a low-voltage detector generates a nonmaskable interrupt as soon as the power begins to decay. The NMI handler routine forces the processor to deal with the interrupt and to perform an orderly shutdown of the system, before the power drops below a critical level and the computer fails completely. The 68000 has a single, level seven, nonmaskable interrupt request called IRQ7*.

Prioritized Interrupts

An environment in which more than one device is able to issue an interrupt request requires a mechanism to distinguish between an important interrupt and a less important one. For example, if a disk drive controller generates an interrupt because it has some data ready to be read by the processor, the interrupt must be serviced before the data is lost and replaced by new data from the disk drive. On the other hand, an interrupt generated by a keyboard interface probably has from 200 ms to several seconds before it must be serviced. Therefore, a request for attention from a keyboard can be forced to wait if interrupts from devices requiring immediate servicing are pending.

For these reasons, microprocessors often support *prioritized interrupts* (i.e., the chip has more than one interrupt request input). Each interrupt has a predefined priority, and a new interrupt with a priority lower than or equal to the current one cannot interrupt the processor until the current interrupt has been dealt with. Equally, an interrupt with a higher priority can interrupt the current interrupt. The 68000 provides seven levels of interrupt priority (IRQ1* = lowest priority and IRQ7* = highest priority).

Vectored Interrupts

A vectored interrupt is one in which the device requesting the interrupt automatically identifies itself to the processor. Typical 8-bit microprocessors lack a vectored interrupt facility and have only a single interrupt request input, IRQ*, shared by all peripherals. When IRQ* is asserted, the processor recognizes an interrupt but not its source — the processor does not know which device interrupted. The processor's interrupt handling routine must examine, in turn, each of the peripherals that may have initiated the interrupt. The interrupt handling routine interrogates an interrupt status bit in each of the peripherals.

More sophisticated processors have an *interrupt acknowledge* output line, IACK, that is connected to all peripherals. Whenever the CPU has accepted an interrupt and is about to service it, the CPU asserts its interrupt acknowledge output. An interrupt acknowledge from the CPU informs the peripheral that its interrupt is about to be serviced. The peripheral then generates an identification number and puts it on the data bus. The CPU reads this number and uses it to calculate the address of the interrupt handling routine appropriate to the peripheral. This process is called a *vectored interrupt*. The 68000 provides the designer with both vectored and non-vectored interrupt facilities. Now that we have introduced the hardware interrupt, the next step is to examine the more general form of interrupt, the exception.

Exceptions

Exception is a word that has trickled down from the world of the mainframe computer. Like an interrupt, an exception is a deviation from the normal sequence of actions carried out by a computer. Interrupts are asynchronous exceptions, initiated by hardware. Some processors, like the 68000, support other types of exception originating from errors detected by the system hardware. For example, an exception can be generated when the processor tries to read data from memory and something goes wrong, such as an attempt to read from non-existent memory or the detection of a memory error.

As well as exceptions raised by external hardware, there are also exceptions initiated by software. Some are related to errors such as an attempt to execute an illegal operation code. Others are actually generated by the programmer. For example, the TRAP instruction acts like a hardware interrupt in the sense that it invokes an interrupt handling procedure. We will see later that a TRAP (as well as other exceptions) can be used to implement special instructions not normally part of the processor's instruction set. Figure 10.1 illustrates the exceptions implemented by the 68000 family. The characteristics and applications of these exceptions are the subject of this chapter.

The 68000 deals with both hardware and software exceptions in a consistent and logical fashion. The exception handling facilities of the 68000 are one of the main factors lifting it out of the world of the 8-bit processor. We will now describe what happens when the 68000 processes an exception

Brief Overview of 68000 Family Exception Processing

Once the 68000 has accepted an exception, the processor treats it rather like a subroutine by saving the program counter on the stack. All exceptions save the status register as well as the PC and some exceptions save a larger amount of information on the stack (indeed, the actual amount of information saved on the stack also depends on whether the processor is a 68000 or a 68020, etc.). This information is later used to return the processor to its pre-exception state.

Figure 10.1 The 68000 family's exceptions

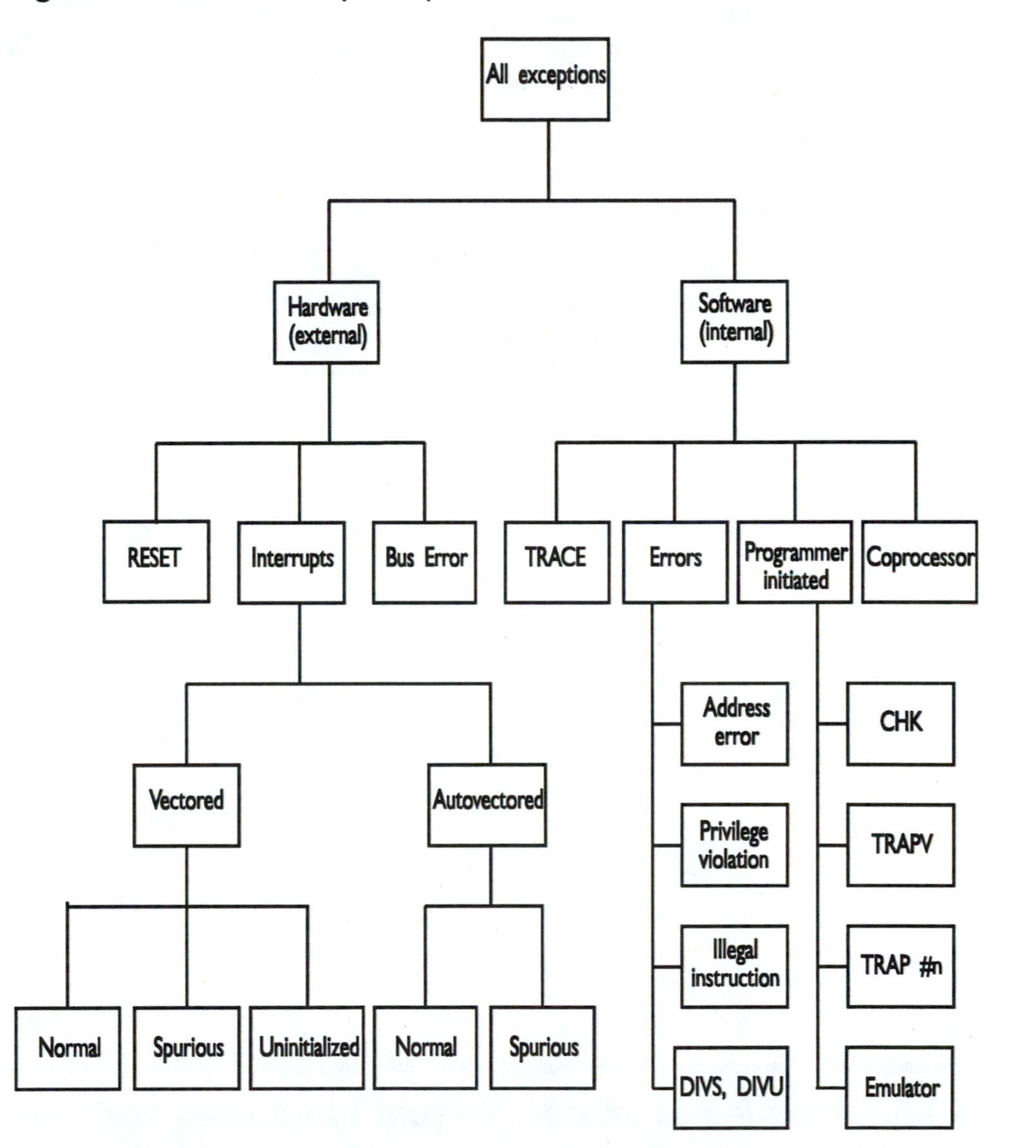

Each exception is associated with a unique longword location in the exception vector table located in memory between $000000 and $0003FF. The processor identifies the type of exception, reads the longword in the vector table, and loads it into the program counter. In other words, the 68000 executes an indirect jump to the exception handling routine via the exception vector table. Once the exception handling routine has been completed, an RTE (return from exception) instruction retrieves the information saved on the stack and the 68000 continues processing normally. Interrupts are treated in a slightly different way. When the 68000 is about to respond to an interrupt, it asks the interrupting device for an identification number and then uses this number to select the desired address in the exception vector table.

The way in which the 68000 processes exceptions can be represented by the following pseudocode. Some of the details are elaborated on later in this chapter.

```
Module Process_exception
  BEGIN
    [TempRegister] ← [SR]        {save status register in temporary location}
    [S] ← 1                      {force supervisor state}
    [T] ← 0                      {turn off trace mode}
    Get VectorNumber             {calculate exception type number}
    Address ← VectorNumber x 4   {exception vector is at 4 x vector number}
    Handler ← [M(Address)]       {read table to get address of exception handler}
    [SSP] ← [SSP] - 4            {pre-decrement stack pointer}
    [M([SSP])] ← [PC]            {push program counter}
    [SSP] ← [SSP] - 2            {pre-decrement stack pointer}
    [M([SSP])] ← [TempRegister]  {push status register}
    [PC]  ←  Handler             {call exception handler}
  END
    .
    .
    .
    ProcessException             {the exception handler processes the exception}
    .                            {the handler is provided by the user}
    .
    .
  BEGIN                          {begin the return from exception sequence - RTE}
    [SR]  ← [M([SSP])]           {restore old SR}
    [SSP] ← [SSP] + 2
    [PC]  ← [M([SSP])]           {restore old PC}
    [SSP] ← [SSP] + 4
  END
```

10.3 Privileged States, Virtual Machines, and the 68000

We are now going to describe the 68000's user and supervisor states and the importance of these states before we look at the fine details of exception handling. The way in which the 68000 processes exceptions is intimately bound up with the notion of privileged states associated with that processor.

At any instant, the 68000 is in one of two states: user or supervisor. By forcing user programs to operate only in the user state and by dedicating the supervisor state to the operating system, it's possible to provide users with a degree of protection against one program corrupting another. The relationship between privileged states and exception processing is quite simple. An exception always forces the 68000 into the supervisor state — of course, if the 68000 was already in the supervisor state prior to the exception, it remains in the supervisor state during exception processing. This means that individual user programs have no direct control over exception processing and interrupt handling.

Privileged States

The supervisor state is the higher state of privilege and is in force whenever the S-bit of the status register is set to one. All the 68000's instructions can be executed while the processor is in the supervisor state. Some of the 68000's instructions are said to be privileged and can be executed only when the 68000 is operating in its privileged supervisor state. The user state is the lower state of privilege, and privileged instructions cannot be executed in this state.

Each of the two states has its own stack pointer, so that the 68000 has two A7 registers. The user-state A7 is called the *user stack pointer*, USP, and the supervisor-state A7 is called the *supervisor stack pointer*, SSP. Note that the SSP cannot be accessed from the user state, whereas the USP can be accessed in the supervisor state by means of the `MOVE USP,An` and `MOVE An,USP` instructions.

Following a hard reset when the 68000's RESET* input is asserted, the S-bit is set and the 68000 begins processing in the supervisor state. All members of the 68000 family enter the supervisor state on initial power-up. The supervisor state register-map of all members of the 68000 family includes at least two registers that cannot be modified from the user state (i.e., the SSP and the status byte of the status register). By the way, no confusion over the two stack pointers normally arises. We do not need to write USP or SSP explicitly for most of the time, as A7 or SP will invariably suffice. There is no confusion because the user program sees only one stack pointer and the supervisor (operating system) also sees only one stack pointer. The user is unaware of the supervisor state, just as the airline passenger is unaware of air traffic control.

Figure 10.2 illustrates the 68000's supervisor state register. The 68000's status register bits are:

S **Supervisor state** When set, S indicates that the 68000 is in its supervisor state. When clear, it indicates that the 68000 is in its user state.

T **Trace bit** When clear, the 68000 operates normally. When T = 1, the 68000 generates a trace exception after the execution of each instruction. We will deal later with the trace exception, which offers a method of debugging programs.

I2,I1,I0 **Interrupt mask bits** The value of I2,I1,I0 (from 0 to 7) indicates the level of the current interrupt mask. An interrupt will not be serviced unless it is of a higher level than that reflected by the interrupt mask. A level seven interrupt is nonmaskable and will be accepted even if I2,I1,I0 = 1,1,1.

Figure 10.2 The 68000's supervisor state register

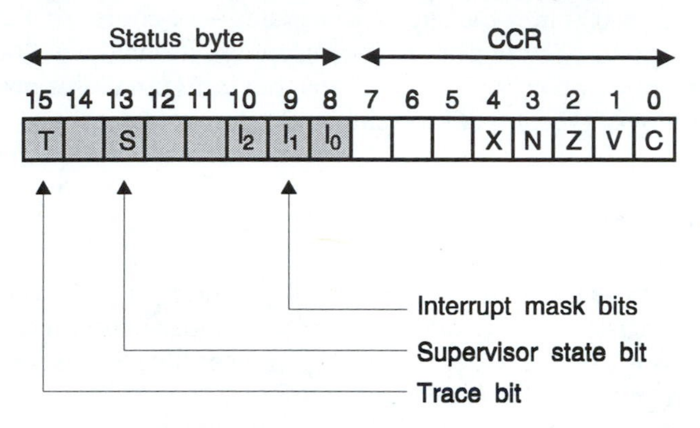

All exception processing is carried out in the supervisor state, because an exception forces a change from user to supervisor state. Indeed, the *only* way of entering the supervisor state from the user state is by means of an exception.

Figure 10.3 shows how a transfer is made between the 68000's two states. An exception causes the S-bit in the 68000's status register to be set, and the supervisor stack pointer to be selected at the start of the exception. Consequently, 68000 saves the return address and pre-exception status register on the *supervisor stack* and not on the user stack.

So, how do we know which state the 68000 is currently operating in? The 68000 has three *function code output pins*, FC2, FC1, FC0, that indicate the type of memory access the 68000 is making. When the processor is in the supervisor state, the function code output from the processor (FC2, FC1, FC0) is 1, 0, 1 if the supervisor is executing a memory access involving data, or 1, 1, 0 if it is accessing an instruction from memory.

It is possible to employ the 68000's function code outputs in address decoding circuits and thereby reserve address space for either supervisor or user applications (or even program or data space). If you wish to use the 68000's function codes in this way, the systems designer has to provide suitable external logic. You can, however, directly make use of function codes in 68030-based systems, because the 68030 has an on-chip memory management unit. The encoding of the 68000's function codes are:

FC2	FC1	FC0	Memory access type
0	0	0	Undefined - reserved
0	0	1	User data
0	1	0	User program
0	1	1	Undefined - reserved
1	0	0	Undefined - reserved
1	0	1	Supervisor data
1	1	0	Supervisor program
1	1	1	IACK space (CPU space)

Figure 10.3 State diagram for user-supervisor state changes

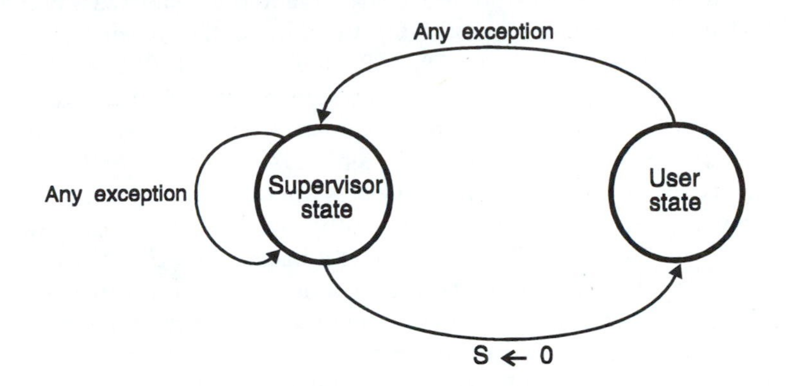

The function code given by FC2, FC1, FC0 = 1, 1, 1 defines an access to CPU space. This space is used by the 68000 for interrupt acknowledge cycles and by the 68020 for accesses to a coprocessor (see Chapter 12).

The change from supervisor to user state is made by clearing the S-bit of the status register, and is carried out by the operating system when it wishes to run a user program. Four instructions are available for this operation: `RTE`, `MOVE.W <ea>,SR`, `ANDI.W #$XXXX,SR`, and `EORI.W #$XXXX,SR`. The `#$XXXX` field represents a 16-bit literal value in hexadecimal form. Note that instructions that modify the status byte of the SR also modify its condition code byte.

The `RTE` (return from exception) instruction terminates an exception handling routine and restores the value of the program counter and the old status register stored on the stack before the current exception was processed. Consequently, if the 68000 was in the user state before the current exception forced it into the supervisor state, an `RTE` restores the processor to its old (i.e., user) state.

A `MOVE.W <ea>,SR` loads the status register with a new value that will force the system into the user state if the S-bit is clear. Similarly, the Boolean operations, `AND` immediate and `EOR` immediate, both may affect the state of the S-bit. For example, we can clear the S-bit by executing an `ANDI.W #$DFFF,SR`.

In the user state, the programmer must not attempt to execute certain privileged instructions. For example, the `STOP` and `RESET` instructions are not available to the user programmer. Why? Because the `RESET` instruction forces the 68000's RESET* output low and resets any peripherals connected to this pin. Suppose a peripheral is being operated by user one. If user two causes RESET* to be pulsed low, user one's program will be adversely affected. Similarly, a `STOP` instruction has the effect of halting the processor until certain conditions are met. Clearly, the user state programmer must not be allowed to bring the entire system to a standstill.

The whole philosophy behind user/supervisor states is to protect the operating system and other user programs from accidental or malicious corruption by any user program. You should not think that being in the user state somehow limits what you can do. The user state simply controls the way in which certain system resources are accessed. A user can always ask the operating system to perform actions that it cannot carry out directly, although the operating system should deny those requests that are destructive to the system as a whole.

Instructions capable of modifying the S-bit in the upper byte of the status register (i.e., RTE, MOVE.W <ea>,SR, ANDI.W #$XXXX,SR, ORI.W #$XXXX,SR, and EORI.W #$XXXX,SR) are *privileged* and are not permitted in the user state. Note that no instruction that performs useful computation is barred from the user state, and therefore the user programmer does not lose any of the 68000's power. Only operations capable of modifying the 68000's operating state are privileged.

Now, suppose the programmer is either wilful or ignorant, and tries to set the S-bit by executing an ORI.W #$2000,SR instruction. If the programmer succeeds in setting the S-bit, he or she will be able to execute privileged instructions. Obviously, there is nothing to prevent the programmer from writing this instruction and running the program containing it. When the program is run, the illegal operation, ORI.W #$2000,SR, causes a privilege violation exception because the instruction may not be executed in the user state. The effect of attempting to execute an ORI.W #$2000,SR is to raise a privilege violation exception, forcing a jump to the routine dealing with this type of exception.

Once the exception handling routine dealing with the privilege violation has been entered, the user no longer controls the processor because the operating system has now taken over. The 68000 is in the supervisor state but it is executing the privilege violation exception handler and not the user program. It is highly probable that the exception handling routine will deal with the privilege violation by terminating the user's program.

10.4 Exceptions and the 68000 Family

We are now going to describe the exception types currently implemented by the 68000. The way in which these exceptions are implemented is described later.

Reset An externally generated reset is caused by forcing the 68000's RESET* pin low for at least ten clock pulses (or for longer than 100 ms on power-up) and is used to place the 68000 in a known state at start-up or following a totally irrecoverable system collapse. A reset loads the program counter and the supervisor stack from memory and sets up the status register (SR is loaded with $2700). The reset is a unique exception, because there is no return from exception following a hard reset. A reset occurs on initial power-up and whenever the "reset" button on the computer is pushed.

Bus error A bus error is an externally generated exception, initiated by hardware driving the 68000's bus error pin, BERR*, active-low. The bus error exception can be used by the systems designer in many different ways, and enables the processor to deal with hardware faults in the system. A typical application of the BERR* input is to indicate either a faulty memory access or an access to a non-existent memory. The bus error is also used in systems with memory management or virtual memory. There are major differences in the way in which the 68000 and later processors implement bus error exception processing.

Interrupt The 68000 implements a conventional, but powerful, hardware interrupt mechanism. A peripheral uses the 68000's three active-low encoded interrupt request inputs, IPL0* to IPL2*, to signal one of seven levels of interrupt. To obtain maximum benefit from the interrupt request inputs, the hardware designer has to employ an external priority encoder chip to convert one of seven interrupt request inputs from peripherals into a three-bit code on IPL0* to IPL2*. The eighth code represents no interrupt request. Fortunately, the assembly language programmer doesn't have to worry about the 68000's interrupt or bus error hardware.

Address error An address error exception occurs when the 68000 attempts to access a 16- or 32-bit longword at an odd address. If you think about it, attempting to read a word at an odd address would require two accesses to memory — one to access the odd byte of an operand, and the other to access the even byte at the next address. Address error exceptions are generated when the programmer does something silly. Consider the fragment of code:

```
LEA     $7000,A0  Load A0 with $00007000
MOVE.B  (A0)+,D0  Load D0 with the byte pointed at by A0 and increment A0 by 1
MOVE.W  (A0)+,D0  Load D0 with the word pointed at by A0 and increment A0 by 2
```

The third instruction results in an address error exception, because the previous operation, `MOVE.B (A0)+,D0`, causes the value in A0 to be incremented by one from $7000 to $7001. Therefore, when the processor attempts to execute `MOVE.W (A0)+,D0`, it is trying to access a word at an odd address.

In many ways, an address error is closer to an exception generated by an event originating in the hardware, than by one originating in the software. The bus cycle leading to the address error is aborted, as the processor cannot complete the operation.

Illegal instruction In the good old days of the 8-bit microprocessor, programmers had much fun finding out what effect *unimplemented* op-codes had on the processor. For example, if the value $A5 did not correspond to a valid op-code, an enthusiast would try to execute it, and then see what happened. This situation was possible because the control unit (i.e., instruction interpreter) of most 8-bit microprocessors was implemented by random logic. Such control units will interpret the bit pattern of any instruction as a sequence of operations (some of which will have no meaningful effect while others might perform a useful operation).

To reduce the number of gates in the control unit of the CPU, some semiconductor manufacturers did not attempt to deal with illegal op-codes. After all (you might erroneously argue), if users try to execute unimplemented op-codes, they deserve everything they get. In keeping with the 68000's approach to programming, an illegal instruction exception is generated whenever an operation code is read that does not correspond to the bit pattern of one of the 68000's legal instructions. A common cause of an illegal instruction exception is a wrongly computed branch, `JMP (Ai)`, that forces a jump to a region of data instead of code.

The 68000 implements a special instruction called `ILLEGAL` that forces an illegal instruction exception. This instruction has been included to allow you to test the illegal instruction exception handler. Some bit patterns that are illegal on the 68000 are legal on the 68020 because the 68020 has several new instructions. The `ILLEGAL` instruction will always be "illegal" on all future members of the 68000 family.

Divide by zero If a number is divided by zero, the result is meaningless and often indicates that something has gone seriously wrong with the program attempting to carry out the division. For this reason, the 68000's designers decided to make any attempt to divide a number by zero an exception-generating event. Good programmers should write their programs so that they never try to divide a number by zero and therefore the divide-by-zero exception should not arise. The divide-by-zero exception is intended as a fail-safe mechanism to avoid the meaningless result that would occur if a number was divided by zero.

CHK instruction The check register against bounds instruction, `CHK`, has the assembly language form `CHK <ea>,Dn`, and compares the contents of the specified data register to the value zero and to the operand at the effective address. If the lower-order word in Dn is less than zero or is greater than the upper bound at the effective address, an exception is generated. For example, when the instruction `CHK D1,D0` is executed, an exception is generated if:

```
[D0(0:15)] < 0 or [D0(0:15)] > [D1(0:15)].
```

Consider the following code fragment that accesses an array pointed at by A0. The array has 26 elements from 0 to 25, and the subscript of the element to be accessed is in D0. Before the element is selected, the subscript in D0 is tested using the `CHK` instruction.

```
MOVE.W  #25,D1          Set up upper bound
.
.
CHK     D1,D0           Check subscript and call o/s if range error
MOVE.B  (A0,D0.W),D7    Now access the array safely
```

Oddly enough, the `CHK` instruction works only with 16-bit signed words, and therefore cannot be used with an address register as an effective address. The `CHK` exception has been included to help compiler writers for languages such as Pascal that have facilities for the automatic checking of array indexes against their bounds.

TRAPV instruction When the trap on overflow instruction, `TRAPV`, is executed, an exception occurs if the overflow bit, V, of the condition code register is set.

Note that an exception caused by dividing a number by zero occurs automatically, while TRAPV is an instruction equivalent to:

```
IF V = 1 THEN exception ELSE continue.
```

The 68020 extends the TRAPV instruction to a general form TRAPcc, where cc is any of the 68020's conditions.

Privilege violation If the processor is in the user state (i.e., the S-bit of the status register is clear) and attempts to execute a privileged instruction, a privilege violation exception occurs. As well as any *logical* operation that attempts to modify the state of the status register (e.g., ANDI #data,SR), the following three instructions cannot be executed in the user state: STOP, RESET, MOVE <ea>,SR.

Trace A popular method of debugging a program is to operate in a *trace mode,* in which the contents of the registers are printed out after each instruction has been executed. The 68000 has a built-in trace facility. If the T-bit of the status register is set, a trace exception is generated after each instruction has been executed. The exception handling routine called by the trace exception can be constructed to display the registers.

Line 1010 emulator Operation codes, whose four most-significant bits (bits 12 to 15) are 1010 or 1111, are unimplemented in the 68000, and therefore represent illegal instructions. However, the 68000 generates a special exception for op-codes whose most-significant nibble is 1010 (also called *line A*, or *line ten*). One of the purposes of this exception is to *emulate* instructions that the 68000 lacks in software. Suppose a version of the 68000 is designed that includes string manipulation operations as well as the normal 68000 instruction set. Clearly, it is impossible to run code intended for the string processor on a normal 68000. But by using 1010 as the four most-significant bits of the new string manipulation instructions, an exception is generated each time the 68000 encounters one of these instructions. The line 1010 exception handler can then be used to allow the 68000 to emulate its more sophisticated brother.

Line 1111 emulator The line 1111 (or line F) emulator behaves in almost exactly the same way as the line 1010 emulator, except that it has a different exception vector number and can therefore call a different exception handling routine. This emulator trap is intended to allow the 68000 to emulate, for example, a floating-point coprocessor. The 68020 uses this exception to communicate with coprocessors.

Uninitialized interrupt vector The 68000 supports vectored interrupts, so that an interrupting device can identify itself and allow the 68000 to execute the appropriate interrupt handling routine without having to poll each device in turn. Before a device can identify itself, the programmer must first configure it by loading its interrupt vector register with the appropriate value. If a 68000 series peripheral is unconfigured and yet generates an interrupt, the 68000 responds by raising an uninitialized interrupt vector exception. 68000 series peripherals are designed to supply the uninitialized interrupt vector number, $0F, during an IACK cycle, if they have not been initialized by software. Their interrupt vector registers are automatically loaded with $0F following a reset operation.

Spurious interrupt If the 68000 receives an interrupt request from external hardware and sends an interrupt acknowledge, but no device responds, the CPU generates a spurious interrupt exception. The spurious interrupt exception prevents the 68000 from hanging up should one of its seven interrupt request lines, IRQ1* to IRQ7*, be asserted and no peripheral respond to the ensuing interrupt acknowledge. To implement the spurious interrupt exception, external hardware is required to assert the 68000's bus error input pin if no peripheral responds to the IACK.

TRAP (software interrupt) The 68000 provides 16 instructions of the form `TRAP #I`, where I = 0, 1,..., 15. These instructions are available to the user programmer. When a `TRAP` instruction is executed, an exception is generated and one of 16 exception-handling routines called. Thus, `TRAP #0` causes `TRAP` exception-handling routine 0 to be called, and so on. The `TRAP #I` instruction is very useful indeed. Suppose we write a program that is to run on all 68000 systems. The greatest problem in designing portable programs is in implementing input or output transactions. One 68000 system may deal with input in a very different way than every other 68000 system. However, if everybody agrees that, for example, `TRAP #0` means input a byte and `TRAP #1` means output a byte, then the software becomes truly portable. All that remains to be done is for an exception handler to be written for each 68000 system to implement the input or output as necessary. The `TRAP` exception is dealt with in more detail later.

Double bus faults A double bus fault is not really an exception in its own right, but a situation in which two exceptions occur in close proximity. Suppose a 68000 system experiences a bus error (or an address error) exception, and the processor begins exception processing by saving the program counter on the stack. Now suppose that a second bus error occurs during the stacking of the PC. The 68000 has nowhere to go. It cannot continue normally because of the original exception, and it cannot enter exception processing because of the second exception. In this case, a double bus fault is said to occur and the 68000 halts. Further instruction execution is stopped and its HALT* pin is asserted active-low. Only a hard reset will restart the 68000 following a double bus fault.

Exception Vectors

Having described the various types of exception supported by the 68000, the next step is to explain how the processor is able to determine the location of the corresponding exception-handling routine. Each exception is associated with a vector, which is the 32-bit absolute address of the appropriate exception handling routine. All the 68000's exception vectors are stored in a table of 256 longwords (i.e., 1024 bytes), extending from address $00 0000 to $00 03FF.

A list of all the exception vectors is given in Table 10.1, and Figure 10.4 shows the physical location of the 256 vectors in memory. The left-hand column of

Table 10.1 The 68000's exception vector table

Vector number	Vector (hex)	Address space	Exception type
0	000	SP	Reset - initial supervisor stack pointer
-	004	SP	Reset - initial program counter value
2	008	SD	Bus error
3	00C	SD	Address error
4	010	SD	Illegal instruction
5	014	SD	Divide by zero
6	018	SD	CHK instruction
7	01C	SD	TRAPV instruction
8	020	SD	Privilege violation
9	024	SD	Trace
10	028	SD	Line 1010 emulator
11	02C	SD	Line 1111 emulator
12	030	SD	(Unassigned - reserved)
13	034	SD	(Unassigned - reserved)
14	038	SD	(Unassigned - reserved)
15	03C	SD	Uninitialized interrupt vector
16	040	SD	(Unassigned - reserved)
.	.	.	.
23	05C	SD	(Unassigned - reserved)
24	060	SD	Spurious interrupt
25	064	SD	Level 1 interrupt autovector
26	068	SD	Level 2 interrupt autovector
27	06C	SD	Level 3 interrupt autovector
28	070	SD	Level 4 interrupt autovector
29	074	SD	Level 5 interrupt autovector
30	078	SD	Level 6 interrupt autovector
31	07C	SD	Level 7 interrupt autovector
32	080	SD	TRAP #0 vector
33	084	SD	TRAP #1 vector
.	.	.	.
47	0BC	SD	TRAP #15 vector
48	0C0	SD	(Unassigned - reserved)
.	.	.	.
63	0FC	SD	(Unassigned - reserved)
64	100	SD	User interrupt vector
.	.	.	.
255	3FC	SD	User interrupt vector

Figure 10.4 Memory map of the 68000's vector table

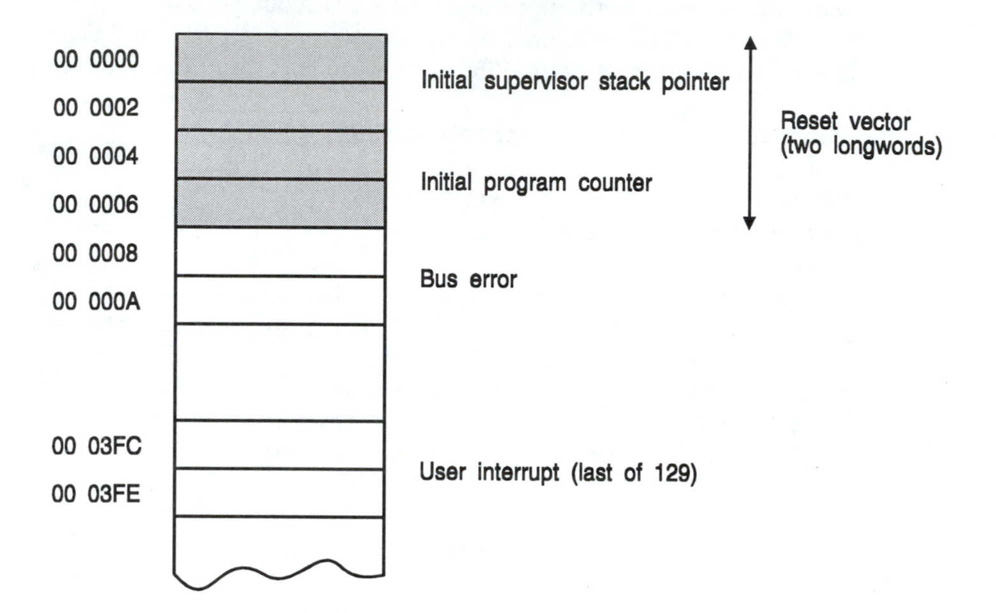

Table 10.1 gives the vector number of each entry in the table. The vector number is a value that, when multiplied by four, gives the address, or offset, of an exception vector. For example, the vector number corresponding to a privilege violation is 8, and the appropriate exception vector is found at memory location 8 x 4 = 32 = $20. Therefore, whenever a privilege violation occurs, the CPU reads the longword at location $20 and loads it into its program counter.

Certain vectors, numbers 12 to 14, 16 to 23, and 48 to 63, have been reserved for possible future enhancements of the 68000. Some of these have been assigned to the 68020 and 68030 processors.

Although we said that two words of memory space are devoted to each 32-bit exception vector, the reset exception (vector number zero) is a special case. The 32-bit longword at address $00 0000 is not the address of the reset-handling routine, but the initial value of the supervisor stack pointer. The actual reset exception vector is at address $00 0004. Thus, the reset exception vector requires four words of memory in the exception vector table instead of the usual two. The 68000's designers have been very clever here. The first operation performed by the 68000 following a reset is to load the system stack pointer. Loading the supervisor stack pointer is important, because until a stack is defined the 68000 cannot deal with any other type of exception. Once the stack pointer has been set up, the reset exception vector is loaded into the program counter and processing continues normally. The reset exception vector is, of course, the initial (or cold-start) entry point into the operating system.

Yet another difference exists between the reset vector and all other exception vectors. Both the reset exception vector and supervisor stack pointer initial value lie in *supervisor program space*, denoted by SP in Table 10.1. Thus, when the 68000 accesses these vectors, it puts out the value 1, 1, 0 on function code pins FC2, FC1, FC0, respectively. All other exception vectors lie in *supervisor data space*, SD, and the function code 1, 0, 1 is put out on FC2, FC1, FC0, when one of these is accessed. The information on FC2 to FC0 is useful to the systems designer.

Exception Processing

We're now going to look at what happens when the 68000 responds to an exception in greater detail. Since the 68020 handles exceptions in a somewhat different way than the 68000, we will not cover the 68020 until Chapter 12.

The 68000 responds to an exception in four phases. In phase one, the processor makes a temporary internal copy of its status register and modifies the current status register ready for exception processing. This process involves setting the S-bit and clearing the T-bit (i.e., Trace-bit). The S-bit is set because all exception processing takes place in the supervisor state. The T-bit is cleared because the trace mode must be disabled during exception processing. Remember that the trace mode forces an exception after the execution of each instruction. If the T-bit were set, an instruction would trigger a trace exception. This would, in turn, cause a trace exception after the first instruction of the trace-handling routine had been executed. In this way, an infinite series of exceptions would be generated.

Two specific types of exception have a further effect on the contents of the status byte of the SR. After a reset, the interrupt mask bits are automatically set to 1,1,1 to indicate an interrupt priority of level-7. That is, all interrupts below level seven are initially disabled. The status byte is also modified by an *interrupt* which causes the interrupt mask bits to be set to the same level as the interrupt currently being processed. The CPU responds only to interrupts with a priority greater than that reflected by the interrupt mask bits.

In phase two, the vector number corresponding to the exception being processed is determined. Apart from interrupts, the vector number is generated internally by the 68000 according to the exception type. If the exception is an interrupt, the interrupting device places the vector number on data lines D_{00} to D_{07} of the processor data bus during the interrupt acknowledge cycle, signified by a function code (FC2, FC1, FC0) of 1,1,1. Under certain circumstances, described when we deal with interrupts, an external interrupt can generate a vector number internally in the 68000, in which case the interrupting device does not supply a vector number. Once the processor has determined the vector number, it multiplies that number by four to calculate the location of the exception-processing routine within the exception vector table.

In phase three, the current CPU context is saved on the stack pointed at by the supervisor stack pointer, A7. The CPU context is the information required by the CPU to return to normal processing after an exception. A reset does not, of course, cause anything to be saved on the stack, as the state of the system is undefined prior to a reset. Phase three of the exception processing is complicated by the fact that the 68000 divides exceptions into two categories, and saves different amounts of information according to the nature of the exception. The information saved by the 68000 is called *the most volatile portion of the current processor context*, and is saved in a data structure called an exception stack frame.

Figures 10.5 and 10.6 show the structure of the 68000's two exception stack frames. The 68000's exceptions are classified into three groups. We will return to this point shortly. The information saved during Group 1 or Group 2 exceptions, Figure 10.5, is only the program counter (two words) and the system status register, temporarily saved during phase one. The PC and the SR are the minimum information required by the processor to restore itself to the state it was in prior to the exception.

Figure 10.5 Stack frame for Group 1 and Group 2 exceptions

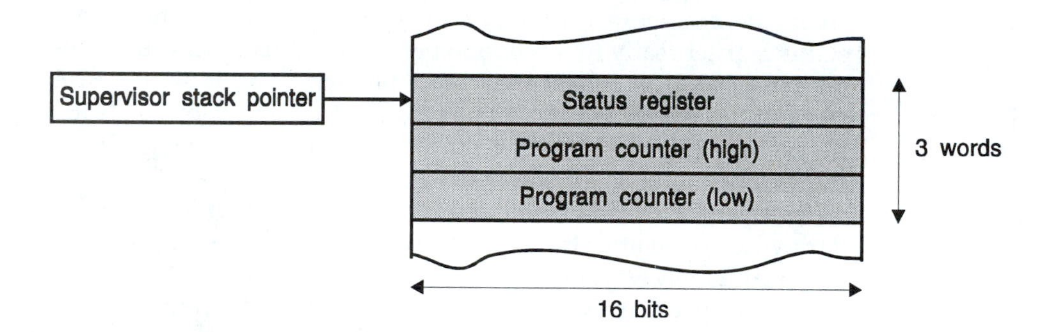

The 68000's exceptions are divided into three groups according to their priority and characteristics, and are categorized in Table 10.2. Basically, a Group 0 exception originates from hardware errors and often indicates that something has gone seriously wrong with the system. Because of this, the information saved in the stack-frame corresponding to a Group 0 exception is more detailed than that for

Table 10.2 68000 exception grouping according to type and priority

Group	Exception	Time at which processing begins
0	Reset Bus error Address error	Exception processing begins within two clock cycles
1	Trace Interrupt Illegal op-code Privilege	Exception processing begins before the next instruction
2	TRAP TRAPV CHK Divide by zero	Exception processing is started by normal instruction execution

Figure 10.6 The 68000's stack frame for Group 0 exceptions

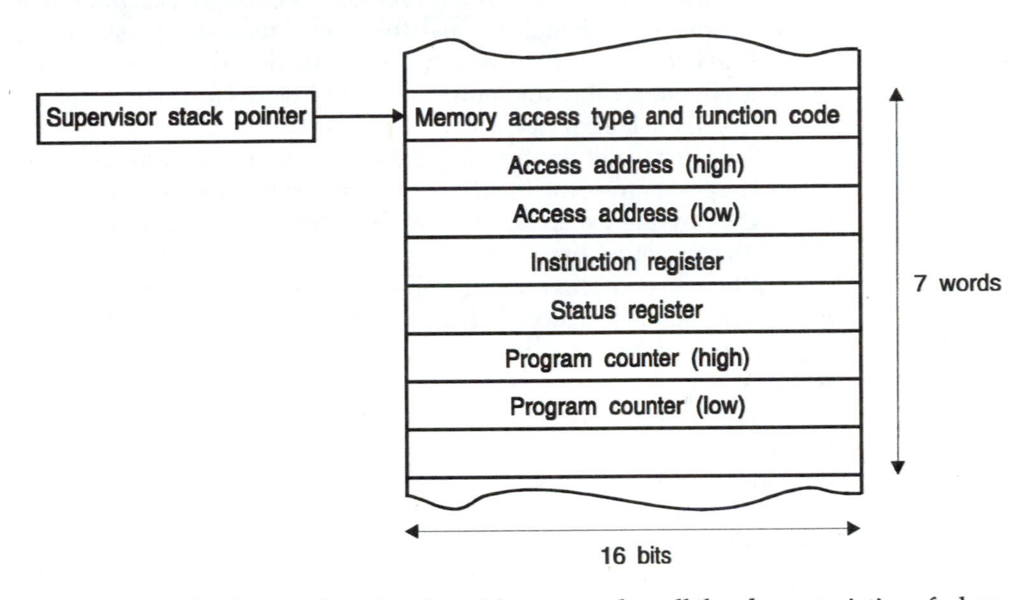

Groups 1 and 2. Remember that the address error has all the characteristics of a bus error but is generated internally by the 68000. Figure 10.6 shows the stack frame for Group 0 exceptions (except for a reset).

Additional items saved in the stack frame by a Group 0 exception are a copy of the first word of the instruction being processed at the time of the exception, and the 32-bit address that was being accessed by the aborted memory access cycle. The third new item saved at the top of the stack is a five-bit code in bits 4:0 of the top word on the exception frame. This gives the function code displayed on FC2, FC1, FC0 at the time the exception occurred, together with an indication of whether the processor was executing a read or a write cycle (Bit 4: 1 = read, 0 = write), and whether it was processing an instruction (I/N bit 3: 0 = instruction, 1= data). For example, if the top of the stack is $00012, corresponding to bits 4:0 = 1 0 010, the faulted bus cycle is interpreted as: a read cycle (bit 4 = 1), an instruction processing cycle (bit 3 = 0), and a user program access (bits 2:0 = 010).

The information saved on a Group 0 exception stack frame is largely intended for diagnostic purposes and may be used by the operating system when dealing with the cause of the exception. In Chapter 12 we shall see that this statement does not apply to the 68020 or later models.

By the way, the value of the program counter saved on a Group 0 stack frame is the address of the first word of the instruction that leads to the bus fault plus a value between two and ten. That is, the program counter value is *indeterminate* and does not point at the next instruction following the exception (as it does in the case of Group 1 and 2 exceptions). This uncertainty arises because a bus error can happen at any point during the execution of a long instruction, and the 68000 does not store enough internal information to deal correctly with a bus error. For example, a `MOVE.L $1234,$3334` instruction might generate a bus error during the instruction fetch, the operand fetch, or the operand store phases. Although the 68000 cannot itself return from a Group 0 exception, you can write a bus error exception handler to use the information on the stack to create a new stack frame from which a return can be made. This procedure is not recommended. If you want to implement a return from a Group 0 exception, you should use a 68020 or a later processor.

The fourth, and final, phase of the exception processing sequence consists of a single operation — the loading of the program counter with the 32-bit address pointed at by the exception vector. Once this has been done, the processor executes the exception-handling routine.

When an exception-handling routine has been run to completion, the return from exception instruction, RTE, is executed to restore the processor to the state it was in prior to the exception. RTE is a privileged instruction, and has the effect of restoring the status register and program counter from the values saved on the system stack. The contents of the program counter and status register, just prior to the execution of the RTE, are lost.

It is important to stress that an RTE instruction cannot be used after a Group 0 exception to execute a return. There are two reasons why you cannot use an RTE instruction following a bus error exception. The first is that the RTE pulls the program counter and status register off the stack. Since the Group 0 stack frame has a different structure than Group 1 and 2 frames, an RTE would just not work. The programmer could always try to modify the Group 0 stack frame and make it look like a Group 1 stack frame before executing the RTE. The second reason is the one we mentioned earlier, the value of the program counter saved on the stack frame after a bus error is not reliable. Some programmers have managed to return from a bus error by means of a clever trick. Since the Group 0 stack frame contains the value of the instruction at the time of the bus error, you can read the program counter from the stack frame, and then search for the instruction that caused the bus error.

Figure 10.7 graphically summarizes the way in which the 68000 processes exceptions.

10.5 Hardware Initiated Exceptions

Three exceptions are initiated by events taking place outside the 68000 and are communicated to the CPU via its input pins: the *reset*, the *bus error*, and the *interrupt*. Each of these exceptions has a direct effect on the hardware design of a 68000-based microcomputer. We now examine these exceptions in more detail.

Reset

A reset is a rather special exception, because it takes place only under two circumstances: a power-up, or a total and irrecoverable system collapse. For this

Figure 10.7 The 68000's exception processing sequence

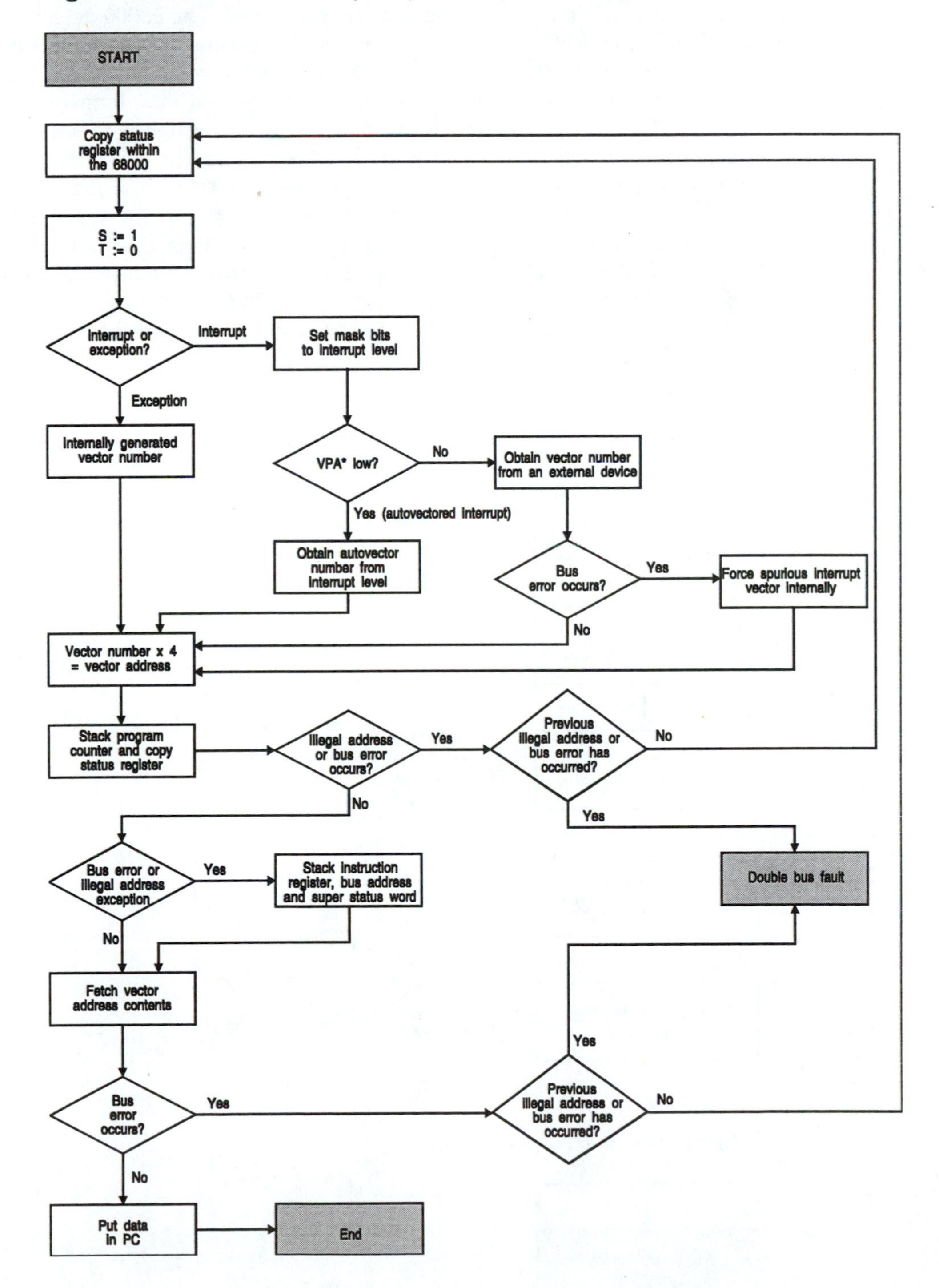

reason, the reset has the highest priority and will be processed before any other exception that is either pending or being processed. The 68000 detects a reset by a low level at its RESET* pin. Following the detection of a reset, the 68000 sets the S-bit, clears the T-bit, and sets the interrupt mask level to seven (i.e., SR = $2700). The 68000 then loads the supervisor stack pointer with the longword at memory location $00 0000, and then loads the program counter with the longword at memory location $00 0004. Once this has been done, the 68000 begins to execute the start-up routine. Figure 10.8 illustrates the 68000's reset sequence.

Although the 68000's supervisor stack pointer is set up during the reset, the user stack pointer is not. Systems designers must take care not to switch from supervisor state to user state and then forget to preset the user stack pointer. The privileged instruction `MOVE An,USP` can be employed to set up the USP while still in the supervisor state. Equally, the user stack pointer can be set up in the user state. The following code demonstrates how you might invoke a user task from the supervisor mode.

Figure 10.8 The 68000's reset sequence

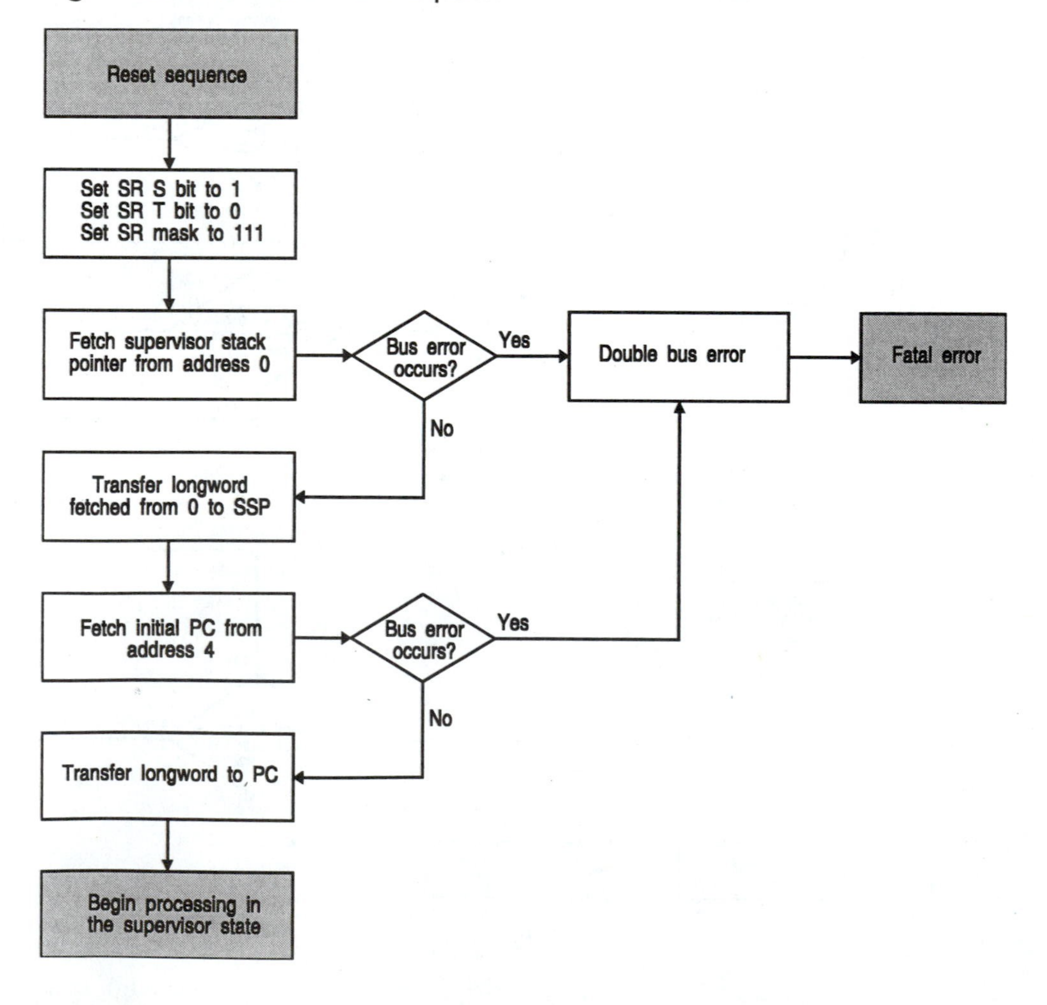

```
MOVEA.L #USER_Stack,A0      Load initial user stack into A0
MOVE.L  A0,USP              Set up the user stack pointer
MOVE.L  User_Start,-(A7)    Push start address of user code
MOVE.W  #$0700,-(A7)        Push initial status - S-bit clear
RTE                         Return by loading new PC and SR
```

Bus Error

A bus error is an exception raised in response to a *failure* by the system to complete a bus cycle. There are many possible failure modes and the details of each mode depend on the hardware used to implement the 68000 system. Therefore, the detection of a bus error has been left to the systems designer, rather than to the 68000 chip itself. All the 68000 provides is an active-low input, BERR*, that, when asserted, initiates a bus-error exception. Typical applications of BERR* are as follows:

- **Illegal memory access** If the processor tries to access memory at an address not populated by memory, the BERR* pin may be asserted. Equally, BERR* may be asserted if an attempt is made to write to a memory address that is read-only. The decision whether to assert BERR* in these cases is a design decision. It is not mandatory. All 8-bit microprocessors are quite happy to access non-existent memory or to write to ROM. The philosophy of 68000 systems design is to trap events that may lead to unforeseen circumstances. If the processor tries to write to ROM, the operating system could intervene because of the exception raised by BERR*.
- **Faulty memory access** If error-detecting memory is employed, a read access to a memory location at which an error is detected can be used to assert BERR*. In this way the processor will never try to process data that is in error due to a fault in the memory.
- **Failure to assert VPA*** The 68000 uses a special hardware interface to communicate with 8-bit peripherals intended for use with the 8-bit 6800 microprocessor. These peripherals use one of the 68000's pins, VPA*, valid peripheral address, to indicate the completion of an access. If the processor accesses a 6800-series peripheral and VPA* is not asserted after some timeout period, BERR* may be asserted to stop the system from hanging up and waiting for VPA* forever.
- **Memory privilege violation** When the 68000 is used in a system with some form of memory management, BERR* may be asserted to indicate that the current memory access is violating a privilege. A privilege violation may be caused by an access by one user to another user's program space, or by a user to supervisor space. In a system with virtual memory, a memory privilege violation may result from a page-fault, indicating that the data being accessed is not currently in read/write memory.

Bus Error Sequence

When the 68000 detects a bus error, the processor begins a normal exception processing sequence for a Group 0 exception. Figure 10.6 shows the information is pushed on the system stack by the 68000 to facilitate recovery from the bus error. Once all phases of the exception-processing sequence have been completed, the 68000 begins to deal with the problem of the bus error in the BERR* exception handling routine.

It must be emphasized that the treatment of the hardware problem that led to the bus error takes place at a software level within the operating system. For example, if a user program generates a bus error, the exception handling routine may abort the user's program and provide him or her with diagnostic information to help deal with the problem that caused the exception. The information stored on the stack by a bus error exception (or an address error) is to be regarded as diagnostic information only, and should not be used to institute a return from exception, as we have already stated.

Interrupts

As we have seen, an interrupt is a request for service generated by a peripheral. In keeping with the 68000's general versatility, it offers two schemes for dealing with interrupts. One is intended for modern peripherals designed for 16-bit processors, while the other is more suited to earlier 8-bit 6800-series peripherals.

An external device signals its need for attention by placing a 3-bit code on the 68000's interrupt request inputs, IPL0*, IPL1*, IPL2*. The code corresponds to the priority of the interrupt and is numbered 0 to 7. A level-7 code indicates the highest priority, level-1 the lowest priority, and level-0 indicates the default state of no interrupt request. While it would be perfectly possible to design peripherals with three interrupt request output lines on which they put a 3-bit interrupt priority code, it is easier to have a single interrupt request output and to design external hardware to convert its priority into a suitable 3-bit code for the 68000. Figure 10.9 illustrates the 68000's hardware interrupt interface.

Processing the Interrupt

All interrupt inputs to the 68000 are latched and made pending. Group 0 exceptions (reset, bus error, address error) take precedence over an interrupt in Group 1. Therefore, if a Group 0 exception occurs, it is serviced before the interrupt. A trace exception in Group 1 takes precedence over the interrupt. Consequently, if an interrupt request occurs during the execution of an instruction while the T-bit is asserted, the trace exception has priority and is serviced first. Assuming that none of the above exceptions has been raised, the 68000 compares the level of the interrupt request with the value recorded in the interrupt mask bits of the processor status word.

Figure 10.9 The 68000's interrupt interface

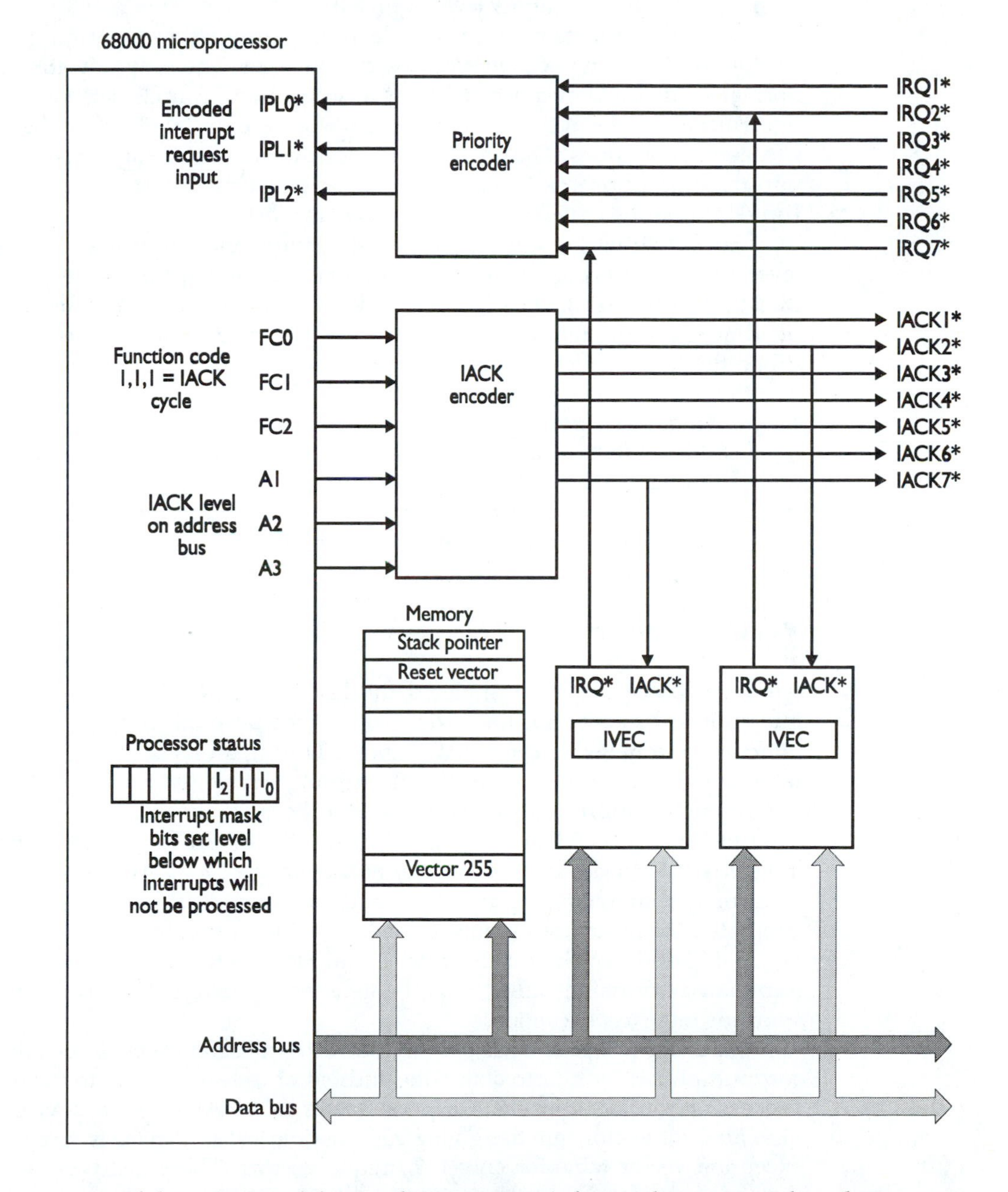

If the priority of the pending interrupt is lower than, or equal to, the current processor priority denoted by the interrupt mask, the interrupt request remains pending and the next instruction in sequence is executed. Interrupt level-7 is treated slightly differently, as it is always processed regardless of the value of the interrupt mask bits. In other words, a level-7 interrupt always interrupts a level-7 interrupt, if one is currently being processed. Any other level of interrupt can be interrupted only by a higher level of priority.

Once the processor has made a decision to process an interrupt, it begins an exception processing sequence as described earlier. The only deviation from the normal sequence of events dictated by a Group 1 or Group 2 exception is that the interrupt mask bits of the processor status word are updated before the exception processing continues. The level of the interrupt request being serviced is copied into the current processor status. This means that the interrupt cannot be interrupted unless the new interrupt has a higher priority.

Suppose that the current (i.e., pre-interrupt) interrupt mask is level-3. If a level-5 interrupt occurs, it is processed, and the interrupt mask set to level-5. If, during the processing of this interrupt, a level-4 interrupt is requested, it is made pending, even though it has a higher priority than the original interrupt mask. When the level-5 interrupt has been processed, a return from exception is made and the former processor status word restored. As the old interrupt mask was 3, the pending interrupt of level-4 is then serviced.

Unlike other exceptions, an interrupt may obtain its vector number externally from the device that made the interrupt request. As stated earlier, there are two ways of identifying the source of the interrupt, one *vectored* and one *autovectored*. A vectored interrupt is dealt with first.

Vectored Interrupt

After the processor has completed the last instruction before recognizing the interrupt and stacked the low-order word of the program counter, it executes an interrupt acknowledge cycle (IACK cycle). During an IACK cycle, the 68000 obtains the vector number from the interrupting device, which it will later use to determine the appropriate exception vector. Because the 68000 puts out the special function code 1,1,1 on FC2, FC1, FC0, during an IACK cycle, the interrupting device is able to detect the interrupt acknowledge cycle. At the same time, the level of the interrupt is put out on address lines A_{01} to A_{03}. The device that generated the interrupt at the specified level then provides a vector number on D_{00} to D_{07} and terminates the read cycle. However, if the IACK cycle is not terminated normally, BERR* must be asserted by external hardware to force a spurious interrupt exception.

After the peripheral has provided a vector number on D_{00} to D_{07}, the processor multiplies it by four to obtain the address of the entry point to the exception-processing routine from the exception vector table. Although a device can provide an 8-bit vector number giving 256 possible values, space is reserved in the exception vector table for only 192 unique vectors. These 192 vectors are more than adequate for the vast majority of applications. But note that a peripheral can put out vector numbers 0 to 63 during an IACK cycle, as there is nothing to stop these numbers being programmed into the peripheral and the processor does not guard against this situation. In other words, if a peripheral is programmed to respond to an IACK cycle with, for example, a vector number 5, then an interrupt from this device would cause an exception corresponding to vector number 5 —

the value also appropriate to a divide-by-zero exception. While I can see times that this might be useful, it seems an oversight to allow interrupt vector numbers to overlap with other types of exceptions.

A possible arrangement of the hardware required to implement a vectored interrupt scheme is given in Figure 10.10. A peripheral indicates that it needs attention by asserting its level-5 interrupt request output, IRQ5*. This interrupt request is encoded by IC3 to provide the 68000 with a level-5 interrupt request on its IPL0* to IPL2* input pins. When the processor acknowledges this request, it places 1,1,1 on its function code output, which is decoded by the 3-line to 8-line decoder IC1. The interrupt acknowledge output (IACK*) from IC1 enables a second 3-line to 8-line decoder, IC2, that decodes address lines A_{01} to A_{03} into seven levels of interrupt acknowledge. In this case, IACK5* from IC2 is fed back to the peripheral, which then responds by placing its vector number, IVEC, onto

Figure 10.10 Implementing the vectored interrupt

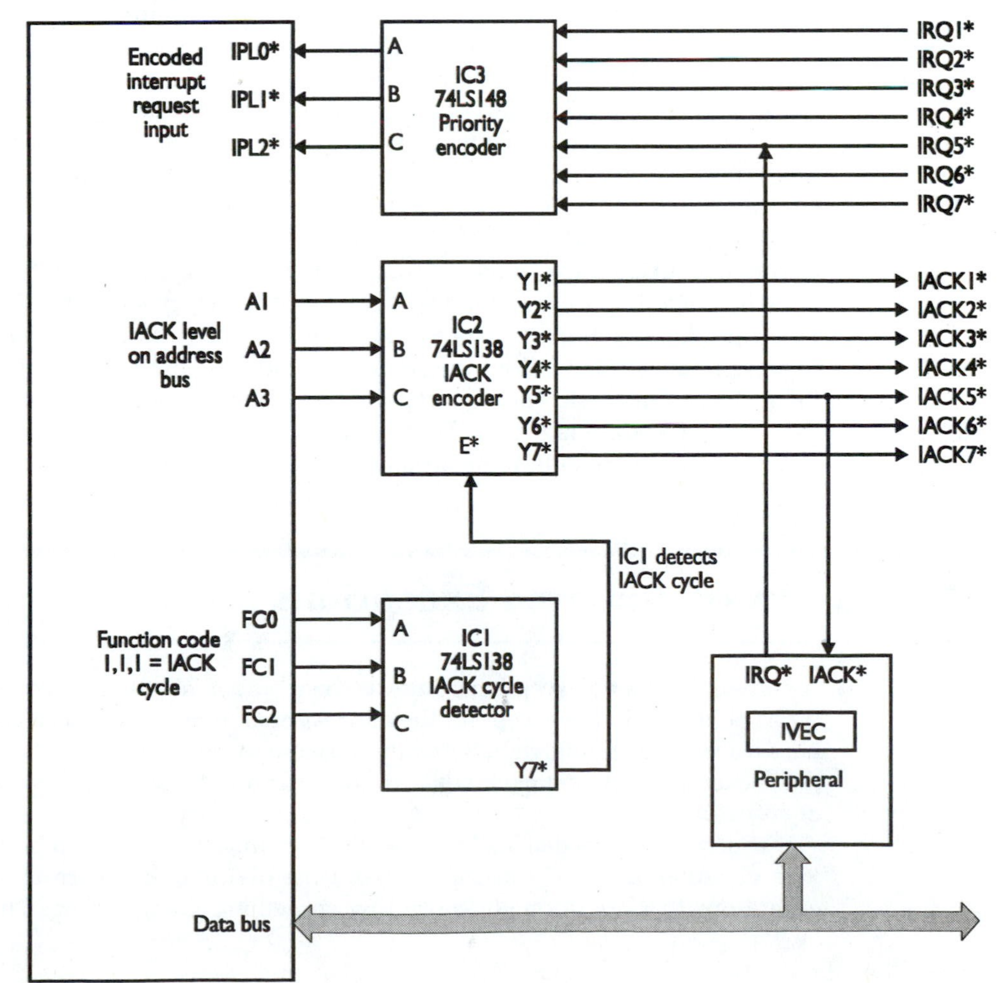

the low-order byte of the system data bus. If the peripheral has not been programmed to supply an interrupt vector number, it should place $0F on the data bus, corresponding to an uninitialized interrupt vector exception.

Autovectored Interrupt

As we have just seen, a device that generates an interrupt request must be capable of identifying itself when the 68000 carries out an interrupt acknowledge sequence. This requirement presents no problem for modern 68000-based peripherals such as the 68230 PI/T.

Unfortunately, older peripherals originally designed for 8-bit processors do not have interrupt acknowledge facilities, and are unable to respond with the appropriate vector number on D_{00} to D_{07} during an IACK cycle. The systems designer could overcome this problem by designing a subsystem that supplied the appropriate vector as if it came from the interrupting peripheral. Such an approach is valid but a little messy. Who wants a single-chip peripheral that needs a handful of components just to provide a vector number in an IACK cycle?

An alternative scheme is available for peripherals that cannot provide their own vector number. If a peripheral asserts the 68000's valid peripheral address (VPA*) input, the 68000 carries out an autovectored interrupt. When asserted, VPA* informs the 68000 that the appropriate vector number should be generated internally. The 68000 reserves vector numbers 25 to 31 (decimal) for its autovector operation (see Table 10.1). Each of these autovectors is associated with an interrupt on IRQ1* to IRQ7*. For example, if IRQ2* is asserted followed by VPA* during the IACK cycle, vector number 26 is generated by the 68000 and the interrupt-handling routine address is read from memory location $000068.

Should several autovectored interrupt requesters assert the same interrupt request line simultaneously, the 68000 will not be able to distinguish between them. The appropriate autovectored interrupt-handling routine must poll each of the possible requesters in turn. That is, the interrupt status register of each peripheral must be read to determine the source of the interrupt.

10.6 Software Initiated Exceptions

A software initiated exception occurs as the result of an attempt to execute certain types of instruction — except for the address error that can really be classified as a hardware initiated interrupt. Software initiated interrupts fall into two categories: those executed deliberately by the programmer and those caused by certain types of software error.

Software errors that lead to exceptions include: the illegal op-code; the privilege violation; the `TRAPV` instruction; and the divide-by-zero error. These are all exceptions that are normally caused by something going wrong. Therefore, the operating system needs to intervene and sort things out. The nature of this intervention is very much dependent on the structure of the operating system. Often, in a multiprogramming environment, the individual task that raised the exception will be aborted, leaving all other tasks unaffected.

Consider the illegal op-code exception. This exception occurs when the 68000 attempts to execute an operation code that does not form part of its instruction set. The only way that this can happen unintentionally is when something has gone seriously wrong. For example, an op-code might have been corrupted in memory by a memory error, or a jump made to a region containing a non-valid 68000 code. The latter event frequently results from wrongly computed `GOTO`s.

Clearly, once an illegal op-code exception has occurred, it is futile to continue trying to execute further instructions, as they have no real meaning. By generating an illegal op-code exception, the operating system can inform users of the problem and invite them to do something about it.

Software exceptions deliberately initiated by the programmer are the trace, the trap, and the emulator. We will deal with each of these in turn.

Trace Exceptions

The trace exception mode is in force whenever the T-bit of the status word is set. After each instruction has been executed, a trace exception is automatically generated, permitting the user to monitor the execution of a program. We have to set the T-bit in the user state in order to trace a user program. However, since it is impossible to modify the status register while the 68000 is operating in the user state, we have a bit of a problem. The solution is to call a supervisor state function (e.g., by means of a `TRAP` instruction) that sets the T-bit of the status register on the supervisor stack before executing an `RTE`. All the `TRAP` handler has to do is:

```
ORI.W #$8000,(SP)    Set trace bit of SR on stack
RTE                  Return from exception
```

The simplest trace facility would allow the user to dump the contents of all registers on the CRT terminal after the execution of each instruction. Unfortunately, such a simple use of tracing leads to the production of vast amounts of utterly useless information. For example, if the 68000 were executing an operation to clear an array by executing a `CLR.L (A4)+` instruction 64K times, the human operator would not wish to see the contents of all registers displayed after each `CLR.L (A4)+` had been executed.

A better approach is to display only the information needed. Before the trace mode is invoked, the user informs the operating system of the conditions under which the results of a trace exception are to be displayed. Some of the events that can be used to trigger the display of registers during a trace exception are:

- The execution of a predefined number of instructions — the contents of registers may be displayed after, say, 50 instructions have been executed.

- The execution of an instruction at a given address. This is equivalent to a *breakpoint*.
- The execution of an instruction falling within a given range of addresses, or the access of an operand falling within the same range.
- As above, but the contents of the register are displayed only when an address generated by the 68000 falls outside the predetermined range.
- The execution of a particular instruction. For example, the contents of the registers may be displayed following the execution of a `TAS` instruction.
- Any memory access that modifies the contents of a memory location — that is, any write access.

Several of the above conditions may be combined to create a composite event. For example, the contents of registers may be displayed whenever the 68000 executes write accesses to the region of memory space between $3A 0000 and $3A 00FF.

Note that the `STOP #data` instruction does not perform its function when it is traced. The `STOP` instruction causes the 68000 to load the 16-bit literal into its status register, and then to stop further processing until an interrupt (or reset) occurs. However, if the T-bit is set, an exception is forced after the status register has been loaded with the literal following the `STOP` instruction. Upon a return from the trace handler routine, execution continues with the instruction following the `STOP`, and the processor never enters the stopped condition.

Example of Application of Trace Mode Exceptions

Let's consider the design of a skeleton generic trace exception handler. The component parts of the trace handler are:

1. A longword at the trace exception vector location pointing to the actual trace handler itself. That is, you have to set up the exception vector table.
2. A mechanism for switching on the trace facility. The switching must be carried out in the supervisor state, since any attempt to modify the status register while in the user state would result in a privilege violation.
3. A mechanism for returning to (or *activating*) the user program once the trace mode has been turned on. You can't use a `JMP` as that would, itself, cause a trace exception once the T-bit of the status register has been set.
4. A trace handler that deals with the trace exception. In this case we will assume that the trace handler simply dumps all registers on the screen.
5. A subroutine, `Print_regs`, that can be called by the trace handler to perform the actual printing of the registers. We assume that the subroutine takes the registers off the stack.

6. A mechanism that permits the tracing to continue or to be suspended. We will assume that after each instruction has been executed, the system waits for a character from the keyboard. If the character is a 'T', the next instruction is printed. If it is not, execution continues without further tracing.

7. A subroutine to input a character from the keyboard and deposit it in D1.B.

The following fragments of code implement a trace exception handler. The fragment labelled Go is responsible for running the code being traced. Before Go is executed, the supervisor stack pointer must be set up with the status register at the top of the stack, and the program counter immediately below that.

```
          ORG      $00000024          Location of trace handler vector
          DC.L     TraceH             The trace handler exception vector
          .
          .
Go        ORI.W    #$8000,(SP)        Set the trace bit in the SR on the stack
          RTE                         Now run the (user) program
          .
          .
          .
*         The trace handler
TraceH    MOVEM.L D0-D7/A0-A7,-(SP) Save all registers on stack (A7=SSP=dummy register)
          MOVE.L   USP,A0             Grab the user's stack pointer and put it on
          MOVE.L   A0,(60,SP)          the stack (overwriting the saved SSP).
          BSR      Print_regs         Display all registers on stack
          BSR      Get_char           See if we want to continue tracing
          CMP.B    #'T',D1            Is the character a 'T' ?
          BEQ      Continue           IF it is THEN continue
          ANDI.W   #$7FFF,(64,SP)              ELSE turn off trace mode
Continue  MOVEM.L (A7)+,D0-D7/A0-A6 Restore registers (except A7 which is the USP)
          LEA      (4,SP),SP          Move past dummy A7 left on the stack
          RTE                         Return from exception
```

A simple way of running the user program is to deposit the entry point to the program and its initial status register on the stack pointed at by the supervisor stack pointer. In this example, we set the trace bit on the stack by means of the ORI.W #$8000,(SP) instruction. Executing an RTE loads the program counter and the status register with the values from the stack. Consequently, the RTE will activate the user program with its T-bit in the status register set.

When the 68000 executes an instruction in the *target* program, a trace exception will be forced, since the T-bit is set. At the start of the exception processing routine, the values of the program counter and the status register are pushed onto the supervisor stack. The program counter points to the instruction after the instruction that caused the trace exception (i.e., the next instruction), and the status register corresponds to the contents of the status register immediately prior to beginning exception processing.

The trace handler, `TraceH`, pushes all the 68000's registers onto the supervisor stack by means of a `MOVEM.L D0-D7/A0-A7,(SP)` instruction. Saving these registers means that we can now use the 68000's registers without worrying about corrupting their pre-exception values. You will appreciate, of course, that the value of A7 saved on the stack is the *supervisor stack pointer*. We can replace it (on the stack) with the value of the user stack pointer. The trace handler first loads the USP into A0 by means of a `MOVEM.L A0,USP` instruction, and then overwrites the old value of 'A7' on the stack with the USP. Thus, when the register display routine is called, it will print D0 to D7, A0 to A6, and the USP. Figure 10.11 describes the state of the stack during the stacking of the registers.

After the display routine has been called, a character is input into D1. If it is a 'T', trace exception continues normally. If not, we access the status register pushed on the stack by the trace exception and clear its T-bit.

In the next step, the copies of the registers saved on the stack at the time of the exception are reloaded into the 68000's registers, except for A7. Since A7 is the supervisor stack pointer, we do not have to load it. In any case, the value of A7 on the supervisor stack frame is the USP. Having reloaded all registers except A7, it is necessary to tidy up the stack with a `LEA (4,SP),SP` instruction to *step past* the USP on the stack.

At this stage the stack is in the same state it was in at the start of exception handling. Executing an `RTE` instruction restores the program counter and the

Figure 10.11 Use of the `MOVEM` instruction to save registers on the stack

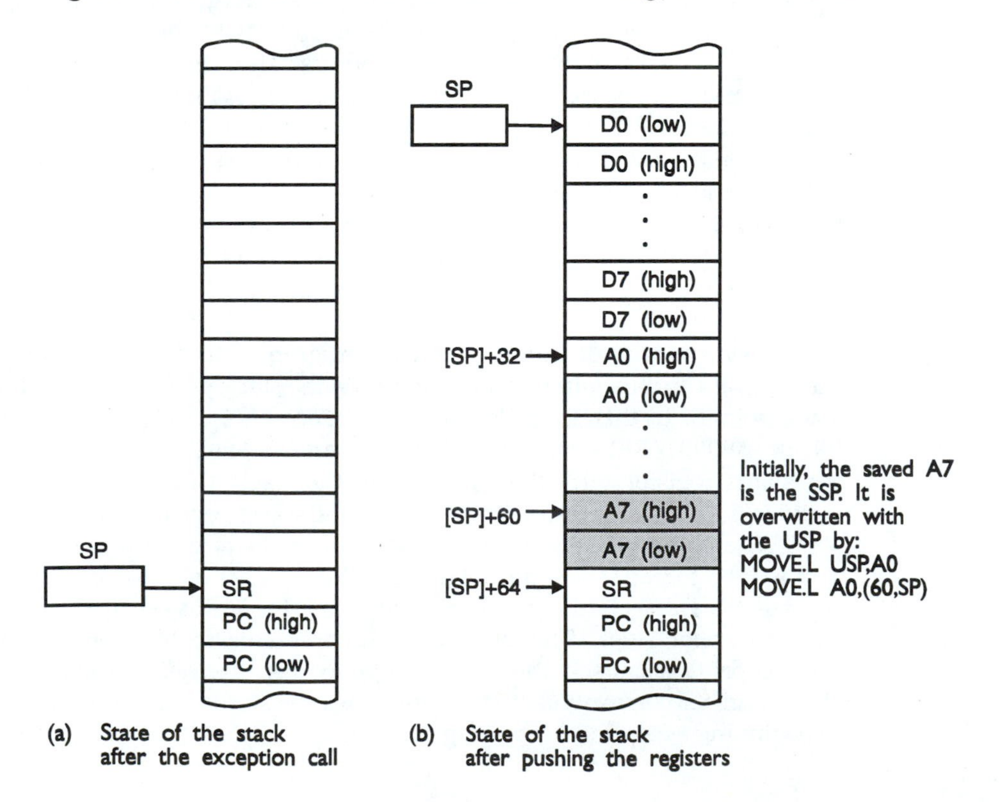

status register to the values they had immediately before the trace exception. The next instruction in the target program will now be executed. If the trace bit is still set, a trace exception will be raised and the entire sequence will be repeated. If the trace bit is clear (because we cleared it during the last trace handling), the 68000 will continue normally.

Emulator Mode Exceptions

The emulator mode exceptions provide the systems designer with tools to develop software for new hardware, before that hardware has been fully realized. Suppose a company is working on a coprocessor to generate the sine of a 16-bit fractional operand. For commercial reasons, it may be necessary to develop software for this hardware long before the coprocessor is in actual production.

By inserting an emulator op-code at the point in a program at which the sine is to be calculated by the hardware, the software can be tested as if the coprocessor were actually present. When the emulator op-code is encountered, a jump is made to the appropriate emulator-handling routine. In this routine, the sine is calculated by conventional techniques. Essentially, the line-A trap is dedicated to user applications and the line-F trap to coprocessor applications — although there is nothing to stop you using them in any way you wish.

A line-A trap has the op-code form $AXXX, where the 12 bits represented by the Xs are user-definable. Line-A traps are vectored to the location pointed at by the contents of location $28 and are generally employed to synthesize new instructions. Note that when a line-A exception occurs (or any other Group 1 or Group 2 exception), the exception handler can read the instruction that caused the exception (i.e., the $AXXX) and use the XXX-field to interpret the instruction.

From the line-A exception handler, you can access the actual line-A instruction by means of the following code.

```
LEA      (2,SP),A0      A0 points at address of the instruction following the exception
MOVEA.L  (A0),A0        Read this address
MOVE.W   (-2,A0),D0     Read the instruction that caused the exception (the $AXXX)
```

Figure 10.12 illustrates the relationship between the line-A trap and the above fragment of code. The first instruction, `LEA (2,SP),A0`, sets A0 to point to the saved program counter on the supervisor stack (the PC is under the status register which is at the top of the stack). The second instruction, `MOVEA.L (A0),A0`, loads A0 with the saved PC. The third instruction, `MOVE.W (-2,A0),D0`, copies the line-A instruction that caused the exception into D0. Note that this instruction uses an offset of -2 because the program counter saved on the exception stack frame points to the instruction after the instruction that caused the trap.

The line-F emulator exception has the format $FXXX and behaves like the corresponding line-A trap. The only difference is that the line-F trap locates the vector to its handler at address $2C. However, the line-F instruction is used to implement coprocessor instructions (in an environment that includes a 68020 or a

Figure 10.12 Accessing a line-A instruction

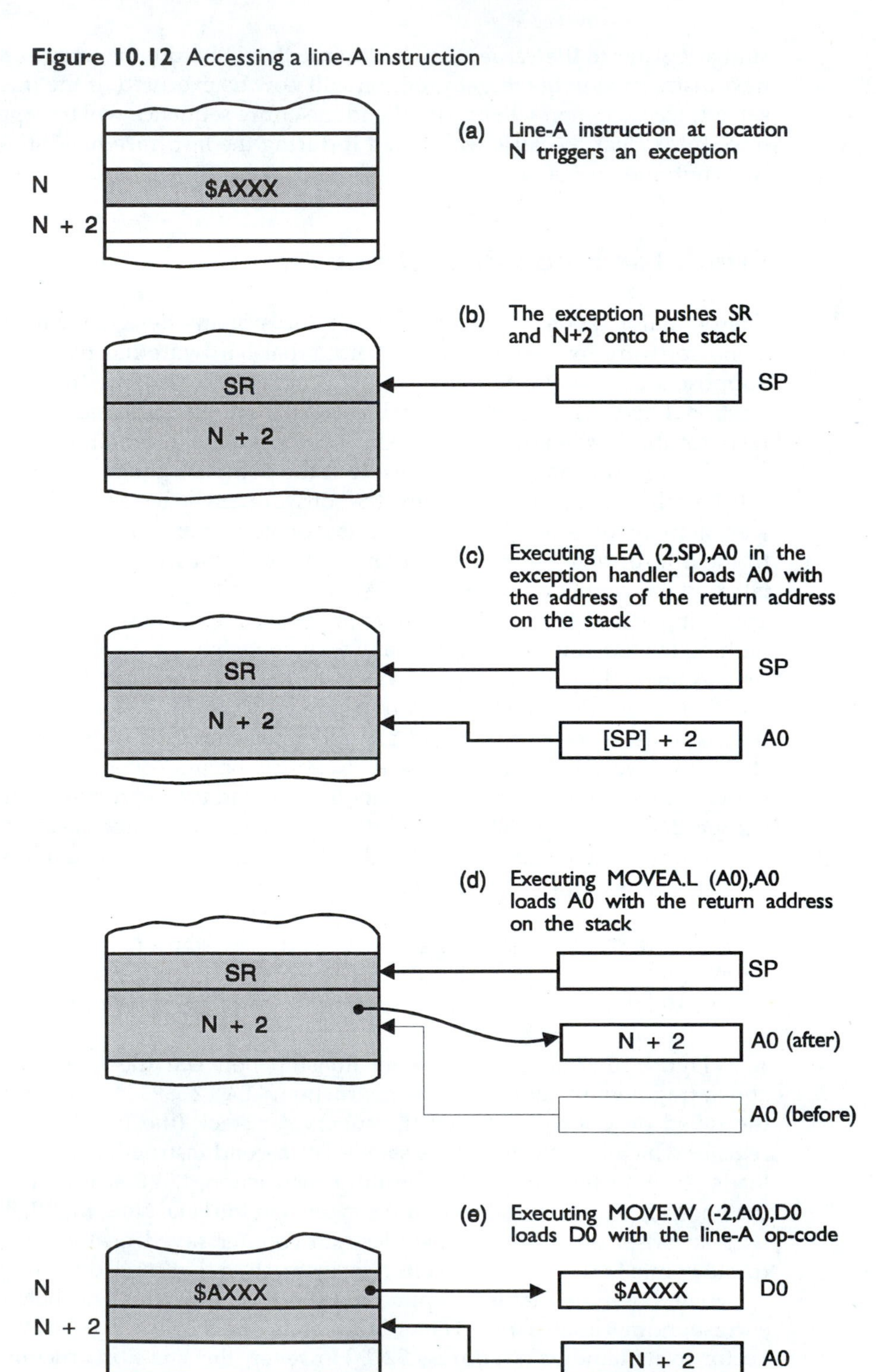
N
$AXXX
N + 2
(a) Line-A instruction at location N triggers an exception
(b) The exception pushes SR and N+2 onto the stack
SR
N + 2
SP
(c) Executing LEA (2,SP),A0 in the exception handler loads A0 with the address of the return address on the stack
SR
N + 2
SP
[SP] + 2
A0
(d) Executing MOVEA.L (A0),A0 loads A0 with the return address on the stack
SR
N + 2
SP
N + 2
A0 (after)
A0 (before)
(e) Executing MOVE.W (-2,A0),D0 loads D0 with the line-A op-code
N
$AXXX
N + 2
$AXXX
D0
N + 2
A0

68030). When the 68000 encounters a line-F trap, it begins exception processing in the normal fashion. We look at the 68020's coprocessors in Chapter 12.

TRAP Exception

The trap is almost certainly the most useful software user-initiated exception available to the programmer. Indeed, it is one of the more powerful functions provided by the 68000. To be honest, there are no real differences between traps and line-A and line-F emulator exceptions. They differ only in their applications. Sixteen traps, TRAP #0 to TRAP #15, are associated with exception vector numbers 32 to 47 decimal (i.e., longword vectors $80 to $BF), respectively.

Just as emulator exceptions provide functions in software that will later be implemented in hardware, trap exceptions create new operations not provided directly by the 68000 itself. However, the real purpose of the trap is to separate the details of housekeeping functions from the applications-level program.

Consider input/output transactions that involve real hardware devices. The precise nature of an input operation on system A may be very different than that on system B, even though both systems put the input to the same use. System A may operate a 6850 ACIA in an interrupt-driven mode, while system B may use an Intel 8055 parallel port in a polled mode to carry out the same function. Clearly, the device drivers (i.e., the software that controls the ports) in these systems differ greatly in their structures.

Applications programmers do not want to know about the fine details of I/O transactions when writing their programs. One solution is to use a *jump table* and to thread all I/O through this table. Figure 10.13 demonstrates how the applications programmer deals with all device-dependent transactions by indirect jumps through a jump table. The applications programmer writes a program and makes all subroutine calls to a table of subroutine addresses (i.e., the jump table). This table contains jumps to the actual subroutines. The programmer does not need to know the actual address of the subroutines.

For example, suppose that all console input at the applications level is carried out by the instruction BSR GETCHAR At the address "GETCHAR" in the jump table, the systems programmer inserts a link (i.e., JMP INPUT) to the actual routine used in their own system.

```
          ORG   $001000              Jump Table
GETCHAR   JMP   INPUT
OUTCHAR   JMP   OUTPUT               Each procedure is vectored to
GETSECTOR JMP   DISK_IN              the actual procedure that carries
PUTSECTOR JMP   DISK_OUT             out the appropriate task
          .
          BSR   GETCHAR
          .                          Application program
          BSR   PUTSECTOR            (address of subroutines not
          .                          system dependent)
```

Figure 10.13 The jump table

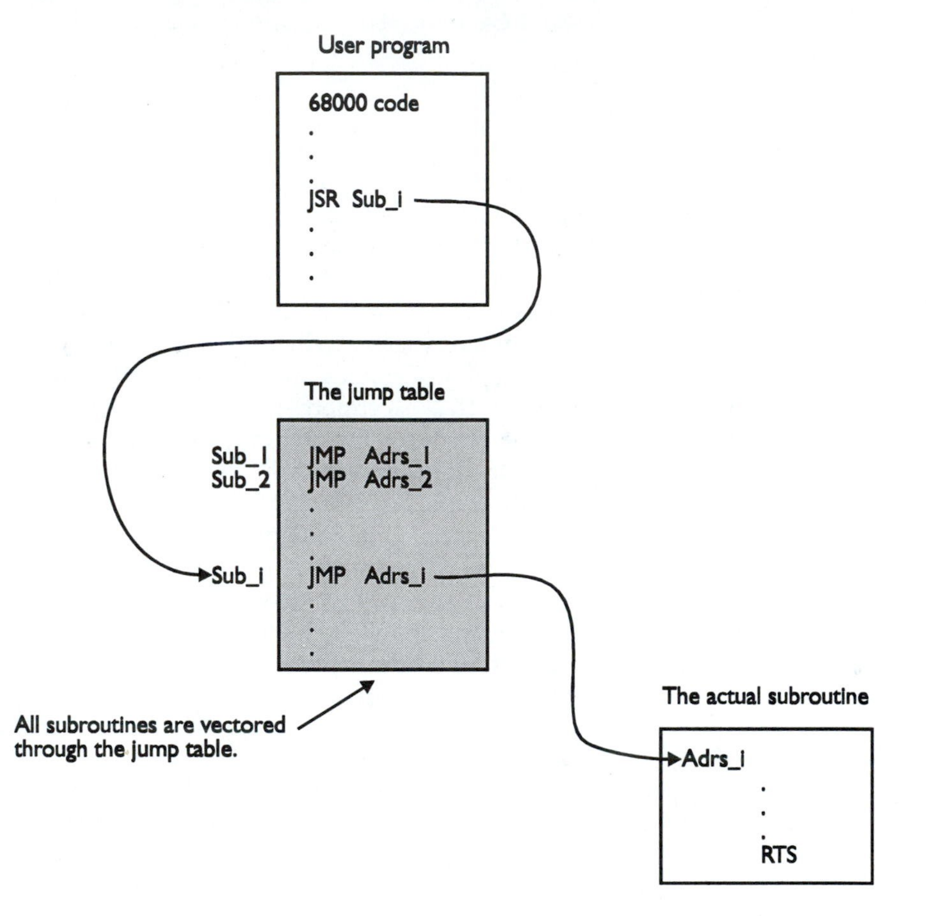

This approach to the problem of device-independency via jump tables is perfectly respectable. Unfortunately, it suffers from the limitation that the applications program must be tailored to fit on to the target system. We can do this by providing the program with a suitable jump table. An alternative approach, requiring no modification whatsoever to the applications software, is provided by the TRAP exception. The TRAP instruction leads to truly system-independent software.

When a trap is encountered, the appropriate vector number is generated and the exception vector table interrogated to obtain the address of the trap-handling routine. Note that the exception-vector table fulfils the same role as the jump table in Figure 10.13. The difference is that the jump table forms part of the applications program while the exception vector table is part of the 68000's operating system.

An example of a trap handler is found on the Motorola ECB computer and is known as the TRAP #14 handler. The TRAP #14 exception provides you with a method of accessing functions within the ECB's monitor software without having to know their addresses. The versatility of a trap exception can be increased by passing parameters from the user program to the trap handler. The TRAP #14 handler of TUTOR (the monitor on the Motorola ECB) provides for up to 256 different functions to be associated with TRAP #14. Before the trap is invoked, the programmer must load the required function code into the least-significant byte of D7. For example, we can use the following calling sequence to transmit a single ASCII character to port 1.

```
OUTCH   EQU   248            Equate the trap function to name of activity
        .
        .
        .
        MOVE.B #OUTCH,D7     Load trap function in D7
        TRAP   #14           Invoke the TRAP #14 handler
```

Table 10.3 gives a list of the functions provided by the TRAP #14 exception handler of the TUTOR monitor on the ECB. Note that the Teesside 68000 simulator uses TRAP #15 to handle input and output operations. There is no standard mechanism used by all 68000-based systems to deal with I/O.

Summary

- An exception is an operating system call that may be made explicitly by the programmer, by certain software events, or by external hardware.
- The 68000 operates either in the supervisor state (S=1) or in the user state, (S=0). The operating system runs in the supervisor state and user programs in the user state. Simple 68000-based systems run only the supervisor state.
- When in the user state, the programmer cannot set the S-bit in the status register, and therefore cannot change the 68000's state. The supervisor state is entered only via an exception. You can read the 68000's status register while in the user state, but not modify it.
- Instructions that modify the S-bit are called *privileged* and cannot be executed in the user state.
- The user state has its own stack pointer, USP. You cannot access the supervisor state stack pointer, SSP, from the user state. Both stack pointers can be written as A7 without ambiguity.
- The 68000 responds to an exception by:

  a. Saving the status register
  b. Setting S = 1, T = 0 of the current status register
  c. Pushing the PC and the saved status register onto the stack
  d. Jumping to the appropriate exception handler.

Table 10.3 The functions provided by the TRAP #14 handler on the ECB

Function value	Function name	Function description
255	-	Reserved functions - end of table indicator
254	-	Reserved function - used to link tables
253	LINKIT	Append user table to TRAP14 table
252	FIXDAOD	Append string to buffer
251	FIXBUF	Initialize A5 and A6 to BUFFER
250	FIXDATA	Initialize A6 to BUFFER and append string to BUFFER
249	FIXDCRLF	Move "CR", "LF", string to buffer
248	OUTCH	Output single character to Port 1
247	INCHE	Input single character from Port 1
246	-	Reserved function
245	-	Reserved function
244	CHRPRNT	Output single character to Port 3
243	OUTPUT	Output string to Port 1
242	OUTPUT21	Output string to Port 2
241	PORTIN1	Input string from Port 1
240	PORTIN20	Input string from Port 2
239	TAPEOUT	Output string to Port 4
238	TAPEIN	Input string from Port 4
237	PRCRLF	Output string to Port 3
236	HEX2DEC	Convert hex values to ASCII-encoded decimal
235	GETHEX	Convert ASCII character to hex
234	PUTHEX	Convert 1 hex digit to ASCII
233	PNT2HX	Convert 2 hex digits to ASCII
232	PNT4HX	Convert 4 hex digits to ASCII
231	PNT6HX	Convert 6 hex digits to ASCII
230	PNT8HX	Convert 8 hex digits to ASCII
229	START	Restart TUTOR; perform initialization
228	TUTOR	Go to TUTOR; print prompt
227	OUT1CR	Output string plus "CR", "LF", to Port 1
226	GETNUMA	Convert ASCII-encoded hex to hex
225	GETNUMD	Convert ASCII-encoded decimal to hex
224	PORTIN1N	Input string from Port 1; no automatic line feed
223-128	-	Reserved
127-0	-	User-defined functions

- The T-bit of the status register can be set to force the 68000 to generate a trace exception at the end of each instruction. This process is used to trace (i.e., step through) a user program line-by-line.
- Hardware exceptions are: reset, interrupt, and bus error. A bus error exception is caused by asserting the BERR* input and indicates a faulty memory access. The spurious interrupt exception is a special case of the interrupt.

- Internally generated software exceptions are: the address error, illegal instruction, divide-by-zero, privilege violation, and the uninitialized interrupt exception. Programmer generated software errors are: CHK, TRAPV, trace, line-A, line-F, and TRAP.
- If the 68000 encounters a bus error or an address error exception while dealing with a bus error or an address error exception, it stops and asserts its HALT* output. The 68000 can be re-started only by asserting its RESET* input. This event is called a double bus fault, and is a fatal error.
- The 68000 maintains a table of 256 longword vectors in memory between $00 0000 and $00 03FF. This is the exception vector table, and the systems programmer loads it with pointers to the appropriate exception handlers. The first two locations in the table are special (i.e., the initial value of the supervisor stack pointer and the initial cold entry point, respectively).
- The 68000 saves the PC and SR on the stack for all exceptions, apart from the bus error and address error (in which case it saves additional information). You return from an exception with the instruction RTE. In general, an address error exception or a bus error exception causes a fatal error — that is, you do not return from one of these exceptions. The operating system must then sort things out.
- When the 68000 is reset, its S-bit is set, its T-bit is cleared, and its interrupt mask is set to 1,1,1 (i.e., only level seven interrupts are recognized).
- The 68000 has seven levels of prioritized interrupt (IRQ1* to IRQ7*). An interrupt is processed only if it has a greater priority than that reflected by the interrupt mask bits in the SR. When an interrupt is processed, the mask bits are set to the level of the current interrupt.
- The 68000 supports vectored interrupts. When the interrupt is processed, the 68000 enters an interrupt acknowledge cycle, IACK, and the interrupting device puts its interrupt vector number on the data bus. The 68000 uses this to locate the appropriate interrupt vector in the exception vector table.
- The line-A emulator mode exception generates an exception if an op-code begins with 1010. This exception permits you to design new instructions that are processed in the supervisor state.
- The line-F emulator mode exception generates an exception if the op-code begins with 1111. This exception is similar to the line-A exception, but is used to support an external coprocessor in conjunction with the 68020 or other later members of the 68000 family.
- In Chapter 12 we will see that the 68020 extends the 68000's exception handling mechanism. The 68020 implements a true *virtual memory system* (in conjunction with external hardware), and you can return from a bus error exception. The 68020 can also relocate the exception vector table anywhere in memory. Finally, the 68020 has yet another supervisor state stack pointer.

Problems

1. Distinguish between hardware and software exceptions, and between internally generated software exceptions and programmer generated software exceptions.

2. An exception handling capability is not essential. So, why does the 68000 implement exception handling?

3. Why does the 68000 have a supervisor state (in contrast to many 8-bit microprocessors)?

4. Why does the 68000 have two stack pointers? How does the programmer know which to use?

5. Write 68000 instructions to perform the following operations:

   a. Set the trace bit.
   b. Set the interrupt level to five.
   c. Put the 68000 in the user state.
   d. Clear the trace bit, set the interrupt level to 6, and set user state.

6. What is the effect of each of the following on the status of the 68000?

   a. `MOVE.W #$0000,SR`
   b. `MOVE.W #$2700,SR`

7. What instructions are privileged, and how do they differ from non-privileged instructions?

8. What are the four actions that the 68000 carries out during exception processing?

9. The 68000 has a user and a supervisor state. When the 68000 is in the user state, it is impossible to execute privileged instructions. If you were a systems hacker, how would you attempt to get around this restriction? If you were a systems designer, what would you do to attempt to make your system hacker-proof?

10. What is the difference between an uninitialized interrupt and a spurious interrupt exception?

11. You are thinking of designing a special version of the 68000 for use in high-speed word processing. A new instruction is to have the specific form `MATCH <source buffer>,<target buffer>`. Its action is to match the char-

acter string starting at address <source buffer> with the character string starting at <target buffer>. Both strings are terminated by a carriage return. If the source string does not occur within the target string, the carry bit of the CCR is cleared. If the source string occurs within the target string, the carry bit of the CCR is set, and the address of the start of the first occurrence of the source string within the target string is pushed on the stack. In order to test the new processor before the *first silicon*, you decide to use the 68000's line 1010 emulator trap. Show how you would do this. Remember that the instruction will have the form:

```
<16-bit opcode>,<32-bit source address><32-bit target address>.
```

12. The 68000 has an instruction `ILLEGAL` (with the bit pattern 0100101011111100). When encountered by the 68000, the CPU carries out the operation:

```
[SSP]        ←  [SSP] - 4
[M([SSP])] ←  [PC]
[SSP]        ←  [SSP] - 2
[M([SSP])] ←  [SR]
[PC]         ←  [M(16)]
```

Explain the action of the above sequence of RTL (register transfer language) operations in plain English. Why do you think that such an instruction was implemented by the designers of the 68000?

13. Why is the reset different from all other 68000 exceptions?

14. What is the exception with the highest priority?

15. Why is the 68000's BERR* exception so important?

16. Describe the 68000's BERR* exception sequence.

17. What is a double bus fault, and why is it described as fatal?

18. What is the difference between a vectored and an autovectored interrupt? Under what circumstances are autovectored interrupts employed?

19. What does context switching mean, and how can it be implemented (making best use of the 68000's features)?

20. What is the effect of the `STOP #d16` instruction, and how do you think it might be used?

21. Write a trace exception handling routine that will display the contents of registers whenever an instruction is executed that falls between two ad-

dresses, `Address_low` and `Address_high`. *Hint*: where can you find the instruction that actually causes the trace exception?

22. A print spooler prints one or more files as background jobs while the processor is busy executing a foreground job. Design a print spooler that will print a file. Assume the existence of `GETCHAR` that reads a character from the disk drive, and `PUTCHAR` that sends a character to the printer. The spooler operates in conjunction with a real-time clock that periodically generates a level-5 interrupt. Clearly state any other assumptions you use in solving this problem.

23. Suppose you require more than 192 interrupt vectors. Can you locate some of the additional vectors in the exception vector table at locations marked "Unimplemented, reserved"?

24. What are the differences between `TRAP`s, illegal instruction exceptions, and line-A and line-F exceptions?

25. A 68000 bus error exception takes place. The top 7 words on the supervisor stack are:

```
$0009
$0100
$123A
$4247
$0701
$00A1
$1C40
```

Interpret the meaning of this data.

CHAPTER 11

Programming Examples

In this chapter we provide two extended examples of assembly language programs. When this book was in its planning stage, several reviewers suggested that we include a few non-trivial examples of assembly language programs. They said that some books on assembly language provide minimal examples, often less than ten or so lines of code. On the other hand, one of my colleagues suggested that there is no need to provide more complex examples, since few programmers ever use an assembly language to do more than write a device driver.

We have decided to provide two extended examples of assembly language programs. A few readers might indeed have to implement programs in assembly language for the sake of speed (they may even have to write programs for unusual or experimental machines). However, we feel that these examples should be provided to illustrate the use of assembly language. Finally, these particular examples have been selected because they are interesting and informative in their own right.

The first example looks at the design of a real-time kernel that switches between tasks in a multitasking system. We describe the software that can be used to end one task and restart another. The task-switching kernel complements Chapter 10 on exception handling. This is a realistic example of the use of assembly language, because a task switcher is often optimized for speed.

The second example is a bootstrap monitor for a single-board computer. This is the type of minimal program you might put in the EPROM of a single-board microprocessor. Essentially, it enables you to test and debug the system with the aid of a terminal. It also lets you down-load software into the system from a host computer. This example provides a library of basic input and output routines.

11.1 Task-switching Kernel

There are those who believe that hardware is hardware, software is software, and never the twain shall meet. While this maxim may be applied to certain aspects of a microcomputer, it cannot be applied to hardware-initiated interrupts. An interrupt is a request for service from some device requiring attention. The request has its origin in *hardware*, but the response (the servicing of the interrupt) is at the *software* level. It is, therefore, difficult to deal with one aspect without at least some consideration of the other. We have already examined how a device physically signals an interrupt request and how the 68000 begins executing an interrupt-handling routine. Now we shall look at the use of the interrupt mechanism to implement a multitasking system.

Interrupts are closely associated with input/output transactions. Without an interrupt mechanism, an input or output device must be polled in order to determine whether or not it is busy. During the polling, the CPU is performing no useful calculations. Permitting a peripheral to indicate its readiness for input or output by asserting an interrupt request line frees the processor to do other work while the peripheral is busy. Implicit in this statement is the assumption that the processor has something else to do. This leads us to the concept of *multitasking* (the executing of a number of programs or tasks apparently simultaneously) and the operating system (the mechanism that controls the execution of the tasks).

Multitasking

Multitasking (or multiprogramming) is a method of squeezing greater performance out of a processor (i.e., a CPU) by chopping the programs into tiny slices and executing slices of different programs one after the other, rather than by executing each program to completion before starting the next. The concept of multitasking should not be confused with *multiprocessing*, which is concerned with the subdivision of a task between several processors.

Figure 11.1 illustrates two tasks (or processes), A and B. Each of these tasks requires several different resources (i.e., input/output via a video display terminal (VDT), disk access, and CPU time) during its execution. One way of executing the tasks would be *end to end*, with task A running to completion before task B begins. Such serial execution is clearly inefficient, as, for much of the time, the processor is not actively involved with either task. Figure 11.2 shows how the system can be made more efficient by scheduling the activities performed by tasks A and B in such a way as to make best use of the resources. For example, in time-slot 3, task A is accessing the disk while task B is using the CPU.

If we examine the idealized picture presented by Figure 11.2, it is immediately apparent that a multitasking system needs two components: a *scheduler* that allocates activities to tasks and a mechanism that *switches* between tasks. The first is called the operating system, and the second the interrupt mechanism.

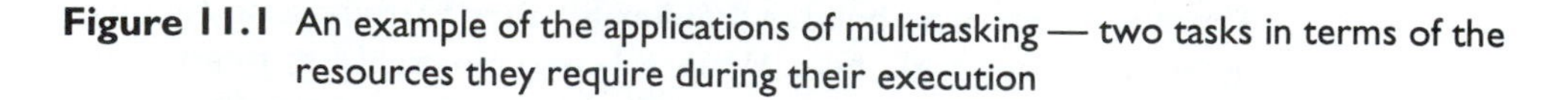

Figure 11.1 An example of the applications of multitasking — two tasks in terms of the resources they require during their execution

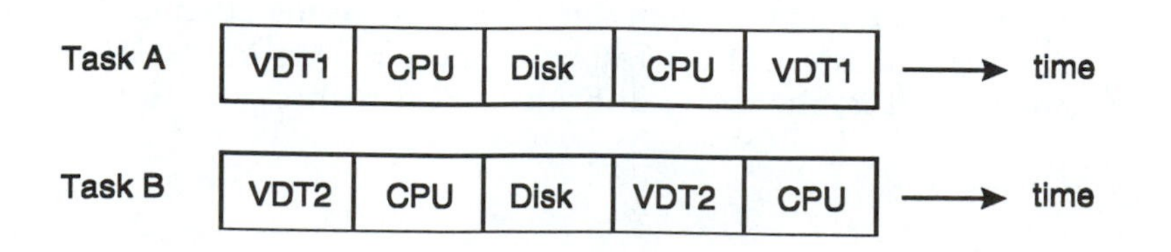

Figure 11.2 Scheduling the two tasks of Figure 11.1

Resource	Activity					
	Slot 1	Slot 2	Slot 3	Slot 4	Slot 5	Slot 6
VDT1	Task A				Task A	
VDT2	Task B				Task B	
Disk			Task A	Task B		
CPU		Task A	Task B	Task A		Task B

⟶ time

Real-time Operating System

It's difficult to define a real-time system precisely, as real-time means different things to different people. The simplest definition is that a real-time system responds to a change in its circumstances within a meaningful period. For example, if a number of users are connected to a multitasking computer, its operation can be called real-time if it responds to the users almost as if each of them had sole access to the machine. Therefore, a maximum response time of no more than 2 seconds must be guaranteed. A real-time system controlling a chemical plant must respond to changes in chemical reactions fast enough to control them. Here, the maximum guaranteed response time may be of the order of milliseconds.

Real-time and multitasking systems are closely related but are not identical. The former optimizes the response time to events while trying to use resources efficiently. The latter optimizes resource utilization while trying to provide a reasonable response time. If this is confusing, consider the mail system. Here we have an example of a real-time process. It offers a (nominally) guaranteed response time (i.e., speed of delivery), and attempts to use its resources well (i.e., pick-up, sorting, and delivery take place simultaneously). Suppose the mail service were made purely multitasking at the expense of its real-time facilities. In that case, the attempt to optimize resources might lead to the following argument. Transport costs can be kept down by carrying the largest load with the least vehicles. Therefore, all vehicles wait on the East Coast until they are full and then travel to the West Coast, and so on. This would increase efficiency (i.e., reduce costs) but at the expense of degrading response time.

Real-time Kernel

Operating systems, real-time or otherwise, can be very complex pieces of software with sizes greater than 10 Mbytes. Here, we are concerned only with the heart (kernel or nucleus) of a real-time operating system, its scheduler. The kernel of a real-time operating system has three functions.

1. The kernel deals with interrupt requests. More precisely, it is a *first-level interrupt handler* that determines how a request should be treated initially. These requests include timed interrupts that switch between tasks after their allocated time has been exhausted, interrupts from peripherals seeking attention, and software interrupts or exceptions originating from the task currently running.
2. The kernel provides a *dispatcher* or *scheduler* that determines the sequence in which tasks are executed.
3. The kernel provides an interprocess (intertask) communication mechanism. Tasks often need to exchange information or to access the same data structures. The kernel provides a mechanism to do this. We will not discuss this topic further, other than to say that a message is often left in a mailbox by its originator and is then collected by the task to which it was addressed.

A task to be executed by a processor may be in one of three states: *running*, *ready*, or *blocked*. A task is *running* when its instructions are currently being executed by the processor. A task is *ready* if it is able to enter a running state when its turn comes. It is *blocked* or dormant if it cannot enter the running state when its turn comes. Such a task is waiting for some event to occur (such as a peripheral becoming free) before it can continue. Figure 11.3 gives the state diagram for a task in a real-time system.

Figure 11.3 State diagram of a real-time system

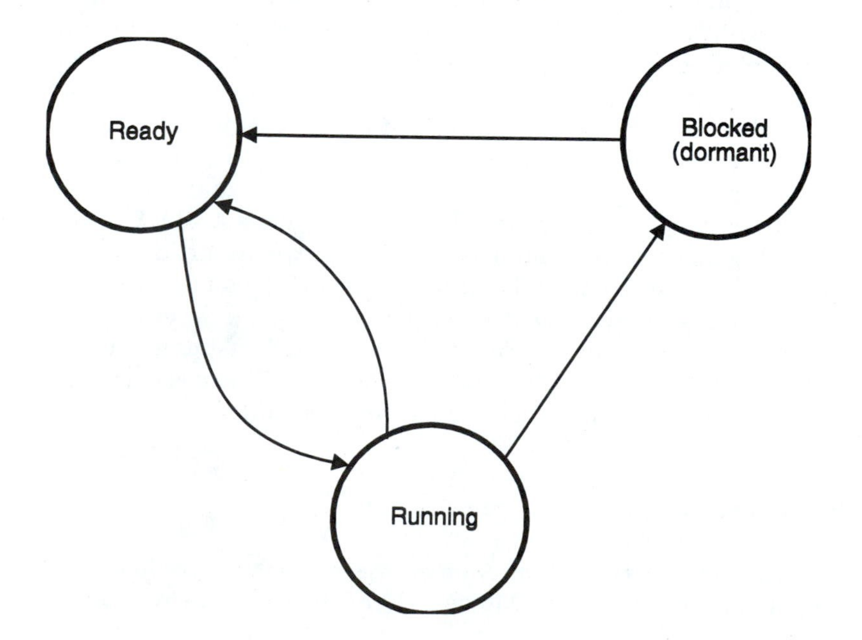

The difference between a running task and a waiting or blocked task lies in the task's *volatile portion*. The volatile portion of a task is the information needed to execute the instructions of that task. This information includes the identity of the next instruction to be executed, the processor status word, PSW, and the contents of any registers being used by the task. In a 68000 system, the processor status word is the processor's 16-bit status register, SR.

When a task is actually running, the CPU's registers, PC, PSW, data, and address registers define the task's *volatile environment*. But when a task is waiting or dormant, this information must be stored elsewhere. All real-time kernels maintain a data structure called a task control block, TCB, that stores a pointer to the volatile portion of each task. The TCB is also called: a run queue, task table, task list, and task status table.

Figure 11.4 provides an example of a possible task control block. Each task is associated with an identifier, which may be a name or simply its number in the task control block. The task block pointer, TBP, contains a vector that points to the location of the task's volatile portion. As you can see, it is not necessary to store the actual volatile portion of a task in the TCB itself.

The task status entry defines the status of the task and marks it as running, ready, or blocked. A task is activated (marked as ready to run) or suspended (blocked) by modifying its task status word in the TCB. The task priority indicates the task's level of priority, and the task time allocation is a measure of how many time slots are devoted to the task every time it runs.

Figure 11.4 The task control block

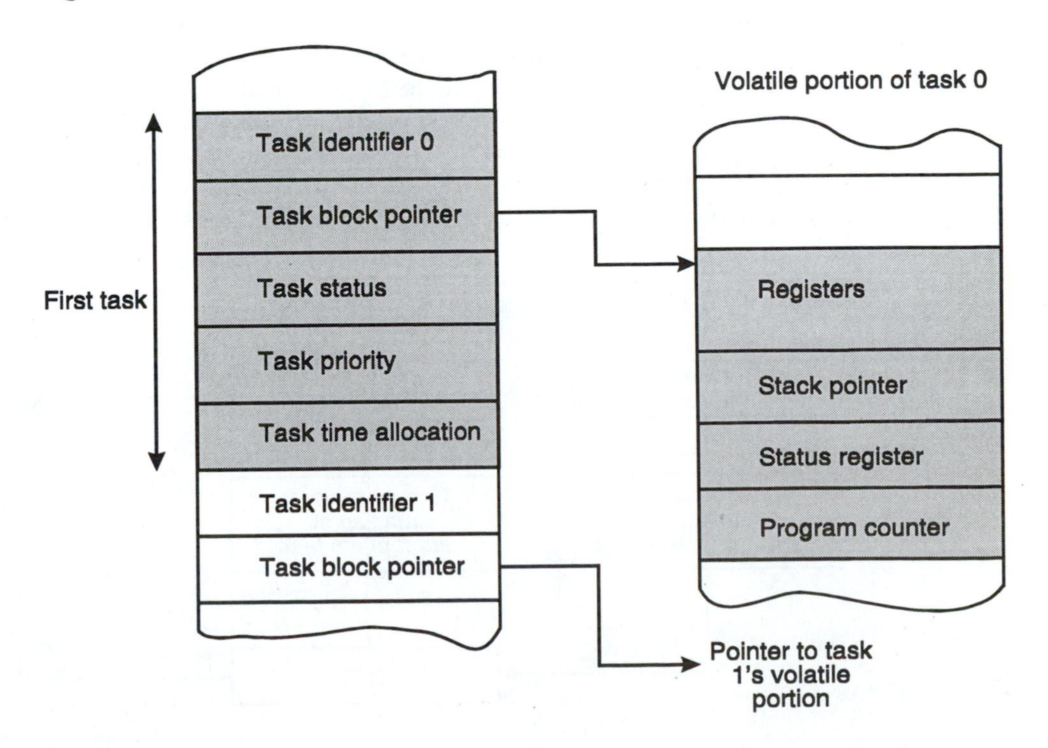

Interrupt Handling and Tasks

The mechanism employed to switch tasks is called a real-time clock, RTC. The RTC generates the periodic interrupts used by the kernel to determine when to locate the next runnable task and to run it. But how do we deal with interrupts originating from sources other than the RTC?

There are two ways of dealing with general interrupts and exceptions. One is to regard them as being outside the scope of the real-time task scheduling system, and to service them as and when they occur (subject to any constraints of priority). An alternative and much more flexible approach is to integrate them into the real-time task structure and to regard them as tasks just like any user task. We will adopt this latter approach in our examination of a real-time kernel.

Figure 11.5 describes a possible arrangement of an interrupt handler in a real-time system. When either an interrupt or an exception occurs, the appropriate interrupt-handling routine is executed via the 68000's exception vector table. However, this interrupt-handling routine does not service the interrupt request itself. It simply locates the appropriate interrupt-handling routine in the TCB and changes its status from blocked to runnable. The next time that this task is encountered by the scheduler, it may be run. We say *may be* because, for example, a `TRAPV` (trap on overflow) exception may have a very low priority and will not be dealt with until all the more urgent tasks have been run. Such an arrangement is called a first-level interrupt handler. The strategy of Figure 11.5 can be modified by permitting the first-level interrupt handler to take pre-emptive action. That is, the interrupt handler not only marks its own task as runnable but suspends the currently running task, as if there had been a real-time clock interrupt.

The real-time interrupt is physically implemented by connecting the output of a pulse generator to one of the 68000's interrupt request inputs. A relatively high priority interrupt (e.g., IRQ5* or IRQ6*) can be reserved for this function. The highest priority interrupt, IRQ7*, is frequently dedicated to the abort (i.e., soft reset) button and is used to restart a system when it has crashed. However, a good argument can be made for using IRQ7* to switch tasks. Clearly, if we use IRQ6*

Figure 11.5 Interrupt handling in a real-time kernel

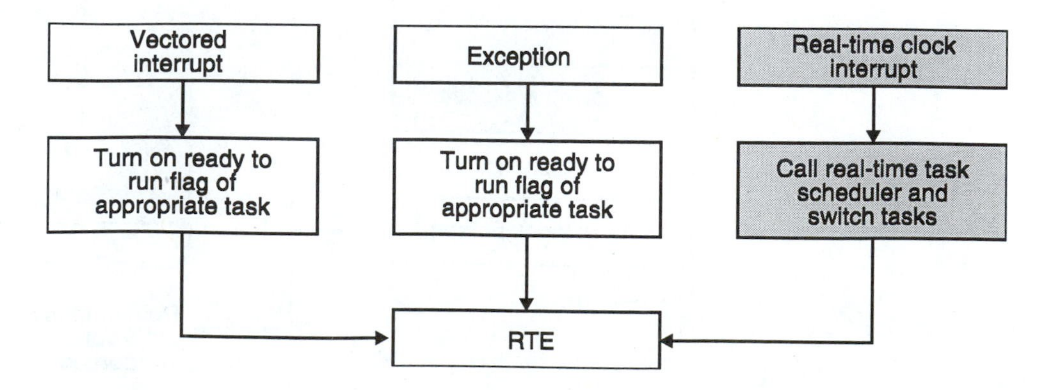

and the interrupt mask gets set to level seven (for whatever reason), a task-switching interrupt at level six will never be recognized and the system could hang-up. Whenever an RTC interrupt is detected, its first-level handler invokes the scheduler part of the real-time kernel.

There are two major approaches to task scheduling, a *round-robin* scheme and a *priority* scheme. The round-robin scheme runs tasks in the order in which they are stored in the TCB. When the task at the end of the TCB has been run, the next task to be run is the task at the start of the TCB. If, for example, the runnable tasks are 2,3,4,7,9, they will be run in the order 2,3,4,7,9,2,3,4,7,9,2,3,4,7,9 and so on. A round-robin scheme is called *fair*, because each task gets an equal chance of running. In a prioritized scheme, entries in the TCB are examined sequentially, but only a task whose priority is equal to the current highest priority may be run.

All interrupts and exceptions from sources other than the real-time clock simply mark the associated task as runnable. This action changes the task's status from blocked to runnable (i.e., ready) and takes only a few microseconds.

Designing a Real-time Kernel for the 68000

We now look at the skeleton design of a simple real-time kernel for a 68000-based microcomputer. Tasks running in the user mode have eight levels of priority from 0 to 7. Priority seven is the highest, and no task with a priority P_j may run if a task with a priority P_i (where $i > j$) is runnable. Each task has a time-slice allocation and may run for that period before a new task is run.

We can assume that the timed interrupt is generated by a hardware timer which pulses IRQ7* low every 20ms. All other interrupts and exceptions are dealt with as in Figure 11.5.

The highest level of abstraction in dealing with task-switching is illustrated in Figure 11.6. A timed interrupt causes the time_to_run allocation of the current task to be decremented. When this reaches zero, the task table is searched for the next runnable task and that task is run. The time_to_run counter is reset to its maximum value before the next task is run. Any interrupt or exception other than `TRAP #15` causes the appropriate task to be activated. A `TRAP #15` instruction allows user programs to access the kernel and to carry out certain actions. Note that a `TRAP #15` exception is not asynchronous — it is executed under program control in the user task currently being run.

A suitable *task control block* structure for this problem is defined in Figure 11.7. A separate task control block is dedicated to each task. The TCBs are arranged as a circular linked list with each TCB pointing to the next one in the chain. The last entry points to the first. Each TCB description occupies 88 bytes: a longword pointing to the next TCB, a 2-byte task number, an 8-byte name, a 2-byte task status word (TSW), a 70-byte task volatile portion, and two reserved bytes. Note that in this arrangement, a task's volatile environment is part of the TCB itself. The task volatile portion is a copy of all the working registers belonging to the task at the moment it was interrupted (A0 to A7, and D0 to D7), plus its program counter and status register. The stored value of A7 is, of course, the user

Figure 11.6 Real-time kernel — this diagram is an elaboration of that of Figure 11.5. TRAP #15 has been reserved to provide user access to the real-time system.

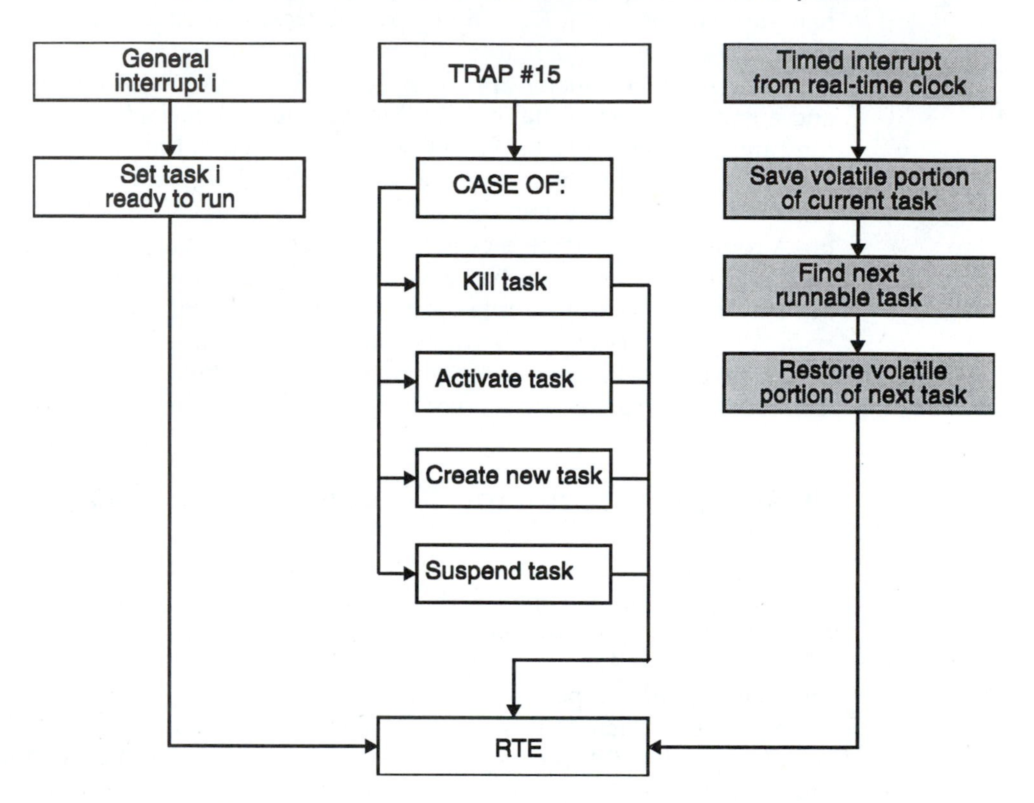

stack pointer (USP) and not the supervisor stack pointer. The format of the TSW is also given in Figure 11.7. The TSW contains a time_to_run field, a priority field, plus a 2-bit activity field. The first-level pseudocode program to implement the real-time kernel of Figure 11.6 is as follows.

```
Module  Interrupt_i
        Activate task for interrupt_i
        RTE
End module

Module  TRAP_#15
        CASE I OF
                1: Kill_task
                2: Activate_task
                3: Create_new task
                4: Suspend_task
        END_CASE
        RTE
End module
```

```
Module  Timed_interrupt
        Decrement time_to_run
        IF time_to_run = 0 THEN
                             BEGIN
                             Move working registers to TCB
                             Save USP in TCB
                             Transfer saved PSW from stack to TCB
                             Transfer saved PC from stack to TCB
                             Find next active task in task table
                             Load new USP from TCB
                             Restore new working registers from TCB
                             Transfer new PSW, PC from TCB to stack
                             Reset time_to_run
                             END
        END_IF
        RTE
End module.
```

The Timed_interrupt module requires a little further explanation. When the 68000 responds to the interrupt, it sets the S-bit in the SR to force the supervisor mode, and then pushes the program counter and status register on the stack. The stacked PC and SR are the values at the time the interrupt took place. The module copies the 68000's registers (apart from A7, which is the supervisor stack pointer) to the appropriate TCB. These registers are part of the state of the task that was interrupted. Finally, the task's own stack pointer is copied to the TCB. At this point, the TCB contains all the information required to rerun the interrupted task. Finally, the software loads all the data from another TCB to the processor. Some registers are transferred directly from the TCB, and some are pushed on the supervisor stack and pulled when the 68000 returns from the exception.

We do not intend to deal with multitasking in great detail, and therefore the level 1 PDL is only partially elaborated to produce the following level 2 PDL.

```
Module Interrupt_i: {An interrupt forces a task into its waiting
                     state by setting the least-significant bits
                     of its task status word, TSW, to 0,1}
                    Calculate address of associated task
                    Get address TSW for this task (TSW_address)
                    Clear bit 1 of TSW at TSW_address
                    Set bit 0 of TSW at TSW_address
                    RTE
End module

Suspend_task: {A task is suspended by setting the least-significant
               bits of its Task Status Word to 1,0}
              Calculate address of associated task
              Get TSW_address for this task
```

Figure 11.7 The task control block arranged as a linked list

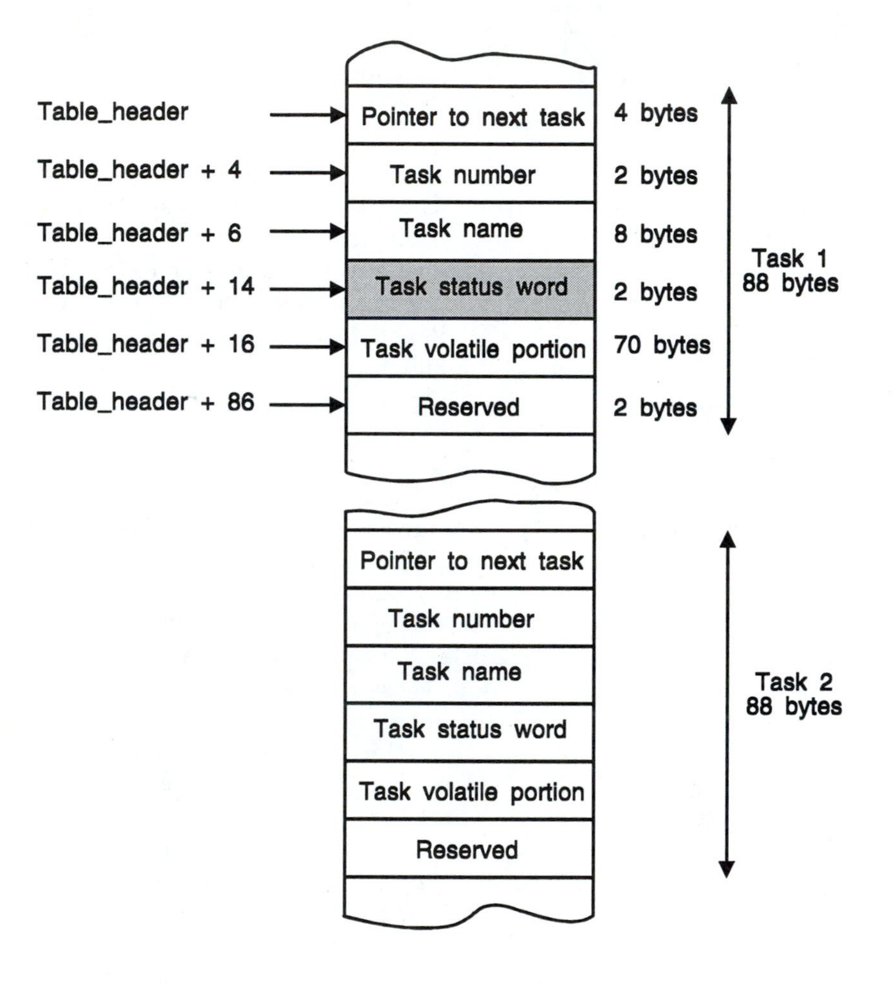

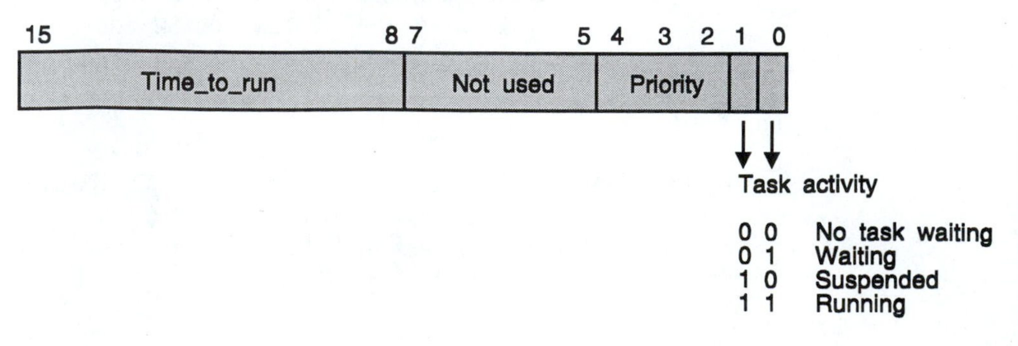

```
            Clear bit 0 of TSW at TSW_address
            Set bit 1 of TSW at TSW_address
            RTE
End suspend_task

Module Timed_interrupt:
                      Global variable: Current_pointer
                                     : Current_priority
                                     : Time_to_run
                      Time_to_run := Time_to_run - 1
                      IF Time_to_run = 0 THEN
                                        BEGIN
                                        Time_to_run := Max_time
                                        Newtask
                                        END
                      END_IF
                      RTE
End module

Newtask: {This swaps task volatile environments}
        Push A0-A6, D0-D7 on the system stack
        Task_volatile_pointer := Current_pointer + 16
        Copy A0-A6, D0-D7 from stack to [Task_volatile_pointer]
        Copy PC, SR from stack to [Task_volatile_pointer + 64]
        Copy USP to [Task_volatile_pointer + 60]
        Mark current task as waiting
        Next_task {Find the next runnable task}
        Task_volatile_pointer := Current_pointer + 16
        Transfer A0-A6, D0-D7 from TCB to stack
        Transfer USP from TCB to USP
        Transfer PC, SR from TCB to supervisor stack
        Transfer A0-A6, D0-D7 from stack to registers
End Newtask

Next_task: {This locates the next runnable task in the TCB}
          Temp_pointer  := Current_pointer
          Temp_priority := Current_priority
          Next_pointer  := Current_pointer
          Next_priority := Current_priority
          REPEAT
            IF Next_priority ¤ Temp_priority AND Next_TSW(0:1) = waiting
                   THEN
                       BEGIN
                       Temp_priority := Next_priority
                       Temp_pointer  := Next_pointer
                   END
            END_IF
```

```
        Next_pointer   := [Next_pointer]
        Next_TSW       := [Next_pointer + 14]
        Next_priority  := Next_TSW(2:4)
      UNTIL Next_pointer = Current_pointer
      Current_pointer   := Temp_pointer
      Current_ priority := Temp_priority
End Next_task
```

Having outlined the PDL required to implement part of the RTL task switching kernel, the next step is to convert the PDL to 68000 assembly language form (assuming the absence of a suitable high-level language). The only fragments of code provided here are `NewTask` that switches tasks, and `NextTask` that locates the next runnable task in the list of TCBs. These subroutines are for illustrative purposes only and are too basic for use in a real multitasking system. Because an understanding of the structure of the stack and the TCB is so essential, Figure 11.8 provides pictures of the supervisor stack and the task control block. The stack is displayed after the exception (i.e., the interrupt) and after the task's registers have been stacked. The TCB displays the task's volatile portion, and shows the location of data with respect to A2, which points to the top of the TCB.

```
*****************************************************************************
*
* Newtask switches tasks by saving the volatile portion of the current task in
* its TCB, transferring the volatile portion of the next task to run to the
* supervisor stack, and then copying all registers on the stack to the 68000's
* registers. The new task is then run by copying the new PC and the SR from the
* TCB to the supervisor stack, and then executing an RTE. Newtask runs in the
* supervisor mode and the supervisor stack is active. Note that all tasks are
* assumed runnable - bits 1,0 of the TSW are not used in this example.
*
* Note: All registers (including A7 = SSP) are saved on the supervisor stack at
* the beginning of the exception processing routine. The saved value of the SSP
* is later overwritten with the USP. Doing this, rather than leaving a "space"
* for the USP on the stack, simplifies the coding.
*
* A2 = CrntPtr        (points to the TCB of the current task)
* A1 = TempPtr        (temporary pointer to a TCB during the search)
* A0 = NextPtr        (pointer to TCB on next task in chain of TCBs)
* D2 = CrntPrty       (the priority of the current task)
* D1 = TempPrty       (the priority of a task during the search)
* D0 = NextPrty       (the priority of the next task in the chain of TCBs)
*
NewTask MOVEM.L A0-A7/D0-D7,-(SP) Save all registers on supervisor stack
        MOVE.L  CrntPtr,A2        Pick up the pointer to this (the current
        LEA     (16,A2),A2        task) and point at its volatile portion.
        MOVE.W  #34,D0
```

Figure 11.8 The supervisor stack and the TCB

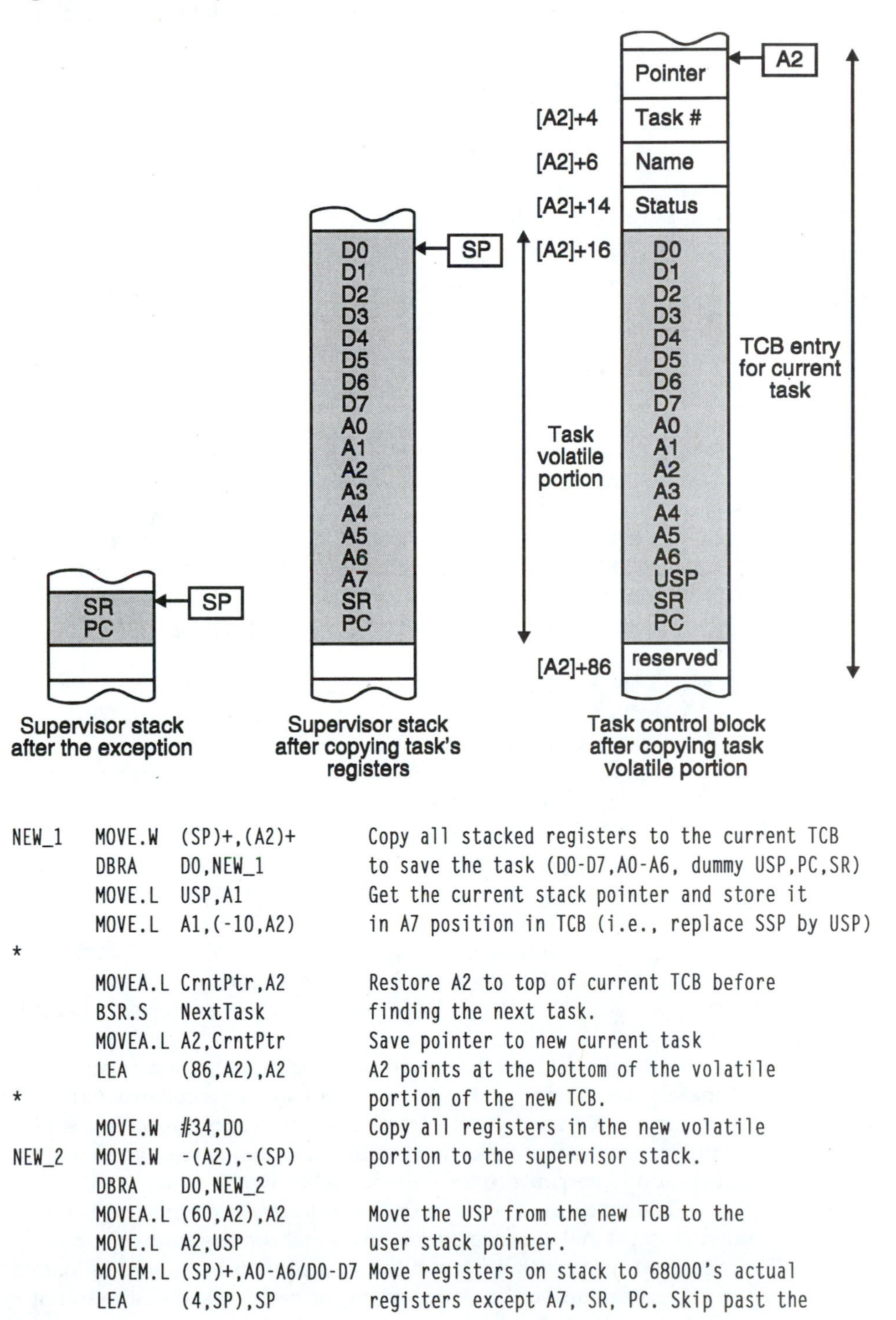


```
NEW_1   MOVE.W   (SP)+,(A2)+        Copy all stacked registers to the current TCB
        DBRA     D0,NEW_1           to save the task (D0-D7,A0-A6, dummy USP,PC,SR)
        MOVE.L   USP,A1             Get the current stack pointer and store it
        MOVE.L   A1,(-10,A2)        in A7 position in TCB (i.e., replace SSP by USP)
*
        MOVEA.L  CrntPtr,A2         Restore A2 to top of current TCB before
        BSR.S    NextTask           finding the next task.
        MOVEA.L  A2,CrntPtr         Save pointer to new current task
        LEA      (86,A2),A2         A2 points at the bottom of the volatile
*                                   portion of the new TCB.
        MOVE.W   #34,D0             Copy all registers in the new volatile
NEW_2   MOVE.W   -(A2),-(SP)        portion to the supervisor stack.
        DBRA     D0,NEW_2
        MOVEA.L  (60,A2),A2         Move the USP from the new TCB to the
        MOVE.L   A2,USP             user stack pointer.
        MOVEM.L  (SP)+,A0-A6/D0-D7  Move registers on stack to 68000's actual
        LEA      (4,SP),SP          registers except A7, SR, PC. Skip past the
```

```
*                                  A7 position on the stack by loading PC, SR
         RTE                       from the stack into the PC and SR.
*
*
*
*******************************************************************************
*
*  Next_task locates the next runnable task
*  A2 = Current_pointer
*  A1 = Temp_pointer
*  A0 = Next_pointer
*  D2 = Current_priority
*  D1 = Temp_priority
*  D0 = Next_priority
*  A2 imports Current_pointer and exports new Current_pointer
*
NextTask MOVEM.L A0-A1/D0-D2,-(SP) Save working registers
         LEA     (A2),A1           Preset ptrs: Temp_pointer := Current_pointer
         LEA     (A2),A0                        Next_ptr:=Current_ptr
         MOVE.W  (14,A2),D2                     D2 := Current_TSW
         ANDI.W  #$001C,D2         Mask D2 (TSW) to priority bits TSW(2:4)
         MOVE.W  D2,D1             D1 := Temp_priority
         MOVE.W  D2,D0             D0 := Next_priority
*
NEXT_1   CMP.W   D1,D0             REPEAT IF Next_priority < Temp_priority
         BMI.S   NEXT_2                   THEN locate next TCB in list
         MOVE.W  D0,D1                    ELSE Temp_priority := Next_priority
         LEA     (A0),A1                       Temp_pointer := Next_pointer
NEXT_2   MOVEA.L (A0),A0                  Locate next TCB: Next_ptr := [Next_ptr]
         MOVE.W  (14,A0),D0               Get TSW of next task in list
         ANDI.W  #$001C,D0                Mask D0 (new TSW) to priority bits
         CMPA.L  A0,A2             UNTIL Current_ptr = Temp_ptr
         BNE     NEXT_1
         LEA     (A1),A2           A2 = new current task = temp task
         MOVEM.L (SP)+,A0-A1/D0-D2 Restore working registers
         RTS                       Return with A2 pointing at new TCB
```

Having sketched the structure of the kernel, the next step is to test it. Unfortunately, we cannot perform a test until we have constructed an *environment* for the kernel. We need a set of task control blocks (initially loaded with dummy values), plus a set of tasks that can be run in sequence. We also need some method of creating interrupts to force the kernel to switch tasks.

The following code demonstrates a simple environment for the kernel. Task switching is not performed by an asynchronous interrupt but by the TRAP #0 exception. Each task control block occupies 88 bytes and is loaded with the following information: the entry point for each task (i.e., the initial PC value), the initial status register for each task, and the task status word containing the priority (in this case, 4). The address of each of these items is given by the offsets with respect to the pointer to the start of the TCBs, A0.

Tasks 2 and 3 simply display sequences of 2s and 3s on the VDT, respectively. However, task 1 asks for a priority level and loads it into the task status word for task 3. This feature allows us to modify the priority level of task 3 while the system is running. All tasks use the Teesside 68000 simulator for input/output transaction (i.e., via TRAP #15).

Initially, all three tasks run. If you enter a priority level of 4 after the prompt, the tasks continue to run in sequence. If you enter 3 or less, task 3 does not run. If you enter 5 or more, only task 3 runs.

We set the initial value of the stack pointer to $2006. The system is initialized by pushing the PC and SR of the first task on the stack, and then jumping to the the task switcher. These two push operations move up the stack pointer by a total of 6 bytes to $2000 (remember that the stack grows *up* to *lower* addresses). This is a nice easy figure to remember and simplifies tracing through the program.

```
Source file: REALTEST.X68
Assembled on: 92-09-04 at: 00:45:29
        by: X68K PC-1.8 Copyright (c) University of Teesside 1989,92

 1                          *       Program to test the real-time kernel
 2 000080                           ORG     $80             ;Location of TRAP #0 vector
 3 000080 00000510                  DC.L    KERNEL          ;Pointer to exception handler
 4 000400                           ORG     $400            ;Program origin
 5 000400 4FF82006                  LEA     $2006,A7        ;Set up the supervisor stack pointer
 6 000404 41F900001000              LEA     TCB,A0          ;Point to the area of three TCBs
 7 00040A 323C0041                  MOVE.W  #65,D1          ;Clear memory space for three TCBs
 8 00040E 4298              CLEAR:  CLR.L   (A0)+
 9 000410 51C8FFFC                  DBRA    D0,CLEAR
10 000414 41F900001000              LEA     TCB,A0
11 00041A 217C000004B4              MOVE.L  #TASK1,82(A0)   ;Set up the initial PCs in the three
          0052
12 000422 217C000004E4              MOVE.L  #TASK2,170(A0)  ;TCBs. A PC is offset by 82 bytes
          00AA
13 00042A 217C000004FA              MOVE.L  #TASK3,258(A0)
          0102
14 000432 317C27000050              MOVE.W  #$2700,80(A0)   ;Set up the initial SRs in the three
15 000438 317C270000A8              MOVE.W  #$2700,168(A0)  ;TCBs. An SR is offset by 80 bytes
16 00043E 317C27000100              MOVE.W  #$2700,256(A0)
17 000444 317C0013000E              MOVE.W  #%010011,14(A0) ;Set up three initial Task Status Words
18 00044A 317C00130066              MOVE.W  #%010011,102(A0) ;in the TCBs. A TSW is offset by 14 bytes
19 000450 317C001300BE              MOVE.W  #%010011,190(A0)
20 000456 20BC00001058              MOVE.L  #TCB+88,(A0)    ;Set up each Pointer to Next Task in the
21 00045C 217C000010B0              MOVE.L  #TCB+176,88(A0) ;TCBs. This makes a ring of TCBs.
          0058
```

```
22 000464 217C00001000          MOVE.L  #TCB,176(A0)
          00B0
23 00046C 23FC00001000          MOVE.L  #TCB,CRNTPTR      ;Make sure that the kernel is pointing to a task
          001108
24 000476 2F3C000004B4          MOVE.L  #TASK1,-(A7)      ;Create an artificial exception by pushing
25 00047C 3F3C2700              MOVE.W  #$2700,-(A7)      ;the PC and SR for Task1
26 000480 6000008E              BRA     KERNEL            ;Force a first jump to the kernel
27 000484 4E722700              STOP    #$2700
28 000488 2A2A2A544153 MESS1:   DC.B    '***TASK 1*** Input priority for task 3  ',0
          4B20312A2A2A
          20496E707574
          207072696F72
          69747920666F
          72207461736B
          2033202000
29 0004B2 00000002              DS.W    1                 ;Force an even address!
30                      *
31 0004B4 41F80488      TASK1:  LEA     MESS1,A0          ;Ask for a task3 priority
32 0004B8 1218          TASK1A: MOVE.B  (A0)+,D1
33 0004BA 6700000A              BEQ     TASK1B
34 0004BE 103C0006              MOVE.B  #6,D0
35 0004C2 4E4F                  TRAP    #15
36 0004C4 60F2                  BRA     TASK1A
37 0004C6 103C0005      TASK1B: MOVE.B  #5,D0             ;Read the priority
38 0004CA 4E4F                  TRAP    #15
39 0004CC 04010030              SUB.B   #$30,D1           ;Convert ASCII to priority level
40 0004D0 E509                  LSL.B   #2,D1             ;Move it to the priority field
41 0004D2 023900E30000          ANDI.B  #$E3,TCB+191      ;Clear old priority
          10BF
42 0004DA 8339000010BF          OR.B    D1,TCB+191        ;Store new priority for task3
43 0004E0 4E40                  TRAP    #0                ;Call kernel to switch tasks
44 0004E2 60D0                  BRA     TASK1             ;and loop back on return
45                      *
46 0004E4 343C0031      TASK2:  MOVE.W  #49,D2            ;Task 2 just prints a string of 50 "2"s
47 0004E8 123C0032      TASK2A: MOVE.B  #'2',D1
48 0004EC 103C0006              MOVE.B  #6,D0
49 0004F0 4E4F                  TRAP    #15
50 0004F2 51CAFFF4              DBRA    D2,TASK2A
51 0004F6 4E40                  TRAP    #0                ;Call kernel to switch tasks
52 0004F8 60EA                  BRA     TASK2             ;and loop back on return
53                      *
54 0004FA 343C0064      TASK3:  MOVE.W  #100,D2           ;Task 3 just prints 100 "3"s
55 0004FE 123C0033      TASK3A: MOVE.B  #'3',D1
56 000502 103C0006              MOVE.B  #6,D0
57 000506 4E4F                  TRAP    #15
58 000508 51CAFFF4              DBRA    D2,TASK3A
```

```
 59 00050C 4E40             TRAP    #0
 60 00050E 60EA             BRA     TASK3
 61                   *
 62        00000510   KERNEL:  EQU     *              ;Entry point for the exception handler
 63                   *
 64                   * Newtask switches tasks by saving the volatile portion of the current task in
 65                   * its TCB, transferring the volatile portion of the next task to run to the
 66                   * supervisor stack, and then copying all registers on the stack to the 68000's
 67                   * registers. The new task is then run by copying the new PC and the SR from the
 68                   * TCB to the supervisor stack, and then executing an RTE. Newtask runs in the
 69                   * supervisor mode and the supervisor stack is active. Note that all tasks are
 70                   * assumed runnable - bits 1,0 of the TSW are not used in this example.
 71                   * Note: All registers (including A7 = SSP) are saved on the supervisor stack at
 72                   * the beginning of the exception processing routine. The saved value of the SSP
 73                   * is later overwritten with the USP. Doing this, rather than leaving a "space"
 74                   * for the USP on the stack, simplifies the coding.
 75                   *
 76                   * A2 = CrntPtr        (points to the TCB of the current task)
 77                   * A1 = TempPtr        (temporary pointer to a TCB during the search)
 78                   * A0 = NextPtr        (pointer to TCB on next task in chain of TCBs)
 79                   * D2 = CrntPrty       (the priority of the current task)
 80                   * D1 = TempPrty       (the priority of a task during the search)
 81                   * D0 = NextPrty       (the priority of the next task in the chain of TCBs)
 82                   *
 83 000510 48E7FFFF   NEWTASK: MOVEM.L A0-A7/D0-D7,-(SP) ;Save all registers on supervisor stack
 84 000514 247900001108        MOVE.L  CRNTPTR,A2       ;Pick up the pointer to this (the current
 85 00051A 45EA0010            LEA     16(A2),A2        ;task) and point at its volatile portion.
 86 00051E 303C0022            MOVE.W  #34,D0
 87 000522 34DF       NEW_1:   MOVE.W  (SP)+,(A2)+      ;Copy all stacked registers to the current TCB
 88 000524 51C8FFFC            DBRA    D0,NEW_1         ;to save the task (D0-D7,A0-A6, dummy USP,PC,SR)
 89 000528 4E69                MOVE.L  USP,A1           ;Get the current stack pointer and store it
 90 00052A 2549FFF6            MOVE.L  A1,-10(A2)       ;in A7 position in TCB (i.e., replace SSP by USP)
 91 00052E 247900001108        MOVEA.L CRNTPTR,A2       ;Restore A2 to top of current TCB before
 92 000534 6124                BSR.S   NEXTTASK         ;finding the next task.
 93 000536 23CA00001108        MOVE.L  A2,CRNTPTR       ;Save pointer to new current task
 94 00053C 45EA0056            LEA     86(A2),A2        ;A2 points at the bottom of the volatile
 95                   *                                  portion of the new TCB.
 96 000540 303C0022            MOVE.W  #34,D0           ;Copy all registers in the new volatile
 97 000544 3F22       NEW_2:   MOVE.W  -(A2),-(SP)      ;portion to the supervisor stack.
 98 000546 51C8FFFC            DBRA    D0,NEW_2
 99 00054A 246A003C            MOVEA.L 60(A2),A2        ;Move the USP from the new TCB to the
100 00054E 4E62                MOVE.L  A2,USP           ;user stack pointer.
101 000550 4CDF7FFF            MOVEM.L (SP)+,A0-A6/D0-D ;Move registers on stack to 68000's actual
102 000554 4FEF0004            LEA     4(SP),SP         ;registers except A7, SR, PC. Skip past the
103                   *                                  A7 position on the stack by loading PS, SR
104 000558 4E73                RTE                      ;and return from exception
```

```
105                            *
106                            * Next_task locates the next runnable task
107                            * A2 = Current_pointer
108                            * A1 = Temp_pointer
109                            * A0 = Next_pointer
110                            * D2 = Current_priority
111                            * D1 = Temp_priority
112                            * D0 = Next_priority
113                            * A2 imports Current_pointer and exports new Current_pointer
114                            *
115 00055A 48E7E0C0   NEXTTASK: MOVEM.L A0-A1/D0-D2,-(SP) ;Save working registers
116 00055E 43D2                 LEA     (A2),A1         ;Preset ptrs: Temp_pointer := Current_pointer
117 000560 41D2                 LEA     (A2),A0         ;Next_ptr:=Current_ptr
118 000562 342A000E             MOVE.W  14(A2),D2       ;D2 := Current_TSW
119 000566 0242001C             ANDI.W  #$001C,D2       ;Mask D2 (TSW) to priority bits TSW(2:4)
120 00056A 3202                 MOVE.W  D2,D1           ;D1 := Temp_priority
121 00056C 3002                 MOVE.W  D2,D0           ;D0 := Next_priority
122 00056E B041       NEXT_1:   CMP.W   D1,D0           ;REPEAT IF Next_priority < Temp_priority
123 000570 6B04                 BMI.S   NEXT_2          ;THEN locate next TCB in list
124 000572 3200                 MOVE.W  D0,D1           ;ELSE Temp_priority := Next_priority
125 000574 43D0                 LEA     (A0),A1         ;Temp_pointer := Next_pointer
126 000576 2050       NEXT_2:   MOVE.L  (A0),A0         ;Locate next TCB: Next_ptr := [Next_ptr]
127 000578 3028000E             MOVE.W  14(A0),D0       ;Get TSW of next task in list
128 00057C 0240001C             ANDI.W  #$001C,D0       ;Mask D0 (new TSW) to priority bits
129 000580 B5C8                 CMPA.L  A0,A2           ;UNTIL Current_ptr = Temp_ptr
130 000582 66EA                 BNE     NEXT_1
131 000584 45D1                 LEA     (A1),A2         ;A2 = new current task = temp task
132 000586 4CDF0307             MOVEM.L (SP)+,A0-A1/D0-D2 ;Restore working registers
133 00058A 4E75                 RTS                     ;Return with A2 pointing at new TCB
134                            *
135 001000                      ORG     $1000           ;Set up the three TCBs at location $1000
136 001000 00000108   TCB:      DS.B    3*88
137 001108 00000004   CRNTPTR:  DS.L    1               ;Reserve a longword for the global variable
138                            *
139        000400               END     $400
```

11.2 A Bootstrap Monitor

When microprocessors first became widely available, semiconductor manufacturers introduced single-board computers, SBCs, for both educational and evaluation purposes. An SBC contains a ROM-based program that enables it to control a serial port connected to a display. This program is sometimes called a *monitor*, and permits the user to examine and modify memory locations, transfer blocks of data between a host computer and the SBC, and execute programs once they have been loaded. More sophisticated monitors include facilities for single-stepping through a program, setting breakpoints, and assembling programs. Modern 68000-based single-board computers can be interfaced to a host computer, such as a PC, via a serial interface. Software can be written on the PC, cross-assembled or cross-compiled, and then downloaded to the target system (i.e., the SBC).

The monitor presented here is very basic with only memory examine/modify, load/dump, and program execution facilities. Data is loaded into memory or transmitted from it in the form of *S-records*. A Motorola S-record is a block of hexadecimal characters beginning with the letter "S" that contains a byte-count, an address, an optional sequence of data elements, and a checksum that provides a limited error detection facility. There are several types of S record — an S1 record is used to transfer data to a 16-bit address, an S9 record is used to transfer a single address, an S0 record is used to transfer a character string or name, and an S8 record is used to transfer data to 24-bit addresses. An S1 record has the format:

```
S1<length><load address><data field><checksum>.
```

The header "S1" indicates the type of record (i.e., a data record with a 16-bit load address). The length field defines the number of hexadecimal pairs (i.e., bytes) in the record excluding the header and record length. This field is therefore the number of bytes in the data field plus 3. The data field is the data to be loaded in memory beginning at the load address. The checksum is the 1's complement of all bytes in the record length, address, and data fields. We will look at the way in which the Teesside cross-assembler generates S-records. Consider the following block of meaningless code, S_RECORD.X68.

```
*         S record test
          ORG     $400
          NOP
          ADD.L   #5,D2
          STOP    #$2700
          DC.B    $12,$34,$56
          END     $400
```

When this code is assembled using the command `X68K S_RECORD -L` the output of the cross-assembler is:

```
Source file: S_RECORD.X68
Assembled on: 93-02-27 at: 11:54:45
          by: X68K PC-2.0 Copyright (c) University of Teesside 1989,93
Defaults: ORG $0/FORMAT/OPT A,BRL,CEX,CL,FRL,MC,MD,NOMEX,NOPCO

    1                                *        S record test
    2  00000400                                ORG       $400
    3  00000400 4E71                           NOP
```

```
4  00000402 5A82              ADD.L     #5,D2
5  00000404 4E722700          STOP      #$2700
6  00000408 123456            DC.B      $12,$34,$56
7           00000400          END       $400
```

```
Lines: 7, Errors: 0, Warnings: 0.
```

The output of the assembler can be converted into S-record format by using the command S68K S_RECORD. This produces the following three records.

```
S0030000FC
S10F04004E715A824E72270012345600CE
S9030400F8
```

The first record is an S0 record that is null because the program does not have a name. The second record is an S1 record containing the data produced by the cross-assembler. The length of the S1 record is $0F = 15 bytes. This means that there are two address bytes ($0400), 12 data bytes, and a checksum ($F8). If you examine the data field of this record, you will see that it contains the code produced by the cross-assembler.

The third record is an S9 record and contains the *transfer address*. The transfer address is the point at which the program is to begin executing (i.e., $400).

Structure of the Monitor

The monitor is, however, extensible and can be modified by adding new facilities. We have included this monitor because it contains a library of useful assembly language routines, such as a hexadecimal input/output.

The monitor can be expressed at its highest level of abstraction in PDL as follows:

```
Module Monitor
      Initialize
      Print banner heading
      REPEAT
            NewLine
            Read command
            CASE OF
                 MEMORY: Display or modify memory
                 JUMP:   Execute a program
                 LOAD:   Load data into memory
                 DUMP:   Send data block from memory
            END_CASE
      FOREVER
End Monitor
```

In many ways the monitor is an easy program to design because it has a naturally modular structure. All we require is a loop that inputs a command and a CASE structure that calls the appropriate module.

The first design decision concerns the way in which the input is processed. Some monitors input a character to determine the command, and then request any further input in the appropriate module. An alternative approach is to assemble a command together with any parameters in a buffer before determining the function to be called. We have adopted this approach because of its generality and because it permits you to edit commands. That is, you can erase mistyped parameters by means of a backspace.

Another initial design decision is to define parameters that modify the operation of the I/O system. For example, we can force lowercase input into uppercase form by setting an automatic case conversion flag. Since an error may occur in the monitor's routines (e.g., the entry of a non-hex character when a hex character is expected), a global error flag is defined and initially cleared. We can now expand the Initialize routine and the monitor's main loop.

```
Module Monitor
     Set automatic input-echo flag
     Set automatic lowercase to uppercase conversion flag
     Print banner headline
     Read longword at location $C000
     IF [$C000] = 'ROM2' THEN Call routine at $C008
     NOP {3 NOPs allow for a future call to auxiliary routine}
     NOP
     NOP
     REPEAT {main loop}
           Clear error flag
           NewLine
           GetLine {read command}
           Tidy    {clean up the command}
           Execute {call the CASE interpreter}
      FOREVER
End Monitor
```

When I built a single-board computer, I wanted to be able to extend the functionality of the monitor by plugging in a supplementary ROM. Although I could have used the monitor's JUMP <address> facility, I decided to add a mechanism that would automatically recognize an extension monitor. The monitor initialization routine reads location $C000_{16}$, which is the starting point of the extension ROM (if any). If the monitor finds that the first longword in the extension ROM contains the ASCII string "ROM2", a subroutine at location $C008_{16}$ is called. This subroutine can be used to activate any of the extension functions.

I included three NOPs in the initialization code to allow me to call an additional routine at a later date. These NOPs take six bytes, allowing me to replace them by a JSR <address> without modifying later addresses.

The `GetLine` routine assembles up to 64 ASCII-encoded characters in a buffer. This routine stores characters one by one as they are entered. However, a backspace moves the pointer back one character and a *control A* character rejects the current line and reruns the routine.

The `Tidy` routine accepts the data in the character buffer and cleans it up. This routine removes any leading spaces before the command, and any multiple spaces between parameters are converted into a single space. For example, the string

```
    MEMORY      1234        1300
```

would be transformed into the string

```
MEMORY 1234 1300.
```

Once the command string has been entered and tidied up, the subroutine `Execute` searches a command table for the first word in the input buffer, locates the address of the appropriate function, and transfers control to it. An important design decision at this stage was the structure of the command table. I decided to make the structure of the command flexible. First, the length of a command is not fixed. Second, the number of characters in a command that need be matched can be specified. Consider the following two possible entries in the table:

```
DC.B   6,4
DC.B   'MEMORY'
DC.L   <Address_of_Memory_handler>
DC.B   4,4
DC.B   'JUMP'
DC.L   <Address_of_Jump_handler>
```

The first two constants are 6 and 4, and indicate that the following string is six characters long and that four characters of `MEMORY` need to be entered to guarantee recognition of the command. The longword following the string is the actual entry point of the command. You can step through successive entries in the command table by means of:

```
     LEA      (ComTab,PC),A3   A3 points at the start of the table
     CLR.W    D0               Clear the offset
Next MOVE.B   (A3),D0          Read length of string in this entry
     BEQ      Exit             If length = 0 THEN end of table
     LEA      (6,A3,D0.W),A4   Calculate the address of the next entry
     .
     .                         Do the string matching here
     .
     LEA      (A4),A3          Update pointer to next entry
     BRA      Next
```

This code well illustrates the use of indexed addressing. The operation MOVE.B (A3),D0 loads D0 with the first byte of the current entry, which is the length of the string of characters that defines the entry. The instruction LEA (6,A3,D0.W),A4 loads A4 with the address of the current entry (i.e., A3) plus the length of the character string (i.e., D0) plus 6. The constant 6 is made up of the two bytes that define the string length at the start of the entry and the longword transfer address at the end of the entry.

In order to make the Execute routine as flexible as possible, an additional user-supplied table in RAM can be searched for a command before searching the command table within the monitor. The Execute routine tests a longword pointer, UserTable, that is initially cleared by the initialization routine. If UserTable is not zero (i.e., it has been loaded by the programmer with a pointer), the command table at the address in UserTable is first searched for the command. We can now express the Execute routine in PDL.

```
Execute
      IF UserTable ≠ 0
          THEN Search Table pointed at by UserTable
                 IF Command found THEN Execute it END_IF
          ELSE Search CommandTable
                 IF Command found THEN Execute it
                                       ELSE Print error message and return
                 END_IF
      END_IF
End Execute
```

We will not go into the detailed design of the functions provided by the monitor, as the commented code should be self-explanatory. The most useful routines are those like HEX and BYTE that input hexadecimal data, and OUT1X and OUT2X that output hexadecimal data in ASCII form.

```
*
*  This is a simple monitor for a single-board computer
*  27 August 1992
*                                     Symbol equates
BS         EQU      $08               Back_space
CR         EQU      $0D               Carriage_return
LF         EQU      $0A               Line_feed
SPACE      EQU      $20               Space
WAIT       EQU      'W'               Wait character (to suspend output)
CTRL_A     EQU      $01               Control_A forces return to monitor
STACK      EQU      $00002000         Stack_pointer
MAXCHR     EQU      64                Length of input line buffer
           ORG      $00000400         Data origin
DATA       EQU      *
LNBUFF     DS.B     MAXCHR            Input line buffer
```

```
BUFFEND  EQU     LNBUFF+MAXCHR-1 End of line buffer
BUFFPT   DS.L    1               Pointer to line buffer
PARAMTR  DS.L    1               Last parameter from line buffer
ECHO     DS.B    1               When clear this enables input echo
U_CASE   DS.B    1               Flag for uppercase conversion
UTAB     DS.L    1               Pointer to user command table
*
*************************************************************************
*
*  This is the main program that assembles a command in the line buffer,
*  removes leading or embedded spaces, and interprets the command by
*  matching it with a command in the user table or the internal table
*  COMTAB. The locations of variables are specified with respect to A6.
*
         ORG     $00001000       Monitor origin
RESET    EQU     *               Cold entry point for monitor
         CLR.L   UTAB            Reset pointer to user extension table
         CLR.B   ECHO            Set automatic character echo
         CLR.B   U_CASE          Clear case conversion flag (lower to upper)
         BSR.S   SetPeris        Set up peripherals (required for an ACIA, etc)
         LEA     BANNER(PC),A4   Point to banner
         BSR.S   HEADING         and print heading
         MOVEA.L #$0000C000,A0   Check for extension EPROM at $C000
         MOVE.L  (A0),D0         IF   first longword in extension ROM
         CMP.L   #'ROM2',D0           begins with 'ROM2'
         BNE.S   NO_EXT               THEN call the subroutine at EXT_ROM+8
         JSR     (8,A0)                ELSE continue
NO_EXT   NOP                     Three NOPs to allow for a future
         NOP                     call to an initialization routine
         NOP
WARM     CLR.L   D7              REPEAT Warm entry point - clear error flag
         BSR.S   NEWLINE                Print a new line
         BSR.S   GETLINE                Get a command line
         BSR     TIDY                   Tidy up the input buffer contents
         BSR     EXECUTE                Interpret command
         BRA     WARM            FOREVER
*
*************************************************************************
*
*  Some initialization and basic routines
*
SetPeris NOP                     Set up peripherals
         NOP                     If you are going to use an ACIA or a DUART
         NOP                     for I/O, you will have to configure them
         RTS                     before you use them. This is dummy routine.
*
```

```
NEWLINE  EQU      *               Move the cursor to start of a new line
         MOVE.L   A4,-(A7)        Save A4
         LEA      CRLF(PC),A4     Point to CR/LF string
         BSR.S    PSTRING         Print it
         MOVEA.L  (A7)+,A4        Restore A4
         RTS                      Return
*
PSTRING  EQU      *               Display the ASCII string pointed at by A4
         MOVE.L   D0,-(A7)        Save D0
PS1      MOVE.B   (A4)+,D0        WHILE character to be printed
         BEQ.S    PS2                   is not a null
         BSR      PUTCHAR               print it
         BRA      PS1             END_WHILE
PS2      MOVE.L   (A7)+,D0        Restore D0 and exit
         RTS
*
HEADING  BSR      NEWLINE         Same as PSTRING but with a new line
         BSR      PSTRING
         BRA      NEWLINE
*
*************************************************************************
*
*  GETLINE  inputs a string of characters into a line buffer
*           A3 points to next free entry in line buffer
*           A2 points to end of buffer
*           A1 points to start of buffer
*           D0 holds character to be stored
*
GETLINE  LEA      LNBUFF,A1       A1 points to the start of the line buffer
         LEA      (A1),A3         A3 points to the start (initially)
         LEA      MAXCHR(A1),A2   A2 points to the end of the buffer
GETLN2   BSR      GETCHAR         Get a character
         CMP.B    #CTRL_A,D0      IF control_A THEN reject this line
         BEQ.S    GETLN5          and get another line
         CMP.B    #BS,D0          IF back_space THEN move back pointer
         BNE.S    GETLN3                         ELSE skip past wind-back routine
         CMP.L    A1,A3           First check for empty buffer
         BEQ      GETLN2          IF buffer empty THEN continue
         LEA      -1(A3),A3                         ELSE decrement buffer pointer
         BRA      GETLN2                            and continue with next character
GETLN3   MOVE.B   D0,(A3)+        Store character and update pointer
         CMP.B    #CR,D0          Test for command terminator
         BNE.S    GETLN4          IF not CR THEN skip past exit procedure
         BRA      NEWLINE                    ELSE new line before next operation
GETLN4   CMP.L    A2,A3           Test for buffer overflow
         BNE      GETLN2          IF buffer not full THEN continue
```

```
GETLN5   BSR      NEWLINE                                  ELSE move to next line and
         BRA      GETLINE                                     repeat this routine
*
**************************************************************************
*
*  TIDY cleans up the line buffer by removing leading spaces and multiple
*       spaces between parameters. At the end of TIDY, BUFFPT points to
*       the first parameter following the command.
*       A0 = pointer to line buffer. A1 = pointer to cleaned-up buffer
*
TIDY     LEA      LNBUFF,A0         A0 points to the line buffer
         LEA      (A0),A1           A1 points to the start of line buffer
TIDY1    MOVE.B   (A0)+,D0          REPEAT Read character from line buffer
         CMP.B    #SPACE,D0
         BEQ      TIDY1             UNTIL non-space character is found
         LEA      -1(A0),A0         Move pointer back to first character
TIDY2    MOVE.B   (A0)+,D0          Move the string left to remove
         MOVE.B   D0,(A1)+          any leading spaces
         CMP.B    #SPACE,D0         Test for embedded space
         BNE.S    TIDY4             IF not space THEN test for end_of_line
TIDY3    CMP.B    #SPACE,(A0)+      IF space THEN skip multiple embedded spaces
         BEQ      TIDY3
         LEA      -1(A0),A0         Move back pointer
TIDY4    CMP.B    #CR,D0            Test for end_of_line
         BNE      TIDY2             IF not end_of_line THEN read next char
         LEA      LNBUFF,A0         Restore buffer pointer
TIDY5    CMP.B    #CR,(A0)          Test for end_of_line
         BEQ.S    TIDY6             IF end_of_line THEN exit
         CMP.B    #SPACE,(A0)+      Test for delimiter
         BNE      TIDY5             REPEAT UNTIL delimiter or end_of_line
TIDY6    MOVE.L   A0,BUFFPT         Update buffer pointer
         RTS
*
**************************************************************************
*
*  EXECUTE matches the first command in the line buffer with the commands
*  in a command table. An external table pointed at by UTAB is searched
*  first and then the built-in table, COMTAB.
*
EXECUTE  TST.L    UTAB              Test pointer to user table
         BEQ.S    EXEC1             IF clear THEN try built-in table
         MOVE.L   UTAB,A3                    ELSE pick up pointer to user table
         BSR.S    SEARCH            Look for command in user table (use C-bit as flag)
         BCC.S    EXEC1             IF not found THEN try internal table
         MOVE.L   (A3),A3                    ELSE get address of command
         JMP      (A3)                            from user table and execute it
```

```
*
EXEC1   LEA     COMTAB(PC),A3   Try built-in command table, COMTAB
        BSR.S   SEARCH          Look for command in COMTAB
        BCS.S   EXEC2           IF found THEN execute command
        LEA     ERMES2(PC),A4           ELSE print "invalid command"
        BRA.L   PSTRING                      and return
EXEC2   MOVE.L  (A3),A3         Get the relative command address
        LEA     COMTAB(PC),A4   pointed at by A3 and add it to
        ADD.L   A4,A3           the PC to generate the actual
        JMP     (A3)            command address. Then execute it
*
SEARCH  EQU     *               Match the command in the line buffer with the
        CLR.L   D0              commands in the command table pointed at by A3
        MOVE.B  (A3),D0         Get the first character in the current entry
        BEQ.S   SRCH7           IF zero THEN exit
        LEA     6(A3,D0.W),A4           ELSE calculate address of next entry
        MOVE.B  1(A3),D1        Get the number of characters to match
        LEA     LNBUFF,A5       A5 points to the command in the line buffer
        MOVE.B  2(A3),D2        Get the first character in this entry
        CMP.B   (A5)+,D2        from the table and match with buffer
        BEQ.S   SRCH3           IF match THEN try the rest of the string
SRCH2   MOVE.L  A4,A3                   ELSE get the address of the next entry
        BRA     SEARCH                  and try the next entry in the table
SRCH3   SUB.B   #1,D1           One less character to match
        BEQ.S   SRCH6           IF match counter is zero THEN all done
        LEA     3(A3),A3                ELSE point to next character in table
SRCH4   MOVE.B  (A3)+,D2        Now match a pair of characters
        CMP.B   (A5)+,D2
        BNE     SRCH2           IF no match THEN try next entry
        SUB.B   #1,D1                   ELSE decrement match counter and
        BNE     SRCH4           repeat until no chars left to match
SRCH6   LEA     -4(A4),A3       Calculate address of command entry
        OR.B    #1,CCR          point. Mark carry flag as success (C-bit = 1)
        RTS                     and return
SRCH7   AND.B   #$FE,CCR        Fail - clear carry flag to indicate
        RTS                     command not found and return
*
*************************************************************************
*
*  Basic input routines
*  HEX    =  Get one   hexadecimal character  into D0
*  BYTE   =  Get two   hexadecimal characters into D0
*  WORD   =  Get four  hexadecimal characters into D0
*  LONGWD =  Get eight hexadecimal characters into D0
*  PARAM  =  Get a longword from the line buffer into D0
*  Bit 0 of D7 is set to indicate a hexadecimal input error
```

```
HEX      BSR     GETCHAR        Get a character N from input device
         SUB.B   #$30,D0        N := N - $30 to convert to binary
         BMI.S   NOT_HEX        IF N < $30 THEN exit with error
         CMP.B   #$09,D0                  ELSE test for number (0 to 9)
         BLE.S   HEX_OK                   IF N ≤ 9 THEN EXIT OK
         SUB.B   #$07,D0                           ELSE convert letter to hex
         CMP.B   #10,D0         IF N < 10 THEN EXIT to not hex
         BMI.S   NOT_HEX
         CMP.B   #$0F,D0        IF character in range "A" to "F"
         BLE.S   HEX_OK            THEN exit successfully
NOT_HEX  OR.B    #1,D7             ELSE set error flag
HEX_OK   RTS                            and return
*
BYTE     MOVE.L  D1,-(A7)       Save D1
         BSR     HEX            Get first hex character
         ASL.B   #4,D0          Move it to most-significant nibble position
         MOVE.B  D0,D1          Save most-significant nibble in D1
         BSR     HEX            Get second hex character
         ADD.B   D1,D0          Merge most- and least-significant nibbles
         MOVE.L  (A7)+,D1       Restore D1
         RTS
*
WORD     BSR     BYTE           Get upper-order byte
         ASL.W   #8,D0          Move it to MS position
         BRA     BYTE           Get LS byte and return
*
LONGWD   BSR     WORD           Get upper-order word
         SWAP    D0             Move it to most-significant position
         BRA     WORD           Get lower-order word and return
*
*  PARAM reads a parameter from the line buffer and puts it in both
*  PARAMTR and D0. Bit 1 of D7 is set on error.
*
PARAM    MOVE.L  D1,-(A7)       Save D1
         CLR.L   D1             Clear input accumulator
         MOVE.L  BUFFPT,A0      A0 points to the parameter in the buffer
PARAM1   MOVE.B  (A0)+,D0       REPEAT Read character from the line buffer
         CMP.B   #SPACE,D0         Test for delimiter
         BEQ.S   PARAM4            The permitted delimiter is a
         CMP.B   #CR,D0            space or a carriage return
         BEQ.S   PARAM4            EXIT on either a space or C/R
         ASL.L   #4,D1             Shift accumulated result 4 bits left
         SUB.B   #$30,D0           Convert the new character to hex
         BMI.S   PARAM5            IF less than $30 THEN not_hex
         CMP.B   #$09,D0           IF less than 10
         BLE.S   PARAM3               THEN continue
```

```
        SUB.B   #$07,D0           ELSE assume $A to $F
        CMP.B   #$0F,D0         IF more than $F
        BGT.S   PARAM5            THEN exit to error on not_hex
PARAM3  ADD.B   D0,D1           Add latest nibble to total in D1
        BRA     PARAM1        UNTIL delimiter found
PARAM4  MOVE.L  A0,BUFFPT     Save the pointer in memory
        MOVE.L  D1,PARAMTR    Save the parameter in memory
        MOVE.L  D1,D0         Put the parameter in D0 for return
        BRA.S   PARAM6        Return without error
PARAM5  BSET    #0,D7         Set the error flag before return
PARAM6  MOVE.L  (A7)+,D1      Restore working register
        RTS                   Return with error
*
*************************************************************************
*
*  Output routines
*  OUT1X   = print one   hexadecimal character
*  OUT2X   = print two   hexadecimal characters
*  OUT4X   = print four  hexadecimal characters
*  OUT8X   = print eight hexadecimal characters
*  In each case, the data to be printed is in D0
*
OUT1X   MOVE.B  D0,-(A7)      Save D0
        AND.B   #$0F,D0       Mask off the most significant nibble
        ADD.B   #$30,D0       Convert to ASCII
        CMP.B   #$39,D0       ASCII = HEX + $30
        BLS.S   OUT1X1        IF ASCII ≤ $39 THEN print and exit
        ADD.B   #$07,D0                  ELSE ASCII := HEX + 7
OUT1X1  BSR     PUTCHAR       Print the character
        MOVE.B  (A7)+,D0      Restore D0
        RTS
*
OUT2X   ROR.B   #4,D0         Get most-significant nibble in LS position
        BSR     OUT1X         Print most-significant nibble
        ROL.B   #4,D0         Restore least-significant nibble
        BRA     OUT1X         Print least-significant nibble and return
*
OUT4X   ROR.W   #8,D0         Get most-significant byte in LS position
        BSR     OUT2X         Print most-significant byte
        ROL.W   #8,D0         Restore least-significant byte
        BRA     OUT2X         Print least-significant byte and return
*
OUT8X   SWAP    D0            Get most-significant word in LS position
        BSR     OUT4X         Print most-significant word
        SWAP    D0            Restore least-significant word
        BRA     OUT4X         Print least-significant word and return
```

```
*************************************************************************
*
* JUMP causes execution to begin at the address in the line buffer.
* Note that you must terminate your routine with RTS to re-run to the command
* line interpreter
*
JUMP     BSR      PARAM          Get target address from the buffer
         BTST     #0,D7          Test for input error
         BNE.S    JUMP1          IF error flag not zero THEN exit
         TST.L    D0                ELSE test for missing address field
         BEQ.S    JUMP1          IF no address THEN exit
         MOVE.L   D0,A0          Put jump address in A0 and call the
         JMP      (A0)           subroutine. User to supply RTS!!
JUMP1    LEA      ERMES1(PC),A4  Here for error - display error
         BRA      PSTRING        message and return
*
*************************************************************************
*
*  Display the contents of a memory location and modify it
*
MEMORY   BSR      PARAM          Get the start address from the line buffer
         BTST     #0,D7          Test for input error
         BNE.S    MEM3           IF error THEN exit
         MOVE.L   D0,A3          A3 points to the location to be opened
MEM1     BSR      NEWLINE
         BSR.S    ADR_DAT        Print the current address and contents,
         BSR.S    PSPACE          update pointer, A3, and print a space
         BSR      GETCHAR        Input character to decide next action
         CMP.B    #CR,D0         IF carriage return then EXIT
         BEQ.S    MEM3
         CMP.B    #'-',D0        IF "-" THEN move back
         BNE.S    MEM2                  ELSE skip wind-back procedure
         LEA      -4(A3),A3      Move pointer back 2+2 bytes
         BRA      MEM1           REPEAT UNTIL carriage return
MEM2     CMP.B    #SPACE,D0      Test for space (= new entry)
         BNE.S    MEM1           IF not space THEN repeat
         BSR      WORD                          ELSE get new word to store
         TST.B    D7             Test for input error
         BNE.S    MEM3           IF error THEN exit
         MOVE.W   D0,-2(A3)              ELSE store new word
         BRA      MEM1           REPEAT UNTIL carriage return
MEM3     RTS
*
ADR_DAT  MOVE.L   D0,-(A7)       Print the contents of A3 and the
         MOVE.L   A3,D0          word pointed at by A3.
         BSR      OUT8X           and print current address
```

```
        BSR.S   PSPACE          Insert delimiter
        MOVE.W  (A3),D0         Get data at this address in D0
        BSR     OUT4X            and print it
        LEA     2(A3),A3        Point to next address to display
        MOVE.L  (A7)+,D0        Restore D0
        RTS
*
PSPACE  MOVE.B  D0,-(A7)        Print a single space
        MOVE.B  #SPACE,D0
        BSR     PUTCHAR
        MOVE.B  (A7)+,D0
        RTS
*
************************************************************************
*
*  LOAD  Loads data formatted in hexadecimal "S" format
*        S1 or S2 records accepted
*
LOAD    BSR     NEWLINE         Send a new line
        MOVE.L  BUFFPT,A4       Any string in the line buffer is
LOAD1   MOVE.B  (A4)+,D0        transmitted to the host computer
        BSR     PUTCHAR         before the loading begins
        CMP.B   #CR,D0          Read from the buffer until EOL
        BNE     LOAD1
        BSR     NEWLINE         Send a new line before loading records
LOAD2   BSR     GETCHAR         Records from the host must begin
        CMP.B   #'S',D0         with S1/S2 (data) or S9/S8 (term)
        BNE.S   LOAD2           REPEAT GETCHAR UNTIL character = "S"
        BSR     GETCHAR         Get character after "S"
        CMP.B   #'9',D0         Test for the two terminators S9/S8
        BEQ.S   LOAD3           IF S9 record THEN exit ELSE test
        CMP.B   #'8',D0         for S8 terminator. Fall through to
        BNE.S   LOAD6           exit on S8 else continue search
LOAD3   EQU     *               Exit point from LOAD
        CLR.B   ECHO            Restore input character echo
        BTST    #0,D7           Test for input errors
        BEQ.S   LOAD4           IF no I/P error THEN look at checksum
        LEA     ERMES1(PC),A4                           ELSE point to error message
        BSR     PSTRING                                 Print it
LOAD4   BTST    #3,D7           Test for checksum error
        BEQ.S   LOAD5           IF clear THEN exit
        LEA     ERMES3(PC),A4               ELSE point to error message
        BSR     PSTRING                              Print it and return
LOAD5   RTS
LOAD6   CMP.B   #'1',D0         Test for S1 record
        BEQ.S   LOAD6A          IF S1 record THEN read it
```

```
         CMP.B    #'2',D0                     ELSE test for S2 record
         BNE.S    LOAD2            REPEAT UNTIL valid header found
         CLR.B    D3               Read the S2 byte count and address
         BSR.S    LOAD8            Clear the checksum
         SUB.B    #4,D0            Calculate the size of the data field
         MOVE.B   D0,D2            D2 contains the number of data bytes to read
         CLR.L    D0               Clear the address accumulator
         BSR.S    LOAD8            Read most-significant byte of address
         ASL.L    #8,D0            Move it one byte left
         BSR.S    LOAD8            Read the middle byte of the address
         ASL.L    #8,D0            Move it one byte left
         BSR.S    LOAD8            Read the least-significant byte of the address
         MOVE.L   D0,A2            A2 points to destination of record
         BRA.S    LOAD7            Skip past S1 header loader
LOAD6A   CLR.B    D3               S1 record found - clear the checksum
         BSR.S    LOAD8            Get byte and update checksum
         SUB.B    #3,D0            Subtract 3 from record length
         MOVE.B   D0,D2            Save byte count in D2
         CLR.L    D0               Clear address accumulator
         BSR.S    LOAD8            Get MS byte of load address
         ASL.L    #8,D0            Move it to MS position
         BSR.S    LOAD8            Get LS byte in D2
         MOVE.L   D0,A2            A2 points to the destination of the data
LOAD7    BSR.S    LOAD8            Get a byte of data for loading
         MOVE.B   D0,(A2)+         Store it
         SUB.B    #1,D2            Decrement the byte counter
         BNE      LOAD7            REPEAT UNTIL count = 0
         BSR.S    LOAD8            Read the checksum
         ADD.B    #1,D3            Add 1 to the total checksum
         BEQ      LOAD2            IF zero THEN start next record
         OR.B     #%00001000,D7           ELSE set the checksum error bit
         BRA      LOAD3            Restore I/O devices and return
*
LOAD8    BSR      BYTE             Get a byte
         ADD.B    D0,D3            Update checksum
         RTS                       and return
*
***************************************************************************
*  DUMP   Transmit S1 formatted records
*         A3 = Starting address of data block
*         A2 = End address of data block
*         D1 = Checksum, D2 = current record length
*
DUMP     BSR      RANGE            Get start and end address
         TST.B    D7               Test for input error
         BEQ.S    DUMP1            IF no error THEN continue
```

```
         LEA      ERMES1(PC),A4            ELSE point to the error message
         BRA      PSTRING                       print it and return
DUMP1    CMP.L    A3,D0            Compare start and end addresses
         BPL.S    DUMP2            If positive THEN start < end
         LEA      ERMES4(PC),A4            ELSE print the error message
         BRA      PSTRING                       and return
DUMP2    BSR      NEWLINE          Send newline to host and wait
         MOVE.L   BUFFPT,A4        Before dumping, send any string
DUMP3    MOVE.B   (A4)+,D0         in the input buffer to the host
         BSR      PUTCHAR          REPEAT
         CMP.B    #CR,D0                Transmit character from buffer to host
         BNE      DUMP3            UNTIL character = C/R
         BSR      NEWLINE
         ADD.L    #1,A2            A2 contains the length of the record + 1
DUMP4    MOVE.L   A2,D2            D2 points to the end address
         SUB.L    A3,D2            D2 contains the number of bytes left to print
         CMP.L    #17,D2           IF this is not a full record of 16
         BCS.S    DUMP5               THEN load D2 with the record size
         MOVE.L   #16,D2              ELSE preset byte count to 16
DUMP5    LEA      HEADER(PC),A4    Point to the record header
         BSR      PSTRING          Print the header
         CLR.B    D1               Clear the checksum
         MOVE.B   D2,D0            Move the record length to output register
         ADD.B    #3,D0            The length includes address + count
         BSR.S    DUMP7            Print the number of bytes in the record
         MOVE.L   A3,D0            Get the start address to be printed
         ROL.W    #8,D0            Get most-significant byte in least-sig position
         BSR.S    DUMP7            Print the most-significant byte of address
         ROR.W    #8,D0            Restore the least-significant byte
         BSR.S    DUMP7            Print the least-significant byte of the address
DUMP6    MOVE.B   (A3)+,D0         Get the data byte to be printed
         BSR.S    DUMP7            Print it
         SUB.B    #1,D2            Decrement the byte count
         BNE      DUMP6            REPEAT UNTIL all this record has been printed
         NOT.B    D1               Complement the checksum
         MOVE.B   D1,D0            Move the checksum to the output register
         BSR.S    DUMP7            Print the checksum
         BSR      NEWLINE
         CMP.L    A2,A3            Have all records been printed?
         BNE      DUMP4            REPEAT UNTIL all done
         LEA      TAIL(PC),A4      Point to the message tail (S9 record)
         BSR      PSTRING          Print it
         RTS
DUMP7    ADD.B    D0,D1            Update the checksum, transmit a byte
         BRA      OUT2X            to the host and return
RANGE    EQU      *                Get the range of addresses to be
```

```
         CLR.B   D7              transmitted from the buffer
         BSR     PARAM           Get the starting address
         MOVE.L  D0,A3           Set up the start address in A3
         BSR     PARAM           Get the end address
         MOVE.L  D0,A2           Set up the end address in A2
         RTS
*
*  GETCHAR gets a character from the console device
*  The I/O system uses the Teesside 68000 simulator's TRAP #15 to
*  input a character. In a real system you would have to modify this
*  routine to control an ACIA or DUART.
*
GETCHAR  MOVE.L  D0,-(A7)        Save the old value of D0 on stack
         MOVE.L  D1,-(A7)        and the old D1
         NOP                     Leave space for branch to alternative
         NOP                     input routine
         NOP
         MOVE.B  #5,D0           Set up parameter 5 for input operation
         TRAP    #15             Call trap 15 to input a byte into D1
         MOVE.L  (A7)+,D0        Restore the old value of D0
         MOVE.B  D1,D0           Load a new byte in D0.B
         AND.B   #$7F,D0         Strip the most-significant bit from input
         TST.B   U_CASE          Test for upper- to lowercase conversion
         BNE.S   GETCH2          IF flag not zero THEN do not convert case
         BTST    #6,D0           Test input for lowercase character
         BEQ.S   GETCH2          IF uppercase THEN skip conversion
         AND.B   #%11011111,D0       ELSE clear bit 5 for uppercase conversion
GETCH2   TST.B   ECHO            Do we need to echo the input?
         BNE.S   GETCH3          IF ECHO flag not zero THEN no echo
         BSR.S   PUTCHAR                         ELSE echo the input
GETCH3   MOVE.L  (A7)+,D1        Restore the old D1
         RTS                     Return
*
************************************************************************
*
*  PUTCHAR sends a character to the console device
*  The I/O system uses the Teesside 68000 simulator's TRAP #15 to
*  output a character. In a real system you would have to modify this
*  routine to control an ACIA or DUART.
*
PUTCHAR  MOVEM.L D0-D1,-(A7)     Save working registers
         NOP                     Leave space for future branch to
         NOP                     alternative output handler
         NOP
         MOVE.B  D0,D1           The simulator outputs the byte in D1
         MOVE.B  #6,D0           Parameter 6 in D0 for output
```

```
        TRAP    #15
        MOVEM.L (A7)+,D0-D1
        RTS
*
**********************************************************************
*
*  All strings and other fixed parameters are located here
BANNER  DC.B    'SIMPLE MONITOR of 26 August 92',0
CRLF    DC.B    CR,LF,'?',0
HEADER  DC.B    CR,LF,'S','1',0
TAIL    DC.B    'S9  ',0
ERMES1  DC.B    'Non-valid hexadecimal input',0
ERMES2  DC.B    'Invalid command',0
ERMES3  DC.B    'Loading error',0
ERMES4  DC.B    'Range error',0
*
*
*  COMTAB is the built-in command table. All entries are made up of: the length
*        of the string length + the number of characters to match + the string
*        + the address of the command relative to COMTAB
*
        DS.W    1               Force an even address
COMTAB  DC.B    4,4             JUMP <address> causes execution to
        DC.B    'JUMP'          begin at <address>
        DC.L    JUMP-COMTAB
        DC.B    6,3             MEMORY <address> examines contents of
        DC.B    'MEMORY'        <address> and allows them to be changed
        DC.L    MEMORY-COMTAB
        DC.B    4,2             LOAD <string> loads S1/S2 records
        DC.B    'LOAD'          from the host. <string> is sent to host
        DC.L    LOAD-COMTAB
        DC.B    4,2             DUMP <string> sends S1 records to the
        DC.B    'DUMP'          host and is preceded by <string>.
        DC.L    DUMP-COMTAB
        DC.B    0,0
*
*  Data area
*
UTAB    DS.L    1               Location of vector to the user command table
BUFFPT  DS.L    1               The pointer to the current element in the buffer
PARAMTR DS.L    1               A parameter read from the line buffer
ECHO    DS.B    1               Input_echo flag
U_CASE  DS.B    1               Automatic lower- to uppercase conversion flag
LNBUFF  DS.B    64              The line buffer
*
        END     $1000
```

Problems

1. The monitor presented in this chapter is actually part of a monitor I wrote for my first 68000 microprocessor system. The assembler used to construct the monitor lacked macro facilities, etc. Rewrite the above monitor using the facilities of a modern assembler.

2. The monitor has very rudimentary facilities. Expand it by adding facilities that you think might be useful. For example, you might want to clear a block of memory or preset it with a given value. You might want to move a block of memory from one place to another. Some monitors include a simple hexadecimal calculator that can evaluate hexadecimal expressions (e.g., $1234*$0F+23).

3. The monitor's S1-record load and dump routines receive data from the console and display it on the console. This mode of operation is not realistic. A typical single-board computer has a secondary port (i.e., an ACIA or a DUART) that can be connected to a disk-based computer. How would you modify the monitor to provide a more realistic download and upload facility?

4. The monitor is not able to return to the Teesside 68000 simulator's command level (i.e., the only way to terminate the monitor is via the escape key). Implement an escape function.

5. The monitor does not correctly implement character input because it was designed to be used in conjunction with a DUART. What is the problem caused by the `TRAP# 15` function and how would you eliminate it?

CHAPTER 12

The 68020 and 68030 Microprocessors

The differences between the 68000 and the 68020 (or the 68030/68040) are both subtle and radical. We can clarify this seemingly contradictory statement by saying that we can either regard the 68020 as just a faster version of the 68000, or as a super 68000 with some very powerful new instructions and addressing modes. For our present purposes, the 68020 and 68030 are identical, since the 68030 is essentially a 68020 plus a sophisticated memory management unit, MMU, and a data cache. An MMU is employed by the operating system to map the memory allocated to programs onto the actual hardware memory. The 68040 is a high performance 68020 with two MMUs, two 4Kbyte caches and an on-chip floating point unit.

The 68020 is compatible with the 68000's *object code* (i.e., its binary machine code) and its source code (i.e., its assembly language). Indeed, the 68020 will execute all but one of the 68000's instructions. The one exception is `MOVE SR,<ea>` that is *privileged* on the 68020 but not on the 68000. The 68020 implements a new non-privileged instruction, `MOVE CCR,<ea>`, that allows the user-mode programmer to access the CCR.

You don't have to make use of the 68020's special instructions and new addressing modes to achieve a significant increase in its performance over the 68000. The 68020 represents an entirely new and more efficient implementation of the 68000 core machine. That is, it shares the 68000's *architecture* but not its *implementation*. However, you can achieve an even greater improvement if you employ the 68020's new facilities when writing a program.

12.1 Basic Differences between the 68000 and 68020

An immediately obvious difference between the 68020 and the 68000 is the 68020's address and data buses. The 68020's address bus is 32 bits wide, extending the logical address space to 4G bytes, and its data bus is 32 bits wide, making it possible to access an entire longword in a single bus cycle. Unlike the 68000, the 68020 has *internal* 32-bit buses and a 32-bit execution unit (i.e., ALU).

The 68020 implements *dynamic bus sizing,* which means that programs and data can be stored in 8-bit wide, 16-bit wide, or 32-bit wide memory. The 68000 is constrained to use 16-bit wide memory. Dynamic bus sizing means that the systems designer can select 8-bit wide memory for cost, or 32-bit wide memory for performance. This aspect of the 68020's implementation is largely transparent to the programmer who doesn't really care where information is stored. However, you should be aware of it, because you might need to relocate data from a slow 8-bit EPROM in 32-bit wide RAM in order to increase speed.

The 68020 is able to access data operands at *misaligned* boundaries. You can now locate a word or a longword at an odd address without experiencing an address error exception whenever you attempt to access the misaligned operand. You could write:

```
        ORG     $00001001
Test    DS.L    Pointer
        .
        .
        LEA     Pointer,A0
```

The 68020 reads the longword, `Pointer`, in three stages, a byte, a word, and a byte. This process is not efficient — but it is legal. Note that the 68020 does not permit *instructions* to be misaligned.

The 68020 has an on-chip instruction cache that keeps a copy of some of the most recently used instructions, so they don't have to be read from memory when they are next required. The 68030 has both an *instruction* and a *data* cache. These caches are quite small, 256 bytes, but they do provide a useful improvement in performance. In general, the instruction and data caches are invisible to the user programmer, although the operating systems designer must be aware of some of the side effects of cache memory. We briefly return to this point later.

The 68020's enhancements applied with a 16 MHz clock provide a four- to six-fold increase in speed with respect to a 68000 clocked at 8 MHz.

One attribute of the 68000 that makes it so much more sophisticated than its predecessors is its ability to support more operating system functions. Before the advent of the 68000 family, microprocessors were designed to execute a particular instruction set with little regard to the needs of operating systems. To be fair, this approach to microprocessor design was not unreasonable, since before the mid-1980s most microprocessors were employed in systems with no operating system or with minimal operating systems.

Why is an operating system different from an applications program (i.e., a user program running under an operating system)? A simple answer to this question is that the operating system must control the computer and protect it from certain types of harm that a user program may inflict. Suppose an applications program is being tested and the programmer intends to clear a block of 1024 bytes of memory. The programmer makes a mistake in the loop terminator (perhaps by writing `BEQ` instead of `BNE`) and the program does not terminate. Instead, it proceeds to clear all memory. A good operating system will prevent this error causing a total system crash by making certain that user programs cannot access regions of memory not allocated to them. That is, a user program is prevented from interfering with either the operating system or with other user programs. However, full protection requires a memory management unit similar to the type built into the 68030 or the 68040.

Some microprocessors like the 68000 provide only limited help to an operating system, while others like the 68020/30/40 provide a much greater degree of help. In this introduction we are interested only in the 68020's user mode, and will describe the 68020's supervisor mode and its operating system facilities later.

The 68020 and the 68030 have more registers than the 68000. However, all these new registers are dedicated to *operating system* functions. As far as we are concerned, there are still eight data and eight address registers. Even the 68020's condition code register is exactly the same as the 68000's.

The 68020 is housed in a larger package than the 68000 to accommodate its full 32-bit data bus and address bus. The 32-bit data bus enables the 68020 to run faster because it can access a longword in a single machine cycle. The 32-bit address bus means that each of the 32 bits of an address register or the program counter are connected to pins. Consequently, the 68020 can address 2^{32} bytes (i.e., 4G bytes). You might think that these modifications have no effect on someone who wishes to run an existing 68000 program on a 68020 microprocessor (apart from causing the program to run faster). Well, some 68000 programmers used the following argument.

"If the 68000 employs only address bits A_{01} to A_{23} to specify a word location (since the 68000 lacks address pins A_{24} to A_{31}, although it does have a 32-bit PC), it does not matter what we do with address bits A_{24} to A_{31}. For example, the addresses $00123456 and $40123456 access the same physical location, since the only difference between them is the state of A_{30} which is not connected to an address pin. Consequently, we can employ address bits A_{24} to A_{31} as *tag bits* that define the type of address being accessed. That is, we can 'label' an address as pointing to a byte or a word, or a vector or a matrix, or to any other type of object. Such a facility is useful in a language like LISP. These tag bits may be used by software to check the type of the address being accessed but have no effect on the actual address leaving the 68000."

Of course, programs written for the 68000 relying on this convention will not run on a 68020 system that makes use of address pins A_{24} to A_{31}. We make this point here because it demonstrates how even apparently harmless differences between two processors can cause problems.

We begin our description of the 68020 by introducing the 68020's new and enhanced instructions. In this context, *enhanced* implies a modest improvement to an existing 68000 facility, while *new* implies a facility that the 68000 lacks.

12.2 Enhanced 68000 Instructions

Designers of the 68020 have gently extended the 68000's instruction set and have not provided a plethora of new instructions. Sometimes they have improved existing instructions by making them more flexible, and sometimes they have added entirely new classes of instructions. We will look at some of the enhanced instructions before tackling the new ones. Although we are concerned with the 68020's *architecture* rather than its internal *organization* or *implementation*, we should point out that the 68020 is not just a 68000 core with new bits tacked on. The 68020 is a new design and is faster than the 68000 even when executing 68000-type instructions. For example, the 68020 implements a fast *barrel shifter* that reduces the time required to perform shift operations.

Multiplication and Division

The 68000's multiplication and division instructions (MULU, MULS, DIVU, and DIVS) have been extended to handle both 16- and 32-bit operations. We will consider only the 68020's unsigned multiplication and division here, as the signed instructions perform corresponding operations on signed two's complement numbers.

The 68000 has just a single multiply instruction, MULU <ea>,Dn, that multiplies the 16-bit word at the specified effective address with the low-order word of Dn and deposits the 32-bit result in Dn. The 68020 provides three versions of the multiplication instruction. The assembly language formats of the 68020's multiplication instructions are as follows:

Assembler syntax	Multiplier	Multiplicand	Result
MULU.W <ea>,Dn	16	16	32
MULU.L <ea>,Dn	32	32	32
MULU.L <ea>,Dh:Dl	32	32	64

These operations can be expressed in RTL form by:

```
MULU.W <ea>,Dn       [Dn(0:31)]              ← [Dn(0:15)]*[ea(0:15)]
MULU.L <ea>,Dn       [Dn(0:31)]              ← [Dn(0:31)]*[ea(0:31)]
MULU.L <ea>,Dh:Dl    [Dh(0:31),Dl(0:31)]     ← [Dl(0:31)]*[ea(0:31)]
```

The 68020's word multiplication instruction, MULU.W <ea>,Dn, is identical to the corresponding 68000 multiplication. The 68020 extends the 68000 multiplication instruction by the inclusion of the two longword multiplications, both of which perform a 32-bit by 32-bit multiplication. One yields a 32-bit result (i.e., the least-significant 32 bits of the 64-bit product), and the other yields a true 64-bit (i.e., quadword) product. In order to handle a 64-bit product, two registers must be specified by the instruction. These are written as Dh:Dl, where Dh is the most-significant 32-bits of the product and Dl the least-significant. For example, the operation MULU.L #$12345678,D0:D1 multiplies the 32-bit contents of D1 by the 32-bit literal $12345678 and deposits the 64-bit result in the register pair D0,D1.

Let's look at an example that uses a new multiplication instruction. Suppose two 32-bit signed numbers are located on the stack and we wish to replace them with their 64-bit product. The following code performs this operation.

```
MOVE.L  X,-(SP)        Push the first number
MOVE.L  Y,-(SP)        Push the second number
.
.
MOVE.L  (SP)+,D0       Pull Y
MULU.L  (SP)+,D0:D1    Pull X and multiply by Y
MOVE.L  D1,-(SP)       Push the least-significant 32 bits
MOVE.L  D0,-(SP)       Push the most-significant 32-bits
```

Division

The 68020 has four unsigned division instructions that carry out the basic operation: dividend/divisor = quotient + remainder. As in the case of the 68000's `MULU` instruction, the 68000's `DIVU` instruction is extended to cater for 64-bit dividends, 32-bit quotients, and 32-bit remainders. The assembly language forms of these operations are as follows:

Assembler syntax	Dividend	Divisor	Quotient	Remainder
`DIVU.W <ea>,Dn`	32	16	16	16
`DIVU.L <ea>,Dq`	32	32	32	none stored
`DIVU.L <ea>,Dr:Dq`	64	32	32	32
`DIVUL.L <ea>,Dr:Dq`	32	32	32	32

These operations can be expressed in RTL form by:

```
DIVU.W  <ea>,Dn        [Dn(0:15)]  ← [Dn(0:31)]/[ea(0:15)]
                       [Dn(16:31)] ← remainder.

DIVU.L  <ea>,Dq        [Dq(0:31)]  ← [Dq(0:31)]/[ea(0:31)]

DIVU.L  <ea>,Dr:Dq     [Dq(0:31)]  ← [Dr(0:31),Dq(0:31)]/[ea(0:31)]
                       [Dr(0:31)]  ← remainder.

DIVUL.L <ea>,Dr:Dq     [Dq(0:31)]  ← [Dr(0:31)]/[ea(0:31)]
                       [Dr(0:31)]  ← remainder.
```

The 68000 provides three `.L` variations of the division instruction in addition to the 68000's `.W` version. In contrast to the multiplication instructions that permit a maximum source operand of 32 bits, the operation `DIVU.L <ea>,Dr:Dq` specifies a 64-bit quadword *dividend*. The most-significant 32-bits are in data register Dr and the least-significant bits in Dq. After the division, the 32-bit quotient is loaded in Dq and the 32-bit remainder in Dq.

Note that if overflow is detected during the execution of a division instruction, the V-bit is set and the operand is left unchanged. Consider:

```
MOVE.L    #$12345678,D0
DIVU.W    #4,D0
```

Since we are dividing the 32-bit value $12345678 by 4 using an instruction that yields a 16-bit operand, overflow will occur. Consequently, the value left in D0 after the division will be $12345678. Division operations on both the 68000 and the 68020 clear the CCR's C-bit and leave its X-bit unmodified.

Branch Instructions

The 68000's Bcc (branch on condition cc = true) is extended by the 68020 to cater for *32-bit* displacements. Instead of being able to branch a maximum of 32K bytes from the current instruction, the 68020 programmer can now execute a relative jump of up to 2G bytes either side of the current instruction. The same extension is applied to unconditional branch, BRA, and branch to subroutine, BSR, instructions. The 68020's LINK instruction also supports a 32-bit displacement as well as a 16-bit displacement.

Check and Compare Instructions

We have already encountered the 68000's check instruction, (CHK <ea>,Dn), that compares the value in data register Dn to zero and to an upper bound specified by the data at the effective address, <ea>. The upper bound is a two's complement integer, X, which means that the check can be carried out either on a negative number in the range -X to 0, or on a positive number in the range 0 to +X. If the value in register Dn is less than zero or greater than the upper bound, an exception occurs and the operating system is invoked.

The 68020 extends the 68000's CHK instruction by including two variations: CHK2 and CMP2 (i.e., check 2 and compare 2). The CHK2 <ea>,Rn instruction compares the value in address/data register Rn with *upper* and *lower* bounds, just like the CHK instruction. The principal difference between CHK and CHK2 is that the effective address specified by CHK2 points to a *pair* of bounds: the lower bound is followed by the upper bound (remember that CHK has a fixed lower bound of zero). A CHK2 instruction can take .B, .W, and .L extensions, so that the boundary value may be a byte, word, or longword.

As in the case of the CHK instruction, a CHK2 instruction causes an exception if the specified value is outside the boundary ranges. Another difference between CHK and CHK2 is that the former tests only a *data* register, while the latter may test a data register or an *address* register. This is a most sensible extension, since the programmer can now easily test whether a pointer is within the range of permitted addresses.

For example, the CHK2.L $1234,A2 instruction compares the contents of address register A2 with the longwords found at address locations $1234 (i.e., the lower bound) and $1238 (i.e., the upper bound). The lower bound and upper bounds are separated by one byte for byte operands, two bytes for word operands, and four bytes for longword operands.

Although the CHK2 instruction can be employed in a variety of applications, its principal use is in the testing of array subscripts in a high-level language. Suppose a programmer writes the code TIME(J) := 12. If the array subscript J has been incorrectly evaluated, the processor will attempt to access data outside the space allocated to the array, TIME. However, by employing a CHK2 instruction to test the array address against the bounds of the array at *run time*, the danger of array bound errors can be removed. We provide an example of bounds testing when we introduce the CMP2 instruction.

The relationship between the upper and lower bounds (i.e., U and L) specified by a CHK2 instruction is illustrated in Figure 12.1. As you can see, the contents of the specified register may be less than the lower bound, greater than the upper bound, equal to one of the bounds, or fall between the bounds.

Figure 12.1 The CHK2 instruction

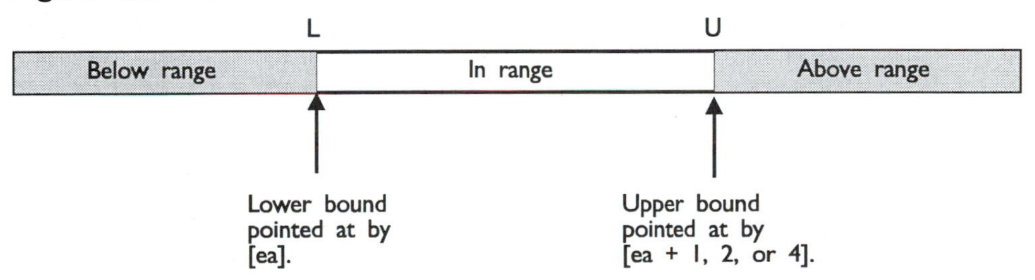

If the CHK2 instruction is used to test an address register and the size is a word or a *byte* (yes — a byte), the bounds are sign-extended to 32 bits and the resultant operands tested against the full 32 bits of the specified address register. However, if the contents of a data register are tested and the operation size is a byte or a word, only the appropriate low-order bits of the data register are tested.

If the upper and lower bounds are the same, the valid range is a single value. For signed comparisons, the *arithmetically smaller* value should be the lower bound, while for unsigned comparisons, the *logically smaller* value should be at the lower bound. This statement is not as confusing as it seems. For example, in unsigned arithmetic the lower bound might be 2 and the upper bound 6, corresponding to 00000010_2 and 00000110_2, respectively. However, in signed two's complement arithmetic, the lower bound might be -2 and the upper bound +4, corresponding to 11111110_2 and 00000100_2, respectively.

The 68020's CMP2 Instruction

The CMP2 <ea>,Rn instruction is identical to the CHK2 instruction with one difference. When the contents of Rn are compared with its lower and upper bounds, the Z- and C-bits of the condition code register are updated. The Z-bit is set if the contents of Rn are equal to either bound and cleared otherwise. The C-bit is set if the contents of Rn are out of bounds and cleared otherwise. The relationship between the contents of the register being tested and the Z- and C-bits can be better appreciated from Figure 12.2.

Figure 12.2 The CMP2 instruction

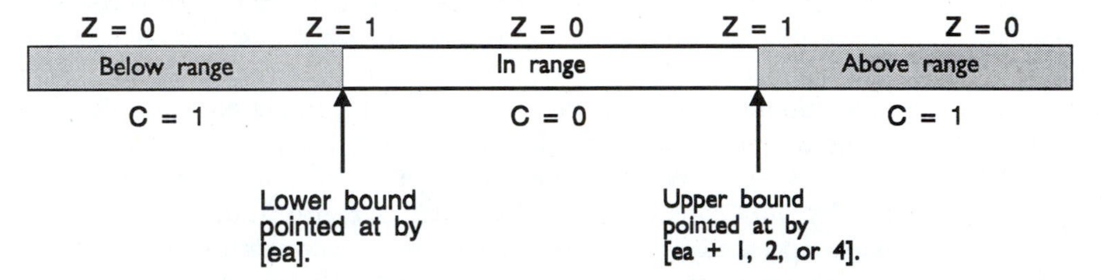

Consider an application of a CMP2 instruction. We have chosen CMP2 rather than CHK2 because a CMP2 doesn't cause a trap, which requires the construction of an appropriate exception handler. A programmer is using a one-dimensional array of bytes, TABLE, and wishes to access the jth element, where the value of j is stored in data register D0.W. The following fragment of code tests for an array bound error at run time.

```
SIZE  EQU     <size>            The size of the array TABLE in bytes
TABLE DS.B    SIZE              Save SIZE bytes for the array TABLE
LOWER DS.L    1                 Reserve longword for lower bound of TABLE
UPPER DS.L    1                 Reserve longword for upper bound of TABLE
      .
      .
*                               Set up the array bounds
      LEA     TABLE,A0          A0 points to the start of the array
      MOVE.L  A0,LOWER          Store the lower bound of the array
      LEA     (SIZE-1,A0),A0    Calculate the upper bound of the array
      MOVE.L  A0,UPPER          Store the upper bound of the array
      .
      .
*                               Perform boundary tests
      LEA     TABLE,A0          A0 points to the start of the array
      LEA     (A0,D0.W),A0      A0 points to the jth element
      CMP2    LOWER,A0          Test for out-of-bounds element
      BCS     ERROR             IF carry set THEN error
      .                                       ELSE continue
      .
ERROR .                         Deal with error condition
```

The upper bound is calculated by LEA (SIZE-1,A0),A0 because the array extends from 0 to SIZE-1. By the way, we could have set up the upper and lower bounds by:

```
MOVE.L #TABLE,LOWER
MOVE.L #TABLE+SIZE-1,UPPER
```

The practical difference between a CHK2 and a CMP2 instruction is that a CHK2 instruction uses an operating system call (i.e., a trap) to deal with the out-of-range condition, while a CMP2 instruction simply modifies the condition codes and leaves any recovery action to the programmer.

Before we leave the CHK2 and CMP2 instructions, we provide an example of how they treat both signed and unsigned bounds. Figure 12.3 illustrates five triplets, (a) to (e). The first element in each triplet is a lower bound, the second element an upper bound, and the third element the value to be compared against the bounds. For example, triplet (a) has a lower bound of $0000, an upper bound of $1000, and a value of $0800. The value is in range because it is between the bounds.

a. 0000	b. 0000	c. F800	d. F000	e. E800
1000	1000	0000	F800	1000
0800	1800	0800	E800	F000

Each of these cases is represented graphically in Figure 12.3. The heavy lines represent the range, lower to upper bounds defined by the first two members of the triplet. The triangles represent the elements to be compared against the bounds.

Figure 12.3 The CMP2 instruction and upper and lower bounds

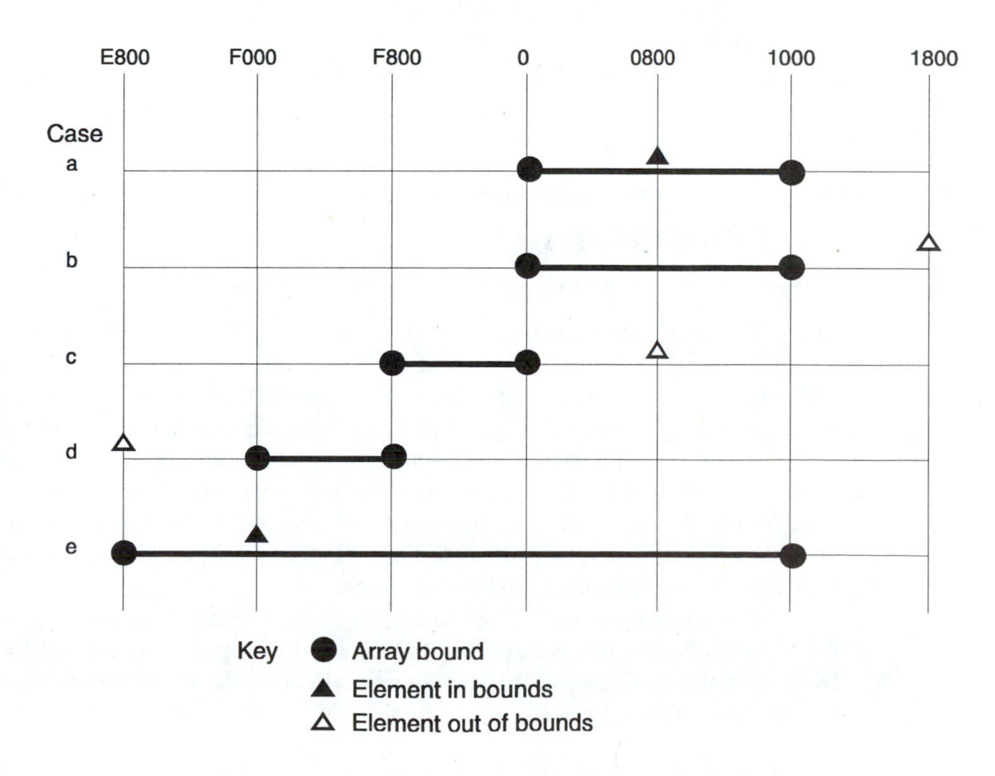

The EXT Instruction

The 68000 has a sign-extend instruction, EXT, that sign-extends an 8-bit byte to a word (i.e., EXT.W Dn), or a 16-bit word to a longword (i.e., EXT.L Dn). The EXT is useful when you have a byte (or word) operand and have to use an instruction that requires a word (or longword) operand. For example, signed multiplication and division requires 16- and 32-bit operands, respectively. You can use the EXT instruction to sign-extend operands before multiplication or division.

The 68020 implements a new instruction, EXTB.L Dn, that sign-extends a byte to a longword by taking bit 7 of Dn and copying it into bits 8 to 31 of Dn. Without the EXTB.L instruction, you would have to sign-extend the byte in Dn by:

```
EXT.W   Dn      Sign-extend byte to a word
EXT.L   Dn      Sign-extend word to longword
```

If D0 contains $1234ABCD, executing these two instructions yields:

```
EXT.W   D0      Sign-extend byte to a word    - $1234FFCD
EXT.L   D0      Sign-extend word to longword - $FFFFFFCD
```

By the way, suppose that the 68000 didn't support an EXT instruction. What would we do to sign-extend a byte in D1 to a longword? You could always write:

```
LSL.W   D0      Move ls byte 8 bits left (and fill with zeros)
MOVEA.W D0,A0   Force to word in D0 to be sign-extended in A0
MOVE.   A0,D0   Copy back the sign extended word to D0
ASR.L   #8,D0   Shift 8 bits right (preserving sign)
```

12.3 New 68020 Instructions

The 68020 provides a modest set of new instructions. Some of these are related to the use of external coprocessors and are not dealt with here. One instruction pair, call module (CALLM) and return from module (RTM), requires an external memory management system and is used only by operating systems. These two instructions are not covered in this text because they were not widely used and have not been implemented by the 68030. Another very specialized pair of instructions are compare and swap, CAS, and compare and swap 2, CAS2. These two instructions are intended for use in sophisticated applications such as multiprocessor systems and interrupt-driven multitasking systems.

The two new groups of 68020 instructions of immediate interest to us are the BCD pack and unpack group, and the bit field group. If you are not interested in BCD operations, we suggest that you skip ahead to the 68020's much more important bit field instructions.

PACK and UNPK Instructions

The PACK and UNPK instructions are used in conjunction with BCD arithmetic and are intended to simplify the conversion between *characters* input in coded form (e.g., ASCII 7-bit code) and the *internal* representation of BCD data. For example, the ASCII code for the character '7' is 37_{16} or 00110111_2, while the BCD representation of the decimal number 7 is 0111_2.

The PACK instruction takes two characters and converts them into two BCD digits; its syntax is:

```
       PACK  -(Ax),-(Ay),#<adjustment>
or     PACK  Dx,Dy,#<adjustment>.
```

We can note two things immediately. First, the PACK instruction has *three* operands and, second, it takes only two addressing modes (data register direct or register indirect with pre-decrementing).

The effect of the PACK instruction is to *translate* the source data into the destination data. Consider the data register form of the PACK instruction, PACK D0,D1,#<adjustment> — see Figure 12.4. The constant, adjustment, is first added to the value in source register D0, and then bits (11:8) and (3:0) of the sum are *concatenated* and placed in bits (7:0) of the destination. Note that the literal, adjustment, used by the PACK instruction is zero for both ASCII and EBCDIC characters. The effect of a PACK D0,D1,#<adjustment> in RTL is:

```
[Temp(0:15)] ←  [D0(0:15)] + <adjustment>
[D1(0:3)]    ←  [Temp(0:3)]; [D1(4:7)] ← [Temp(8:11)]
```

Figure 12.4 Effect of a PACK instruction

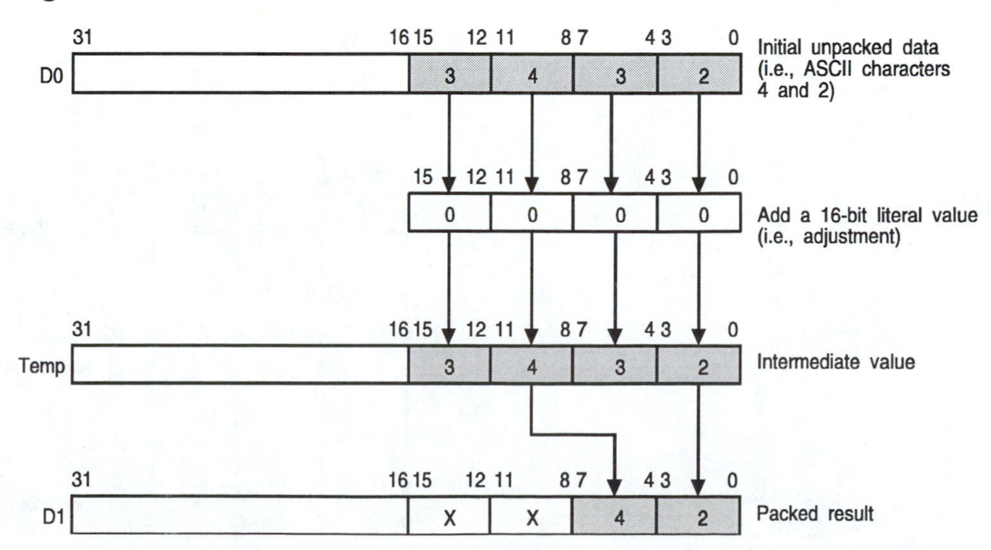

Let's look at an application of the PACK D0,D1,#0, instruction in Figure 12.4, where the source register D0(0:15) contains the ASCII characters for '4' and '2' (i.e., [D0] = $3432). The PACK instruction adds the literal (i.e., 0) to D0 to produce a result of $3432. In this case, the literal is zero and there is no change. In the next step, the least-significant nibbles of this result (i.e., 4 and 2) are extracted and concatenated in the destination register D1 to give D1 = $XX42. The Xs indicate that the most-significant byte of D1.W is unaffected by this instruction.

The UNPK instruction performs the inverse operation of the PACK instruction. The syntax of UNPK can be expressed as either UNPK Dx,Dy,#<adjustment> or UNPK -(Ax),-(Ay),#<adjustment>. The UNPK instruction takes two BCD digits from the source byte, separates them, and uses them to generate the least-significant nibble of two bytes (i.e., they are unpacked). The 16-bit literal specified by <adjustment> is added to these two bytes to give the 16-bit destination operand. The effect of UNPK Dx,Dy,#<adjustment> expressed in RTL form is:

```
[Dy(0:3)]    ← [Dx(0:3)]; [Dy(8:11)]  ← [Dx(4:7)]
[Dy(0:15)]   ← [Dy(0:15)] + adjustment
```

Consider now the action of UNPK D0,D1,#$3030, where D0 = $25, in Figure 12.5. In the first step, the contents of data register D0 are unpacked to give the 16-bit value $0205. In the second step, the literal $3030 is added to the $0205 to provide $3235 in D1, which corresponds to the ASCII characters for '2' and '5'.

Consider the following example of the PACK and UNPK instructions. Data register D0 contains the two ASCII-coded characters '1' and '2' (i.e., [D0] = $3132). Similarly, data register D1 contains $3334, corresponding to the characters '3' and '4'. We wish to add together the BCD numbers represented by these values and deposit the result, in ASCII form, in D0. We cannot add the ASCII characters

Figure 12.5 Effect of an UNPK instruction

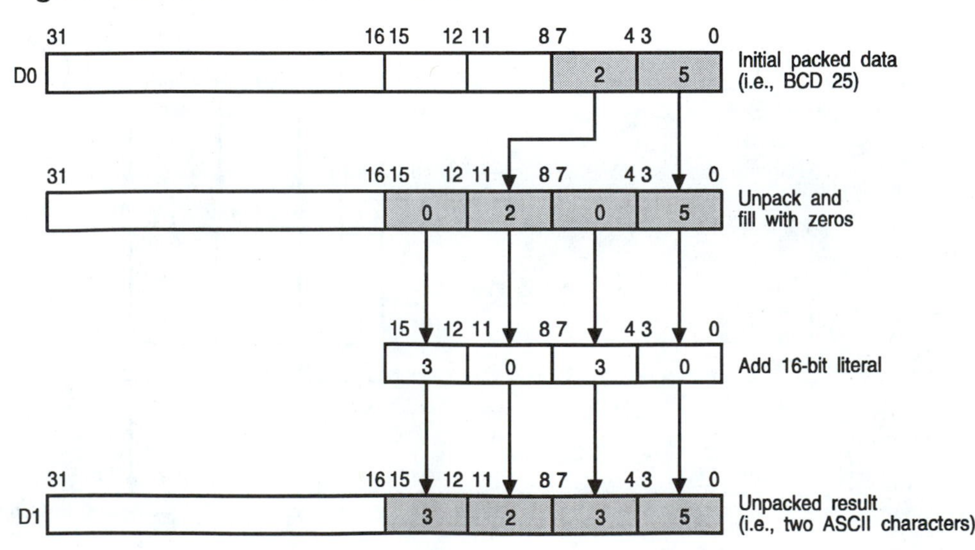

themselves and must first convert them into BCD numbers. The following code will perform the desired action:

```
PACK  D0,D0,#0       Convert 2 ASCII chars into 2 BCD digits
PACK  D1,D1,#0       Do the same for D1
ABCD  D0,D1          Add the pair of BCD digits
UNPK  D1,D1,#$3030   and convert then into ASCII characters
```

Bit Field Instructions

The 68020's bit field group of instructions represents the most significant enhancement to the 68000's instruction set from the point of view of the programmer writing applications programs. Conventional microprocessors are byte-, word-, or longword-oriented, and operate on data located at a byte boundary. Being restricted to one of these boundaries doesn't usually cause a problem, because most of the data structures employed by programmers fit within these boundaries quite naturally. For example, ASCII-encoded characters (which may be 7-bit or 8-bit values) fit within byte boundaries, just as 64-bit IEEE-format floating point numbers fit into word or longword boundaries.

Sometimes programmers have to deal with data structures that do not fall within these *natural* boundaries. Suppose a programmer is working with 17-bit data structures. The 68000 programmer must fit each item into two consecutive words (16 bits in one and a single bit in the other), or into a byte and a word. Not only is this arrangement inefficient in terms of storage, it is cumbersome in terms of operations on the 17-bit items. The 68020's bit field operations permit you to forget about byte, word, and longword boundaries and to handle data items falling across these boundaries, that have been imposed by the hardware.

Since a bit field is an arbitrary group of bits falling anywhere within the 68020's address space, it is tempting to wonder just what bit field instructions might be designed to do. Because a bit field is an arbitrary data structure, conventional arithmetic operations on a bit field are meaningless. Bit field instructions are designed largely to copy strings of bits between registers and arbitrary locations in memory, and to test the bit fields. Once bit fields have been transferred from memory to a register, they can be processed in the normal way by the 68020's conventional instructions. Figure 12.6 demonstrates the idea underlying a bit field. As you can see, the bit field straddles three consecutive bytes.

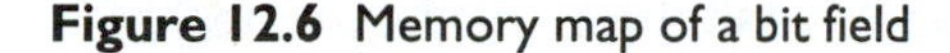

Figure 12.6 Memory map of a bit field

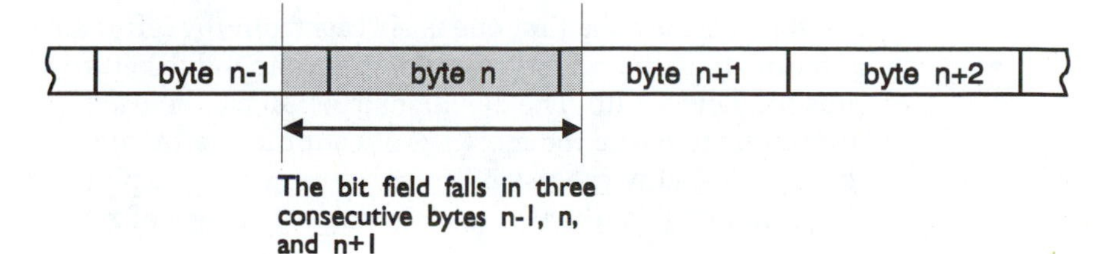
byte n-1
byte n
byte n+1
byte n+2
The bit field falls in three consecutive bytes n-1, n, and n+1

We now briefly describe the bit field instructions provided by the 68020, and then look in more detail at how they are arranged in memory. The 68020 implements the following eight bit field instructions.

BFEXTU The bit field extract unsigned instruction reads a bit field and deposits it in a data register. This instruction is the bit field equivalent of the MOVE <source>,Dn. If the size of the bit field is less than 32 bits, it is loaded into the low-order bits of the data register, and the high-order bits are set to zero.

BFEXTS The bit field extract signed instruction reads a bit field, sign-extends it to 32 bits, and deposits it in a register. For example, if the bit field is 1100101, it is sign-extended to 11111111111111111111111111100101.

BFINS The bit field insert instruction copies a bit field in a data register into a location in memory. This instruction is the bit field equivalent of a MOVE Dn,<destination>.

BFTST The bit field test instruction tests the specified bit field and sets the flag bits in the CCR accordingly. The N-bit is set if the most-significant bit of the bit field is one, and the Z-bit is set if all the bits of the bit field are zeros.

BFCLR The bit field clear instruction tests the bit field exactly like BFTST, and then clears all the bits of the bit field.

BFSET The bit field set instruction tests the bit field exactly like BFTST, and then sets the bits of the bit field to one.

BFCHG The bit field test and change instruction behaves exactly like a BFTST instruction, except that the bits of the bit field are all inverted after the test.

BFFFO The bit field find first one instruction is the only instruction that actually performs a calculation on the bits of a bit field (although the bit field itself is not modified). The bit field at the specified address is read and scanned by the processor. The location of the first one bit in the bit field is then loaded into the specified data register. We interpret the *first one* as the most-significant bit of the bit field that is set to the value one. The location of the first one, which is loaded into the destination register, is specified as the offset of the bit field itself plus the offset of the first one. The precise meanings of *offset* and *field* width are defined later. However, if the offset is 7, and the bit field equal to 00000011011010, the BFFFO instruction will return the value 7 + 6 = 13 in the specified data register (because the first one is six bits from the left-most bit). If no one is found (i.e., the bit field is all zeros), the value returned is the offset plus the field width. The BFFFO instruction can be used in floating point arithmetic to locate the most-significant bit of a mantissa without going through the slow process of shifting and testing. Equally, it can be used in bit-mapped data structures to scan past strings of zeros.

Figure 12.7 illustrates how a bit field extract instruction, BFEXTU, operates. The bit field is copied from memory, even though it falls wholly or partially in three consecutive bytes, into the low-order bits of data register D0. The higher-order bits of D0 are padded with zeros.

Figure 12.7 Effect of a bit field extract instruction on data register D0

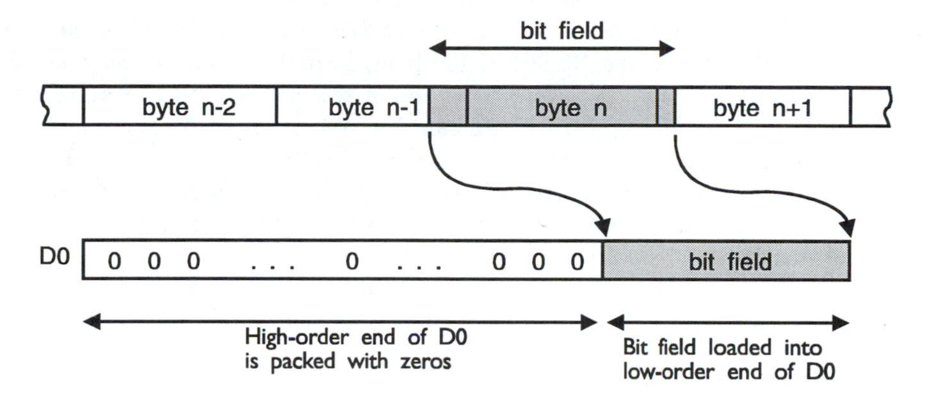

The assembler syntax of the 68020's bit field instructions is:

```
BFCHG  <ea>{offset:width}         Bit field change
BFCLR  <ea>{offset:width}         Bit field clear
BFEXTS <ea>{offset:width},Dn      Bit field signed extract
BFEXTU <ea>{offset:width},Dn      Bit field unsigned extract
BFFFO  <ea>{offset:width},Dn      Bit field find first one
BFINS  Dn,<ea>{offset:width}      Bit field insert
BFSET  <ea>{offset:width}         Bit field set
BFTST  <ea>{offset:width}         Bit field test.
```

Bit field instructions take either *three* or *four* operands — the bit field clear instruction has three operands and the syntax BFCLR <ea>{offset:width}, while the insert bit field instruction, BFINS, has four operands and has the syntax BFINS Dn,<ea>{offset:width}. Since a bit field is a user-defined data structure that is not aligned on a byte boundary, three pieces of information are needed to define it: its *size* (i.e., its width), its *location* (i.e., a byte address and an offset from that position), and its *value.*

The *base address* of a bit field is specified in the conventional way by means of an effective address. The actual bit field is located at *offset* bits from the base byte. For example, the instruction BFCLR $1000{10:8} refers to a bit field with a base address $1000. The *offset,* in this case 10, is measured from field-bit zero of the base address. The *width* of the field, in this case 8, is a positive integer in the range 1 to 32 that tells us how many bits there are in the field. Bit fields wider than 32 bits are not supported by the 68020. The instruction BFCLR $1000{10:8} is interpreted as *clear the 8 bits of the bit field whose location is 10 bits from base address*

$1000. However, there is one rather confusing point that we must come to terms with — the way in which the bits of the bit field are numbered.

Figure 12.8 illustrates the relationship between base address, base bit, field offset, and field width. The memory locations are numbered consecutively *0, 1, ..., i-1, i, i+1, ..., n,* and the individual bits of each byte are numbered 0 to 7, where 0 is the least-significant bit (this numbering is fundamental to the 68000 family). The effective address in the instruction points to the *base address* of the bit field, which is, in Figure 12.8, byte *i*. The first bit (called the *base bit*) of the bit field is located 10 bits (i.e., the bit field offset) away from the *base address.* Note that the bit field is located at 10 bits from the *most-significant* bit of the base address and not 10 bits from its least-significant bit, as you might expect. The bit field itself extends from the base bit to the base bit plus the field width minus one.

Having stated the above, we need to look at Figure 12.8 a little more closely. Once again it is necessary to stress that the field offset, i.e., 10 bits, is measured from the most-significant bit of the base byte. The bits of the actual bit field itself are numbered *0* to *width-1* in the same sense as the field offset (i.e., in the opposite direction to the numbering of the bits of individual bytes).

Figure 12.8 How field data is located in memory

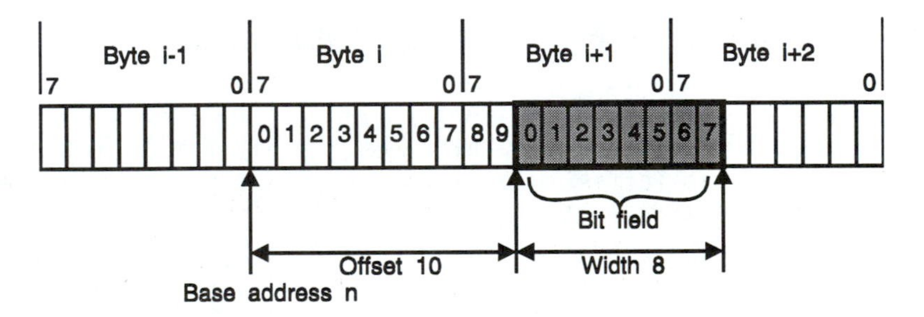

The effective address (i.e., the base address) of each of the bit field instructions is specified in the *normal* way by any of the 68020's addressing modes (apart from those employing auto-incrementing and auto-decrementing). The offset may be specified by either a literal in the range 0 to 31 or by a data register which permits an offset in the range -2^{31} to $2^{31}-1$. The bit field width may be specified by a literal in the range 1 to 31 (or 0, which specifies 32 bits), or by the contents of a data register modulo 32. Note again that a value of zero for the bit field width is interpreted as a width of 32. Typical legal bit field instructions are:

```
BFCLR (A0){5:7}
BFCLR (D2,A3){D6:12}
BFCLR ($F4,PC,A4){30:D2}
BFCLR $1234{D5:D6}
```

When a bit field offset is specified by the contents of a data register, the range is -2^{31} to $2^{31}-1$, which means that the offset can be on either side of the base byte (i.e., at a lower address or a higher address).

What then is the purpose of bit field instructions? Bit field operations make it easy to operate on data structures that are not byte-oriented. Such structures are associated with graphics and with information stored on disks. Without bit field operations it would be very tedious to manipulate arbitrary data structures using only byte operations and shifting. For example, consider the bit field operation `BFCLR $1000{23:13}`. This instruction clears the 13 bits of the specified bit field that lies 23 bits from bit 7 of the base byte $1000, as illustrated in Figure 12.9.

Figure 12.9 The effect of a `BFCLR $1000{23:13}` instruction

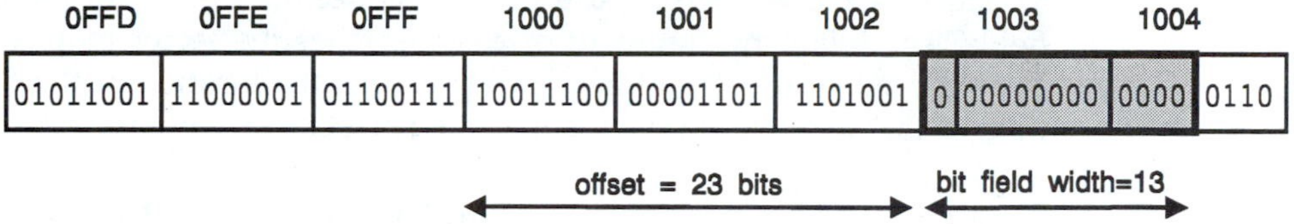

0FFD	0FFE	0FFF	1000	1001	1002		1003	1004	
01011001	11000001	01100111	10011100	00001101	1101001	0	00000000	0000	0110

offset = 23 bits

bit field width=13

Bit Field Applications

A typical application of bit field instructions is found in the realm of disk file systems. A disk is made up of a number of concentric *tracks*, each of which is composed of a series of *sectors*. When a file is created, new sectors are allocated to it, and when a file is deleted, its sectors are released. A simple way of keeping track of sectors is to create a *bit map* that indicates the availability of the sectors. This bit map is often located in the first sector of the first track. Consider an arrangement with 256-byte sectors. The first sector contains 8 x 256 = 2048 bits, each of which is associated with a sector, as illustrated in Figure 12.10. If a sector is free, the corresponding bit in the bit map is set, and if it is allocated, the corresponding bit is clear.

Now suppose the operating system wishes to create a new file. It must read the bits of the bit map sequentially until it finds the first bit set to a logical one. The operating system then allocates the corresponding sector to the file, clears the bit to mark the sector as allocated, and searches for the next unallocated sector, and so on, until the file has been created.

Figure 12.10 The free-sector bit map of a disk

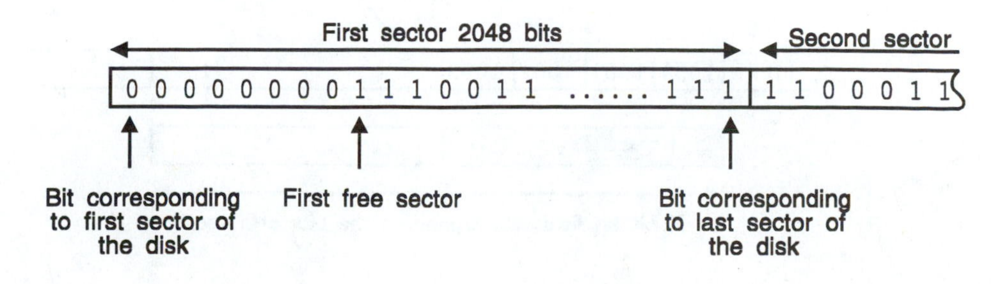

In practice, the first sector containing the free-sector bit map is copied into main store to make processing easier. Reading the image of a bit map in memory is not difficult in 68000 code. You can read a byte at a time and check the eight bits of the byte in sequence using, for example, BTST instructions. The 68020's bit field instructions make this task much simpler.

It would be nice if we could turn the memory-map of the entire sector containing the free-sector bit-map into a single 2048-bit-wide bit field. Then we could use the *find first one in bit field* instruction, BFFFO, to locate the first free sector. However, since the maximum bit field width supported by the 68020 is 32 bits, we must regard the free sector bit map as a sequence of 64 bit fields of length 32 (since 64 x 32 = 2048). Figure 12.11 describes the free-sector map in terms of bit fields. The following fragment of code searches the sector map for a free sector. Remember that the BFFFO instruction returns the location of the first 1 in the bit field or the field width plus the offset if a bit set to 1 is not found.

```
FIELDS DS.L    64                 Reserve space for the bit-map
*
*      Assume that the sector map is made up of 64 longwords
*      (this assumption means that we can use 32-bit bit fields)
*
       LEA     FIELDS,A0          A0 points to the sector map
       MOVE.W  #63,D7             Up to 64 bit fields to search
LOOP   BFFFO   (A0){0:32},D1      Look for a free sector
       LEA     (4,A0),A0          Increment pointer to next bit field
       CMP.B   #32,D1             IF D1 = 32 THEN no free sector found
       BNE     FOUND                         ELSE process free sector
       DBRA    D7,LOOP            REPEAT UNTIL all 64 fields tested
       BRA     FULL               IF here THEN no free sector located
FOUND  LEA     (-4,A0),A0         Wind back the pointer
       BFCLR   (A0){D1:1}         Clear bit to claim sector
       .
       .
FULL   deal with disk full
```

Figure 12.11 Using bit fields to implement a free-sector bit-map

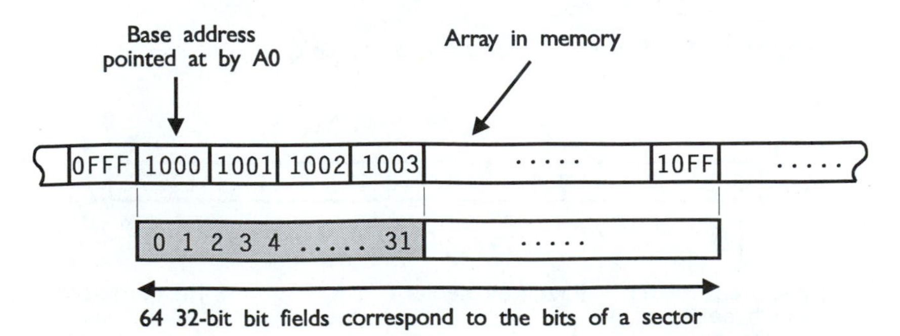

This code could be slightly simplified by combining the two lines

```
BNE     FOUND
DBRA    D7,LOOP
```

into the single line

```
DBNE    D7,LOOP.
```

When I showed the above program to John Hodson who was then Motorola's Northern Europe Training Manager, John suggested rewriting it to make even better use of the BFFFO instruction.

```
        CLR.L   D0                Initial bit field offset = 0
        LEA     FIELDS,A0         A0 points at the sector bitmap
        MOVE.W  #64-1,D7          Up to 64 bit fields to test
LOOP    BFFFO   (A0){D0:32},D0    Look for free sector, IF found Z = 0 and
*                                  bit field offset from (A0) is loaded
*                                  into D0, ELSE Z = 1 and bit field offset
*                                  plus 32 is loaded into D0.
        DBEQ    D7,LOOP           Decrement counter until Z = 0 or all 64
                                   fields searched without success.
        BEQ     FULL              IF Z = 1, no 1's found and disk is full
        BFCLR   (A0){D0:1}        ELSE claim sector. D0 = sector number
```

In John's version of the example, the BFFFO (A0){D0:32},D0 instruction looks for the first 1 in the 32-bit-wide bit field that is offset by the contents of data register D0 from the base byte pointed at by address register A0. If a 1 is found, its total offset from the base byte is loaded into D0. If a 1 is not found, the initial offset plus the field width (i.e., 32) is loaded into D0. That is, D0 is the offset pointer which is incremented by 32 until either a 1 is found or until all sectors have been checked. Note also that the BFFFO instruction sets the Z-bit of the CCR if a 1 is not found in the bit field. We use this fact in the DBEQ D7,LOOP instruction to terminate the loop if a 1 is found.

Bit Fields and Graphics

Another example of the application of bit field instructions can be taken from the world of computer graphics. Consider the bit plane of Figure 12.12 which contains an image made up of individual *pixels*. A pixel is a picture element that may be on or off to create a dot or a no-dot. In this example, a rectangular part of the image, denoted by heavy shading in Figure 12.12, is to be copied from one place to another. The image is 64 pixels (8 bytes) wide. A 15-bit by 6-line field is to be moved down by ten lines and to the right by 3 pixels. Assume that the address of the source image is in address register A0.

Figure 12.12 Use of bit field operations in bit-mapped graphics

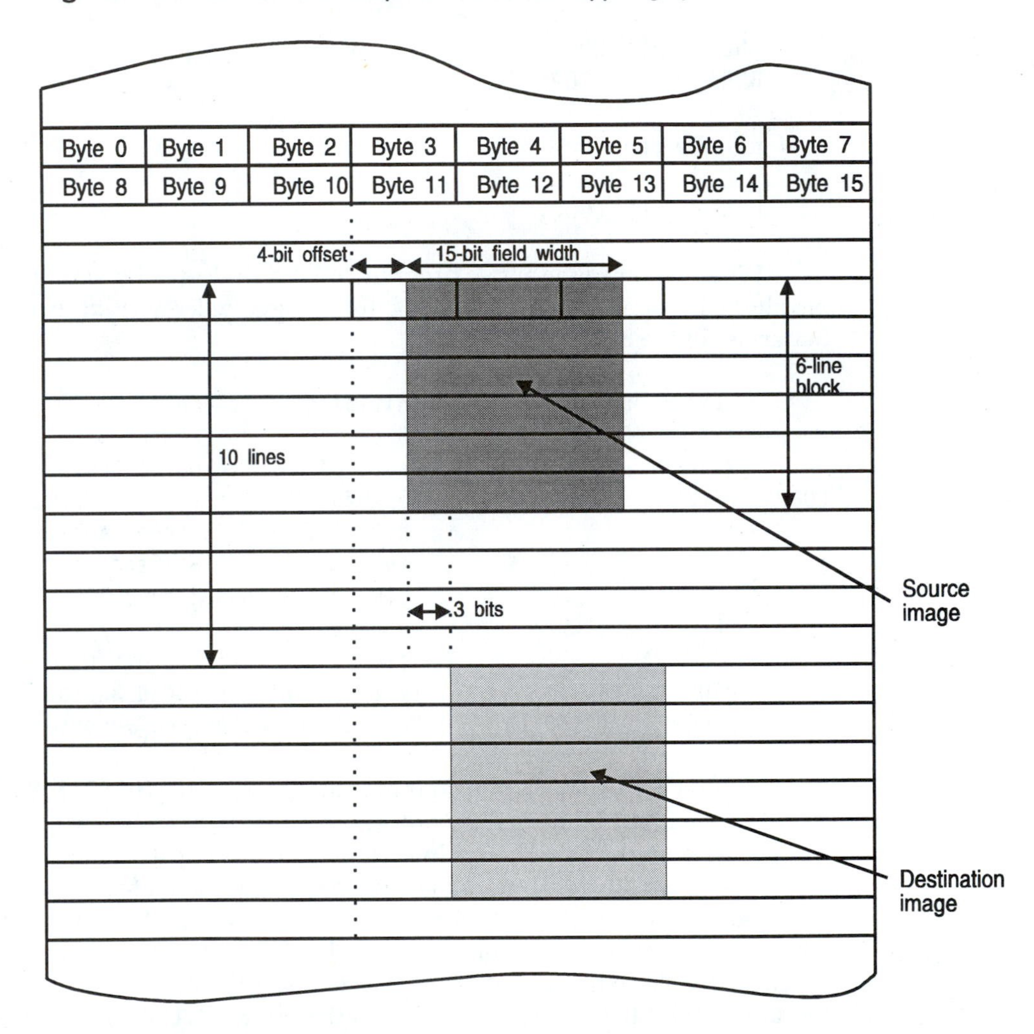

We can carry out the pixel translation by reading a line of the image using the BFEXTU instruction and then copying it to its destination with a BFINS instruction. Figure 12.13 demonstrates the action of the bit field extract and bit field insert instructions. Note that the offset in the effective address of BFEXTU (A0){4:15},D1 is *80 bytes* because the destination of the bit field is 10 lines of 8 bytes on from the source.

```
      MOVE.W #5,D0              Six lines to move
LOOP  BFEXTU (A0){4:15},D1      Copy a line of the image to D1
      BFINS  D1,(80,A0){7:15}   Copy it to its destination
      LEA    (8,A0),A0          Update the pointer by 8 bytes (one line)
      DBRA   D0,LOOP            Repeat until all lines moved
```

Figure 12.13 Copying a bit field from one point to another

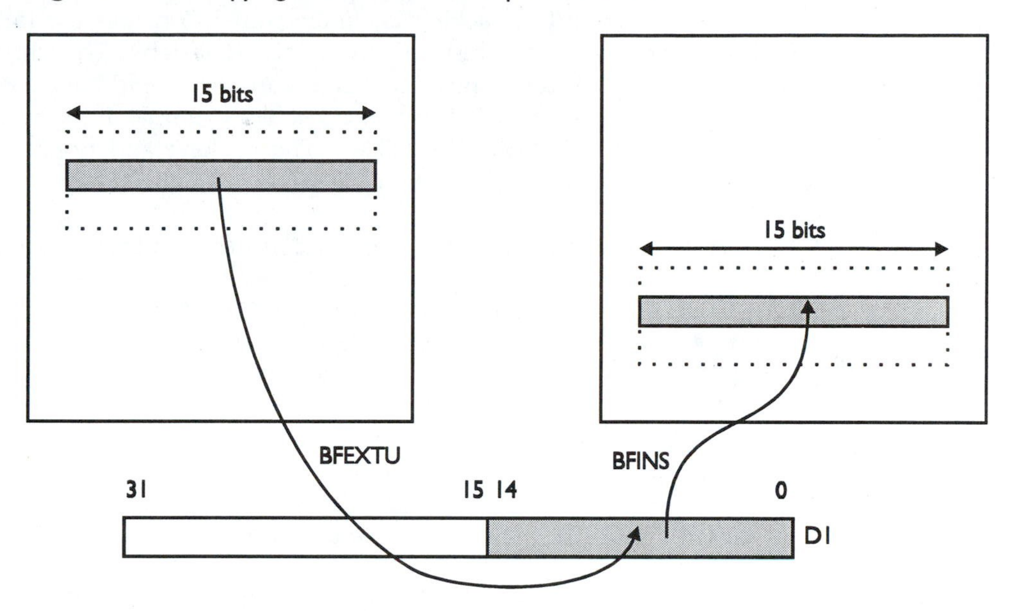

Bit Fields and Packed Data

A Motorola application report by Bob Beims, AR219, provides an excellent example of the use of bit field instructions. Beims points out that high-level languages such as C often pack several small variables into a single word. For example, a single 16-bit word can hold a 6-bit variable and two 5-bit variables. Packing variables saves memory space at the expense of the time taken to access the individual variables. Figure 12.14 illustrates how these variables can be packed into a word and then unpacked and stored in three words.

Figure 12.14 Packing and unpacking data

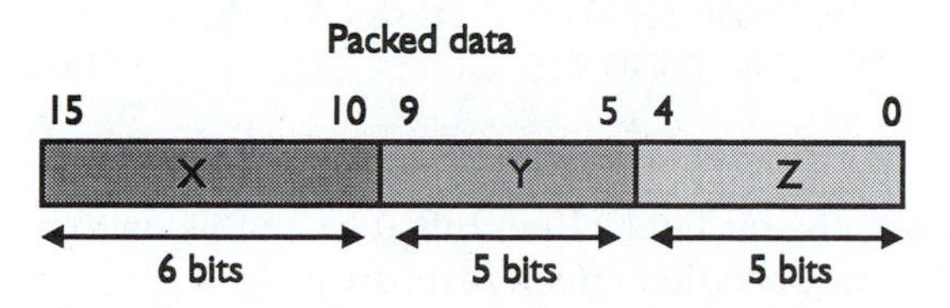

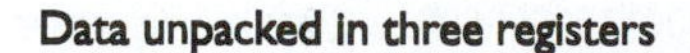

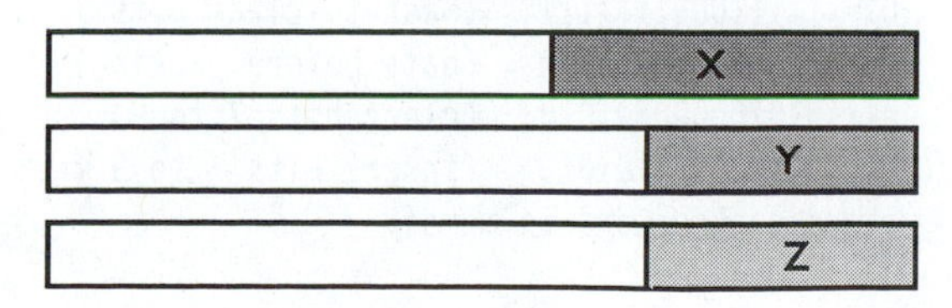

Since most microprocessors lack specific instructions to manipulate packed variables, packing and unpacking is cumbersome. Consider the following routines that retrieve and store, respectively, a packed variable. This variable is three bits wide and is packed into bits 7 to 9 of a word. We could have used the terms *stored* or *loaded* instead of *packed* and *unpacked*. For example, if the bit field is $\mathtt{PQR}_2$, the packed word is $\mathtt{XXXXXXPQRXXXXXXX}_2$. The packing and unpacking are performed first in 68000 code and then in 68020 code using its bit field instructions.

```
*     Read the packed variable from memory and store it in bits 0 to 2 of D0
*
LOAD  MOVE.W #$0380,D0   Load the mask word 0000001110000000
      AND.W  <ea>,D0     Packed word in D0 is masked to bits 7 to 9
      LSR.W  #7,D0       Justify D0 to get bits in least-significant position
      RTS
*
*     The 3-bit variable to pack is initially in bits 0 to 2 of D0 and is to be
*     packed in bits 7 to 9 of the specified memory location. The store operation
*     must not modify any other bits of the packed word - bits 0 to 6 and 10 to 15
*
STORE LSL.W  #7,D0       Shift bit field to be stored to bits 7 to 9
      MOVE.W #$FC7F,D1   Load mask word 1111110001111111 in D1
      AND.W  <ea>,D1     Get packed word in D1, clear bits 7 to 9
      OR.W   D0,D1       Insert bit field from D0 in bits 7 to 9 of D1
      MOVE.W D1,<ea>     Store packed word in memory
      RTS
```

Since the action of the above code might not be immediately clear, consider an example in which the bit pattern 010 is packed into bits 7 to 9 of the word at the specified effective address and is to be unpacked and loaded into D0.

```
*                     [<ea>]=0101010101010100   (initial packed string)
LOAD  MOVE.W #$0380,D0   D0=0000001110000000   (mask bits = bits 7 to 9)
      AND.W  <ea>,D0     D0=0000000100000000   (mask D0 to bits 7 to 9)
      LSR.W  #7,D0       D0=0000000000000010   (left justify bit field)
      RTS
```

In the next example, the bit field 110 in bits 0 to 2 of D0 is to be packed into bits 7 to 9 of the word at the specified effective address.

```
STORE LSL.W  #7,D0       D0=0000001100000000   (move variable to bits 7 to 9)
      MOVE.W #$FC7F,D1   D1=1111110001111111   (mask to clear bits 7 to 9)
      AND.W  <ea>,D1        1010101010101010   (data before packing)
*                        D1=1010100000101010   (clear bits 7 to 9)
      OR.W   D0,D1       D1=1010101100101010   (insert bits 7 to 9 from D0)
      MOVE.W D1,<ea>     Store packed word in memory
      RTS
```

As you can see, there is nothing remarkable about the above fragments of code. They are simply rather long-winded. Now consider the use of the bit field instructions BFEXTU and BFINS. BFEXTU is a bit field extract unsigned instruction that extracts a bit field from the specified effective address, zero-extends the result to 32 bits, and loads it into the destination data register. Its assembly language form is BFEXTU <ea>{offset:width},Dn. We can therefore recode the above LOAD (i.e., pack) operation as the single instruction:

```
BFEXTU <ea>{6,3},D0
```

Note that the field width is 3 bits and that the offset is 6 because bit field operations number bits from left to right, whereas the bits of a word are numbered from right to left (i.e., the bit field is 6 bits from bit 15 of the effective address). Figure 12.15 shows how the bits are numbered in this example.

Figure 12.15 Inserting a bit field

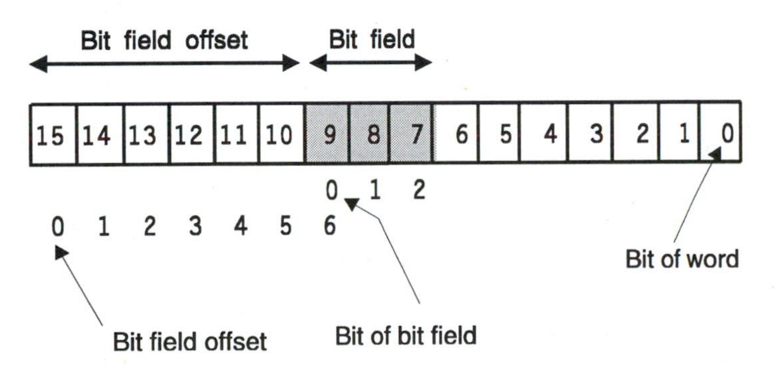

Similarly, the BFINS instruction can be used to insert a bit field. Its assembly language form is BFINS Dn,<ea>{offset:width}, and its effect is to take the bit field from the low-order bits of the specified data register and insert them into the bit field at the effective address. In terms of the above example, we can re-code the STORE operation as:

```
BFINS D0,<ea>{6:3}
```

Beims provides a second example of an application of bit field instructions that manages to combine both the power of the 68020's new addressing modes and its bit field operations in a single instruction. The reader who is unfamiliar with the 68020's indirect addressing modes should read the next section before working through this example.

Beims constructs a system in which a set of records is stored in memory and a pointer to the records is pushed onto the stack at an offset FILEPTR below the top of the stack. This situation might arise when a subroutine is called to process the records and the address of the records is passed on the stack. Figure 12.16 illustrates the data structure relevant to this example.

Figure 12.16 Using bit field instructions to access a complex data structure: executing a `BFEXTU ([FILEPTR,SP],REC5,D3.W*4){D4:24},D0` instruction

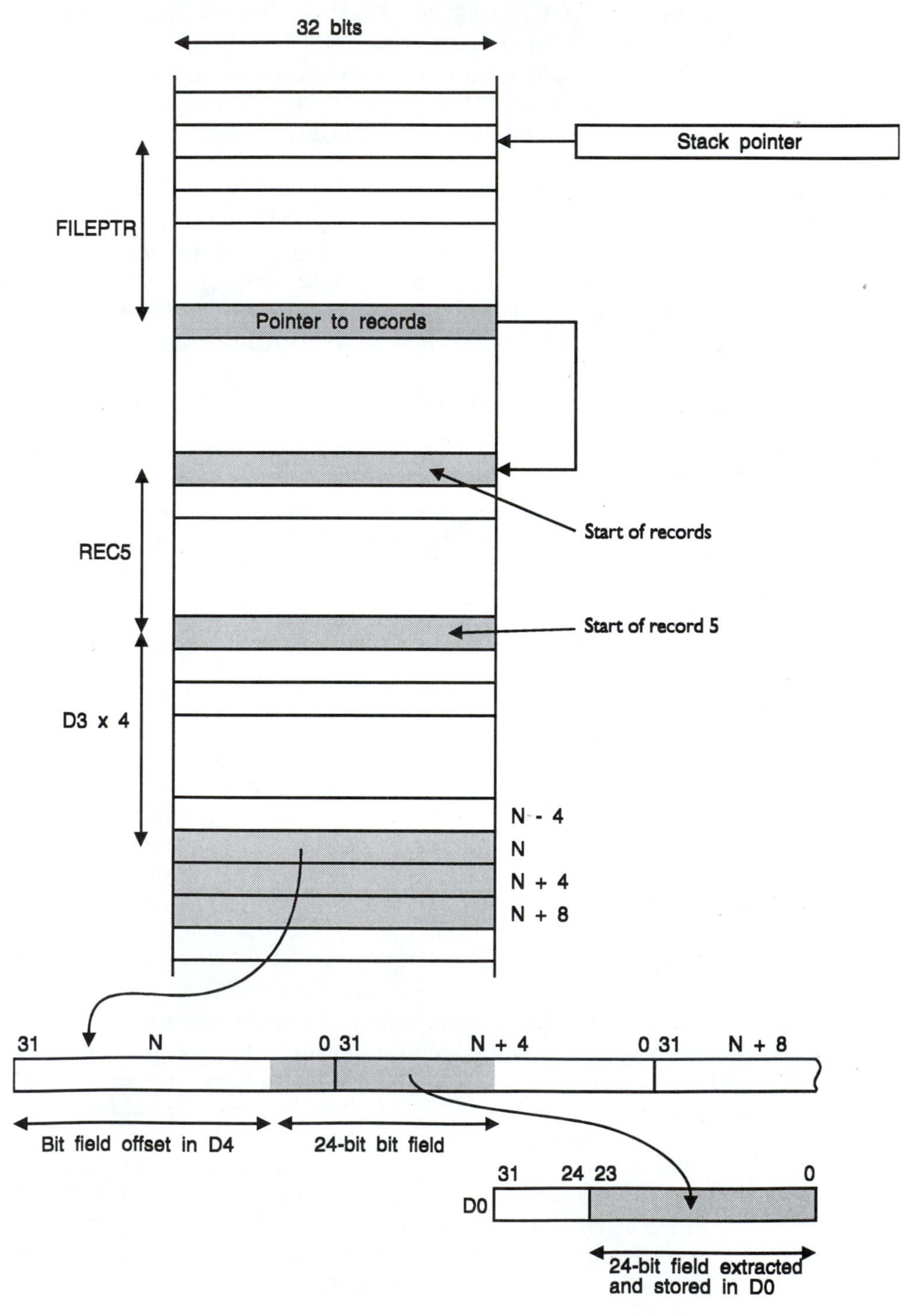

In order to access the required data, you first have to read the longword on the stack at [SP] + FILEPTR. This 32-bit address is a *pointer* to the record structure, and the actual record can be found at some offset from the beginning of the records. Assume that we wish to access record five at offset REC5. Once record five has been located, it is necessary to extract an item within the record. Data register D3.W provides an offset to the longword containing the required bit field. Data register D4 contains the offset from the start of the bit array to the most-significant bit of the longword pointed at by D3. Assume that the bit field to be accessed is 24 bits wide. We can access the bit field by the single instruction:

```
BFEXTU ([FILEPTR,SP],REC5,D3.W*4){D4:24},DO
```

This instruction contains eight operands and performs a remarkable amount of computing in one instruction. Imagine performing an equivalent operation in pure 68000 code.

12.4 The 68020's New and Extended Addressing Modes

The 68000 has all the basic addressing modes you would expect to find on most 8- or 16-bit microprocessors. However, the 68020 introduces powerful new addressing modes with several variations, making it very much more sophisticated than most other microprocessors. The 68020's new addressing modes are also implemented by the 68030 and the 68040. When we talk about the 68020's addressing modes in this section, we mean also the 68030 and the 68040.

You might almost say that the 68020 represents the high point of microprocessor development — in the sense that future developments will probably be in different directions. For example, I don't expect to see new generations of microprocessors similar to the 68020, but with more and more complex addressing modes and instruction sets. I expect to see microprocessors that have streamlined and simplified addressing modes (e.g., like the RISC processors), or microprocessors that have been designed to execute high-level languages directly, or microprocessors that incorporate system functions such as memory management and memory caches (e.g., like the 68030 and the 68040).

Extended Addressing Modes

Some of the 68000's addressing modes have been enhanced by the 68020, just as the 68000's operation set has been enhanced. These enhancements are the extension of the 68000's addressing modes to include 32-bit offsets. For example, the 68000's register indirect with index addressing mode expresses the effective address as (d8,An,Xn). The 68020 permits the constant offset to be either d8, d16, or d32 (i.e., a byte, word, or longword).

Memory Indirect Addressing

Although a glance at the 68020's instruction manual might lead you to believe that the 68020 has quite a few new addressing modes, a purist could argue that there is really only one new addressing mode. This is *memory indirect addressing*, in which the effective address generated by an instruction points to a memory location that points to the actual operand to be accessed. The reason that the 68020 seems to have a lot of new addressing modes is that memory indirect addressing is implemented with a large number of options. Memory indirect addressing has been included to make it easier to access arrays and similar data structures.

Consider the most basic form of memory indirect addressing, that is synthesized from the 68020's more complex general memory indirect addressing modes. In 68020 assembler syntax, the effective address is written `([address])` and is calculated by first reading the contents of memory location `address` and then using the 32-bit value at that address to access the actual location of the operand. An instruction of the form `MOVE D0,([$1234])` expressed in RTL has the effect:

$$[D0] \leftarrow [M([M(1234_{16})])].$$

Note that the conventions of RTL and the 68020's assembler syntax differ. We can understand this RTL expression better if it is split into two parts:

```
Temp_ea ← [M(1234)]
[D0]    ← [M(Temp_ea)].
```

Figure 12.17 illustrates the action of memory indirect addressing. One advantage of this addressing mode is immediately clear. It provides the same functionality as address register indirect addressing. However, instead of having just eight address registers, we have an *index register* for each longword of the processor's memory space. By operating on the contents of the effective address in memory, we can skip about an array or similar data structure, just as we could by operating on the contents of an address register.

The 68020 provides two memory indirect addressing modes: *memory indirect postindexed* and *memory indirect preindexed*. 68020 literature also refers to two other memory indirect addressing modes: *program counter memory indirect postindexed* and *program counter memory indirect preindexed*. These are variations on memory indirect addressing in which the program counter expresses a relative displacement rather than an address register that expresses an absolute displacement.

The syntax of the 68020's new indirect addressing modes is:

Addressing mode	**Assembler syntax**
Memory indirect postindexed	`([bd,An],Xn,od)`
Memory indirect preindexed	`([bd,An,Xn],od)`
PC memory indirect postindexed	`([bd,PC],Xn,od)`
PC memory indirect preindexed	`([bd,PC,Xn],od)`

Figure 12.17 Example of indirect addressing `MOVE.W ([$1234]),D0`

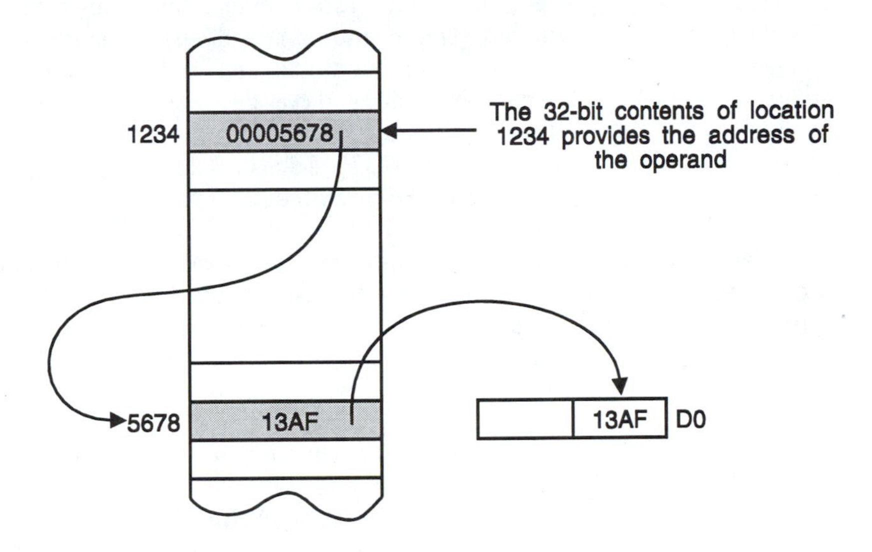

In this assembler syntax, bd is a 16-bit or a 32-bit signed constant called a *base displacement* or an *inner displacement*. Xn is an address or data register whose contents may be scaled by multiplying them by 1, 2, 4, or 8, as we shall soon see. Finally, od is a second 16-bit or 32-bit signed literal, called an *outer displacement*. We can use these addressing modes in simpler forms by *suppressing* the displacements bd and/or od, and by *omitting* An or Xn.

When a constant is *suppressed*, it is omitted from the encoding of the instruction rather than simply setting it to zero. Suppressing a constant makes the assembled code more compact. The 68020's numerous indirect addressing modes provide the programmer (or compiler) with a series of options. For example, both the following two instructions are legal examples of memory indirect postindexed addressing:

```
    MOVE  D0,([$12345678,A0],D4,$FF000000)
and MOVE  D0,([A0]).
```

Using the 68020's ability to omit constants and to suppress registers in addressing modes, we can take the full effective address `([d32,An,Rx],d32)` and create a number of options by simply *taking an eraser* and rubbing out any unnecessary components of the effective address. Below are some of the possible legal options of memory indirect postindexed addressing (we use the term *legal* here, since some of the options are not sensible). The meaning of these addressing modes will become clearer when you have read the remainder of this section. At the moment we are simply interested in demonstrating the wide range of options available.

```
([d32,An,Rx],d32) Full effective address
([An,Rx])         Rub out inner and outer displacements
([An],d32)        Rub out inner displacement and index register
([Rx])            Rub out inner and outer displacements and An
(An,Rx)           Rub out displacements and indirection
(Rx)              We can even have data register indirect
[(d32)]           Address of address
()                Strange but still legal. Address zero.
[()]              Even stranger - address zero.
```

Before continuing we demonstrate one simple advantage of memory indirect addressing. A table of longword addresses, each of which corresponds to a subroutine entry point, is pointed at by A0 (see Figure 12.18). Suppose that the number in D0 indicates which subroutine is to be called. We can call the appropriate subroutine (in 68000 code) by the following action:

```
LSL.L    #2,D0         Multiply subroutine number in D0 by 4
MOVEA.L (A0,D0),A1     Get subroutine address in A1
JSR      (A1)          Call the subroutine.
```

Note that the contents of D0 must be multiplied by 4 because a subroutine address occupies a longword, but the subroutine identifier in D0 is a one-byte value. We can now make use of the 68020's indirect addressing mode and write:

```
JSR     ([A0,D0*4])  Call the subroutine specified by D0
                     That is, [PC] ← [M([A0] + 4 x [D0])]
```

Figure 12.18 Accessing a jump table

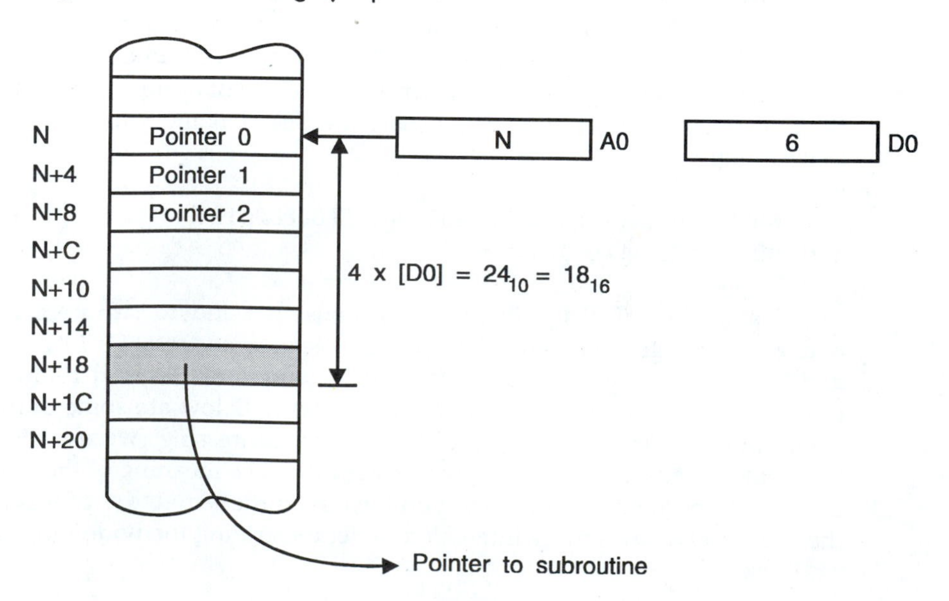

In this example the contents of A0 are added to the contents of D0 multiplied by 4. The resulting effective address is used to access the table. The longword at this address is read and loaded into the program counter to call the subroutine. The 68020's memory indirect addressing mode has allowed us to replace three 68000 instructions by a single 68020 instruction. Moreover, the scaling does not modify the contents of the register scaled.

Constants Used in Memory Indirect Addressing

Three constants or literals are employed by the 68020's new memory indirect addressing modes and are written bd, od, and sc in 68020 literature. The base displacement, bd, corresponds approximately to the 68000's d8 and d16 offsets, except that bd may also be a full 32-bit literal. The base displacement is added to an address register to give the location in memory containing the pointer to the actual operand.

The outer displacement, od, is a 16-bit or 32-bit literal and has no 68000 equivalent. The constant od is added to the value read from memory to give the actual address of the operand. The index register used in memory indirect addressing can be written `Xn.S*sc`. The `.S` indicates size (i.e., `.W` or `.L`). The scale factor, sc, is a constant whose value is 1, 2, 4, or 8, and is used to scale the contents of an index register. The scale factor has no explicit 68000 equivalent, although it is used implicitly in certain operations.

Consider the 68000 operation `MOVE.W (A0)+,D0`. The contents of address register A0 are incremented by 2 after it has been used as an index register. Had the instruction been `MOVE.L (A0)+,D0`, A0 would have been incremented by 4. In other words, the contents of an auto-incremented address register are incremented by 1 scaled by 1, 2, or 4.

The need for a scale factor is related to the 68020's ability to support byte, word, longword, and quadword operands. Clearly, successive bytes differ by one location, successive words differ by two locations, successive longwords differ by four locations, and successive quadwords differ by eight locations because the 68000 family's memory is byte addressed, irrespective of the type of data being accessed. Suppose that A0 points to an array of elements, *Y*, and D0 contains an index *i*. The address of the *i*th element is given by [A0] + [D0]*sc, where sc is 1, 2, 4, or 8 for byte, word, longword, or quadword (i.e., 64-bit) data elements. Unfortunately, since the 68000 has no automatic mechanism for scaling, the 68000 programmer must carry out the calculation. For example, in 68000 code we would write:

```
LSL.L #s,D0          Scale ith element (s = 0,1,2 for sc = 1,2,4)
LEA   (A0,D0.L),A1   Calculate address of ith element.
```

The 68020 provides explicit scaling in conjunction with memory indirect addressing. The index register, Xn, may be an address or data register and is written Xn*sc in assembler form. Typical effective addresses might be written:

```
MOVE D0,([A0,D4*1])  Scale factor = 1
MOVE D0,([A0,D4*4])  Scale factor = 4.
```

Figure 12.19 demonstrates how the scale factor relates to the size of data objects. Of course, the built-in scale factors of the 68020 cannot be used with data objects of arbitrary size.

Example of Memory Indirect Addressing

The effect of the base displacement is best illustrated by means of an example. Suppose we have a list of students and each student has a record consisting of six elements. Each of these elements corresponds to the student's results in that subject. We can organize a suitable data structure in several ways. One is to choose a structure consisting of the student's name followed by the six results. An alternative is to create a list of pointers, one per student, where each pointer points to the student's results (see Figure 12.20). This example takes the latter approach and demonstrates how the three constants, bd, sc, and od, are related.

Register A0 in Figure 12.20 points to the base of a region of memory devoted to the students' records. This region may include other items of related data. The base displacement, bd, points to the start of the list of students with respect to the start of the region of data. That is, the first student's entry is at address [A0] + bd. Of course, if A0 had been loaded with the address of the first student's entry, the base displacement could have been set to zero and the addressing mode simplified by omitting bd. Note that the students are stored as a list of 32-bit pointers.

Data register D0 contains the index of the student to be accessed (i.e, if [D0] = *i*, the *i*th student is to be accessed). Since each entry in the table of pointers is a longword, we have to scale the contents of D0 by 4. The effective address of the

Figure 12.19 The scale factor (only scale factors of 1, 2, and 4 illustrated)

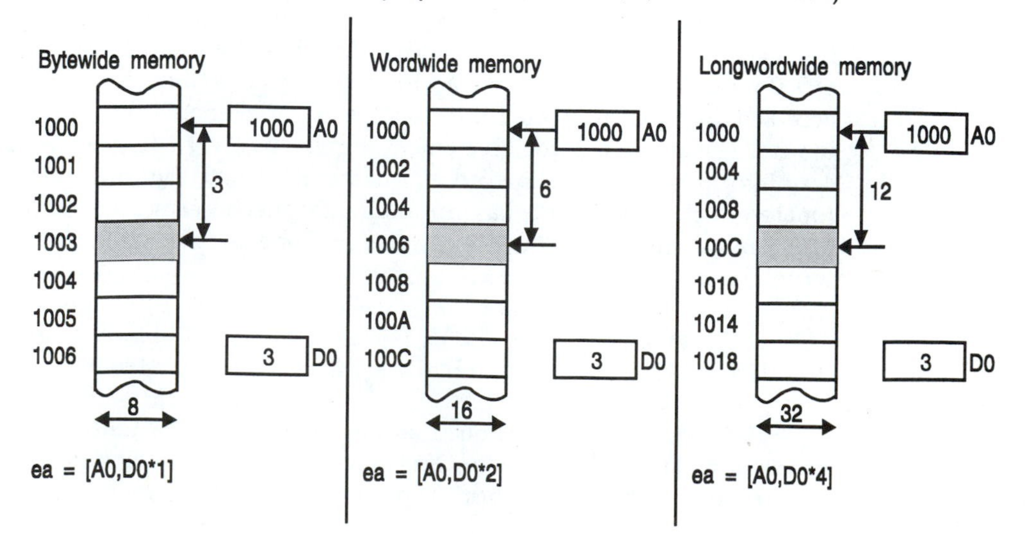

Figure 12.20 Example of memory indirect addressing with preindexing

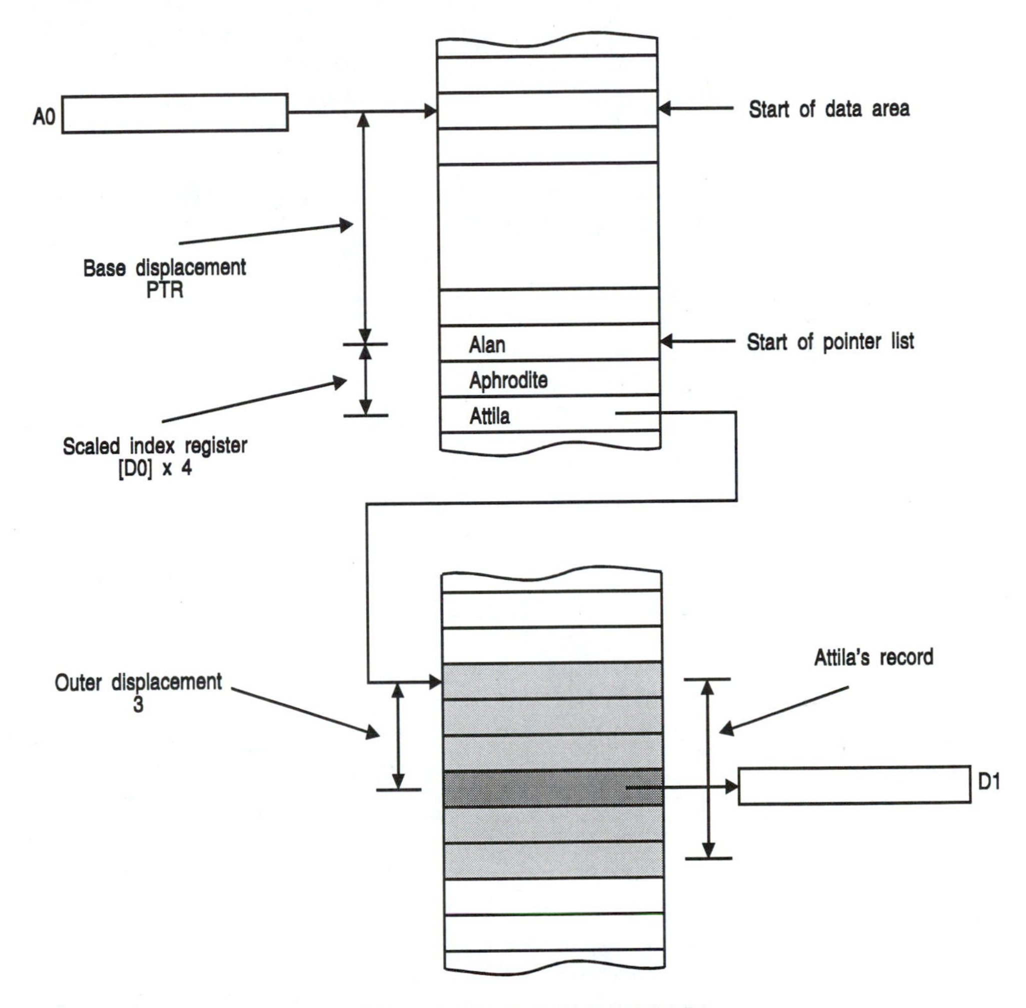

pointer to the selected student's record is therefore [A0] + bd + 4 x [D0]. The 68020 reads this pointer which points to the start of the student's record.

Suppose we want to know how the student performed in computer science, which is the fourth out of the six results. We need to access the fourth item in the table (i.e., item number 3 because the first item is numbered zero). The outer displacement, od, provides us with a facility to do this. When the processor reads the pointer from memory, it adds the outer displacement to it to calculate the actual effective address of the desired operand.

If this example were to be coded for the 68000, the assembly form might look like:

```
PTR     EQU      <pointer to data>
        LSL.L    #2,D0                    Multiply the student index by 4
        LEA      (PTR,A0,D0.L),A1         Calculate address of pointer to record
        MOVEA.L  (A1),A1                  Read the actual pointer
        ADDA.L   #3,A1                    Calculate address of CS result
        MOVE.B   (A1),D1                  Read the result.
```

The same calculation can be carried out by the 68020 using memory indirect addressing with preindexing:

```
MOVE.B   ([PTR,A0,D0.L*4],3),D1
```

Before looking at another example of the 68020's two memory indirect addressing modes, it is worthwhile looking at how they differ. Figure 12.21 illustrates the effect of postindexing and Figure 12.22 the effect of preindexing.

Figure 12.21 The 68020's postindexing memory indirect addressing

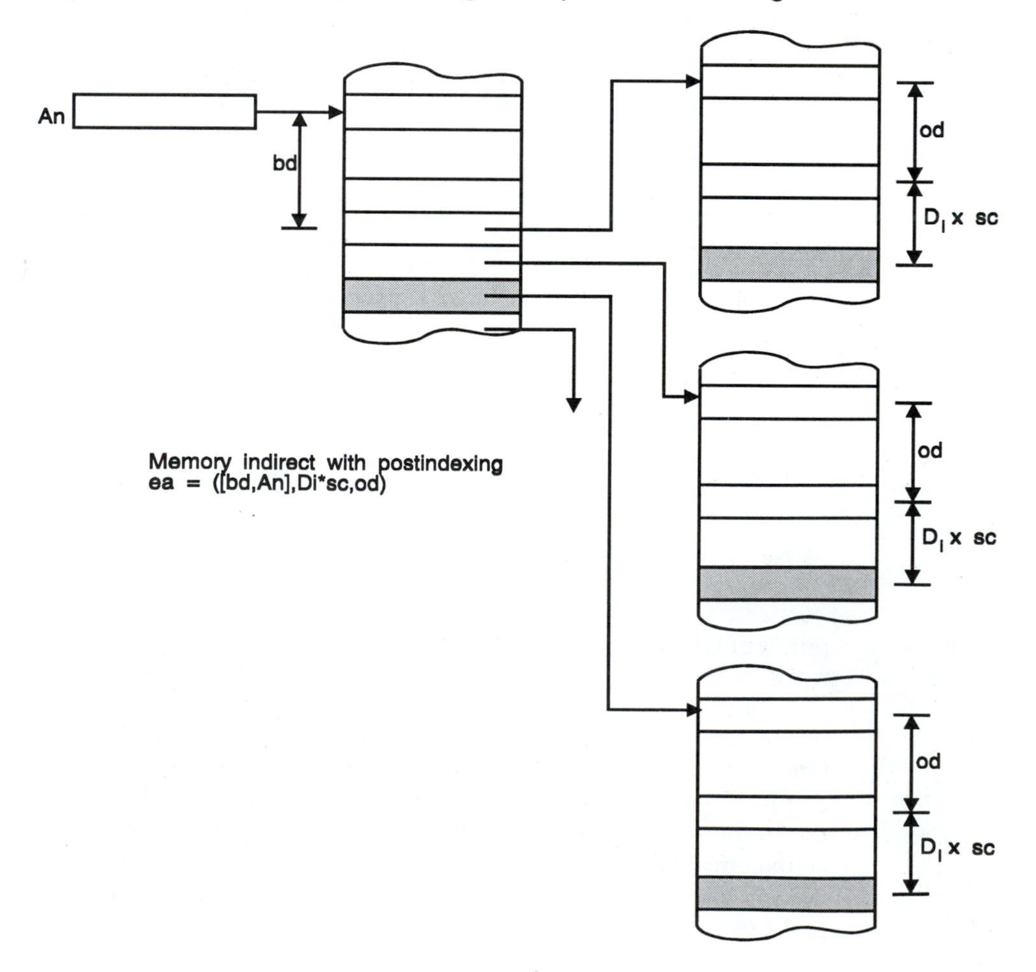

Figure 12.22 The 68020's preindexing memory indirect addressing

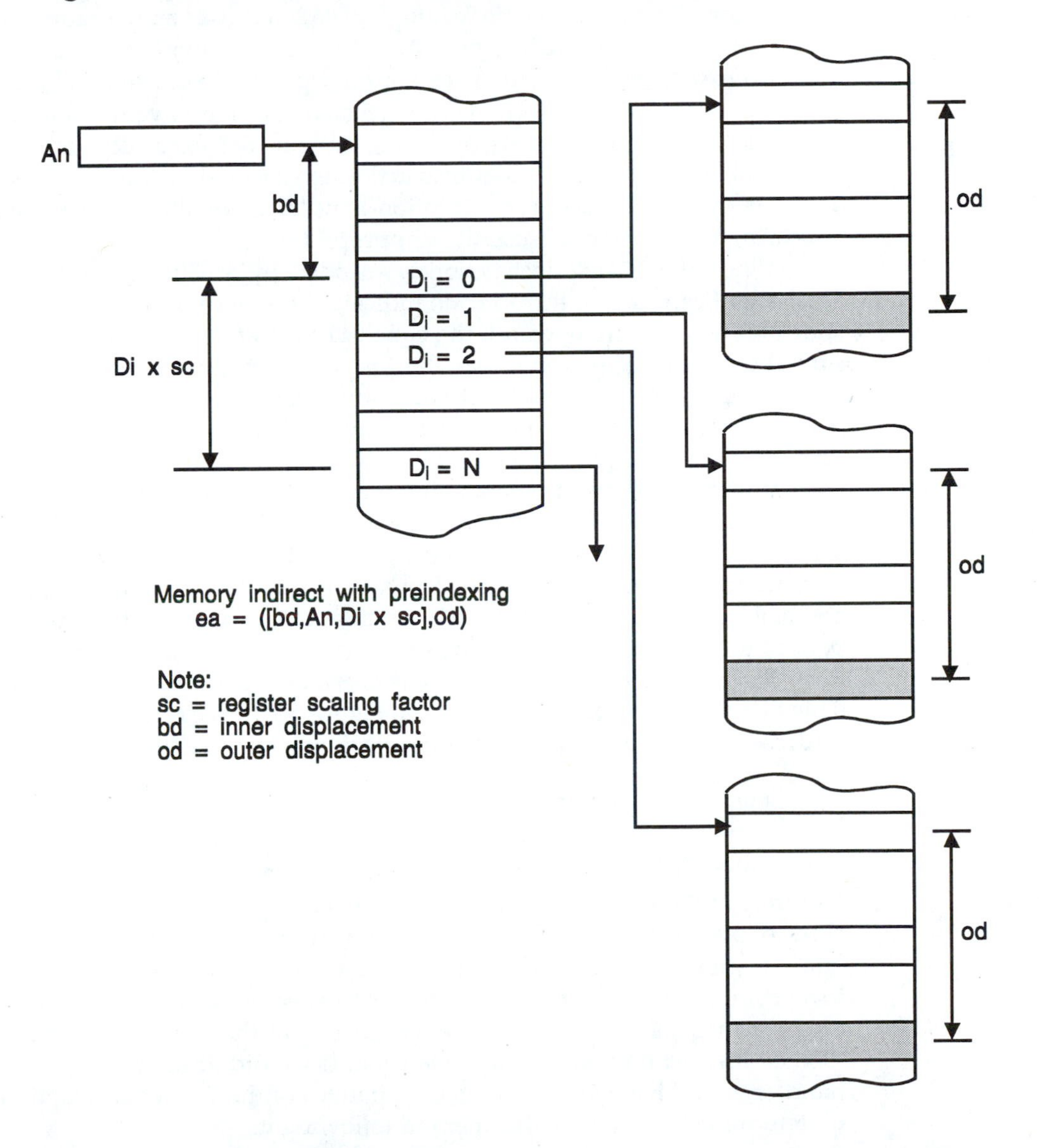

Using Memory Indirect Addressing with a TRAPcc Instruction

It is not easy to provide both simple and realistic applications of memory indirect addressing, since many examples of memory indirect addressing involve complex high-level language data structures and are inappropriate in this text. The following example demonstrates how memory indirect addressing can be used even by the assembly language programmer. This example uses two new concepts: the 68020's `TRAPcc` instruction and the 68020's exception stack frame. In brief, a TRAPcc instruction causes a trap or call to the operating system to take place if condition cc is true. For example, TRAPCS causes an operating system call if the carry bit in the CCR is one. A TRAPcc instruction has three formats, one with no extension, one with an extension word (TRAPcc.W), and one with two extension words (TRAPcc.L). These extensions are literal values and are parameters that can be read by the operating system. For example, you might write TRAPCS.W #MyData. When this instruction is executed, a trap exception will be taken if the C-bit is set. The exception handler reads the 16-bit value, MyData, following the op-code and uses it as appropriate.

When the TRAPcc instruction is encoded by a 68020 assembler, the three least-significant bits of the op-code indicate the size of instruction. The code 010 indicates a word-size operand, the code 011 indicates a long-word-size operand, and the code 100 indicates that the instruction has no operand. The systems programmer can read these bits to determine the type of TRAPcc instruction.

When a TRAPcc instruction is encountered and the specified condition cc is true, a call to the operating system (i.e., the TRAPcc handler) is made and certain information saved on the stack pointed at by the stack pointer, A7. Figure 12.23 illustrates the state of the system immediately after a TRAPCS.W #d16 instruction has been executed and the trap taken. You don't have to understand the details of exception handlers — all you need know for the purpose of this example is that the stack frame pointed at by A7 contains the address of the instruction that caused the exception (i.e., the address of the TRAPCS.W #d16) at location [A7] + 8.

Suppose the TRAPcc handler needs to examine the literal following the TRAPcc. This literal is stored at address N + 2 in Figure 12.23. We can use memory indirect addressing to access this literal via the stack pointer, A7. That is:

```
MOVE.W ([8,A7],2),D0.
```

Eight is added to the contents of the stack pointer to get the address of the "address of the TRAPcc exception". The contents of this location (i.e., [A7] + 8) are read to give the address of the TRAPcc. The outer displacement, 2, is added to this value to give the address of the literal following the TRAPCS.W. This literal is then loaded into D0. As we have seen, this entire sequence is carried out by a single 68020 instruction. By the way, we assumed that the trap had been called by a TRAPcc instruction with a word extension. If we didn't know this, the exception handler would have to examine the bit pattern of the TRAPcc instruction in order to determine the length of the operand following it.

12.5 The 68020's Supervisor State

The first enhanced member of the 68000 family was the 68010. This microprocessor has essentially the same architecture as the 68000, and is physically compatible with it (i.e., the 68010 fits in the same socket as the 68000). The real difference

Figure 12.23 Effect of executing a `TRAPCS.W #data` instruction

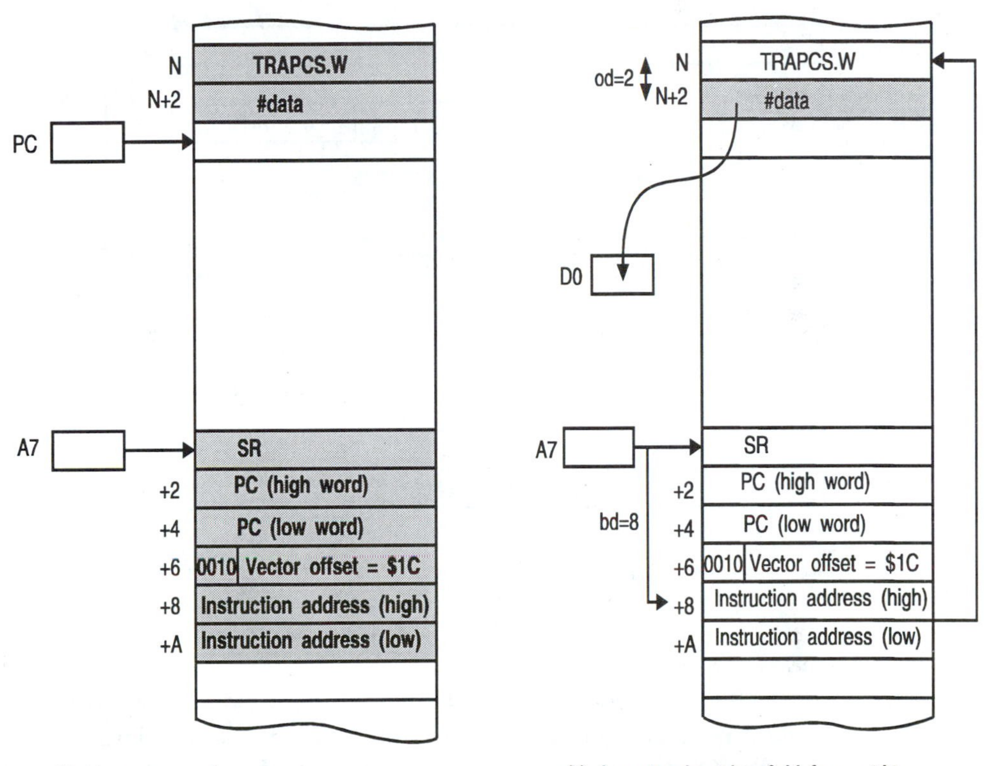

(a) Memory map after executing TRAPCS.W #data

(b) Accessing the #data field from within the TRAPCS.W exception handler by means of MOVE.W ([8,A7],2),D0

between the 68010 and the 68000 is that the 68010 has enhanced exception handling capabilities. In particular, the 68010 implements a true *virtual* machine; it can return from a bus error exception; and its exception vector table is relocatable.

The 68020 extends the 68010's exception handling capabilities even further by implementing two supervisor state stack pointers and a variable stack frame format. The 68030 is essentially a 68020 with an on-chip memory management unit, and its exception handling facilities are only marginally different from those of the 68020. Therefore, in this section we concentrate on the 68020. Figure 12.24 illustrates the 68020's new supervisor state registers of interest to us here.

The *alternate function code registers,* SFC and DFC, allow the systems programmer to force a particular value on the 68020's three function code output pins, FC0, FC1, and FC2, during a memory access. These three pins tell an external memory manager whether a member of the 68000 family is accessing data or executable code and whether it is accessing supervisor memory space or user memory space. The memory management unit can use the information on FC0, FC1, FC2 to decide whether the current memory access is legal.

Figure 12.24 Supervisor state registers of the 68000 and the 68020

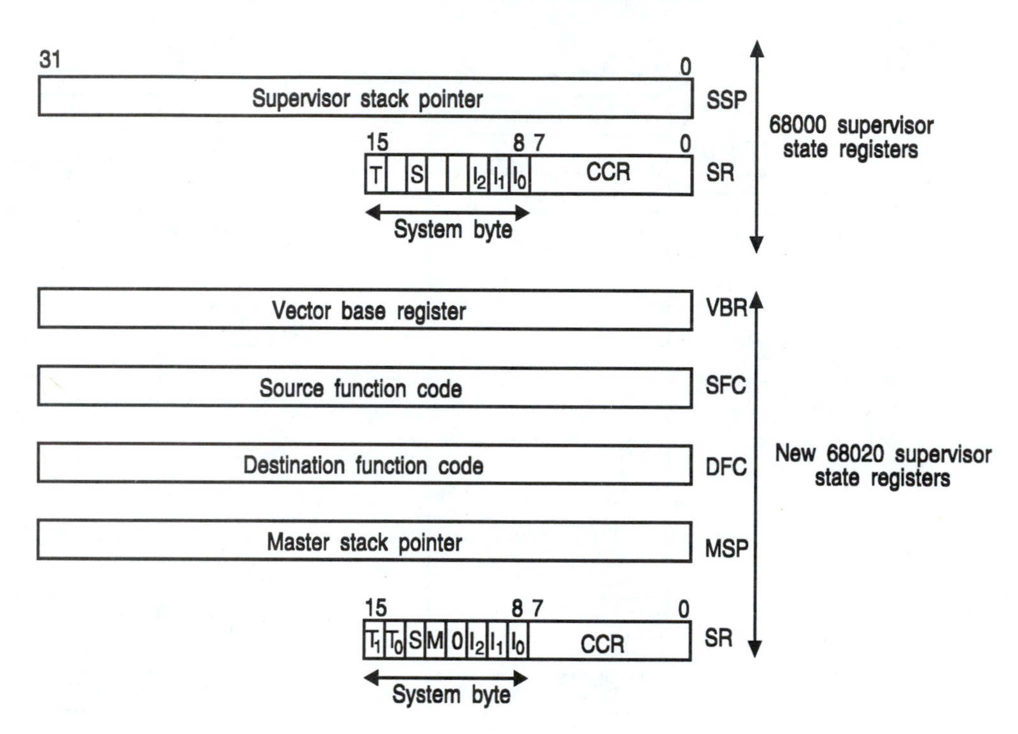

The *vector base register*, VBR, is used to relocate the 68020's exception vector table anywhere within the processor's memory space. This table contains the pointers to each of the exception handlers, and therefore the 68020 can support multiple exception vector tables. We shall look at the VBR in more detail shortly.

The 68020 has a third stack pointer called the *master stack pointer*, MSP, in addition to the USP and SSP. 68020 literature renames the 68000's supervisor stack pointer the ISP (*interrupt stack pointer*) or A7' (*A7 prime*). The MSP is sometimes referred to as A7'' (*A7 double prime*). As we shall see, the ISP can be dedicated to interrupt handling, and the MSP used for all other exception handling.

Accessing Supervisor-state Registers

These new registers require a new instruction to access them. The 68020 employs a move to/from control register instruction, `MOVEC`, to access the VBR, SFC, and DFC registers. `MOVEC` is a privileged instruction with the assembly language form:

```
      MOVEC  Rc,Rn
or    MOVEC  Rn,Rc
```

where Rn is a general register (A0 to A7 or D0 to D7) and Rc is a control register (i.e., VBR, SFC, or DFC). The `MOVEC` instruction is also used to access the cache control register in the 68020/68030. Note that MOVEC takes a longword operand (even though only three bits of the SFC and DFC registers are defined). When, for example, a MOVEC SFC,D0, is executed, bits 0 to 2 of the source function code register are moved to data register D0, and bits 3 to 31 of D0 are padded with zeros. Since MOVEC permits only register-to-register operations, data must be first loaded into a data or address register before it can be loaded into a control register. We can relocate the 68020's exception vector table from its default location $00 0000 to $00 8000 by executing the following instructions.

```
MOVE.L  #$8000,D0
MOVEC   D0,VBR
```

The *alternate function code registers*, SFC and DFC, are used in conjunction with the 68020 instruction MOVES (move to or from address space). Before continuing, we have to make it clear that the alternate function code registers have a meaning only in systems with memory management units that distinguish between *user address space* and *supervisor address space* (Figure 12.25). The 68000 employs the function code outputs on its FC0, FC1, FC2 pins to indicate the type of address space being accessed (user or supervisor) to an MMU (memory management unit). If the type of address space accessed does not match the type of address space indicated by FC0, FC1, FC2, the MMU issues a bus error exception by asserting the 68020's BERR* (bus error) input. Consequently, the MMU can prevent a user from accessing memory space that has been allocated to the operating system.

This arrangement of processor and MMU suffers from a subtle flaw. When the 68000 is operating in the *supervisor mode,* it cannot make a memory access to *user* address space, because all its accesses are to supervisor address space (by definition — that's what being in the supervisor state means). What we need is a method of fooling the MMU into thinking that the microprocessor is operating in the user mode when it is, in fact, operating in the supervisor mode. Why should we wish to do this? Such a facility permits the operating system to transfer data to

Figure 12.25 The 68x00, the function code, and the memory management unit

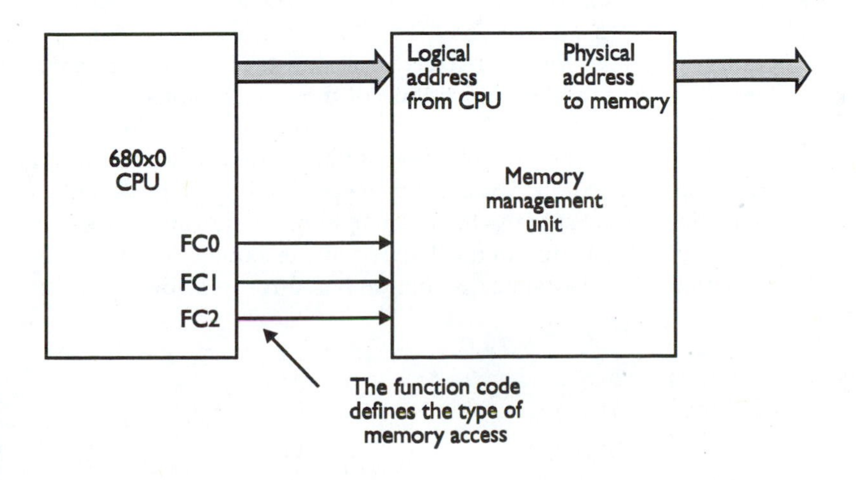

user data space. Equally, it allows the operating system to perform diagnostic tests on user address space.

The 68020 can access any address space when it is in the supervisor mode by means of the privileged instruction MOVES, which has the assembly language forms

```
    MOVES Rn,<ea>
and MOVES <ea>,Rn
```

Rn is a general register (i.e., A0 to A7 or D0 to D7), and <ea> is an effective address. Note that MOVES permits byte, word, and longword operands. The address space used by the MOVES instruction is determined by the source function code register, SFC, if the source operand is in memory. Similarly, the address space is determined by the destination function code register, DFC, if the MOVES instruction specifies a destination operand. For example, suppose the operating system wishes to read the contents of location $40000 which lies in user address space. The following sequence of operations will perform this task:

```
MOVE.L  #%001,D0    Load D0 with the user data function code 0,0,1
MOVEC.L D0,SFC      Copy the user space code into the SFC register
MOVES.W $40000,D1   Read data from the user data space
```

Again we must stress that the MOVES instruction would be used only in a system with some form of memory management. Since Motorola defines address space 3 (i.e., FC2,FC1,FC0 = 0,1,1) as *user reserved*, you could use this address space to trigger your hardware in some way. That is, you can employ the MOVES instruction to put the special code 0,1,1 on the function code pins. What you do with this code is up to you.

New 68020 and 68030 Exceptions

Apart from exceptions associated with the 68000's new instructions (CHK2 and TRAPcc), the 68020's and the 68030's new exceptions are concerned with the coprocessor interface, the format error, and the 68030's memory management unit. We will provide a brief description of these exceptions.

TRAPcc The TRAPcc exception causes an exception if condition cc is true when the instruction is executed. The condition specified by cc represents one of the 16 conditions that are the same as the branch, Bcc, conditions. The exception routine called by the TRAPcc instruction is located at the same address as that called by a TRAPV instruction. The TRAPcc instruction has three possible formats:

```
TRAPcc
TRAPcc.W    #<d16>
TRAPcc.L    #<d32>
```

A TRAPcc can take no extension, a word, or a longword extension. These optional extensions have no effect on the execution of the TRAPcc exception itself, and can be used to pass a parameter to the TRAPcc exception handling routine. Since the TRAPV and TRAPcc exceptions share the *same* exception vector, it is up to the writer of the exception handlers for these instructions to provide the appropriate course of action.

For example, the instruction TRAPEQ.W #$00FC has the effect of calling the TRAPcc exception handler if the Z-bit of the CCR is set to one. The trap handler can read the constant $00FC and use it as required. The 68020 shares the TRAPcc exception handler with the 68000's TRAPV exception handler (i.e., memory location $00 001C). You can determine the type of trap and the size of the data following the TRAPcc by reading its op-code. The least-significant bits of the op-code are set to 010 for a word-size operand, 011 for a longword size operand, and 100 for no operand.

CHK2 The CHK2 instruction is an extension of the 68000's existing CHK. CHK2 has the assembly language form CHK2 <ea>,Rn and can take byte, word, and longword operands. The value in register Rn is compared with the lower and upper bounds at the address specified by <ea>. An exception is called if:

```
       Rn < lower bound
or if  Rn > upper bound.
```

Note that the CHK and the CHK2 instructions share the same exception vector. That is, they share the same exception handling routine. Suppose you are accessing an array of bytes and the array extends from $201200 to $2012FF in memory. You might employ the following code:

```
      LEA     (A3,D4.W),A5       A5 points to the next element
      CHK2.L  Bound,A5           Check address against bounds
      .                          Continue if no violation
      .
      .
Bound DS.L    $201200,$2012FC    Lower and upper bounds
```

Format Error Exception Members of the 68000 family save different amounts of information on an exception stack frame, depending on the nature of the exception. For example, the 68000 has two possible stack frames and the 68020 has seven. Each of these stack frames has a format that defines its size and the type of information stored in it. When a return from exception is made by an RTE instruction, it is necessary to restore information from the stack frame to the processor. The processor (i.e., 68020 or 68030) determines the type of stack frame by reading the format number on the stack frame (which is held in bits 12 to 15 of the word that contains the vector offset). A format error exception takes place when the processor encounters an RTE instruction, and the information saved in the stack frame does not match that specified by the frame's format number.

cpTRAPcc Exceptions The cpTRAPcc exception is employed in conjunction with an external coprocessor and causes a trap if the selected condition code of the coprocessor is true. All we need say here is that you can force an exception on the coprocessor's condition code register (which is not the same as the 68020's own CCR). We take a brief look at the 68020's coprocessors later in this chapter.

Privilege Violation Exception In addition to the 68000's privileged instructions, the 68020 and 68030 generate privilege violation exceptions for the following privileged instructions:

Supervisor mode instructions	Coprocessor instructions	MMU instructions
MOVEC	cpRESTORE	PFLUSH
MOVES	cpSAVE	PLOAD
		PMOVE
		PTEST

Exception Vectors and the 68020

The 68020's exception vector table is the same as the 68000's but with the addition of vectors to deal with the 68020's new exceptions. Table 12.1 illustrates the 68020's and 68030's new additions to the vector table.

As we discovered in Chapter 10, whenever the 68000 responds to an exception, it reads an exception vector from the appropriate location in the exception vector table in the region $00 0000 to $00 03FF. The 68010 and later members of the 68000 family calculate the address of an exception vector in exactly the same

Table 12.1 Additions to the 68000's exception vector table

Vector	Vector address	Address space	Exception type
6	018	SD	CHK2 instruction
7	01C	SD	TRAPcc, cpTRAPcc instructions
13	034	SD	Coprocessor protocol violation
14	038	SD	Format error
48	0C0	SD	FPCP Bcc or Scc incorrect
49	0C4	SD	FPCP inexact result
50	0C8	SD	FPCP divide by zero
51	0CC	SD	FPCP underflow
52	0D0	SD	FPCP operand error
53	0D4	SD	FPCP overflow
54	0D8	SD	FPCP signalling no number (i.e., NaN)
56	0E0	SD	PMMU configuration
57	0E4	SD	PMMU illegal operation
58	0E8	SD	PMMU access level violation

way as the 68000, but then add the address of the exception vector to the contents of the 32-bit vector base register, VBR, to provide the actual address of the exception vector. Figure 12.26 illustrates how the exception vector table can be remapped. As Figure 12.26 demonstrates, the VBR permits the exception vector table to be relocated anywhere within the processor's memory space. The VBR is cleared following a hardware reset, so that the processor initially behaves just like a 68000 with the exception vector table mapped in the region \$000000 to \$0003FF.

The 68020's VBR supports multiple exception vector tables. Simply reloading the VBR selects a new exception vector table anywhere in the 68020's address space. Such a facility might be useful in certain classes of multitasking systems, in which each task maintains its own copy of the exception vector table. For example, the various tasks may treat, say, a divide-by-zero exception in different ways. If each task shared the same exception vector table, the exception handler would have to determine which task generated the current divide-by-zero exception, and then call the corresponding procedure. By changing the exception vector table each time a new task is run, exceptions are automatically vectored to the handler appropriate to the task now running.

Exception Processing and the 68020

Broadly speaking, the 68020 and 68030 handle exceptions in the same way as the 68000. These newer microprocessors build on the exception handling ability of the 68000. It's reasonable to say that as the applications of microprocessors have grown increasingly sophisticated, the 68000 family has been extended to match the requirements of these new applications. This statement is probably more true

Figure 12.26 Using the VBR to remap the exception vector table

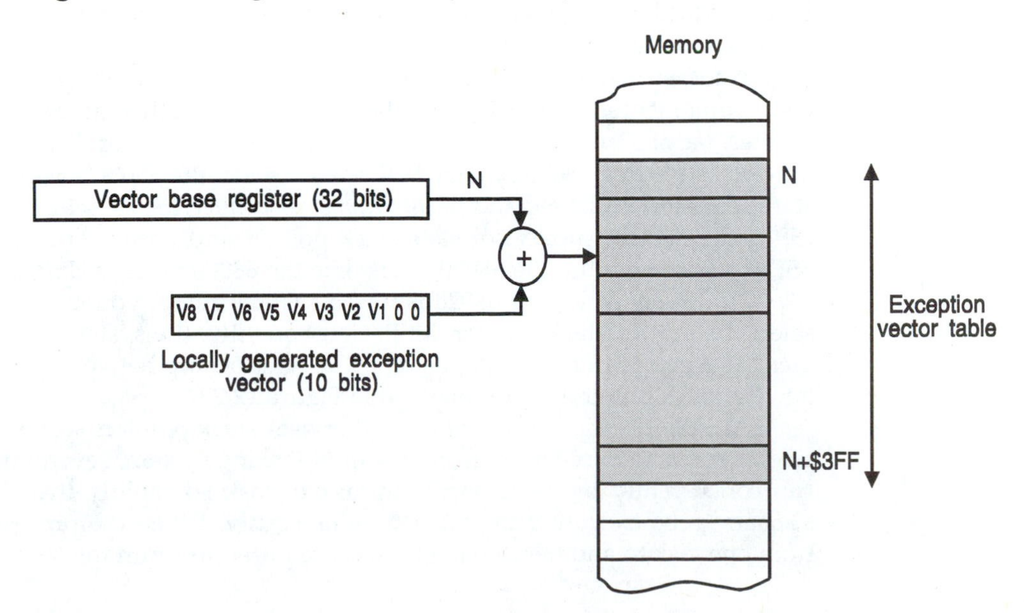

when applied to exception handling than to any other aspect of microprocessor architecture. Consider the way in which the 68000 family has progressed.

Traditional 8-bit microprocessors	**68000 microprocessor**
Simple interrupt handling with fixed vectors to interrupt handlers and one or two interrupt request inputs. Very limited software exception handling capabilities.	Seven levels of prioritized and vectored interrupts. Extensive range of software exceptions. User and supervisor modes permit protected operating system.
The 68010 microprocessor	**The 68020/30 microprocessors**
Same as 68000 but with the ability to recover from a bus error. This makes it possible to implement virtual memory systems. Addition of a vector base register improves its multitasking performance. Making `MOVE from SR` a privileged instruction transforms the 68010 into a true virtual machine.	Same as 68010 but with two supervisor state stack pointers. One of these can be devoted to interrupts and the other to maintaining task control blocks in a multitasking environment. This enhancement makes the 68020 ideal for environments supporting both multitasking and extensive interrupt handling.

The 68020's Stack Pointers

We have already seen that the 68020 improves on the 68000's exception handling mechanism, because it is able to remap the exception vector table by means of its VBR. The addition of a VBR is not the only change in the way the 68020 processes exceptions. A 68020 saves more information on the exception stack frame than the 68000. Indeed, six different stack frame formats are necessary to facilitate a return from exception under all circumstances.

Another major extension to the 68000 family's exception processing mechanism implemented by the 68020 is the inclusion of another supervisor state stack pointer, *the master stack pointer*, MSP. At any instant, one of three stack pointers may be active in a 68020 system. Following a reset, the 68020's M-bit (bit 12 of its status register) is cleared and its *interrupt stack pointer*, ISP, is selected. The 68020's ISP is the default supervisor state stack pointer and corresponds to the 68000's SSP. You can now use the 68020 *exactly* like the 68000 with its USP and SSP.

However, when the 68020 is in the system mode, you can set its M-bit to select the master stack pointer, MSP. Consequently, the systems mode programmer has a choice of *two* stack pointers. The relationship between the S and M bits and the stack pointers is summarized in Figure 12.27.

Although the provision of two supervisor stack pointers seems a little excessive, there is an excellent reason. In a multitasking system, several user tasks are run concurrently by switching from task to task so rapidly that the processor appears to be executing the tasks simultaneously. When the processor switches from one task to another, the information required to rerun the task just switched

Figure 12.27 Relationship between the 68020's S-bit and M-bit

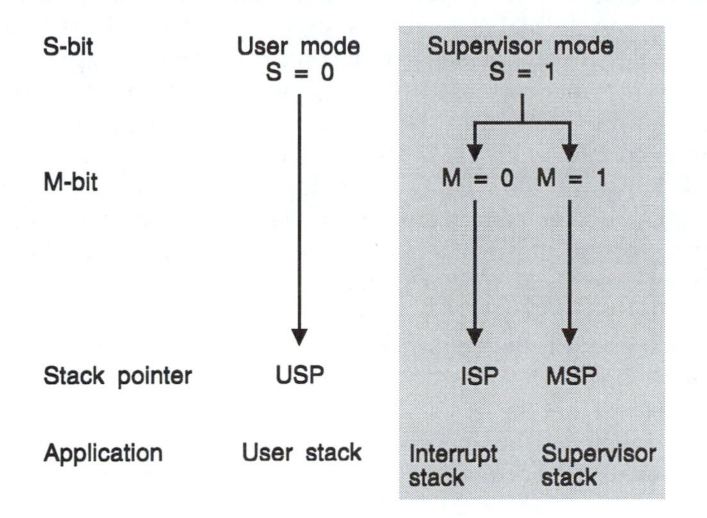

off must be saved. This information is a copy of the processor's registers, its CCR, and its stack pointer (i.e., the USP), and is saved in a structure called a *task control block* or TCB. The TCB is often saved on the supervisor mode stack. The real-time kernel example in Chapter 11 discusses the TCB in greater detail.

Now suppose that an interrupt occurs during the time a given task is being processed. The interrupt's stack frame is stored on the task's (supervisor state) stack. If interrupts are heavily used, a task might require an enormous TCB, just to store interrupt stack frames. Moreover, each task must have an equally large TCB to store interrupt stack frames generated while the task was active. A much more sensible solution is to implement two entirely separate supervisor-state stacks. One stack is dedicated to multitasking (and general purpose supervisor-state applications) and the other dedicated to interrupt processing. That is, interrupts store information on an entirely separate stack from that used to handle all other exceptions. In this way, the systems designer does not have to worry about saving interrupt information within a task's TCB.

The 68020 uses its master stack pointer, MSP, to maintain a general purpose stack, and its interrupt stack pointer, ISP, to maintain a stack dedicated to interrupt handling. When M = 0 (following a reset), the 68020 behaves exactly like the 68000 and uses a single supervisor mode stack pointer, the ISP, for all exception processing. When the M-bit of the status register is set to one (by the operating system — `ORI #$1000,SR`), both the 68020's supervisor-state stacks are activated. By switching the master stack pointer each time a new task is activated, a separate master stack pointer value can be assigned to each task in a multitasking environment. Once the M-bit has been set, interrupts use the interrupt stack pointer rather than the master stack pointer.

When the 68020 processes an *interrupt* (but not any other exception), it tests the status of the M-bit after the processor context has been saved on the currently active supervisor stack. If the M-bit is clear, exception processing continues normally (since there is only one supervisor-state stack pointed at by the ISP).

If the M-bit in the status register is set, the processor clears it and creates a so-called *throwaway* exception stack frame on top of the interrupt stack. Note that there are now *two* interrupt stack frames, one on each of the supervisor-state stacks. This second throwaway stack frame on the interrupt stack contains the same program counter value and vector offset as the frame created on top of the master stack. However, the stack frame on top of the interrupt stack has a format number 1 instead of 0 or 9 (stack formats are described later).

The copy of the status word on the throwaway frame is the same as the version in the frame on the master stack, except that the S-bit is set in the version placed on the interrupt stack. The version of the status word on the master stack may have S = 0 or S = 1, depending on whether the processor was in the user or supervisor state before the interrupt. Interrupt processing then takes place in the normal way, except that the ISP is the active stack pointer.

As we've just said, at the end of the interrupt processing sequence, the processor's S-bit is set and its M-bit cleared. When a return from exception is executed, the processor reads the SR from the throwaway frame on the interrupt stack, updates the SR, increments the active stack pointer by eight, and begins RTE processing again. This repetition may seem strange, but remember that there are two stack frames (one on the ISP stack and one on the MSP stack).

The 68020's Stack Frames

We already know that the 68000 has two types of stack frame, one for both Group 1 and Group 2 exceptions and a more complex stack frame for Group 0 exceptions (i.e., bus errors and address errors). The 68020 and 68030 employ a total of six stack frames according to the type of exception being processed. No simple *68000-type* stack frame is implemented, because all stack frames include at least four words. In fact, the first four words of all six stack frames are identical and comprise (starting at the top-of-stack): status register, program counter, stack frame format, and vector offset.

The *format* of the stack frame is stored in bits 12 to 15 of the word at address [SP] + 6. Bits 0 to 11 of the same word contain the *vector offset*, which is the exception number multiplied by four. That is, the vector offset is the index into the exception vector table and permits the exception handler to determine the nature of the exception that invoked it. The format is read by the processor itself when it executes a return from exception (otherwise, how would the processor know how much information to restore when it encounters an RTE?).

Figures 12.28 to 12.33 describe the 68020's six stack frames. Note that the *name* of the stack frame is determined by the binary value of the frame's format. For example, if the format code is 1010, the stack frame is called *stack frame ten*. Stack frame 0, Figure 12.28, has a four-word structure and is used by the following exceptions:

Figure 12.28 The 68020's stack frame 0

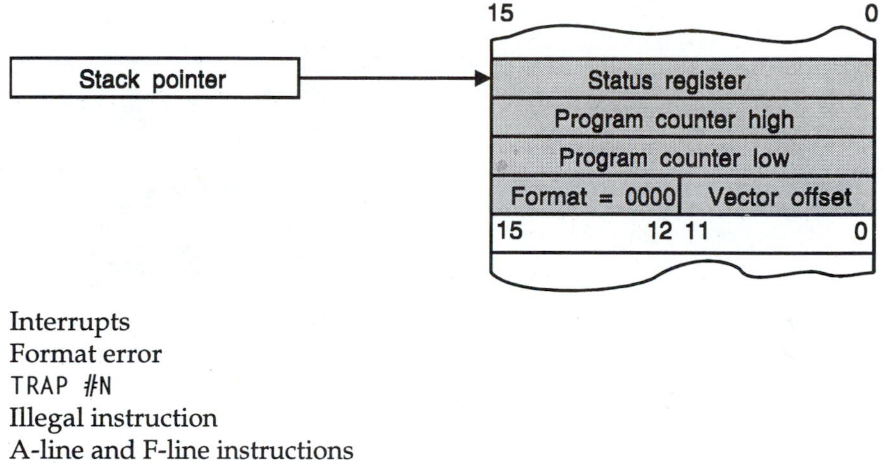

Interrupts
Format error
TRAP #N
Illegal instruction
A-line and F-line instructions
Privilege violation
Coprocessor pre-instruction

Stack frame 1, Figure 12.29, is a *throwaway frame* used during interrupt processing when a change from master state to interrupt state is made. When the M-bit is set, the stack frame is saved on the stack pointed at by the master stack pointer. If the exception is an interrupt, a second copy of the stack frame, a type 1 stack frame, is pushed onto the stack pointed at by the ISP, and the M-bit cleared.

Figure 12.29 The 68020's stack frame 1

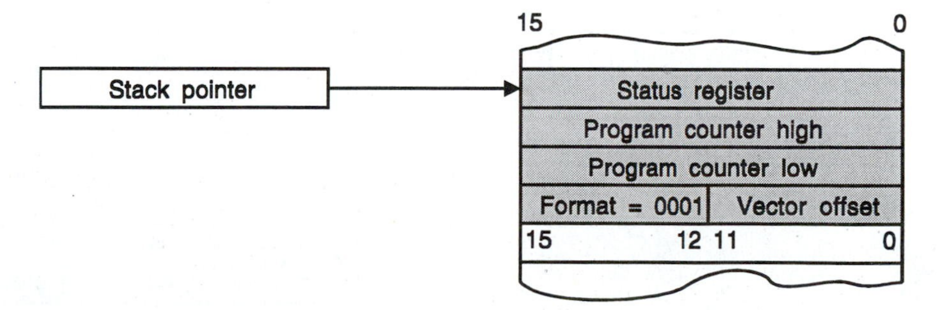

Stack frame 2, figure 12.30, uses a 6-word structure and is used by the following exceptions:

CHK, CHK2
cpTRAPcc, TRAPcc, TRAPV
Trace
Divide by zero
MMU configuration
Coprocessor post-instruction

Figure 12.30 The 68020's stack frame 2

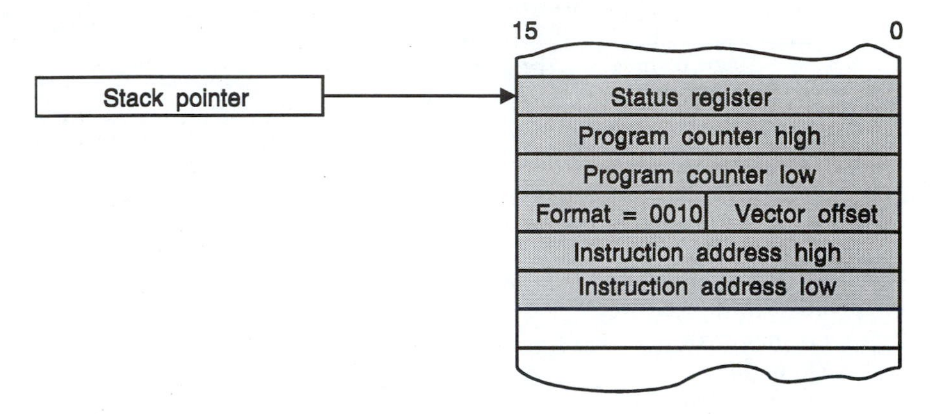

Stack frame 9, Figure 12.31, stores 10 words and is used by the following exceptions:

Coprocessor mid-instruction
Main-detected protocol violation
Interrupt detected during coprocessor instruction

Figure 12.31 The 68020's stack frame 9

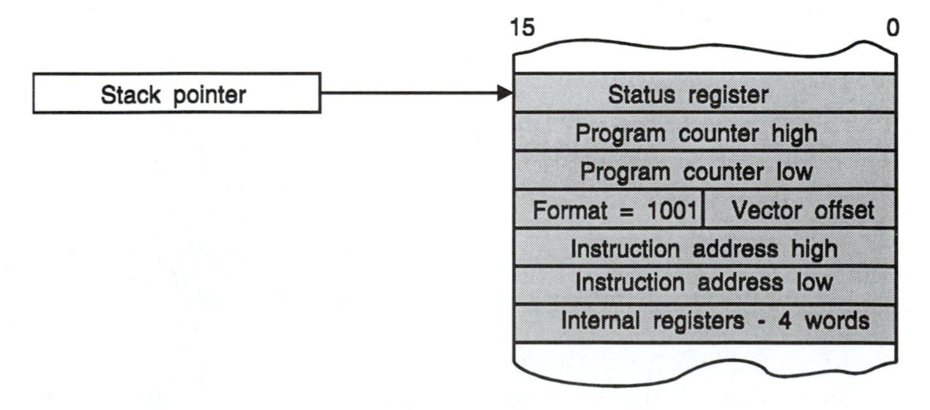

Stack Frame 10, Figure 12.32, stores 16 words on the stack and is used when an address error or bus error exception occurs at an instruction boundary. This is called a *short bus cycle fault format* because it is relatively easy to return from an exception at an instruction boundary.

Stack frame 11, Figure 12.33, is the longest stack frame and is composed of 46 words. It is used when an address error or a bus error exception occurs during the execution of the faulted instruction.

Figure 12.32 The 68020's stack frame 10

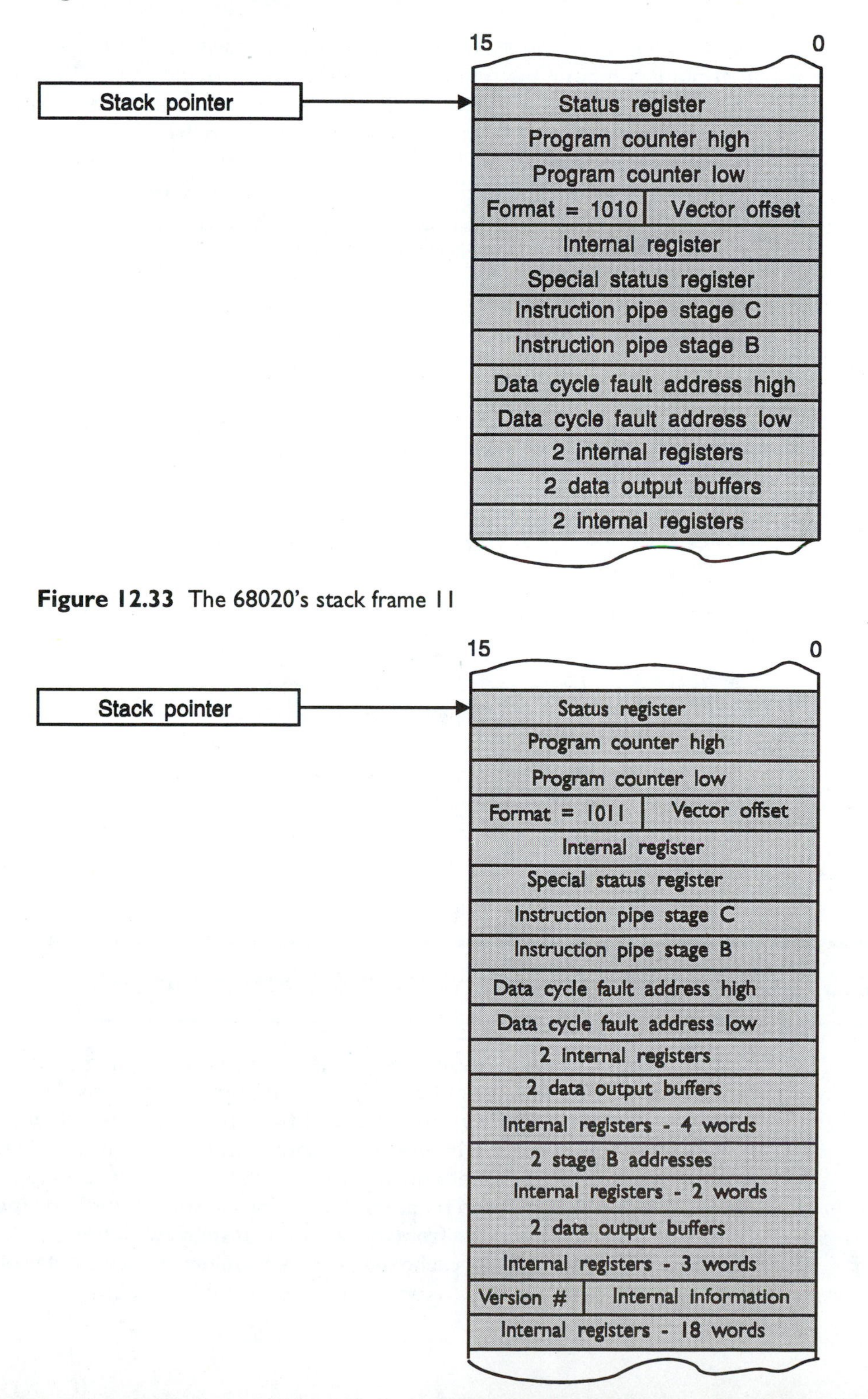

Figure 12.33 The 68020's stack frame 11

Multiple Exceptions and the 68020/30

Since two or even more exceptions can occur at the same time, the processor must prioritize competing exceptions. The 68020/30 provides a fixed prioritization for exceptions, as listed in Table 12.2. Note that exceptions that include the prefix 'Cp' are exceptions related to the 68020's coprocessor interface. When multiple simultaneous exceptions occur, the processor deals with the highest priority exception, then processes the exception with the next highest priority, and so on. High priority exceptions such as bus and address errors are processed immediately — even if another exception is currently being processed.

68020 Trace Mode

The 68020 implements a modest improvement in the 68000's trace mode by including a simple trace filter. The 68020's status register has two trace bits, T_1 and T_0, which are bits 15 and 14 of the status register, respectively. The effect of these bits is as follows:

Trace bit		Trace function
T_1	T_0	
0	0	No tracing
0	1	Trace on change of program flow
1	0	Trace on any instruction
1	1	Undefined state — reserved

The new mode provided by the 68020 is activated when $T_1, T_0 = 0, 1$. A trace exception is generated only when a change of flow takes place, caused by the execution of, for example, a `BRA`, `JMP`, `TRAP`, and return instruction. The 68020 also considers a change of flow to take place when its status register is modified.

12.6 Cache Memory and the 68000 Family

Cache memory provides designers with a way of exploiting high-speed processors without incurring the cost of large high-speed memory systems. The word *cache* is pronounced *cash* or *cash-ay* and is derived from the French word meaning *hidden*. Cache memory is hidden from the programmer and appears as part of the system's memory space. There's nothing mysterious about cache memory. It's simply a quantity of very high-speed memory that can be accessed rapidly by the processor. The element of magic comes from the ability of systems with tiny amounts of cache memory (e.g., 64K bytes of cache memory in a system with 4 Mbytes of DRAM) to make over 95% of accesses to cache rather than slower DRAM.

Table 12.2 Exception Priority Groups for the 68020/30 (Group 0 highest priority)

Group	Exception	Characteristics
0.0	Reset	Aborts all processing and does not save the old machine context.
1.0 1.1	Address error Bus error	Suspends processing and saves internal machine context.
2.0	`BKPT #n`, `CHK`, `CHK2`, Cp mid-instruction, Cp protocol violation, `CpTRAPcc`, Divide by zero, `RTE`, `TRAP #n`, `TRAPV`, MMU configuration	Exception processing is part of the instruction execution.
3.0	Illegal instruction, Line A Unimplemented line F, Privilege violation, Cp pre-instruction	Exception processing begins before the instruction is executed.
4.0 4.1 4.2	Cp post-instruction Trace Interrupt	Exception processing begins when the current instruction or previous exception processing is completed.

The structure of a cache memory is given in Figure 12.34. A block of high-speed cache memory sits on the processor's address and data buses in parallel with the much larger main memory. When the CPU accesses memory, the cache controller determines whether the data is in the cache. If it is, a *hit* is declared and the data fetched from the cache. If the data is not cached, a *miss* is declared and the data is fetched from the main store and the cache updated. Because of the nature of programs and their attendant data structures, the data required by a processor is often highly clustered throughout memory. This aspect of memories is called the *locality of reference* and makes the use of cache memory possible.

Cache memories are usually organized in one of three ways: direct mapped, associative mapped, and set associative mapped. Each of these systems has its own advantages and disadvantages. However, since this is a text on assembly language programming, we do not cover the way in which cache memories are actually implemented. We are more interested in the impact on the programmer.

Cache memory systems are both expensive and complex, and until recently they were not associated with low-cost microprocessor systems. Modern microprocessors like the 68020 and 68030 operate at speeds great enough to make cache memory worthwhile. The addition of a large cache memory subsystem can substantially increase the cost of a microcomputer and can sometimes push its price into the minicomputer range.

Figure 12.34 Structure of cache memory

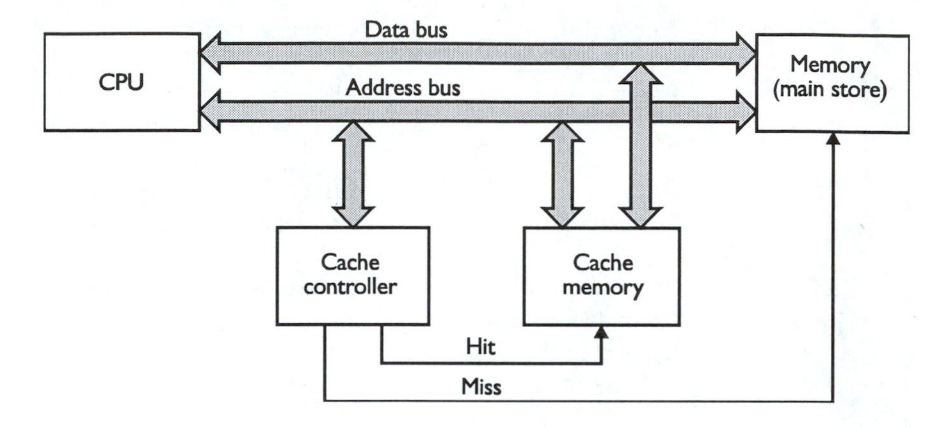

Designers of the 68020 and 68030 have placed a modest quantity of cache on-chip, and have therefore eliminated all the design and implementation overheads normally associated with cache memories. On-chip cache gives these processors a boost in performance at no cost to the user. Modern technology has enabled the 68040 to implement an impressive 8 Kbyte on-chip cache system.

The 68020's Cache

The 68020 implements a 64-longword, direct-mapped instruction cache. That is, instructions are cached, but not data. By not caching data, the design of the cache is greatly simplified, since the problem of updating memory is eliminated (you never perform data writes to the cache). The 256 bytes permit small program loops to be run entirely from cache, and thereby remove the need for instruction fetches from main store. As you can imagine, the 68020's cache has relatively little effect on the execution of pure in-line code with no loops.

When the 68020 pre-fetches an instruction it first looks for the instruction in the cache. If it hasn't been cached, the instruction is read from the external memory and the cache updated. Each entry in the cache includes a single control bit, the V-bit. The V-bit validates entries in the cache. During a processor reset (e.g., after power-up), all 64 V-bits are cleared to indicate that the cache is empty.

The 68020's instruction cache is almost, but not quite, invisible to the systems programmer. The 32-bit cache control register, CACR, in the 68020's supervisor space can be accessed by the privileged instructions `MOVEC CACR,Rn` and `MOVEC Rn,CACR` to control the operation of the cache. Although CACR, Figure 12.35, is a 32-bit register, only four bits are defined — C, CE, F, and E.

During a reset, the cache is cleared by resetting all V-bits to invalidate the entries in the cache, and the cache enable and cache freeze bits of the CACR are also cleared. The cache enable bit determines whether or not the cache is to be used. If CACR(E) = 0, the cache is disabled — it does not exist. You might wonder

Figure 12.35 The 68020's cache control register

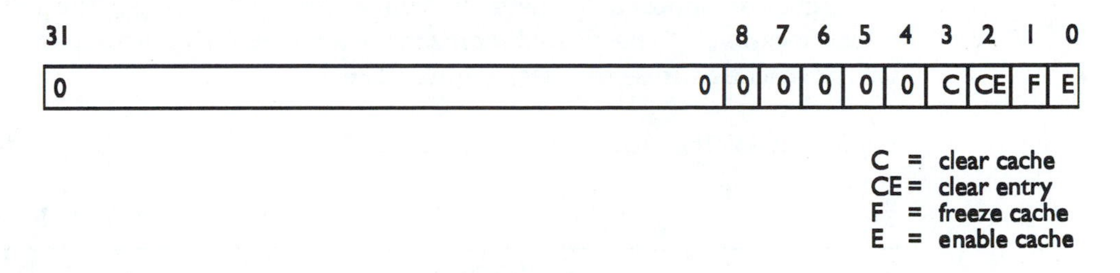

why it may be necessary to disable the cache. Suppose you are debugging the 68020 by observing the flow of data on the data and address buses. The internal cache suppresses the fetching of instructions that are already cached, and therefore might make it difficult to debug the 68020.

Since the cache enable bit is cleared on reset, the supervisor program must explicitly set it to enable the cache. We can do this by:

```
MOVE.L  #$01,D0     Set bit zero (i.e., the E bit)
MOVEC   D0,CACR     Load cache control register.
```

If you disable the cache and then re-enable it, the previously valid entries remain valid and can be used again. The cache freeze bit, when set to 1, suspends the instruction cache's update mechanism. If a miss occurs, the line in the instruction cache is not updated. When the F-bit is cleared, the cache updates instructions normally. Do not confuse the E- and F-bits. The E-bit simply switches off the cache and prevents the 68020 from using it. The F-bit permits the cache to be accessed but not to be updated. Since the 68020's cache is rather small, it can hold relatively few instructions and will be refilled every time a section of in-line code is executed. Suppose you have a short task switcher that must be fast. You can unfreeze the cache on entry to the code and then freeze it again at the end of the code block. In this way, the cache is not updated and the task switcher's code will be cached next time it is called. Consider the example,

```
SWITCH MOVE.L  #%0001,D0  Enable and unfreeze cache
       MOVEC   D0,CACR    Set up cache control register
       .
       .                  Code of task switcher
       .
       MOVE.L  #%0011,D0  Enable and freeze cache
       MOVEC   D0,CACR    Set up cache control register
       RTS                Return
```

The clear cache bit, CACR(3), is used to clear or to *flush* all entries in the cache. Setting the C-bit of the CACR clears all V-bits. If the operating system performs a context switch or a new program is loaded into main store, it is necessary to clear (flush) the cache to remove old (i.e., stale) instructions.

The clear entry bit, CACR(2), can be used to clear a single entry (i.e., line) in the cache (as opposed to the C-bit which clears all entries). We tell the 68020 which location is to be cleared by means of another supervisor-space register, the CAAR (cache address register), Figure 12.36.

Figure 12.36 The cache address register, CAAR

31 … 8	7 6 5 4 3 2	1 0
Cache function address	Index	

The index field of the CAAR determines which line of the cache is to be cleared when the CE-bit is set (the cache function address is not used by the 68020). The CE-bit is automatically reset to zero after it has been loaded with 1.

As you can see, it is easy to use the 68020's cache memory — you enable it and then forget about it. The use of cache memory has big implications and problems caused by cache memories are some of the hardest to track down. The real danger lies in modifying data in main memory without changing the data in the cache. Suppose you load a new program from disk into main memory and begin executing it. The cache will contain old instructions even though corresponding information in main memory has now been overwritten by new instructions. You can avoid this problem by flushing the cache each time you load new code. If you forget to flush the cache, you will spend weeks looking at the code byte by byte and wondering why the 68020 does not do what it was told. By the way, there is no explicit way in which you can examine the contents of the 68020's (or 68030's) cache.

The 68030's Cache

The 68030 doubles the size of the 68020's cache by supporting a 256-byte data cache in addition to the 68020's 256-byte instruction cache. It is important to note that the 68030's two caches are logical caches, since they cache logical addresses from the 68030. This statement is necessary because the 68030 contains an on-chip memory management unit, and therefore the address at its address pins is a *physical* address, rather than a *logical* address. The point we must appreciate here is that if the mapping performed by the 68030's memory management unit is changed, both the 68030's caches must be flushed.

Because the 68030's instruction and data caches are entirely independent, they can operate autonomously. That is, the 68030 can perform an instruction fetch and a data fetch to its internal caches, perform a data fetch to external memory, and execute an instruction — all simultaneously. Before looking at the details of the 68030's cache, we will comment briefly on the hardware interface used to support the cache.

Since the 68000 family uses memory-mapped I/O, imagine the effect of caching data on a peripheral. The first time the CPU reads a peripheral, the data will be read correctly and then cached. The next time it reads the peripheral, it will read the old data from the cache even though the peripheral may have new data for the CPU. The 68030 has a cache disable input, CDIS*, that can be asserted to disable both caches. A separate cache inhibit input, CIIN*, can be asserted to disable the cache during certain memory accesses (e.g., I/O accesses to peripherals). CIIN* is ignored in write cycles.

A cache inhibit output signal, CIOUT*, indicates to any external cache that the bus cycle should be ignored. Two signals are provided to permit an entire line of the cache to be filled in a burst of data transfers. Cache burst request, CBREQ*, is an output that requests a line of data, and cache burst acknowledge, CBACK*, is an input informing the cache that one more longword can be supplied.

Read and write data accesses are treated differently by the 68030. Data reads are treated exactly like instruction reads. If a miss occurs, the operand is read from main store and the cache updated. When data is written to memory, it is written to the cache and it is also written in parallel to the external main store. This approach means that the memory and cache always remain in step and that a line of the cache can be updated or flushed without having to write it back to memory. Such a cache is called a write-through cache.

Figure 12.37 describes the 68030's cache control register. Although the 68030's CACR looks much more complex than that of the 68020's CACR, it is not. All the 68020's instruction cache control bits have been duplicated to refer to the instruction or the data cache explicitly. That is, you can enable or freeze either the instruction cache or the data cache independently.

The real additions to the 68030 are a write allocate bit and data/instruction cache burst enable bits. The burst enable bits, DBE and IBE, are cleared after a reset, and can be set to permit the 68030 to fill a line in its cache using the burst mode described above.

The write allocate bit can be set to select the 68030's data cache write-allocate mode. The WA bit is cleared after a reset and is ignored if the data cache is frozen. When WA = 0, *a write around policy* is implemented and write cycles resulting in a miss do not alter the data cache's contents. That is, the main store is updated but not the cache, as the cache is updated only during a write hit.

When WA = 1, the 68030 operates in its *write allocation mode* and the processor always updates the data cache on (cachable) write cycles, but validates an updated entry only during hits or when the entry is a longword aligned at a longword boundary.

Figure 12.37 The 68030's cache control register, CACR

31–14	13	12	11	10	9	8	7	6	5	4	3	2	1	0
	WA	DBE	CD	CED	FD	ED	0	0	0	IBE	CI	CEI	FI	EI

WA = write allocate
DBE = data burst enable
CD = dear data cache
CED = clear entry in data cache
FD = freeze data cache
ED = enable data cache

IBE = instruction burst enable
CI = clear instruction cache
CEI = clear instruction cache entry
FI = freeze instruction cache
EI = enable instrutt on cache

Once again, it is necessary to stress that the 68030's cache memory is invisible to the user programmer. Care has to be taken to inhibit the 68030's data cache by asserting CIIN* when a memory-mapped I/O port is accessed. If the supervisor mode implements task switching (using the on-chip MMU), the caches must be flushed before the new task runs, since the relationship between the cache and main memory will have altered.

The 68040's Cache

The 68040 represents a giant leap forward in cache technology over the 68020 and 68030. Like the 68030, the 68040 has independent and autonomous instruction and data caches. However, these are both four-way set-associative caches with 64 sets of four 16-byte lines (i.e., 4 Kbytes each), that are located after the on-chip MMU to provide a *physical* cache. That is, the 68030 caches data *before* its logical address is passed to the MMU, while the 68040 caches data *after* the MMU has calculated the physical address.

The 68040's cache operates in conjunction with the on-chip MMU. Each page (i.e., unit of memory) is assigned a two-bit code that determines how data on that page is to be cached. For example, pages containing I/O space can be marked as non-cachable, or pages can be marked as write-through or copyback pages.

Each entry in the instruction cache contains a V-bit, whereas each entry in the data cache contains both a V-bit and a D-bit (dirty-bit). Instead of relying on the writeback algorithm, the 68040 marks a line in the data cache as dirty if it has been written to. Lines with their D-bits set must be written back to memory. Writeback occurs when the line is accessed and a miss results. The 68040 permits an explicit writeback by means of the new `CPUSH` instruction, which pushes the selected dirty data cache lines to memory and then invalidates the lines in cache. All data transfers between the 68040's cache and external memory operate in a burst mode and transfer an entire line.

One of the major problems caused by cache memory in sophisticated microprocessor systems is called *cache coherency*. Consider an item that is in both the cache and main memory. If the processor modifies the value in the cache, the processor knows that it must also modify the value in memory. However, if the system has DMA or uses multiple processors, the data in main memory might be modified *without* the cache being updated. A cached system must ensure that when memory is updated, the cached copy of the data is also updated and not left *stale*. The term *stale* when applied to cache systems indicates old non-updated data.

The 68040's cache solves the problem of cache coherency by means of a remarkable technique called *bus snooping*. Instead of only *actively* accessing the system bus like other members of the 68000 family, the 68040 also monitors the bus in a passive fashion — that is, the 68040 monitors bus traffic even when it is not accessing the bus itself. If an alternative bus master accesses the bus and writes data to memory (when the 68040's snooping is enabled), the 68040 may then either invalidate the same line in its own cache or it may update its cache (depending on its programming).

Consider now a read cycle by an alternate bus master. Suppose the alternate master performs a read cycle and accesses a location cached by the 68040 that has its dirty bit set. In this case the alternate master will be in danger of reading stale data from memory. The 68040 can be programmed to prevent the memory from responding to the read access and to supply the correct data itself.

It should now be clear why the 68040 has a physical cache rather than a logical cache. If it had a logical cache it would be able to deal only with addresses generated by the processor. By using a physical cache, the 68040 caches addresses that appear on the system address bus and are meaningful to the system memory. Without physical address caching, the 68040 would not be able to perform bus snooping and ensure cache coherency.

12.7 The Coprocessor

The designer of any microprocessor would like to extend its instruction set almost infinitely but is limited by the quantity of silicon available (not to mention the problems of complexity and testability). Consequently, a real microprocessor represents a compromise between what is desirable and what is acceptable to the majority of the chip's users. Having said this, there are many applications for which a given microprocessor lacks sufficient power. For example, even the powerful 68020 is not optimized for applications that require a large volume of scientific (i.e., floating point) calculations. We are now going to look at one way in which the power of an existing microprocessor can be considerably enhanced by means of an external coprocessor.

An ideal coprocessor appears to the programmer as an extension of the CPU itself. For example, the 68882 floating-point coprocessor, FPC, can be employed in a 68020-based system to provide an extended 68020 instruction set that is rich in floating point operations. As far as the programmer is concerned the architecture of the 68020 has just been expanded. A coprocessor like the FPC enhances not only the 68000's instruction set but also other components of its architecture. For example, the FPC adds eight 80-bit floating point registers to the 68000's existing complement of registers. Moreover, the 68020's extended instruction set can cope with branches on FPC conditions and even handle exceptions initiated by the FPC. The 68040 doesn't need a coprocessor — it has an internal floating point unit.

The designers of 68000-family coprocessors decided to implement coprocessors that could work with existing and future generations of microprocessors with a minimal hardware and software overhead. The actual approach adopted by 68000-family coprocessors is to tightly couple the coprocessor to the host microprocessor and to treat the coprocessor as a memory-mapped peripheral lying in CPU address space. In effect, the microprocessor fetches instructions from memory, and, if an instruction is a coprocessor instruction, the microprocessor passes it to the coprocessor by means of its data transfer bus. By adopting this approach, the coprocessor does not have to fetch or interpret instructions itself. If a coprocessor requires data from memory, the host processor must fetch it. A corollary of this is that the coprocessor does not have to deal with, for example, bus errors, since all memory accesses are performed by a 68000-series processor.

68000-series coprocessors work efficiently with the 68020/68030 because the microprocessor-coprocessor communication protocol is built into the firmware of the microprocessor itself. In what follows, we will regard the 68020 as the host processor when describing coprocessors, because coprocessors are not normally used in 68000 systems. By the way, the 68040 doesn't have a hardware coprocessor interface — not least because it has its own MMU and floating point unit.

So, how can new coprocessor instructions be mapped onto the 68020's existing instruction set? In order to provide new instructions that can be interpreted by a coprocessor, a 68020 must allocate suitable op-code bit patterns. That is, a certain bit pattern must be interpreted by the processor as a coprocessor instruction. The 68020 uses its F-line op-codes to communicate with coprocessors. You will remember from Chapter 10 that F-line op-codes are also software exceptions or traps. The 68020 first treats an F-line bit-pattern as a coprocessor instruction, and then, if a coprocessor doesn't respond, treats it as an F-line exception.

The coprocessor is not located in conventional memory space, but is memory-mapped in *CPU address space* for which the function code on FC2, FC1, FC0 = 1, 1, 1. Although it is theoretically possible to locate a coprocessor anywhere within the 68020's 2^{32}-byte address space, each coprocessor is restricted to a specific slice of CPU memory space. The 68020 family supports up to eight coprocessors in the region \$20000 to \$2E01F. Each coprocessor is allocated a 32-byte slice of memory space. The address of a coprocessor within CPU space is given by Figure 12.38.

When the 68020 communicates with a coprocessor, address bits A_{31} to A_{20} are set to zero (along with bits A_{12} to A_{05}), and the coprocessor identification code 0010 is placed on address bits A_{19} to A_{16}. This code indicates to the system hardware that a coprocessor access is taking place. Address bits A_{15} to A_{13} comprise the Cp-ID field and define one of eight possible coprocessor types from 000 to 111. For example, the code A_{15}, A_{14}, A_{13} = 0, 0, 1 indicates a 68882 IEEE floating-point coprocessor. Currently, only two coprocessor types are defined by Motorola — the FPC and the MMU with a cp-ID of 0, 0, 0. You can, of course, design your own coprocessor.

Figure 12.38 Address of a coprocessor within CPU space

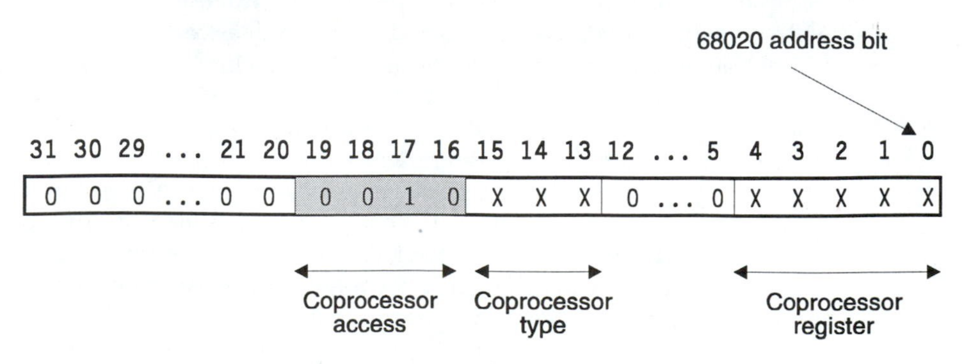

Finally, the least-significant five address bits A_{04} to A_{01} are used to access memory-mapped registers within the selected coprocessor.

The way in which a coprocessor is memory-mapped into a specific region of CPU address space is of interest to designers who have to interface coprocessors to microprocessors. It is of no interest to programmers. Why? Because the coprocessor appears as an extension of the microprocessor's instruction set. All programmers have to know about are the new instructions and registers provided by the coprocessor.

The Coprocessor Instruction Set

All coprocessor instructions have an *F-line format* that must begin with the bit-pattern 1, 1, 1, 1. Coprocessor instructions must be at least one word long and multi-word instructions are provided. A generic coprocessor instruction has the format described in Figure 12.39.

When the processor reads a coprocessor instruction, it uses the *Cp-ID field* to determine the particular coprocessor and the *type field* to determine the class of the instruction (see Table 12.3). If the coprocessor field contains all zeros and the type field is non-zero, the processor treats the instruction as an F-line exception. The field labelled *type-dependent* is determined by the nature of the actual instruction, and other words may follow the first if the instruction requires it.

Figure 12.39 Generic coprocessor instruction format

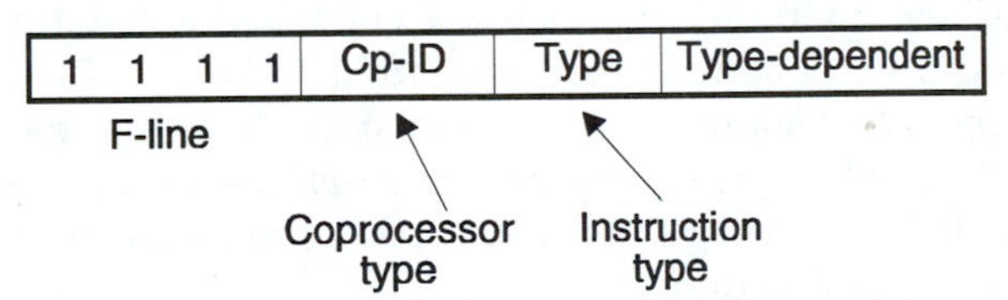

Table 12.3 Interpreting the coprocessor instruction type field

Type bits	Mnemonic	Meaning
8 7 6		
0 0 0	`cpGEN`	General instruction
0 0 1	`cpDBcc, cpScc, cpTRAPcc`	DBcc, set, and TRAP on condition
0 1 0	`cpBcc.W`	Branch on condition cc
0 1 1	`cpBcc.L`	Branch on condition cc
1 0 0	`cpSAVE`	Save context
1 0 1	`cpRESTORE`	Restore context
1 1 0	not defined	
1 1 1	not defined	

Although these mnemonics look strange, they are really generic mnemonics and can be applied to any type of coprocessor, irrespective of its actual function. For example, when writing actual instructions 'cp' is replaced by the appropriate mnemonic for the coprocessor, and 'GEN' is replaced by the actual general instruction mnemonic. Thus, an MC68882 FPC represents the instruction to calculate a tangent as FTAN (the 'F' indicates floating point coprocessor and the 'TAN' is the general instruction to calculate a tangent). Similarly, an FPC cpTRAPcc instruction might be written as FTRAPEQ (i.e., floating point trap on zero). Incidentally, the coprocessor condition codes represented by 'cc' in the mnemonics are not necessarily the same as the 68020's conditions. A floating point coprocessor can trap or branch on 32 conditions (e.g., *unordered condition*) while a memory management unit can trap on 16 conditions (e.g., *write-protected*).

Coprocessor instructions are used in exactly the same way as real 68020 instructions. For example, the FPC instruction FBEQ NEXT causes a branch to the line labelled 'NEXT' if the zero-bit of the condition code register of the FPC (not the 68020's CCR) is set. Of course, you can write programs for coprocessors only if your assembler *knows* about coprocessors. Otherwise, you could always write your own macros.

A general coprocessor operation, cpGEN, may be monadic or dyadic and take one or two arguments, respectively. Typical 68882 cpGEN instructions are: FCOSH (floating point hyperbolic cosine), FACOS (floating point arc cosine), and FADD (floating point addition).

The cpSAVE and cpRESTORE instructions transfer an internal coprocessor *state frame* between a coprocessor and external memory. All this means is that you can save a coprocessor's status by means of a cpSAVE instruction and then restore it with a cpRESTORE. You would use these instructions in a multitasking or an interrupt-driven environment to save a coprocessor's status before it is used by another task. For example, the instruction FSAVE -(A7) would save a floating-point coprocessor's status on the system stack. Note that the FSAVE saves only the *invisible* status of the FPC, and not its *visible* status made up of FP0 to FP7 and system control/status registers. You can save registers FP0 to FP7 with a FMOVEM FP0-FP7,-(A7) instruction.

As we have already stated, the coprocessor is allocated 32 bytes of CPU memory space, which is, effectively, arranged as eight contiguous longwords. All coprocessors must have a specific arrangement of registers in order to communicate with the 68020 host. Figure 12.40 defines the register structure of a generic coprocessor. It is important to note that these registers are *invisible* to the programmer and are required to implement the 68020-coprocessor protocol.

Coprocessor Operation

When a 68020 detects an F-line instruction, it writes a coprocessor command word to the coprocessor's memory-mapped command register in CPU space. Remember, once again, that the programmer does not have to provide an address because the coprocessor instruction contains a cp-ID field that identifies the type of coprocessor, and each type of coprocessor has its own unique region of CPU memory space. Having sent a command to the coprocessor, the processor reads the response from the coprocessor's response register. At this stage the coprocessor may use its response to request further actions such as *fetch an operand from the calculated effective address and store it in my operand register*. Once the host processor has complied with the coprocessor's demands, it is free to continue normal computing. That is, processor and coprocessor may overlap their operation.

Figure 12.40 The coprocessor interface register map

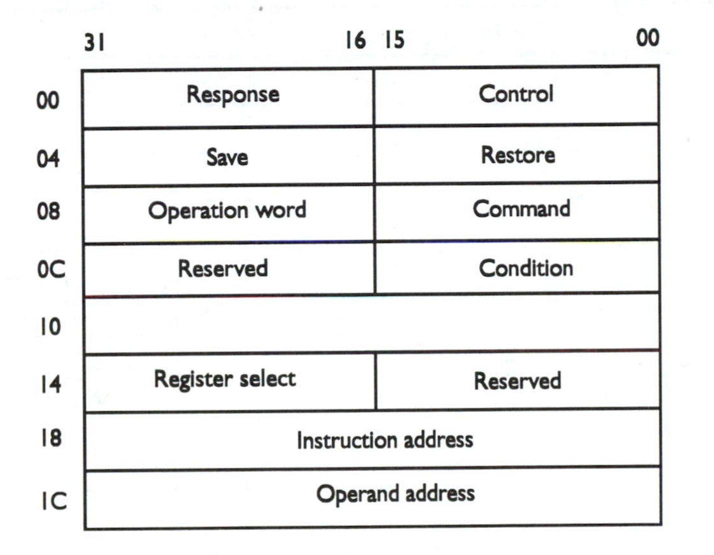

Introduction to the MC68882 Floating Point Coprocessor

In principle, the 68882 floating point coprocessor is a very simple device, although in practice its full details are rather complex, as a glance at its extensive data manual will indicate — the 68882's manual is as large as that of the 68000 itself. Much of this complexity arises from the nature of IEEE floating-point arithmetic, rather than from the nature of the FPC.

The 68882 extends the 68020's architecture to include the eight 80-bit floating point data registers, FP0 to FP7, described in Figure 12.41. It is important to understand that as far as the programmer is concerned, these registers magically appear within the 68020's register space. In addition to the standard byte (`.B`), word (`.W`), and longword (`.L`) operands, the FPC supports four new operand sizes: single precision real (`.S`), double precision real (`.D`), extended precision real (`.X`), and packed decimal string (`.P`). Figure 12.42 illustrates the structure of these operand formats. All on-chip calculations take place in extended precision format and all floating point registers hold extended precision values. The single real and double real formats are used to input and output operands.

Figure 12.41 The 68882 FPC register model

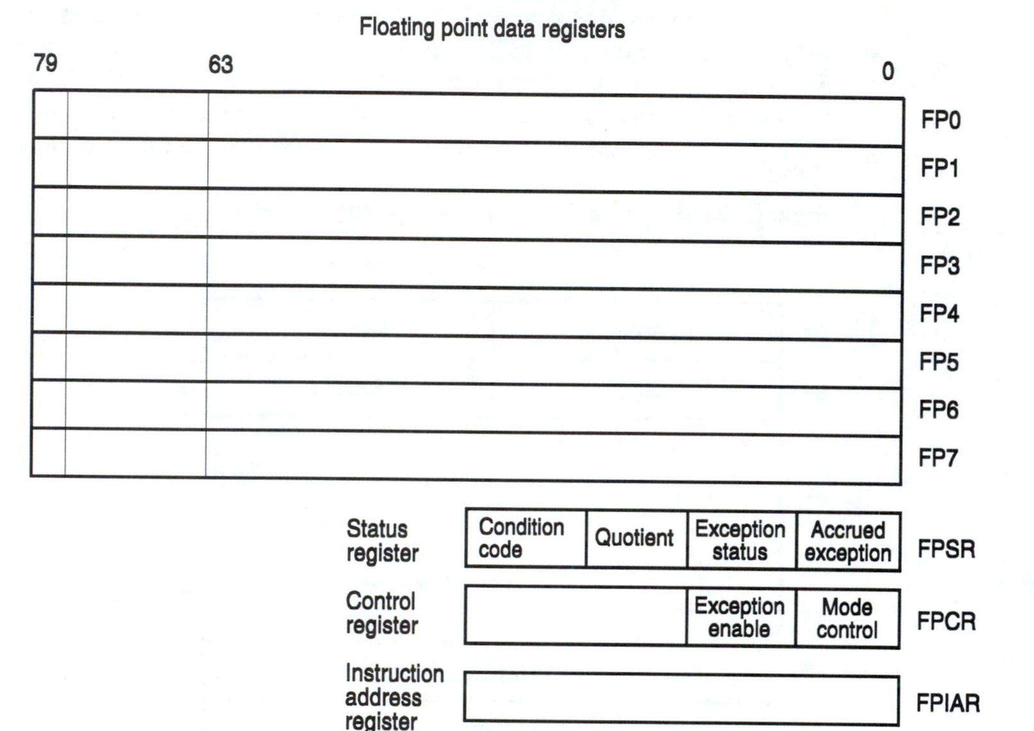

The three real floating point formats support the corresponding IEEE floating point numbers (single, double, and extended precision). The packed decimal real format holds a number in the form of a packed BCD 3-digit base-10 exponent and a 17-digit base-10 mantissa, as illustrated in Figure 12.42. A complete packed decimal string value is 96 bits (packed in three longwords when stored in external memory), and is used to convert between decimal and binary floating point values. For example, you can write the instruction:

```
FADD.P    #-4.123456E+17,FP0
```

to add the literal decimal number -4.123456 x 10^{17} to the contents of floating point register FP0. Note that the FPC *automatically* converts a packed decimal value to an extended precision value.

You can convert from binary to decimal form by executing an instruction with a packed decimal real as a destination operand. For example, the instruction

```
FMOVE.P   FP6,Result{#6}
```

has the effect of taking the value in FP6, converting it into packed decimal form, and moving it to main store at address `Result`. Note that the `{#6}` field after the

Figure 12.42 Data types supported by the 68882 FPC

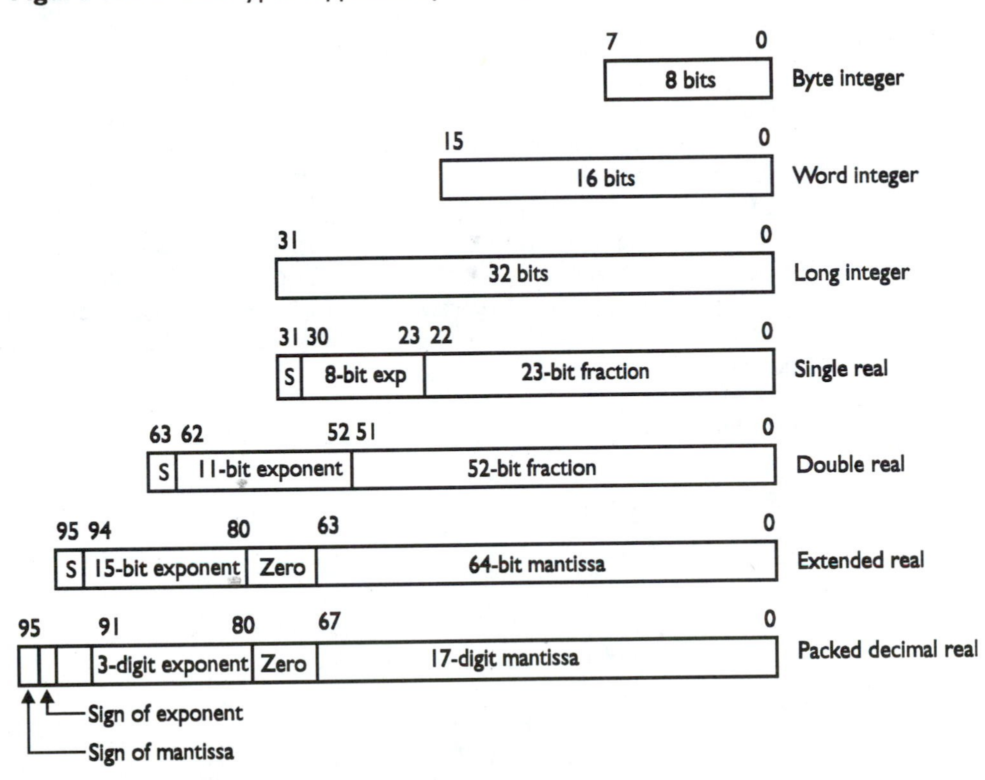

destination operand tells the floating point coprocessor to store the number with six digits to the right of the decimal point.

The FPC's `FMOVE <source>,<destination>` instruction not only copies an operand from its source to its destination, but also forces a data conversion. If the source is memory, the operand is converted to the internal extended precision format inside the FPC, and if the source is the FPC, the operand is converted into its destination format before storing in memory. The FPC also supports a move multiple register instruction, `FMOVEM`, that permits any set of the eight 80-bit floating point registers to be moved to or from memory. Each of these floating-point registers is stored as three longwords, and no data conversion takes place when a `FMOVEM` is executed. A *new* move instruction is `FMOVECR.X #data,FPn` which can move one of 22 floating point constants to a floating point register. These constants include 0, 1, e, $\log_2 e$, and π. Table 12.4 illustrates some of the FPC's monadic and dyadic instructions.

In addition to the eight floating point registers, the 68882 has a 32-bit control register, FPCR, and a 32-bit status register, FPSR. The FPCR is used by the programmer to determine which events cause floating point exceptions and to specify how rounding is to be carried out (the rounding of inexact numbers in floating point arithmetic is a very important consideration in numerical methods). The 32-bit FPSR provides a floating point status (like a conventional CCR), the sign and

Table 12.4 Some of the 68882's monadic and dyadic instructions

Monadic instructions

FCOSH	Hyperbolic cosine	FNEG	Negate
FETOX	e to the power x	FSIN	Sine
FETOXM1	e to the power x-1	FSINCOS	Simultaneous sine and cosine
FGETEXP	Get exponent	FSINH	Hyperbolic sine
FGETMAN	Get mantissa	FSQRT	Square root
FINT	Integer part	FTAN	Tangent
FINTRZ	Integer part (truncated)	FTANH	Hyperbolic tangent
FLOG10	Log base 10	FTENOX	10 to the power x
FLOG2	Log base 2	FTST	Test
FLOGN	Log base e	FTWOTOX	2 to the power x
FLOGNP1	Log bas e of x+1		

Dyadic instructions

FADD	Add
FCMP	Compare
FDIV	Divide
FMOD	Modulo remainder
FMUL	Multiply
FREM	IEEE remainder
FSCALE	Scale exponent
FSGLDIV	Single precision divide
FSGLMUL	Single precision multiply
FSUB	Subtract

least-significant seven bits of a quotient, the exception status, and the accrued exception status. Figure 12.43 illustrates the 68882's FPSR.

In order to show you what coprocessor code looks like, consider the following fragment of code.

```
*       Calculate a vector times a constant plus a vector
*       FOR I = 1 to N
*               X(i) := Y(i) x C + X(i)
*       END_FOR
*       C = address of constant
*       XVec = address of vector X
*       YVec = address of vector Y
*
        MOVE.W      #N-1,D0          D0 contains the loop counter
        FMOVE.D     C,FP0            FP0 contains the constant
        LEA         XVec,A0          A0 points to XVec
        LEA         YVec,A1          A1 points to YVec
```

Figure 12.43 Structure of the 68882's status register

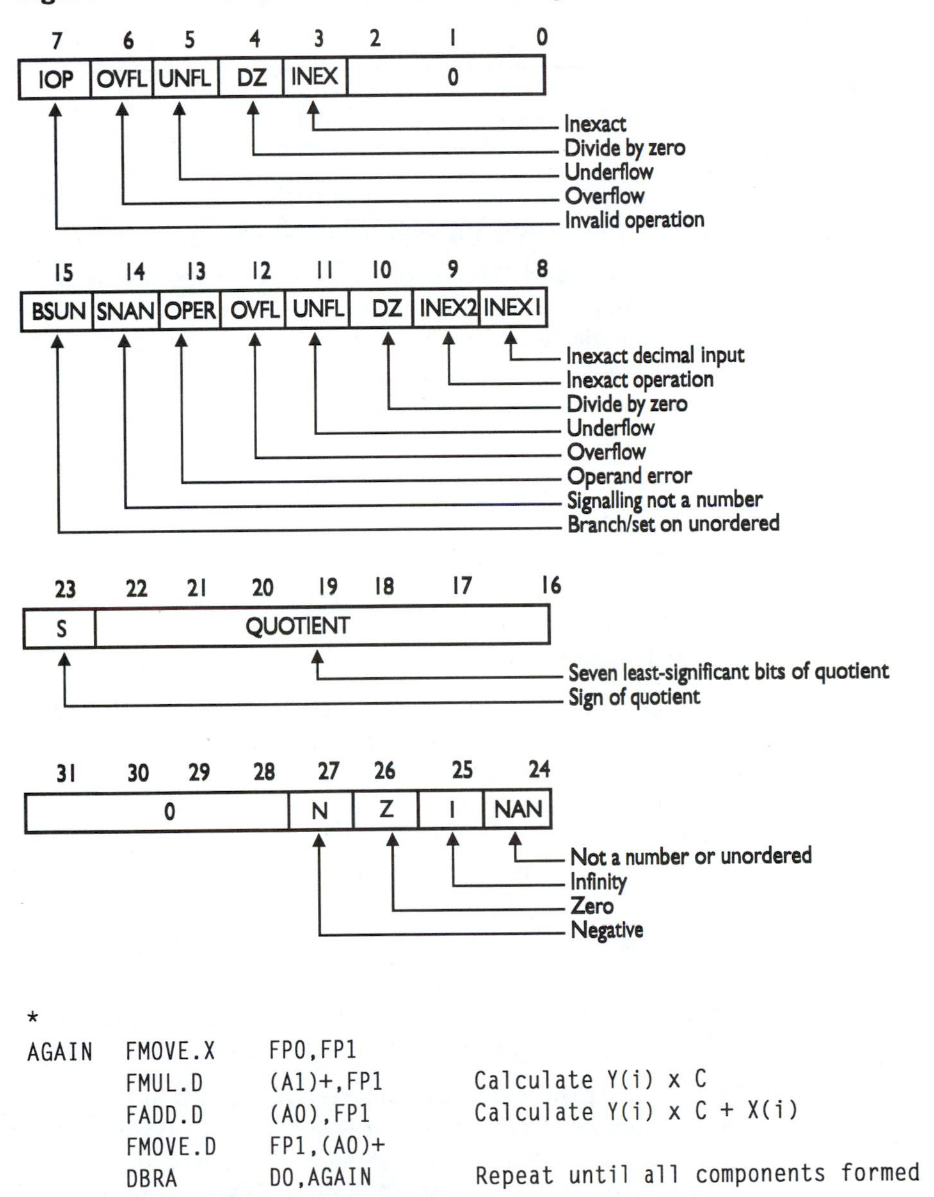


```
*
AGAIN   FMOVE.X    FP0,FP1
        FMUL.D     (A1)+,FP1       Calculate Y(i) x C
        FADD.D     (A0),FP1        Calculate Y(i) x C + X(i)
        FMOVE.D    FP1,(A0)+
        DBRA       D0,AGAIN        Repeat until all components formed
```

Exceptions and the FPC

The 68882 handles exceptions via the 68020 host processor. That is, 68020 and 68882 exceptions are treated in the same way — by the 68020. However, since the FPC can generate new exceptions (e.g., inexact result), the 68000's exception vector table is extended to include new exception vectors as described in Table 12.1.

Summary

- The 68020 and the 68030 are both extended versions of the 68000. These processors have full 32-bit address and data buses plus a 32-bit architecture.
- The 68020 has enhanced multiply and divide instructions. 68020 branch and LINK instructions now support 32-bit offsets.
- The new CMP2 and CHK2 instructions permit the contents of a data or an address register to be compared with both upper and lower bounds. The bounds may be signed or unsigned. The CMP2 instruction performs the test and sets the CCR accordingly, whereas the CHK2 instruction performs the test and calls a CHK exception if the bounds are violated.
- The new PACK and UNPK instructions facilitate the conversion between ASCII-encoded characters and the BCD representation of integers. These instructions also permit the conversion between BCD and less common codes such as EBCDIC.
- The 68020's most significant new instructions are the bit field operations. A bit field is a group of 1 to 32 consecutive bits that may start at any bit location in memory and cross byte, word, or longword boundaries.
- Three parameters define a bit field. The base byte is the effective address of the start of the bit field in memory. The field width is the width of the bit field and must be in the range 1 to 32. The offset is the location of the first bit of the bit field with respect to the most-significant bit of the base byte. Both the bit field offset and bit field bit number are numbered from the left. This arrangement is the reverse of the normal bit numbering convention used by the 68000 family.
- The instruction BFEXTU (A0){6,12},D2 has the effect of accessing the byte pointed at by A0 and then taking the 12-bit bit field that starts 6 bits from the most-significant bit of the base byte. This bit field is loaded into bits 0 to 11 of D2, and bits 12 to 31 of D2 are all set to zero.
- One of the most interesting bit field instructions is BFFFO (find first 1 in a bit field). The effect of BFFFO <ea>{offset:width},Dn is to access the bit field and locate the position of the first one within the bit field. For example, if the bit field is 000100011110, the value 3+offset would be returned in Dn (because the first 1 is 3 bits from bit zero of the bit field). If the bit field is all zeros, the offset plus the bit field width is loaded into the data register.
- The 68020 supports four radically new addressing modes: memory indirect (with postindexing and preindexing) and program counter memory indirect (with postindexing and preindexing). The memory indirect and PC memory indirect modes differ in that the former uses Ai as a pointer and the latter uses the program counter.

- The assembly language form of the effective address for memory indirect postindexed addressing is `([bd,An],Xn,od)`. The contents of address register An are added to a base displacement, bd, and the resulting address used to access memory. The contents of this location are added to register Xn (address or data) together with an outer displacement, od, to give the actual effective address. Furthermore, register Xn my be scaled by 1, 2, 4, or 8.
- The assembly language form of the effective address for memory indirect preindexed addressing is given by `([bd,An,Xn],od)`. The contents of address register An are added to a base displacement, bd, together with the contents of Xn, and the resulting address used to access memory. The contents of this location are added to an outer displacement, od, to give the actual effective address.
- These memory indirect addressing modes are very powerful and are used to support complex data structures in which a table of addresses in memory is used to access the required data.
- The 68020's supervisor state includes four additional registers: a second supervisor state stack pointer (MSP), a vector base register (VBR), and two alternate function code registers (SFC, DFC). These four registers would almost certainly be used only by the designers of operating systems and can be ignored by most programmers.
- The 68000's two supervisor-state stack pointers can be used to provide two stacks: one for exceptions due to hardware interrupts and one for all other exceptions.
- The 68020 supports a return from exception following a bus error. Therefore, the 68020 can be used in systems with virtual memory.
- There is one incompatibility between the 68000 and the 68020. The 68000 instruction `MOVE SR,<ea>` may be executed when the 68000 is in the user mode (i.e., the S-bit = 0). This instruction is privileged on the 68020. However, the new non-privileged 68020 instruction `MOVE CCR,<ea>` can be used to access the CCR. The incompatibility with the `MOVE SR,<ea>` may easily be handled by the 68020's privilege violation handler.
- The 68020's hardware interface and internal microcode (i.e., firmware) allow a coprocessor to look like an extension of the 68020's architecture, as far as the programmer is concerned.
- The 68882 floating point coprocessor adds eight 80-bit floating point registers to the 68020's architecture. The programmer can use new instructions to manipulate these registers and transfer data between them and memory. Moreover, new branch instructions can be used to branch on floating point conditions as well as the normal 68020 branches.

Problems

1. What are the major architectural differences between the 68000 and the 68020? If you were part of the 68020 design team, are there any other changes you would have included?

2. Describe how the 68020 has enhanced: (a) the 68000's multiply instructions, and (b) the 68000's divide instructions.

3. Why do bit field instructions number bits from the most-significant bit of the effective address of the bit field?

4. Why do some of the 68020's new addressing modes employ a scale factor?

5. What is the difference between postindexing and preindexing (when applied to the 68020's memory indirect addressing)?

6. *The 68020 microprocessor has almost the same instruction set as its predecessor, the 68000, and yet represents a giant leap forward in terms of its architecture. Indeed, you might reasonably say that the 68020 represents a high point in terms of the development of mainstream von Neumann architectures.*

   Discuss the truth, or otherwise, of the above statement. You should consider the broad range of applications for which a microprocessor might be used.

7. A 68020-based single-board microprocessor system is to be used to keep track of items in a shop. The shop sells up to N items. Each item is characterized by its 28-byte ASCII-encoded name followed by a longword pointer. This pointer points to the actual record that describes the item.

   A record consists of a 256-byte text field that describes the item, followed by four consecutive 4-byte fields. These fields define the item's cost price, selling price, number bought, and number sold.

   Suppose you need to access field x of item y. Before accessing the field, you may assume that the 68000's registers have been set up as follows:

   - A0 points at the first item in the list of N items
   - D0 contains item number y and is in the range 0 to N-1
   - D1 contains the field number x and is in the range 0 to 3

   a. Draw a map to illustrate the above data structure.

   b. Show how you would use the **68000** microprocessor to access field x of item y from the data structure.

   c. Show how you would use the memory indirect addressing mechanism of the 68020 to access the same field in the data structure.

8. A 68000 subroutine is designed to search a table for a longword. On entry, A0 contains the starting address of the table to be searched, D0 contains the longword to be matched, and D2 contains the total number of entries in the table. First write a 68000 subroutine to perform the search, and then write a 68020 subroutine to perform the same action (making best use of the 68020's facilities). Your subroutine must not modify the contents of A0.

9. Identify the bit fields accessed by the following instructions. Assume that the sequence of data stored in memory from address \$0FFD onward is: \$59C1, \$679C, \$0DD2, \$01E6. Register D4 contains \$FFFFFFF2 = -14_{10}.

   a. `BFCLR $1000{23:3}`
   b. `BFCLR $1000{5:12}`
   c. `BFCLR $1000{30:5}`
   d. `BFCLR $1000{D4:1}`
   e. `BFCLR $1000{D4:23}`

10. D3 = \$000F0124, D2 = \$0000F2AC, A1 = \$00033740; what effective addresses are generated by the instructions:

   a. `MOVE.W ($40,A1,D3.L),D1`
   b. `MOVE.W ($40,A1,D3.W),D1`
   c. `MOVE.W ($70,A1,D3.W),D1`
   d. `MOVE.W ($80,A1,D2.W),D1`

11. The first two lines of a 68000 subroutine might be:

```
        PEA       Param2       Push pointer to Param2 on the stack
        BSR       Sub          Call the subroutine
        .
Sub     LEA       (4,A7),A0    Get pointer to parameter 2 on stack
*                              (buried under the return address)
        MOVEA.L   (A0),A0      Get the pointer itself in A0
        MOVE.W    (A0),D2      Now read the actual parameter
        .
        RTS
```

How can the retrieval of the parameter in the subroutine be improved by using 68020 code?

12. For the same data block as question 9, explain the action of the following instructions. Assume that D1 initially contains \$12345678, and that D4 contains \$FFFFFFF2 = -14.

   a. `BFEXTU $1000{4:14},D0`
   b. `BFEXTS $1000{4:14},D0`
   c. `BFINS  D1,$1000{D4:23}`
   d. `BFFFO  $1000{23:8},D0`
   e. `BFFFO  $1000{6:11},D0`

APPENDIX A

Glossary

This appendix provides a glossary of some of the terms, symbols, and expressions used in this book. You can refer to the glossary whenever you find a term with which you are unfamiliar and do not wish to spend time searching for it through the text. Of course, this presupposes that I thought of putting the term in the glossary in the first place. Another reason for including a glossary is to provide an aid to revision.

The # symbol indicates that the following value is an immediate or literal value; for example, #1234, #Monday. See *immediate*.

$ The $ symbol indicates that the following number is a hexadecimal value; for example, $12, $12345678, and $00FE. Note that #$0AF2 indicates a *literal* hexadecimal value. See *hexadecimal*.

% The % symbol indicates that the following number is a binary value; for example %00001111. See *binary*.

* The asterisk is used in 68000 assembly language programs in three circumstances. When located in the *first* column of a line, the asterisk indicates a comment field. Anything on this line is ignored by the assembler. A second use of the asterisk is to indicate the current value of the location counter (i.e., the current memory location into which code or data is to be loaded by the assembler). The third use is as the conventional multiplier symbol. Consider the following example which illustrates all three applications.

```
*       This demonstrates the three-way use of the asterisk
*
        ORG     $400
Here    EQU     *
Table   DS.B    Rows*Cols
```

The 68000 assembler directive ORG (which resets the location counter) cannot take a label. However, we can use the directive EQU to equate a symbolic name to a value. In this case, the symbolic name Here is equated to the asterisk which is, of course, the current value of the location counter (i.e., $400).

ABSOLUTE ADDRESSING The address field of the instruction provides the actual or absolute address of the operand. For example, the MOVE 1000,D0 instruction means copy the contents of memory location 1000 into data register D0. Similarly, `ADD Val6,D3` means *add the contents of the memory location called* `Val6` *to the contents of D3*. Absolute addressing is used to access *variables* (e.g., expressions in the form A := B + C in a high-level language). Since absolute addressing does not yield position independent code, it is often avoided by the programmer.

ACTIVE-LOW A signal is called active-low if it is asserted by putting it into a low voltage state. An active-low signal is sometimes indicated by an asterisk as a suffix (e.g., RESET*, BERR*). For example, the signal RESET* is active-low and must be approximately 0V to cause a reset to take place. An active-high signal is one that is asserted by putting it into high voltage state.

ACTUAL PARAMETER See *formal parameter*.

ADDRESS ERROR EXCEPTION The 68000 must read word and longword operands from an *even* address boundary. If you attempt to read a .W or a .L operand at an odd boundary, an address error exception will occur. A word or a longword located at an odd boundary is called a *misaligned* operand. Note that the 68020 is able to handle misaligned operands.

ADDRESS REGISTER The 68000 has eight 32-bit special purpose address registers, A0 to A7. An address register is so-called because it is used to hold the address of an operand. An operation on an address register always yields a 32-bit result. Even if a 16-bit operation is applied to the contents of an address register, the result is *sign-extended* to 32 bits. Operations on the contents of an address register (with the exception of `CMPA`) do not modify the CCR. Address registers are intended to be used in conjunction with the address register indirect addressing mode; for example, `MOVE D0,(A2)`. Address register A7 is also used as a stack pointer by the 68000.

ADDRESSING MODE An addressing mode represents the way in which the location of an operand is expressed. Typical addressing modes are literal, absolute (memory direct), register indirect, and indexed.

ARRAY An array (also called a matrix) is a type of data structure. A two-dimensional n-column by m-row array consists of n x m elements. The 68000's indexed addressing mode makes it easy to access array elements.

ASCII The American Standard Code for Information Interchange provides a means of representing the common alphanumeric symbols by a 7-bit code. The ASCII code also includes special codes for nonprinting characters (such as carriage return or back space), and for communications control characters (such as ENQ, SYN, and NUL).

ASSEMBLE TIME Assemble time describes events that take place during the assembly of a program. That is, *assemble time* contrasts with *run time*. For example, symbolic expressions in the assembly language are evaluated at assemble time and not run time. Note that an assemble time operation is performed once only. For example, the expression `ARRAY DS.B ROWS*COLS+1` is evaluated by the assembler, and the result of `ROWS*COLS+1` is used to reserve the appropriate number of bytes of storage. By way of contrast, the address of the source operand `8+[A0]+[D3]` in the instruction `MOVE (8,A0,D3),D6` is evaluated at *run time*, because the contents of registers A0 and D0 are not known at assemble time and because the contents of these registers may change as the program is executed.

ASSEMBLER DIRECTIVE An assembler directive is a statement in an assembly language program that provides the assembler with some information it requires to assemble a source file into an object file. Typical assembler directives tell the assembler where to locate a program and data in memory and equate symbolic names to actual values. The most important assembler directives are: `ORG`, `EQU`, `DS`, and `DC`.

ASSEMBLY LANGUAGE An assembly language is a symbolic or human-readable form of a computer's machine code. An assembly language is normally the most primitive language in which the programmer can write programs.

ASSERTED Digital systems use a high voltage and a low voltage to represent the two logical states. Sometimes a high level represents the active or true state, and sometimes a low level represents the active state. To avoid confusion, we use the term *asserted* to mean that a signal is put in a true or active state.

ATOMIC OPERATION An atomic operation is one that cannot be interrupted or subdivided in any way. Such an operation is also called indivisible.

AUTO-DECREMENTING A register is said to be auto-decrementing if its contents are automatically *decremented* before or after it is used. The 68000 implements address register indirect addressing with *pre-decrementing*. The effective address is expressed as `-(Ai)`, and the contents of address register Ai are decremented *before* the address register is employed to access an operand. The size of the decrement depends on the operand (1 for a byte, 2 for a word, and 4 for a longword). For example, if address register A0 initially contains the value 12, the operation `MOVE.B -(A0),D0` would decrement the contents of register A0 by 1 to get 11, read the byte in memory location 11, and deposit this byte in data register D0. This addressing mode makes it easy to step through tables and to implement stacks.

AUTO-INCREMENTING A register is said to be auto-incrementing if its contents are automatically *incremented* before or after it is used. The 68000 implements address register indirect addressing with *post-incrementing*. The effective address is expressed as `(Ai)+`, and the contents of address register Ai are incremented *after* the address register is employed to access an operand. The size of the increment depends on the operand (1 for a byte, 2 for a word, and 4 for a longword). For example, if address register A0 initially contains the value 12, the operation `MOVE.B (A0)+,D0` would copy the byte in location 12 into D0, and then increase the contents of A0 by 1 to get 13. This addressing mode makes it easy to step through tables and to implement stacks. The 68000 uses address register indirect with pre-decrementing to implement *push* operations and address register indirect with post-incrementing to implement *push* operations.

BASE In positional notation each digit in a number has a *weighting* determined by the *location* of the digit within the number. We can express the integer N in base b as: $N = d_{m-1} b^{m-1} + d_{m-2}b^{m-2} + ... + d_1b^1 + d_0b^0$, where the d_i represents the ith digit of the number. Assembly language programmers work with three number bases: 10 (decimal), 2 (binary), and 16 (hexadecimal). Base 8 (octal) was once common but is rarely used today. By convention, a number on its own (e.g., 1234) is regarded as a decimal number. If it is prefixed by '%' it is a binary number, and if it is prefixed by '$' it is a hexadecimal number. Programmers choose the most suitable base when writing programs. For example, you might write:

```
     MOVE.W  #365,D0         Set up this year's array
     LEA     Year,A0
LOOP ANDI.W  #$FF00,(A0)     Clear the most-significant byte of each entry
     ORI.W   #%1010,(A0)+    Set bits 1 and 3 of the entry
     SUBQ.W  #1,D0
     BNE     LOOP            Repeat until all days processed
```

In this example, we use three bases (decimal 365, hexadecimal $FF00, binary %1010). Although there is no *technical* reason for choosing these bases, we use decimal to express the number of days in a year, hexadecimal to show that we wish to clear the upper byte of a word, and binary to show that we wish to clear bits 1 and 3 of a byte.

BAUD RATE The baud rate of a signal is a measure of the number of times at which it can change state per second. For example, a modem that transmits at 2400 baud causes the data to change state (i.e., to switch between two frequencies) 2400 times a second. If the signal being transmitted is binary (i.e., two-state), the data is being transmitted at 2400 bits per second. However, if the signal is transmitted as 4 levels, each signal element carries two bits of information, and therefore the data is transmitted at 4800 bits per second.

BINARY A binary number is expressed in the base two. A binary value is indicated to the assembler by prefixing it with "%". For example, we can write `AND.B %00010110,D0` to mask out bits 1, 2, and 4 of D0.

BINARY FILE An assembler takes a source file and produces two new files. One is an optional listing file and the other is the binary file, containing object code, that can be executed by the target processor. Some operating systems apply the extension `.BIN` to binary files.

BINDING When a symbolic name is attached to a memory location it is said to be *bound* to that location. For example, the operations

```
       ORG   $1000
Vector DS.B  6
```

bind the name `Vector` to the memory location 1000_{16}.

BIT FIELD The 68020 can operate on a group of consecutive bits, called a *bit field,* falling anywhere in memory (i.e., the bit field need not lie on a byte-boundary). A bit field may be 1 to 32 bits long and is specified by <ea>{offset:width}, where `offset` is the distance of the bit field from the most-significant bit of the effective address, and `width` is the size of the bit field. Consider the three consecutive bytes in locations $2000, $2001, and $2002 with the values 11001100 10101001 11011001, respectively. The 12-bit bit field represented by the address $2000{7:12} crosses two byte boundaries and is: 1100110**0 10101001 11**011001.

BOUND An array of n elements extends from address N to address N+n-1. These two addresses are the upper and lower bounds of the array, respectively. When accessing an array you can (should?) check that the element does not lie outside these bounds.

BRANCH A branch instruction has the format `Bcc Branch_target_address`, and forces the processor to take one of two courses of action: either to execute the next instruction immediately following the branch instruction (i.e., branch not taken), or to execute the instruction at the branch target address (i.e., branch taken). The branch is taken or not taken depending on the outcome of the condition specified by cc (e.g. `BEQ` = branch on carry clear). Note that the 68000's branch instructions employ *relative addressing* — when the branch instruction is encoded by the assembler, its operand is not the actual or *absolute* branch target address but the location of the branch target address relative to the current contents of the program counter.

BREAKPOINT A breakpoint is a point in a program at which the execution of the program is stopped and the contents of the processor's registers displayed on the console. Breakpoints are used to debug programs. Software breakpoints are often implemented by replacing an instruction in a program with a code that forces an exception (e.g., `ILLEGAL` or `TRAP #i`). You can also use the 68000's trace facility to implement a breakpoint (i.e., generate a trace exception after each instruction and then break when a certain point is reached).

BUFFER In the context of serial and parallel input/output systems, a buffer is a storage system arranged as a first-in-first-out queue. The buffer is part of the interface device and allows data to enter and leave the chip at different rates (at least over a short period of time). For example, a serial interface may receive a burst of data and store the data in its buffer until the processor is ready to read it.

BUS CYCLE The 68000 executes three types of cycle (clock cycle, bus cycle, and instruction cycle). A *clock cycle* is the smallest event in which the 68000 can participate. Only hardware systems designers are interested in clock cycles. A *bus cycle* involves a read or write access to external memory. The 68000 requires a minimum of four clock cycles to perform a bus cycle. An *instruction cycle* involves the execution of an instruction and may require several bus cycles.

BUS ERROR The 68000 has a BERR* (i.e., bus error) input pin. When this pin is asserted by user-supplied hardware, the 68000 executes a bus error exception. The bus error exception is used to handle accesses to illegal memory locations or accesses to protected regions of memory.

BYTE A byte is a unit of 8 bits, and is the smallest quantity of data that can be transferred into and out of the 68000. The 68000 can also perform operations on bytes. An operation on a byte is generally indicated by appending the suffix `.B` to a mnemonic (e.g., `ADD.B`, `MOVE.B`). The 68000 is byte-addressed in the sense that consecutive bytes have consecutive addresses (i.e., 0,1,2,3,...).

C-BIT The C-bit is a carry bit in the condition code register. It is the least-significant bit of the CCR. The C-bit is set if an operation results in a carry-out from the most-significant bit position of the operand. The C-bit is also set by a shift out of the least-significant bit during right shift operations. The 68000 provides two instructions that branch on the value of the C-bit (i.e., `BCC` = branch if C = 0 and `BCS` = branch if C = 1). There are two other branch conditions that test the carry bit, but these also test the Z-bit.

CASE-SENSITIVE The letters of the Roman alphabet may be uppercase or lowercase. Programming languages are said to be case-sensitive if they distinguish between upper- and lowercase letters. For example, a case-sensitive language would regard the symbolic names `Day_One` and `DAY_ONE` as different. A case-insensitive language would regard these two names as identical. The 68000 assembler used in this text is case-insensitive.

CCR The condition code register records the state of status flags after an operation is carried out by the 68000. The 68000's CCR has a C-bit (carry), an N-bit (negative, i.e., the most-significant bit), a Z-bit (zero), a V-bit (overflow), and an X-bit (extend bit). Some instructions set or clear one or more of these bits, some instructions leave one of more bits unchanged, and some instructions have an indeterminate effect on one or more of these bits. You have to look up the 68000's instruction set to determine how a particular instruction affects the CCR. The order of the CCR's bits is 000XNZVC (the three most-significant bits are zeros).

COMMENT A comment field in either a low-level or a high-level language is a text string that is ignored by the assembler or compiler, respectively. The only purpose of the comment field is to permit the programmer to describe some aspect of the program (i.e., to make a comment). In 68000 assembly language, any text following the instruction or assembly directive is regarded as a comment. Any line that begins with an asterisk, *, in column one is also treated as a comment. The following example demonstrates two types of comment.

```
*   This is a comment because the line starts with an "*"
    MOVE.B D1,D2    The MOVE instruction gets executed - but this comment doesn't
```

COMMON INSTRUCTIONS You can write 68000 assembly language programs with very few instructions. The following list provides a subset of 68000 basic instructions.

```
ADD     SUB
AND     OR
NOT     BEQ
BNE     ADDA
BRA     CMPA
CMP     SUBA
LSL     MOVEA
LSR     BSR
RTS     MOVE
```

COMPILER A compiler is a program that translates or *maps* a high-level language into machine code.

COPROCESSOR A coprocessor is an auxiliary processor that performs certain functions not implemented by the CPU itself. Typical coprocessors are floating point units and memory management units. You can also design coprocessors to manage graphics operations or string handling. Members of the 68000 family use line F exceptions to communicate with the coprocessor. When a coprocessor is fitted, it appears to the programmer as an extension of the CPU itself.

CPU SPACE The 68000 divides memory into three types: user space, supervisor space, and CPU space. The type of memory being accessed is indicated by a three-bit value on the 68000's function code output pins, FC2, FC1, FC0. CPU space is used by the 68000 to indicate an interrupt acknowledge cycle or a coprocessor access. That is, a CPU access is a special operation that should not be interpreted by the rest of the system as a normal memory access.

DATA REGISTER A data register is a 32-bit on-chip register used to hold information being processed by the 68000. Most of the 68000's instructions operate on the contents of a data register and the contents of a memory location (e.g., `ADD D0,$1234`). The 68000 has eight data registers, D0 to D7, and they all behave identically. An operation on a data register may be applied to bits 0 to 7 (`.B`), bits 0 to 15 (`.W`), or 0 to 31 (`.L`). An operation on a byte does not modify bits of the register not taking part in the operation. For example, `CLR.B D0` sets bits 0 to 7 to zero and does not affect bits 8 to 31.

DATA TYPE All information stored in a digital computer is stored in binary form. However, it is possible to assign a type to a data element. For example, an element may be a 16-bit integer, an IEEE format floating point number, or a two-dimensional array of vectors. In a *strongly typed* language, a data element may take part only in operations appropriate to its type.

DEBUGGING Debugging is the process of removing bugs (i.e., errors) from a program. Debugging is normally performed by using breakpoints and single stepping through a program to locate errors.

DFC The destination function code, DFC, indicates the *destination* address space being currently accessed. Members of the 68000 family put out a code on function control pins, FC0 to FC2, whenever they access memory. The function code indicates the type of memory access (user/supervisor, program/data). The 68020 has a destination function code register, DFC, that permits a user-determined code to be placed on FC0 to FC2 when the privileged instruction `MOVES Rn,<ea>` is executed. This operation is used by the operating system (running in the supervisor mode) when it needs to access user space.

DISASSEMBLER A disassembler reads object code (i.e., binary code) and converts it into mnemonic form. That is, it performs the inverse function of an assembler. Most disassemblers are unable to provide symbolic addresses, and the disassembled code uses absolute addresses expressed in hexadecimal form.

DEQUE A deque, pronounced "deck," is a *double-ended queue*. Items can be added to or removed from a deque at either end.

DIRECT MEMORY ACCESS (DMA) Direct memory access is a mechanism whereby information is transferred directly between a peripheral like a disk drive and the system memory without the active intervention of the processor. DMA requires special purpose hardware and a dedicated logic element called a DMA controller, DMAC. The DMAC keeps track of the source or destination of the data in memory and controls the port through which the transfer is taking place.

DYADIC An operator is said to be dyadic if it requires two operands. The following operators are dyadic: AND, OR, EOR, addition, subtraction, multiplication, division. Operations with a single operand such as 1/x, -x, sqrt(x) are said to be *monadic*.

EDGE-SENSITIVE Digital systems operate with signals that have two states: electrically low or electrically high. For most of the time, digital systems are concerned only with whether a signal is in a high state or a low state. However, the input circuit of some systems can be configured to detect the *change of state* of a signal rather than its *actual value*. Such an input is called edge-sensitive. For example, you might wish a peripheral to interrupt the processor only when the voltage it is monitoring changes.

EFFECTIVE ADDRESS The *effective address* of an operand is a generic term that stands for all the various ways of calculating an address. We often write an effective address as <ea> to stand for all the legal addressing modes that can be used. For example, we might express the clear instruction as CLR <ea>. When we use this instruction, we might write CLR 1000 to clear the location at the effective address 1000, or CLR (12,A0) to clear the location at the address given by the contents of A0 plus 12, and so on.

EQUIVALENT INSTRUCTIONS The following operations demonstrate instruction "equivalences." Note that some operations are not entirely identical because they affect the CCR in different ways, or they handle sign-extension differently.

```
CLR  Di       SUB      Di,Di
TST  Di       CMP      #0,Di
BTST #i,Di    ANDI     #n,Di
SWAP Di       ROL.L    Dj,Di   (Dj = 16)
MULS #4,Di    ASL      #2,Di
LEA  10,A0    MOVEA.L  #10,A0
```

ERROR Programmers can make many types of error. An assembly error (or *syntax* error) occurs when you write an illegal instruction (e.g., CLR A2), or employ the same label twice, etc. Even if your program is syntactically correct, it may not do what you wish it to do. In this case it has a *semantic* error; that is, the program does not carry out the actions you intended. This type of error is normally found by debugging the program.

EXCEPTION An exception is an event that causes a change in the flow of control of a program by calling the operating system. An interrupt is an externally generated exception whose origin lies in hardware. A software exception is caused by certain events — some of them are programmed and some are due to certain types of error. The 68000 responds to an exception by entering its supervisor state, pushing the program counter and status word on the stack, and jumping to the appropriate exception handler.

EXCEPTION VECTOR TABLE The 68000 maintains a table of 256 longwords from 000000_{16} to $0003FF_{16}$. Each longword is a pointer to an exception handler. When an exception occurs, the appropriate vector is loaded into the program counter to start exception processing. Note that the first two longwords in this table are used by a special exception — the reset (the first longword is the initial value of the stack pointer, and the second is the location of the reset vector).

EXPONENT The decimal number 1.234×10^{12} has an exponent of 12. The exponent is the power or index of the base used to scale the mantissa. A 32-bit IEEE floating point number is scaled by 2^{E-127}. The 8-bit stored exponent is said to be *biased* because it is 127 larger than the true exponent. For example, the actual scale factor 2^{15} would use a stored exponent of $15 + 127 = 142 = 10001110_2$.

FIELD When a unit of data can be subdivided, the individual subdivisions are known as *fields*. For example, an instruction might have an op-code field and an operand field. An IEEE floating point number has a sign field, an exponent field, and a mantissa field.

FIFO A first-in-first-out buffer, FIFO, is a form of queue in which new items join the queue at its back and oldest items leave the queue at its front. That is, the next item to leave is the oldest item in the queue. A FIFO is the opposite of a LIFO (i.e., a last-in-first-out buffer or *stack*).

FLOATING POINT Large and small numbers are represented in floating point format. A decimal floating point number might be represented by 1.234×10^{-3}. Most computers now represent floating point numbers in the IEEE format. A 32-bit floating point value is represented by $(-1)S \times 1.F \times 2^{E-127}$, where S is the sign bit, F the fractional mantissa, and E the biased exponent. Members of the 68000 family cannot operate on floating point numbers directly. You must either write appropriate software or use a floating point coprocessor.

FORMAL PARAMETER When a procedure (or a macro) is written, the actual parameters it uses might not be known by the programmer. The programmer therefore employs *formal* parameters that are later replaced by the *actual* parameters used when the procedure is executed. For example, the high-level language procedure `PutChar(x)` may display a character called `x` on the display device. When the programmer calls the procedure with the statement `PutChar(p)`, the actual value of `p` replaces the formal parameter `x` during the execution of the procedure. The concept of a formal parameter is encountered by the assembler programmer when dealing with macros. The parameters used by the programmer when writing the body of the macro are the formal parameters (i.e., `\1`, `\2`, `\3` ... in the case of the 68000) that are replaced by the actual parameters when the macro is called.

FRAME POINTER A subroutine might require work space for any temporary values it creates. By using the top of the stack as a work space, it is possible to make the program re-entrant (i.e., it can be interrupted and re-used without corrupting the current work space). This work space is called a *stack frame*. You can use an address register to point to this frame and then access all data with respect to this register (called a frame pointer). The 680000's `LINK` instruction is used to create a stack frame, and an `UNLK` instruction to remove it.

FUNCTION CODE Whenever the 68000 accesses memory it puts out a three-bit code on function control outputs FC0, FC1, and FC2. The function code indicates whether the 68000 is accessing user memory space or supervisor memory space, and whether it is accessing program space or data space. The function code also indicates accesses to CPU space when the processor executes an interrupt acknowledge cycle or a coprocessor access. Basic 68000 systems do not make use of the function code outputs (other than to detect an interrupt acknowledge).

The function code outputs are normally used in systems with memory management — e.g., workstations running Unix.

GLOBAL A global variable is one that can be accessed from all parts of a program. Contrast this with a *local* variable that is private to a particular subroutine. The range over which a variable or a parameter is used is called its *scope.*

GRAMMAR The grammar of a language comprises the set of rules that allow the construction of a legal sentence in the language.

HARD RESET A hard reset occurs when the 68000's RESET* input is asserted (i.e., pulled low). A hard reset takes place when the system is switched on. It is also performed after a total system failure. Full system initialization (re-booting) is performed during a hard reset. A soft reset shares some of the characteristics of a hard reset, except that only part of the initialization process is performed.

HANDSHAKE Data may be transferred from one device to another synchronously or asynchronously. In a synchronous transfer (or open-loop transfer), data is transmitted and its reception assumed. In an asynchronous data transfer (or closed-loop transfer) a signal, called a *strobe,* is transmitted from the data source to the data destination. This signal indicates that a data transfer is underway. The data destination responds to the strobe by sending a reply indicating that the data has been received. This reply or *acknowledgment* is called a handshake.

HEXADECIMAL The base 16 is called the hexadecimal base and uses the digits 0,1,2,3,4,5,6,7,8,9,A,B,C,D,E,F to represent the decimal values 0 to 15. For example, the hexadecimal value 234_{16} = 2x256 + 3x16 + 4 = 564. Hexadecimal representation is used because a binary number can be converted into a hexadecimal value by dividing the number into groups of four bits (starting at the right), and then converting each binary group into the corresponding hexadecimal character. For example, 0010111000111010 is equivalent to $2E3A_{16}$.

IDENTIFIER An identifier is a name given to a variable or a label by the programmer. A legal 68000 assembly language identifier must begin with an alphabetic character. Only the first eight alphanumeric characters of the identifier are recognized by the assembler. For example, both `DateOfMonth` and `DateOfMo` are regarded by the 68000 assembler as the same identifier.

IEEE FORMAT See *floating point.*

ILLEGAL INSTRUCTION The 68000 uses 16-bit operation codes. Theoretically, there are 65,536 possible op-codes. Not all these possible op-codes are assigned to actual 68000 instructions. If the 68000 reads a 16-bit op-code that does not correspond to a legal op-code, the 68000 generates an illegal op-code exception. The 68000 has a special instruction, `ILLEGAL`, that can be used to force an illegal instruction exception.

IMMEDIATE An immediate operand is one that forms part of an instruction and is indicated by the prefix '#'. For example, the immediate operand 5 in the instruction `MOVE.W #5,D0` is part of the instruction. When the instruction is executed, the operand 5 is immediately available, since the CPU does not have to read memory again to obtain it. Immediate addressing is also referred to as *literal addressing*, because the operand is a literal or a constant.

INDIRECT ADDRESSING In indirect addressing, the address of the required operand is not provided directly by the instruction. Instead, the instruction tells the processor where to find the address. That is, indirect addressing gives the *address of the address*. The 68000 uses *address register* indirect addressing in which the address of an operand is in an address register. This addressing mode is specified by enclosing the address register in round brackets. For example, `MOVE (A3),D2` means move the contents of the memory location whose address is in address register A3 into D2. The computer must then use A3 to locate the actual operand that is to take part in the instruction. The 68000 does not support memory indirect addressing — but the 68020 does.

INDIVISIBLE An operation is said to be indivisible if it must be completed without any form of interruption. A 68000 instruction is indivisible in the sense that it cannot be interrupted by either a hardware or a software exception until after it has been executed. A segment of code can be made indivisible by turning off the 68000's interrupt handling mechanism until it has been executed (although level-7 interrupts cannot be turned off). The 68000 has a special indivisible test and set, `TAS`, instruction that reads an operand, tests it, and sets it in one complete indivisible action. This instruction is used to reliably synchronize processors in a multiprocessor system.

INTERLOCKED A sequence of actions is said to be interlocked if each action can continue only when the previous action has been completed. For example, a data transfer may employ an interlocked handshake. That is, the transmitter first sends a *data available strobe* to the receiver. The receiver detects this strobe and responds by sending its own *data acknowledge strobe* back to the transmitter. The transmitter then negates its data available strobe when it detects the receiver's data acknowledge strobe, and so on.

INTERRUPT An interrupt is a request for service from an external device seeking attention. The external device requests service by asserting an interrupt request line connected to the processor. The processor may or may not deal with the interrupt depending on whether the interrupt is masked (i.e., ignored). If the interrupt is not masked, the processor deals with it by executing a piece of code called an interrupt handler. Once this handler has been executed, the processor returns to the point which it had reached immediately before the interrupt.

INTERRUPT ACKNOWLEDGE CYCLE The 68000 executes an IACK or interrupt acknowledge cycle after it accepts an interrupt request on one of the seven interrupt request lines, IRQ1* to IRQ7*, from an external device. During the IACK cycle the 68000 informs peripherals that an interrupt acknowledge is in progress and indicates the level of the interrupt request (1 to 7) by placing the number on the address bus. The device that requested the interrupt at the same level as the interrupt acknowledge places its 8-bit interrupt vector number on the data bus. The 68000 reads the interrupt vector number, multiplies it by four, and uses the result to index into a table of interrupt vectors in the 256-longword region of memory 000000_{16} to $0003FF_{16}$. The longword accessed in this table is the address of the interrupt handler and is loaded into the 68000's program counter.

INTERRUPT-DRIVEN I/O In interrupt-driven I/O, the processor normally carries out a background task until an input or output device is ready to be serviced. When the device is ready, it generates an interrupt and the processor deals with it by performing the appropriate data transfer. Interrupt-driven I/O is much more efficient than programmed I/O using polling, because the processor takes part in the data transaction only when the peripheral is ready.

INTERRUPT HANDLER An interrupt handler is a piece of code, rather like a subroutine or procedure, that deals with the cause of an interrupt. Interrupt handlers are invariably part of a computer's operating system.

INTERRUPT LATENCY The period between the generation of an interrupt and the time at which the processor begins to process the interrupt.

INTERRUPT MASK When the 68000 receives an interrupt request on IRQ1* to IRQ7*, it checks the interrupt level against the three bits of its interrupt mask in the status byte of its status register. If the incoming interrupt has a level i and the interrupt mask is set to level j, the interrupt will be serviced if $i > j$. That is, the interrupt must be at a higher level than that reflected in the interrupt mask. When an interrupt is serviced, the value of the interrupt mask is set to the level of the current interrupt. Consequently, if an interrupt at level i is serviced, only an interrupt at level $i+1$ or greater can interrupt the servicing of the interrupt at level i. Note that a level-7 interrupt is a special case. A new level-7 interrupt can always interrupt the processing of an existing level-7 interrupt.

I/O STRATEGY An input/output strategy describes the way in which a system goes about performing I/O transactions. The three basic I/O strategies are programmed I/O, interrupt-driven I/O, and DMA (direct memory access).

JUMP In 68000 terminology, a jump is a change of flow of program (like an unconditional branch, an `RTS`, or an `RTE`). The jump, `JMP`, instruction is equivalent to a `GOTO` in a high-level language. An important form of `JMP` (or `JSR`) is `JMP (Ai)` (or `JSR (Ai)`) that permits you to compute a variable destination at run time.

LABEL A label is a symbolic name that identifies a particular line in an assembly language program. In 68000 assembly language programs, a label is any valid identifier that begins in column one. A label permits a programmer to refer to a line without having to know its actual (i.e., numeric) address. It is illegal to employ the same label to identify two different lines in the same program.

LINK The 68000's `LINK` instruction is used to support stack frames by creating a region of work space on the top of the current stack. This instruction is widely used by compilers. The `UNLK` instruction undoes the work of a link instruction by collapsing the stack.

LINKED LIST A (singly) linked list is a data structure in which each element has two parts, the actual data and a pointer to the next element in the list.

LINKER An assembly language program may be written as a number of separately assembled modules. A linker combines the modules into a single program.

LIST A list is a sequence of elements that are arranged according to some algorithm. For example, a queue, a deque, and a stack are all types of list. A linked list is a list in which each element points to the next element in the list.

LISTING FILE When an assembler assembles a source file into object code (i.e., binary code), it may also produce a listing file. The listing file is the assembled source file complete with line numbers, the instruction opcodes (in hexadecimal form), plus any error messages or warnings. You use the listing file to help you to debug a program if it does not function correctly. Since the listing file contains the address of both instructions and operands, you can readily trace through the program and set breakpoints.

LITERAL A literal is an operand that is *directly* used by the computer, as opposed to a *reference to a memory location.* In an arithmetic expression like `X := Y + 5`, the '5' is a literal value. In 68000 assembler language a literal is indicated by prefixing it with a '#' symbol. `ADD #5,D0` means *add the literal value 5 to the contents of data register D0.* The terms *literal* and *immediate* are interchangeable.

LOCAL A local variable is a variable that *belongs* to a subroutine. and cannot be accessed from any other part of the program. The variable is therefore private to the subroutine and is used for the subroutine's working storage. A local variable can be contrasted with a *global* variable that can be accessed from other parts of the program. The 68000 implements the special instructions `LINK` and `UNLK` to manage a subroutine's local variables.

LOCATION COUNTER The location counter is a variable maintained by an assembler. When an assembly language program is assembled, the location counter keeps track of where information is to be located in memory when the assembled program is loaded in a real computer. For example, if the location counter is currently pointing at memory location $001234, the effect of the assembler directive `DS.B $100` is to change the location counter to $001334.

LOGICAL ADDRESS The address of an operand generated by the CPU is called a logical address. The logical address is mapped onto the actual address (*physical address*) of an operand by a *memory management unit*. In systems without an MMU, there is no difference between a logical and a physical address.

LOGICAL OPERATION A logical operation operates on the bits of one or two operands but does not produce carry bits from one position (i.e., column) into another. That is, an operation involving bits a_i and b_i does not affect the i+1th bit position. The 68000's logical operations are AND, OR, NOT, EOR, and logical shifts left and right.

LONGWORD A longword, in the context of the 68000 family, is a 32-bit unit of data. The 68000's address and data registers can all hold a longword. The 68000 is able to perform operations on longwords (that is, it is a 32-bit computer).

MACHINE LANGUAGE The binary language that is actually executed by a computer. The human-readable form of this language is called *assembly language*.

MACRO A macro is a symbolic name given to a sequence of instructions. If you are going to use a certain sequence of instructions frequently, you can create a named macro that is equivalent to these instructions. Whenever you use this macro in the assembly language program, the assembler expands it into the instructions it represents.

MANTISSA A floating point number represents a number in the form of mantissa x $2^{exponent}$. In the IEEE floating point format, the mantissa lies in the range 1.000...0 to 1.111...1. Note that the IEEE mantissa is stored in memory in a fractional form (i.e., the leading 1 is not stored).

MASK In computer terms the word *mask* means to hide or to cover one or more bits of a word. The logical `AND` is frequently used as a mask operator. For example, if we have 8 bits in D0 and are only interested in bits 4, 5, and 6, we can mask all other bits to zero by `AND.B #%01110000,D0`.

MEMORY MANAGEMENT UNIT A memory management unit, MMU, is used in sophisticated computer systems to map a *logical* address generated by the CPU onto the address of data in the *physical* (i.e., real) memory system. MMUs support virtual memory systems and also free the programmer from worrying about where to locate programs and data. The operating system uses the MMU to map programs onto whatever physical (i.e., actual) memory is currently available.

MEMORY MAP The 68000's memory can be represented as a linear list of memory elements from $00 0000 to $FF FFFF. This arrangement is called a memory map. It is used by programmers to show how data structures are arranged, and by hardware designers to show how the various memory elements are arranged. Throughout this text the memory map is drawn conventionally with low addresses at the top of the diagram and high addresses at the bottom. This splendidly confusing convention means that *up* on the page means *down* (i.e., toward lower addresses) in memory.

MEMORY-MAPPED Since the 68000 has no special hardware or software devoted to performing I/O operations, all I/O must be performed via the 68000's existing memory interface. I/O ports are arranged by the hardware system's designer so that they look like normal memory locations. These ports are said to be memory-mapped because they appear as part of the system's memory space.

METALANGUAGE A metalanguage is one used to describe another language. Most of this text uses English as a metalanguage to describe the 68000 assembly language. However, since English sometimes lacks precision, special-purpose metalanguages have been designed. A popular metalanguage is BNF (Backus-Naur Form).

MISALIGNED OPERAND The 68000 supports byte, word, and longword operands. A byte may be stored at any location. A word or a longword must be located at an even address. If you attempt to read a word or longword operand at an odd address, the 68000 generates an address error exception. The 68020 and the 68030 can access misaligned operands.

MNEMONIC The symbolic representation of an assembly level instruction. For example, the mnemonic for the operation that adds two values is `ADD`. The mnemonic for the operation that forces a change of program flow if the overflow bit in the CCR is set is `BVS` — Branch on oVerflow bit Set.

MODEM A modem is a MODulator DEModulator and is used to transmit digital signals over the public switched telephone network, PSTN.

MONADIC An operator is said to be monadic if it operates on a single value. The following operators are monadic: logical NOT, negate, sign-extension, reciprocal, square root.

N-BIT The N-bit is the *negative* bit of the CCR and is set if the most-significant bit of the result is 1 (i.e., the operand is negative when interpreted as a two's complement number). Programmers can use the most-significant bit of an operand as a general flag bit. If you set this bit (or clear it), you can later employ a branch on negative (or branch on positive) instruction to test the state of the flag.

NONMASKABLE INTERRUPT An interrupt that cannot be turned off or otherwise disabled is called nonmaskable. Nonmaskable interrupts (IRQ7*, in the case of the 68000) are used for interrupts that must never be missed.

OBJECT CODE The machine code output of an assembler or a compiler is called object code. An object file holds this code. Object code can be executed.

OFFSET In 68000 terminology an offset is a literal value that forms part of an instruction used in conjunction with address register indirect addressing. For example, the operation `CLR.B (12,A0)` clears the memory location that is offset by 12 bytes from the address pointed at by address register A0.

OP-CODE The op-code or *operation code* is the binary pattern that represents an instruction.

OPERAND The operand of an instruction is the data used by that instruction. For example, the instruction `MOVE.B #5,(A0)+` has two operands. The source operand is the literal value 5, and the destination operand is the memory location pointed at by address register A0.

OVERFLOW Arithmetic overflow occurs when a two's complement number goes outside its permitted range. This results in the V-bit of the CCR being set. Note that exponent overflow occurs in floating point arithmetic when the exponent becomes too large or small to represent. The 68000's division instruction can give rise to overflow if the result is too large to fit in the operand.

PARALLEL In the context of input/output operations, a *parallel* data transfer moves two or more bits between a peripheral and the CPU (or its memory) in a single operation. Most parallel interface devices move 8 or 16 bits at a time.

PARITY A parity bit is an additional bit appended to a word to make the total number of 1's in the word including the parity bit even (even parity) or odd (odd parity). The purpose of the parity bit is to detect errors (usually in storage or transmission). If one or an odd number of bits in a word is corrupted and the parity bit is re-computed, it will differ from the actual parity bit.

PATCH A patch is a modification to object code (i.e., machine code) made by hand. For example, a program might use two `NOP`s (no operation) op-codes that can later be replaced by a patch to fix an error. The `NOP`s might later be replaced by a `BSR FIX` instruction to perform the necessary operation. The use of a patch to modify a program is as near to a criminal offense as you can get. Only other programmers employ a patch to correct an error. A program should be modified by rewriting its source code and then reassembling it.

PERIPHERAL The term *peripheral* is somewhat ambiguous. It is used to describe external devices like disk drives, keyboards, mouses (mice?), and displays. It is also used to describe the *hardware* that interfaces these devices to the processor. For example, both the *floppy disk controller chip* that interfaces a floppy disk drive to a processor and the *floppy disk drive* itself are often called peripherals. Fortunately, the context of the word *peripheral* usually indicates its actual meaning.

PHYSICAL ADDRESS The address of an operand generated by the CPU is called a logical address. The logical address is mapped onto the actual, or physical, address of an operand by a memory management unit. In systems without an MMU, there is no difference between a logical and a physical address.

POINTER A pointer is a variable that provides the location of an operand. The 68000 uses its address registers as pointers because they point to operands in memory. Pointers are necessary to implement data structures.

POLLING LOOP A polling loop is a programmed loop that continually asks a question until it gets an answer. For example, a polling loop in a program performing programmed I/O continually tests whether a device is ready or not. Such a loop usually takes the form:

```
REPEAT
      Test device status
UNTIL device is ready
```

However, it is also possible to have polling loops in other situations. For example, suppose that several peripherals share the same level of interrupt and the processor cannot directly identify the device that generated the interrupt. A polling loop can be used to ask each of the devices in turn:

```
i = 0
REPEAT
      i := i + 1
      Ask device i if it interrupted
UNTIL an interrupter is found
```

PORT A port is a device that interfaces the processor to external systems. Since the 68000 uses memory-mapped I/O, a port usually refers to a location in memory used to write data to an external peripheral or to read data from the peripheral.

POSITION INDEPENDENT CODE If you write a program or a subroutine in such a way that all the addresses of operands are expressed by means of relative addressing, the code is said to be position independent. That is, all addresses are independent of the actual location of the code in memory. Position independent code is written by using relative branches and program counter relative addressing, rather than jump instructions and absolute addresses. Position independent code, PIC, is normally regarded as a good thing.

PRIORITIZED INTERRUPTS The 68000 supports seven levels of interrupt request, IRQ1* to IRQ7*. These interrupt requests are said to be prioritized because a higher-numbered interrupt will always be accepted in preference to a lower-numbered interrupt (if they occur at the same time).

PRIVILEGED INSTRUCTION A privileged instruction is an instruction that cannot be executed when the 68000 is running in the user mode. Attempting to execute a privileged instruction causes an exception. Privileged instructions are those that operate on the 68000's supervisor state registers, plus STOP and RESET.

PROGRAM COUNTER The program counter contains the address of the next instruction to be executed. The program counter is incremented after each instruction is executed. It is altered by jump and branch instructions, RTS and RTE instructions, and by all exception calls.

PROCEDURE See *subroutine.*

PSEUDOCODE Pseudocode provides a semi-formal method of expressing algorithms before they are translated into assembly language. A typical pseudocode borrows the constructs of a high-level language like Pascal, without importing all its complexities. Some of the pseudocode conventions used in this text are as follows.

```
IF L THEN S

IF L THEN S1 ELSE S2

CASE I OF
       I1: S1
       I2: S2
        .
        .
END_CASE

FOR I := N1 TO N2 DO
       S
END_FOR

WHILE L
       DO S
END_WHILE

REPEAT
       S
UNTIL L
```

Pseudocode Notation

[ ] Indicates an index into an array — e.g., Day[Monday]
( ) Indicates a parameter — e.g., GetChar(Char)
{ } Indicates a comment field — e.g., {Clear the array}
<, >, =, ≠, ≤, ≥ Indicate logical comparison operators — e.g., X ≤ Y
:= Assignment — e.g., X := Y + 3

QUEUE A type of two-ended list, or data structure, whose elements are added at one end and removed from the other. It is sometimes called a FIFO (first-in-first-out) queue.

RECURSION Recursion means defining something in terms of itself. For example, the value of 2^n can be defined recursively by stating that $2^n = 2 \times 2^{n-1}$ (you also need to state that $2^0 = 1$ to terminate the recursive procedure). A subroutine can be written in such a way that it can be called recursively. That is, it can call itself. Recursive procedures make use of the stack to store their return addresses and any working data.

RE-ENTRANT Code is said to be re-entrant if it can be interrupted and called by another procedure.

RELATIVE ADDRESS An absolute address provides the actual location of an operand. A relative address (e.g., `MOVE.B (Data,PC),D3`) is an address that is specified with respect to the program counter or to an address register. Relative addressing is used to create position independent code in which the program can be relocated in memory without having to recalculate the addresses of operands.

RELOCATABLE CODE Code is relocatable if it can be moved from one location in memory to another without modifying the address of operands. In order to write relocatable code, you have to use program counter relative addressing for both branch instructions and operands. The following code is relocatable.

```
      LEA    (Table,PC),A0  A0 points to the data source
      LEA    (Dest,PC),A1   A1 points to the data destination
      MOVE.W #5,D1          Six bytes to move
Next  MOVE.B (A0)+,D2       FOR i = 0 to 5
      ADD.B  #2,D2            Get element from TABLE and add 2
      MOVE.B D2,(A1)+         Store the element in DEST
      DBRA   D1,Next        END_FOR
```

RETURN ADDRESS The address of the next instruction after a subroutine call instruction is called the *return address*, because this is the address of the instruction that will be executed after the subroutine has been executed. This address is saved on the stack (automatically) when the subroutine is called, and is removed from the stack at the end of a subroutine by an `RTS` instruction.

RS232C This is a standard that specifies the interface between DCE (data communications equipment) and DTE (data terminal equipment). It is frequently used to connect a computer's serial port to a modem.

RUN TIME Run time describes the period during which a program is being executed (i.e., when it is running). A run time operation is performed each time it is encountered during the execution of the program. For example, the source address in the assembly language instruction `MOVE.B (12,A6,D3.W),D4` is evaluated at run time. The source address `(12,A6,D3.W)` is evaluated by adding 12 to the contents of A6 plus the low-order word of D3. The actual result obtained depends on the contents of A6 and D3 at the time the calculation is performed and can change as the program is executed. Contrast run time with assemble time. The source address in the operation `MOVE.B 3*TEMP+6,D0` is evaluated once during the assembly process. For example, if the symbolic name `TEMP` has the value 20, the assembler treats the instruction as `MOVE.B 66,D0`.

SCOPE A symbolic name may be bound to a variable or a constant. The region over which the name is *visible* is called its scope. For example, a name may be declared in a subroutine and its scope may extend only to the subroutine (i.e., it is a *local* name). The scope of a name may be *global* and be visible to all subroutines.

SELF-MODIFYING CODE Programs that actually change or modify their own code at run-time are said to be self-modifying. For example,

```
MODLINE ADD.W   #TEMP,D3
        MOVE.W  D6,MODLINE+2
```

is an example of self-modifying code. The first instruction appears to add a literal to D3. However, the second instruction writes data to the constant field of the first instruction (it modifies the instruction itself). Programmers have done things like this to save storage space. Although occasionally employed to overcome the restrictions of first-generation microprocessors, self-modifying code should never ever be used today — it is difficult to read and nearly impossible to debug.

SEPARATE ASSEMBLY When writing a large program, it is possible to divide the task into several modules. These modules can be assembled independently of each other (i.e., separate assembly). In order to do this, the assembler has to be told which variables are defined outside a module and which variables have to be exported from the module. The modules are then put together by a *linker*.

SERIAL In the context of input/output operations, a serial data transfer moves data between a microprocessor system and a peripheral a bit at a time. Serial data transfer is associated with communications links.

SERIALLY REUSABLE A serially reusable procedure may be called by a program and then called again at any time after it has been executed. That is, it must be executed to completion without interruption. Any procedure is serially reusable as long as its code is not modified when it is run. A serially reusable procedure is to be contrasted with reusable procedures and re-entrant procedures.

SFC The source function code, SFC, defines the type of address space from which the 68000 is currently accessing an operand — user program, user data, supervisor program, or supervisor data. The 68020 puts out the source function code on its function control output pins, FC0 to FC2, whenever it reads from memory. The 68020 has a source function code register, SFC, that permits a user-determined code to be placed on FC0 to FC2 when the privileged instruction `MOVES <ea>,Rn` is executed. This function code override the 68020's *normal* function code. For example, you can load the SFC with a suitable source function code using the `MOVEC D0,SFC` instruction and then access an operand using `MOVES (A0),D2`.

SHIFT A shift operation moves the bits of a memory location or a data register one or more places left or right. There are three types of shift (logical, arithmetic, and rotate). In a logical shift, a zero enters the bit at the end that is vacated. In an arithmetic operation, the sign bit is replicated during a shift right. In a rotate operation, the bit that falls off one end is copied to the vacated bit.

SIGN-EXTENSION The 68000 represents negative numbers by their two's complement. One of the properties of a two's complement number is that a number in m bits can be represented in m+1 bits by replicating the most-significant bit. For example, if the two 4-bit two's complement numbers 0011 and 1011 are converted to a 6-bit format, they become 000011 and 111011, respectively. The process of increasing the number of bits while preserving the correct two's complement value is called sign-extension. In the context of the 68000, sign-extension means extending an 8-bit value to 16 bits or a 16-bit value to 32 bits. Sign-extension is important because the 68000 treats addresses as signed values.

SOURCE CODE The program as written by the programmer is in source code. If the source code is written in assembly language, an assembler translates it into machine code. If the source code is in high-level language, a compiler translates it into machine code.

S-RECORD Binary (object) data is often stored or transmitted in the form of hexadecimal code in the range $00 to $FF. One such code is represented by S-records. There are several types of S-record (S0, S3, S8). An S3 record consists of the header S3, a four-character address, a byte count, up to 16 bytes of data, and a check sum.

STATUS REGISTER The 68000 has a 16-bit status register, subdivided into an 8-bit system byte and an 8-bit condition code register. The system byte includes the S-bit (user/supervisor mode), the T-bit (trace on/off), and the interrupt mask level (0 to 7). Programs running in the user mode can't modify the status register.

STACK A stack is a data structure in the form of a first-in-last-out queue. A stack is so-called because it behaves like a stack of papers on a desk. Items are added to the top of the stack and bury those below. When items are removed from the top of the stack, they are removed in the reverse order to which they were entered. For example, the last sheet of paper added to the stack is the first sheet removed from it. Data stacks are used to implement a subroutine return mechanism. The 68000 can support up to eight stacks simultaneously using A0 to A7 as a stack pointers. Although a true stack can be accessed only at one end (its top), you can access any element in a real stack. For example, the word below the top of the stack pointed at by A7 can be accessed by: `MOVE.W (2,A7),D0`.

STACK POINTER The 68000 implements a stack in external memory. The top item on the stack is pointed at by the stack pointer, which may be any one of address registers A0 to A7. However, address register A7 is always used by the 68000 to point at the stack used to store subroutine return addresses. Since the 68000's stack pointer points at the item on top of the stack, it must be pre-decremented to put a new item on the stack, and post-incremented to remove one. For example, `MOVE.W (A7)+,D0` pulls a word off the stack, and `MOVE.W D0,-(A7)` pushes a word on the stack.

STRONG TYPING A language is said to be strongly typed when the operations that can be performed on data are strictly governed by their type. For example, floating point operations might not be permitted on numbers of type integer. Strong typing helps to prevent the programmer performing inappropriate or illegal operations on data. Assembly languages are not strongly typed.

SUBROUTINE A unit of code (or a fragment of a program) that can be called from a point in a program and executed, and a return made to the instruction after the calling point. You call a subroutine to perform some task or function that is used frequently in the program.

SUPERVISOR STATE The 68000 operates in its supervisor state when the S-bit is 1, and in its user state when S = 0. User programs run in the user state and the operating system runs in the supervisor state. The supervisor state has its own stack pointer, A7 (or SSP). All exception and interrupt handling takes place in the supervisor state.

SYMBOL TABLE An assembler uses a source file to produce a listing file and a binary file. The listing file may contain an optional symbol table giving the address of each symbol (i.e., symbolic name) in the program. The symbol table is helpful in debugging an assembly language program (in particular, large programs). Most assemblers provide an option that suppresses the listing of the symbol table. The 68000 assembler can be forced to produce a cross-referenced symbol table by including the assembler directive `OPT CRE`.

SYMBOLIC NAME A symbolic name is the name given to a variable, a constant, or a line of a program by the programmer.

SYNTAX The set of rules that govern what constitutes a legal program is the syntax of a language. If a statement in a program breaks one of these rules, it is said to have a syntax error. Syntax errors are discovered by the assembler before the program can be run, since a program with syntax errors cannot be translated into machine code. Note that a program can be syntactically correct and yet not function correctly. The meaning of a program is called its semantics.

SYSTEMS PROGRAMMER For the purpose of this text we define the *systems programmer* as someone who writes the operating system (including the exception processing routines). *Systems programmer* is to be contrasted with *applications programmer* (i.e., someone who writes applications programs running under the control of an operating system). This distinction is important in 68000-based systems because the operating system runs in the supervisor mode, and applications programs run in the user mode.

TRAP In the context of a 68000, a trap is a call to the exception handling system. 16 trap instructions `TRAP #0` to `TRAP #15` are provided. Systems designers may employ the trap in any way they wish (e.g., for device-independent I/O).

TWO'S COMPLEMENT A method of representing a negative number is by means of its two's complement. In n-bit arithmetic, a number -N is represented by 2^n - N. You can also form the two's complement of N by inverting all the bits of N and adding 1. When using two's complement arithmetic, the most-significant bit of a number is a *sign bit* (0 = positive and 1 = negative). If two positive numbers are added and the resulting sign bit is 1, arithmetic overflow has occurred and the result is out of range. Similarly, if two negative numbers are added and the resulting sign bit is 0, arithmetic overflow has occurred. Addition and subtraction operations are the same for signed and unsigned arithmetic. However, you have to use the appropriate multiplication or division instructions for signed and unsigned arithmetic.

TWO-PASS ASSEMBLER Many assemblers operate in a two-pass (or two-phase) mode. During the first pass, the source code is read and mnemonics and addressing modes checked against all legal combinations. Symbolic names (variables and labels) are stored in an internal table. On the second pass, the op-codes and symbols are translated into binary form. A two-pass assembler is needed to resolve forward references that cannot be translated on the first pass.

USER STATE The 68000 operates in its supervisor state when the S-bit in the status register is 1, and in its user state when S = 0. In general, user programs run in the user state and the operating system runs in the supervisor state. The user state has its own stack pointer, A7 (or USP). Instructions that modify the state of the 68000's status byte cannot be executed in the user state.

V-BIT The V-bit or overflow bit of the 68000's condition code register is set if the result of an operation leads to arithmetic overflow. For example, if the most-significant bits of the two source operands taking part in an addition are both 0 and the most-significant bit of the result is 1, the V-bit is set to indicate arithmetic overflow. Similarly, the V-bit is set if the two most-significant bits are 1 and the most-significant bit of the result is 0.

VBR The 68020 and later processors have a vector base register, VBR, that can be used to relocate the processor's exception vector table anywhere in memory. When the 68020 is reset, the VBR is cleared and the exception vector table is located in the address range $000000 to $0003FF. If the VBR is loaded with the value N, the exception vector table is relocated to the address range N to N+$3FF. The VBR is loaded in the supervisor state by means of the privileged instruction `MOVEC Ri,VBR`.

VECTORED-INTERRUPT When an external device interrupts the 68000, the CPU executes an IACK (interrupt acknowledge) cycle. During the IACK cycle, the interrupting device supplies a *vector number* to the 68000 which is used to identify the appropriate interrupt handling routine. The 8-bit vector number is multiplied by four and used to index into the exception vector table. The 68000 is said to have a vectored-interrupt mechanism because the interrupting device identifies itself to the processor.

VOLATILE PORTION When a task (i.e., program) is running, its state is described by the program counter, status register, and address and data registers. This information is said to comprise a task's volatile portion or *context*.

WEIGHTING In the positional notation, the value of each digit depends on its weighting. Consider the decimal number 276. The weighting of the 6 is one, the weighting of the 7 is ten, and the weighting of the 2 is one hundred. The weightings used in this text are 10 (decimal), 2 (binary), and 16 (hexadecimal). These weightings are sometimes referred to as *natural weightings* because they define the positional notation system employed in conventional arithmetic. However, you can define other weightings. For example, if a 3-digit number has a weighting of 6, 3, 5, the number 123 would be equal to $1 \times 6 + 2 \times 3 + 3 \times 5 = 27_{10}$.

WORD A word is the basic unit of data processed by a computer. 68000 texts use the word to refer to a 16-bit value, which is the width of the 68000's data bus.

X-BIT The X-bit, or extend bit, of the condition code register is, essentially, the same as the C-bit (carry bit). Some operations that generate a carry bit also set the X-bit to the same value as the carry bit. Many operations that clear the carry bit do not affect the X-bit. Unlike other bits in the CCR, the X-bit is not used to determine the outcome of a conditional branch. The X-bit is used in chain arithmetic with the `ADDX`, `SUBX`, `NEGX`, `ABCD`, and `SBCD` instructions. Note that the X-bit is used as an extension bit in `ROXL` and `ROXR` instructions. That is, the X-bit is considered to be part of the operand and is shifted into one end of the operand, while the bit moved out of the other is shifted into the X-bit.

Z-BIT The Z-bit is the *zero bit* in the 68000's condition code register. The Z-bit is set to 1 if an operation yields a zero result, and cleared to 0 if the result is nonzero.

APPENDIX B

The 68000's Instruction Set

We have included this appendix to save you the task of having to turn to secondary material when writing 68000 assembly language programs. Since most programmers are not interested in the encoding of instructions, details of instruction encoding have been omitted (i.e., the actual op-code bit patterns). Applications of some of the instructions have been provided to demonstrate how they can be used in practice.

Instructions are listed by mnemonic in alphabetical order. The information provided about each instruction is: its assembler syntax, its attributes (i.e., whether it takes a byte, word, or longword operand), its description in words, the effect its execution has on the condition codes, and the addressing modes it may take. The effect of an instruction on the CCR is specified by the following codes:

- `U` The state of the bit is undefined (i.e., its value cannot be predicted)
- `-` The bit remains unchanged by the execution of the instruction
- `*` The bit is set or cleared according to the outcome of the instruction.

Unless an addressing mode is implicit (e.g., `NOP`, `RESET`, `RTS`, etc.), the legal source and destination addressing modes are specified by their assembly language syntax. The following notation is used to describe the 68000's instruction set.

`Dn, An`	Data and address register direct.
`(An)`	Address register indirect.
`(An)+, -(An)`	Address register indirect with post-incrementing or pre-decrementing.
`(d,An), (d,An,Xi)`	Address register indirect with displacement, and address register indirect with indexing and a displacement.
`ABS.W, ABS.L`	Absolute addressing with a 16-bit or a 32-bit address.
`(d,PC), (d,PC,Xi)`	Program counter relative addressing with a 16-bit offset, or with an 8-bit offset plus the contents of an index register.
`imm`	An immediate value (i.e., literal) which may be 16 or 32 bits, depending on the instruction.

Two notations are employed for address register indirect addressing. The notation originally used to indicate address register indirect addressing has been superseded. However, the Teesside 68000 simulator supports only the older form.

Old notation	Current notation
`d(An), d(An,Xi)`	`(d,An), (d,An,Xi)`
`d(PC), d(PC,Xi)`	`(d,PC), (d,PC,Xi)`

ABCD Add decimal with extend

Operation: $[\text{destination}]_{10} \leftarrow [\text{source}]_{10} + [\text{destination}]_{10} + [X]$

Syntax:

```
ABCD Dy,Dx
ABCD -(Ay),-(Ax)
```

Attributes: Size = byte

Description: Add the source operand to the destination operand along with the extend bit, and store the result in the destination location. The addition is performed using BCD arithmetic. The only legal addressing modes are data register direct and memory to memory with address register indirect using pre-decrementing.

Application: The `ABCD` instruction is used in chain arithmetic to add together strings of BCD digits. Consider the addition of two nine-digit numbers. Note that the strings are stored so that the least-significant digit is at the high address.

```
      LEA  Number1,A0    A0 points at first string
      LEA  Number2,A1    A1 points at second string
      MOVE #8,D0         Nine digits to add
      MOVE #$04,CCR      Clear X-bit and set Z-bit of CCR
LOOP  ABCD -(A0),-(A1)   Add a pair of digits
      DBRA D0,LOOP       Repeat until 9 digits added
```

Condition codes:

X	N	Z	V	C
*	U	*	U	*

The Z-bit is cleared if the result is non-zero, and left unchanged otherwise. The Z-bit is normally set by the programmer before the BCD operation, and can be used to test for zero after a chain of multiple-precision operations. The C-bit is set if a decimal carry is generated.

ADD Add binary

Operation: [destination] ← [source] + [destination]

Syntax:
```
ADD <ea>,Dn
ADD Dn,<ea>
```

Attributes: Size = byte, word, longword

Description: Add the source operand to the destination operand and store the result in the destination location.

Condition codes:

X	N	Z	V	C
*	*	*	*	*

Source operand addressing modes

Dn	An	(An)	(An)+	-(An)	(d,An)	(d,An,Xi)	ABS.W	ABS.L	(d,PC)	(d,PC,Xn)	imm
✔	✔	✔	✔	✔	✔	✔	✔	✔	✔	✔	✔

Destination operand addressing modes

Dn	An	(An)	(An)+	-(An)	(d,An)	(d,An,Xi)	ABS.W	ABS.L	(d,PC)	(d,PC,Xn)	imm
✔		✔	✔	✔	✔	✔	✔	✔			

ADDA Add address

Operation: [destination] ← [source] + [destination]

Syntax:
```
ADDA <ea>,An
```

Attributes: Size = word, longword

Description: Add the source operand to the destination address register and store the result in the destination address register. The source is sign-extended before it is added to the destination. For example, if we execute `ADDA.W D3,A4` where A4 = 00000100_{16} and D3.W = 8002_{16}, the contents of D3 are sign-extended to $FFFF8002_{16}$ and added to 00000100_{16} to give $FFFF8102_{16}$, which is stored in A4.

Application: To add to the contents of an address register and not update the CCR. Note that `ADDA.W D0,A0` is the same as `LEA (A0,D0.W),A0`.

Condition codes:

X	N	Z	V	C
-	-	-	-	-

An `ADDA` operation does not affect the state of the CCR.

Source operand addressing modes

Dn	An	(An)	(An)+	-(An)	(d,An)	(d,An,Xi)	ABS.W	ABS.L	(d,PC)	(d,PC,Xn)	imm
✔	✔	✔	✔	✔	✔	✔	✔	✔	✔	✔	✔

ADDI Add immediate

Operation: `[destination] ← <literal> + [destination]`

Syntax: `ADDI #<data>,<ea>`

Attributes: Size = byte, word, longword

Description: Add immediate data to the destination operand. Store the result in the destination operand. `ADDI` can be used to add a literal directly to a memory location. For example, `ADDI.W #$1234,$2000` has the effect $[M(2000_{16})] \leftarrow [M(2000_{16})] + 1234_{16}$.

Condition codes:

X	N	Z	V	C
*	*	*	*	*

Destination operand addressing modes

Dn	An	(An)	(An)+	-(An)	(d,An)	(d,An,Xi)	ABS.W	ABS.L	(d,PC)	(d,PC,Xn)	imm
✔		✔	✔	✔	✔	✔	✔	✔			

ADDQ Add quick

Operation: `[destination] ← <literal> + [destination]`

Syntax: `ADDQ #<data>,<ea>`

Sample syntax: `ADDQ #6,D3`

Attributes: Size = byte, word, longword

Description: Add the immediate data to the contents of the destination operand. The immediate data must be in the range 1 to 8. Word and longword operations on address registers do not affect condition codes. Note that a word operation on an address register affects all bits of the register.

Application: ADDQ is used to add a small constant to the operand at the effective address. Some assemblers permit you to write ADD and then choose ADDQ *automatically* if the constant is in the range 1 to 8.

Condition codes:

Z	N	Z	V	C
*	*	*	*	*

Note that the CCR is not updated if the destination operand is an address register.

Destination operand addressing modes

Dn	An	(An)	(An)+	-(An)	(d,An)	(d,An,Xi)	ABS.W	ABS.L	(d,PC)	(d,PC,Xn)	imm
✔	✔	✔	✔	✔	✔	✔	✔	✔			

ADDX Add extended

Operation: `[destination] ← [source] + [destination] + [X]`

Syntax:

```
ADDX Dy,Dx
ADDX  -(Ay),-(Ax)
```

Attributes: Size = byte, word, longword

Description: Add the source operand to the destination operand along with the extend bit, and store the result in the destination location. The only legal addressing modes are data register direct and memory to memory with address register indirect using pre-decrementing.

Application: The ADDX instruction is used in chain arithmetic to add together strings of bytes (words or longwords). Consider the addition of two 128-bit numbers, each of which is stored as four consecutive longwords.

```
        LEA  Number1,A0     A0 points at first number
        LEA  Number2,A1     A1 points at second number
        MOVE #3,D0          Four longwords to add
        MOVE #$04,CCR       Clear X-bit and set Z-bit of CCR
LOOP    ADDX -(A0),-(A1)    Add pair of numbers
        DBRA D0,LOOP        Repeat until all added
```

Condition codes:

X	N	Z	V	C
*	*	*	*	*

The Z-bit is cleared if the result is non-zero, and left unchanged otherwise. The Z-bit can be used to test for zero after a chain of multiple precision operations.

AND AND logical

Operation: `[destination] ← [source].[destination]`

Syntax:

```
AND <ea>,Dn
AND Dn,<ea>
```

Attributes: Size = byte, word, longword

Description: AND the source operand to the destination operand and store the result in the destination location.

Application: AND is used to mask bits. If we wish to clear bits 3 to 6 of data register D7, we can execute `AND #%10000111,D7`. Unfortunately, the AND operation cannot be used with an address register as either a source or a destination operand. If you wish to perform a logical operation on an address register, you have to copy the address to a data register and then perform the operation there.

Condition codes:

X	N	Z	V	C
-	*	*	0	0

Source operand addressing modes

Dn	An	(An)	(An)+	-(An)	(d,An)	(d,An,Xi)	ABS.W	ABS.L	(d,PC)	(d,PC,Xn)	imm
✔		✔	✔	✔	✔	✔	✔	✔	✔	✔	✔

Destination operand addressing modes

Dn	An	(An)	(An)+	-(An)	(d,An)	(d,An,Xi)	ABS.W	ABS.L	(d,PC)	(d,PC,Xn)	imm
✔		✔	✔	✔	✔	✔	✔	✔			

ANDI AND immediate

Operation: `[destination] ← <literal>.[destination]`

Syntax: `ANDI #<data>,<ea>`

Attributes: Size = byte, word, longword

Description: AND the immediate data to the destination operand. The ANDI permits a literal operand to be ANDed with a destination other than a data register. For example, ANDI #$FE00,$1234 or ANDI.B #$F0,(A2)+.

Condition codes:

X	N	Z	V	C
-	*	*	0	0

Destination operand addressing modes

Dn	An	(An)	(An)+	-(An)	(d,An)	(d,An,Xi)	ABS.W	ABS.L	(d,PC)	(d,PC,Xn)	imm
✔		✔	✔	✔	✔	✔	✔	✔			

ANDI to CCR AND immediate to condition code register

Operation: `[CCR] ← <data>.[CCR]`

Syntax: `ANDI #<data>,CCR`

Attributes: Size = byte

Description: AND the immediate data to the condition code register (i.e., the least-significant byte of the status register).

Application: ANDI is used to clear selected bits of the CCR. For example, ANDI #$FA,CCR clears the Z- and C-bits, i.e., XNZVC = X N 0 V 0.

Condition codes:

```
X  N  Z  V  C
*  *  *  *  *
```

X: cleared if bit 4 of data is zero
N: cleared if bit 3 of data is zero
Z: cleared if bit 2 of data is zero
V: cleared if bit 1 of data is zero
C: cleared if bit 0 of data is zero

ANDI to SR AND immediate to status register

Operation:

```
IF [S] = 1
   THEN
        [SR] ← <literal>.[SR]
   ELSE TRAP
```

Syntax:

```
ANDI #<data>,SR
```

Attributes: Size = word

Description: AND the immediate data to the status register and store the result in the status register. All bits of the SR are affected.

Application: This instruction is used to clear the interrupt mask, the S-bit, and the T-bit of the SR. ANDI #<data>,SR affects both the status byte of the SR and the CCR. For example, ANDI #$7FFF,SR clears the trace bit of the status register, while ANDI #$7FFE,SR clears the trace bit and also clears the carry bit of the CCR.

Condition codes:

```
X  N  Z  V  C
*  *  *  *  *
```

ASL, ASR Arithmetic shift left/right

Operation:

```
[destination] ← [destination] shifted by <count>
```

Syntax:

```
ASL Dx,Dy
ASR Dx,Dy
ASL #<data>,Dy
ASR #<data>,Dy
ASL <ea>
ASR <ea>
```

Attributes: Size = byte, word, longword

Description: Arithmetically shift the bits of the operand in the specified direction (i.e., left or right). The shift count may be specified in one of three ways. The count may be a literal, the contents of a data register, or the value 1. An immediate (i.e., literal) count permits a shift of 1 to 8 places. If the count is in a register, the value is modulo 64 (i.e., 0 to 63). If no count is specified, one shift is made (i.e., `ASL <ea>` shifts the contents of the *word* at the effective address one place left).

The effect of an arithmetic shift left is to shift a zero into the least-significant bit position and to shift the most-significant bit out into both the X- and the C-bits of the CCR. The overflow bit of the CCR is set if a sign change occurs during shifting (i.e., if the most-significant bit changes value during shifting).

The effect of an arithmetic shift right is to shift the least-significant bit into both the X- and C-bits of the CCR. The most-significant bit (i.e., the sign bit) is *replicated* to preserve the sign of the number.

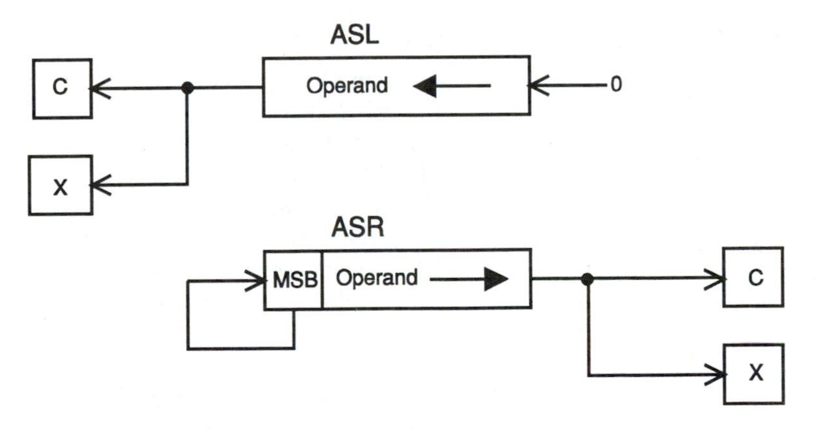

Application: ASL multiplies a two's complement number by 2. ASL is almost identical to the corresponding logical shift, LSR. The only difference between ASL and LSL is that ASL sets the V-bit of the CCR if overflow occurs, while LSL clears the V-bit to zero. An ASR divides a two's complement number by 2. When applied to the contents of a memory location, all 68000 shift operations operate on a word.

Condition codes:

X	N	Z	V	C
*	*	*	*	*

The X-bit and the C-bit are set according to the last bit shifted out of the operand. If the shift count is zero, the C-bit is cleared. The V-bit is set if the most-significant bit is changed at any time during the shift operation and cleared otherwise.

Destination operand addressing modes

Dn	An	(An)	(An)+	-(An)	(d,An)	(d,An,Xi)	ABS.W	ABS.L	(d,PC)	(d,PC,Xn)	imm
✔		✔	✔	✔	✔	✔	✔	✔			

Bcc Branch on condition cc

Operation: If cc = 1 THEN [PC] ← [PC] + d

Syntax: Bcc <label>

Sample syntax: BEQ Loop_4
BVC *+8

Attributes: BEQ takes an 8-bit or a 16-bit offset (i.e., displacement).

Description: If the specified logical condition is met, program execution continues at location [PC] + displacement, d. The displacement is a two's complement value. The value in the PC corresponds to the current location plus two. The range of the branch is -126 to +128 bytes with an 8-bit offset, and -32K to +32K bytes with a 16-bit offset. A short branch to the next instruction is impossible, since the branch code 0 indicates a long branch with a 16-bit offset. The assembly language form BCC *+8 means branch to the point eight bytes from the current PC if the carry bit is clear.

BCC	branch on carry clear	$\overline{C}$
BCS	branch on carry set	C
BEQ	branch on equal	Z
BGE	branch on greater than or equal	$N.V + \overline{N}.\overline{V}$
BGT	branch on greater than	$N.V.\overline{Z} + \overline{N}.\overline{V}.\overline{Z}$
BHI	branch on higher than	$\overline{C}.\overline{Z}$
BLE	branch on less than or equal	$Z + N.\overline{V} + \overline{N}.V$
BLS	branch on lower than or same	$C + Z$
BLT	branch on less than	$N.\overline{V} + \overline{N}.V$
BMI	branch on minus (i.e., negative)	N
BNE	branch on not equal	$\overline{Z}$
BPL	branch on plus (i.e., positive)	$\overline{N}$
BVC	branch on overflow clear	$\overline{V}$
BVS	branch on overflow set	V

Note that there are two types of conditional branch instruction:

those that branch on an unsigned condition and those that branch on a signed condition. For example, $FF is greater than $10 when the numbers are regarded as unsigned (i.e., 255 is greater than 16). However, if the numbers are signed, $FF is less than $10 (i.e., -1 is less than 16).

The signed comparisons are:

BGE	branch on greater than or equal
BGT	branch on greater than
BLE	branch on lower than or equal
BLT	branch on less than

The unsigned comparisons are:

BHS	BCC	branch on higher than or same
BHI		branch on higher than
BLS		branch on lower than or same
BLO	BCS	branch on less than

The *official* mnemonics BCC (branch on carry clear) and BCS (branch on carry set) can be renamed as BHS (branch on higher than or same) and BLO (branch on less than), respectively. Many 68000 assemblers support these alternative mnemonics.

Condition codes:

X	N	Z	V	C
-	-	-	-	-

BCHG Test a bit and change

Operation:

$[Z] \leftarrow \overline{\langle\text{bit number}\rangle\ \text{OF [destination]}}$

$\langle\text{bit number}\rangle\ \text{OF [destination]} \leftarrow \overline{\langle\text{bit number}\rangle\ \text{OF [destination]}}$

Syntax:

```
BCHG Dn,<ea>
BCHG #<data>,<ea>
```

Attributes: Size = byte, longword

Description: A bit in the destination operand is tested and the state of the specified bit is reflected in the condition of the Z-bit in the CCR. After the test operation, the state of the specified bit is changed in the destination. If a data register is the destination, then the bit numbering is modulo 32, allowing bit manipulation of all bits in a data register. If a memory location is the destination, a byte is read from that location, the bit operation performed using the bit number modulo 8, and the byte written back to the location. Note that bit zero refers to the least-significant bit. The bit number for this operation may be specified either *statically* by an immediate value or *dynamically* by the contents of a data register.

Application: If the operation `BCHG #4,$1234` is carried out and the contents of memory location \$1234 are 10101010_2, bit 4 is tested. It is a 0 and therefore the Z-bit of the CCR is set to 1. Bit 4 of the destination operand is changed and the new contents of location 1234_{16} are 10111010_2.

Condition codes:

X	N	Z	V	C
-	-	*	-	-

Z: set if the bit tested is zero, cleared otherwise.

Destination operand addressing modes

Dn	An	(An)	(An)+	-(An)	(d,An)	(d,An,Xi)	ABS.W	ABS.L	(d,PC)	(d,PC,Xn)	imm
✔		✔	✔	✔	✔	✔	✔	✔			

Note that data register direct (i.e., Dn) addressing uses a longword operand, while all other modes use a byte operand.

BCLR Test a bit and clear

Operation:

```
[Z] ← NOT <bit number> OF [destination]
<bit number> OF [destination] ← 0
```

Syntax:

```
BCLR Dn,<ea>
BCLR #<data>,<ea>
```

Attributes: Size = byte, longword

Description: A bit in the destination operand is tested and the state of the specified bit is reflected in the condition of the Z-bit in the condition code. After the test, the state of the specified bit is cleared in the destination. If a data register is the destination, the bit numbering is modulo 32, allowing bit manipulation of all bits in a data register. If a memory location is the destination, a byte is read from that location, the bit operation performed using the bit number modulo 8, and the byte written back to the location.

Bit zero refers to the least-significant bit. The bit number for this operation may be specified either by an immediate value or dynamically by the contents of a data register.

Application: If the operation `BCLR #4,$1234` is carried out and the contents of memory location $1234 are 11111010_2, bit 4 is tested. It is a 1 and therefore the Z-bit of the CCR is set to 0. Bit 4 of the destination operand is cleared and the new contents of $1234 are: 11101010_2.

Condition codes:

X	N	Z	V	C
-	-	*	-	-

Z: set if the bit tested is zero, cleared otherwise.

Destination operand addressing modes

Dn	An	(An)	(An)+	-(An)	(d,An)	(d,An,Xi)	ABS.W	ABS.L	(d,PC)	(d,PC,Xn)	imm
✔		✔	✔	✔	✔	✔	✔	✔			

Note that data register direct (i.e., Dn) addressing uses a longword operand, while all other modes use a byte operand.

BRA Branch always

Operation: `[PC] ← [PC] + d`

Syntax:

```
BRA <label>
BRA <literal>
```

Attributes: Size = byte, word

Description: Program execution continues at location [PC] + d. The displacement, d, is a two's complement value (8 bits for a short branch and 16 bits for a long branch). The value in the PC corresponds to the current location plus two. Note that a short branch to the next instruction is impossible, since the branch code 0 is used to indicate a long branch with a 16-bit offset.

Application: A `BRA` is an unconditional relative jump (or goto). You use a `BRA` instruction to write position independent code, because the destination address (*branch target address*) is specified with respect to the current value of the PC. A `JMP` instruction does not produce position independent code.

Condition codes: X N Z V C
- - - - -

BSET Test a bit and set

Operation: [Z] ← <bit number> OF [destination]
<bit number> OF [destination] ← 1

Syntax: BSET Dn,<ea>
BSET #<data>,<ea>

Attributes: Size = byte, longword

Description: A bit in the destination operand is tested and the state of the specified bit is reflected in the condition of the Z-bit of the condition code. After the test, the specified bit is set in the destination. If a data register is the destination then the bit numbering is modulo 32, allowing bit manipulation of all bits in a data register. If a memory location is the destination, a byte is read from that location, the bit operation performed using bit number modulo 8, and the byte written back to the location. Bit zero refers to the least-significant bit. The bit number for this operation may be specified either by an immediate value or dynamically by the contents of a data register.

Condition codes: X N Z V C
- - * - -

Z: set if the bit tested is zero, cleared otherwise.

Destination operand addressing mode for BSET Dn,<ea> form

Dn	An	(An)	(An)+	-(An)	(d,An)	(d,An,Xi)	ABS.W	ABS.L	(d,PC)	(d,PC,Xn)	imm
✔		✔	✔	✔	✔	✔	✔	✔			

Note that data register direct (i.e., Dn) addressing uses a longword operand, while all other modes use a byte operand.

BSR Branch to subroutine

Operation: [SP] ← [SP] - 4; [M([SP])] ← [PC]; [PC] ← [PC] + d

Syntax: `BSR <label>`
`BSR <literal>`

Attributes: Size = byte, word

Description: The longword address of the instruction immediately following the BSR instruction is pushed onto the system stack pointed at by A7. Program execution then continues at location [PC] + displacement. The displacement is an 8-bit two's complement value for a short branch, or a 16-bit two's complement value for a long branch. The value in the PC corresponds to the current location plus two. Note that a short branch to the next instruction is impossible, since the branch code 0 is used to indicate a long branch with a 16-bit offset.

Application: BSR is used to call a procedure or a subroutine. Since it provides relative addressing (and therefore position independent code), its use is preferable to JSR.

Condition codes:

```
X  N  Z  V  C
-  -  -  -  -
```

BTST Test a bit

Operation: $[Z] \leftarrow \overline{\text{<bit number> OF [destination]}}$

Syntax: `BTST Dn,<ea>`
`BTST #<data>,<ea>`

Attributes: Size = byte, longword

Description: A bit in the destination operand is tested and the state of the specified bit is reflected in the condition of the Z-bit in the CCR. The destination is not modified by a BTST instruction. If a data register is the destination, then the bit numbering is modulo 32, allowing bit manipulation of all bits in a data register. If a memory location is the destination, a byte is read from that location, the bit operation performed. Bit 0 refers to the least-significant bit. The bit number for this operation may be specified either statically by an immediate value or dynamically by the contents of a data register.

Condition codes:

```
X  N  Z  V  C
-  -  *  -  -
```

Z: set if the bit tested is zero, cleared otherwise.

Destination operand addressing modes for BTST Dn,<ea> form

Dn	An	(An)	(An)+	-(An)	(d,An)	(d,An,Xi)	ABS.W	ABS.L	(d,PC)	(d,PC,Xn)	imm
✔		✔	✔	✔	✔	✔	✔	✔	✔	✔	

Note that data register direct (i.e., Dn) addressing uses a longword operand, while all other modes use a byte operand.

CHK Check register against bounds

Operation: IF [Dn] < 0 OR [Dn] > [<ea>] THEN TRAP

Syntax: CHK <ea>,Dn

Attributes: Size = word

Description: The contents of the low-order word in the data register specified in the instruction are examined and compared with the upper bound at the effective address. The upper bound is a two's complement integer. If the data register value is less than zero or greater than the upper bound contained in the operand word, then the processor initiates exception processing.

Application: The CHK instruction can be used to test the bounds of an array element before it is used. By performing this test, you can make certain that you do not access an element outside an array. Consider the following fragment of code:

```
MOVE.W  subscript,D0      Get subscript to test
CHK     #max_bound,D0     Test subscript against 0 and upper bound
*                         TRAP on error ELSE continue if ok
```

Condition codes:
```
X  N  Z  V  C
-  *  U  U  U
```
N: set if [Dn] < 0; cleared if [Dn] > [<ea>]; undefined otherwise.

Source operand addressing modes

Dn	An	(An)	(An)+	-(An)	(d,An)	(d,An,Xi)	ABS.W	ABS.L	(d,PC)	(d,PC,Xn)	imm
✔		✔	✔	✔	✔	✔	✔	✔			

CLR Clear an operand

Operation: [destination] ← 0

Syntax: CLR <ea>

Sample syntax: CLR (A4)+

Attributes: Size = byte, word, longword

Description: The destination is cleared — loaded with all zeros. The CLR instruction can't be used to clear an address register. You can use SUBA.L A0,A0 to clear A0. Note that a side effect of CLR's implementation is a *read* from the specified effective address before the clear (i.e., write) operation is executed. Under certain circumstances this might cause a problem (e.g., with write-only memory).

Condition codes:

X	N	Z	V	C
-	0	1	0	0

Source operand addressing modes

Dn	An	(An)	(An)+	-(An)	(d,An)	(d,An,Xi)	ABS.W	ABS.L	(d,PC)	(d,PC,Xn)	imm
✔		✔	✔	✔	✔	✔	✔	✔			

CMP Compare

Operation: [destination] - [source]

Syntax: CMP <ea>,Dn

Sample syntax: CMP (Test,A6,D3.W),D2

Attributes: Size = byte, word, longword

Description: Subtract the source operand from the destination operand and set the condition codes accordingly. The destination must be a data register. The destination is not modified by this instruction.

Condition codes:

X	N	Z	V	C
-	*	*	*	*

Source operand addressing modes

Dn	An	(An)	(An)+	-(An)	(d,An)	(d,An,Xi)	ABS.W	ABS.L	(d,PC)	(d,PC,Xn)	imm
✔	✔	✔	✔	✔	✔	✔	✔	✔	✔	✔	✔

CMPA Compare address

Operation: `[destination] - [source]`

Syntax: `CMPA <ea>,An`

Sample syntax:
```
CMPA.L #$1000,A4
CMPA.W (A2)+,A6
CMPA.L D5,A2
```

Attributes: Size = word, longword

Description: Subtract the source operand from the destination address register and set the condition codes accordingly. The address register is not modified. The size of the operation may be specified as word or longword. Word length operands are sign-extended to 32 bits before the comparison is carried out.

Condition codes:
```
X  N  Z  V  C
-  *  *  *  *
```

Source operand addressing modes

Dn	An	(An)	(An)+	-(An)	(d,An)	(d,An,Xi)	ABS.W	ABS.L	(d,PC)	(d,PC,Xn)	imm
✔	✔	✔	✔	✔	✔	✔	✔	✔	✔	✔	✔

CMPI Compare immediate

Operation: `[destination] - <immediate data>`

Syntax: `CMPI #<data>,<ea>`

Attributes: Size = byte, word, longword

Description: Subtract the immediate data from the destination operand and set the condition codes accordingly — the destination is not modified. CMPI permits the comparison of a literal with memory.

Condition codes:
```
X N Z V C
- * * * *
```

Destination operand addressing modes

Dn	An	(An)	(An)+	-(An)	(d,An)	(d,An,Xi)	ABS.W	ABS.L	(d,PC)	(d,PC,Xn)	imm
✔		✔	✔	✔	✔	✔	✔	✔	✔	✔	

CMPM Compare memory with memory

Operation: `[destination] - [source]`

Syntax: `CMPM (Ay)+,(Ax)+`

Attributes: Size = byte, word, longword

Sample syntax: `CMPM.B (A3)+,(A4)+`

Description: Subtract the source operand from the destination operand and set the condition codes accordingly. The destination is not modified by this instruction. The only permitted addressing mode is address register indirect with post-incrementing for both source and destination operands.

Application: Used to compare the contents of two blocks of memory. For example:

```
*       Compare two blocks of memory for equality
        LEA      Source,A0         A0 points to source block
        LEA      Destination,A1    A1 points to destination block
        MOVE.W   #Count-1,D0       Compare Count words
RPT     CMPM.W   (A0)+,(A1)+       Compare pair of words
        DBNE     D0,RPT            Repeat until all done
        .
        .
```

Condition codes:
```
X N Z V C
- * * * *
```

DBcc Test condition, decrement, and branch

Operation:

```
IF(condition false)
  THEN [Dn] ← [Dn] - 1 {decrement loop counter}
   IF [Dn] = -1 THEN [PC] ← [PC] + 2 {fall through to next instruction}
                ELSE [PC] ← [PC] + d {take branch}
  ELSE [PC] ← [PC] + 2 {fall through to next instruction}
```

Syntax: `DBcc Dn,<label>`

Attributes: Size = word

Description: The DBcc instruction provides an automatic looping facility and replaces the usual decrement counter, test, and branch instructions. Three parameters are required by the DBcc instruction: a branch condition (specified by 'cc'), a data register that serves as the loop down-counter, and a label that indicates the start of the loop. The DBcc first tests the condition 'cc', and if 'cc' is true the loop is terminated and the branch back to <label> not taken. The 14 branch conditions supported by Bcc are also supported by DBcc, as well as DBF and DBT (F = false, and T = true). Note that many assemblers permit the mnemonic DBF to be expressed as DBRA (i.e., decrement and branch back).

It is important to appreciate that the condition tested by the DBcc instruction works in the *opposite* sense to a Bcc, conditional branch, instruction. For example, BCC means branch on carry clear, whereas DBCC means continue (i.e., exit the loop) on carry clear. That is, the DBcc condition is a loop terminator. If the termination condition is not true, the low-order 16 bits of the specified data register are decremented. If the result is -1, the loop is not taken and the next instruction is executed. If the result is not -1, a branch is made to 'label'. Note that the label represents a 16-bit signed value, permitting a branch range of -32K to +32K bytes. Since the value in Dn decremented is 16 bits, the loop may be executed up to 64K times.

We can use the instruction DBEQ, decrement and branch on zero, to mechanize the high-level language construct REPEAT...UNTIL.

```
LOOP ...                   REPEAT
     ...
     ...                     [D0] := [D0] - 1
     ...
     DBEQ   D0,REPEAT  UNTIL [D0] = - 1 OR [Z] = 1
```

Application: Suppose we wish to input a block of 512 bytes of data (the data is returned in register D1). If the input routine returns a value zero in D1, an error has occurred and the loop must be exited.

```
      LEA     Dest,A0    Set up pointer to destination
      MOVE.W #511,D0    512 bytes to be input
AGAIN BSR     INPUT      Get the data in D1
      MOVE.B D1,(A0)+   Store it
      DBEQ    D0,AGAIN   REPEAT until D1 = 0 OR 512 times
```

Condition codes:

X	N	Z	V	C
-	-	-	-	-

Not affected

DIVS, DIVU Signed divide, unsigned divide

Operation: [destination] ← [destination]/[source]

Syntax:

```
DIVS <ea>,Dn
DIVU <ea>,Dn
```

Attributes: Size = longword/word = longword result

Description: Divide the destination operand by the source operand and store the result in the destination. The destination is a longword and the source is a 16-bit value. The result (i.e., destination register) is a 32-bit value arranged so that the quotient is the lower-order word and the remainder is the upper-order word. DIVU performs division on unsigned values, and DIVS performs division on two's complement values. An attempt to divide by zero causes an exception. For DIVS, the sign of the remainder is always the same as the sign of the dividend (unless the remainder is zero).

Attempting to divide a number by zero results in a divide-by-zero exception. If overflow is detected during division, the operands are unaffected. Overflow is checked for at the start of the operation and occurs if the quotient is larger than a 16-bit signed integer. If the upper word of the dividend is greater than or equal to the divisor, the V-bit is set and the instruction terminated.

Application: Consider the division of D0 by D1, DIVU D1,D0, which results in:

```
[D0(0:15)]  ← [D0(0:31)]/[D1(0:15)]
[D0(16:31)] ← remainder
```

Condition codes: X N Z V C
- * * * 0

The X-bit is not affected by a division. The N-bit is set if the quotient is negative. The Z-bit is set if the quotient is zero. The V-bit is set if division overflow occurs (in which case the Z- and N-bits are undefined). The C-bit is always cleared.

Source operand addressing modes

Dn	An	(An)	(An)+	-(An)	(d,An)	(d,An,Xi)	ABS.W	ABS.L	(d,PC)	(d,PC,Xn)	imm
✔		✔	✔	✔	✔	✔	✔	✔	✔	✔	✔

EOR Exclusive OR logical

Operation: `[destination] ← [source] ⊕ [destination]`

Syntax: `EOR Dn,<ea>`

Sample syntax: `EOR D3,-(A3)`

Attributes: Size = byte, word, longword.

Description: EOR (exclusive or) the source operand with the destination operand and store the result in the destination location. Note that the source operand must be a data register and that the operation `EOR <ea>,Dn` is not permitted.

Application: The EOR instruction is used to *toggle* (i.e., change the state of) selected bits in the operand. For example, if [D0] = 00001111, and [D1] = 10101010, the operation `EOR.B D0,D1` toggles bits 0 to 3 of D1 and results in [D1] = 10100101.

Condition codes: X N Z V C
- * * 0 0

Destination operand addressing modes

Dn	An	(An)	(An)+	-(An)	(d,An)	(d,An,Xi)	ABS.W	ABS.L	(d,PC)	(d,PC,Xn)	imm
✔		✔	✔	✔	✔	✔	✔	✔			

EORI EOR immediate

Operation: [destination] ← <literal> ⊕ [destination]

Syntax: EORI #<data>,<ea>

Attributes: Size = byte, word, longword

Description: EOR the immediate data with the contents of the destination operand. Store the result in the destination operand.

Condition codes:

X	N	Z	V	C
-	*	*	0	0

Destination operand addressing modes

Dn	An	(An)	(An)+	-(An)	(d,An)	(d,An,Xi)	ABS.W	ABS.L	(d,PC)	(d,PC,Xn)	imm
✔		✔	✔	✔	✔	✔	✔	✔			

EORI to CCR EOR immediate to CCR

Operation: [CCR] ← <literal> ⊕ [CCR]

Syntax: EORI #<data>,CCR

Attributes: Size = byte

Description: EOR the immediate data with the contents of the condition code register (i.e., the least-significant byte of the status register).

Application: Used to toggle bits in the CCR. For example, EORI #$0C,CCR toggles the N- and Z-bits of the CCR.

Condition codes:

X	N	Z	V	C
*	*	*	*	*

X:= toggled if bit 4 of data = 1; unchanged otherwise
N:= toggled if bit 3 of data = 1; unchanged otherwise
Z:= toggled if bit 2 of data = 1; unchanged otherwise
V:= toggled if bit 1 of data = 1; unchanged otherwise
C:= toggled if bit 0 of data = 1; unchanged otherwise

EORI to SR EOR immediate to status register

Operation:
```
IF [S] = 1
   THEN
       [SR] ← <literal> ⊕ [SR]
   ELSE TRAP
```

Syntax: `EORI #<data>,SR`

Attributes: Size = word

Description: EOR (exclusive OR) the immediate data with the contents of the status register and store the result in the status register. All bits of the status register are affected.

Condition codes:
```
X  N  Z  V  C
*  *  *  *  *
```
X:= toggled if bit 4 of data = 1; unchanged otherwise
N:= toggled if bit 3 of data = 1; unchanged otherwise
Z:= toggled if bit 2 of data = 1; unchanged otherwise
V:= toggled if bit 1 of data = 1; unchanged otherwise
C:= toggled if bit 0 of data = 1; unchanged otherwise

EXG Exchange registers

Operation: `[Rx] ← [Ry]; [Ry] ← [Rx]`

Syntax: `EXG Rx,Ry`

Sample syntax:
```
EXG D3,D4
EXG D2,A0
EXG A7,D5
```

Attributes: Size = longword

Description: Exchange the contents of two registers. The size of the instruction is a longword because the entire 32-bit contents of two registers are exchanged. The instruction permits the exchange of address registers, data registers, and address and data registers.

Application: One application of EXG is to load an address into a data register and then process it using instructions that act on data registers. Then the reverse operation can be used to return the result to the address register. Doing this preserves the original contents of the data register.

Condition codes:
```
X N Z V C
- - - - -
```

EXT Sign-extend a data register

Operation: `[destination] ← sign-extended[destination]`

Syntax:
```
EXT.W Dn
EXT.L Dn
```

Attributes: Size = word, longword

Description: Extend the least-significant byte in a data register to a word, or extend the least-significant word in a data register to a longword. If the operation is word sized, bit 7 of the designated data register is copied to bits (8:15). If the operation is longword sized, bit 15 is copied to bits (16:31).

Application: If `[D0] = $12345678, EXT.W D0` results in 12340078_{16}
If `[D0] = $12345678, EXT.L D0` results in 00005678_{16}

Condition codes:
```
X N Z V C
- * * 0 0
```

ILLEGAL Illegal instruction

Operation:
```
[SSP] ← [SSP] - 4; [M([SSP])] ← [PC];
[SSP] ← [SSP] - 2; [M([SSP])] ← [SR];
[PC] ← Illegal instruction vector
```

Syntax: `ILLEGAL`

Attributes: None

Description: The bit pattern of the illegal instruction, $4AFC_{16}$ causes the illegal instruction trap to be taken. As in all exceptions, the contents of the program counter and the processor status word are pushed onto the supervisor stack at the start of exception processing.

Application: Any *unknown* pattern of bits read by the 68000 during an instruction read phase will cause an illegal instruction trap. The ILLEGAL instruction can be thought of as an *official* illegal instruction. It can be used to test the illegal instruction trap and will always be an illegal instruction in any future enhancement of the 68000.

Condition codes:

X	N	Z	V	C
-	-	-	-	-

JMP Jump (unconditionally)

Operation: [PC] ← destination

Syntax: JMP <ea>

Attributes: Unsized

Description: Program execution continues at the effective address specified by the instruction.

Application: Apart from a simple unconditional jump to an address fixed at compile time (i.e., JMP label), the JMP instruction is useful for the calculation of *dynamic* or *computed* jumps. For example, the instruction JMP (A0,D0.L) jumps to the location pointed at by the contents of address register A0, offset by the contents of data register D0. Note that JMP provides several addressing modes, while BRA provides a single addressing mode (i.e., PC relative).

Condition codes:

X	N	Z	V	C
-	-	-	-	-

Source operand addressing modes

Dn	An	(An)	(An)+	-(An)	(d,An)	(d,An,Xi)	ABS.W	ABS.L	(d,PC)	(d,PC,Xn)	imm
		✔			✔	✔	✔	✔	✔	✔	

JSR Jump to subroutine

Operation: [SP] ← [SP] - 4; [M([SP])] ← [PC]
[PC] ← destination

Syntax: JSR <ea>

Attributes: Unsized

Description: JSR pushes the longword address of the instruction immediately following the JSR onto the system stack. Program execution then continues at the address specified in the instruction.

Application: JSR (Ai) calls the procedure pointed at by address register Ai. The instruction JSR (Ai,Dj) calls the procedure at the location [Ai] + [Dj] which permits dynamically computed addresses.

Condition codes:

X	N	Z	V	C
-	-	-	-	-

Source operand addressing modes

Dn	An	(An)	(An)+	-(An)	(d,An)	(d,An,Xi)	ABS.W	ABS.L	(d,PC)	(d,PC,Xn)	imm
		✔			✔	✔	✔	✔	✔	✔	

LEA Load effective address

Operation: [An] ← <ea>

Syntax: LEA <ea>,An

Sample syntax:

```
LEA Table,A0
LEA (Table,PC),A0
LEA (-6,A0,D0.L),A6
LEA (Table,PC,D0),A6
```

Attributes: Size = longword

Description: The effective address is computed and loaded into the specified address register. For example, LEA (-6,A0,D0.W),A1 calculates the sum of address register A0 plus data register D0.W sign-extended to 32 bits minus 6, and deposits the result in address register A1. The difference between the LEA and PEA instructions is that LEA calculates an effective address and puts it in an address register, while PEA calculates an effective address in the same way but pushes it on the stack.

Application: LEA is a very powerful instruction used to calculate an effective address. In particular, the use of LEA facilitates the writing of position independent code. For example, LEA (TABLE,PC),A0 calculates the effective address of 'TABLE' with respect to the PC and deposits it in A0.

```
      LEA  (Table,PC),A0    Compute address of Table with respect to the PC
      MOVE (A0),D1          Pick up the first item in the table
      .                     Do something with this item
      MOVE D1,(A0)          Put it back in the table
      .
      .
Table DS.B  100
```

Source operand addressing modes

Dn	An	(An)	(An)+	-(An)	(d,An)	(d,An,Xi)	ABS.W	ABS.L	(d,PC)	(d,PC,Xn)	imm
		✔			✔	✔	✔	✔	✔	✔	

Condition codes:

X	N	Z	V	C
-	-	-	-	-

LINK Link and allocate

Operation: [SP] ← [SP] - 4; [M([SP])] ← [An];
[An] ← [SP]; [SP] ← [SP] + d

Syntax: LINK An,#<displacement>

Sample syntax: LINK A6,#-12

Attributes: Size = word

Description: The contents of the specified address register are first pushed onto the stack. Then, the address register is loaded with the updated stack pointer. Finally, the 16-bit sign-extended displacement is added to the stack pointer. The contents of the address register occupy two words on the stack. A *negative displacement* must be used to allocate stack area to a procedure. At the end of a LINK instruction, the old value of address register An has been pushed on the stack and the new An is pointing at the base of the stack frame. The stack pointer itself has been moved up by d bytes and is pointing at the top of the stack frame. Address register An is called the *frame pointer* because it is used to reference data on the stack frame. By convention, programmers often use A6 as a frame pointer.

Application: The LINK and UNLK pair are used to create local workspace on the top of a procedure's stack. Consider the code:

```
Subrtn LINK A6,#-12     Create a 12-byte workspace
       .
       MOVE D3,(-8,A6) Access the stack frame via A6
       .
       .
       UNLK A6          Collapse the workspace
       RTS              Return from subroutine
```

Condition codes:

X	N	Z	V	C
-	-	-	-	-

The LINK instruction does not affect the CCR.

LSL, LSR Logical shift left/right

Operation: [destination] ← [destination] shifted by <count>

Syntax:

```
LSL Dx,Dy
LSR Dx,Dy
LSL #<data>,Dy
LSR #<data>,Dy
LSL <ea>
LSR <ea>
```

Attributes: Size = byte, word, longword

Description: Logically shift the bits of the operand in the specified direction (i.e., left or right). A zero is shifted into the input position and the bit shifted out is copied into both the C- and the X-bit of the CCR. The shift count may be specified in one of three ways. The count may be a literal, the contents of a data register, or the value 1. An immediate count permits a shift of 1 to 8 places. If the count is in a register, the value is modulo 64 — from 0 to 63. If no count is specified, one shift is made (e.g., LSL <ea> shifts the *word* at the effective address one position left).

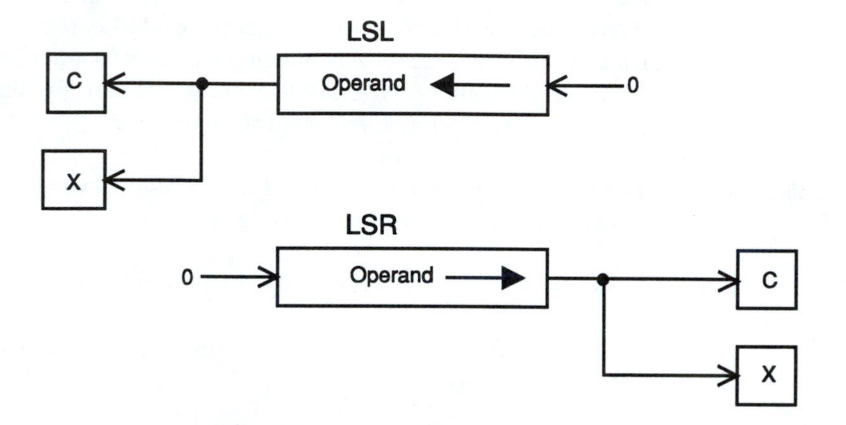

Application: If [D3.W] = 1100110010101110_2, the instruction `LSL.W #5,D3` produces the result 1001010111000000_2. After the shift, both the X-and C-bits of the CCR are set to 1 (since the last bit shifted out was a 1).

Condition codes:

X	N	Z	V	C
*	*	*	0	*

The X-bit is set to the last bit shifted out of the operand and is equal to the C-bit. However, a zero shift count leaves the X-bit unaffected and the C-bit cleared.

Destination operand addressing modes

Dn	An	(An)	(An)+	-(An)	(d,An)	(d,An,Xi)	ABS.W	ABS.L	(d,PC)	(d,PC,Xn)	imm
✔		✔	✔	✔	✔	✔	✔	✔			

MOVE Copy data from source to destination

Operation: [destination] ← [source]

Syntax: `MOVE <ea>,<e>`

Sample syntax:

```
MOVE (A5),-(A2)
MOVE -(A5),(A2)+
MOVE #$123,(A6)+
MOVE Temp1,Temp2
```

Attributes: Size = byte, word, longword

Description: Move the contents of the source to the destination location. The data is examined as it is moved and the condition codes set accordingly. Note that this is actually a *copy* command because the source is not affected by the move. The move instruction has the widest range of addressing modes of all the 68000's instructions.

Condition codes:

X	N	Z	V	C
-	*	*	0	0

Source operand addressing modes

Dn	An	(An)	(An)+	-(An)	(d,An)	(d,An,Xi)	ABS.W	ABS.L	(d,PC)	(d,PC,Xn)	imm
✔	✔	✔	✔	✔	✔	✔	✔	✔	✔	✔	✔

Destination operand addressing modes

Dn	An	(An)	(An)+	-(An)	(d,An)	(d,An,Xi)	ABS.W	ABS.L	(d,PC)	(d,PC,Xn)	imm
✔		✔	✔	✔	✔	✔	✔	✔			

MOVEA Move address

Operation: [An] ← [source]

Syntax: MOVEA <ea>,An

Attributes: Size = word, longword

Description: Move the contents of the source to the destination location. The destination is an address register. The source must be a word or longword. If it is a word, it is sign-extended to a longword. The condition codes are not affected.

Application: The MOVEA instruction is used to load an address register (some assemblers simply employ the MOVE mnemonic for both MOVE and MOVEA). Note that the instruction LEA can often be used to perform the same operation (e.g., MOVEA.L #$1234,A0 is the same as LEA $1234,A0).

Take care because the MOVEA.W #$8000,A0 instruction sign-extends the source operand to $FFFF8000 before loading it into A0, whereas LEA $8000,A0 loads A0 with $00008000.

You should appreciate that the MOVEA and LEA instructions are not interchangeable. The operation MOVEA (Ai),An cannot be implemented by an LEA instruction, since MOVEA (Ai),An performs a memory access to obtain the source operand, as the following RTL demonstrates.

```
LEA   (Ai),An   =  [An] ← [Ai]
MOVEA (Ai),An   =  [An] ← [M([Ai])]
```

Condition codes:

X	N	Z	V	C
-	-	-	-	-

Source operand addressing modes

Dn	An	(An)	(An)+	-(An)	(d,An)	(d,An,Xi)	ABS.W	ABS.L	(d,PC)	(d,PC,Xn)	imm
✔	✔	✔	✔	✔	✔	✔	✔	✔	✔	✔	✔

MOVE to CCR Copy data to CCR from source

Operation: [CCR] ← [source]

Syntax: MOVE <ea>,CCR

Attributes: Size = word

Description: Move the contents of the source operand to the condition code register. The source operand is a *word*, but only the low-order *byte* contains the condition codes. The upper byte is neglected. Note that MOVE <ea>,CCR is a word operation, but ANDI, ORI, and EORI to CCR are all byte operations.

Application: The move to CCR instruction permits the programmer to preset the CCR. For example, MOVE #0,CCR clears all the CCR's bits.

Condition codes:

X	N	Z	V	C
*	*	*	*	*

Source operand addressing modes

Dn	An	(An)	(An)+	-(An)	(d,An)	(d,An,Xi)	ABS.W	ABS.L	(d,PC)	(d,PC,Xn)	imm
✔		✔	✔	✔	✔	✔	✔	✔	✔	✔	✔

MOVE from SR Copy data from SR to destination

Operation: `[destination] ← [SR]`

Syntax: `MOVE SR,<ea>`

Attributes: Size = word

Description: Move the contents of the status register to the destination location. The source operand, the status register, is a word. This instruction is not privileged in the 68000, but is privileged in the 68010, 68020, and 68030. Executing a `MOVE SR,<ea>` while in the user mode on these processors results in a privilege violation trap.

Condition codes:

X	N	Z	V	C
-	-	-	-	-

Destination operand addressing modes

Dn	An	(An)	(An)+	-(An)	(d,An)	(d,An,Xi)	ABS.W	ABS.L	(d,PC)	(d,PC,Xn)	imm
✔		✔	✔	✔	✔	✔	✔	✔			

MOVE to SR Copy data to SR from source

Operation:

```
IF [S] = 1
   THEN [SR] ← [source]
ELSE TRAP
```

Syntax: `MOVE <ea>,SR`

Attributes: Size = word

Description: Move the contents of the source operand to the status register. The source operand is a word and all bits of the status register are affected.

Application: The MOVE to SR instruction allows the programmer to preset the contents of the status register. This instruction permits the trace mode, interrupt mask, and status bits to be modified. For example, MOVE #$2700,SR moves 00100111 00000000 to the status register which clears all bits of the CCR, sets the S-bit, clears the T-bit, and sets the interrupt mask level to 7.

Condition codes:

X	N	Z	V	C
*	*	*	*	*

Source operand addressing modes

Dn	An	(An)	(An)+	-(An)	(d,An)	(d,An,Xi)	ABS.W	ABS.L	(d,PC)	(d,PC,Xn)	imm
✔		✔	✔	✔	✔	✔	✔	✔	✔	✔	✔

MOVE USP Copy data to or from USP

Operation 1:

```
IF [S] = 1                        {MOVE USP,An form}
   THEN [USP] ← [An]
ELSE TRAP
```

Operation 2:

```
IF [S] = 1                        {MOVE An,USP form}
   THEN [An] ← [USP]
ELSE TRAP
```

Syntax 1: MOVE USP,An
Syntax 2: MOVE An,USP

Attributes: Size = longword

Description: Move the contents of the user stack pointer to an address register or vice versa. This is a privileged instruction and allows the operating system running in the supervisor state either to read the contents of the user stack pointer or to set up the user stack pointer.

Condition codes:

X	N	Z	V	C
-	-	-	-	-

MOVEM Move multiple registers

Operation 1:
```
REPEAT
   [destination_register] ← [source]
UNTIL all registers in list moved
```

Operation 2:
```
REPEAT
   [destination] ← [source_register]
UNTIL all registers in list moved
```

Syntax 1:
```
MOVEM <ea>,<register list>
```
Syntax 2:
```
MOVEM <register list>,<ea>
```

Sample syntax:
```
MOVEM.L D0-D7/A0-A6,$1234
MOVEM.L (A5),D0-D2/D5-D7/A0-A3/A6
MOVEM.W (A7)+,D0-D5/D7/A0-A6
MOVEM.W D0-D5/D7/A0-A6,-(A7)
```

Attributes: Size = word, longword

Description: The group of registers specified by `<register list>` is copied to or from consecutive memory locations. The starting location is provided by the effective address. Any combination of the 68000's sixteen address and data registers can be copied by a single `MOVEM` instruction. Note that either a word or a longword can be moved, and that a word is sign-extended to a longword when it is moved (even if the destination is a data register).

When a group of registers is transferred to or from memory (using an addressing mode other than pre-decrementing or post-incrementing), the registers are transferred starting at the specified address and up through higher addresses. The order of transfer of registers is data register D0 to D7, followed by address register A0 to A7.

For example, `MOVEM.L D0-D2/D4/A5/A6,$1234` moves registers `D0,D1,D2,D4,A5,A6` to memory, starting at location $1234 (in which D0 is stored) and moving to locations $1238, $123C,... Note that the address counter is incremented by 2 or 4 after each move according to whether the operation is moving words or longwords, respectively.

If the effective address is in the pre-decrement mode (i.e., `-(An)`), only a register to memory operation is permitted. The registers are stored starting at the specified address minus two (or four for longword operands) and down through lower addresses. The order of storing is from address register A7 to address register A0, then from data register D7 to data register

D0. The decremented address register is updated to contain the address of the last word stored.

If the effective address is in the post-increment mode (i.e., (An)+), only a memory to register transfer is permitted. The registers are loaded starting at the specified address and up through higher addresses. The order of loading is the inverse of that used by the pre-decrement mode and is D0 to D7 followed by A0 to A7. The incremented address register is updated to contain the address of the last word plus two (or four for longword operands).

Note that the MOVEM instruction has a side effect. An extra bus cycle occurs for memory operands, and an operand at one address higher than the last register in the list is accessed. This extra access is an 'overshoot' and has no effect as far as the programmer is concerned. However, it could cause a problem if the overshoot extended beyond the bounds of physical memory. Once again, remember that MOVEM.W sign-extends words when they are moved to data registers.

Application: This instruction is invariably used to save working registers on entry to a subroutine and to restore them at the end of a subroutine.

```
                BSR      Example
                .
                .
Example         MOVEM.L D0-D5/A0-A3,-(SP)  Save registers
                .
                .
                Body of subroutine
                .
                .
                MOVEM.L (SP)+,D0-D5/A0-A3  Restore registers
                RTS                        Return
```

Condition codes:

X	N	Z	V	C
-	-	-	-	-

Source operand addressing modes (memory to register)

Dn	An	(An)	(An)+	-(An)	(d,An)	(d,An,Xi)	ABS.W	ABS.L	(d,PC)	(d,PC,Xn)	imm
		✔	✔		✔	✔	✔	✔	✔	✔	

Destination operand addressing modes (register to memory)

Dn	An	(An)	(An)+	-(An)	(d,An)	(d,An,Xi)	ABS.W	ABS.L	(d,PC)	(d,PC,Xn)	imm
		✔		✔	✔	✔	✔	✔			

MOVEP Move peripheral data

Operation: `[destination] ← [source]`

Syntax:
```
MOVEP Dx,(d,Ay)
MOVEP (d,Ay),Dx
```

Sample syntax:
```
MOVEP D3,(Control,A0)
MOVEP (Input,A6),D5
```

Attributes: Size = word, longword

Description: The MOVEP operation moves data between a data register and a byte-oriented memory mapped peripheral. The data is moved between the specified data register and *alternate bytes* within the peripheral's address space, starting at the location specified and incrementing by two. This instruction is designed to be used in conjunction with 8-bit peripherals connected to the 68000's 16-bit data bus. The high-order byte of the data register is transferred first and the low-order byte transferred last. The memory address is specified by the address register indirect mode with a 16-bit offset. If the address is even, all transfers are to or from the high-order half of the data bus. If the address is odd, all the transfers are made to the low-order half of the data bus.

Application: Consider a memory-mapped peripheral located at address $08 0001 which has four 8-bit internal registers mapped at addresses $08 0001, $08 0003, $08 0005, and $08 0007. The longword in data register D0 is to be transferred to this peripheral by the following code.

```
LEA       $080001,A0
MOVEP.L   D0,0(A0)
```

This code results in the following actions:

```
[M(080001)] ← [D0(24:31)]
[M(080003)] ← [D0(16:23)]
[M(080005)] ← [D0(8:15)]
[M(080007)] ← [D0(0:7)]
```

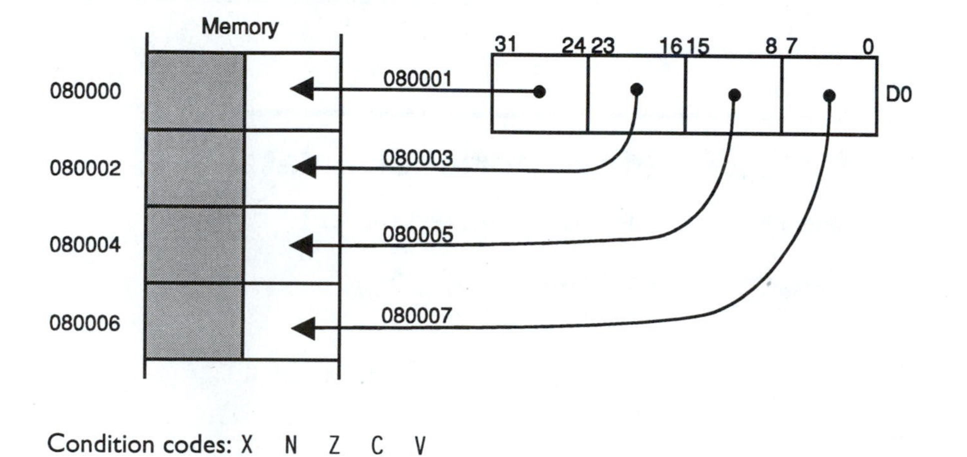

Condition codes: X N Z C V
- - - - -

MOVEQ Move quick (copy a small literal to a destination)

Operation: `[destination] ← <literal>`

Syntax: `MOVEQ #<data>,Dn`

Attributes: Size = longword

Description: Move the specified literal to a data register. The literal data is an eight-bit field within the `MOVEQ` op-code and specifies a signed value in the range -128 to +127. When the source operand is transferred, it is sign-extended to 32 bits. Consequently, although only 8 bits are moved, the `MOVEQ` instruction is a *longword* operation.

Application: `MOVEQ` is used to load small integers into a data register. Beware of its sign-extension. The two operations `MOVE.B #12,D0` and `MOVEQ #12,D0` are not equivalent. The former has the effect [D0(0:7)] ← 12, while the latter has the effect [D0(0:31)] ← 12 (with sign-extension).

Condition codes:

X	N	Z	V	C
-	*	*	0	0

MULS, MULU Signed multiply, unsigned multiply

Operation: [destination] ← [destination] * [source]

Syntax:

```
MULS <ea>,Dn
MULU <ea>,Dn
```

Attributes: Size = word (the product is a longword)

Description: Multiply the 16-bit destination operand by the 16-bit source operand and store the result in the destination. Both the source and destination are 16-bit word values and the destination result is a 32-bit longword. The product is therefore a correct product and is not truncated. `MULU` performs multiplication with unsigned values and `MULS` performs multiplication with two's complement values.

Application: `MULU D1,D2` multiplies the low-order words of data registers D1 and D2 and puts the 32-bit result in D2. `MULU #$1234,D3` multiplies the low-order word of D3 by the 16-bit literal $1234 and puts the 32-bit result in D3.

Condition codes:

X	N	Z	V	C
-	*	*	0	0

Source operand addressing modes

Dn	An	(An)	(An)+	-(An)	(d,An)	(d,An,Xi)	ABS.W	ABS.L	(d,PC)	(d,PC,Xn)	imm
✔		✔	✔	✔	✔	✔	✔	✔	✔	✔	✔

NBCD Negate decimal with sign extend

Operation: $[\text{destination}]_{10} \leftarrow 0 - [\text{destination}]_{10} - [X]$

Syntax: `NBCD <ea>`

Attributes: Size = byte

Description: The operand addressed as the destination and the extend bit in the CCR are subtracted from zero. The subtraction is performed using binary coded decimal (BCD) arithmetic. This instruction calculates the ten's complement of the destination if the X-bit is clear, and the nine's complement if X = 1. This is a byte-only operation. Negating a BCD number (with X = 0) has the effect of subtracting it from 100_{10}.

Condition codes:

```
X  N  Z  V  C
*  U  *  U  *
```

The Z-bit is cleared if the result is non-zero and is unchanged otherwise. The C-bit is set if a decimal borrow occurs. The X-bit is set to the same value as the C-bit.

Destination operand addressing modes

Dn	An	(An)	(An)+	-(An)	(d,An)	(d,An,Xi)	ABS.W	ABS.L	(d,PC)	(d,PC,Xn)	imm
✔		✔	✔	✔	✔	✔	✔	✔			

NEG Negate

Operation: `[destination] ← 0 - [destination]`

Syntax: `NEG <ea>`

Attributes: Size = byte, word, longword

Description: Subtract the destination operand from 0 and store the result in the destination location. The difference between `NOT` and `NEG` instructions is that `NOT` performs a bit-by-bit logical complementation, while a `NEG` performs a two's complement arithmetic subtraction. All bits of the condition code register are modified by a `NEG` operation. For example, if D3.B = 11100111_2, the logical operation `NEG.B D3` results in D3 = 00011001_2 (XNZVC=10001) and `NOT.B D3` = 00011000_2 (XNZVC=-0000).

Condition codes:

```
X  N  Z  V  C
*  *  *  *  *
```

Note that the X-bit is set to the value of the C-bit.

Destination operand addressing modes

Dn	An	(An)	(An)+	-(An)	(d,An)	(d,An,Xi)	ABS.W	ABS.L	(d,PC)	(d,PC,Xn)	imm
✔		✔	✔	✔	✔	✔	✔	✔			

NEGX Negate with extend

Operation: `[destination] ← 0 - [destination] - [X]`

Syntax: `NEGX <ea>`

Attributes: Size = byte, word, longword

Description: The operand addressed as the destination and the extend bit are subtracted from zero. NEGX is the same as NEG except that the X-bit is also subtracted from zero.

Condition codes:

X	N	Z	V	C
*	*	*	*	*

The Z-bit is cleared if the result is non-zero and is unchanged otherwise. The X-bit is set to the same value as the C-bit.

Destination operand addressing modes

Dn	An	(An)	(An)+	-(An)	(d,An)	(d,An,Xi)	ABS.W	ABS.L	(d,PC)	(d,PC,Xn)	imm
✔		✔	✔	✔	✔	✔	✔	✔			

NOP No operation

Operation: None

Syntax: `NOP`

Attributes: Unsized

Description: The no operation instruction, NOP performs no *computation*. Execution continues with the instruction following the NOP instruction. The processor's state is not modified by a NOP.

Application: NOPs can be used to introduce a *delay* in code. Some programmers use them to provide space for *patches* — two or more NOPs can later be replaced by branch or jump instructions to fix a bug. This use of the NOP is seriously frowned upon, as errors should be corrected by re-assembling the code rather than by patching it.

Condition codes:

X	N	Z	V	C
-	-	-	-	-

NOT Logical complement

Operation: `[destination] ← [destination]`

Syntax: `NOT <ea>`

Attributes: Size = byte, word, longword

Description: Calculate the logical complement of the destination and store the result in the destination. The difference between NOT and NEG is that NOT performs a bit-by-bit logical complementation, while a NEG performs a two's complement arithmetic subtraction. Moreover, NEG updates all bits of the CCR, while NOT clears the V- and C-bits, updates the N- and Z-bits, and doesn't affect the X-bit.

Condition codes:

X	N	Z	V	C
-	*	*	0	0

Source operand addressing modes

Dn	An	(An)	(An)+	-(An)	(d,An)	(d,An,Xi)	ABS.W	ABS.L	(d,PC)	(d,PC,Xn)	imm
✔		✔	✔	✔	✔	✔	✔	✔			

OR OR logical

Operation: `[destination] ← [source] + [destination]`

Syntax:

```
OR <ea>,Dn
OR Dn,<ea>
```

Attributes: Size = byte, word, longword

Description: OR the source operand to the destination operand, and store the result in the destination location.

Application: The OR instruction is used to set selected bits of the operand. For example, we can set the four most-significant bits of a longword operand in D0 by executing:

```
OR.L #$F0000000,D0
```

Condition codes:

X	N	Z	V	C
-	*	*	0	0

Source operand addressing modes

Dn	An	(An)	(An)+	-(An)	(d,An)	(d,An,Xi)	ABS.W	ABS.L	(d,PC)	(d,PC,Xn)	imm
✓		✓	✓	✓	✓	✓	✓	✓	✓	✓	✓

Destination operand addressing modes

Dn	An	(An)	(An)+	-(An)	(d,An)	(d,An,Xi)	ABS.W	ABS.L	(d,PC)	(d,PC,Xn)	imm
✓		✓	✓	✓	✓	✓	✓	✓			

ORI OR immediate

Operation: [destination] ← <literal> + [destination]

Syntax: ORI #<data>,<ea>

Attributes: Size = byte, word, longword

Description: OR the immediate data with the destination operand. Store the result in the destination operand.

Condition codes:

X	N	Z	V	C
-	*	*	0	0

Application: ORI forms the logical OR of the immediate source with the effective address, which may be a memory location. For example,

```
ORI.B #%00000011,(A0)+
```

Destination operand addressing modes

Dn	An	(An)	(An)+	-(An)	(d,An)	(d,An,Xi)	ABS.W	ABS.L	(d,PC)	(d,PC,Xn)	imm
✔		✔	✔	✔	✔	✔	✔	✔			

ORI to CCR Inclusive OR immediate to CCR

Operation: `[CCR] ← <literal> + [CCR]`

Syntax: `ORI #<data>,CCR`

Attributes: Size = byte

Description: OR the immediate data with the condition code register (i.e., the least-significant byte of the status register). For example, the Z flag of the CCR can be set by `ORI #$04,CCR`.

Condition codes:

```
X N Z V C
* * * * *
```

X is set if bit 4 of data = 1; unchanged otherwise
N is set if bit 3 of data = 1; unchanged otherwise
Z is set if bit 2 of data = 1; unchanged otherwise
V is set if bit 1 of data = 1; unchanged otherwise
C is set if bit 0 of data = 1; unchanged otherwise

ORI to SR Inclusive OR immediate to status register

Operation:

```
IF [S] = 1
   THEN
      [SR] ← <literal> + [SR]
   ELSE TRAP
```

Syntax: `ORI #<data>,SR`

Attributes: Size = word

Description: OR the immediate data to the status register and store the result in the status register. All bits of the status register are affected.

Application: Used to set bits in the SR (i.e., the S, T, and interrupt mask bits). For example, ORI #$8000,SR sets bit 15 of the SR (i.e., the trace bit).

Condition codes:

```
X  N  Z  V  C
*  *  *  *  *
```

X is set if bit 4 of data = 1; unchanged otherwise
N is set if bit 3 of data = 1; unchanged otherwise
Z is set if bit 2 of data = 1; unchanged otherwise
V is set if bit 1 of data = 1; unchanged otherwise
C is set if bit 0 of data = 1; unchanged otherwise

PEA Push effective address

Operation: [SP] ← [SP] - 4; [M([SP])] ← <ea>

Syntax: PEA <ea>

Attributes: Size = longword

Description: The longword effective address specified by the instruction is computed and pushed onto the stack. The difference between PEA and LEA is that LEA calculates an effective address and puts it in an address register, while PEA calculates an effective address in the same way but pushes it on the stack.

Application: PEA calculates an effective address to be used later in address register indirect addressing. In particular, it facilitates the writing of position independent code. For example, PEA (TABLE,PC) calculates the address of TABLE with respect to the PC and pushes it on the stack. This address can be read by a procedure and then used to access the data to which it points. Consider the example:

```
           PEA      Wednesday     Push the parameter address on the stack
           BSR      Subroutine    Call the procedure
           LEA      (4,SP),SP     Remove space occupied by the parameter
           .
Subroutine MOVEA.L  (4,SP),A0     A0 points to parameter under return address
           MOVE.W   (A0),D2       Access the actual parameter - Wednesday
           .
           RTS
```

Condition codes:

```
X  N  Z  V  C
-  -  -  -  -
```

Source operand addressing modes

Dn	An	(An)	(An)+	-(An)	(d,An)	(d,An,Xi)	ABS.W	ABS.L	(d,PC)	(d,PC,Xn)	imm
		✔			✔	✔	✔	✔	✔	✔	

RESET Reset external devices

Operation:
```
IF [S] = 1 THEN
            Assert RESET* line
        ELSE TRAP
```

Syntax: `RESET`

Attributes: Unsized

Description: The reset line is asserted, causing all external devices connected to the 68000's RESET* output to be reset. The RESET instruction is privileged and has no effect on the operation of the 68000 itself. This instruction is used to perform a programmed reset of all peripherals connected to the 68000's RESET* pin.

Condition codes:

X	N	Z	V	C
-	-	-	-	-

ROL, ROR Rotate left/right (without extend)

Operation: `[destination] ← [destination] rotated by <count>`

Syntax:
```
ROL Dx,Dy
ROR Dx,Dy
ROL #<data>,Dy
ROR #<data>,Dy
ROL <ea>
ROR <ea>
```

Attributes: Size = byte, word, longword

Description: Rotate the bits of the operand in the direction indicated. The extend bit, X, is not included in the operation. A rotate operation is circular in the sense that the bit shifted out at one end is shifted into the other end. That is, no bit is lost or destroyed by a rotate operation. The bit shifted out is also copied into the C-bit of the CCR, but not into the X-bit. The shift count may be specified in one of three ways: the count may be a literal, the contents of a data register, or the value 1. An immediate count permits a shift of 1 to 8 places. If the count is in a register, the value is modulo 64, allowing a range of 0 to 63. If no count is specified, the *word* at the effective address is rotated by one place (e.g., ROL <ea>).

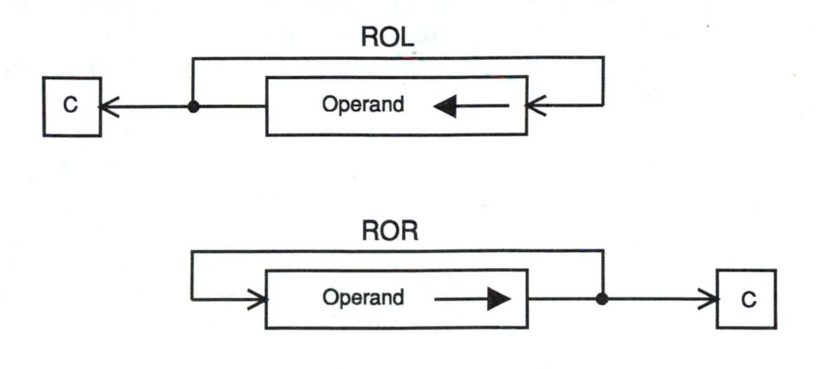

Condition codes:

```
X N Z V C
- * * 0 *
```

The X-bit is not affected and the C-bit is set to the last bit rotated out of the operand (C is set to zero if the shift count is 0).

Destination operand addressing modes

Dn	An	(An)	(An)+	-(An)	(d,An)	(d,An,Xi)	ABS.W	ABS.L	(d,PC)	(d,PC,Xn)	imm
✔		✔	✔	✔	✔	✔	✔	✔			

ROXL, ROXR Rotate left/right with extend

Operation: `[destination] ← [destination] rotated by <count>`

Syntax:

```
ROXL Dx,Dy
ROXR Dx,Dy
ROXL #<data>,Dy
ROXR #<data>,Dy
ROXL <ea>
ROXR <ea>
```

Attributes: Size = byte, word, longword

Description: Rotate the bits of the operand in the direction indicated. The extend bit of the CCR is included in the rotation. A rotate operation is circular in the sense that the bit shifted out at one end is shifted into the other end. That is, no bit is lost or destroyed by a rotate operation. Since the X-bit is included in the rotate, the rotation is performed over 9 bits (.B), 17 bits (.W), or 33 bits (.L). The bit shifted out is also copied into the C-bit of the CCR as well as the X-bit. The shift count may be specified in one of three ways: the count may be a literal, the contents of a data register, or the value 1. An immediate count permits a shift of 1 to 8 places. If the count is in a register, the value is modulo 64 and the range is from 0 to 63. If no count is specified, the word at the specified effective address is rotated by one place (i.e., ROXL <ea>).

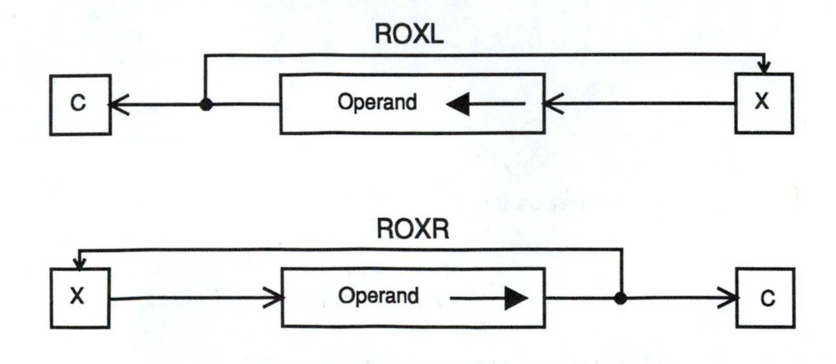

Condition codes:

```
X  N  Z  V  C
*  *  *  0  *
```

The X- and the C-bit are set to the last bit rotated out of the operand. If the rotate count is zero, the X-bit is unaffected and the C-bit is set to the X-bit.

Destination operand addressing modes

Dn	An	(An)	(An)+	-(An)	(d,An)	(d,An,Xi)	ABS.W	ABS.L	(d,PC)	(d,PC,Xn)	imm
✔		✔	✔	✔	✔	✔	✔	✔			

RTE Return from exception

Operation:

```
IF [S] = 1 THEN
      [SR] ← [M([SP])]; [SP] ← [SP] + 2
      [PC] ← [M([SP])]; [SP] ← [SP] + 4
    ELSE TRAP
```

Syntax: RTE

Attributes: Unsized

Description: The status register and program counter are pulled from the stack. The previous values of the SR and PC are lost. The RTE is used to terminate an exception handler. Note that the behavior of the RTE instruction depends on the nature of both the exception and processor type. The 68010 and later models push more information on the stack following an exception than the 68000. The processor determines how much to remove from the stack.

Condition codes:

```
X  N  Z  V  C
*  *  *  *  *
```

The CCR is restored to its pre-exception state.

RTR Return and restore condition codes

Operation:

```
[CCR] ← [M([SP])]; [SP] ← [SP] + 2
[PC] ← [M([SP])];  [SP] ← [SP] + 4
```

Syntax: RTR

Attributes: Unsized

Description: The condition code and program counter are pulled from the stack. The previous condition code and program counter are lost. The supervisor portion of the status register is not affected.

Application: If you wish to preserve the CCR after entering a procedure, you can push it on the stack and then retrieve it with RTR.

```
        BSR      Procl      Call the procedure
        .                   .
        .                   .
Procl   MOVE.W   SR,-(SP)   Save old CCR on stack
        .                   .
        .                   Body of procedure
        .                   .
        RTR                 Return and restore CCR (not SR!)
```

Condition codes:

```
X  N  Z  V  C
*  *  *  *  *
```

The CCR is restored to its pre-exception state.

RTS Return from subroutine

Operation: [PC] ← [M([SP])]; [SP] ← [SP] + 4

Syntax: RTS

Attributes: Unsized

Description: The program counter is pulled from the stack and the previous value of the PC is lost. RTS is used to terminate a subroutine.

Condition codes:

X	N	Z	V	C
-	-	-	-	-

SBCD Subtract decimal with extend

Operation: $[\text{destination}]_{10} \leftarrow [\text{destination}]_{10} - [\text{source}]_{10} - [X]$

Syntax:

```
SBCD Dy,Dx
SBCD -(Ay),-(Ax)
```

Attributes: Size = byte

Description: Subtract the source operand from the destination operand together with the X-bit, and store the result in the destination. Subtraction is performed using BCD arithmetic. The only legal addressing modes are data register direct and memory to memory with address register indirect using auto-decrementing.

Condition codes:

X	N	Z	V	C
*	U	*	U	*

Z: Cleared if result is non-zero. Unchanged otherwise. The Z-bit can be used to test for zero after a chain of multiple precision operations.

Scc Set according to condition cc

Operation: IF cc = 1 THEN [destination] ← 11111111_2
ELSE [destination] ← 00000000_2

Syntax: Scc <ea>

Attributes: Size = byte

Description: The specified condition code is tested. If the condition is true, the bits at the effective address are all set to one (i.e., $FF). Otherwise, the bits at the effective address are set to zeros (i.e., $00).

SCC	set on carry clear	$\overline{C}$
SCS	set on carry set	C
SEQ	set on equal	Z
SGE	set on greater than or equal	$N.V + \overline{N}.\overline{V}$
SGT	set on greater than	$N.V.\overline{Z} + \overline{N}.\overline{V}.\overline{Z}$
SHI	set on higher than	$\overline{C}.\overline{Z}$
SLE	set on less than or equal	$Z + N.\overline{V} + \overline{N}.V$
SLS	set on lower than or same	$C + Z$
SLT	set on less than	$N.\overline{V} + \overline{N}.V$
SMI	set on minus (i.e., negative)	N
SNE	set on not equal	$\overline{Z}$
SPL	set on plus (i.e., positive)	$\overline{N}$
SVC	set on overflow clear	$\overline{V}$
SVS	set on overflow set	V
SF	set on false (i.e., set never)	0
ST	set on true (i.e., set always)	1

Condition codes:

X	N	Z	V	C
-	-	-	-	-

Destination operand addressing modes

Dn	An	(An)	(An)+	-(An)	(d,An)	(d,An,Xi)	ABS.W	ABS.L	(d,PC)	(d,PC,Xn)	imm
✔		✔	✔	✔	✔	✔	✔	✔			

STOP Load status register and stop

Operation:

```
IF [S] = 1 THEN
           [SR] ← <data>
           STOP
        ELSE TRAP
```

Syntax: `STOP #<data>`

Sample syntax:

```
STOP #$2700
STOP #SetUp
```

Attributes: Unsized

Description: The immediate operand is copied into the entire status register (i.e., both status byte and CCR are modified), and the program counter advanced to point to the next instruction to be executed. The processor then suspends all further processing and halts. That is, the privileged STOP instruction stops the 68000.

The execution of instructions resumes when a trace, an interrupt, or a reset exception occurs. A trace exception will occur if the trace bit is set when the STOP instruction is encountered. If an interrupt request arrives whose priority is higher than the current processor priority, an interrupt exception occurs, otherwise the interrupt request has no effect. If the bit of the immediate data corresponding to the S-bit is clear (i.e., user mode selected), execution of the STOP instruction will cause a privilege violation. An external reset will always initiate reset exception processing.

Condition codes:

X	N	Z	V	C
*	*	*	*	*

Set according to the literal.

SUB Subtract binary

Operation: [destination] ← [destination] - [source]

Syntax:

```
SUB <ea>,Dn
SUB Dn,<ea>
```

Attributes: Size = byte, word, longword

Description: Subtract the source operand from the destination operand and store the result in the destination location.

Condition codes:

X	N	Z	V	C
*	*	*	*	*

Source operand addressing modes

Dn	An	(An)	(An)+	-(An)	(d,An)	(d,An,Xi)	ABS.W	ABS.L	(d,PC)	(d,PC,Xn)	imm
✔	✔	✔	✔	✔	✔	✔	✔	✔	✔	✔	✔

Destination operand addressing modes

Dn	An	(An)	(An)+	-(An)	(d,An)	(d,An,Xi)	ABS.W	ABS.L	(d,PC)	(d,PC,Xn)	imm
✔		✔	✔	✔	✔	✔	✔	✔			

SUBA Subtract address

Operation: [destination] ← [destination] - [source]

Syntax: SUBA <ea>,An

Attributes: Size = word, longword

Description: Subtract the source operand from the destination operand and store the result in the destination address register. Word operations are sign-extended to 32 bits prior to subtraction.

Condition codes:

X	N	Z	V	C
-	-	-	-	-

Source operand addressing modes

Dn	An	(An)	(An)+	-(An)	(d,An)	(d,An,Xi)	ABS.W	ABS.L	(d,PC)	(d,PC,Xn)	imm
✔	✔	✔	✔	✔	✔	✔	✔	✔	✔	✔	✔

SUBI Subtract immediate

Operation: [destination] ← [destination] - <literal>

Syntax: SUBI #<data>,<ea>

Attributes: Size = byte, word, longword

Description: Subtract the immediate data from the destination operand. Store the result in the destination operand.

Condition codes:

X	N	Z	V	C
*	*	*	*	*

Destination operand addressing modes

Dn	An	(An)	(An)+	-(An)	(d,An)	(d,An,Xi)	ABS.W	ABS.L	(d,PC)	(d,PC,Xn)	imm
✔		✔	✔	✔	✔	✔	✔	✔			

SUBQ Subtract quick

Operation: `[destination] ← [destination] - <literal>`

Syntax: `SUBQ #<data>,<ea>`

Attributes: Size = byte, word, longword

Description: Subtract the immediate data from the destination operand. The immediate data must be in the range 1 to 8. Word and longword operations on address registers do not affect condition codes. A word operation on an address register affects the entire 32-bit address.

Condition codes:

X	N	Z	V	C
*	*	*	*	*

Destination operand addressing modes

Dn	An	(An)	(An)+	-(An)	(d,An)	(d,An,Xi)	ABS.W	ABS.L	(d,PC)	(d,PC,Xn)	imm
✔	✔	✔	✔	✔	✔	✔	✔	✔			

SUBX Subtract extended

Operation: `[destination] ← [destination] - [source] - [X]`

Syntax:

```
SUBX Dx,Dy
SUBX -(Ax),-(Ay)
```

Attributes: Size = byte, word, longword

Description: Subtract the source operand from the destination operand along with the extend bit, and store the result in the destination location.

The only legal addressing modes are data register direct and memory to memory with address register indirect using auto-decrementing.

Condition codes:
```
X N Z V C
* * * * *
```
Z: Cleared if the result is non-zero, unchanged otherwise. The Z-bit can be used to test for zero after a chain of multiple precision operations.

SWAP Swap register halves

Operation: `[Register(16:31)] ← [Register(0:15)];`
`[Register(0:15)] ← [Register(16:31]`

Syntax: `SWAP Dn`

Attributes: Size = word

Description: Exchange the upper and lower 16-bit words of a data register.

Application: The `SWAP Dn` instruction enables the higher-order word in a register to take part in word operations by moving it into the lower-order position. `SWAP Dn` is effectively equivalent to `ROR.L Di,Dn`, where [Di] = 16. However, `SWAP` clears the C-bit of the CCR, whereas `ROR` sets it according to the last bit to be shifted into the carry bit.

Condition codes:
```
X N Z V C
- * * 0 0
```
The N-bit is set if most-significant bit of the 32-bit result is set and cleared otherwise. The Z-bit is set if 32-bit result is zero and cleared otherwise.

TAS Test and set an operand

Operation: `[CCR] ← tested([operand]); [destination(7)] ← 1`

Syntax: `TAS <ea>`

Attributes: Size = byte

Description: Test and set the byte operand addressed by the effective address field. The N- and Z-bits of the CCR are updated accordingly. The high-order bit of the operand (i.e., bit 7) is set. This operation is *indivisible* and uses a read-modify-write cycle. Its principal application is in multiprocessor systems.

Application: The TAS instruction permits one processor in a multiprocessor system to test a resource (e.g., shared memory) and claim the resource if it is free. The most-significant bit of the byte at the effective address is used as a semaphore to indicate whether the shared resource is free. The TAS instruction reads the semaphore bit to find the state of the resource, and then sets the semaphore to claim the resource (if it was free). Because the operation is indivisible, no other processor can access the memory between the testing of the bit and its subsequent setting.

Condition codes:

X	N	Z	V	C
-	*	*	0	0

Source operand addressing modes

Dn	An	(An)	(An)+	-(An)	(d,An)	(d,An,Xi)	ABS.W	ABS.L	(d,PC)	(d,PC,Xn)	imm
✔		✔	✔	✔	✔	✔	✔	✔			

TRAP Trap

Operation:

```
S ← 1;
[SSP] ← [SSP] - 4; [M([SSP])] ← [PC];
[SSP] ← [SSP] - 2; [M([SSP])] ← [SR];
[PC] ← vector
```

Syntax: TRAP #<vector>

Attributes: Unsized

Description: This instruction forces the processor to initiate exception processing. The vector number used by the TRAP instruction is in the range 0 to 15 and, therefore, supports 16 traps (i.e., TRAP #0 to TRAP #15).

Application: The TRAP instruction is used to perform operating system calls and is system independent. That is, the effect of the call depends on the particular operating environment. For example, the University of Teesside 68000 simulator uses TRAP #15 to perform

I/O. The ASCII character in D1.B is displayed by the following sequence.

```
MOVE.B  #6,D0  Set up the display a character parameter in D0
TRAP    #15    Now call the operating system
```

Condition codes:
```
X  N  Z  V  C
-  -  -  -  -
```

TRAPV Trap on overflow

Operation:
```
IF V = 1  THEN:
          [SSP] ← [SSP] - 4;  [M([SSP])] ← [PC];
          [SSP] ← [SSP] - 2;  [M([SSP])] ← [SR];
          [PC] ← [M($01C)]
          ELSE no action
```

Syntax: `TRAPV`

Attributes: Unsized

Description: If the V-bit in the CCR is set, then initiate exception processing. The exception vector is located at address $01C_{16}$. This instruction is used in arithmetic operations to call the operating system if overflow occurs.

Condition codes:
```
X  N  Z  V  C
-  -  -  -  -
```

TST Test an operand

Operation: `[CCR] ← tested([operand])`
i.e., `[operand] - 0; update CCR`

Syntax: `TST <ea>`

Attributes: Size = byte, word, longword

Description: The operand is compared with zero. No result is saved, but the contents of the CCR are set according to the result. The effect of `TST <ea>` is the same as `CMPI #0,<ea>` except that the `CMPI` instruction also sets/clears the V- and C-bits of the CCR.

Condition codes:

X	N	Z	V	C
-	*	*	0	0

Source operand addressing modes

Dn	An	(An)	(An)+	-(An)	(d,An)	(d,An,Xi)	ABS.W	ABS.L	(d,PC)	(d,PC,Xn)	imm
✔		✔	✔	✔	✔	✔	✔	✔	✔	✔	

UNLK Unlink

Operation: `[SP] ← [An]; [An] ← [M([SP])]; [SP] ← [SP] + 4`

Syntax: `UNLK An`

Attributes: Unsized

Description: The stack pointer is loaded from the specified address register and the old contents of the stack pointer are lost (this has the effect of collapsing the stack frame). The address register is then loaded with the longword pulled off the stack.

Application: The `UNLK` instruction is used in conjunction with the `LINK` instruction. The `LINK` creates a stack frame at the start of a procedure, and the `UNLK` collapses the stack frame prior to a return from the procedure.

Condition codes:

X	N	Z	V	C
-	-	-	-	-

APPENDIX C

Selected 68020 Instructions

In this appendix we describe some of the 68020's instructions. Instructions that have not been covered elsewhere in this text have been omitted — for example, the CALLM instruction and the CAS instruction. In fact, the CALLM instruction is not supported by the 68030 and later processors. The CAS instruction has been omitted because it is used to support multiprocessor systems.

BFCHG Test a bit field and change it

Operation: <bit field> OF <destination> ← $\overline{\langle\text{bit field}\rangle\ \text{OF}\ \langle\text{destination}\rangle}$

Syntax: `BFCHG <ea>{offset:width}`

Sample syntax:

```
BFCHG (A0){5:6}
BFCHG Table_3{D0:D5}
```

Attributes: Unsized

Description: The condition code is set according to the state of the bit field at the selected destination address. The contents of the bit field are then inverted (i.e., complemented). The effective address points to the base byte of the instruction and the offset determines the location of the bit field with respect to the base byte. The offset may either be a literal in the range 0 to 31 or the contents of a data register in the range -2^{31} to $+2^{31}-1$. The field width determines the width of the bit field and may be in the range 1 to 32. Note that the width is from 1 to 32 and not from 0 to 31, because a width of zero is meaningless. The offset and width fields may be literals or data registers.

Condition codes:

X	N	Z	V	C
-	*	*	0	0

The N-bit is set if the most-significant bit of the bit field is set and cleared otherwise. The Z-bit is set if all bits of the bit field are zero and cleared otherwise.

Source operand addressing

Dn	An	(An)	(An)+	-(An)	(d,An)	(d,An,Xi)	ABS.W	ABS.L	(d,PC)	(d,PC,Xn)	imm
✔		✔			✔	✔	✔	✔			

([bd,An,Xn],od)	([bd,An],Xn,od)	([bd,PC,Xn],od)	([bd,PC],Xn,od)
✔	✔		

BFCLR Test a bit field and clear it

Operation: `<bit field> OF <destination> ← 0`

Syntax: `BFCLR <ea>{offset:width}`

Sample syntax:

```
BFCLR Vector{5:D7}
BFCLR (12,A3){D2:D3}
```

Attributes: Unsized

Description: The condition code is set according to the state of the bit field at the selected destination address. The contents of the bit field are then cleared (i.e., set to zero).

Condition codes:

```
X  N  Z  V  C
-  *  *  0  0
```

The N-bit is set if the most-significant bit of the bit field is set and cleared otherwise. The Z-bit is set if all bits of the bit field are zero and cleared otherwise.

Source operand addressing

Dn	An	(An)	(An)+	-(An)	(d,An)	(d,An,Xi)	ABS.W	ABS.L	(d,PC)	(d,PC,Xn)	imm
✔		✔			✔	✔	✔	✔			

([bd,An,Xn],od)	([bd,An],Xn,od)	([bd,PC,Xn],od)	([bd,PC],Xn,od)
✔	✔		

BFEXTS Extract a bit field signed

Operation: [Dn] ← <bit field> OF <source>

Syntax: BFEXTS <ea>{offset:width},Dn

Attributes: Unsized

Description: Extract a bit field from memory, sign-extend it to 32 bits, and put it in the destination data register.

Condition codes:

X	N	Z	V	C
-	*	*	0	0

The N-bit is set if the most-significant bit of the bit field is set and cleared otherwise. The Z-bit is set if all bits of the bit field are zero and cleared otherwise.

Source operand addressing

Dn	An	(An)	(An)+	-(An)	(d,An)	(d,An,Xi)	ABS.W	ABS.L	(d,PC)	(d,PC,Xn)	imm
✔		✔			✔	✔	✔	✔	✔	✔	

([bd,An,Xn],od)	([bd,An],Xn,od)	([bd,PC,Xn],od)	([bd,PC],Xn,od)
✔	✔	✔	✔

BFEXTU Extract a bit field unsigned

Operation: [Dn] ← <bit field> OF <source>

Syntax: BFEXTU <ea>{offset:width},Dn

Sample syntax:

```
BFEXTU $1234{12:15}
BFEXTU (12,A3){D2:D3}
```

Attributes: Unsized

Description: Extract a bit field from memory and put it in the destination data register. The bit field is zero-extended to 32 bits before it is loaded into Dn.

Condition codes:

```
X  N  Z  V  C
-  *  *  0  0
```

The N-bit is set if the most-significant bit of the bit field is set and cleared otherwise. The Z-bit is set if all bits of the bit field are zero and cleared otherwise.

Source operand addressing

Dn	An	(An)	(An)+	-(An)	(d,An)	(d,An,Xi)	ABS.W	ABS.L	(d,PC)	(d,PC,Xn)	imm
✔		✔			✔	✔	✔	✔	✔	✔	

([bd,An,Xn],od)	([bd,An],Xn,od)	([bd,PC,Xn],od)	([bd,PC],Xn,od)
✔	✔	✔	✔

BFFFO Find first one in a bit field

Operation: `[Dn] ← <bit offset> OF source bit scan`

Syntax: `BFFFO <ea>{offset:width},Dn`

Sample syntax: `BFFFO Sector{D2:16},Dn`

Attributes: Unsized

Description: The `BFFFO` instruction searches the source operand bit field for the most-significant bit that is set to a one. That is, it locates the leading one. The bit offset of that bit (the bit offset in the instruction plus the offset of the first one bit) is placed in data register Dn. If no bit in the bit field is set to one, the value in Dn is set to the field offset plus the field width. The condition codes are set according to the bit field value. The following diagram demonstrates the effect of `BFFFO $1000{10:8},D0`. Note that data register D0 is loaded with 10 (the offset) + 4 (location of first 1) = 14.

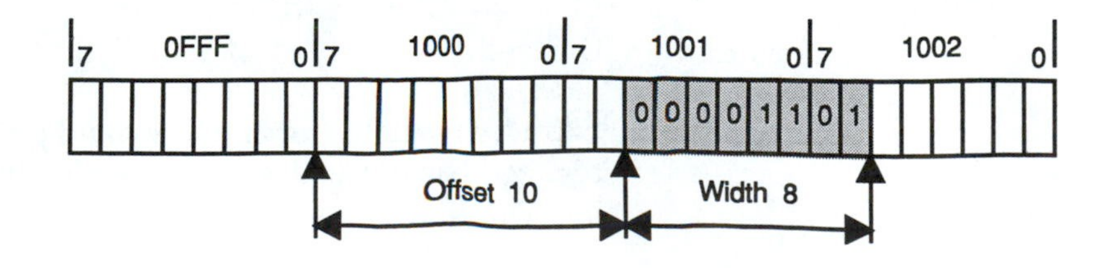

Condition codes: X N Z V C
- * * 0 0

The N-bit is set if the most-significant bit of the bit field is set and cleared otherwise. The Z-bit is set if all bits of the bit field are zero and cleared otherwise.

Source operand addressing

Dn	An	(An)	(An)+	-(An)	(d,An)	(d,An,Xi)	ABS.W	ABS.L	(d,PC)	(d,PC,Xn)	imm
✔		✔			✔	✔	✔	✔	✔	✔	

([bd,An,Xn],od)	([bd,An],Xn,od)	([bd,PC,Xn],od)	([bd,PC],Xn,od)
✔	✔	✔	✔

BFINS Insert a bit field

Operation: <bit field> OF <destination> ← [Dn]

Syntax: `BFINS Dn,<ea>{offset:width}`

Sample syntax:
```
BFINS D4,(A0){23:D4}
BFINS D0,([Date,A0,D7],Time){23:D4}
```

Attributes: Unsized

Description: Take the bit field in the low-order bits of data register Dn and insert the bit field in memory at the specified effective address. This instruction is the bit field equivalent of `MOVE Dn,<ea>`.

As in the case of all bit field instructions, the offset may be the 32-bit contents of a data register or a literal in the range 0 to 31. The field width is a value in the range 1 to 32 and may be a literal or the contents of a data register.

Condition codes: X N Z V C
- * * 0 0

The N-bit is set if the most-significant bit of the bit field is set and cleared otherwise. The Z-bit is set if all bits of the bit field are zero and cleared otherwise.

Source operand addressing

Dn	An	(An)	(An)+	-(An)	(d,An)	(d,An,Xi)	ABS.W	ABS.L	(d,PC)	(d,PC,Xn)	imm
✔		✔			✔	✔	✔	✔			

([bd,An,Xn],od)	([bd,An],Xn,od)	([bd,PC,Xn],od)	([bd,PC],Xn,od)
✔	✔		

BFSET Test a bit field and set it

Operation: `<bit field> OF <destination> ← 1s`

Syntax: `BFSET <ea>{offset:width}`

Attributes: Unsized

Description: The condition code is set according to the state of the bit field at the selected destination address. The contents of the bit field are then set to all ones.

Condition codes:

```
X  N  Z  V  C
-  *  *  0  0
```

The N-bit is set if the most-significant bit of the bit field is set and cleared otherwise. The Z-bit is set if all bits of the bit field are zero and cleared otherwise.

Source operand addressing

Dn	An	(An)	(An)+	-(An)	(d,An)	(d,An,Xi)	ABS.W	ABS.L	(d,PC)	(d,PC,Xn)	imm
✔		✔			✔	✔	✔	✔			

([bd,An,Xn],od)	([bd,An],Xn,od)	([bd,PC,Xn],od)	([bd,PC],Xn,od)
✔	✔		

BFTST Test a bit field

Operation: `Test <bit field> OF <destination>`

Syntax: `BFTST <ea>{offset:width}`

Attributes: Unsized

Description: The condition code is set according to the state of the bit field at the selected destination address.

Condition codes:

```
X  N  Z  V  C
-  *  *  0  0
```

The N-bit is set if the most-significant bit of the bit field is set and cleared otherwise. The Z-bit is set if all bits of the bit field are zero and cleared otherwise.

Source operand addressing

Dn	An	(An)	(An)+	-(An)	(d,An)	(d,An,Xi)	ABS.W	ABS.L	(d,PC)	(d,PC,Xn)	imm
✔		✔			✔	✔	✔	✔	✔	✔	

([bd,An,Xn],od)	([bd,An],Xn,od)	([bd,PC,Xn],od)	([bd,PC],Xn,od)
✔	✔	✔	✔

BKPT Breakpoint

Operation:

```
Run breakpoint acknowledgment cycle
IF acknowledged
      THEN execute returned operation word
      ELSE TRAP as illegal instruction
```

Syntax: `BKPT #<data>`

Attributes: Unsized

Description: The breakpoint instruction is an aid to hardware debugging. It executes a breakpoint acknowledge cycle with the immediate data (in the range 0 to 7) on bits 2 to 4 of the address bus (bits A_{00} and

A_{01} are set to zero). The breakpoint acknowledge cycle accesses CPU space (i.e., FC0 to FC2 = 1,1,1). If external hardware terminates the bus cycle normally, the data on the bus (i.e., an instruction word) is executed after the breakpoint instruction. If the external logic terminates the breakpoint acknowledge cycle with a bus error, the processor takes an illegal instruction exception (not a bus error exception).

Application: The breakpoint instruction is used to support debug monitors and real-time hardware emulators.

Condition codes:

X	N	Z	V	C
-	-	-	-	-

CHK2 Check register against bounds

Operation: `IF [Rn] < lower bound OR [Rn] > upper bound THEN TRAP`

Syntax: `CHK2 <ea>,Rn`

Sample syntax:

```
CHK2.L Bounds,A4
CHK2.W (A0),D3
```

Attributes: Size = byte, word, longword

Description: The contents of register Rn are compared to each bound (Rn may be an address or a data register). The effective address contains the bounds pair: the lower bound is followed by the upper bound. For signed comparisons, the arithmetically smaller value should be used as the lower bound. For unsigned comparisons, the logically smaller value should be the lower bound.

The size of the data bounds can be specified as a byte, a word, or a longword. If Rn is a data register and the operation size is byte or word, only the appropriate low-order part of Rn is checked. If Rn is an address register and the operand size is a *byte* or a word, the bounds operands are sign-extended to 32 bits and the resultant operands are compared to the full 32 bits in An.

If the upper bound equals the lower bound, the valid range is a single value. If the register value is less than the lower bound or greater than the upper bound, a `CHK` instruction exception (vector number 6) occurs.

Application: The `CHK` instruction can be used to test the bounds of an array element before it is used. By performing this test, you can make certain that you do not access an element outside an array. For example, CHK2.L (A0),D7 has the effect of comparing the longword in D7 with the lower bound at the address pointed at by A0, and with the upper bound at the address pointed at by A0+4.

Condition codes:

X	N	Z	V	C
-	U	*	U	*

Z: set if [Rn] is equal to either bound, cleared otherwise. C: set if [Rn] is out of bounds, cleared otherwise.

Source operand addressing

Dn	An	(An)	(An)+	-(An)	(d,An)	(d,An,Xi)	ABS.W	ABS.L	(d,PC)	(d,PC,Xn)	imm
		✔			✔	✔	✔	✔	✔	✔	

([bd,An,Xn],od)	([bd,An],Xn,od)	([bd,PC,Xn],od)	([bd,PC],Xn,od)
✔	✔	✔	✔

CMP2 Compare register against bounds

Operation: Compare [Rn] with lower and upper bounds and set condition codes

Syntax: CMP2 <ea>,Rn

Attributes: Size = byte, word, longword

Description: The contents of register Rn are compared to each bound. The effective address provides the location of the upper and lower bounds: the lower bound is followed by the upper bound. For signed comparisons, the arithmetically smaller value should be used as the lower bound. For unsigned comparisons, the logically smaller value should be the lower bound. For example, if the bounds pair is expressed as $FFFE, $0006 the bounds are -2 to 6. However, if the bounds pair is $0006, $FFFE the bounds are 6 to 65,534. The CMP2 instruction is identical to the CH2K, except that a trap is never called after the comparison.

The size of the data bounds can be specified as byte, word, or longword. If Rn is a data register and the operation size is .B or .W, only the appropriate low-order part of Rn is checked. If Rn is an address register and the operand size is a byte or a word, the bounds operands are sign-extended to 32 bits and the resultant operands are compared to the full 32 bits in An. If the upper bound equals the lower bound, the valid range is a single value.

Condition codes:

X	N	Z	V	C
-	U	*	U	*

Z: set if [Rn] is equal to either bound, cleared otherwise. C: set if [Rn] is out of bounds, cleared otherwise.

Source operand addressing

Dn	An	(An)	(An)+	-(An)	(d,An)	(d,An,Xi)	ABS.W	ABS.L	(d,PC)	(d,PC,Xn)	imm
		✔			✔	✔	✔	✔	✔	✔	

([bd,An,Xn],od)	([bd,An],Xn,od)	([bd,PC,Xn],od)	([bd,PC],Xn,od)
✔	✔	✔	✔

DIVS, DIVSL Signed divide

Operation: [Destination] ← [Destination]/[Source]

Syntax:

```
DIVS.W   <ea>,Dn      32/16, 16r:16q
DIVS.L   <ea>,Dq      32/32, 32q
DIVS.L   <ea>,Dr:Dq   64/32, 32r:32q
DIVSL.L  <ea>,Dr:Dq   32/32, 32r:32q
```

Sample syntax:

```
DIVS.W   (A0)+,D2     D2(0:15) quotient,
                      D2(16:31) remainder
DIVS.L   (A1),D4      D4(0:31) quotient, no remainder
DIVS.L   D0,D1:D2     D2(0:31) quotient,
                      D1(0:31) remainder
DIVSL.L  D0,D1:D2     D2(0:31) quotient,
                      D1(0:31) remainder
```

Attributes: Size = word, longword

Description: Divide the signed destination operand by the signed source operand and store the result in the destination. This instruction has three forms. The division may be 32 bits by 16 bits, 32 bits by 32 bits, or 64 bits by 32 bits. The result consists of a quotient and a remainder. The `DIVS.W` form stores the quotient in the low-order word of the destination and the remainder in the high-order word. All other cases store the quotient and remainder in separate registers (except `DIVS.L <ea>,Dq` which does not record the remainder). An attempt to divide by zero results in divide-by-zero exception. The `DIVU` and `DIVUL` instructions perform the same operations on unsigned values. Note that in `DIVS.L <ea>,Dr:Dq`, register Dr specifies the destination for the remainder and the high-order byte of the source, and Dq specifies the destination for the quotient and the low-order byte of the source.

Condition codes:

X	N	Z	V	C
-	*	*	*	0

Source operand addressing

Dn	An	(An)	(An)+	-(An)	(d,An)	(d,An,Xi)	ABS.W	ABS.L	(d,PC)	(d,PC,Xn)	imm
✔		✔	✔	✔	✔	✔	✔	✔	✔	✔	✔

([bd,An,Xn],od)	([bd,An],Xn,od)	([bd,PC,Xn],od)	([bd,PC],Xn,od)
✔	✔	✔	✔

MOVE from CCR — Copy data from CCR to destination

Operation: `[destination] ← [CCR]`

Syntax: `MOVE CCR,<ea>`

Attributes: Size = word

Description: Move the contents of the condition code register to the destination location. The source operand is a word, but only the low-order byte contains the condition codes. The upper byte is all zeros.

Application: The `MOVE from CCR` instruction enables you to read the bits of the CCR without using a conditional branch or a bit test instruction. For example, you might want to preserve the current value of the CCR for testing later.

```
SaveCCR     MOVE.W  CCR,D7      Save CCR in D7
            .
            .                   Do operations that change CC
            .
            MOVE.W  D7,CCR      Restore CCR
            BEQ     Test1       Now do a test on the old CCR
            BMI     Test2
```

Condition codes:

X	N	Z	V	C
-	-	-	-	-

Destination operand addressing

Dn	An	(An)	(An)+	-(An)	(d,An)	(d,An,Xi)	ABS.W	ABS.L	(d,PC)	(d,PC,Xn)	imm
✔		✔	✔	✔	✔	✔	✔	✔			

([bd,An,Xn],od)	([bd,An],Xn,od)	([bd,PC,Xn],od)	([bd,PC],Xn,od)
✔	✔		

MOVEC Move to or from control register

Operation:

```
IF [S] = 1
    THEN [Rx] ← [Ry]
    ELSE TRAP
```

Rx is a control register and Ry a general register, or vice versa.

Syntax:

```
MOVEC Rc,Rn
MOVEC Rn,Rc
```

Sample syntax:

```
MOVEC D0,SFC
MOVEC DFC,D2
```

Attributes: Size = longword

Description: Copy the contents of the specified control register to the specified general register (i.e., D0 to D7 or A0 to A7), or vice versa. The transfer is always a 32-bit operation, even though the control register may be implemented with fewer bits. Unimplemented bits are treated as zeros. This is a privileged instruction.

Application: The `MOVEC` instruction permits control registers of the 68010 and later processors to be loaded or read. The control registers of the 68020 are: SFC, DFC, CACR, USP, VBR, CAAR, MSP, and ISP.

Condition codes:

X	N	Z	V	C
-	-	-	-	-

MOVES Move to or from address space

Operation 1:
```
MOVES Rn,<ea>        IF [S] = 1
                        THEN [destination] ← [Rn]
                        ELSE TRAP
```

Operation 2:
```
MOVES <ea>,Rn        IF [S] = 1
                        THEN [Rn] ← [source]
                        ELSE TRAP
```

Syntax:
```
MOVES Rn,<ea>
MOVES <ea>,Rn
```

Sample syntax:
```
MOVES D4,(A6)+
MOVES Table_6,A4
```

Attributes: Size = byte, word, longword

Description: The operation `MOVES Rn,<ea>` moves the byte, word, or longword operand in the specified general register (D0 to D7, A0 to A7) to a location within the address space specified by the destination function code register (DFC). The operation `MOVES <ea>,Rn` performs the inverse operation by moving an operand from a location within the address space specified by the source function code register, SFC. If the destination is a data register, the source operand replaces the corresponding low-order bits of the data register. If the destination is an address register, the source operand is sign-extended to 32 bits and then loaded into the address register.

Application: The function of the `MOVES` instruction is simple — it is used by the operating system running in the supervisor mode to access any of the 68020's possible address spaces. That is, it overrides the current address space. The `MOVES` instruction is necessary in systems with memory management units that do not permit an instruction in supervisor space to access an operand user space. Note that it is necessary to set up the SFC and DFC registers (by means of a `MOVEC`) before the `MOVES` can be used. For example, consider an access by the operating system (running in supervisor space) to data in a user program.

```
MOVEQ    #%001,D0   Function code = user data space
MOVEC    D0,SFC     Source operand lies in user space
MOVEC    D0,DFC     Destination operand is also in user space
MOVES.W (A0),D2     Read byte from user space
.                       .
.                   Process it
.                       .
MOVES.W D2,(A0)     Restore it to user space
```

Condition codes:

X	N	Z	V	C
-	-	-	-	-

Effective address modes

Dn	An	(An)	(An)+	-(An)	(d,An)	(d,An,Xi)	ABS.W	ABS.L	(d,PC)	(d,PC,Xn)	imm
		✔	✔	✔	✔	✔	✔	✔			

([bd,An,Xn],od)	([bd,An],Xn,od)	([bd,PC,Xn],od)	([bd,PC],Xn,od)
✔	✔		

MULS Signed multiply

Operation: [Destination] ← [Destination] x [Source]

Syntax:

```
MULS.W  <ea>,Dn         16x16, 32-bit product
MULS.L  <ea>,Dl         32x32, 32-bit product
MULS.L  <ea>,Dh:Dl      32x32, 64-bit product
```

Sample syntax:

```
MULS.W  (A0)+,D2        D2(0:31) product
MULS.L  (A1),D4         D4(0:31) product
MULS.L  D0,D1:D2        D1(0:31),D2(0:31) product
```

Attributes: Size = word, longword

Description: Multiply the signed destination operand by the signed source operand and store the result in the destination. This instruction has three forms. The multiplication may be 16 bits by 16 bits, 32 bits by 32 bits with a 32-bit product, or 32 bits by 32 bits with a 64-bit product. The result consists of a 32-bit product or a 64-bit product. The `MULS.L <ea>,Dh,Dl` form multiplies the 32-bit source operand at the effective address by the 32-bit contents of `Dl` and puts the 64-bit product in the register pair `Dh:Dl`, with the most-significant 32 bits in `Dh`. The `MULU` instruction performs the same operations on unsigned values.

Condition codes:

X	N	Z	V	C
-	*	*	*	0

Note that overflow can occur only when multiplying two 32-bit operands to yield a 32-bit result.

Source operand addressing

Dn	An	(An)	(An)+	-(An)	(d,An)	(d,An,Xi)	ABS.W	ABS.L	(d,PC)	(d,PC,Xn)	imm
✔		✔	✔	✔	✔	✔	✔	✔	✔	✔	✔

([bd,An,Xn],od)	([bd,An],Xn,od)	([bd,PC,Xn],od)	([bd,PC],Xn,od)
✔	✔	✔	✔

PACK Pack BCD

Operation: Add adjustment to unpacked BCD and move to destination.

Syntax:

```
PACK  -(Ax),-(Ay),#<adjustment>
PACK  Dx,Dy,#<adjustment>
```

Attributes: Unsized

Description: The contents of the source operand are first added to the adjustment. Bits 11:8 of this result are moved into bits 7:4 of the destination, and bits 7:0 of the result are moved into bits 3:0 of the destination. This instruction is used to take two ASCII bytes and convert them to a single BCD character.

Application: If [D0.W] = $3639 (i.e., the ASCII characters for 6 and 9), the operation `PACK D0,D1,#0` loads data register D1.B with $69.

RTD Return and deallocate parameters

Operation: `[PC] ← [M([SP])]; [SP] ← [SP] + 4; [SP] ← [SP] + d`

Syntax: `RTD #<displacement>`

Attributes: Word

Description: The program counter is pulled from the stack and the previous value of the PC is lost. After the program counter has been read from the stack, the 16-bit displacement, d, is sign-extended to 32 bits and added to the stack pointer. This instruction is similar to an RTS, except that the stack pointer is also adjusted.

Application: An RTD instruction moves the stack pointer down by d bytes, and is therefore used to deallocate any parameters that were on the stack while the procedure was active.

```
MOVE.L D0,-(SP)        Push D0 on the stack
MOVE.W D1,-(SP)        Push D1 on the stack
BSR    TEST            Call procedure TEST
       .
       .
       .
TEST   .               Body of TEST
       .
       .
RTD    #6              Return and deallocate
```

Note that the deallocation is six because data register D0 takes up 4 bytes on the stack and data register D1 takes up 2 bytes.

Condition codes:

X	N	Z	V	C
-	-	-	-	-

UNPK Unpack BCD

Operation: Add adjustment to a packed BCD, unpack, and move to destination.

Syntax:

```
UNPK  -(Ax),-(Ay),#<adjustment>
UNPK  Dx,Dy,#<adjustment>
```

Attributes: Unsized

Description: The two BCD digits in the low-order byte of the source operand are unpacked and placed in the low-order two bytes of the destination operand. The literal, adjustment, is added to the unpacked value.

Application: If the low-order byte in D0 is $45 (i.e., the BCD value 45), the operation `UNPK D0,D1,#$3030` loads data register D1.W with $3435. That is, this instruction converts the two-digit BCD value 45 into the two ASCII characters '4' and '5'.

The `PACK` and `UNPK` instructions make it possible to read in BCD as ASCII characters, convert from ASCII to BCD, and then take the result of any BCD calculations and convert the result back to ASCII for printing.

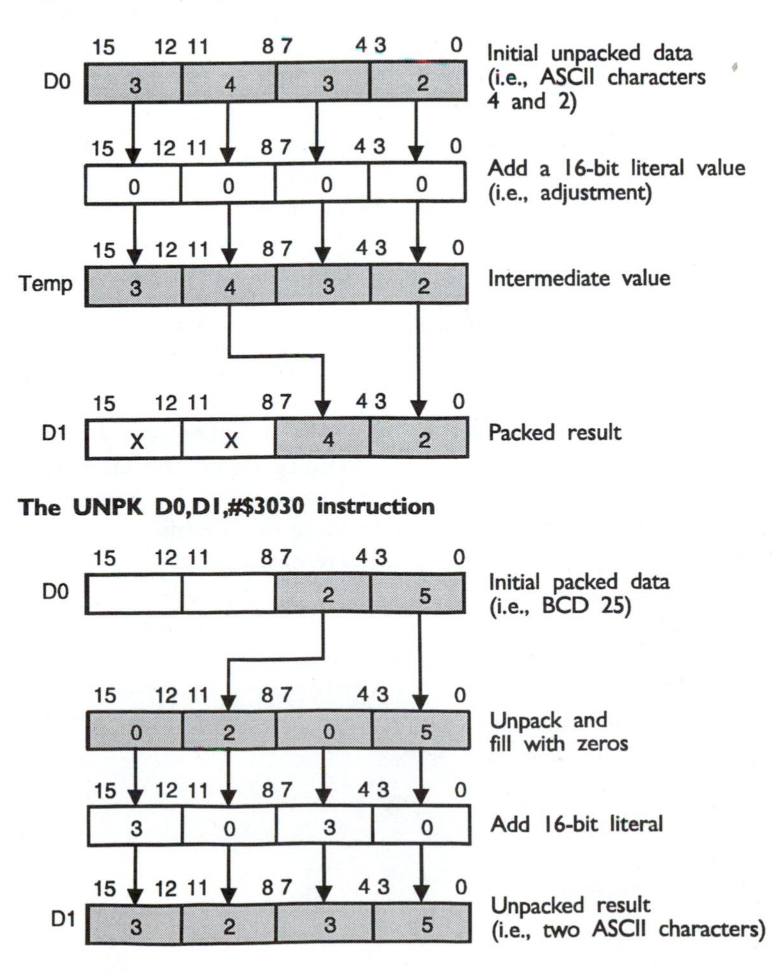

APPENDIX D

The 68000 Cross-assembler and Simulator

This appendix describes the University of Teesside 68000 cross-assembler and 68000 simulator, written by Paul Lambert and Eric Pearson. We first cover the cross-assembler and then the 68000 simulator. The cross-assembler takes a 68000 ASCII source file and produces a binary 68000 code file. The 68000 simulator takes this binary file and executes it by simulating a 68000 microprocessor and its operating system environment. Together, these two pieces of software permit you to develop 68000 programs on a PC.

Using the cross-assembler and simulator is straightforward. You first create a source file (i.e., the assembly language program) with any ASCII text editor. An ASCII editor keeps text in ASCII form and doesn't use special control characters to format the text. For example, you can use Microsoft's EDIT or Window's Notepad to create a source file, but not WordPerfect or Word For Windows. The source file is then cross-assembled with the X68K.EXE cross-assembler to produce a binary file (i.e., the 68000 code to be executed) and an optional listing file. The listing file contains the source program together with the assembled code expressed in hexadecimal format together with any error messages. If errors are found during the assembly phase, you have to return to the source file and edit it to correct the errors.

When the program has been correctly assembled, the simulator, E68K.EXE, can be run using the binary file as an input. If the run is successful and the program behaves as expected, all is well. If the program doesn't run as expected, the facilities of the simulator can be used to locate the source of the problem. Then you have to return to the source file, edit it, assemble it, and run the simulator again. This entire process is repeated until either the program works or you give up and go home.

Debugging a program in this way involves two steps. The first step is to produce a *syntactically* correct program that contains no errors. That is, the instructions and assembler directives all conform to the standard required by the assembler. This step detects typographic errors, illegal instructions and addressing modes, missing subroutines, and the use of the same label more than once. Syntax errors are best corrected by taking a hard copy of the listing and editing the source file until they have been removed.

Once the program is syntax-error free, the simulator can be used to detect *semantic* errors. These are errors of logic in the program *design* (e.g., using an `ADD.B 2,D0` instruction when you intended to write an `ADD.B #2,D0` instruction). By the way, once the program runs in the way you expect, the debugging process has not ended. Bugs can still surface long after a program has passed all its initial testing.

Using the Cross-assembler

The Teesside 68000 cross-assembler runs on any IBM PC or compatible clone. It takes as its input an ASCII-encoded text file and produces a binary file. The source file must have the extension '.X68' in order for the cross-assembler to *recognize* it (e.g., TEST.X68). The binary file generated by the cross-assembler has the same name as the source file but with the extension '.BIN'. This binary code will run on an appropriate 68000 microprocessor system.

The cross-assembler is run by entering the command: X68K <file name>. For example, to cross-assemble TEST.X68, you enter X68K TEST. You don't have to include the full name of the file and can omit the .X68 extension. In what follows, we use a simple program called TTL.X68 to demonstrate the operation of the cross-assembler. If we wish to produce a listing file as well as a binary file, we append the suffix -L to the command and enter the line:

```
X68K <file name> -L
```

The command X68K TTL -L, produces two new files: TTL.BIN and TTL.LIS. If errors are detected during assembly, we will be informed. A binary object-code file is produced only if assembly takes place without errors. Once the source code has been assembled without errors, we can move on to the simulation phase. In this example, the TTL.X68 source code is:

```
TTL     'This is a test program'
ORG     $1000                   It starts here
NOP
NOP                             and does nothing...
NOOP                            It even has an error
END     $1000
```

If we examine the listing file, TTL.LIS, we get:

```
This is a test program
Source file: TTL.X68
Assembled on: 93-03-04 at: 14:20:09
          by: X68K PC-2.0 Copyright (c) University of Teesside 1989,93
Defaults: ORG $0/FORMAT/OPT A,BRL,CEX,CL,FRL,MC,MD,NOMEX,NOPCO
    1                                TTL      'This is a test program'
    2  00001000                      ORG      $1000                  ;It starts here
    3  00001000 4E71                 NOP
    4  00001002 4E71                 NOP                             ;and does nothing...
    5                                NOOP                            ;It even has an error
***** Invalid operation code.
    6           00001000             END      $1000

Lines: 6, Errors: 1, Warnings: 0.
```

As you can see, line 5 of the listing has generated an error message because I typed NOOP, rather than NOP. We now have to edit TTL.X68, correct the error, and re-assemble it.

Details of the Cross-assembler

The University of Teesside X68K cross-assembler is based on the Motorola's M68000 Family Resident Structured Assembler. It assembles all the valid instruction mnemonics as described in Motorola's literature, and accepts most of the assembler directives. The current release of the X68K cross-assembler generates only absolute code and does not include facilities for structured assembly. The following assembler directives are accepted by X68K.

```
DC, DCB, DS, END, ENDC, ENDM, EQU, FAIL, FORMAT, NOFORMAT, IFxx,
LIST, NOLIST, MACRO, MEXIT, PAGE, REG, SET, SPC, TTL, OPT (A, BRL,
BRS, CEX, NOCEX, CL, NOCL, CRE, FRL, FRS, MC, NOMC, MD, NOMD, MEX,
NOMEX, PCO, NOPCO)
```

The use of all these assembler directives is documented in the Motorola assembler manual. However, there are some minor points to be made:

- By default, X68K generates no symbol table information in the listing file. `OPT CRE` displays both symbol table and cross-reference information.

- `OPT NOCL` is used to stop conditional `IFxx` directives being displayed in the listing file. This has been extended to include instructions that are skipped as a result of an `IFxx` directive being false.

- The text string that follows the `TTL` directive *must* be enclosed in single quotes. Only the last `TTL` directive will be used as a title in the listing file.

- The `END` directive must be included and have an operand, as this defines the point at which execution begins in the assembled program. For example, `END $400` causes the program to start execution at location 400_{16} when the program is loaded.

- The `OPT MEX` directive may generate more lines in the listing file than in the original source file. This apparent increase in size is due to macro expansion. In order to make it easier to relate a listing file to the original source file, X68K places two sets of line numbers in the listing file if macros have been called. The first column of line numbers is used by X68K for cross-referencing purposes. The second column of line numbers relates to source lines in the original file, and is displayed in parentheses. Any error messages displayed on the screen during assembly are of the form:

  `Line NN, error message` or `Line NN.XX, error message`

  The number `NN` refers to a line number in the original source file being assembled. The extension `XX` is used when an error is detected in a macro expansion starting at line `NN`.

The assembler was written primarily for use in conjunction with the 68000 E68K simulator. Consequently, you should note the following points. The assembler initializes the program counter to zero before any code is produced, and will allow the PC to be altered only to addresses higher than the PC through the use of ORG directives. This restriction prevents users from causing code overlays. Other than comments, the first line of most programs will probably be an ORG $400 directive, because the 68000 locates its exception vectors in the range from $000000 to $0003FF. The PC is set to zero initially, allowing users to employ an ORG to place the addresses of exception handlers in the vector table before generating code.

Expressions within the assembler are divided into three types: string, arithmetic, and logical. All non-string expressions are treated as arithmetic expressions unless they contain one of the operators !, &, <<, >>, or non-base-ten numbers. An expression may contain a string. For example, the expression 'A'+$80 results in the *logical* hexadecimal value $C1_{16}$, whereas the expression 65+128 results in the *arithmetic* value $C1_{16}$. This has the effect that statements such as AND.B #$FF,D0 are legal, since $FF is an eight-bit quantity. However, statements such as AND.B #255,D0 are not legal since the arithmetic range for byte operations is -128 to 127. Any forward reference to a label defined by a SET directive causes the value of the last SET directive in the program to be returned.

The example below demonstrates the use of three special symbols interpreted by the cross-assembler. Each instruction loads the value 9 (i.e., %00001001) into a data register. In line 2, the expression '9'-$30 causes the assembler to subtract hexadecimal 30 from the ASCII code for the symbol 9. In line, 3 the ASCII value for 9 is logically ANDed with $0F. In line 4, hexadecimal 90 is shifted four places right. Note that Operand>>n means shift the operand right by n places. Similarly, Operand<<n means shift the operand left by n places.

```
Source file: OPS.X68
Assembled on: 93-03-04 at: 17:51:52
          by: X68K PC-2.0 Copyright (c) University of Teesside 1989,93
Defaults: ORG $0/FORMAT/OPT A,BRL,CEX,CL,FRL,MC,MD,NOMEX,NOPCO

    1 00001000                      ORG       $1000
    2 00001000 103C0009             MOVE.B    #'9'-$30,D0        ;Load D0 with 9
    3 00001004 123C0009             MOVE.B    #'9'&$0F,D1        ;Load D1 with 9
    4 00001008 143C0009             MOVE.B    #$90>>4,D2         ;Load D2 with 9
    5          00001000             END       $1000

Lines: 5, Errors: 0, Warnings: 0.
```

Using the Simulator

The simulator permits us to run the 68000 machine code generated by the cross-assembler on a PC, as if the PC were an actual 68000. Simulation is, of course, much slower than running the same code on a true 68000 machine. Furthermore, it does not provide the same *environment* as the target machine on which the code is eventually to run. By *environment* we mean the operating system and input/output devices. However, the simulator enables us to test much of the code before it runs on the target machine. We can therefore design and debug software before the target machine has been built.

Programmers often use a cross-compiler to generate binary code rather than a cross-assembler. For example, you might write the source code in C, instead of assembly language.

The simulator is called by typing: E68K <filename> at the MS-DOS prompt. This command runs the simulator program, E68K.EXE, and then loads <FILENAME>.BIN. For example, we might type E68K TEST — we don't have to explicitly specify the extension .BIN.

After starting the simulator with the command E68K TEST, the computer displays the copyright banner, the filename, the low address, high address, and starting point. The starting point is the address of the first executable instruction supplied by the END directive in the source file (e.g., END $400). The low and high addresses define the region of memory allocated by the simulator to the program.

Once the simulator has been loaded, nothing appears to happen. The computer just sits there and waits for a valid command from the keyboard. Let's looks at some of the simulator's commands. The facilities provided by the simulator fall into two groups — those that enable us to set up or examine the environment, and those that enable us to monitor the execution of a program. We begin by describing functions that let us look at the 68000's environment — its memory and registers.

MD The memory display command allows us to examine the contents of the 68000's memory. This memory is, of course, the simulated memory of the 68000 processor. If we type MD followed by an address, the simulator displays the contents of 16 consecutive bytes in a hexadecimal format. Consider the following example.

```
>MD 44E

00044E  41 6C 61 6E 20 43 6C 65 6D 65 6E 74 73 0A 00 00
00045E  00 00 00 00 00 00 00 00 00 00 00 00 00 00 00 00.
```

This command displays data in memory starting at memory location $44E. A line of data is displayed each time you hit the enter (i.e., return) key. Entering a period, ".", terminates this command process. The memory display function is frequently used to examine a block of data. For example, if you have a table of 10 bytes and run a program to put the bytes into ascending order, you can use the MD function to determine whether the program has worked correctly. Equally, you can use it to examine the state of the stack.

The memory display function can also be used to disassemble memory (i.e., the reverse function of an assembler). Binary code is read from memory and converted into 68000 mnemonics. The simulator's disassembler is not able to convert addresses into symbolic form, and therefore all addresses are expressed numerically.

In the following example, a program has already been loaded into the simulator and we enter the command MD 400 -DI. The simulator displays 16 lines of the program starting from hexadecimal address 400. The extension to the MD command, -DI, tells the simulator to disassemble the contents of memory rather than presenting data in hexadecimal form. You do not need to follow this program, as it is for demonstration purposes only. However, note that addresses are given in hexadecimal form and literals in decimal form.

```
>MD 400 -DI

000400: MOVEA.L  #4096,SP
000406: BSR.L    $00000434
```

```
00040A: LEA      $0000044E,A0
000410: BSR.L    $00000442
000414: MOVE.B   D1,(A0)+
000416: CMPI.B   #10,D1
00041A: BNE.S    $00000410
00041C: BSR.L    $00000434
000420: MOVE.B   -(A0),D1
000422: BSR.L    $00000448
000426: CMPA.W   #1102,A0
00042A: BNE.S    $00000420
00042C: BSR.L    $00000434
000430: STOP     #8192
000434: MOVEM.L  A0/D0-D7,-(SP)
000438: MOVEQ    #0,D0
.
```

We can use the disassembly function to examine code — we might not have an assembled listing of the source file available. It is rather difficult to follow a disassembled program, as the disassembler is not able to bind addresses and data values to their symbolic names. For example, all branch and subroutine addresses are given as absolute values.

`MM` The memory modify function allows you to modify the contents of a memory location. Entering `MM <address>` causes the simulator to display the contents of memory location `<address>`. If you type new data, it overwrites the existing data, and displays the contents of the next location. The memory modify command is terminated by entering a period, ".".

`DF` The `DF`, display formatted registers, command lists the contents of all the 68000's internal registers on the screen — D0 to D7, A0 to A7, the supervisor stack pointer (SS), the user stack pointer (US), the status register (SR), and the condition codes. The instruction currently pointed at by the PC is disassembled and displayed on the screen — this is the next instruction to be executed. An example of the screen you might see after entering `DF` is given below.

```
PC=000400 SR=2000 SS=00A00000 US=00000000          X=0
A0=00000000 A1=00000000 A2=00000000 A3=00000000 N=0
A4=00000000 A5=00000000 A6=00000000 A7=00A00000 Z=0
D0=00000000 D1=00000000 D2=00000000 D3=00000000 V=0
D4=00000000 D5=00000000 D6=00000000 D7=00000000 C=0
———>MOVEA.L  #4096,SP
```

Remember that the information you see on the screen is the contents of the 68000's synthetic registers in the simulator. These are the values that will be used when the 68000 program is run.

Following the register list is the instruction at the address pointed at by the PC. In this example the instruction is `MOVEA.L #4096,SP`, which is the first instruction in the program, because the program counter has been set to 400 hexadecimal. Note that this instruction was actually written in the program as `MOVEA.L #$1000,A7`. The simulator displays literal data in decimal form when code is disassembled (hence, 4096) and has replaced A7 by SP (the stack pointer).

We can modify the contents of any of the 68000's registers by entering the expression: `.<register> <value>`. For example, the program counter is set to $400 by entering `.PC 400`.

`GO` The `GO` command causes the simulator to execute the program previously loaded starting from the value currently in the program counter. Execution continues until:

a. The program reaches a `STOP` instruction
b. The program is interrupted from the keyboard by hitting the ESC (i.e., escape) key
c. The program reaches a breakpoint that was previously set by the user.

If *nothing* appears to happen after you have typed `GO`, the program is probably either waiting for input from the keyboard or it is in a loop from which it cannot exit. In the latter case, the escape key must be used to stop execution.

`TR` The trace command, `TR`, is one of the most useful of the simulator commands and permits the execution of a single instruction at a time. After an instruction has been executed, the contents of all registers are displayed on the screen. Entering further carriage returns steps through the program instruction by instruction. You can use the trace command to walk through a program an instruction at a time until you find the cause of an error. You really need a hard copy of the assembled program listing with you when using the trace function. However, you cannot rely on the trace function alone. Suppose you clear an array of 10,000 bytes in your program by going around a loop made up of two instructions 10,000 times. To trace this loop you have to hit the return button a mere 20,000 times....

`BR` The simulator's breakpoint command, `BR <address>`, places a *marker* in the program at the location specified by `<address>`. When the program is executed, it runs normally until the breakpoint address is reached. The simulator then halts execution and prints the contents of the registers on the screen (i.e., in the same way as a `DF` command). You can use one or more `.<register> <value>` commands (e.g., `.DO 1234`) to modify registers at this stage if necessary. A breakpoint gives you a snapshot of the state of the processor at any desired point in the execution of a program. It is possible to set more than one breakpoint at a time.

The breakpoint function overcomes the limitation of the trace function we described above. By putting a breakpoint after the end of a loop that is executed many times, we can execute the code of the loop and then halt the processor at its termination. Then we can revert to the trace mode. The art of program debugging is to insert breakpoints at judicious points. If you know the expected behavior of a program, it is usually possible to locate the point at which the actual behavior of the program differs from the intended behavior.

Note that we can resume execution after a breakpoint by entering `TR` to step past the breakpoint, followed by `GO` to run the program. Entering a `GO` after a breakpoint will simply re-run the instruction at the breakpoint again.

`QU` The `QUIT` command exits the simulator and returns you to the MS-DOS operating system. Once again, note that if a program is being simulated and the system hangs up for whatever reason, hitting the escape key forces an exit from the simulation mode.

Example of a Session with the Simulator

We're now going to walk through a demonstration of the cross-assembler and simulator. The following example matches a substring against a string. That is, it attempts to determine whether the sequence xxx...x ocurrs in the sequence yyyyyy...y. Having designed the program, the first step is to write the assembly program, called MATCH1.X68.

```
         ORG      $400             Program to compare two strings
START    LEA      $1000,A7         Set up stack pointer
         PEA      Striing          Push address of string
         MOVE.W   String_L,-(SP)   Push length of string
         PEA.B    Sub              Push address of substring
         MOVE.W   Sub_L,-(SP)      Push length of substring
         BSR      MATCH            Call match
         LEA      12(SP),SP        Clean up stack
         STOP     #$2700           End of program
*************************************************************************
*
* A0 = pointer to substring
* D0 = length of substring
* A2 = A0+D0 = end of substring+1
* A1 = pointer to string
* D1 = length of string
* A3 = A1+D1 = end of string+1
*
MATCH    MOVE     4SP),D0          Get substring length
         MOVEA.L  6(SP),A0         Get substring start
         MOVE     10(SP),D1        Get string length
         MOVEA.L  12(SP),A1        Get string start
*
         LEA      (A0,D0.W),A2     Get end of substring+1 in A2
         LEA      (A1,D1.W),A3     Get end of string+1 in A3
         SUBB.L   D0,A3            Get address of last character to match in A3
*
Next     MOVE.B   (A0),D7          Match first substring character
         CMP.B    (A1),D9           with string character
         BNE      No_Match         If no match then continue with next pair
         BSR      Rest             Else match rest of substring
         BCS      Exit_OK          If carry set then submatch found
No_Math  ADDA.L   #1,A1            Point to next character in string
         CMPA.L   A1,A3            Are we at the end of the string?
         BNE      Next             If not then continue
Exit     SUBA.L   A0,A0            Clear A0 for fail
         RTS
Exit_OK  LEA      (A1),A0          Put address of match in A0
         RTS
```

```
*
Rest       LEA     1(A0),A4        A4 points to second character of substring
           LEA     1(A1),A5        A5 points to next character of string
Again      CMP.B   (A4)-,(A5)+     Match a pair of characters
           BNE     Fail            If different exit
           CMPA.L  A2,A4           Test for end of substring
           BNE     Again           Continue
ExitOK     MOVE    #$FF,CCR        Set carry
           RTS
Fail       MOVE    #0,CCR          Clear carry
           RTS
*
           ORG     $600
String     DS.B    'This that then'
String_L   DC.W    14
Sub        DC.B    'that'
Sub_L      DC.W    4
           END     START
```

The next step is to assemble the program using the cross-assembler. We enter the command `X68K MATCH1 -L` to produce a binary file and a listing file.

```
Source file: MATCH1.X68
Assembled on: 93-04-23 at: 18:42:34
        by: X68K PC-2.1 Copyright (c) University of Teesside 1989,93
Defaults: ORG $0/FORMAT/OPT A,BRL,CEX,CL,FRL,MC,MD,NOMEX,NOPCO

  1 00000400                       ORG     $400           ;Program to compare two strings
  2 00000400 4FF81000     START:   LEA     $1000,A7       ;Set up stack pointer
  3 00000404 487900000000          PEA     STRIING        ;Push address of string
***** Undefined symbol - operand 1.
  4 0000040A 3F3900000600          MOVE.W  STRING_L,-(SP) ;Push length of string
  5                                PEA.B                  ;Sub         Push address of substring
***** Invalid operation code.
  6 00000410 3F3900000606          MOVE.W  SUB_L,-(SP)    ;Push length of substring
  7 00000416 6100000A              BSR     MATCH          ;Call match
  8 0000041A 4FEF000C              LEA     12(SP),SP      ;Clean up stack
  9 0000041E 4E722700              STOP    #$2700         ;End of program
 10                       ******************************************************************************
 11                       *
 12                       * A0 = pointer to substring
 13                       * D0 = length of substring
 14                       * A2 = A0+D0 = end of substring+1
 15                       * A1 = pointer to string
 16                       * D1 = length of string
 17                       * A3 = A1+D1 = end of string+1
```

```
   18                       *
   19                       MATCH:   MOVE      4SP),D0              ;Get substring length
***** Missing operand(s).
   20  00000422 206F0006             MOVEA.L   6(SP),A0             ;Get substring start
   21  00000426 322F000A             MOVE      10(SP),D1            ;Get string length
   22  0000042A 226F000C             MOVEA.L   12(SP),A1            ;Get string start
   23                       *
   24  0000042E 45F00000             LEA       (A0,D0.W),A2         ;Get end of substring+1 in A2
   25  00000432 47F11000             LEA       (A1,D1.W),A3         ;Get end of string+1 in A3
   26                                SUBB.L                         ;D0,A3  Get address of last character to match in A3
***** Invalid operation code.
   27                       *
   28  00000436 1E10        NEXT:    MOVE.B    (A0),D7              ;Match first substring character
   29                                CMP.B     (A1),D9              ;with string character
***** Illegal addressing mode - operand 2.
   30  00000438 66000000             BNE       NO_MATCH             ;If no match then continue with next pair
***** Undefined symbol - operand 1.
   31  0000043C 61000018             BSR       REST                 ;Else match rest of substring
   32  00000440 65000010             BCS       EXIT_OK              ;If carry set then submatch found
   33  00000444 D3FC00000001 NO_MATH: ADDA.L   #1,A1                ;Point to next character in string
   34  0000044A B7C9                 CMPA.L    A1,A3                ;Are we at the end of the string?
   35  0000044C 66E8                 BNE       NEXT                 ;If not then continue
   36  0000044E 91C8        EXIT:    SUBA.L    A0,A0                ;Clear A0 for fail
   37  00000450 4E75                 RTS
   38  00000452 41D1        EXIT_OK: LEA       (A1),A0              ;Put address of match in A0
   39  00000454 4E75                 RTS
   40                       *
   41  00000456 49E80001    REST:    LEA       1(A0),A4             ;A4 points to second character of substring
   42  0000045A 4BE90001             LEA       1(A1),A5             ;A5 points to next character of string
   43                       AGAIN:   CMP.B     (A4)-,(A5)+          ;Match a pair of characters
***** Illegal use of a register - operand 1.
***** Illegal addressing mode - operand 1.
   44  0000045E 6600000C             BNE       FAIL                 ;If different exit
   45  00000462 B9CA                 CMPA.L    A2,A4                ;Test for end of substring
   46  00000464 66F8                 BNE       AGAIN                ;Continue
   47  00000466 44FC00FF    EXITOK:  MOVE      #$FF,CCR             ;Set carry
   48  0000046A 4E75                 RTS
   49  0000046C 44FC0000    FAIL:    MOVE      #0,CCR               ;Clear carry
   50  00000470 4E75                 RTS
   51                       *
   52  00000600                      ORG       $600
   53                       STRING:  DS.B      'This that then'
***** Illegal use of a string - operand 1.
   54  00000600 000E        STRING_L: DC.W     14
   55  00000602 74686174    SUB:     DC.B      'that'
   56  00000606 0004        SUB_L:   DC.W      4
```

```
 57          00000400              END      START

Lines: 57, Errors: 9, Warnings: 0.
```

As you can see, the program has nine errors. Unlike most errors in this book, these nine were deliberate. So, we correct the errors and re-assemble the program to get the following error-free listing. Remember that a program free of syntax errors doesn't mean that it is a correct program. We still have to determine whether it will do what we intend it to do. The following listing is the corrected program.

```
Source file: MATCH.X68
Assembled on: 93-04-23 at: 18:42:54
          by: X68K PC-2.1 Copyright (c) University of Teesside 1989,93
Defaults: ORG $0/FORMAT/OPT A,BRL,CEX,CL,FRL,MC,MD,NOMEX,NOPCO

   1 00000400                      ORG      $400
   2 00000400 4FF81000    START:   LEA      $1000,A7           ;Set up stack pointer
   3 00000404 487900000600         PEA      STRING             ;Push address of string
   4 0000040A 3F390000060E         MOVE.W   STRING_L,-(SP)     ;Push length of string
   5 00000410 487900000610         PEA      SUB                ;Push address of substring
   6 00000416 3F3900000614         MOVE.W   SUB_L,-(SP)        ;Push length of substring
   7 0000041C 6100000A             BSR      MATCH              ;Call 'match'
   8 00000420 4FEF000C             LEA      12(SP),SP          ;Clean up stack
   9 00000424 4E722700             STOP     #$2700
  10                      *************************************************************************
  11                      *
  12                      * A0 = pointer to substring
  13                      * D0 = length of substring
  14                      * A2 = A0+D0 = end of substring+1
  15                      * A1 = pointer to string
  16                      * D1 = length of string
  17                      * A3 = A1+D1 = end of string+1
  18                      *
  19 00000428 302F0004    MATCH:   MOVE     4(SP),D0           ;Get substring length
  20 0000042C 206F0006             MOVEA.L  6(SP),A0           ;Get substring start
  21 00000430 322F000A             MOVE     10(SP),D1          ;Get string length
  22 00000434 226F000C             MOVEA.L  12(SP),A1          ;Get string start
  23                      *
  24 00000438 45F00000             LEA      (A0,D0.W),A2       ;Get end of substring+1 in A2
  25 0000043C 47F11000             LEA      (A1,D1.W),A3       ;Get end of string+1 in A3
  26 00000440 97C0                 SUBA.L   D0,A3              ;Get address of last character to match in A3
  27                      *
  28 00000442 1E10        NEXT:    MOVE.B   (A0),D7            ;Match first substring character
  29 00000444 BE11                 CMP.B    (A1),D7            ;with string character
  30 00000446 6600000A             BNE      NO_MATCH           ;If no match then continue with next pair
  31 0000044A 61000018             BSR      REST               ;Else match rest of substring
```

```
 32 0000044E 65000010                BCS      EXIT_OK         ;If carry set then submatch found
 33 00000452 D3FC00000001 NO_MATCH: ADDA.L   #1,A1           ;Point to next character in string
 34 00000458 B7C9                    CMPA.L   A1,A3           ;Are we at the end of the string?
 35 0000045A 66E6                    BNE      NEXT            ;If not then continue
 36 0000045C 91C8         EXIT:      SUBA.L   A0,A0           ;Clear A0 for fail
 37 0000045E 4E75                    RTS
 38 00000460 41D1         EXIT_OK:   LEA      (A1),A0         ;Put address of match in A0
 39 00000462 4E75                    RTS
 40                       *
 41 00000464 49E80001     REST:      LEA      1(A0),A4        ;A4 points to second character of substring
 42 00000468 4BE90001                LEA      1(A1),A5        ;A5 points to next character of string
 43 0000046C BB0C         AGAIN:     CMP.B    (A4)+,(A5)+     ;Match a pair of characters
 44 0000046E 6600000C                BNE      FAIL            ;If different exit
 45 00000472 B9CA                    CMPA.L   A2,A4           ;Test for end of substring
 46 00000474 66F6                    BNE      AGAIN           ;Continue
 47 00000476 44FC00FF     EXITOK:    MOVE     #$FF,CCR        ;Set carry
 48 0000047A 4E75                    RTS
 49 0000047C 44FC0000     FAIL:      MOVE     #0,CCR          ;Clear carry
 50 00000480 4E75                    RTS
 51                       *
 52 00000600                         ORG      $600
 53 00000600 546869732074 STRING:    DC.B     'This that then'
             686174207468
             656E
 54 0000060E 000E         STRING_L:  DC.W     14
 55 00000610 74686174     SUB:       DC.B     'that'
 56 00000614 0004         SUB_L:     DC.W     4
 57          00000400                END      START
Lines: 57, Errors: 0, Warnings: 0.
```

Because this listing has no syntax errors, we can therefore run it using the simulator. The command E68K MATCH1 is entered at the DOS prompt to give the following heading:

```
[68K PC-2.1 Copyright (c) University of Teesside 1989,93]

Address space 0 to ^10485759 (10240 kbytes).

Load binary file "MATCH.BIN".
Start address: 0004000, Low: 00000400, High: 00000615

>
```

We will now demonstrate some of the simulator's features. One of the most useful is the LOG function that records all screen input/output in a file. Entering LOG MATCH1.LOG puts all screen I/O in the file MATCH1.LOG. We have employed this function to reproduce the screen output used throughout this book. We begin by using the DF function to display the initial contents of the registers and then TR to trace the first few instructions.

```
>DF
PC=000400 SR=2000 SS=00A00000 US=00000000          X=0
A0=00000000 A1=00000000 A2=00000000 A3=00000000 N=0
A4=00000000 A5=00000000 A6=00000000 A7=00A00000 Z=0
D0=00000000 D1=00000000 D2=00000000 D3=00000000 V=0
D4=00000000 D5=00000000 D6=00000000 D7=00000000 C=0
———>LEA.L    $1000,SP
```

The simulator initially clears all registers and memory locations to zero, sets the status register to $2000 and the supervisor stack pointer to $00A00000. The PC is set to $000400 by the `END` directive in the source program, and the instruction pointed at by the program counter, `LEA.L $1000,SP`, displayed. When we enter the trace mode, the simulator traces an instruction after each return. To leave the trace mode, just enter a period followed by a return.

```
>TR

PC=000404 SR=2000 SS=00001000 US=00000000          X=0
A0=00000000 A1=00000000 A2=00000000 A3=00000000 N=0
A4=00000000 A5=00000000 A6=00000000 A7=00001000 Z=0
D0=00000000 D1=00000000 D2=00000000 D3=00000000 V=0
D4=00000000 D5=00000000 D6=00000000 D7=00000000 C=0
———>PEA      $00000600

Trace>

PC=00040A SR=2000 SS=00000FFC US=00000000          X=0
A0=00000000 A1=00000000 A2=00000000 A3=00000000 N=0
A4=00000000 A5=00000000 A6=00000000 A7=00000FFC Z=0
D0=00000000 D1=00000000 D2=00000000 D3=00000000 V=0
D4=00000000 D5=00000000 D6=00000000 D7=00000000 C=0
———>MOVE.W   $0000060E,-(SP)

Trace>

PC=000410 SR=2000 SS=00000FFA US=00000000          X=0
A0=00000000 A1=00000000 A2=00000000 A3=00000000 N=0
A4=00000000 A5=00000000 A6=00000000 A7=00000FFA Z=0
D0=00000000 D1=00000000 D2=00000000 D3=00000000 V=0
D4=00000000 D5=00000000 D6=00000000 D7=00000000 C=0
———>PEA      $00000610

Trace>

PC=000416 SR=2000 SS=00000FF6 US=00000000          X=0
A0=00000000 A1=00000000 A2=00000000 A3=00000000 N=0
A4=00000000 A5=00000000 A6=00000000 A7=00000FF6 Z=0
```

```
D0=00000000 D1=00000000 D2=00000000 D3=00000000 V=0
D4=00000000 D5=00000000 D6=00000000 D7=00000000 C=0
——————>MOVE.W   $00000614,-(SP)

Trace>

PC=00041C SR=2000 SS=00000FF4 US=00000000          X=0
A0=00000000 A1=00000000 A2=00000000 A3=00000000 N=0
A4=00000000 A5=00000000 A6=00000000 A7=00000FF4 Z=0
D0=00000000 D1=00000000 D2=00000000 D3=00000000 V=0
D4=00000000 D5=00000000 D6=00000000 D7=00000000 C=0
——————>BSR.L    $00000428
```

At this stage the program is about to call the subroutine MATCH. Note that the simulator displays the instruction as BSR.L $0000428 because it has no knowledge of symbolic names and addresses. Let's look at the state of the stack — the current top of the stack is indicated by A7 (i.e., $0000FF4). The MD function can be used to display the contents of memory. This command displays a line of 16 bytes each time you hit the enter key, and is terminated by typing a period followed by an enter.

```
>MD 4 FF4

000FF4  00 04 00 00 06 10 00 0E 00 00 06 00 00 00 00 00
001004  00 00 00 00 00 00 00 00 00 00 00 00 00 00 00 00.
```

You have to refer to the assembled program or the trace output to understand this display. After setting the stack pointer to $1000, the program pushes four items on the stack — the address and lengths of the string and substring. Working from the *bottom* of the stack, these are $00000600, $000E, $00000610, and $0004, respectively. If you examine the assembled program, you will find that these are all the expected values. Let's continue tracing.

```
>TR
PC=000428 SR=2000 SS=00000FF0 US=00000000          X=0
A0=00000000 A1=00000000 A2=00000000 A3=00000000 N=0
A4=00000000 A5=00000000 A6=00000000 A7=00000FF0 Z=0
D0=00000000 D1=00000000 D2=00000000 D3=00000000 V=0
D4=00000000 D5=00000000 D6=00000000 D7=00000000 C=0
——————>MOVE.W   4(SP),D0

Trace>
PC=00042C SR=2000 SS=00000FF0 US=00000000          X=0
A0=00000000 A1=00000000 A2=00000000 A3=00000000 N=0
A4=00000000 A5=00000000 A6=00000000 A7=00000FF0 Z=0
D0=00000004 D1=00000000 D2=00000000 D3=00000000 V=0
D4=00000000 D5=00000000 D6=00000000 D7=00000000 C=0
——————>MOVEA.L  6(SP),A0
```

```
Trace>

PC=000430 SR=2000 SS=00000FF0 US=00000000          X=0
A0=00000610 A1=00000000 A2=00000000 A3=00000000 N=0
A4=00000000 A5=00000000 A6=00000000 A7=00000FF0 Z=0
D0=00000004 D1=00000000 D2=00000000 D3=00000000 V=0
D4=00000000 D5=00000000 D6=00000000 D7=00000000 C=0
——————>MOVE.W    10(SP),D1

Trace>

PC=000434 SR=2000 SS=00000FF0 US=00000000          X=0
A0=00000610 A1=00000000 A2=00000000 A3=00000000 N=0
A4=00000000 A5=00000000 A6=00000000 A7=00000FF0 Z=0
D0=00000004 D1=0000000E D2=00000000 D3=00000000 V=0
D4=00000000 D5=00000000 D6=00000000 D7=00000000 C=0
——————>MOVEA.L  12(SP),A1

Trace>

PC=000438 SR=2000 SS=00000FF0 US=00000000          X=0
A0=00000610 A1=00000600 A2=00000000 A3=00000000 N=0
A4=00000000 A5=00000000 A6=00000000 A7=00000FF0 Z=0
D0=00000004 D1=0000000E D2=00000000 D3=00000000 V=0
D4=00000000 D5=00000000 D6=00000000 D7=00000000 C=0
——————>LEA.L     0(A0,D0.W),A2
```

At this stage in the subroutine, the address and length of the string and the substring have been pulled off the stack and deposited in registers A1, D1, and A0, D0, respectively. The above display demonstrates that these actions have been correctly carried out.

We could trace the entire program instruction by instruction, but this might prove tedious. Here's where the breakpoint facility comes it. We run the program until a breakpoint is reached and display the current simulator status. To a considerable extent, the art of debugging a program is all about choosing appropriate places to put breakpoints. In this case, we will locate a single breakpoint in the subroutine `MATCH` at the point after two characters have been found equal and the subroutine `REST` called to compare the remainder of the substring. This point is at address $44E and is found by referring to the assembled listing on page 693.

```
>BR 44E
Breakpoints: Address   Count
             00044E    00000000
```

The program can be run from the current value of the PC until its termination or a breakpoint by the `GO` command.

```
>GO

Breakpoint at: 00044E.

PC=00044E SR=20FF SS=00000FF0 US=00000000          X=1
A0=00000610 A1=00000605 A2=00000614 A3=0000060A N=1
A4=00000614 A5=00000609 A6=00000000 A7=00000FF0 Z=1
D0=00000004 D1=0000000E D2=00000000 D3=00000000 V=1
D4=00000000 D5=00000000 D6=00000000 D7=00000074 C=1
———>BCS.L    $00000460
```

At this point you have to match the output from the simulator (i.e., the contents of the 68000's registers) with the values you would expect. If the values are the same, you can continue. If they are not, you have to localize the problem.

The breakpoint occurs after the substring has been compared with the string. The source program tells us that the carry bit is set for a match and cleared for no match. The display tells us that the carry bit C=1 and therefore a match has occurred. At this point, address register A1 should contain the location of the character in the string at which the match begins. This value is $605. Subroutine `REST` tells us that address register A5 should point to the character in the string after the end of the substring — $609.

If we look at the string "This that then" set up by the `DC.B` assembly directive at location $600, we find that the location of the match is $605 and that the location of the first character after the "that" is $609. These values correspond to those expected. To continue executing from this point, it is necessary to first perform a trace operation to step past the current breakpoint.

Details of the E68K Motorola M68000 System Simulator

E68K is a program that simulates the instructions and architecture of the 68000. The simulator supports an addressing range of 10240K bytes, and memory is allocated when first written to. To call E68K we enter the command

```
E68K filename
```

where the full name of the input file is `filename.BIN`. The binary file may be one produced by the X68K cross-assembler, or one produced by the E68K simulator as a memory dump. Memory is allocated in 1K-byte chunks from address 0. You don't need to use memory in a contiguous fashion. If you try to exceed the memory available on the computer, an error message is printed and a bus error exception is generated.

The simulator initializes the exception vector table to zeros before any programs are loaded. If, when an exception occurs, the vector is zero, the simulator will take a default action. Otherwise, normal exception processing takes place. The simulator is always in one of two modes. It is either executing a program or processing commands. Below is a list of the available commands and their meaning.

HELP The HELP command responds with the following display.

```
HELP
 .D0 .D1 .D2 .D3 .D4 .D5 .D6 .D7
 .A0 .A1 .A2 .A3 .A4 .A5 .A6 .A7
 .PC .SR .US .SS .C  .N  .V  .X  .Z

 BF - Block fill.                         GT - Go with temp brkpt.
 BM - Block move.                         HE - Help.
 BR - Set/display brkpts.                 LO - Load binary file.
 BS - Block search.                       LOG - LOG screen output.
 DC - Data conversion.                    MD - Memory dump.
 DF - Display registers.                  MM - Memory modify.
 DU - Dump memory to file.                MS - Memory set.
 GD - Go direct.                          NOBR - Remove brkpts.
 GO - Execute program with brkpts.        TR - Trace program.
                                          QUIT - Return to MS-DOS.

 For more detailed information type: HE command
                                     HE INFO
                                     HE EXCEPTIONS
                                     HE TRAP#15
```

The first three lines of the help command show how register names are specified. The contents of register Rn can be displayed by entering .Rn. The contents of registers can be altered in a similar fashion. For example, entering .PC $1000 assigns the value 1000_{16} to the program counter. We could have entered .PC 1000 to do the same thing, since the simulator assumes that the input will be hexadecimal. The register modify feature is also available for the status register flags — entering .Z 1 sets the Z-bit flag to true.

Before continuing, we will take a look at one of the help commands. If you type HE LOG, you will see the following display:

```
>HE LOG
LOG screen output.
 Format: LOG [filename]

 This command causes all screen output to be copied to the specified
 file. If the file does not exist it is created, else appended too. The
 LOG command with no filename argument terminates logging.
 E.G. LOG MYPROG.LOG
```

The simulator supports full expression processing using the same operators and operator priorities as the assembler. The only difference is that the default base of the simulator is hexadecimal, and decimal numbers must be preceded by a "^". The register names may also be used in expressions; e.g., .PC+2 increments the program counter by 2. We will now look at some more of the simulator's commands.

BF - Block fill

Format: BF <ADDR_1> <ADDR_2> <DATA>

The BF command fills a block of memory starting at ADDR_1 and ending at ADDR_2 with DATA. All arguments to this command are integers. The command uses word operations only. If you need to fill the 4K byte memory space from $1000 to $1FFF with the value $FFFF, you enter BF 1000 1FFF FFFF.

BM - Block move

Format: BM <ADDR_1> <ADDR_2> <ADDR_3>

The byte-oriented BM command moves a block of memory starting at ADDR_1 and ending at ADDR_2 to a new location starting at ADDR_3. The two areas of memory must not overlap.

BR - Set and display breakpoints

Format: BR <brkpt_1> ... <brkpt_7> -<count>

The BR command allows you to insert or remove breakpoints in memory. The parameters are the addresses of the breakpoints (e.g., BR 1234 locates a breakpoint at address $1234). An integer, count, is associated with each breakpoint and specifies how many times the breakpoint location can be referenced before a breakpoint is taken. If no value for count is specified, the default value is zero. Up to seven breakpoints at a time can be inserted with the BR command if a count is specified, eight otherwise. If BR is used with no options, the existing breakpoint table is printed. Up to 16 breakpoints can be in force at once.

Breakpoints should be inserted only in the first word of an instruction. Placing them in the middle of instructions will probably cause an illegal instruction exception when the instruction is executed. The breakpoint count is reset when any of the following commands is executed: GO, GT, TR.

Consider the following example of setting breakpoints. We use BR to list breakpoints (initially, none are set). Then we set three, each with a count of 5. The NOBR, remove breakpoint, command is used to remove the breakpoint at address 1200. Finally, two new breakpoints are added, each with a count of 12.

```
>BR
There are no breakpoints set.

>BR 1000 1200 1400 -5
Breakpoints: Address   Count
             001000    00000005
             001200    00000005
             001400    00000005
```

```
>NOBR 1200
Breakpoints: Address   Count
             001000    00000005
             001400    00000005

>BR 2000 3000 -12
Breakpoints: Address   Count
             001000    00000005
             001400    00000005
             002000    00000012
             003000    00000012
```

BS - Block search

Format: BS <ADDR_1> <ADDR_2> 'string'
BS <ADDR_1> <ADDR_2> <DATA> [<MASK>] [-B] [-W] [-L]

The BS command that searches a block of memory between addresses ADDR_1 and ADDR_2 has two formats. The first format, BS <ADDR_1> <ADDR_2> 'string', searches for a character string. The second format, BS <ADDR_1> <ADDR_2> <DATA> [<MASK>] [-B] [-W] [-L], searches for DATA. The contents of the location being tested may be optionally ANDed with a literal MASK before testing. The block search for non-strings may be byte-, word-, or longword-based. The default is byte. If a match is found, the start address of the match is printed.

DC - Data conversion

Format: DC expression

This command evaluates an expression and prints it in hexadecimal, decimal, octal, and binary formats. Expressions must not contain embedded blanks. The DC command allows you to perform a calculation while in the simulator, without having to look for a pocket calculator. Consider the following two examples of this command.

```
>DC $2*@10+%101010+2
DEC: ^60
HEX: $0000003C
OCT: @00000000074
BIN: %00000000000000000000000000111100

>DC $10*$ABCD
DEC: ^703696
HEX: $000ABCD0
OCT: @00002536320
BIN: %00000000000010101011110011010000
```

`DF - Display formatted registers`

The `DF` command displays all the current register values plus the CCR. The instruction pointed to by the current value of the PC is also disassembled and displayed.

`DU - Dump memory and registers to file`

Format: `DU <ADDR_1> <ADDR_2> dumpfilename`

The `DU` command dumps a block of memory starting at `ADDR_1` and ending at `ADDR_2` to a disk file specified by `dumpfilename`. The full name of the file produced is `dumpfilename.BIN`. The current register values are also dumped. The data dumped is a block of words.

`GO - Execute program`

Format: `GO <ADDR>`

The `GO` command causes execution to begin at address `ADDR`. The program executes instructions until a trap, exception, or breakpoint is reached.

`GD - Go direct (execute program without breakpoints)`

Format: `GD <ADDR>`

The `GD` command causes execution to begin at address `ADDR`. The program executes until a trap or exception is reached. Breakpoints are ignored.

`GT - Execute program with temporary breakpoint`

Format: `GT <ADDR>`

The `GT` command inserts a temporary breakpoint at address `ADDR` and begins execution at the address currently in the PC. The program executes instructions until a trap, exception, or any breakpoint is reached. The temporary breakpoint is removed when execution terminates.

`LO - Load binary file`

Format: `LO filename`

The load binary file command loads the file specified by filename into memory. The full name of the file must be `filename.BIN`, and will either be an E68K dump file (i.e., a file saved by the `DU` command) or one produced by the X68K cross-assembler. The loading is done by word, not byte. If the binary file is an E68K dump file, the saved register contents will replace the current settings.

MD - Memory dump

Format: MD <ADDR> <COUNT> [-DI]

The MD command dumps COUNT bytes of memory (or disassembles COUNT instructions if the -DI option is used) starting at address ADDR. Once the first line has been dumped, entering carriage returns will cause the MD command to display the next line of data. The default value of COUNT is 16. To stop the command, enter any character before the carriage return. The following output was produced by listing a block of memory containing a program (part of the example we used earlier in this appendix) first as data and then as disassembled instructions.

```
>MD 442

000442  1E 10 BE 11 66 00 00 0A 61 00 00 18 65 00 00 10
000452  D3 FC 00 00 00 01 B7 C9 66 E6 91 C8 4E 75 41 D1
000462  4E 75 49 E8 00 01 4B E9 00 01 BB 0C 66 00 00 0C
000472  B9 CA 66 F6 44 FC 00 FF 4E 75 44 FC 00 00 4E 75.

>MD 442 -DI

000442: MOVE.B   (A0),D7
000444: CMP.B    (A1),D7
000446: BNE.L    $00000452
00044A: BSR.L    $00000464
00044E: BCS.L    $00000460
000452: ADDA.L   #1,A1
000458: CMPA.L   A1,A3
00045A: BNE.S    $00000442
00045C: SUBA.L   A0,A0
00045E: RTS
000460: LEA.L    (A1),A0
000462: RTS
000464: LEA.L    1(A0),A4
000468: LEA.L    1(A1),A5
00046C: CMPM.B   (A4)+,(A5)+
00046E: BNE.L    $0000047C
.
```

MM - Memory modify

Format: MM <ADDR> [-B] [-W] [-L] [-O] [-E] [-HEX] [-OCT] [-DEC] [-BIN]

The MM command is used to display and change memory locations. The options determine the operation size and type. The defaults are -B and -HEX. The -O option selects odd bytes, while the -E option selects even bytes. The other options -B, -W, and -L are obvious. The MM command prints address ADDR and displays its contents. You may then specify new data for the location. The new data, if any, is followed by a symbol that determines the next location to be displayed. A "^" character means step backwards to the previous location, and a "=" character means don't move. The default is to step forward to the next memory location. Specifying a "." terminates the MM command. Because this is an important command, we'll look at some examples. First, we will display and modify a word at a time using binary mode.

```
>MM 400 -W -BIN
000400 0010111001111100 ?  10110
000402 0000000000000000 ?  111111100000
000404 0001000000000000 ?  0011
000406 0011000000111100 ? .
```

Next, we'll look at word-mode display and read/write data in decimal form.

```
>MM 400 -W -DEC
000400     0 ?  3
000401    22 ?  6
000402    15 ?  9
000403   -32 ?  27
000404     0 ?  55
000405     3 ? .
```

We can use the -0 option to examine and modify odd bytes. This function might be useful when you are splitting code into odd and even bytes before writing it into EPROM.

```
>MM 401 -0
000401 12 ?  F1
000403 1B ?  23
000405 03 ?  0C
000407 3C ? .
```

Finally, we'll look at two symbols, ^ and =, you can use to alter the default sequence.

```
>MM 400
000400 03 ? 26
000401 F1 ? ^
000400 26 ? 27
000401 F1 ? 34
000402 AB ? =
000402 AB ? 43
000403 23 ? .
```

MS - Memory set

Format:

```
MS <ADDR> 'string'
MS <ADDR> n1 n2 ... n7
```

The MS command presets memory locations starting at ADDR to the values in the arguments. Successive bytes are read until the data is exhausted. The data may be a string value or numbers no greater than a longword; e.g., 'Test_4' takes 4 bytes, and $123 takes 2 bytes.

NOBR - Remove breakpoint

Format: NOBR <BRKPT_1> ... <BRKPT_8>

The NOBR command removes the breakpoints specified by the breakpoint address arguments (BRKPT_n) from the breakpoint table. NOBR also prints the remaining breakpoint table.

TR - Trace program execution

Format: TR <COUNT>

The TR command causes COUNT instructions to be executed starting at the current value of the PC. The program executes until the count expires or a trap, exception, or breakpoint is reached. After COUNT instructions have been executed, the registers are printed and a trace prompt is displayed. Typing an enter causes the next instruction to be executed and the registers are dumped again. Since the trace facility does not use the 68000's own trace mode facility, exceptions may also be traced. The default value of COUNT is 1. Breakpoints are in effect only when the value of COUNT is greater than one.

LOG - Log a session

Format: LOG <Filename>

The LOG command is very useful indeed. It causes everything on the screen (i.e., both keyboard input and simulator output) to be loaded into filename which is stored on disk. This means that you can examine the results of a session with the simulator at your leisure. Equally, you can edit this file and incorporate it in other documents (as I have done). If you type LOG again during the session, the logging function is turned off.

I/O and Other Facilities

The 68000 simulator provides a mechanism to allow you to perform I/O on numbers and characters. This is achieved by using a TRAP #15 instruction. Should you alter the default vector for TRAP #15 to a non-zero value, this feature will cease to work, and normal exception processing will take place for this instruction. You have to put the task number in data register D0.B before calling the trap. The ten I/O operations supported by the simulator are:

0: Print the string pointed at by A1 and D1.W characters long with CRLF.

1: Print the string pointed at by A1 and D1.W characters long without CRLF.

2: Read the string pointed at by A1. Length returned in D1.W (maximum 80).

3: Print the number in D1.L in decimal form in the smallest field.

4: Read a number into D1.L.

5: Read a single character into D1.B.

6: Print a single character in D1.B.

7: Set D1.B to 1 if character input is pending; otherwise, set it to 0.

8: Return the time in hundredths of a second since midnight in D1.L.

9: Call the monitor (i.e., it's like a permanent breakpoint).

For example, if you wish to print the ASCII-encoded character in D1, you use the code:

```
        MOVE.B  #$6,D0              Set up the output code
        TRAP    #15                 Call the operating system
```

We now look at a simple example of the use of two of the above I/O functions. This code fragment prints a message using function 0. Note that address register A1 is set to point to the start of the string `Strg1`. The length of the string is stored in memory by:

```
Strg1E  DC.W    Strg1E-Strg1.
```

Another service provided by `TRAP #15` is function 8, that returns the time since midnight in hundredths of a second in D1. The following fragment of code first scales D1.L by 256 before dividing by 100 to keep the result of the division within a word.

```
        ORG     $400
        LEA     Strg1,A1            Point to string 1
        MOVE.W  Strg1E,D1           Put its size in D1
        MOVE.B  #0,D0               Tell O/S to print string with CR/LF
        TRAP    #15                 Call OS to do printing
*
        MOVE.B  #8,D0               Tell O/S to find time
        TRAP    #15                 Call OS to get time in D1
        LSR.L   #8,D1               Scale D1 by 256 to keep in range
        DIVU    #100,D1             Find seconds
        AND.L   #$0000FFFF,D1       Mask out remainder to get seconds in D1
        LSL.L   #8,D1               Restore time by multiplying by 256
        DIVU    #60,D1              Find minutes by dividing seconds by 60
        AND.L   #$0000FFFF,D1       Mask out remainder
        DIVU    #60,D1              Find hours by dividing minutes by 60
        MOVE.W  D1,D2               Copy hours to D2
        SWAP    D1                  Move remainder to low-order word
        MOVE.W  D1,D3               Copy remainder (minutes to D3)
        STOP    #$2700              End of the program
*
```

```
Strg1   DC.B    'This is a test',
Strg1E  DC.W    Strg1E-Strg1
        END     $400
```

Final Comments

The simulator has a built-in disassembler that is available through the MD and DF commands. Should it encounter an illegal instruction pattern, it will be disassembled as a DC.W $nnnn statement.

A read to an unallocated area of memory causes zeros to be returned as the result. To keep the scheme consistent, newly allocated memory is always initialized to zero. It is, of course, possible to execute programs with infinite loops that cause the simulator to lock up. Pressing the ESC key during program execution will force the simulator to return to command mode.

A new utility, S68K.EXE, has been included on the disk. This utility takes a .BIN file produced by the cross-assembler and creates a .REC file. The .REC file is the binary output of the cross-assembler expressed in Motorola's S record format. You can use the .REC file as an input to an EPROM programmer, etc. All you need do to convert Name.BIN into Name.REC is to enter S68K Name.

BIBLIOGRAPHY

Bacon, J. *The Motorola MC68000: An Introduction to Processor, Memory and Interfacing.*
Englewood Cliffs, N.J.: Prentice-Hall, 1986.

Bramer, B., and Bramer, S. *MC68000 Assembly Language Programming*, 2d ed.
London, England: Edward Arnold, 1991.

Carter, E.M., and Bonds, A.B. "A 68000-based System for only $200."
Byte (January 1984): 403–416

Clements, A. "A Microprocessor for Teaching Computer Technology."
Computer Bulletin Vol. 2, Part 1 (March 1986), 14–16.

Clements, A. *Microprocessor Systems Design.*
Boston, Mass.: PWS Publishing Co., 1992.

Clements, A. *Microprocessor Interfacing and the 68000: Peripherals and Systems.*
New York: John Wiley and Sons, 1989.

Clements, A. *Microprocessor Support Chips Sourcebook.*
New York: McGraw-Hill Book Co., 1992.

Dr. Dobb's Journal. *Dr. Dobb's Toolbook of 68000 Programming.*
Englewood Cliffs, N.J.: Prentice-Hall, 1986.

Eccles, W.J. *Microprocessor Systems: A 16-bit Approach.*
Reading, Mass.: Addison-Wesley, 1985.

Ford, W., and Topp, W. *Assembly Language and Systems Programming for the M68000 Family*, 2d ed.
Lexington, Mass.: D.C. Heath and Company, 1992.

Foster, C.C. *Real-Time Programming — Neglected Topics.*
Reading, Mass.: Addison-Wesley, 1981.

Gillet, W.D. *An Introduction to Engineered Software.*
Orlando, Fla.: Holt, Rinehart and Winston, 1982.

Gorsline, G.W. *Assembly and Assemblers: The Motorola MC68000 Family.*
Englewood Cliffs, N.J.: Prentice-Hall, 1988.

Hall, D.V., and Rood, A.L. *Microprocessors and Interfacing: Programming and Hardware — 68000 Version.*
Westerville, OH: Glencoe, 1993.

Harel, D. *Algorithmics — The Spirit of Computing.*
Reading, Mass.: Addison-Wesley, 1987.

Harman, L.T. *The Motorola MC68020 and MC68030 Microprocessors: Assembly Language, Interfacing and Design.*
Englewood Cliffs, N.J.: Prentice-Hall, 1989.

Harper, K. *A Terminal Interface, Printer Interface, and Background Printing for an MC68000-based System using the 68681 DUART.*
Application note AN899, Motorola Inc., 1984.

Heath, W.S. *Real-Time Software Techniques.*
New York: Van Nostrand Reinhold, 1991.

Jaulent, P. *The 68000 Hardware and Software.*
London, England: Macmillan, 1985.

Johnston, H. *Learning to Program.*
Englewood Cliffs, N.J.: Prentice-Hall, 1985.

Kane, G., et al. *68000 Assembly Language Programming.*
New York: Osborne/McGraw-Hill, 1986.

Kelly-Bootle, S. *680x0 Programming by Example.*
Carmel, Ind.: Howard W. Sams, 1988.

King, T., and Knight, B. *Programming the M68000.*
Reading, Mass.: Addison-Wesley, 1986.

Livadas, P. E., and Ward, C. *Computer Organization and the MC68000.*
Englewood Cliffs, N.J.: Prentice-Hall, 1993.

Leventhal, L., and Cordes, F. *Assembly Language Subroutines for the 68000.*
New York: McGraw-Hill, 1989.

Lipovski, G.J. *16- and 32-bit Microcomputer Interfacing: Programming Examples in C and M68000 Family Assembly Language.*
Englewood Cliffs, N.J.: Prentice-Hall, 1990.

Lipovski, G.J., *Object-Oriented Interfacing to 16-bit Microcontrollers.*
Englewood Cliffs, N.J.: Prentice-Hall, 1993.

MacGregor, D., Mothersole, D.S., and Moyer, B. *The Motorola MC68020.*
Motorola Inc. AR217 [reprinted from *IEEE Micro*, Vol. 4, No. 4: 101–118.

MacGregor, D., and Mothersole, D.S. "Virtual Memory and the MC68010."
IEEE Micro 3 (June 1983), 24–39.

Mimar, T. *Programming and Designing with the 68000 Family.*
Englewood Cliffs, N.J.: Prentice-Hall, 1991.

Morton, M. "68000 Tricks and Traps."
Byte, Vol. 11, No. 9 (September 1986): 163–172.

Motorola Inc. *A Discussion of Interrupts for the MC68000.*
Application Note AN1012, Motorola Inc.

Motorola Inc. *Educational Computer Board User's Manual.*
Motorola Inc., Austin, Texas, 1982.

Motorola Inc. *M68000 vs. iAPX86 Benchmark Performance.*
Note BR150, Motorola Inc.

Motorola Inc. *MC68000 16/32-bit Microprocessors Reference Manual, 4th ed.*
Englewood Cliffs, N.J.: Prentice-Hall, 1986.

Motorola Inc. *MC68000 16-/32-bit Microprocessor.*
Note AD1814R6, Motorola Inc., 1985.

Motorola Inc. *MC68230 Parallel Interface/Timer.*
AD1860R2, Motorola Inc., 1983.

Motorola Inc. *MC68451 Memory Management Unit.*
Motorola Inc., April 1983.

Motorola Inc. *MC68881 Floating-Point Coprocessor as a Peripheral in an M68000 System.*
Application Note AN947, Motorola Inc., 1987.

Ripps, D., and Mushinsky, B. "Benchmarks Contrast 68020 Cache-memory Operations."
EDN (August 8, 1985): 177–202.

Scanlon, L.J. *The 68000: Principles and Programming.*
Carmel, Ind.: Howard W. Sams, 1981.

Scherer, V.A., and Peterson, W.G. *The MC68230 Parallel Interface/Timer Provides an Effective Printer Interface.*
Application Note AN854, Motorola Inc.

Shooman, M.L. *Software Engineering.*
New York: McGraw-Hill, 1983.

Sommerville, I. *Software Engineering*, 4th ed.
Reading, Mass.: Addison-Wesley, 1992.

Starnes, T.W. *Design Philosophy behind Motorola's MC68000.*
Note AR208, Motorola Inc.

Stenstrom, P., *68000 Microcomputer Organization and Programming.*
Englewood Cliffs, N.J.: Prentice-Hall, 1992.

Treibel, W.A., and Singh, A. *The 68000 Microprocessor: Architecture, Software and Interfacing Techniques.*
Englewood Cliffs, N.J.: Prentice-Hall, 1986.

Treibel, W.A., and Singh, A. *The 68000 and 68020 Microprocessors: Architecture, Software and Interfacing Techniques.*
Englewood Cliffs, N.J.: Prentice-Hall, 1991.

Veronis, A. *The 68000 Microprocessor.*
New York: Van Nostrand Reinhold, 1988.

Wakerly, J.F. *Microcomputer Architecture and Programming: The 68000 Family.*
New York: John Wiley, 1989.

Wilcox, A.D. *68000 Microcomputer Systems: Designing and Troubleshooting.*
Englewood Cliffs, N.J.: Prentice-Hall, 1987.

Williams, S. *68030 Assembly Language Reference.*
Reading, Mass.: Addison-Wesley, 1988.

Yu-Cheng Liu *The M68000 Microprocessor Family: Fundamentals of Assembly Language.*
Englewood Cliffs, N.J.: Prentice-Hall, 1991.

Zehr, G. "Memory Management Units for 68000 Architectures."
Byte, Vol. 11, No. 12 (December 1986): 127–135.

INDEX